MATHEMATICS IN THE BEHAVIORAL AND SOCIAL SCIENCES

MATHEMATICS IN THE
BEHAVIORAL AND SOCIAL SCIENCES

JOHN W. BISHIR/DONALD W. DREWES

North Carolina State University

HARCOURT, BRACE & WORLD, INC.

New York / Chicago / San Francisco / Atlanta

To Mary and Betty

ISBN 0-15-555251-1

Library of Congress Catalog Card Number: 70-113704

Printed in the United States of America

PART I. Paul Klee. *Disturbed Equilibrium.* 1922. Paul Klee Foundation. Museum of Fine Arts, Berne. (photo by Museum of Modern Art, New York)

PART II. Paul Klee. *Centrifugal Forces.* 1929. Paul Klee Foundation. Museum of Fine Arts, Berne. (photo by Museum of Modern Art, New York)

PART III. Paul Klee. *Inscription.* 1918. Felix Klee.

PART IV. Paul Klee. *Monument in the Orchard.* 1929. Paul Klee Foundation. Museum of Fine Arts, Berne.

For all the above, permission S.P.A.D.E.M. 1970 by French Reproduction Rights, Inc.

PREFACE

The utility of scientific knowledge is judged according to the degree to which it permits understanding of past events and prediction and control of future events. Description of events, being a historical record of past experience, cannot be used directly for future prediction and control. Rather, expectations of future events are generated within a logical system consisting of two or more abstract variables and the rules for specifying their interdependence. This logical system is called a *model*.

A logical system in which the variables are mathematical symbols and the structural relations between variables consist of a set of equations is called a *mathematical model*. Formulation of models in mathematical rather than literary or verbal form is often advantageous in that (1) mathematical language is often more concise and analytic than verbal language, (2) the relations between variables must be explicitly stated, thereby facilitating public scrutiny and evaluation of explanations, (3) the model builder has access to supportive mathematical theories, and (4) an explanatory structure can be readily generalized to the *n*-variable case.

This book introduces the mathematical methods most frequently used in the behavioral and social sciences. The mathematical content follows closely the recommendations made by the Committee on the Undergraduate Program in Mathematics of the Mathematical Association of America for students in the behavioral, management, and social sciences. Whenever possible, usage of mathematical content is illustrated by models drawn from the scientific literature. In other cases, hypothetical models are offered which attempt to capture the essence, if not the actuality, of the use of mathematical models.

The book is written for both those reasonably competent in mathematics who are seeking to relate mathematics to the social and behavioral sciences and those who have yet to develop their mathematical skills. Admittedly the task is easier for the former. To make it more feasible for the latter, the material is presented so as to facilitate an intuitive understanding of the mathematical concepts that govern useful techniques in the analysis of social and behavioral problems.

For those interested in pursuing applications in specific content areas literature references are footnoted within the textual material. Mathematical readings are listed at the end of each chapter.

Topics in the book are arranged according to a meaningful order of mathematical material rather than a hierarchy of models. Part I considers the finite mathematics relevant to the concept of system as a group of related entities. Matrices and linear algebra are discussed in Part II and applied to the analysis of systems containing n variables. Part III introduces differential and integral calculus and includes a parallel discussion of difference equations and their use in studying the dynamics of system behavior. Part IV deals with probability theory and its application to the development of random models in the social and behavioral sciences.

For each mathematical topic, we have included examples and problems from each discipline in proportion to usage. For instance, there are relatively more examples and problems drawn from classical economics in Part III (Calculus), whereas Part IV (Probability) contains proportionately more material drawn from psychology and management science.

The book contains more than 1500 problems, some within the main textual discussion to allow immediate verification of understanding. In addition, most sections contain problems which call for "proofs." Although these problems do not contain results required for an understanding of later material, it is our feeling that such problems facilitate the development of mathematical skills.

The following diagram shows how the book may be used in one-, two-, or three-semester courses. A solid line indicates dependence on the preceding chapter; a dashed line indicates use of only a portion of the preceding

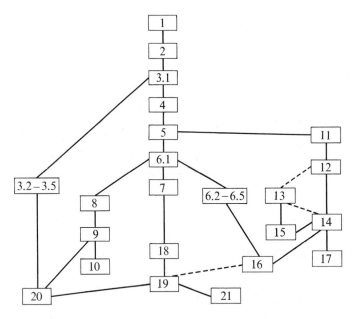

material. For example, a one-semester course emphasizing matrix algebra and its applications would survey the necessary parts of Chapters 1 through 6 before concentrating on Chapters 8 through 10. A one-semester course emphasizing basic discrete probability can be taught using parts of Chapters 1 through 7, together with Chapters 18 and 19. (In each of these cases, some selection is necessary from Chapters 1 through 7. In our experience, a typical class will be able to cover 200 text pages when meeting three times per week for one semester.)

A two-semester course emphasizing the general ideas of "finite mathematics" including matrix algebra and Markov chains would use Chapters 1 through 9 and 18 through 20. (The material need not be covered in order.) A two-semester course emphasizing techniques of calculus consists of Chapters 1 through 6, and 11 through 17, with options of including none, some, or all of Chapters 13, 15, 16, and 17. Finally, the entire book may be covered in three semesters. In this case, the large number of problems and the presence of Chapters 10, 17, and 21 allow considerable latitude for the individual preferences of the instructor.

It is a pleasure to thank the many people who have contributed so much to the completion of this book. Special thanks go to Mrs. Carol Little, Mrs. Debra Currin, and Mrs. Sherry Ford for their many hours of careful typing; to Miss Jane Woodbridge of Harcourt, Brace and World for a superb job of editing; to Dr. Nicholas Rose, Chairman of the Department of Mathematics at North Carolina State University, for his assistance and encouragement; to Professor David Rosen of Swarthmore College for a careful review which led to substantial improvement of Part III on calculus; and especially to our wives for their encouragement and understanding through the seemingly endless job of writing.

John W. Bishir
Donald W. Drewes

CONTENTS

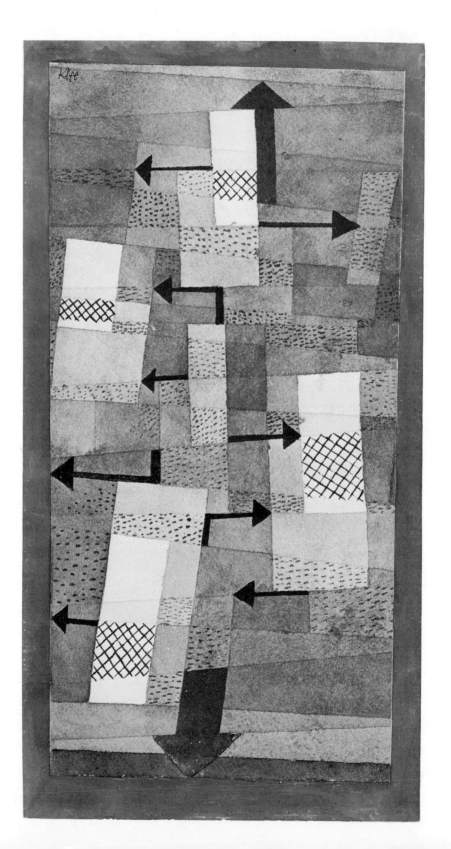

PART I

Finite Mathematics

SET THEORY **1**

1.1 SET MEMBERSHIP

The language and techniques of set theory provide the basic tools for many branches of mathematics of significance to the behavioral and social scientist. Sets also appear in mathematical models developed for the behavioral sciences. Thus we begin our mathematical discussion with a look at sets.

A set is any collection of objects. We may speak, for example, of the set of closing quotations on the New York Stock Exchange for September 23, or the set of possible dominance relations among a group of ten people, or the set of all solutions of the equation $x^2 - 1 = 0$, or the set of all purple cows. Other terms, such as *class* or *collection*, are sometimes used as synonyms for set.

The objects in a set are called *elements*, or *members*, of the set. We say that elements *belong to* the set. The notation $x \in A$ is used to indicate that the object x is a member of the set A. If, for example, P is the set of all psychotics, we might indicate that Mr. X is psychotic by writing $X \in P$.

One way of describing a set is to enclose, in braces, letters or numbers separated by commas to represent the members of the set. Thus, the set S of all solutions of the equation $x^2 - 1 = 0$ could be written as $S = \{1, -1\}$. If R and L denote, respectively, "the rat turns right" and "the rat turns left," then the set of possible choices made in a single run through a T-maze is $C = \{R, L\}$. Similarly, {Adams, Jefferson, Monroe} denotes the set of U.S. Presidents who died on July 4.

PROBLEMS

1. Write the following sets.

    (a) The set of countries lying in both the Eastern and Western hemispheres.

(b) The set of living former Presidents of the United States.

(c) The set of States having only one representative in the House of Representatives.

(d) The set of possible combinations of coins which amount to 41¢.

A set may also be described by stating a criterion which members of the set must satisfy. For instance, the set $S = \{1, -1\}$ may be denoted in the alternative form $S = \{x: x^2 - 1 = 0\}$ read "S is the set of all objects x having the property that $x^2 - 1 = 0$." In general, the notation $A = \{x: p(x)\}$ means that A is the set of all objects x about which the proposition $p(x)$ is true. It is common to use this method of representing a set when the members are not known exactly or are too numerous to list. Thus we represent the set of all millionaires by $M = \{x: x \text{ is a millionaire}\}$ and the set of all real numbers larger than 4 by $G = \{y: y \text{ is a real number and } y > 4\}$. We have, for example, Onassis $\in M$ and $9 \in G$.

PROBLEMS

2. Denote the following sets.

(a) The set of citizens of Canada.

(b) The set of stocks listed on the American Stock Exchange.

(c) The set of positive, even integers.

Some sets, such as the set of all English words that begin with the letters *qa*, contain no elements. Such sets are called *empty sets*, or are referred to as the *null set*, and are denoted by the symbol $\emptyset$. Other empty sets are the set of U.S. Presidents who have lived to be 100 years old and the set of all audible tones greater than 20,000 cps.

PROBLEMS

3. Which of the following sets are empty?

(a) $\{0\}$

(b) The set of integer solutions of the equation $x^2 - 3 = 0$.

(c) $\{\emptyset\}$

The members of a set may themselves be sets. A social club represents a certain set of people, and it may be of interest to speak of the set of social clubs in a given town. However, we never allow a set to be a member of itself. The following example, due to Bertrand Russell, shows why.

Example 1   *Russell's Paradox*   Suppose we wish to divide the collection of all sets into two smaller collections one of which, M, is to contain all those sets which *are* members of themselves while the other, N, is to contain those sets which *are not* members of themselves. Obviously, every set is assigned either to M or to N but no set belongs to both M and N. Question—Where should we place the set N?

Suppose N is placed in M. Then by definition of M, N must be a set which is a member of itself. But this means that N belongs to N, not to M. On the other hand, if N is placed in N, then N is a set which is not a member of itself, and hence must be a member of M. Either assumption is self-contradictory.

We may resolve this paradox by restricting our theory to those sets which are not members of themselves. Unfortunately, this solution of our troubles only leads to further complications. For we now have $M = \emptyset$ and N has become the collection of all sets. But since N is itself a set (and thus $N \in N$, which is now illegal), we must also agree to consider the concept of "the set of all sets" as meaningless. ▶

Example 1 has dealt briefly with some problems in the logical foundations of set theory. Of primary concern is the realization that in order to be properly specified a set must be *well defined*. That is, for each particular object we must be able to decide whether it does or does not belong to the set. We do not pursue this discussion further but refer the reader to other books, such as Wilder, or Luchins and Luchins, listed in the Supplementary Reading at the end of this chapter.

PROBLEMS

    4. Denote each of the following sets in as many ways as possible.

        (a) The set of living Presidents of the United States.

        (b) The set of stocks listed on the New York Stock Exchange.

        (c) The set of all living personal acquaintances of George Washington.

        (d) The set of possible times required by a rat to run down a straight alley to get food.

        (e) The set of 26-year-old male citizens of the United States who have legally voted in three presidential elections.

        (f) The set of real numbers satisfying the equation $x^2 + 3x - 10 = 0$.

    5. Let x and y be coordinates of a point in the plane. Identify the following sets and give a geometric interpretation of your results.

        (a) $\{(x, y): 3x - y = 3\}$

        (b) $\{(x, y): x + y = 5\}$

        (c) $\{(x, y): x + y = 5 \text{ and } 3x - y = 3\}$

        (d) $\{(x, y): x + y = 5 \text{ and } 2x + 2y = 3\}$

        (e) $\{(x, y): x + y = 5 \text{ and } 2x + 2y = 10\}$

    6. In an experiment the same problem is presented to each of four people A, B, C, and D. The *outcome* of the experiment is defined as the set of people who solve the problem.

        (a) How many possible outcomes are there?

        (b) Write the set of possible outcomes.

7. Four paintings, one each by Renoir, Picasso, Klee, and Braque, are presented to a student who is asked to match paintings and names. List the set of possible outcomes. (*Hint:* The set contains 24 elements.)

Problems 8 and 9 relate to Russell's Paradox.

8. The village barber shaves all those males (and only those) who do not shave themselves. Who shaves the barber? (Designate by S the set of males who shave themselves and by B the set of those who do not shave themselves. Which set contains the village barber?)

9. Upon being captured by cannibals a missionary is told that he may make one statement. If the statement is true, he will be boiled, and if the statement is false, he will be roasted. What should the cannibals do if the missionary says "I will be roasted"?

10. In an experiment on concept formation in small children, an experimenter wishes to vary the dimensions of color, size, and number of petals in a flower sketch. If there are three colors, R, G, and Y, three sizes, S, M, and L, and three petal configurations, 4, 6, and 8, list the set of possible stimulus configurations available. (There are 27.)

11. To illustrate the logical inadequacy of displaying a few elements of a set and indicating by three dots that the pattern is to be continued, consider the set A of all numbers of the form

$$n^2 + (n-1)(n-2)(n-3) \tag{1}$$

where n is a positive integer. Show that the first three elements, obtained when $n = 1, 2$, and 3, are 1, 4, and 9 so that one is tempted to write

$$A = \{1, 4, 9, \ldots\}$$

If A were written in this way on an intelligence test, we would not hesitate to write the next element as 16. Show, however, that the next element, obtained by putting $n = 4$ in (1), is 22 rather than 16.

12. Write a defining relation for elements of a set so that its fourth element is, say, 73, while its first three elements are 1, 4, and 9.

1.2 SUBSETS AND SET EQUALITY

If every member of set A is also a member of set B, then A is said to be a *subset* of B. If A is a subset of B, we write $A \subseteq B$. For instance, if B is the set of states which lie east of the Mississippi River, then

$A = \{$Illinois, Indiana, Kentucky, North Carolina, Virginia, West Virginia$\}$

representing the collection of those states whose state bird is the cardinal, is a subset of B. Or, if B is the set of annual gross national products for the years 1900–1960 and A is the set of annual gross national products for the years 1941–1945, then $A \subseteq B$ since every member of set A is also a member of set B.

Our definition of subset may be rephrased by saying that A is a subset of B if there are no elements of A which are not contained in B. Of course, if A is an empty set $\emptyset$ this condition is automatically satisfied and it follows that $\emptyset$, having no members, must be a subset of every set.

In principle, a subset A of a set B is completely determined when we have decided for each element of B whether or not it is to be included in A. When B has only a few elements, the number of decisions required is small and we may easily list all subsets of B. Thus the subsets of the two-element set $B = \{x, y\}$ are B itself, obtained by including both elements, the sets $\{x\}$ and $\{y\}$ obtained by including one element and excluding the other, and $\emptyset$, obtained by excluding both elements. Similarly, the subsets of $B = \{x, y, z\}$ are $\emptyset$, containing no elements, $\{x\}$, $\{y\}$, and $\{z\}$ each containing one element, the two-element sets $\{x, y\}$, $\{x, z\}$, and $\{y, z\}$, and B itself.

PROBLEMS

1. (a) List the subsets of $B = \{x, y, z, w\}$. How many are there?

(b) If B contains n elements, how many subsets are there?

As the number n of elements in B increases, it soon becomes impractical to list all subsets. For each element of B, there are two possible choices (include or exclude) and thus there are $2 \times 2 \times \cdots \times 2 = 2^n$ subsets which may be formed using elements of B. For $n = 2$, 3, or 4 we have seen that it is easy to list all the subsets. But when $n = 10$, there are $2^{10} = 1024$ subsets, while if $n = 20$, the number of subsets exceeds one million!

PROBLEMS

2. If A contains 40 elements, how many subsets does A have?

Two sets are said to be *equal* if each is a subset of the other. Equivalently, two sets are equal if they contain the same elements, or if neither contains an element not also contained in the other.

For simplicity we shall adopt the convention that any listing of elements of a set contains no duplications. Stated another way, we shall agree that listing the same element twice does not change the set. For instance, the sets $A = \{1, 2, 1, 3\}$ and $B = \{1, 2, 3\}$ are considered equal. (Any element contained in one of the sets is also contained in the other.)

As a final point concerning set equality we note that a set having a single member is not to be considered as being identical with that member. If, for example, L is the set of 95 people belonging to the local Lions club, then L and $\{L\}$ are two different sets. For $\{L\}$ is the collection of local Lions clubs and in our example has but one element, while L is a set of people belonging to the local Lions club and contains 95 elements. The two sets must differ since they do not contain the same elements.

PROBLEMS

3. Determine whether $A = B$ or $A \neq B$.

   (a) $A = \{2, 4, 6\}$ and $B = \{4, 6, 2\}$

   (b) $A = \{1, 2, 3\}$, $B = \{$Mercury, Venus, Earth$\}$

   (c) $A = \{T: T$ is a plane equilateral triangle$\}$

   $B = \{U: U$ is a plane equiangular triangle$\}$

4. Which of the following are true?

   (a) $2 = \{2\}$ (b) $2 \in \{2\}$ (c) $0 = \emptyset$ (d) $0 \in \emptyset$

5. Let $A = \{1, 2, 3\}$. Identify the sets B such that $\{1\} \subseteq B$, $B \subseteq A$, and $B \neq A$.

6. Which of the following are correct and why?

   (a) $\{x\} \in \{\{x\}\}$ (b) $\{x\} \subseteq \{\{x\}\}$

   (c) $\{x\} \in \{x, \{x\}\}$ (d) $\{x\} \subseteq \{x, \{x\}\}$

   (e) $a = \{a\}$ (f) $\{a, b\} = \{b, a\}$

7. Seven objects are to be presented in groups of three in such a manner that each object appears with each other object once and only once during the sequence of presentations. How many three-element subsets must be presented in order to meet this restriction?

8. Ivan, Sean, Juan, Maria, and Betty line up for a picture. Naturally, boys and girls alternate. List the set of all possibilities. Then list the following subsets.

   (a) The set in which Ivan is in the middle.

   (b) The set in which a boy is at each end.

   (c) The set in which Ivan is between Maria and Betty.

   (d) The set in which Maria is at the left end of the line.

9. The Tredmore Shoe Company has a top management committee consisting of four vice-presidents and the president. The committee reaches its decision by simple majority vote.

   (a) List the set of all possible outcomes of the vote on a given motion, assuming that no member abstains from voting.

   (b) List the subset of winning outcomes, that is, those outcomes which result in the motion being passed.

   List the subset of winning outcomes if:

   (c) the president has veto power.

   (d) the president has two votes.

   (e) the president votes only in the case of a tied vote.

10. (a) Prove that every set is a subset of itself. Contrast this with the comments preceding Russell's Paradox (Example 1 of Section 1.1).

    (b) Prove that $\emptyset$ has only one subset. What is it?

11. The set A is called a *proper subset* of B if $A \subseteq B$ and $A \neq B$.　$A \subset B$

(a) List the proper subsets of $B = \{L, R\}$.

(b) Show that $\emptyset$ has no proper subsets.

(c) Show that $A = B$ if and only if A is a subset of B but not a proper subset.

12. (a) If $A \subseteq B$, is it necessarily true that $B \subseteq A$?

(b) If $A \in B$, does it follow that $B \in A$?

13. (a) If $A \subseteq B$ and $B \subseteq C$, does it follow that $A \subseteq C$?

(b) Same question with $\subseteq$ replaced by $\in$.

14. (a) If $Z \in A$ and $A \subseteq B$, does it necessarily follow that $Z \in B$?

(b) If $Z \in B$ and $A \subseteq B$, does it necessarily follow that $Z \in A$?

15. Four people A, B, C, and D are waiting for an interview. The interviewer may elect to see them singly or in groups of two, three, or four. List the collection of possible sets of interviewees which the interviewer may see in the first interview session. Compare with Problem 1(a).

16. In a committee of four individuals A, B, C, and D each having one vote, a *majority* is a set of three or more committeemen. Thus the set M of possible majorities is a set having sets as elements.

(a) List the five elements of M.

(b) Argue that the majority $\{A, B, C\}$, although a set, is not a subset, but rather an element, of M.

(c) Is $\{\{A, B, C\}\}$ an element of M? a subset of M?

1.3　COMPLEMENT, UNION, AND INTERSECTION

In building workable models to be used in the behavioral and social sciences we are generally not interested in all possible objects contained in the model, but rather only in those which are contained in some fixed set. In economics, for instance, one often limits the discussion to some set of commodities with the understanding that unless an explicit statement to the contrary is made, all statements refer to this set.

The set of those objects which are of interest in a particular discussion is called the *domain of discourse* or the *universal set* for that discussion. The domain of ordinary discourse is often vague and limits of applicability of a statement are more often assumed than precisely set forth. Thus, when it is remarked that "Everyone knows that the Beatles are a pop singing group," it is to be understood that "everyone" refers to persons, principally of Western cultures, who know something of pop music, but not to infants, Bhantu tribesmen, etc. The scientist, however, is allowed no such latitude. His task is to express information in such a way as to minimize the possibility of misunderstanding. In order to achieve this goal he finds it necessary to limit his choice of language and symbols. In particular, he must begin with a domain of discourse definite enough to allow an objective evaluation as to what does and what does not belong to it.

To see how the choice of a universal set may affect our investigations, consider the problem of "solving" the equation $2x^2 - x - 3 = 0$. Different results are obtained depending on whether we mean integer solutions $\{-1\}$, real solutions $\{-1, \frac{3}{2}\}$, or positive solutions $\{\frac{3}{2}\}$. We must specify beforehand those objects to which the term *solution* is to be applied. That is to say, we must specify the universal set of objects which may be substituted for x. Once a universal set U has been specified the term *set* will always refer to a subset of U.

We now turn to a consideration of how sets may be manipulated to produce other sets. Throughout we shall assume that a universal set U has been specified and that all sets mentioned are subsets of U.

If A is any subset of U, the collection of those elements of U which are not contained in A is called the *complement of A* (sometimes, the term "the complement of A relative to U" is used) and is denoted by A'. Thus, if $U = \{a, b, c, d, e\}$ and $A = \{a, c\}$, then $A' = \{b, d, e\}$. Similarly, the complement of $B = \{a, b, c, e\}$ is $B' = \{d\}$. If U is the set of all integers and E is the set of even integers, then E' is the set of all odd integers. The complement of the set of negative integers is the set of integers which are positive or zero.

PROBLEMS

1. The universal set is $U = \{$Washington, Adams, Jefferson, Monroe$\}$. What is the complement of the set containing

    (a) The names of those Presidents who were Federalists?

    (b) The names which begin with A?

    (c) The names of those Presidents who served after 1834?

    (d) Those names containing nine letters?

Since in our discussion all elements will be contained in U, it follows that the complement of U is the empty set $\emptyset$. In turn, the complement of $\emptyset$ is U. It is apparent from the definition that complements always come in pairs. That is, if A' is the complement of A, then A is the complement of A'. Symbolically,

$$(A')' = A$$

A convenient way of indicating complements is shown in Figure 1, called a *Venn diagram*. The rectangle represents the universal set U, the elements of U being contained within its boundaries. Elements of A are contained within the circle while the elements of A' lie in the shaded region, in U but outside A.

If A and B are any two sets, the collection of elements common to both A and B is called the *intersection* of A and B and is denoted by $A \cap B$. The collection of those elements which belong either to A or to B or to both is called the *union* of A and B and is denoted by $A \cup B$.

If $U = \{a, b, c, d, e\}$, $A = \{a, c\}$, and $B = \{a, b, d, e\}$, then $A \cap B = \{a\}$ and $A \cup B = \{a, b, c, d, e\} = U$. If $C = \{b, d\}$, then $A \cap C = \emptyset$, while $C \cap B = C$ and $C \cup B = B$.

In the Venn diagram of Figure 2, the intersection of A and B is represented by the heavily shaded region common to both sets while the union consists of the entire shaded area.

FIGURE 1

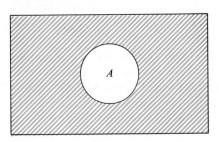

FIGURE 2

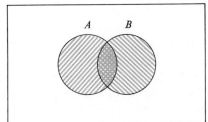

2. If $U = \{1, 2, 3, 4, 5, 6, 7\}$, $A = \{1, 4\}$, $B = \{2, 4, 7\}$, and $C = \{2, 6\}$, what is $A \cap B$? $A \cup B$? $A \cap C$?

Example 1 Luce and Rogow* use the language of set theory in an analysis of the congressional power structure. If F is the set of congressmen who voted in favor of a bill sponsored by the majority party, M is the set of minority congressmen and U is the set of all congressmen, how might we denote

(a) the set of defectors from the minority party?

(b) the set of loyal minority party members?

FIGURE 3

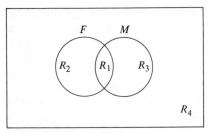

In Figure 3 are labeled four regions corresponding to the four possibilities for any element x (or congressman) in U. We have

Region $R_1 = F \cap M$. Here $x \in F$ and $x \in M$.
Region $R_2 = F \cap M'$. Here $x \in F$ and $x \notin M$.

*Luce, R. D., and Rogow, A. A., "A Game Theoretic Analysis of Congressional Power Distribution for a Stable Two-Party System," *Behavioral Science* **1**, 83–95 (1956).

Region $R_3 = F' \cap M$. Here $x \notin F$ and $x \in M$.

Region $R_4 = F' \cap M'$. Here $x \notin F$ and $x \notin M$.

Since the defectors are those minority congressmen who voted with the majority party, the set of defecting minority congressmen is $F \cap M$, represented by region R_1 in Figure 3.

In (b) we are interested not in set F but rather in the complement of F and specifically in that part of F' which is contained in M. Thus $M \cap F'$, represented by region R_3, is the set of minority members who did not vote with the majority party. ▶

The operations of union and intersection may be extended to any number of sets. We define the intersection of any collection of sets as the set containing those elements which belong to *all* the sets. The union of a collection of sets is the set of those elements which belong to *at least one* of the sets in the collection.

Example 2 Stoyva and Kamiya* have conducted studies in which they attempt to relate dreaming to the subject's verbal reports of dreams and to the rapid eye movement of the subject while sleeping. Let U denote the set of subjects participating in the experiment, D the set of subjects in which dream activity occurs, V the set of subjects who, upon awakening, report having dreamed, and E the set of subjects on whom rapid eye movement is measured. The nonoverlapping regions in Figure 4(a) represent the eight logically possible categories into which a subject may be placed relative to the sets D, V, and E. Region R_1, the intersection $D \cap V \cap E$ of the three sets, represents the set of subjects in whom dreaming occurs, rapid eye movement is measured, and who, upon awakening, report dream activity. Similarly, region $R_4 = D \cap V' \cap E'$ represents the set of subjects in which dreaming occurs which is not subsequently reported, and in which no rapid eye movement is detected. The set of those who make verbal reports is the union $V = R_1 \cup R_2 \cup R_5 \cup R_6$, while those who dream and in whom rapid eye movement is measured is the set $D \cap E = R_1 \cup R_3$. ▶

PROBLEMS

3. Proceeding as in Example 2, express each of the regions R_2, R_3, R_5, R_6, R_7, and R_8 in terms of the sets D, V, and E. In each case give a verbal description of the resulting combination.

Example 3 Suppose that in Example 2 the following statistics are reported on a group of 100 persons tested:

> 55 dreamed (D)
>
> 50 reported dreams (V)

*Stoyva, J., and Kamiya, J., "Electrophysiological Studies of Dreaming as the Prototype of a New Strategy in the Study of Consciousness," *Psychological Review* **75**, 192–205 (1968).

40 exhibited rapid eye movement (*E*)
23 did both *D* and *V*
15 did both *D* and *E*
19 did both *V* and *E*
 7 did all three, *D*, *V*, and *E*

How many persons had dream activity which was neither reported nor indicated by eye movement? That is, how many persons fell in region R_4?

To answer this sort of question we use the data to determine the number of people in each of the eight regions of Figure 4(a). The results are shown in Figure 4(b). The trick is to work backward, using the data in reverse order. Thus the last item indicates that there are seven people in $R_1 = D \cap V \cap E$. Next the 19 persons in $V \cap E$ go either in R_1 or R_5. Since seven are already placed in R_1, this leaves 12 for R_5. Continuing in this way we complete the enumeration one category at a time. In particular, we find 24 persons in region R_4. ▶

FIG 4

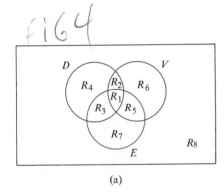

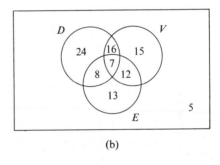

(a)   (b)

PROBLEMS

4. What universal set is understood in the statement "The only states extending farther south than Texas are Florida and Hawaii"?

5. In the Venn diagram of Figure 4 which region or combination of regions does each of the following sets represent?

(a) $D' \cup V$ (b) $E' \cup D'$

(c) $(D \cup V)'$ (d) $E \cap (D \cap V)$

(e) $(E \cap D)'$ (f) $(E \cap D) \cup V$

6. The sets *A*, *B*, and *C* are subsets of a universal set *U*. Arrange the following sets in sequential order so that each set in the sequence is a subset of the next set: $A \cup B$, U, $A \cap B$, $\emptyset$, B, $A \cup (B \cup C)$, $(A \cap B) \cap C$, $(A \cup B) \cup C$, $\emptyset'$, $B \cap A$.

7. Sets *A* and *B* are called *disjoint* if $A \cap B = \emptyset$.

(a) Prove that if *A* and *B* are disjoint, then $(A \cap C) \cap (B \cap C) = \emptyset$.

(b) Show that $A \cap B \cap C = \emptyset$ *does not* imply $A \cap B = \emptyset$ or $A \cap C = \emptyset$ or $B \cap C = \emptyset$.

8. Restle* defines *betweenness* for sets as follows: Set Y is said to be *between* sets X and Z if and only if $X \cap Y' \cap Z = X' \cap Y \cap Z' = \emptyset$.

(a) Give a verbal description of betweenness.

(b) Show that Y is between X and Z if and only if $X \cap Z \subseteq Y \subseteq X \cup Z$.

(c) Prove that if $X \subseteq Y \subseteq Z$, then Y is between X and Z.

(d) Prove that if X is between W and Z, and Y is between X and Z, then Y is between W and Z.

9. Let U denote the set of possible outcomes of an experiment in which three subjects A, B, and C each answer a simple yes–no question. Let T denote the subset of U containing those outcomes in which subject A responds yes, V the set of outcomes in which all three subjects respond yes, and W the set of outcomes in which the number of no's exceeds the number of yes's.

(a) List the elements of U. (There are eight.)

(b) Write the following subsets of U: T', V', $T \cup V$, $T' \cap V'$, $(T \cup V)'$, $(T \cap V)'$, $T' \cup V'$, and $(T \cap W') \cup V$.

10. Which of the following are correct and which incorrect:

(a) $(A \cup B) \cap C' = A \cup (B \cap C')$

(b) $A \cap B \cap C = A \cap B \cap (C \cup B)$

(c) $A \cup B \cup C = A \cup (B \cap A' \cap B) \cup (C \cap (A \cap C)')$

(d) $A \cup B = A \cup (B \cap A' \cap B)$

(e) $(A \cap B \cap C) \subseteq (A \cap B) \cup (B \cap C) \cup (C \cap A)$

(f) $(A \cup B) \cap A' = B$

(g) $(A \cap B' \cap C) \subseteq A \cup B$

(h) $(A \cup B \cup C)' = A' \cap B' \cap C'$

(i) $(A \cup B)' \cap C = A' \cap B' \cap C$

(j) $(A \cup B)' \cap C = (A' \cap C) \cup (B' \cap C)$

11. Let U be the set of all employees of the Behavioral Research, Inc., and let A_1, A_2, and A_3 denote the sets of newly hired employees who score, respectively, below 50, between 50 and 150, and above 150 on the initial placement test. Further, let J_1, J_2, and J_3 denote the sets of employees in respective job classifications I, II, and III. Translate each of the following statements into set notation.

(a) All newly hired employees who score below 50 are placed in job classification I.

(b) No newly hired employee who scores above 150 is placed in either job classification I or II.

(c) All newly hired employees who score between 50 and 150 are placed in either job classification II or III.

(d) Some newly hired employees who score over 150 are placed in job classification III.

*Restle, F., *Psychology of Judgment and Choice: A Theoretical Essay* (John Wiley & Sons, Inc., New York, 1961), p. 46.

(e) No employee of the company is assigned multiple job classifications, there being only three classifications.

12. For each of the following draw a Venn diagram to show that the pairs of sets are represented by the same region.

(a) $(A')'$ and A

(b) $A \cup A'$ and U

(c) $A \cup (B \cap C)$ and $(A \cup B) \cap (A \cup C)$

(d) $(A \cup B)'$ and $A' \cap B'$

13. A convention of insurance agents was attended by 180 agents who sell life insurance, 230 who sell fire and casualty insurance, and 240 who sell auto insurance. Of these agents, 140 sell both auto insurance and fire and casualty insurance, 120 sell both life insurance and fire and casualty insurance, 110 sell both life insurance and auto insurance, and 60 sell all three. All agents in attendance sell at least one of the aforementioned types of insurance.

(a) How many agents attended the convention?

(b) How many agents sell auto insurance but not fire and casualty insurance?

(c) How many agents sell only life insurance?

(d) How many agents specialize in only one type of insurance?

14. According to National Safety Council estimates,* there were 49,000 motor-vehicle fatalities in 1965. Of these, 15,000 occurred in places classi-fied as urban, 23,200 occurred during daylight hours, and 17,500 oc-curred at night in places classified as rural. A further classification re-vealed that 3100 pedestrian fatalities occurred in rural places, 15,900 nonpedestrian fatalities occurred in rural places at night, 2550 pedestrian fatalities occurred in urban places during the daylight hours, and 1600 pedestrian fatalities occurred at night in rural places.

(a) What proportion of the victims of daylight accidents in rural places were pedestrians?

(b) What proportion of the victims of daylight accidents in urban places were nonpedestrian?

(c) What proportion of the total fatalities in rural areas occurred during the daylight hours?

(d) What proportion of the total pedestrian fatalities in rural areas occurred at night?

15. In a survey of the voting habits of 1000 registered voters, it was found that 605 voted in the 1956 presidential election, 595 voted in the 1960 presidential election, and 675 voted in the 1964 presidential election. Of these, 415 voted in both the 1964 and 1960 elections, 385 voted in both the 1956 and 1960 elections, 395 voted in both the 1956 and 1964 elec-tions, and 65 did not vote in any of the three elections.

(a) How many voted in all three elections?

(b) How many voted in exactly one election?

(c) How many voted in two or more elections?

*Accident Facts (National Safety Council, Chicago, Ill.), 1966 ed., p. 41.

16. Rice and Smith* use set theory in dealing with the problems involved in generalizing laboratory results to the real world. Suppose we denote the set of outcomes in the real world as N, the set of outcomes predicted by a model as M, and the set of outcomes of the experiment used to test the model as E. Using a Venn diagram, denote the regions and the related sets that correspond to the following statements.

(a) The model outcomes duplicate exactly those in the real world.

(b) The experimental conditions perfectly duplicate the environment assumed in the model.

(c) The experimental outcomes are common to both the model and the real world outcomes.

(d) The outcomes predicted by the model are found in the real world but do not duplicate the experimental outcomes.

(e) There are model outcomes that are not found in the experimental situation or the real world.

17. Show that

(a) $U' = \emptyset$.

(b) if $A \subseteq B$, then $B' \subseteq A'$.

18. The collection of elements which belong to A or to B, but *not* to both, is called the *symmetric difference*† of the sets A and B and is denoted by $A \bigtriangleup B$. Show that $A \bigtriangleup B = (A' \cap B) \cup (B' \cap A)$.

19. The set of elements contained in A but not in B is sometimes called the *set difference* of A and B and denoted by $A - B$.

(a) Show that $A - B = A \cap B'$.

(b) Show that the complement of A may be written as $U - A$.

(c) Show that $A \bigtriangleup B = (A \cup B) - (A \cap B)$.

(d) Show that $A \bigtriangleup B = (A - B) \cup (B - A)$. This explains why $A \bigtriangleup B$ is called the *symmetric difference*.

(e) What sets in Example 1 may be written in the form $A - B$?

1.4 ALGEBRA OF SETS—COMBINING UNION, INTERSECTION, AND COMPLEMENT

In this section we consider various relations which exist among the set operations union, intersection, and complement. In so doing, we will discover that there are close analogies between the operations of union and intersection for sets and the operations of addition and multiplication for numbers. The basic results are summarized in the following theorem.

*Rice, D. B., and Smith, V. L., "Nature, the Experimental Laboratory, and the Credibility of Hypotheses," *Behavioral Science* **9**, 239–246 (1964).

†For an example of its use see Galanter, E., "An Axiomatic and Experimental Study of Sensory Order and Measure," *Psychological Review* **63**, 218–227 (1956).

Theorem 1 Let A, B, and C be subsets of a universal set U. Then the following relationships hold:

Identity laws:

$$A \cup \emptyset = A \qquad\qquad\qquad\qquad A \cap \emptyset = \emptyset$$
$$A \cup U = U \qquad\qquad\qquad\qquad A \cap U = A$$

Complement laws:

$$A \cup A' = U \qquad\qquad\qquad\qquad A \cap A' = \emptyset$$
$$(A')' = A$$

Idempotent laws:

$$A \cup A = A \qquad\qquad\qquad\qquad A \cap A = A$$

Commutative laws:

$$A \cup B = B \cup A \qquad\qquad\qquad\qquad A \cap B = B \cap A$$

Associative laws:

$$(A \cup B) \cup C = A \cup (B \cup C) \qquad\qquad (A \cap B) \cap C = A \cap (B \cap C)$$

Distributive laws:

$$A \cup (B \cap C) = (A \cup B) \cap (A \cup C)$$
$$A \cap (B \cup C) = (A \cap B) \cup (A \cap C)$$

De Morgan's laws:

$$(A \cup B)' = A' \cap B' \qquad\qquad\qquad (A \cap B)' = A' \cup B' \qquad \blacktriangleright$$

Before proving this theorem, let us note that most of these names and operations are familiar from our knowledge of the algebra of numbers studied in high school. Since adding zero to any number yields that same number as a sum ($a + 0 = a$), zero is called an *identity number* for addition. Similarly, 1 is an identity number for multiplication since multiplying any number by unity yields that number as a product ($a \cdot 1 = a$). A glance at the identity laws for sets shows that the empty set $\emptyset$ is an identity set with respect to union and the universal set U is an identity set for intersection. Because of these analogies to numbers, $A \cup B$ is often called the *set theoretic sum* of A and B and $A \cap B$ the *set theoretic product*.

The analogy may be carried further with the commutative and associative laws. Thus, addition and multiplication of numbers are commutative, that is,

$$a + b = b + a \quad \text{and} \quad a \cdot b = b \cdot a$$

for any numbers a and b. Moreover, the order in which numbers are added or multiplied is irrelevant. That is, the associative laws

$$a + (b + c) = (a + b) + c \quad \text{and} \quad a \cdot (b \cdot c) = (a \cdot b) \cdot c$$

hold for any three numbers a, b, and c. (Parentheses indicate those operations which are to be performed first.)

A distinction between operations on sets and operations on numbers arises with the distributive laws. There are two such laws for sets, one of which distributes union over intersection while the other distributes intersection over union. For numbers, there is only one distributive law, which distributes multiplication over addition. That is,

$$a \cdot (b + c) = a \cdot b + a \cdot c$$

for any three numbers a, b, and c. It is not true that addition distributes over multiplication since, in general, we find

$$a + (b \cdot c) \neq (a + b) \cdot (a + c)$$

For example, $3 + (4 \cdot 2) = 11$, while $(3 + 4) \cdot (3 + 2) = 35$.

A further distinction between numbers and sets arises from the fact that numbers do not satisfy idempotent laws. For instance, if A is a number, then $A + A = 2A$, while if A is a set, $A \cup A = A$.

PROBLEMS

1. Let $U = \{a, b, c, d, e, f, g\}$, $A = \{a, f\}$, $B = \{b, c, f, g\}$, and $C = \{a, b, e, g\}$. Verify the statements of Theorem 1.

PROOF OF THEOREM 1   We shall prove only the first distributive law, leaving the remainder of Theorem 1 for the exercises. Let us first consider the procedure we must follow. By definition two sets L and R are equal if and only if each is a subset of the other. Our procedure for proving equality is thus clearly indicated. We first choose an arbitrary element $x \in L$, and show that $x \in R$. Since the choice of x was arbitrary we conclude that every member of L is also a member of R and hence that $L \subseteq R$. We then reverse the argument by choosing an arbitrary element $x \in R$, show that $x \in L$, and deduce that $R \subseteq L$. From these two results we conclude $L = R$.

In order to prove the first distributive law, then, we must show both

$$A \cup (B \cap C) \subseteq (A \cup B) \cap (A \cup C) \tag{2}$$

and

$$(A \cup B) \cap (A \cup C) \subseteq A \cup (B \cap C) \tag{3}$$

To prove (2), we choose an arbitrary element $x \in A \cup (B \cap C)$, the total shaded area shown in Figure 5(a). There are two cases to consider:

case (i) $x \in A$. Then certainly $x \in A \cup B$ and $x \in A \cup C$ and it follows that $x \in (A \cup B) \cap (A \cup C)$.

case (ii) $x \in B \cap C$. Then $x \in B$ and $x \in C$. Hence $x \in A \cup B$ and $x \in A \cup C$ so again $x \in (A \cup B) \cap (A \cup C)$.

Having considered all possible cases, we have established that $A \cup (B \cap C) \subseteq (A \cup B) \cap (A \cup C)$.

To complete the proof, we choose an arbitrary element x in $(A \cup B) \cap (A \cup C)$, the cross-hatched area shown in Figure 5(b). Then $x \in A \cup B$ and

FIGURE 5 Venn diagrams for Proof of Theorem 1

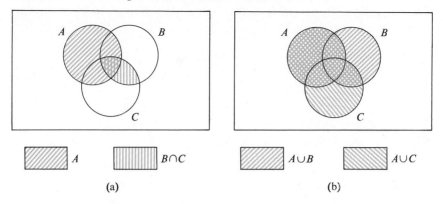

(a)   (b)

$x \in A \cup C$. Again there are two cases:

case (i) $x \in A$. If so, then x must also be an element of $A \cup (B \cap C)$.

case (ii) $x \notin A$. Then in order to have $x \in A \cup B$ and $x \in A \cup C$, we must have $x \in B$ and $x \in C$. Hence $x \in B \cap C$ and again it follows that $x \in A \cup (B \cap C)$.

The proof of (3), and of the first distributive law, is now complete. ▶

PROBLEMS

2. Draw Venn diagrams illustrating the other statements of Theorem 1.

In the Proof of Theorem 1 we repeatedly used the following result, the proof of which is left as an exercise (see Figure 6).

FIGURE 6 Venn diagrams to illustrate Theorem 2

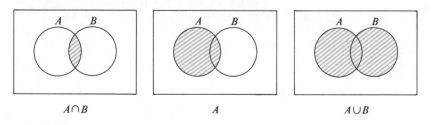

$A \cap B$ A $A \cup B$

Theorem 2 For any two sets A and B,

$$A \cap B \subseteq A \subseteq A \cup B$$

and

$$A \cap B \subseteq B \subseteq A \cup B \qquad \blacktriangleright$$

Example 1 Orcutt, *et al.*,* have developed a simulation model of the labor force in which each person is classified according to each of the following categories:

Place of Residence	Age	Education
R_1 Rural, farm	A_1 25–29	E_1 5–6 years elementary school
R_2 Rural, nonfarm	A_2 30–34	E_2 7–8 years elementary school
R_3 City 2500–25,000	A_3 35–44	E_3 1–3 years high school
R_4 City 25,000–100,000	A_4 45–64	E_4 4 years high school
		E_5 1–3 years college
		E_6 4 years college

Let us take the universal set U to be the set of all persons in the labor force, and let us denote by L the set of all employed persons and by $R_1, R_2, \ldots, E_6$ the sets of persons in the categories listed above.

Assuming that no person with 4 or more years of college is unemployed, then the set of unemployed persons other than those who have neither 4 or more years of college nor only 5 or 6 years of grade school is identical with the set of unemployed persons having only 5 or 6 years of grade school. Translated into set notation we have

$$L' \cap (E_6' \cap E_1')' = L' \cap [(E_6')' \cup (E_1')'] \qquad \text{[De Morgan's Laws]}$$
$$= L' \cap (E_6 \cup E_1) \qquad \text{[Complement Laws]}$$
$$= (L' \cap E_6) \cup (L' \cap E_1) \qquad \text{[Distributive Law]}$$
$$= \emptyset \cup (L' \cap E_1) \qquad \text{[Assumption]}$$
$$= L' \cap E_1 \qquad \text{[Identity Laws]} \qquad \blacktriangleright$$

An argument similar to that given in the Proof of Theorem 1 serves to establish a more general form of the distributive laws for sets. Recall that by the intersection of a collection C of sets A_i we mean the set of those elements which belong to *all* the sets A_i. We denote this set by $\bigcap\limits_{C} A_i$. The union of the collection is the set of elements belonging to *at least one* of the sets A_i, and is denoted by $\bigcup\limits_{C} A_i$.

Theorem 3 *The Generalized Distributive Laws* Let C be an arbitrary collection of sets A_i, and E be any set. Then

$$E \cup \left(\bigcap_{C} A_i \right) = \bigcap_{C} (E \cup A_i) \qquad \qquad (4)$$

*Orcutt, G. H., Greenberger, M., Korbel, J., and Rivlin, A. M., *Microanalysis of Socioeconomic Systems; a Simulation Study* (Harper and Row Publishers, Inc., New York, 1961).

and

$$E \cap \left(\bigcup_C A_i \right) = \bigcup_C (E \cap A_i) \tag{5}$$

PROOF The proof of (4) closely parallels those of (2) and (3) above. If $x \in E \cup \left(\bigcap_C A_i \right)$, then $x \in E$ or x is a member of every A_i. If $x \in E$, then by Theorem 2, $x \in E \cup A_i$ for every A_i, and hence $x \in \bigcap_C (E \cup A_i)$. If x is a member of every A_i, then x is a member of every set $E \cup A_i$ and again $x \in \bigcap_C (E \cup A_i)$. We conclude that $E \cup \left(\bigcap_C A_i \right)$ is a subset of $\bigcap_C (E \cup A_i)$.

Conversely, if $x \in \bigcap_C (E \cup A_i)$, then either $x \in E$ or $x \notin E$. If $x \in E$, Theorem 2 guarantees that $x \in E \cup \left(\bigcap_C A_i \right)$. If $x \notin E$, then x can be a member of $\bigcap_C (E \cup A_i)$ only by being a member of every A_i. Hence $x \in \bigcap_C A_i$ and $x \in E \cup \left(\bigcap_C A_i \right)$. We conclude that $\bigcap_C (E \cup A_i)$ is a subset of $E \cup \left(\bigcap_C A_i \right)$, completing the proof of (4). The proof of (5) is left as an exercise (Problem 14). ▶

Example 2 In the notation of Example 1, the set of persons between ages 25 and 44 is $A_1 \cup A_2 \cup A_3$. Hence the set of these persons who are residents of rural nonfarm areas is $R_2 \cap (A_1 \cup A_2 \cup A_3)$. Using the generalized distributive law (5) this may be rewritten

$$(R_2 \cap A_1) \cup (R_2 \cap A_2) \cup (R_2 \cap A_3) \qquad ▶$$

The De Morgan's laws may be extended to any collection of sets in a manner similar to the distributive law.

Theorem 4 *Generalized De Morgan's Laws* Let C be an arbitrary collection of sets A_i. Then

$$\left(\bigcup_C A_i \right)' = \bigcap_C A_i' \tag{6}$$

and

$$\left(\bigcap_C A_i \right)' = \bigcup_C A_i' \tag{7}$$

In words, this says that the complement of a "union" is the "*intersection*" of the complements, while the complement of an "intersection" is the "*union*" of the complements.

PROOF We shall prove (6), leaving (7) as an exercise (Problem 14). Choose an arbitrary element x in $\left(\bigcup_C A_i \right)'$. Then $x \notin \bigcup_C A_i$ which means that x is not a member of any A_i. That is, x lies outside every A_i, and thus $x \in \bigcap_C A_i'$ (see Figure 7).

FIGURE 7 Venn diagram to illustrate Formula (6)

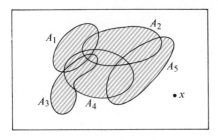

$$\bigcup_C A_i$$

Conversely, if $x \in \bigcap_i A_i'$, then x lies outside every A_i, from which it follows that x is not a member of at least one A_i, or that $x \in \left(\bigcup_C A_i\right)'$. This completes the proof. ▶

Example 3 In Example 1, any person having less than 5 years of schooling is by definition not included in any of the sets $E_1, \ldots, E_6$. That is, any such person is a member of

$$(E_1 \cup E_2 \cup \cdots \cup E_6)' = E_1' \cap E_2' \cap \cdots \cap E_6' \qquad ▶$$

PROBLEMS

3. A sociologist is studying the membership structure of cliques in high school. Let U be the set of all students and let the following be subsets of U:

   M = the set of male students

   F = the set of female students

   D = the set of students who date

   C = the set of students who plan on going to college

   O = the set of students who are members of the "out group" as defined in this study

   I = the set of students who are members of the "in group" as defined in this study

   Translate each of the following empirical observations into either an equation or an inequality using set notation and the symbols $=$, $\neq$, $\emptyset$, $\cap$, $\cup$, $'$, $\subseteq$.

   (a) Some male students do not plan on attending college.

   (b) Not all the female students date.

   (c) Those girls who date or who plan on attending college are members of the "in group."

   (d) No student is a member of both the "in group" and the "out group."

(e) Neither the male students who do not plan on attending college nor the girls who do not date are members of the "in group."

(f) Some boys who do not date but who plan on attending college are members of the "in group."

(g) All students are members of either the "in group" or the "out group."

(h) Given this information, what conclusions can we draw concerning the membership of the "out group"? Express these conclusions in set notation.

4. Prove that the following laws hold for any subsets A, B, and C of a universal set U. In each case draw an appropriate Venn diagram.

(a) $A \cup \emptyset = A$ (b) $A \cap \emptyset = \emptyset$

(c) $(A \cap B) \cap C = A \cap (B \cap C)$ (d) $(A' \cap B')' = A \cup B$

(e) $(A \cap B) \cap (A \cap B') = \emptyset$

(f) $A \cap (B \cup C) = (A \cap B) \cup (A \cap C)$

(g) $A \cap B \subseteq A \subseteq A \cup B$

5. Simplify each of the following by drawing an appropriate Venn diagram.

(a) $A \cap (A \cap B)$ (b) $A \cap [(A \cap B') \cup (A' \cap B)]$

(c) $(A \cup B) \cap (A \cup B')$ (d) $[(A \cap B') \cup (B \cap C')] \cap (A \cap C')$

6. Corcoran* presented cards containing symbols which differed in shape, color, and size to a group of subjects in order to assess the effects of task complexity and practice on performance after sleep loss. The symbols were either squares or circles, black or white, and large or small. Let S, C, B, W, L, and N denote the respective sets of cards containing square, circular, black, white, large, and small symbols. Describe the attributes of the following sets of cards:

(a) $S \cap L$ (b) $W \cap N'$

(c) $(C \cap W) \cup (C \cap B)$ (d) $(B \cap W) \cup S$

(e) $S - (L \cap B)$ (f) $(S \cap B) \cup (S \cap L)$

(g) $(B \cap W)' \cap (C \cup L)$ (h) $C \cap (L - W)$

7. Simplify each of the following:

(a) $A \triangle U$ (b) $A \triangle \emptyset$

(c) $A \triangle A'$ (d) $A \triangle A$

(The symbol $\triangle$ is defined in Problem 18, Section 1.3.)

8. Construct an example to show that, in general,

$$A \triangle (B \cap C) \neq (A \triangle B) \cap (A \triangle C)$$

That is, there are sets A, B, and C for which the relation does not hold.

9. Prove that $[A \cap (B \cup C)] \cup [B \cap C] = [A \cap B] \cup [(A \cup B) \cap C]$.

[*Hint:* The analogous statement for numbers is $a(b + c) + bc = ab + (a + b)c$.]

---

*Corcoran, D. W. J., "The Influence of Task Complexity and Practice on Performance After Loss of Sleep," *Journal of Applied Psychology* **48**, 339–343 (1964).

10. Prove that $(A \cup B) \cap (A \cup C) = A \cup [(B \cup C) \cap A] \cup (B \cap C)$.

11. It is true for numbers that $ab + (a + b)c = (a + c)b + ac$. State and prove an analogous result for sets.

12. Prove the remaining statements in Theorem 1.

13. Prove Theorem 2.

14. Prove Equations (5) and (7).

15. Show by a suitable example that the following statements are true:
   (a) Subtraction is not associative.
   (b) Subtraction is not commutative.
   (c) Division is not associative.
   (d) Division is not commutative.

16. Show by choosing suitable examples that
   (a) the operation of taking differences of sets is not associative.
   (b) set differencing is not commutative.
   (c) the operation of symmetric difference is not distributive over union.
   (d) symmetric difference is not distributive over intersection.
   (The difference of two sets is defined in Problem 19, Section 1.3.)

17. As an example of noncommutative mathematical operations, let $M(x)$ indicate that the number x is to be multiplied by 3 and $A(x)$ that 2 is to be added to x. Show that $A[M(x)]$ is not the same as $M[A(x)]$.

18. Let $a \circ b$ be defined as the average of two numbers, that is, $(a + b)/2$. Show that the operation $\circ$ is commutative but not associative.

19. Suppose that an applicant is evaluated by two separate interviewers. If the applicant fails either or both of the evaluations, he fails to qualify for employment. The selection operation $*$ defined on the set $\{0, 1\}$ where 0 denotes "failure" and 1 denotes "success" may be described by the following table

	evaluation B	
$*$	0	1
0	0	0
1	0	1

evaluation A

Thus $0 * 0 = 0, 0 * 1 = 0, 1 * 0 = 0$, and $1 * 1 = 1$. Determine whether the properties of commutativity and associativity hold for the operation $*$.

20. The operation $\wedge$ defined on the set $\{0, 1\}$ may be described as follows:

$\wedge$	0	1
0	1	0
1	0	1

That is, $0 \wedge 0 = 1, 0 \wedge 1 = 0, 1 \wedge 0 = 0$, and $1 \wedge 1 = 1$. Test the operation $\wedge$ for commutativity and associativity.

SUPPLEMENTARY READING

Goldberg, S., *Probability*, *An Introduction* (Prentice-Hall, Inc., Englewood Cliffs, N. J., 1960), Chapter 1.

Luchins, A. S., and Luchins, E. H., *Logical Foundations of Mathematics for Behavioral Scientists* (Holt, Rinehart and Winston, Inc., New York, 1965).

Suppes, P., *Introduction to Logic* (D. Van Nostrand Company, Princeton, N. J., 1957), Chapter 9.

Wilder, R. L., *Introduction to the Foundations of Mathematics* (John Wiley & Sons, Inc., New York, 1965), 2nd ed.

SYMBOLIC LOGIC **2**

2.1 STATEMENTS AND THEIR TRUTH SETS

Behavioral science like most subjects has its own special vocabulary used with ordinary language. The set theoretic notation of Chapter 1 is only used if it helps to clarify the discussion. In this chapter, we extend the clarity and precision of set notation to the analysis of statements by relating set theory to logic. This allows us to translate behavioral systems into simple sentences and then use logical notation and operations to formalize these systems.

Logic deals with statements which can be classified as being either true (T) or false (F) but not both. The symbols T and F represent the possible *truth values* a statement may have. To be considered within the scope of logic, a statement must be capable of being assigned a specific truth value. Thus, such a paradoxical statement as "This sentence enclosed in quotation marks is false" will not be allowed in our discussion. If the statement is false, that is, if it is false that the sentence is false, then it must be true. On the other hand, if it is true, then by its own admission it is false. Either way a contradiction results. This situation is similar to Russell's Paradox and the requirement that a set be well defined.

It is not required that we actually know the truth value of each statement used. For instance, the sentence "The population of the United States is 190 million" is acceptable for inclusion in a logical discussion, even though we do not know whether it is true or false. The important point is that it is possible to assign a truth value to the statement.

As in the theory of sets, we begin by specifying a domain of discourse (universal set) U. All statements are assumed to pertain to the elements of U. A particular statement will be true for some elements of U and false for others. With any statement p we associate the set P containing those members of U for which p is true. The set P is called the *truth set in U* or, simply, the *truth set* of the statement p.

Example 1   If U is the set of real numbers, the truth set of the statement

$$p: x^2 + 3x = 4$$

is the set $P = \{1, -4\}$ of solutions of this equation. If U is the set of positive numbers, the truth set of p is $Q = \{1\}$.   ▶

Example 2   Let U be the set of all business establishments in the United States. Then the set P of profitable businesses may be denoted

$$P = \{z : z \text{ is a profitable business}\}$$

In this form it is apparent that P is the truth set of the statement

$$p: z \text{ is a profitable business}$$ ▶

PROBLEMS

1. Let $A = \{a, b, c, d\}$ and let B denote a subset of A. What is the truth set of the statement $B \subseteq \{a, c\}$? What universal set U is implied here?

In Section 1.1, we indicated that one way of describing a set is to state a criterion which members of the set must satisfy. The set so described then becomes the truth set of the stated criterion. Since this may be done for any set whatsoever, it follows that there is a complete correspondence between truth sets and statements, and hence between set theory and logic. We shall utilize this correspondence and our knowledge about sets to develop the basic concepts of logic.

A statement which is true for all elements of U is said to be *logically true*, or to be a *tautology*. A statement which is never true is *logically false*, or a *self-contradiction*. The universal set U itself is the truth set of a tautology, while the empty set $\emptyset$ is the truth set of a self-contradiction. For instance, if U is the set of real numbers x, then $2x + 2 = 2(x + 1)$ is a tautology* while $x - 3 = x + 2$ is a self-contradiction.

To take a behavioral example, suppose we define reinforcement as any stimulus which increases the strength of an immediately preceding response R. Then if learning is defined as an increase in the strength of response R, the statement "Reinforcement facilitates learning response R" is a tautology.

With any statement p may be associated another statement not-p, called the *negation* of p, and symbolized by $\sim p$. If, for example, p is the statement "absolute power corrupts," then $\sim p$ is "absolute power does not corrupt."

*A tautology for numbers is usually called an *identity*.

The negation of p is true when p is false, and false when p is true. If P is the truth set for p, then the complementary set P', containing those elements for which p is false (and hence for which $\sim p$ is true), is the truth set for $\sim p$.

PROBLEMS

2. The price x of a certain commodity can be any value in the set $U = \{1, 2, 4, 6, 8, 10\}$. The statement p is "$x^2 + 16 = 10x$." What is the truth set of $\sim p$?

In addition to the relation between negation and complementation, there are other operations on statements which correspond to intersection, union, subset, and equality for sets.

Let p be the statement "Material costs have increased" and q the statement "Labor productivity has declined." The statement "Material costs have increased *and* labor productivity has declined" is called the *conjunction* of statements p and q and is denoted by $p \wedge q$. The *disjunction* of p and q, denoted by $p \vee q$, is "Material costs have increased *or* labor productivity has declined."

An expression of the form "*If* material costs have increased, *then* labor productivity has declined" is a *conditional statement* (denoted $p \Rightarrow q$) connecting p and q. Finally, the expression "Material costs have increased *if and only if* labor productivity has declined" is a *biconditional statement* ($p \Leftrightarrow q$) relating p and q.

The symbols $\sim$, $\wedge$, $\vee$, $\Rightarrow$, and $\Leftrightarrow$ representing, respectively, negation, conjunction, disjunction, conditional, and biconditional, are the five basic *sentential connectives* of symbolic logic. We shall adopt the point of view that a connective is *defined* by the way in which its truth set is related to the truth sets of the statements being connected. For instance, the following definition formalizes our discussion of the negation of a statement.

Definition 1 Let p be a statement having truth set P. Any statement whose truth set is P' is called the negation of p.   ▶

In deciding upon the truth sets of the remaining connectives, we shall be guided by the way in which these connectives are used in ordinary discourse. In asserting the conjunction $p \wedge q$, for instance, we normally mean to imply that p and q are both true. If either p or q is false then the conjunction is false. If P and Q are the respective truth sets for p and q, it follows that $p \wedge q$ is true only for those elements of U which are contained in both P and Q. In other words, the truth set of $p \wedge q$ is $P \cap Q$.

The word *or* is used in ordinary language in two different senses. In the *exclusive* sense p or q means either p or q is true, *but not both*. If you say "The treatment is either effective or ineffective," everyone knows that the exclusive sense is intended. On the other hand, the *inclusive* sense (the legal and/or) is apparent in "Extensive negotiation may weaken our bargaining position or prolong the conflict." Of course, both might happen.

In mathematics and logic *p or q* is always used in the *inclusive* sense, *p* or *q, or both*. If the exclusive *p* or *q*, but not both, is intended, it will be explicitly stated. Thus the truth set of $p \lor q$ is $P \cup Q$, the set of elements contained in either P or Q, or in both.

Example 3 If x denotes a real number, p is the statement $x^2 + 3x = 4$, and q the statement $2x^2 - 6x = -4$, then the truth set of $p \land q$ is $\{1\}$, the common solution of the two equations. The truth set of $p \lor q$ is

$$\{1, 2, -4\} \qquad \blacktriangleright$$

PROBLEMS

   3. Verify the results stated in Example 3.

When stating the conditional "If p, then q" we are claiming that if p is true, then q is also true. Hence, if e is an element of U for which p is true—that is, if $e \in P$—then we must also have $e \in Q$. To say that the conditional $p \Rightarrow q$ is true, then, is to say that P is a *subset* of Q.

Looking at it another way, the conditional $p \Rightarrow q$ is false only for those elements e of U which violate the condition $P \subseteq Q$, that is, only for an e which is a member of P, but not a member of Q. These are the elements in $P \cap Q'$, shown unshaded in Figure 1. The remaining elements, those for which $p \Rightarrow q$ is true, are members of the set $(P \cap Q')' = P' \cup Q$.

FIGURE 1

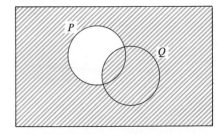

The truth set $P' \cup Q$ is represented by the shaded region of Figure 1. Note that within this truth set, an element e for which p is true ($e \in P$) is also an element for which q is true ($e \in Q$). However, the truth of p is not necessary for the truth of $p \Rightarrow q$. In fact, since P' is entirely shaded it follows that $p \Rightarrow q$ is considered true whenever p is false, regardless of the truth or falsity of q.

As most persons initially find this point of view at variance with their intuitive feelings, there may be some objections to our procedure. In particular, it might be maintained that in order to express the conditional "if p, then q" the statements p and q should be related in some way.

But how is one to decide which statements are related and which are not? Certainly any intuitive feelings about relatedness will vary from one person to another and will depend on the context in which the statements appear. Our approach has been deliberately chosen to avoid the complications which arise from questions of relatedness, the definitions being made in such a way that the truth set of a logical form depends *only on the form itself* and not on the specific statements involved.

Granting, then, that truth sets should depend only on the logical form, have we made the proper decision for the conditional form $p \Rightarrow q$ when p is false? For the case when q is also false, consider the statement:

F T

> If Canada has 300 million inhabitants and the United States only 200 million, then the population of Canada is greater than that of the United States.

For the case when q is true, consider:

F F

> If Canada has 300 million inhabitants and the United States has 400 million, then the population of the United States exceeds that of Canada.

It is hard to imagine anyone denying the truth of these statements. But if this is the case, then the truth set for the conditional form is determined, and must be as we have chosen it.

The biconditional $p \Leftrightarrow q$ is intended to mean $(p \Rightarrow q) \wedge (q \Rightarrow p)$. Since the conjunction of two statements is true only when both statements are true, the truth set of $p \Leftrightarrow q$ must be the intersection of $P' \cup Q$ (the truth set of $p \Rightarrow q$) and $P \cup Q'$ (the truth set of $q \Rightarrow p$). This intersection is the set $(P \cap Q) \cup (P' \cap Q')$ shaded in Figure 2. We see that the biconditional is true either when p and q are both true or when they are both false.

FIGURE 2

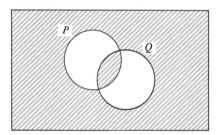

Example 4 In Example 3, the truth set of $p \Rightarrow q$ is the set containing all real numbers except the integer -4. The truth set of $p \Leftrightarrow q$ contains all real numbers except -4 and 2. ▶

PROBLEMS

   4. Work out the details of Example 4.

Definition 2 summarizes our discussion.

Definition 2 Let p and q be statements having truth sets P and Q, respectively. Then

(a) the conjunction $p \wedge q$ is a statement whose truth set is $P \cap Q$;

(b) the disjunction $p \vee q$ is a statement whose truth set is $P \cup Q$;

(c) the conditional $p \Rightarrow q$ is a statement whose truth set is $P' \cup Q$;

(d) the biconditional $p \Leftrightarrow q$ is a statement whose truth set is

$$(P \cap Q) \cup (P' \cap Q') \qquad \blacktriangleright$$

Definitions 1 and 2 may be used to determine truth sets of statements involving combinations of the basic connectives.

Example 5 To find the truth set of the statement

$$\sim[p \wedge (\sim p \vee q)] \wedge q \qquad (1)$$

we work from the innermost parentheses outward, as follows:

1. The truth set of $\sim p$ is P' [Definition 1]

2. The truth set of $\sim p \vee q$ is $P' \cup Q$ [Definition 2(b)]

3. The truth set of $p \wedge (\sim p \vee q)$ is   [Definition 2(a) and distributive law for sets]

$$P \cap (P' \cup Q) = P \cap Q$$

4. The truth set of $\sim[p \wedge (\sim p \vee q)]$ is   [Definition 1 and De Morgan's law for sets]

$$(P \cap Q)' = P' \cup Q'$$

5. The truth set of (1) is   [Definition 2(a) and distributive law for sets]

$$(P' \cup Q') \cap Q = P' \cap Q \qquad \blacktriangleright$$

PROBLEMS

    5. Find the truth sets of the following:

      (a) $p \wedge p$ (b) $\sim p \Rightarrow q$ (c) $p \wedge \sim p$

The next example shows how to determine truth sets of logical forms which combine three or more distinct statements.

Example 6 The truth set of

$$[(p \wedge q) \vee \sim p] \Rightarrow (r \Rightarrow q) \qquad (2)$$

is

$$P \cup Q \cup R' \qquad (3)$$

determined as follows:

1. The truth set of $(p \wedge q) \vee (\sim p)$ is   [Definitions 1, 2(a), 2(b), and dis-

$$(P \cap Q) \cup P' = P' \cup Q$$ tributive law for sets]

2. The truth set of $r \Rightarrow q$ is $R' \cup Q$ [Definition 2(c)]

3. The truth set of (2) is

$$[P' \cup Q]' \cup [R' \cup Q]$$ [Definition 2(c)]

This latter expression may be reduced to (3) by use of Theorem 1 of Chapter 1, or by studying the diagrams shown in Figure 3. We have shown that the logical form (2) is true if either p or q is true, or if r is false. That is, (2) has the same truth set as $p \vee q \vee \sim r$.   ▶

FIGURE 3   The truth set of (2) is, indeed, given by (3) since the total shaded regions of both Venn diagrams are equal.

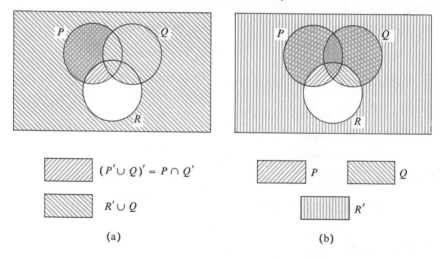

(a)   (b)

PROBLEMS

6. Translate the following compound statements into symbolic notation using letters to stand for the component statements.

(a) Either person a will cooperate with his opponent or he will not.

(b) If an employer attempts to hire that amount of labor such that the marginal revenue of the last man hired approximates his wage, then he will obtain maximum profits.

(c) Either the product is faulty and is accepted by the quality control department or the product is not faulty and is not accepted by the quality control department.

(d) The state legislature will determine the winner if no candidate receives a majority of the popular vote.

(e) If John takes the exam and scores over 70, he does not have to take the course for credit.

(f) John Smith is not an employee of the ABC Company or the XYZ Company.

(g) Theories either explain events or they are eventually discarded and replaced by those which do.

7. In the following example, determine the truth value of the compound statements (a)–(c) given the truth values of the following component statements:

(i) "Mr. Black has a more favorable attitude than Mr. Blue" is true. *P*

(ii) "Mr. Greene has a less favorable attitude than Mr. White" is false. *q*

(iii) "Mr. Blue has a neutral attitude" is true. √

(iv) "Mr. Brown has equally as favorable an attitude as Mr. Black" is false. *s*

(a) If Mr. Black has a more favorable attitude than Mr. Blue, then Mr. Greene does not have a less favorable attitude than Mr. White.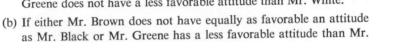

(b) If either Mr. Brown does not have equally as favorable an attitude as Mr. Black or Mr. Greene has a less favorable attitude than Mr. White, then Mr. Blue has a neutral attitude.

(c) If Mr. Brown does not have equally as favorable an attitude as Mr. Black, then either Mr. Greene has a less favorable attitude than Mr. White or Mr. Blue does not have a neutral attitude.

8. Suppose we assume that the statements:

p = consumption is a function of total real profits and total real wages

q = net investment is a function of profits and capital stock

r = private wage bill is a function of net national income and indirect taxes

are true, and that the statement

s = net national income plus indirect taxes is a function of investment

is false. Find the truth values of the following statements:

(a) $(p \Leftrightarrow q) \Rightarrow (s \Leftrightarrow r)$

(b) $(p \wedge q) \wedge r$

(c) $s \Rightarrow p$

(d) $p \Rightarrow [q \Leftrightarrow (r \Rightarrow s)]$

(e) $(p \wedge q) \Leftrightarrow (r \wedge \sim s)$

(f) $(\sim p \Rightarrow q) \Rightarrow (s \Rightarrow r)$

(g) $p \Rightarrow [q \Rightarrow (q \Rightarrow p)]$

9. Let p, q, and r be statements having associated truth sets P, Q, and R, respectively. Which of the following statements are tautologies?

(a) $(p \vee q) \Rightarrow (q \vee p)$

(b) $p \Rightarrow (p \vee q) \vee r$

(c) $p \Rightarrow [q \Rightarrow (q \Rightarrow p)]$

(d) $[(p \Rightarrow q) \Leftrightarrow q] \Rightarrow p$

(e) $(p \wedge q) \Rightarrow (p \vee r)$

(f) $[(p \wedge q) \Rightarrow \{(p \wedge \sim p) \Rightarrow (q \vee \sim q)\}] \wedge (q \Rightarrow q,$

(g) $(p \Rightarrow q) \Leftrightarrow (q \Rightarrow p)$

10. Let $\underline{\vee}$ denote the exclusive "or." That is,

$$p \underline{\vee} q = (p \vee q) \wedge \sim (p \wedge q)$$

Finish filling Table 1.

TABLE 1.　Truth table for $\underline{\vee}$

Truth value of p	Truth value of q	Truth value of $p \underline{\vee} q$
T	T	F
T	F	
F	T	
F	F	

11. Write truth tables for the five basic connectives. Table 2 is a partial answer to this problem.

TABLE 2.　Truth table for conjunction $\wedge$

Truth value of p	Truth value of q	Truth value of $p \wedge q$
T	T	T
T	F	F
F	T	F
F	F	F

12. Show that the statement $\sim p \wedge [q \wedge (\sim q \vee p)]$ is a self-contradiction.

2.2 EQUIVALENCE AND IMPLICATION

Of the five sentential connectives, two—the biconditional and the conditional—are of fundamental importance in mathematics. However, these connectives are ordinarily used in a somewhat restricted form. In mathematics, as in everyday discourse, when we state the biconditional $p \Leftrightarrow q$, we are claiming that it is true. We wish to rule out the possibility that one statement, p or q, is true and the other false. A biconditional statement $p \Leftrightarrow q$ which cannot be false, that is, whose truth set is the universal set U, is called an *equivalence* between p and q. A glance at Figure 2 shows that equivalence of a pair of statements may be deduced from *equality* of their respective truth sets.

R implies S if S is true whenever r is true i.e. if S is true in all the logically possible cases in which r is true.

Two relations are equivalent if they have the same truth tables.

Since equivalent statements have identical truth values, they may be substituted for one another in a logical form. The same considerations arise as in algebra where, for example, it is useful to know that $(x + y)^2$ may always be replaced by $x^2 + 2xy + y^2$ (or vice versa). Substitutions in logical forms, as in algebraic expressions, are made with an eye toward reducing the complexity of an expression or making it more amenable to the solution of the problem at hand. Table 3 lists some important equivalences which correspond to the set theoretic equalities detailed in Theorem 1 of Chapter 1.

TABLE 3

Logical terminology	Equivalent statements		Corresponding set theoretic result
	Statement 1	Statement 2	
Law of double negation	$\sim(\sim p)$	p	$(P')' = P$
Idempotent laws	$p \vee p$	p	Idempotent laws
	$p \wedge p$	p	for sets
Commutative laws	$p \vee q$	$q \vee p$	Commutative laws
	$p \wedge q$	$q \wedge p$	for sets
Associative laws	$(p \vee q) \vee r$	$p \vee (q \vee r)$	Associative laws
	$(p \wedge q) \wedge r$	$p \wedge (q \wedge r)$	for sets
Distributive laws	$p \vee (q \wedge r)$	$(p \vee q) \wedge (p \vee r)$	Distributive laws
	$p \wedge (q \vee r)$	$(p \wedge q) \vee (p \wedge r)$	for sets
De Morgan's laws	$\sim(p \vee q)$	$(\sim p) \wedge (\sim q)$	De Morgan's laws
	$\sim(p \wedge q)$	$(\sim p) \vee (\sim q)$	for sets
Law of the excluded middle	$p \vee \sim p$ is always true (is a tautology)		$P \cup P' = U$
Law of contradiction	$p \wedge \sim p$ is never true (is a self-contradiction)		$P \cap P' = \emptyset$

Example 1 The statement $(q \vee \sim q) \Rightarrow \sim p$ is equivalent to $\sim p$ since its truth set is

$$(Q \cup Q')' \cup P' = U' \cup P'$$
$$= \emptyset \cup P'$$
$$= P'$$

which is the truth set of $\sim p$.

Since the two statements $(q \vee \sim q) \Rightarrow \sim p$ and $\sim p$ have the same truth set P', *either* of these statements may be called the negation of statement p. This explains what may have seemed to be odd wording in Definitions 1 and 2, which allowed a statement to have many negations or a pair of statements to have more than one conjunction, disjunction, conditional, or biconditional.

▶

    1. Show that $P \cup Q$ is the truth set of $\sim p \Rightarrow q$ and hence that this statement is a disjunction of p and q.

Like the biconditional, the conditional also is usually stated only when it is true. If the conditional statement $p \Rightarrow q$ cannot be false, then it is called an *implication* and we say that p *implies q*.

    A look at Figure 1 shows that p implies q if and only if there are no elements x such that $x \in P$ and $x \notin Q$. That is to say, p implies q whenever P is a *subset* of Q. In this case, q is true whenever p is true.

Example 2   Does the conjunction of the statements "Interest rates do not increase or taxes increase" and "Interest rates increase" imply the statement "Taxes increase"?

    Let p be "Interest rates increase" and q be "Taxes increase." The problem is whether $(\sim p \vee q) \wedge p$ implies q. Using Definitions 1 and 2 we find that the truth set of $(\sim p \vee q) \wedge p$ is

$$(P' \cup Q) \cap P = (P' \cap P) \cup (Q \cap P)$$

$$= Q \cap P$$

Since $Q \cap P \subseteq Q$, it follows that q is indeed implied by $(\sim p \vee q) \wedge p$.

    Incidentally, $Q \cap P$ is also a subset of P, of $Q \cap P$, and of $Q \cup P$. Hence the statements p, $p \wedge q$, and $p \vee q$ are also implied by $(\sim p \vee q) \wedge p$. In fact, any statement is implied whose truth set contains $P \cap Q$.   ▶

    2. Does the conjunction of the statements "If interest rates increase, then taxes decrease" and "Taxes decrease" imply the statement "Interest rates increase"?

Implication differs from other sentential connectives in that it lacks symmetry. While $p \wedge q$ is equivalent to $q \wedge p$, $p \vee q$ is equivalent to $q \vee p$, and $p \Leftrightarrow q$ is equivalent to $q \Leftrightarrow p$, $p \Rightarrow q$ is not equivalent to $q \Rightarrow p$.

    For instance, if p "The Dodgers win the pennant" and q "The final standings show the Dodgers ahead of the Mets" are two statements about this year's National League race, then p implies q but q certainly does not imply p. There are many possible outcomes of the pennant race in which q is true and p false.

    The statement $q \Rightarrow p$ is called the *converse* of $p \Rightarrow q$. Many common fallacies in thinking arise from confusion of a statement with its converse. For example, if p is the statement "The enemy halts his aggressive acts" and q

the statement "U.S. forces will be withdrawn," it is obvious that $p \Rightarrow q$ is not equivalent to $q \Rightarrow p$.

The implication $p \Rightarrow q$ is true if the corresponding truth sets are related by $P \subseteq Q$. But if $P \subseteq Q$, then it must be that $Q' \subseteq P'$, and conversely. That is, $P \subseteq Q$ is equivalent to $Q' \subseteq P'$. Since Q' and P' are the respective truth sets of $\sim q$ and $\sim p$, it follows that $p \Rightarrow q$ is equivalent to $\sim q \Rightarrow \sim p$. The form $\sim q \Rightarrow \sim p$ is called the *contrapositive* of the implication $p \Rightarrow q$.

Example 3 A well-known perceptual law states that if an observer looks at a colored object for a sufficiently long period of time and immediately afterward looks at an illuminated neutral surface, he will see an after-image of the object in the complementary color.

If we let

  p = an observer looks at a colored object for a sufficiently long period of time

  q = the observer looks at an illuminated neutral surface immediately afterward

  r = the observer sees an after-image in the complementary color

the law can be expressed in symbolic notation as

$$p \wedge q \Rightarrow r$$

The contrapositive of the implication is $\sim r \Rightarrow \sim(p \wedge q)$ which can be rewritten as $\sim r \Rightarrow \sim p \vee \sim q$. This states that if an observer did not see an after-image of complementary color, then either he did not look at the object for a sufficiently long period of time or he did not look at an illuminated neutral surface immediately afterward. ▶

An implication $p \Rightarrow q$, then, is equivalent to its contrapositive $\sim q \Rightarrow \sim p$, but is not equivalent to its converse $q \Rightarrow p$. Of course, the converse $q \Rightarrow p$ *is* equivalent to *its* contrapositive $\sim p \Rightarrow \sim q$. Thus, associated with any implication are three other implications, the four forming two pairs of equivalent statements. These statements are summarized in Table 4.

The language used to express implications takes a variety of forms. For instance, the statement "Interest rates will rise only if there is excessive in-

TABLE 4. Original implication $p \Rightarrow q$ and associated implications

Original statement	$p \Rightarrow q$	
Converse	$q \Rightarrow p$	
Contrapositive	$\sim q \Rightarrow \sim p$	
Contrapositive of converse	$\sim p \Rightarrow \sim q$	

These statements are equivalent

These statements are not equivalent

flation" (*p* only if *q*) expresses the same thought as the implication "If interest rates rise, then there is excessive inflation" (if *p*, then *q*). If it is true that the only time interest rates rise is in the presence of rapid inflation, then an observed rise in interest rates allows the conclusion that such inflation must have occurred.

Another way commonly used to express the implication "if *p*, then *q*" is to say that *p* is a *sufficient condition* for *q*. That is, in order to conclude *q*, it is sufficient to know *p*. The statement "*p* is a *necessary condition* for *q*" is equivalent to "if *q*, then *p*" and to "*q* only if *p*." If *q* occurs, it necessarily follows that *p* also occurs. The assertion of a necessary condition is the converse of the assertion of a sufficient condition.

Equivalence between two statements *p* and *q* may be expressed either by "*p* *if and only if q*" or by "*p* is both *a necessary and a sufficient condition* for *q*."

A source of confusion between implication and equivalence arises from the way in which definitions are stated. For instance, the concept of power is one of central importance to the political scientist. A definition often used is "If *A* can influence the behavior of *B*, then *A* has power over *B*." In logical form this is a conditional statement $(p \Rightarrow q)$ claiming that influence is a sufficient condition for power. If influence is a sufficient but not a necessary condition for power, then it is possible that *A* has power over *B* but no influence. In this case there are conditions of power other other than those specified and the definition is deficient in that it fails to specify the necessary conditions.

Properly stated, a definition specifies an *equivalence* between two statements. However, it is almost universal in science and mathematics to state definitions in conditional terms $(p \Rightarrow q)$, as above, it being *understood* that equivalence ($q \Rightarrow p$ as well as $p \Rightarrow q$) is intended. Thus, in the illustration above the possession of power is intended to be equivalent to the ability to influence behavior.

PROBLEMS

3. Let *p*, *q*, and *r* be statements with truth sets *P*, *Q*, and *R*, respectively. Show that the following are pairs of equivalent statements.

(a) $\sim(\sim p \wedge \sim q)$ and $p \vee q$

(b) $(p \wedge q) \vee (p \wedge q')$ and p

(c) $\sim[\sim p \wedge (p \vee q)]$ and $p \vee \sim q$

(d) $\sim[\sim p \wedge (q \wedge r)]$ and $p \vee \sim q \vee \sim r$

(e) $\sim[p \wedge (p \vee q)]$ and $\sim p$

(f) $q \wedge \sim[p \wedge (\sim p \vee q)]$ and $\sim p \wedge q$

4. Which of the following statements are equivalent to the statement "Either there are government subsidies of agriculture or there is an overproduction."

(a) It is not the case that there are no government subsidies and no overproduction.

(b) Either there are government subsidies and overproduction or there is overproduction.

(c) There are government subsidies or there is overproduction and no government subsidies.

(d) There are neither government subsidies nor overproduction.

(e) There are government subsidies and no overproduction.

(f) If there are no government subsidies, then there is overproduction.

5. Which of the following conditional statements are implications?

(a) $\sim(\sim p) \Rightarrow p$

(b) $p \wedge q \Rightarrow q$

(c) $(p \Rightarrow \sim p) \Rightarrow p$

(d) $(p \wedge q) \Rightarrow (p \vee r)$

(e) $(p \wedge q) \Rightarrow (p \Leftrightarrow q \vee r)$ yes with any $(p \Leftrightarrow q) \vee r$ or $p \Leftrightarrow (q \vee r)$

(f) $p \wedge q \Rightarrow p \vee q$

(g) $(q \vee \sim q) \Rightarrow p$

(h) $[\sim q \wedge (p \Rightarrow q)] \Rightarrow \sim p$

6. If $\underline{\vee}$ is as defined in Problem 10, Section 2.1, show that

(a) $p \underline{\vee} p$ is a self-contradiction

(b) $p \underline{\vee} q$ is equivalent to p when q is false

(c) $p \underline{\vee} q$ is equivalent to $\sim p$ when q is true

(d) $p \underline{\vee} q$ is equivalent to $q \underline{\vee} p$

(e) $p \underline{\vee} (q \underline{\vee} r)$ and $(p \underline{\vee} q) \underline{\vee} r$ are equivalent

(f) $p \wedge (q \underline{\vee} r)$ and $(p \wedge q) \underline{\vee} (p \wedge r)$ are equivalent

7. Find truth values for p, q, and r such that

$$p \underline{\vee} (q \wedge r) \text{ and } (p \underline{\vee} q) \wedge (p \underline{\vee} r)$$

have different truth values. What does this say concerning the equivalence of these two statements?

8. How is each of the following statements related to the statement "If increased housing expenditures do not accompany demolition, then slums will not disappear"?

(a) Increased housing expenditures accompany demolition only if slums disappear.

(b) A necessary condition for increased housing expenditures to accompany demolition is the disappearance of slums.

(c) Slums disappear only if increased housing expenditures accompany demolition.

(d) A sufficient condition for the continuation of slums is that increased housing expenditures do not accompany demolition.

(e) If increased housing expenditures accompany demolition, then slums will disappear.

(f) If slums disappear, then increased housing expenditures accompany demolition.

2.3 VALID ARGUMENTS

Let us suppose that you wake up on the morning of your big tennis match and find that it is raining. Recalling that you and your opponent have agreed that in case of rain the match will be cancelled, you conclude that there will be no match and so roll over and go back to sleep. In this hypothetical situation you have arrived at your conclusion by using a simple logical argument called the *Law of Detachment*. If p is the statement "It is raining" and q the statement "There will be no tennis match," then you have reasoned that the conjunction of the premises p and $p \Rightarrow q$ allows you to conclude q. This line of reasoning is tabulated in Table 5.

TABLE 5.   The Law of Detachment

	Statement	Truth set
First premise	p	P
Second premise	$p \Rightarrow q$	$P' \cup Q$
Conclusion	q	Q

An *argument* consists of two major components, a collection $p_1, p_2, \ldots, p_n$ of statements, called *premises*, and a statement q, the *conclusion*. The argument is said to be *valid* if the conjunction of the premises implies the conclusion. Since a conditional statement $p \Rightarrow q$ is an implication if and only if the truth set of p is a subset of the truth set of q, it follows that an argument is valid if and only if the intersection $P_1 \cap P_2 \cap \cdots \cap P_n$ of the truth sets of the respective premises is a subset of the truth set Q of the conclusion.

In our example, the intersection of the truth sets P and $P' \cup Q$ of the premises is $P \cap (P' \cup Q) = P \cap Q$. Since this latter set is a subset of Q it follows that the conjunction of the premises does indeed imply the conclusion. Hence the argument is valid.

Note that the validity of an argument depends only upon its logical form and not upon the truth or falsity of the particular statements involved. The fact that a valid argument is an implication means that the conclusion must be true whenever the premises are all true. However, if the premises are not all true, then the truth of the conclusion is neither necessary nor sufficient for the validity of the argument.

Example 1

If a person is a leader, then he has power to influence others. $p \Rightarrow q$

If a person has power to influence others, then he is respected. $q \Rightarrow r$

Therefore, if a person is a leader, then he is respected by others. $p \Rightarrow r$

Obviously, the conclusion is false. History is replete with examples of despotic leaders who were feared but not respected. However, the argument is valid since the intersection of the truth sets $P' \cup Q$ and $Q' \cup R$ of the premises is a subset of the truth set $P' \cup R$ of the conclusion.

The seeming paradox of a valid argument resulting in a false conclusion is resolved if we observe that the second premise in the argument is obviously false. It is not surprising that a false premise should lead to a false conclusion, even though the conclusion is correctly derived from the premises.

▶

PROBLEMS

1. Draw a Venn diagram to verify the validity of the argument in Example 1.

Example 2

If Lincoln was born in Quebec, then Lincoln was born in Canada.	$p \Rightarrow q$
Lincoln was not born in Quebec.	$\sim p$
Therefore, Lincoln was not born in Canada.	$\sim q$

The conclusion, of course, is true. But the argument is not valid, the truth set of the conjunction of the premises being $(P' \cup Q) \cap P' = P'$, which is not a subset of Q', the truth set of the conclusion. ▶

An argument which is invalid is called a ~~fallacy~~. Example 2 illustrates one of the most common fallacies in which the "if" part, or *antecedent*, of a conditional statement (p in the example above) is denied in an attempt to deny the "then" part, or *consequent* (q above).

A form of *valid* argument closely related to the Law of Detachment is Modus Tollendo Tollens, the method of denying (tollendo) the consequent of a conditional statement in order to deny (tollens) the antecedent. This type of argument is displayed in Table 6. Replacing the premise $p \Rightarrow q$ by its equivalent contrapositive form $\sim q \Rightarrow \sim p$ shows that this is merely the Law of Detachment in disguised form. Alternatively, we may note that the conjunction $Q' \cap (P' \cup Q) = Q' \cap P'$ of the truth sets of the premises is a subset of P'. Hence this is a valid form of argument.

TABLE 6.   Modus Tollendo Tollens

	Statement	Truth set
First premise	$\sim q$	Q'
Second premise	$p \Rightarrow q$	$P' \cup Q$
Conclusion	$\sim p$	P'

Other common forms of valid arguments are listed in Table 7 and their use is illustrated in the following examples. For simplicity the conjunction

of the premises is written on the same line with the conclusion and the name
of the form.

TABLE 7. Useful forms of valid argument

	Premises	Conclusion
Law of Detachment	$p \land (p \Rightarrow q)$	q
Modus Tollendo Tollens	$\sim q \land (p \Rightarrow q)$	$\sim p$
Modus Tollendo Ponens	$\sim p \land (p \lor q)$	q
Law of Simplification	$p \land q$	p
Law of Addition	p	$p \lor q$
→ Law of Exportation	$(p \land q) \Rightarrow r$	$p \Rightarrow (q \Rightarrow r)$
→ Law of Importation	$p \Rightarrow (q \Rightarrow r)$	$(p \land q) \Rightarrow r$
Law of Hypothetical Syllogism	$(p \Rightarrow q) \land (q \Rightarrow r)$	$p \Rightarrow r$
Law of Absurdity	$p \Rightarrow (q \land \sim q)$	$\sim p$

PROBLEMS

    2. Prove the validity of the Law of Exportation and give an example of its
use.

    In Section 2.2 it was shown that the conditional $p \Rightarrow q$ may always be
replaced by the equivalent contrapositive form $\sim q \Rightarrow \sim p$. Applied to arguments this means that showing at least one of the premises must be false
when the conclusion is false is equivalent to making the original argument.
The next examples illustrate this method of *indirect argument*.

Example 3   Suppose we wish to prove the theorem: If n is a positive integer and n^2 is odd, then n is also odd. The equivalent contrapositive form
is: If n is not odd (is even), then n^2 is not odd (is even). Now if n is even,
then $n = 2k$ where k is some integer. Hence $n^2 = (2k)^2 = 4k^2 = 2(2k^2)$
is also even, being twice the integer $2k^2$. This establishes the theorem. ▶

PROBLEMS

    3. Try to find a direct proof of the theorem of Example 3.

    Usually an indirect proof is attempted when it is easier than a direct proof
(as in Example 3) or when it is difficult to know just where to begin an
argument from the premises given. There are a variety of forms in which
one may cast an indirect proof. A little checking will show that the following
conditional forms are all equivalent and hence provide alternative forms for
argument:

    (1) $p \Rightarrow q$

    (2) $\sim q \Rightarrow \sim p$

(3) $(p \wedge \sim q) \Rightarrow \sim p$

(4) $(p \wedge \sim q) \Rightarrow q$

(5) $(p \wedge \sim q) \Rightarrow (r \wedge \sim r)$

In cases (3)–(5) we see that in venturing an indirect proof we are permitted to introduce the negation of the conclusion as another premise. This adds to the available information and is especially convenient in cases where it is difficult to see how to begin a direct proof. There are then a variety of conclusions, any one of which establishes the proof.

Case (5) is a particularly useful form closely related to the Law of Absurdity. If from the conjunction $p \wedge \sim q$, we are able to imply the false statement $r \wedge \sim r$, then $p \wedge \sim q$ must itself be false. Hence if p is true, $\sim q$ must be false. That is, if p is true, then so is q, which means $p \Rightarrow q$.

Example 4 If the President appointed a candidate (p), then the Senate confirmed the appointment (c). Either the candidate was acceptable to the Senate (a), or the Senate did not confirm the appointment. It is not the case that the President appointed a candidate and the candidate was confirmed by the Senate. Therefore, the President did not appoint a candidate.

Since the argument involves an implication and two disjunctions, it is difficult to begin a direct proof. However, by introducing the negation of the conclusion as another premise we arrive at the following indirect proof:

Premise(s) Used	Statement	Derived from
1	(1) $p \Rightarrow c$	[Premise 1]
2	(2) $a \vee \sim c$	[Premise 2]
3	(3) $\sim(p \wedge c)$	[Premise 3]
4	(4) p (negation of conclusion)	[Premise 4]
1, 4	(5) c	[Statements (1) and (4)]
1, 2, 4	(6) a	[Statements (2) and (5)]
3	(7) $\sim p \vee \sim c$	[Statement (3); De Morgan's Laws]
1, 2, 3, 4	(8) $\sim p$	[Statements (5) and (7)]
1, 2, 3, 4	(9) $p \wedge \sim p$	[Statements (4) and (8)]
1, 2, 3	(10) $\sim p$	[Statements (4) and (9); using case (5) above]

Since introducing the negation of the conclusion results in a contradiction, we conclude that the conclusion is indeed implied by the stated premises. ▶

PROBLEMS

4. Draw a Venn diagram to show that in Example 4 the truth set of the conjunction of premises (1)–(3) is a subset of P', the truth set of the conclusion $\sim p$.

Example 5 Davis* has listed 56 formal propositions drawn from the writings of sociologists and social psychologists. By slightly modifying some of the propositions for the purpose of simplification, we can advance the following general argument.

Person P_i and Person P_j are friends (P_iFP_j) if and only if P_i likes P_j (P_iLP_j) and P_j likes P_i. If P_i and P_j are friends, then they are similar in attitude (P_iSP_j). If P_i likes P_j and P_j dislikes P_k, then P_i will dislike P_k. If P_i likes P_j and P_j likes P_k, then P_i will like P_k. If P_i dislikes P_j and P_j likes P_k, then P_i will dislike P_k. Suppose it is the case that P_1 and P_3 are friends and P_2 likes P_3. Therefore, P_1 and P_2 are similar in attitude.

The validity of the above argument is verified in the following analysis. It is assumed that P_i either likes or dislikes P_j, but not both.

Premise(s) Used	Statement	Derived from
1	(1) $P_iFP_j \Leftrightarrow P_iLP_j \wedge P_jLP_i$	[Premise 1]
2	(2) $P_iFP_j \Rightarrow P_iSP_j$	[Premise 2]
3	(3) $P_iLP_j \wedge \sim(P_jLP_k) \Rightarrow \sim(P_iLP_k)$	[Premise 3]
4	(4) $P_iLP_j \wedge P_jLP_k \Rightarrow P_iLP_k$	[Premise 4]
5	(5) $\sim(P_iLP_j) \wedge P_jLP_k \Rightarrow \sim(P_iLP_k)$	[Premise 5]
6	(6) P_2LP_3	[Premise 6]
7	(7) P_1FP_3	[Premise 7]
1, 2	(8) $P_iLP_j \wedge P_jLP_i \Rightarrow P_iSP_j$	[Statements (1) and (2)]
3	(9) $P_1LP_3 \Rightarrow P_1LP_2 \vee \sim(P_2LP_3)$	[Statement (3)]
3, 6	(10) $P_1LP_3 \Rightarrow P_1LP_2$	[Statements (6) and (9)]
3, 6, 7	(11) P_1LP_2	[Statements (7) and (10)]
1	(12) $P_1FP_3 \Rightarrow P_3LP_1$	[Statement (1)]
1, 7	(13) P_3LP_1	[Statements (7) and (12)]
4	(14) $P_2LP_3 \wedge P_3LP_1 \Rightarrow P_2LP_1$	[Statement (4)]
1, 4, 6, 7	(15) P_2LP_1	[Statements (6), (13), and (14)]
1, 2, 3, 4, 6, 7	(16) P_1SP_2	[Statements (8), (11), and (15)]

In contrast to Example 4 where there did not seem to be enough premises for a direct proof and where the negation of the conclusion was offered as another premise, the number of premises in this example may be reduced by one since the conclusion may be obtained without use of premise (5). ▶

PROBLEMS

In Problems 5–10 determine whether the arguments are valid.

5. If work methods are uneconomical, then they are not socially desirable. If work methods are boring, they are harmful to initiative. If work meth-

*Davis, J. A., "Structural Balance, Mechanical Solidarity, and Interpersonal Relations," *American Journal of Sociology* **68**, 444–462 (1963).

ods are harmful to initiative, they are uneconomical. Merely mechanical labor is boring. Therefore, merely mechanical labor is not socially desirable.

6. Either a social institution satisfies some need, or it will not survive. A social institution satisfies some need. Therefore, it will survive.

7. If an animal is not satiated, the animal will enter the goal box on the subsequent trial. If the animal enters the goal box on the subsequent trial, then it is not the case that the strength of the goal-seeking response was reduced on the previous trial. The strength of the animal's goal-seeking response on the previous trial is either reduced or increased. It is known that the strength of the response was not increased on the previous trial. Therefore, the animal is satiated.

8. If the fossil remains are human, then if the dating process is accurate these remains are the oldest human remains yet discovered. Either the fossil remains are human or the finding will not receive world wide acclaim. It is a fact that the dating process is accurate. Therefore, if the finding is given world wide acclaim, the remains are the oldest human remains yet discovered.

9. If the Bilge Company is able to purchase raw material at a favorable price or if the sales of the company increase, the company will not suffer a loss. If there is a shortage of material, the Bilge Company will not be able to purchase raw material at a favorable price. The fact is that there is no shortage of material. Therefore, the Bilge Company will not suffer a loss.

10. If the candidate receives the support of the liberals and the support of labor, then he will win. Either the candidate will not win or he will introduce a bill favorable to labor. The candidate will not introduce a bill favorable to labor. Furthermore, the candidate will receive the support of the liberals. Therefore, the candidate will not receive the support of labor.
[*Hint:* Try an indirect proof.]

Example 6 Classical, or Aristotelian, logic is concerned with relations among classes (sets) of objects. A typical example is

> No Fascist is a Communist
>
> All Chinese are Communists
>
> Therefore, no Chinese are Fascists

By introducing the sets F, C_o, and C_h of Fascists, Communists, and Chinese, respectively, these statements may be translated into set notation as

$$F \subseteq C_o'$$
$$C_h \subseteq C_o$$

Therefore,

$$C_h \subseteq F'$$

Recalling that $A \subseteq B$ and $B' \subseteq A'$ are equivalent statements about sets, we see that $F \subseteq C_o'$ and $C_h \subseteq C_o$ are equivalent to $C_h \subseteq C_o$ and $C_o \subseteq F'$. Together these imply the conclusion $C_h \subseteq F'$ and the argument is valid. ▶

PROBLEMS

Use the method of Example 6 to determine whether the arguments in Problems 11–14 are valid.

11. In some forms of government, the individual is subservient to the state. All governmental systems which make the individual subservient to the state are nondemocratic. Therefore, some forms of governments are nondemocratic.

12. Television programs are designed for maximum audience appeal. Nothing designed for maximum appeal can be of high quality. Therefore, no television program is of high quality.

13. Some Democrats are liberals. No liberals are conservative. Hence, no Democrats are conservative.

14. All subjects are male or overachievers or anxious. All males are students or anxious. All anxious males are students and overachievers. Some overachieving males are students. Therefore, some subjects are students.

15. A young man was imprisoned in a cell having two doors, one of which led to freedom while the other led to certain death. The jailor, being a sporting man, agreed to give the prisoner his choice of doors. The cell was guarded by two guards, one of whom, the prisoner was told, could tell no lies, and the other could tell no truth. The prisoner was permitted to ask one question which each guard would answer with a simple "yes" or "no." Our prisoner, having been a student at Professor Foole's logic school, asked a question and then walked out, a free man. How did he know which door to choose?

16. A personnel manager is seeking an individual who is adept in problem solving. From a number of applicants, he selects three promising candidates and uses the following test to eliminate all but one. The three individuals are blindfolded and seated in a triangle. They are instructed that a black or white mark is to be placed on each forehead. When the blindfold is removed, they are told to look at each of the others. If they see a black mark, they are to raise their hand until they can reach a decision as to the color of the mark on their forehead. Then they are to stand and give the reason for the decision. Unknown to the applicants, the personnel manager places a black mark on each forehead. After a period of time, one applicant rises, and indicates that the mark on his forehead is black. How did he arrive at that decision?

17. A set of premises is said to be an *inconsistent* set if it is impossible that
all be true at the same time; that is, if the truth set of their conjunction
is empty.

(a) Which of the following sets of premises are inconsistent?

(i) $\sim(p \lor \sim q)$ (ii) $p \Rightarrow q$ (iii) $p \Rightarrow q$

    $p \lor \sim r$ $p \Leftrightarrow r$ $q \Rightarrow r$

    $q \Rightarrow r$ $(r \lor s) \Leftrightarrow \sim q$ $s \Rightarrow \sim r$

                                                     $p \land s$

(b) Prove that from an inconsistent set of premises one may validly conclude any statement whatsoever.

18. Prove the validity of the arguments listed in Table 7.

SUPPLEMENTARY READING

Kemeny, J. G., Snell, J. L., and Thompson, G. L., *Introduction to Finite Mathematics* (Prentice-Hall, Inc., Englewood Cliffs, N. J., 1966), 2nd ed., Chapter 1.

Luchins, A. S., and Luchins, E. H., *Logical Foundations of Mathematics for Behavioral Scientists* (Holt, Rinehart and Winston, Inc., New York, 1965).

Langer, S. K., *An Introduction to Symbolic Logic* (Dover Publications, Inc., New
York, 1953), 2nd ed.

Suppes, P., *Introduction to Logic* (D. Van Nostrand Company, Inc., Princeton,
N. J., 1957), Chapters 1 and 2.

An ordered pair consists of two element, say a and b, in which one of them, say a, is designated as the first element and the other as the second element.

$$(a,b) = (c,d) \longleftrightarrow a = c \quad \text{and} \quad b = d$$

RELATIONS 3

A binary relation or, simply, relation R from a set A to a set B assigns to each pair (a,b) in A×B exactly one of the statements (i) "a is related to b" written a R b (ii) "a is not related to b" " a R̸ b

3.1 BINARY RELATIONS

In ordinary discourse we think of relations as holding between two things, or among several things. We speak of John Adams as being in the relation of father to John Quincy Adams. We express the relative standings of the National League baseball teams by indicating the number of games each team is behind the leader.

Relations which hold between pairs of objects are called *binary* relations, those involving three objects are *ternary* relations and, in general, an *"n-ary"* relation connects n objects. As well as being the simplest relations, binary relations are the type most used in the behavioral sciences. We shall therefore devote most of our discussion to these relations.

From your everyday experiences you are familiar with various kinds of relations connecting two people. Examples are family relations such as mother, father, uncle, sister, and cousin, and emotional relations such as loving and hating. Relations involving pairs of numbers have appeared in your study of mathematics. To say that a real number x has the relation *greater than* to a real number y is to say that the pair (x, y) lies in the shaded region of Figure 1. Similarly, x is *equal* to y if and only if the pair (x, y) lies on the line passing through the origin and inclined 45° from the horizontal.

A word about pairs is perhaps in order here. A pair of objects in which we distinguish one of the objects as the first and the other (which need not be different) as the second is called an *ordered pair*. The most familiar use of ordered pairs is the representation, as in Figure 2, of points on a plane by a pair of numbers, the first of which denotes the horizontal distance and the second the vertical distance from a fixed point O, called the origin. In this interpretation it is apparent that the pairs (1, 5) and (5, 1) represent different points and should be considered to be different pairs of numbers.

FIGURE 1

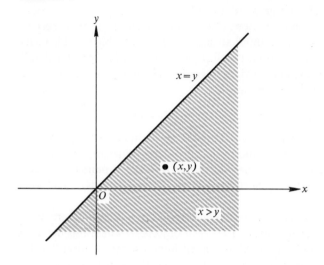

FIGURE 2

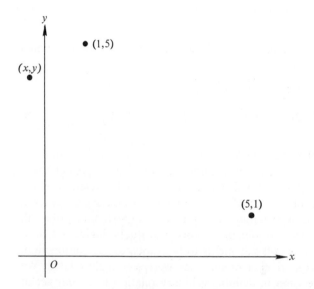

In general, ordered pairs are considered different unless they contain the same entries, in the same order. That is, the pairs (x, y) and (r, s) are equal if and only if $x = r$ and $y = s$. This property distinguishes ordered pairs from sets having two elements. For, while the sets $\{x, y\}$ and $\{y, x\}$ are identical, since they have the same members, the ordered pairs (x, y) and (y, x) are not equal unless $x = y$.

The objects in an ordered pair need not be numbers. For instance, the list of grades for a class is a set of ordered pairs in which the first element is a student's name, and the second element is his grade. Obviously, we want to distinguish the pair (Jones, A), Jones received an A in the course, from the pair (A, Jones), A received a grade of Jones.

From a mathematical point of view, it is useful to view relations as special kinds of sets, namely sets containing all ordered pairs of objects which satisfy the description of the relation. For instance, the relation "greater than" is the set G of all pairs (x, y) of real numbers for which $x > y$. Then $x > y$ if and only if $(x, y) \in G$. In set notation

$$G = \{(x, y): x \text{ and } y \text{ are real numbers and } x > y\}$$

Any other binary relation may be similarly represented. The relation of "dominating" may be viewed as the set of all ordered pairs (a, b) such that a dominates b; the relation of "father" as the set of ordered pairs (f, c) such that f is the father of c; etc. In short, for our purposes, *a binary relation is a set of ordered pairs.*

In order to give precise meaning to the description of a relation there must be a previously agreed upon universal set U of objects to which this description refers. If, for example, U is the set of all humans, the relation "is the brother of" contains the pair (Cesare, Lucretia), but this pair is not included if U is the set of all males. In general, a binary relation defined relative to a set U is a set of ordered pairs of elements of U.

Example 1 If U is the set $\{0, 1, 2\}$, then the relation "is less than or equal to" defined on U is the set

$$L = \{(x, y): x \in U, y \in U, \text{ and } x \leq y\}$$

$$= \{(0, 0), (0, 1), (0, 2), (1, 1), (1, 2), (2, 2)\} \qquad \blacktriangleright$$

When we speak of relations in ordinary discourse we usually have in mind some sort of natural connection (for example, father, cousin, greater than) between the objects being related which serves to identify the relation. However, this intuitive idea of connectedness is difficult to describe in general, although it may have a clear meaning in a particular context. You will recall that a similar situation arose in our discussion of symbolic logic. Our decision there was to allow any two statements to be combined, by conjunction, implication, and so forth, whether or not they were seemingly related. We have made the same decision in defining a binary relation to be *any* set of ordered pairs, regardless of how unconnected the members of the pairs may seem intuitively. Thus, for example, the set

$\{(\textit{War and Peace}, 6\frac{3}{4}), (\emptyset, \text{John Adams}), (\text{Sandy Koufax, Suez Canal})\}$

is a relation although any intuitive significance it may have is not readily apparent.

Relations involving more than two objects may also be represented by sets. Thus, *a ternary relation is a set of ordered triples* (x, y, z), and, in general, *any set of ordered n-tuples is an n-ary relation.*

Example 2   The political scientist interested in voting behavior relative to a specific elective office might define the relation "votes for" as a set of all ordered pairs (x, y) such that y is the candidate chosen by voter x.

If he were interested in a slate of four offices, the quinary relation "votes for candidates" consisting of the set of ordered 5-tuples (x, y_1, y_2, y_3, y_4), such that y_1, y_2, y_3, and y_4 are the respective candidates chosen by x for the four offices, would be of greatest value to his work.   ▶

PROBLEMS

1. Which of the following are relations?
   - (a) $\{(x, y): x \text{ and } y \text{ are real numbers}\}$
   - (b) (Red card, 10 seconds)
   - (c) $\{(x, y): x \text{ and } y \text{ are real numbers and } y = 5\}$
   - (d) $\{(J. \text{ Oates}, 27 \text{ years})\}$
   - (e) $\{10 \text{ years seniority, } 120 \text{ I.Q., machinist}\}$
   - (f) $\{(5, \text{IBM, Standard Oil}), (120, \text{Du Pont, GMC})\}$

2. Each of the following is intended to describe a binary relation connecting elements of the set $U = \{1, 2, 3, 4\}$. Write each relation by listing all its elements.
   - (a) "is equal to" $\{(1,1), (2,2), (3,3), (4,4)\}$
   - (b) "is less than"
   - (c) "is not equal to" $A = \{1, 2, 3, 4\}$
   - (d) "is a multiple of"
   - (e) "differs by at least two from" $B = \{1, 2, 3, 4\}$
   - (f) "is equal to or greater than"

3. Let the universal set U be the collection of all subsets of the set $P = \{a, b\}$. List the elements of each of the following relations.
   - (a) "is a subset of" $\{\phi, \{a\}, \{b\}, \{a, b\}\} = U$
   - (b) "is equal to"

4. Let R be the relation "is one less than" defined on the set $U = \{1, 2, 3, 5\}$. List the elements of R.

Whenever we have two sets we can form ordered pairs by choosing the first object in a pair from the first set and the second object from the second set. The collection of all pairs which can be formed in this way is called the *Cartesian product** of the two sets. The Cartesian product of sets A and B (in that order) is denoted $A \times B$.

*After René Descartes, a seventeenth-century French philosopher and mathematician.

Cartesian Product is a set y all ordered pairs such that

$A \times B = \{(a, b): a \in A, b \in B\}$

Example 3 If $A = \{a, b\}$ and $B = \{1, 2, 3\}$, then

$A \times B = \{(a, 1), (a, 2), (a, 3), (b, 1), (b, 2), (b, 3)\}$

$B \times A = \{(1, a), (1, b), (2, a), (2, b), (3, a), (3, b)\}$

$A \times A = \{(a, a), (a, b), (b, a), (b, b)\}$

$B \times B = \{(1, 1), (1, 2), (1, 3), (2, 1), (2, 2), (2, 3), (3, 1), (3, 2), (3, 3)\}$

Note that $A \times B$ is not equal to $B \times A$. ▶

PROBLEMS

5. Let $A = \{a, b\}$, $B = \{$subject A, subject $B\}$, and $C = \{$yes$\}$. Write the following sets.

   (a) $A \times A$ (b) $C \times C$ (c) $A \times B$ (d) $B \times A$

   (e) $A \times (B \cup C)$ (f) $(A \times B) \cup (A \times C)$ (g) $A \times B \times C$

6. Argue that a binary relation defined relative to a set U is a subset of the Cartesian product $U \times U$.

7. Let U denote the universal set.

   (a) Show that $X = U \times U$ is a relation, called the *universal relation*.

   (b) Show that $\emptyset$ can be regarded as a relation, called the *empty relation*.

8. Given $A = \{1, 2\}$ and $B = \{2, 4\}$, write the following relations in (subsets of) $A \times B$. $A \times B = \{(1, 2)\ (1, 4)\ , (2, 2)\ , (2, 4)\}$

   (a) "is less than"

   (b) "is not equal to" $\cup = \{1, 2, 4\}$

   (c) "is at least as great as"

   (d) "is half as large as"

9. In Problem 8, how many distinct relations can be defined in $A \times B$?

10. Prove the distributive laws

    (a) $A \times (B \cup C) = (A \times B) \cup (A \times C)$

    (b) $A \times (B \cap C) = (A \times B) \cap (A \times C)$

11. Let A and B be subsets of some universal set U. Prove that $(A \times U) \cap (U \times B) = A \times B$.

12. The concept of ordered pair may be defined in terms of sets by writing $(a, b) = \{\{a\}, \{a, b\}\}$. Prove that with this definition $(a, b) = (r, s)$ if and only if $a = r$ and $b = s$.

3.2 EQUIVALENCE RELATIONS

It is convenient to introduce letters to designate relations. For instance, we indicate that c stands in the relation of *friend* to b by writing $c\,F\,b$. Similarly, $x\,G\,y$ might be used to indicate that x is greater than y.

A relation R is said to be *reflexive* in (relative to) a set U if it is true for every $x \in U$ that xRx; that is, that the pair (x, x) is an element of R, or that each element $x \in U$ bears the relation R to itself. The relation "is at least as tall as" is reflexive in the set of all people, every person being at least as tall as himself. The relation

$$G = \{(1, 1), (\text{liberty, liberty}), (1, 2), (3, 3)\}$$

is not reflexive in the set $W = \{1, 2, 3\}$ since $(2, 2)$ is not a member of G, but is reflexive in the set $V = \{\text{liberty}, 3\}$.

A relation R is *symmetric* in a set U if for all x and y in U, it is true that yRx whenever xRy. If U consists of a set of five people lined up for a picture, then the relation "stands next to" is symmetric since if A stands next to B, then B also stands next to A. The relation "is the brother of" is not symmetric in the set of all humans, but is symmetric in the set of all males. Sadly, the relation "loves" is not symmetric in the set of all people. The relation G, above, is not symmetric in W since $(2, 1)$ is not a member of G. However, G is symmetric in V since neither (liberty, 3) nor (3, liberty) belongs to G.

A relation R is *transitive* in a set U if for all x, y, and z in U, it is true that whenever xRy and yRz, then xRz. The relation $\leq$ is transitive in the set of real numbers since $x \leq y$ and $y \leq z$ together imply $x \leq z$. The relation "stands next to" is not transitive in a line of more than two people while the relation "is a sibling (brother or sister) of" is transitive in the set of all people. The relation G above is transitive in both W and V.

PROBLEMS

1. Classify the relations in Problem 2, Section 3.1 as to whether they are reflexive, symmetric, and transitive.

A relation which is reflexive, symmetric, and transitive is called an *equivalence relation*. The most familiar example is the relation "equality" between real numbers. Other equivalence relations are "is the same height as" and "has the same color hair as." The relation "is a sibling of" is not an equivalence relation since it is not reflexive. The relation $\leq$ fails because it is not symmetric.

Example 1 Bavelas* in his work on small-group communication introduces a relation "touching" about which the following assumptions are made.

(a) The group consists of a collection of subgroups called *cells*. We shall take the universal set U to be the set of all cells.

(b) A given cell may or may not be touching another cell.

(c) If cell C_1 is touching C_2, then C_2 is touching C_1.

(d) No cell touches itself.

*Bavelas, A., "A Mathematical Model for Group Structures," *Applied Anthropology* **7**, 16–30 (1948).

Property (c) states that "touching" (T) is symmetric while (d) implies that it is not reflexive. Also, it is apparent that T is not transitive since it is not required that we must have $C_1 T C_3$ whenever $C_1 T C_2$ and $C_2 T C_3$. "Touching" is not an equivalence relation.

Suppose, however, that we define another relation "linked" by:

Cells A and B are linked if either $A = B$ or there exist cells $C_1, C_2, \ldots, C_n$ such that A touches C_1, C_1 touches C_2, $\ldots$, C_{n-1} touches C_n, and C_n touches B.

The symmetry of "touches" guarantees that if A is linked to B, then B is linked to A. Further, if A is linked to B and B to C, then A must be linked to C. The relation "linked" is therefore an equivalence relation, being reflexive, symmetric, and transitive. ▶

PROBLEMS

2. Give details of the proofs of the statements in the last paragraph of Example 1.

Suppose that E is an equivalence relation defined on a set U and let x be an element of U. The set of all those elements of U which have the relation E with x is called the *equivalence class* of x. That is, the equivalence class of x is the set

$$E_x = \{y \in U : yEx\}$$

Since E is reflexive we must have xEx, and thus each element of U is a member of its own equivalence class. Moreover, the symmetry of E guarantees that if yEx, then xEy. Hence if y is a member of the equivalence class E_x, then x is in turn a member of E_y. In fact, if yEx, the equivalence classes E_x and E_y must be identical. Let us suppose z is an element of E_y. Then zEy, and this together with yEx and the transitivity of E gives zEx or $z \in E_x$. It follows that E_y is a subset of E_x. A similar argument shows $E_x \subseteq E_y$ and hence that $E_y = E_x$. We conclude that if xEy, then E_x and E_y are the *same* equivalence class. Actually, the equality of E_y and E_x follows if they have *any* element in common. For suppose $z \in E_x \cap E_y$ (see Figure 3). Then zEx and zEy. By symmetry zEx becomes xEz which together with zEy gives xEy. This, as we have seen, implies $E_y = E_x$.

FIGURE 3

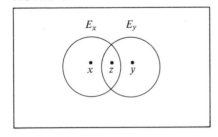

$E_x \qquad E_y$

Equivalence classes, then, either have no elements in common or are identical. Every element of U lies in at least one equivalence class (namely, its own) but no element is a member of two different classes. The distinct equivalence classes determined by E thus partition U in the sense of the following definition.

Definition 1 A collection of sets A_i constitutes a *partition* of a set U if

(1) $\bigcup_i A_i = U$; every element in U is a member of at least one A_i.

(2) $A_i \cap A_j = \emptyset$ whenever $i \neq j$; no element of U lies in more than one A_i. ▶

PROBLEMS

3. Write two partitions of the set $U = \{a, b, c, d, e\}$.

4. Let U be the set of all U. S. Presidents and B be the relation "was born the same month as." List the equivalence classes of B in U.

We have seen that each equivalence relation E defined on a set U classifies the elements of U into various disjoint subsets forming a partition of U. Conversely, any partition of U determines an equivalence relation E in U. We simply take xEy if and only if x and y belong to the same set in the partition.

Classifications utilizing equivalence relations are widely used in the behavioral sciences. A sociologist may classify human beings according to the region of the country in which they reside and may then compare regions according to various attributes such as average income, amount of education, etc. An economist may classify firms as distributing, manufacturing, and so forth.

Such classification is not always successful. In grouping humans into generations a natural rule might be xGy if x and y were born within 25 years of each other. However, this relation is not transitive, and hence is not an equivalence relation. For instance, if x was born in 1920, y in 1935, and z in 1955, we have xGy and yGz but not xGz. We must look for another criterion for classification.

Example 2 Economists are concerned with consumer demand for commodities. As a means of explaining consumer demand, a *preference relation* P is defined on the set U of commodities by xPy if commodity x is preferred to commodity y. It is assumed that

(a) P is not reflexive. No commodity is preferred to itself.

(b) P is not symmetric. If x is preferred to y then y is not preferred to x.

(c) P is transitive. If x is preferred to y and y to z, then x is preferred to z.

Thus P is not an equivalence relation. ▶

5. The *indifference relation I* is defined by xIy if and only if x is not preferred to y and y is not preferred to x. That is,

$$xIy \Leftrightarrow \sim xPy \wedge \sim yPx$$

Show that I is an equivalence relation. The relation I is called the equivalence relation *determined by* the preference relation P. What are the equivalence classes for I?

6. Determine which of the following are equivalence relations. For those which are not, state which properties $(R, S, \text{or } T)$ fail to hold.

(a) "votes for"   (b) "costs more than"

(c) "is equivalent to"   (d) "is at war with"

(e) "is biologically related to"   (f) "causes"

(g) "communicates with"   (h) "has at least as much utility as"

(i) "influences"

7. A subject is presented with a set U of 64 color patches differing in brightness. He is requested to sort the patches according to the magnitude of the brightness differential between the patches and the standard. Show that the relation "differs in brightness from the standard by the same amount as" induces a partition of U and is therefore an equivalence relation.

8. Let R_1 be the equivalence relation "belongs to the same social group as" and R_2 the equivalence relation "belongs to the same religious group as." Let xRy mean $xR_1y \wedge xR_2y$. Prove that R is an equivalence relation.

9. We know that any equivalence relation partitions the universal set into equivalence classes. In Problem 8, how are the equivalence classes for R related to those for R_1 and R_2?

10. Let R_1 and R_2 be as defined in Problem 8 and let xQy be defined as $xR_1y \vee xR_2y$. Show that Q need not be an equivalence relation.

11. The *product* of two relations A and B is defined by $AB = \{(x, y): \text{there exists } z \text{ such that } xAz \text{ and } zBy\}$. Show that if A is an equivalence relation, then $AA \subseteq A$.

12. Suppose that an experimenter is studying the communication pattern of a task oriented group. The communication network is so organized that for any two members, neither can contact the other directly. All communication between members is directed to the experimenter who functions as a relay. If A is the relation "can send a message to" and B is the relation "can receive a message from," prove that AB is an equivalence relation.

13. Prove that if $U \subseteq A$ and $V \subseteq B$ where $U, V, A,$ and B are relations, then $UV \subseteq AB$.

14. For the relations A, B, and U, prove that if $A \subseteq B$ then $UA \subseteq UB$ for all U.

15. *Idempotency* of a relation is defined as $AA = A$. Prove that if A is an equivalence relation, then it must be an idempotent relation.

3.3 WEAK ORDERING RELATIONS AND PREFERENCE RELATIONS

An equivalence relation provides the mathematical counterpart to the operation of classifying a group of objects. Equally important in the behavioral sciences is the problem of ordering a collection of objects.

Mathematically, orderings arise by omitting one or more of the three properties possessed by equivalence relations. For example, the relation $\leq$ is reflexive and transitive, but not symmetric. While it is possible that we have both $x \leq y$ and $y \leq x$, it is not in general true that $x \leq y$ implies $y \leq x$. The relation $<$ is transitive but not symmetric or reflexive. However, in this case more can be said since $x < x$ is always false and $x < y$ implies that $y < x$ is false.

A relation R for which xRx is always false is said to be *irreflexive*. If R is such that xRy always implies that yRx is false, then R is called *asymmetric*. Thus the relation $<$ is transitive, asymmetric, and irreflexive. The relation $\leq$ is reflexive and transitive, but neither symmetric nor asymmetric.

A relation R which is reflexive and transitive on a set U is called a *weak ordering relation* on U. A relation R which is transitive and asymmetric on U is called a *preference relation* on U. Thus the relation $\leq$ is a weak ordering, while $<$ is a preference relation. In Example 2 of Section 3.2, P is a preference relation in this technical sense.

Example 1 Let U be the set of all families in a community and R be the relation "is socially more prominent than." Then R is:

(1) irreflexive—no family is socially more prominent than itself

(2) asymmetric—if family F is more prominent than family G, then G is not more prominent than F.

(3) transitive—if family F is more prominent than family G, and G is more prominent than H, then F is more prominent than H.

Hence R is a preference relation. ▶

PROBLEMS

1. Let A denote the relation "is at least as socially prominent as." Prove that A is a weak ordering.

2. Prove that a relation which is asymmetric must also be irreflexive. In particular, a preference relation is irreflexive.

From any weak ordering relation it is possible to obtain a corresponding preference relation. For instance, from the relation $\leq$ we may obtain the relation $<$ by writing $x < y$ if and only if $x \leq y$ *and not* $y \leq x$.

Theorem 1 If W is a weak ordering relation and P is a relation defined by

$$xPy \Leftrightarrow (xWy) \wedge \sim(yWx) \tag{1}$$

then P is a preference relation.

PROOF We must show that (1) means P is asymmetric and transitive. First, xPy means xWy and not yWx, while yPx means yWx and not xWy. The two statements are obviously contradictory, and it follows that P is asymmetric.

Transitivity of P follows from transitivity of W. For, if xPy and yPz, then, among other things, we have xWy and yWz. Hence xWz. But if we also had zWx, then this, together with xWy, would imply zWy, contradicting yPz. We have proved xPz and thus P is transitive. ▶

The relation P defined in (1) is called the *preference relation determined by W*.

Example 2 Behavioral scientists are becoming increasingly interested in the process by which an individual, when confronted with the necessity of making a decision, chooses one course of action rather than another from a set of possible alternatives. The alternatives may range from various bundles of commodities for the housewife to alternative strategies of international politics for national leaders.

The notions of preference relation and indifference relation (Example 2 and Problem 5, of Section 3.2) apply here. Thus the relation W defined by xWy if an individual either prefers x to y or is indifferent between the two is a weak ordering relation. The relation P defined by xPy if x is preferred to y is the preference relation determined by W.

One way which is often used to formalize problems involving preference or choice is to assign to each alternative x a number $u(x)$, called the *utility* of x. A basic assumption of utility theory, the *ordinal assumption*, states that if an individual prefers x to y or is indifferent between them (that is, if xWy) then $u(x) \geq u(y)$. Based on the ordinal assumption and certain other assumptions,* the notion of utility serves as a numerical characterization of preference and indifference. ▶

Equivalence relations as well as preference relations may be obtained from weak ordering relations. For example, if W is the relation $\leq$, the relation E of equality is obtained by writing

$$xEy \Leftrightarrow xWy \wedge yWx$$

*See, for example, the discussion in Chapter 2 of Luce, R. D., and Raiffa, H., *Games and Decisions* (John Wiley & Sons, Inc., New York, 1957).

That is

$$x = y \Leftrightarrow x \le y \text{ and } y \le x$$

Theorem 2   If W is a weak ordering relation and E is a relation defined by

$$xEy \Leftrightarrow xWy \wedge yWx \tag{2}$$

then E is an equivalence relation.

PROOF   We must show that E is reflexive, symmetric, and transitive. By (2), xEx if and only if xWx. But xWx for all x since W is reflexive. Hence the same is true of E.

The symmetry of E is immediate since xEy and yEx have the same definition. To show that E is transitive we must argue that $xEy \wedge yEz$ implies xEz. By definition, $xEy \wedge yEz$ is equivalent to

$$xWy \wedge yWx \wedge yWz \wedge zWy$$

Because W is transitive, the first and third conjunctive forms yield xWz, while the fourth and second give zWx. Hence E is transitive.   ▶

Example 3   Continuing Example 2, let us define the indifference relation I by

$$xIy \quad \text{if neither} \quad xPy \text{ nor } yPx$$

It follows from the definitions of W and P that

$$xIy \Leftrightarrow xWy \text{ and } yWx$$

By Theorem 2 then, I is an equivalence relation, a fact previously verified directly in Problem 5 of Section 3.2.

The ordinal assumption of utility theory requires that $u(x) \ge u(y)$ whenever xWy. Thus xIy implies $u(x) = u(y)$. Equality of utility values reflects indifference between alternatives.   ▶

PROBLEMS

3. For each of the following relations, state whether it is (i) reflexive, (ii) irreflexive, (iii) symmetric, (iv) asymmetric, or (v) transitive:

  (a) "negotiates with"

  (b) "is superior to"

  (c) "is the brother of" in the set of all humans

  (d) "is the brother of" in the set of all males

  (e) "is the brother of" in the set of all females

  (f) "is as bright as"

  (g) "merges with"

  (h) "differs from"

  (i) "is subsumed by"

(j) "is no worse than"

(k) "is the same as"

4. From the relations in Problem 3, select the weak ordering, the preference, and the equivalence relations.

5. Find the preference relation determined by the following weak ordering relations.

(a) "is no wealthier than"

(b) "is at least as liberal as"

(c) "is no more popular than"

6. Let U be a domain of discourse. Prove that

(a) the relation I of logical implication is a weak ordering relation.

(b) the relation E of logical equivalence is an equivalence relation.

(c) the relation E is the equivalence relation determined by the weak ordering relation I.

7. Let U be a universal set. Prove that

(a) the relation S of *subset* is a weak ordering.

(b) the relation E of *set equality* is an equivalence relation.

(c) E is the equivalence relation determined by the weak ordering S.

(d) What is the preference relation determined by S?

8. For each of the following relations R, find the weak ordering relation that determines R.

(a) xRy if schools x and y have the same racial balance.

(b) xRy if cities x and y have equal crime rates.

(c) xRy if decision x is equally as risky as decision y.

9. Prove that every equivalence relation is also a weak ordering relation.

10. Prove that a relation which is irreflexive and transitive is a preference relation.

Problems 11 and 12 refer to Examples 2 and 3.

11. Prove that for each pair x, y of elements of U, exactly one of the statements xIy, xPy, or yPx is true.

12. Prove that for all $x, y, z \in U$, if xPy and yIz, then xPz, and if xPy and xIz, then zPy.

13. Complete the following table concerning the R, S, and T properties of the set relations membership, inclusion (subset), and equality:

	R	S	T
Membership	no		
Inclusion			
Equality		yes	

▶

3.4 PARTIAL AND SIMPLE ORDERINGS

As we have seen, an equivalence relation E partitions the universal set U into disjoint subsets called equivalence classes. The elements of a particular equivalence class are alike relative to E. If, for instance, U is a set of people and E is the relation "has the same aptitude as," an equivalence class contains all people having the same aptitude. If aptitude is the only criterion upon which a person is to be selected, then members of the same equivalence class may be freely substituted for one another.

An equivalence relation E is said to be *consistent* with another relation S if elements in the same equivalence class (relative to E) may be substituted for each other in statements involving S. A familiar example is the relation E of equality among numbers. The statement that "equals may be substituted for equals" is a statement concerning the consistency of E with any other relation involving numbers.

Definition 2 An equivalence relation E and another relation S are said to be *consistent* if

$$(x \, E \, y) \wedge (y \, S \, z) \Rightarrow x \, S \, z$$

and

$$(x \, E \, y) \wedge (z \, S \, y) \Rightarrow z \, S \, x$$

for all x, y, and z in U. ▶

Example 1 Measurement models in the behavioral sciences generally include transitivity as one of the postulates necessary for the construction of a measurement scale. Intuitively speaking, if a is in some sense "greater than" b and b is in turn "greater than" c, then a should be "greater than" c.

Unfortunately, measurement relations which are defined by experimental procedures do not always exhibit the desired transitivity. For instance, suppose that we wish to scale a set of tones with respect to loudness. It is possible, of course, to measure mechanically (in decibels) the loudness of each tone and to arrange them according to the relations L, aLb means a is louder than b, and E, aEb means a and b are equally loud. Assuming no measurement error, it is apparent that L is irreflexive and transitive, and hence is a preference relation, while E is an equivalence relation.

More interesting to the behavioral scientist than the mechanical measurement of loudness is the set of responses given by a subject who is presented with pairs of tones and asked in each case to choose the louder one. If tones x, y, and z are of almost equal loudness but differ in pitch, it is quite conceivable that the responses xLy, yLz, and zLx might be obtained, yielding an obvious violation of transitivity.

One way of partially overcoming this difficulty is to present each pair of tones a large number of times. Then aLb whenever a is judged louder than b more times than b is judged louder than a, and aEb if each is judged to be louder in exactly half of the presentations.

We arbitrarily require that E be reflexive. And it is apparent from the way in which the relations are defined that L is irreflexive and E symmetric. What is not apparent, and indeed may fail to occur, is the transitivity of L and E. The determination of experimental conditions or subject types for which this condition does or does not hold is of particular interest.

We cannot pursue this situation further here* but will content ourselves with proving (in the case where transitivity obtains) that L and E are consistent relations.

We must show first that aLb and bEc together imply aLc. It will be simpler to proceed indirectly and prove the equivalent contrapositive statement:

$$\sim aLc \quad \text{implies either} \quad \sim aLb \quad \text{or} \quad \sim bEc$$

From the way in which the relations L and E are defined, we must have either aLc or cLa or cEa. Thus $\sim aLc$ is equivalent to $cLa \lor cEa$. If cLa and aLb, then (transitivity of L) cLb from which we obtain $\sim cEb$. If cEa and bEc, then bEa, which implies $\sim aLb$. Since all cases have been considered, this completes the proof of the first criterion of consistency. The second is established in a similar manner. ▶

PROBLEMS

1. Complete the proof that L and E are consistent; that is, prove that aLb and aEc together imply that cLb.

2. Are the following relations consistent? Why or why not?
   (a) S "lives next to" and E "is in the same precinct as."
   (b) D "pays the same dividend as" and E "is listed on the same exchange as."
   (c) S "is shorter than" and E "is the same height as."
   (d) F "is a friend of" and C "lives in the same community as."

3. List three pairs of consistent relations.

4. The statement "L is irreflexive and E is symmetric" appears in the fifth paragraph of Example 1. Prove that this is a true statement.

It is easily seen that if E is an equivalence relation whose equivalence classes each contain only a single element and R is *any* relation, then E and R are consistent. For, in this case, since each equivalence class contains only one element, we have xEy if and only if $x = y$. The requirements of consistency are automatically satisfied when $x = y$.

An equivalence relation whose equivalence classes consist of single elements is called an *identity relation*. Familiar examples are the relations of equality for numbers and for sets. Roughly speaking, no two distinct elements of the universal set are alike relative to an identity relation.

*Methods of scaling which use information gained from paired comparisons are discussed at length in Torgerson, W. S., *Theory and Methods of Scaling* (John Wiley & Sons, Inc., New York, 1958).

Suppose W is a weak ordering and E is the equivalence relation determined by W. If E is an identity relation, then W is called a *partial ordering*. A good example is the ordering of the real numbers.

The intended interpretation of a weak ordering W is that it arranges the elements of U in a certain order. We may think of xWy as expressing that y is at least as far ahead in the order as x. However, in using the ordering of the real numbers as an example of a weak ordering, one may easily be misled into thinking that every pair of elements in the universal set may be compared concerning their relative position. That this is not the case for all weak orderings is shown in the next example.

Example 2   Let A be the set $\{a, b, c\}$ and let U be the collection of all subsets of A. The relation W defined by xWy if and only if x is a subset of y is easily shown to be reflexive and transitive and hence to be a weak ordering. (Since the equivalence relation determined by W is the relation of equality, W is also a partial ordering.) Then if $x = \{c\}$ and $y = \{b, c\}$, we would have xWy and would say that y is at least as far ahead in the ordering as x. Continuing this analogy suggests that A itself should be at the top of the ordering and $\emptyset$ at the bottom since for every $x \in U$ we have xWA $(x \subseteq A)$ and $\emptyset Wx$ $(\emptyset \subseteq x)$. However, some pairs are not capable of comparison. For example, if $x = \{a, b\}$ and $y = \{b, c\}$, we have neither xWy nor yWx.

Figure 4 shows the position in the order of each element of U. If xWy, we place y above x and connect them by a line. Not all lines have been drawn in. For instance, we have $\emptyset Wx$ for all $x \in U$ but only three of the seven possible lines emanating from $\emptyset$ appear on the diagram. However, from the diagram we read, for example, $\emptyset W\{c\}$ and $\{c\} W\{b, c\}$. The transitivity of W then yields $\emptyset W\{b, c\}$. Other cases may be similarly obtained by "reading" transitivity into the diagram.   ▶

FIGURE 4

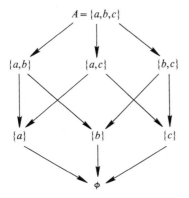

Two elements x and y of a set U are said to be *comparable* with respect to an ordering relation O if either xOy or yOx. Otherwise x and y are not comparable. In Figure 4, for instance, sets $\{c\}$ and $\{b, c\}$ are comparable while $\{a\}$ and $\{b, c\}$ are not, neither being a subset of the other.

An ordering relation W is said to be *connected* in U if every pair of elements of U is a comparable pair. Thus the ordering relation $\leq$ for the real numbers is connected, while the relation $\subseteq$ for sets, considered in Example 2, is not connected.

A partial ordering which is connected in U is called a *simple ordering* of U. Simple ordering is one of the most important ordering relations and the one which is usually intended when one speaks of ordering a collection of objects. The most familiar example is the relation $\leq$ for real numbers. The diagram of every simple ordering consists of a single vertical line connecting the elements of U, as shown in Figure 5. This is because in a simple ordering all pairs of elements are comparable and no two elements are alike relative to the ordering.

FIGURE 5

Example 3 A recurring problem in behavioral studies is that of assigning positions on some *hypothetical scale* which reflect the preference orderings of a group of subjects for a certain set of stimuli objects.

One theory of preference ordering is the "unfolding technique" developed by Coombs.* In case (I) of Coombs' theory each stimulus object j is assumed to have a fixed scale position Q_j and each individual i is assumed to have chosen (perhaps subconsciously) a fixed scale position C_i representing his ideal stimulus value. The basic postulate which attempts to explain observed behavior is: Given the choice between objects j and k, a subject will respond jPk (object j is preferred to object k) if and only if the scale value Q_j is closer than the scale value Q_k to the ideal value C_i.

The usual restrictions imposed require that P be a simple ordering relation. Hence, presentation of all pairs of stimuli will result in a rank ordering of the stimulus objects for a given individual. Such a rank ordering is called an individual or I scale. A scale on which both stimuli and individuals are assigned positions is called a joint continuum or a J scale. An example of a J scale with stimuli positions A, B, C, D, and E and a subject's ideal value X

---

*Coombs, C. H., "Psychological Scaling without a Unit of Measurement," *Psychological Review* **57**, 145–158 (1950). For a more comprehensive description, see Coombs, C. H., *A Theory of Data* (John Wiley & Sons, Inc., New York, 1964).

is shown in Figure 6. Given such a *J* scale, the individual's *I* scale can be obtained as follows. Consider the *J* scale to be hinged at the point *X*. If we fold the scale so that the part of the scale to the left of the hinge is merged with the part to the right, we obtain Figure 7. It is now apparent that the *I* scale for this individual must be *ABCDE*, reflecting the relative nearness of the stimuli positions to his ideal stimulus value *X*. Any *I* scale, then, may be regarded as a folded *J* scale. Since the experimental data consist of observed *I* scales, the *J* scale is recovered by unfolding the *I* scales, hence the reason for the name of the scaling technique.

FIGURE 6   Joint continuum or *J* scale

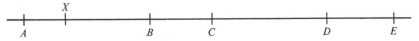

FIGURE 7   Individual or *I* scale

Consider all individuals whose ideal point is located to the left of the stimulus position *A* on the *J* scale in Figure 6. Since the location of the ideal point determines the distance to the respective stimuli positions, the *I* scales cannot, in general, be regarded as being quantitatively equivalent for all such individuals. However, the preference order of the stimuli objects will be the same for all such individuals. Thus, the *I* scales can be regarded as being qualitatively equivalent.

In fact, the order will not change until we consider individuals whose ideal points are immediately to the right of the midpoint between *A* and *B*. For these individuals, the qualitative *I* scale is *BACDE*. In passing the midpoint between *A* and *C*, stimuli *A* and *C* change positions and the *I* scale is *BCADE*. Hence the midpoint between *A* and *B* and the midpoint between *A* and *C* bound an interval on the *J* scale such that the *I* scales are the same for any *X* within the interval.

Continuing to move across the *J* scale, we see that the midpoints between all possible pairs of positions partition the entire *J* scale, as shown in Figure 8, into intervals such that the *I* scales are qualitatively the same for all individuals whose ideal point is located within the same interval.

Before the *J* scale can be recovered by unfolding the *I* scales, it is necessary to order the *I* scales. From Figure 8, we note that a *J* scale with 5 stimulus positions is partitioned into 11 segments. Since each segment generates a

FIGURE 8

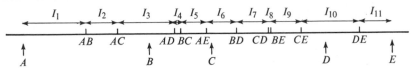

unique I scale, the J scale in our example yields a set of 11 different I scales. All I scales must of necessity end with the first or last stimulus position on the J scale. The I scale associated with interval I_1 establishes the simple ordering of the stimuli on the J scale. The remaining scales can be ordered according to the rule that a stimulus must first move to the left in an I scale before it can move to the right. Adjacent stimuli in I scales simply change order when we move from one adjacent interval to another on the J scale.

Ordering of the I scales induces a simple ordering on the midpoints and as such provides information about relative distances between stimuli positions on the J scale. For example, the ordering of the I scales resulting from the J scale of Figure 6 as presented in Table 1 indicates that the midpoint BC precedes AE on the J scale. This can only be the case if the distance between A and B, denoted $\overline{AB}$, is less than the distance $\overline{CE}$ between C and E.

TABLE 1. Ordering of the I scales of Figure 8

I scale	Arrangement of I scale	Order of midpoints	Relation of distances
I_1	ABCDE	AB	
I_2	BACDE	AC	
I_3	BCADE	AD	AD precedes $BC \Rightarrow \overline{AB} > \overline{CD}$
I_4	BCDAE	BC	BC precedes $AE \Rightarrow \overline{CE} > \overline{AB}$
I_5	CBDAE	AE	AE precedes $BD \Rightarrow \overline{AB} > \overline{DE}$
I_6	CBDEA	BD	AE precedes $CD \Rightarrow \overline{AC} > \overline{DE}$
I_7	CDBEA	CD	CD precedes $BE \Rightarrow \overline{DE} > \overline{BC}$
I_8	DCBEA	BE	
I_9	DCEBA	CE	
I_{10}	DECBA	DE	
I_{11}	EDCBA		

Information on the relative distances between stimuli positions on the J scale of our example is illustrated by the partial ordering in Figure 9. The utility of the unfolding procedure is that, given only the ordering of the observable I scales, we can infer not only the stimuli ordering on the hypothetical unobservable J scale but also information about the relative distances between the stimuli positions on the scale. ▶

FIGURE 9

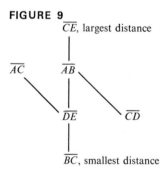

$\overline{CE}$, largest distance

$\overline{AC}$ $\overline{AB}$

$\overline{DE}$ $\overline{CD}$

$\overline{BC}$, smallest distance

5. Let xFy denote the relation "x voted in favor of all the bills that y did" defined on the set U of all congressmen.

   (a) Suppose that no two congressmen voted for exactly the same bills. Show that the relation F is a partial ordering.

   (b) Consider the possibility that x and y may have voted for exactly the same bills. Does F still give a partial ordering of U? Give a reason for your answer.

6. Let U be a set of machines. Let T be the weak ordering relation for which xTy means x is at least as efficient and costs no more than y. Show that:

   (a) T is not connected on the set of all machines.

   (b) T is connected on a set of machines all of the same cost.

   (c) T is connected on a set of machines all of the same efficiency.

7. Suppose that you wish to determine a subject's preferences for a set U of selected foods. You present all possible pairs of foods to the subject. When confronted with a pair (x, y), the subject is required to make one of three responses: (1) I prefer x to y (xPy), (2) I prefer y to x (yPx), or (3) I cannot make a choice. Argue that the relation P may not be connected on the set U even if it is transitive.

8. Suggest a modification of the experimental procedure for Problem 7 which will change the relation P into a simple ordering relation.

9. Prove that the relations P and I as defined in Examples 2 and 3, Section 3.3, are consistent.

10. Prove that if E is the equivalence relation determined by a weak ordering relation W, and P is the preference relation determined by W, then P and E are consistent.

11. Prove that any simple ordering is determined by its preference relation.

12. Referring to Example 1, define the relation S by xSy if and only if $xEy \lor xLy$.

    (a) Prove that S is a weak ordering relation and that L and E are, respectively, the preference and equivalence relations determined by S.

    (b) Prove that S and E are consistent relations.

13. Prove that if W is any weak ordering relation and E is the equivalence relation determined by it, then W and E are consistent. Compare with Problem 12 of Section 3.3.

3.5 GRAPHIC REPRESENTATION OF ORDER RELATIONS

Diagrams, such as Figures 4 and 5 of the preceding section, facilitate the application of the theory of relations to specific problems. In this section we shall consider some important notions concerning relations which are most easily presented in graphic form.

One of the most common representations of a weak ordering is the *flow diagram* (see Figure 10). Consider, for instance, the set of employees of a utility providing service for a city. The most efficient communication network is one which allows any person to contact any other person in the organization, but for practical reasons this may not be possible. For any two employees in the network one of the following four cases exists:

(1) Each can contact the other.

(2) The first can contact the second, but not conversely. This might be the case for a repairman working in a home and the central dispatcher.

(3) The second can contact the first, but not conversely.

(4) Neither can contact the other. This might be the case for two repairmen working in different parts of the city.

FIGURE 10

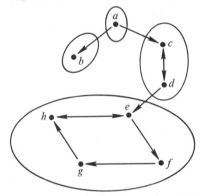

Suppose the relation C is defined by xCy if and only if x can *contact y* directly or indirectly. This relation is reflexive since anyone can contact himself. And C is transitive since if x can contact y and y in turn can contact z, then x, by using y as an intermediary, can contact z. The relation C is thus a weak ordering.

The flow diagram of Figure 10 illustrates a possible contact structure for eight persons in the firm. The arrows indicate the direction in which a contact can be made. However, the diagram has been simplified by not showing all possible contacts. For instance, each person can contact himself. In addition, transitivity means, for example, that eCf and fCg together imply eCg. Only the essential arrows are shown and reflexivity and transitivity must be read into the graph.

The equivalence relation E determined by C is such that xEy if and only if $xCy \wedge yCx$; that is, xEy if and only if x and y can each contact the other. We shall say that x and y can *communicate* if xEy. The four equivalence (communication) classes for E are indicated by the circles in Figure 10. Two employees communicate when they are members of the same communication class. In the most efficient setup there would be a single communication class containing all eight persons.

A simpler version of Figure 10 is shown in Figure 11, in which arrows are drawn only to indicate contacts between equivalence classes, it being understood that contact is always possible between members of the same communication class. Note that $\{a\}$ is ahead of each of the other classes in the ordering and that $\{c, d\}$ is above $\{e, f, g, h\}$. However, $\{b\}$ is comparable with $\{a\}$ but not comparable with either of the other two classes. As an ordering of equivalence classes, then, it follows that Figure 11 represents a partial ordering but not a simple ordering.

FIGURE 11

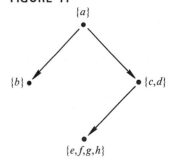

The transition from an ordering of elements (Figure 10) to an ordering of equivalence classes (Figure 11) may be made for any weak ordering W. The new relation $\widetilde{W}$ which orders the equivalence classes is called the *induced ordering relation* corresponding to W. An induced ordering relation $\widetilde{W}$ is always a partial ordering. To see why this is so, we note first that if E_1 and E_2 are two different equivalence classes determined by a weak ordering W, exactly one of the following three cases must occur:

I. There are no two elements $x \in E_1$ and $y \in E_2$ which are comparable. That is, for every such choice of x and y we have neither xWy nor yWx. As an example, take $E_1 = \{b\}$ and $E_2 = \{c, d\}$ in Figure 11.

II. For every choice of two elements $x \in E_1$ and $y \in E_2$, we have xWy. That is, every element of E_1 is at least as far ahead in the ordering as any element of E_2.* For example, let $E_1 = \{a\}$ and $E_2 = \{e, f, g, h\}$ in Figure 11.

III. For every choice of two elements $x \in E_1$ and $y \in E_2$ we have yWx. That is, every element in E_2 is at least as far ahead as any element in E_1.

Of course, elements in the same equivalence class are equally far ahead in the ordering since

$$xEy \Leftrightarrow xWy \wedge yWx$$

---

*Actually, each element of E_1 must be farther ahead in the ordering. If not, that is, if $x \in E_1$ and $y \in E_2$ were equally far ahead, we would have xEy and hence $E_1 = E_2$.

The fact that W places every member of an equivalence class in the same position relative to the ordering means that $\widetilde{W}$ may be considered as imposing the same ordering on the equivalence classes themselves. It is apparent from I–III above that $\widetilde{W}$ is transitive and reflexive because W is. Thus $\widetilde{W}$ is itself a weak ordering. However, since no two equivalence classes are "alike" relative to W (if they were, they would be lumped together to form a single class), it follows that $\widetilde{W}$ is actually a partial ordering.

PROBLEMS

1. Let U be the set of members of your class and define the relation W by xWy if x is at least as tall and has hair at least as dark as y. List the members of each equivalence class determined by W. Draw a flow diagram for W and another for the induced ordering $\widetilde{W}$.

A glance at Figure 10 shows that each communication class forms a circle or a *loop* in the ordering around which contacts may be made successively. Through such a loop every person may contact every other.

Definition 3 The elements $a_1, a_2, \ldots, a_n$ are said to form a *loop* of the relation R if $n \geq 2$ and

$$a_1 R a_2 \wedge a_2 R a_3 \wedge \cdots \wedge a_{n-1} R a_n \wedge a_n R a_1 \qquad (3)$$

A relation is *loop free* if no loop exists. ▶

Loops are common in weak orderings but cannot occur at all in preference relations or in partial orderings. To see why, suppose P is a preference relation and that (3) holds for P. Then the first statement in (3) is $a_1 P a_2$, while the remaining statements, by the transitivity of P, imply $a_2 P a_1$. These two statements together violate the asymmetry of P. We conclude that (3) cannot be true for P. Hence P is loop free.

Similarly, if O is a partial ordering and (3) determines a loop of O, then we have

$$a_1 O a_2 \wedge a_2 O a_1 \qquad (4)$$

Recall that the equivalence relation E determined by O is such that $a_1 E a_2$ if and only if (4) holds. But for a partial ordering, $a_1 E a_2$ if and only if $a_1 = a_2$. Similar arguments show that if (3) holds, then all the elements $a_1, a_2, \ldots, a_n$ must be the same.

Example 1 Suppose the U. S. Employment Service is interested in establishing a skill hierarchy for 10 job classifications. They are interested in the relation S defined by xSy if and only if the skills required to perform job x suffice also for the performance of job y. It is apparent that S is reflexive and transitive and hence a weak ordering.

On the basis of past experience, the following skill relations have been established:

$$dSj \quad cSf \quad aSh \quad iSh$$
$$fSe \quad dSa \quad gSa \quad jSh$$
$$jSb \quad hSc \quad fSi \quad aSg$$

This information is summarized in the flow diagram of Figure 12, in which arrows indicate the direction of skill transference. Again reflexivity and transitivity must be read into the graph.

FIGURE 12

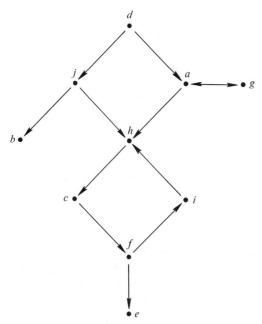

Absence of a chain of arrows linking two jobs indicates that the relative skills of these jobs are not comparable and that there is not sufficient skill transference from one to the other. On the other hand a loop in the graph indicates a bilateral transference of skill between any two of its members and these jobs can be regarded as being alike in terms of skill demands.

The induced partial ordering of equivalence classes shown in Figure 13 represents the skill hierarchy for the 10 jobs. We note that job d requires the greatest skill whereas jobs b and e make minimal skill demands.

It seems reasonable to interpret any simple ordering as representing an ordering of the jobs in terms of the relative amounts required of a single skill or a homogeneous group of skills. From this view there appear to be three separate skill classes, one represented by the simple ordering

$$\{d\} \rightarrow \{j\} \rightarrow \{b\}$$

FIGURE 13

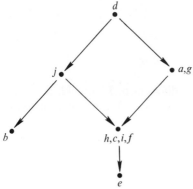

a second by $\{d\} \rightarrow \{j\} \rightarrow \{h, c, i, f\} \rightarrow \{e\}$, and the third by

$$\{d\} \rightarrow \{a, g\} \rightarrow \{h, c, i, f\} \rightarrow \{e\}$$

One would suspect that those equivalence classes which are members of more than one simple ordering contain jobs which require several distinct types of skills. ▶

PROBLEMS

2. The salary classification for an employee of the Burke Company is determined by years of formal education and amount of experience. Let

$U = \{(16, 3), (14, 1), (14, 5), (12, 1), (20, 5), (14, 3), (16, 10), (12, 15), (12, 3)\}$

be the set of ordered pairs of numbers of years formal education and experience, respectively, for 9 employees. Draw the flow diagram for the relation T where xTy if and only if x has at least as many years formal education and at least as many years experience as y.

3. Redraw the flow diagram of Problem 2 under the assumption that

(a) one year of experience is equivalent to 2 years of formal education.

(b) each year of experience is equivalent to $\frac{1}{2}$ year of formal education.

4. Explain why the rules in Problem 3 changed T from a partial ordering to a simple ordering.

5. Let U be a set of corporations. State a rule for each of the following weak ordering relations which will change it into a simple ordering relation on its equivalence classes.

(a) T is defined as "has at least as many employees as and no greater sales than."

(b) T is defined as "has no more assets than and at least as good a profit margin as."

(c) T is defined as "has at least as much net income, and no greater per share earnings, and no less dividends per share than."

6. Prove that a weak ordering relation T defined on a set U is a weak ordering on every subset of U.

7. Let W be a weak ordering relation and P the preference relation determined by W. An element $x \in U$ is termed "maximal" if there is no $y \in U$ for which yPx. If x is the unique maximal element, it is termed "the maximum." In a similar manner, if there is no y such that xPy, then x is termed "minimal"; and if x is unique, it is called "the minimum." Prove that every weak ordering defined for finite U has at least one maximal element and at least one minimal element.

8. In the communication example illustrated by the flow diagram of Figure 10, prove that x is minimal if he can contact only members of his equivalence class.

9. Let T be a weak ordering of the elements of a set U, and $\widetilde{T}$ the corresponding induced ordering of the equivalence classes determined by T. Prove that if $\widetilde{T}$ is a simple ordering, then every subset of U has both a maximum and a minimum relative to T.

10. According to Suppes and Zinnes,* a relation P defined on a set U is a *semiorder* whenever the following axioms are satisfied for all x, y, u, v in U:

    (1) P is irreflexive; that is, xPx is always false.

    (2) If xPy and uPv, then either xPv or uPy.

    (3) If xPy and yPu, then either xPv or vPu.

    (a) If an indifference relation I is defined as xIy if and only if not xPy and not yPx, show that I is not an equivalence relation.

    (b) If a relation E is defined as xEy if and only if for every u in U, xIu if and only if yIu, show that E is an equivalence relation.

11. In Problem 10, prove that

    (a) If xPu and xEy, then yPu.

    (b) If uPx and xEy, then uPy.

    (c) Given the results in (a) and (b), show how a relation $\widetilde{P}$ can be defined which will order the E equivalence classes of U.

12. Let U be the set of all positive integers and define the relation R by xRy if x and y both give the same remainder when divided by the positive integer m. (We express this by saying that x equals y, modulo m.) Let the relation S be defined by xSy if x/m has a greater remainder than y/m.

    (a) Show that R is an equivalence relation.

    (b) Show that S is a preference relation.

    (c) Show that R and S are consistent.

13. In Problem 12, prove that if xRy and zRw, then

$$(x + z) R (y + w) \quad \text{and} \quad (xz) R (yw)$$

*Suppes, P., and Zinnes, J. L., "Basic Measurement Theory," in *Handbook of Mathematical Psychology*, Luce, R. D., Bush, R. R., and Galanter, E., eds. (John Wiley & Sons, Inc., New York, 1963), Vol. I, Chapter 1.

Problem 13 may be used in the following way. The numbers 10 and 2 both yield the same remainder (2) when divided by 8. Thus $100 = 10 \cdot 10$ yields the same remainder as $2 \cdot 2$ (that is, 4) when divided by 8. Similarly, $1000 = 10^3$ gives the same remainder as $2^3 = 8$ (that is, zero remainder) when divided by 8.

14. What is the remainder when

(a) 1000 is divided by 7?

(b) 1,000,000 is divided by 8?

(c) 3^{50} is divided by 13?

15. Suppose you have 12 objects, indistinguishable in outward appearance, of which 11 have the same weight, while the 12th is either heavier or lighter than the rest. Let O_iEO_j denote that objects O_i and O_j are of equal weight, while O_iHO_j means that O_i is heavier than O_j.

(a) Argue that the relations E and H are consistent.

(b) Prove that three weighings on a simple balance scale are sufficient to determine the two equivalence classes of the relation E.

SUPPLEMENTARY READING

Cogan, E. J., Kemeny, J. G., Norman, R. Z., Snell, J. L., and Thompson, G. L., *Modern Mathematical Methods and Models* (Mathematical Association of America, 1958), Vol. 2, Unit II: Order Relations.

Suppes, P., *Introduction to Logic* (D. Van Nostrand Company, Princeton, N. J., 1957), Chapter 10.

Suppes, P., and Zinnes, J. L., "Basic Measurement Theory," in *Handbook of Mathematical Psychology*, Luce, R. D., Bush, R. R., and Galanter, E., eds. (John Wiley & Sons, Inc., New York, 1963), Vol. I, Chapter 1.

FUNCTIONS AND REAL NUMBERS 4

4.1 FUNCTIONS

The concept of function, or the dependence of one quantity upon another, pervades social and behavioral science. The economist attempts to describe and to predict the manner in which economic processes change over time. Quantities, such as *GNP*, the prime interest rate, the national debt, and so forth, are thought of as functions whose values change as time passes. To the sociologist, the growth of a population depends on many variables such as time, the present size of the population and migration rates.

From a mathematical point of view, a *function* is a special kind of binary relation. Specifically, a function is a set of ordered pairs in which no two pairs have the same first element. Let us look at some examples.

Example 1 Table 1 lists the number of known cases of robbery per 100,000 population for selected states for the year 1965. This table contains ordered

TABLE 1*

State	Rhode Island	Illinois	North Dakota	Kansas	Louisiana	Hawaii	Oklahoma	California
Cases of robbery per 100,000 population	19	165	5	24	51	19	38	113

*Source: *Statistical Abstract of the United States* (U. S. Department of Commerce, 1967), 88th ed., p. 151.

pairs of the form (x, y) where x is a state and y a number. As such, it constitutes a binary relation. This relation is a function since each state (first element) appears only once. Note that the same number may correspond to different states. ▶

Example 2 (a) The relation

$$D = \{(\text{Robert, James}), (\text{Sam, Teddy}), (\text{Robert, Teddy}), (\text{James, Sam})\}$$

representing the dominance relations among four men, $(a, b) \in D$ if a dominates b, is *not* a function since two pairs, (Robert, James) and (Robert, Teddy), have the same first element.

(b) The relation

$$E = \{(\text{Robert, James}), (\text{Sam, Teddy}), (\text{James, Sam}), (\text{Teddy, James})\}$$

is a function. The fact that the pairs (Robert, James) and (Teddy, James) have the same second element is of no consequence. ▶

Example 3 The pair (x, y) of real numbers x and y indicates the point lying x units to the right and y units above the origin $(0, 0)$. The distance D from $(0, 0)$ to (x, y) is $D = \sqrt{x^2 + y^2}$, computed using the Pythagorean Theorem for right triangles (see Figure 1(a)). For instance, if $x = 4$ and $y = 3$, then $D = \sqrt{4^2 + 3^2} = 5$.

FIGURE 1

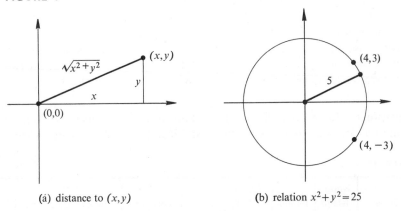

(a) distance to (x,y) (b) relation $x^2 + y^2 = 25$

To each point $P = (x, y)$ in the plane, there corresponds a distance $D = \sqrt{x^2 + y^2}$. The set of all pairs (P, D) is a function since to each point P there corresponds a unique distance D. Note that in this case the first elements P are themselves ordered pairs. ▶

Example 4 The points which all lie the same distance r from the origin form the circle with radius r and center at $(0, 0)$. If (x, y) is a point on this circle, then $\sqrt{x^2 + y^2} = r$ or $x^2 + y^2 = r^2$. A circle with radius 5 is a set of ordered pairs of real numbers and as such it constitutes a relation

$$C = \{(x, y): x^2 + y^2 = 25\} \tag{1}$$

defined on the set R of real numbers. Figure 1(b) is called the *graph* of this relation. Note that the relation C is *not* a function since it contains different pairs, for instance (4, 3) and (4, −3), having the same first elements.   ▶

When it is understood that the universal set is the set R of real numbers, we shall normally use the shorthand of writing only the description of a relation rather than the relation itself. For instance, in Figure 1(b) we speak of the relation $x^2 + y^2 = 25$ rather than writing the full statement (1).

Example 5   The points (x, y) on the horizontal line through the origin (called the horizontal axis or the x axis) all have the property that $y = 0$. No other points have this property. Hence this line is the graph of the relation $y = 0$ (see Figure 2). This relation is also a function since no two pairs have the same first element and different second elements.

FIGURE 2

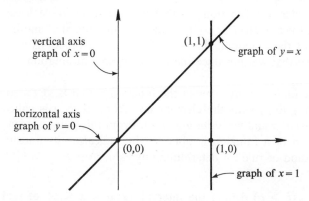

Similarly, the vertical line through the origin (the vertical or y axis) represents the graph of the relation $x = 0$, while the vertical line passing one unit to the right of the origin is the graph of the relation $x = 1$. Neither of these relations is a function, since in each case all pairs have the same first element, but different second elements.

The line passing through the origin at an angle of 45° with the positive end of the horizontal axis represents the graph of the relation $y = x$; that is, of the set of points $E = \{(x, y): y = x\}$. This relation is a function.   ▶

The following definition formalizes our discussion of functions and introduces some useful terminology.

Definition 1 A *function f* is a set of ordered pairs having the property that if (a, b) and (a, c) are both elements of f, then $b = c$. The set of all quantities which occur as first members of elements of f is called the *domain* of f and is denoted by $D(f)$. The set of all quantities which occur as second members of elements of f is called the *range* of f and is denoted by $R(f)$. ▶

Example 6 (a) The domain of the function determined by Table 1 is the set of states

{Rhode Island, Illinois, North Dakota, Kansas, Louisiana, Hawaii,
Oklahoma, California}

while the range is the set

{5, 19, 24, 38, 51, 113, 165}

(b) The domain of the function $y = 0$ graphed in Figure 2 is the set R of all real numbers. The range is the set {0} containing the single element zero. ▶

PROBLEMS

    1. What are the domain and range of the function in Example 2(b)? Of the function $y = x$ in Example 5?

The essential character of a function f is that it associates with each element of a certain set, the domain $D(f)$ of the function, an element of another set, the range $R(f)$ of the function. This way of looking at functions is so convenient in practice that we introduce the following special notation and terminology.

If the ordered pair (a, b) is a member of the function f, that is, if f associates the element $b \in R(f)$ with the element $a \in D(f)$, then b is called the *value of f at the point a* and we write $b = f(a)$. From this point of view, a complete description of a function is obtained by indicating its domain and describing a method or rule for determining its values.

Example 7 Let $I_{a,b}$ $(b \geq a)$ denote the interval $\{x: a \leq x \leq b\}$ of real numbers lying between a and b, inclusive (see Figure 3). Let L be the function which assigns to each interval its length. Then

$$L(I_{a,b}) = b - a$$

The domain of L is the collection of all intervals on the real line while the range of L is the set of all non-negative real numbers. ▶

FIGURE 3

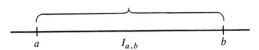

a $I_{a,b}$ b

2. What is the value of $L(I_{4,10})$? of $L(I_{-3,6})$?

Example 8 The *complement function* C is defined on the subsets of a universal set U by

$$C(A) = A'$$

The domain and range of C are the same, each being the collection of all subsets of U. ▶

PROBLEMS

3. Let $U = \{a, b, c, d\}$, $A = \{a, c\}$, $B = \{b, c, d\}$, and $D = \{b, d\}$. Find $C(A)$, $C(B)$, $C(A \cup B)$, and $C(A \cap B)$. What is the domain of the function C?

Example 9 The *identity function* I is defined on any set A by

$$I(a) = a \quad \text{for each } a \in A$$

That is, an identity function pairs every element with itself. The function $y = x$ in Example 5 is the identity function on the set of real numbers R.

▶

Example 10 Let U be a given universal set and let n be the function which assigns to any subset A of U the number $n(A) =$ number of elements in A. For instance, $n(\{a, b, c\}) = 3$, $n(\{7, 2\}) = 2$, and $n(\emptyset) = 0$. ▶

PROBLEMS

4. (a) Suppose the set U in Example 10 contains k elements. Prove that for any subset A of U, $n(A') = k - n(A)$.

   (b) Prove that if subsets A and B have no elements in common, that is, if $A \cap B = \emptyset$, then $n(A \cup B) = n(A) + n(B)$.

A geometrical way of viewing a function is shown in Figure 4. An element c in the domain of f is *mapped*, or *transformed*, by f into the corresponding element $b = f(c)$ in the range of f. In this way of looking at functions, b is sometimes called the *image* of c relative to the function f.

FIGURE 4 A function as a mapping

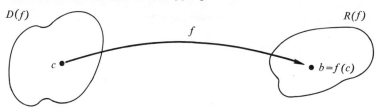

Still another way of visualizing a function is as a machine or system which accepts elements of $D(f)$ as inputs and which produces corresponding elements of $R(f)$ as outputs (see Figure 5). If we insert an element $a \in D(f)$ into the system, the corresponding value $f(a)$ comes out. If another element $c \in D(f)$ is inserted, we obtain another (not necessarily different) value $f(c)$. If we try to insert something not contained in the domain of f, it is rejected, for f operates only on elements belonging to its domain. Interpreting a function in this way makes clear the distinction between a function (the machine) and its values (outputs of the machine). A function should no more be confused with its values than a vending machine should be confused with a soft drink.

FIGURE 5 A function as a machine

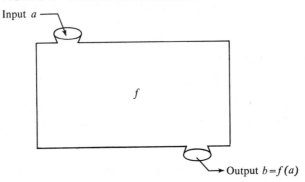

Input a

f

Output $b = f(a)$

PROBLEMS

5. Which of the following are functions?

   (a) $\{(6, 1), (5, 4), (4, 1), (10, 5)\}$

   (b) $\{(\text{Red}, 10), (\text{Green}, 9), (\text{Red}, 8)\}$

   (c) $\{\text{Jones}, 1934\}$

   (d) $\{(\text{Corn}, 34\cancel{c}), (\text{Beans}, 16\cancel{c}), (\text{Bread}, 25\cancel{c})\}$

6. Define a function whose domain is the set

$$W = \{\text{John, red, plays, I, softly, Iowa}\}$$

and whose range is

$$P = \{\text{verb, noun, adverb, pronoun, adjective}\}$$

7. Plot the points (ordered pairs) in each of the following tables and then draw a graph which passes through the points.

(a)

x	1	2	3	4	5	6
y	5	7	9	11	13	15

(b)

x	1	2	3	4	5	6
y	3	6	11	18	27	38

8. Which of the following relations defined on the set R of all real numbers also qualify as functions?

(a) $y = x^2 - x - 6$

(b) $y^2 - y = x + 4$

(c) $y^2 - x^2 = 25$

(d) $y^2 = x^4$

(e) $9x^2 + 25y^2 = 225$

(f) $9x^2 - 16y^2 = 144$

9. The total cost C of a certain commodity is given in terms of the quantity Q as

$$C = f(Q) = 3 - 0.2Q + 0.1Q^2$$

For the values $Q = 1, 2, 3, 4, 5$, calculate the corresponding values of C, plot the points (Q, C), and draw the graph of the function f through the points.

10. Deutsch* has proposed the following index of government stability:

$$S_t = f(g, L, \text{pol}, y, y_{10}) = \frac{g}{L \cdot (\text{pol})} \frac{y}{y_{10}}$$

where S_t indicates stability, g is the ratio of government income to total national income, L is the percentage of literacy, "pol" the percentage of political participation, y is the per capita national income, and y_{10} is the percentage of the total national income received by the top 10% of income receivers.

(a) Assuming that $g = \frac{2}{5}$, $L = \frac{4}{5}$, $y = 2000$, and $y_{10} = \frac{1}{5}$, compute S_t when pol $= \frac{1}{10}, \frac{1}{5}, \frac{1}{2}, \frac{3}{4}$.

(b) Solve the above equation for pol, thereby obtaining a formula which expresses pol as a function of g, L, S_t, y, and y_{10}.

11. It has been empirically demonstrated† that if $N(i)$ is the number of cities with population greater than i million, then

$$N(i) \approx ki^{-\rho}$$

where k and ρ are constants (the symbol $\approx$ means approximately equal). The 1960 U. S. census lists 5 metropolitan areas with population in excess of 3 million and 24 metropolitan areas with population in excess of 1 million. Use these figures to compute estimates of the constants k and ρ. Then draw a curve representing the function N.

12. The following is a simple system of equations in macroeconomic variables:

$$Y = C + I$$
$$C = a + bY$$
$$I = u + vY$$

*Deutsch, Karl, "Towards an Inventory of Basic Trends and Patterns in Comparative and International Politics," *American Political Science Review* **54**, 34–57 (1960).

†See Zipf, G. K., *Human Behavior and the Principle of Least Effort* (Addison-Wesley Publishing Company, Reading, Mass., 1949), Chapters 9 and 10.

where C is the money value of aggregate consumption, I is the money value of aggregate investment, and a, b, u, and v are real-valued constants. Solve the above system of equations for Y, C, and I thereby representing these quantities as functions of the constants only.

13. Harrah* has a model of how a rational human receiver behaves in certain communication situations. In his model, Harrah postulates the existence of a *semantic information function I* such that if one declarative sentence in the receiver's language L implies another, the I values of the two sentences are connected in a specified manner. The statement "F is L-true" is defined by Harrah to mean that F is a true statement according to the conventions of the language system L. The statement "F L-implies G" is defined to mean that $(F \to G)$ is a theorem of L and similarly for "F and G are L-equivalent" and "F is L-false". The statement "F is L-consistent" means that F is not L-false.

The information function I assigns real numbers to statements of L and is characterized by the following assumptions:

(1) $0 \leq I(F) \leq \infty$.

(2) $I(F) = 0 \Leftrightarrow F$ is L-true.

(3) $I(F) = \infty \Leftrightarrow F$ is an L-false statement.

(4) If F L-implies G, then $I(F) \geq I(G)$.

(5) The information of F, given G, is defined by

$$I(F \mid G) = I(F \wedge G) - I(G)$$

From these assumptions, prove the following theorems:

(a) If F and G are L-equivalent, then $I(F) = I(G)$.

(b) If F L-implies G, then $I(F \wedge G) = I(F)$.

(c) $I(F \wedge G) \geq I(F) \geq I(F \vee G)$.

(d) $I(F \mid G) \geq 0$.

(e) If F L-implies G, then $I(G \mid F) = 0$.

(f) If F L-implies G, then $I(F \mid H) \geq I(G \mid H)$.

(g) $I[(F \wedge G) \mid (G \wedge H)] = I[F \mid (G \wedge H)]$.

14. Let n be the function defined in Example 10. Prove that for any subsets A and B
$$n(A \cup B) = n(A) + n(B) - n(A \cap B)$$

15. Let $D(x, y)$ denote the distance between quantities x and y.† The distance function D is said to be a *metric* if it satisfies four axioms:

(i) $D(x, x) = 0$

(ii) $D(x, y) \geq 0$

(iii) $D(x, y) = D(y, x)$

(iv) $D(x, y) + D(y, z) \geq D(x, z)$

*Harrah, D., "A Model of Semantic Information and Message Evaluation," in *Mathematical Explorations in Behavioral Science*, Massarik, F., and Ratoosh, P., eds. (Irwin, Inc., Homewood, Ill., 1965), Chapter 6.

†Adapted from Restle, F., *Psychology of Judgement and Choice: A Theoretical Essay* (John Wiley & Sons., Inc., New York, 1961).

Let s_1 and s_2 be two situations from the universal set of situations that a person might encounter. Let S_1 and S_2 be the sets of aspects corresponding to situations s_1 and s_2, respectively, and let n be the counting function defined in Example 10.

The *distance* between two situations s_1 and s_2 is a function of the common aspects and is defined as

$$\begin{aligned}
D(s_1, s_2) &= n[(S_1 \cap S_2') \cup (S_1' \cap S_2)] \\
&= n(S_1 \cap S_2') + n(S_1' \cap S_2) \\
&= n[(S_1 \cup S_2) - (S_1 \cap S_2)] \\
&= n(S_1) + n(S_2) - 2n(S_1 \cap S_2)
\end{aligned}$$

Show that this distance function is a metric. (*Hint:* See Problem 14.)

4.2 ABSOLUTE VALUE FUNCTION

There are many special mathematical functions that will be of interest to us. Among them is the absolute value function.

Definition 2   The *absolute value* of a real number x, denoted $|x|$, is defined by

$$\begin{aligned}
|x| &= x & \text{when } x > 0 \\
|x| &= 0 & \text{when } x = 0 \\
|x| &= -x & \text{when } x < 0
\end{aligned} \tag{2}$$

Example 1 $|5| = 5$, $|\frac{3}{2}| = \frac{3}{2}$, $|0| = 0$, $|-1| = -(-1) = 1$, $|-\frac{7}{4}| = -(-\frac{7}{4}) = \frac{7}{4}$, and $|-5| = -(-5) = 5$. ▶

The definition of absolute value determines a function A which associates with each real number another number $A(x) = |x|$, called its absolute value. The graph of the absolute value function A is shown in Figure 6. Note that

FIGURE 6   Graph of the absolute value function

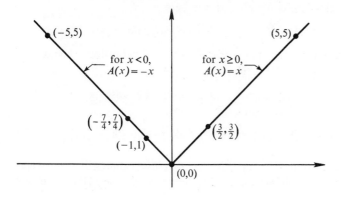

$|x|$ is never negative, no matter what the value of x may be. In fact, $|x| = 0$ if and only if $x = 0$; otherwise $|x|$ is positive.

PROBLEMS

1. Let $x = 2$ and $y = -1$. Then compare:
   (a) $|x \cdot y|$ with $|x| \cdot |y|$
   (b) $\left|\dfrac{x}{y}\right|$ with $\dfrac{|x|}{|y|}$
   (c) $|x|$ with $|-x|$
   (d) $|x + y|$ with $|x| + |y|$
   (e) $|x - y|$ with $|x| - |y|$

2. Let $x = -2$ and $y = -3$ in Problem 1.

3. Let $x = 3$ and $y = 4$ in Problem 1. What general conclusions do you draw from Problems 1–3?

The following list contains important properties of the absolute value function which we shall use frequently. We shall provide some of the proofs and leave the rest to you. In each case you should compare these results with your calculations and conclusions in Problems 1–3.

Property 1 Absolute value is never negative. That is, $|x| \geq 0$ no matter what the value of x may be. Also $|x| = 0$ if and only if $x = 0$.

PROOF The definition of absolute value takes a different form depending on whether the quantity whose absolute value is desired is positive or negative. Thus the proof of each property must consider positive and negative cases separately.

For this particular property, we note from the definition that when x is positive, $|x| = x$, so that $|x|$ is positive. If x is negative, then $-x$ is positive. By definition, $|x| = -x$, so again $|x|$ is positive. The fact that $|0| = 0$ completes the proof. ▶

Property 2 The absolute value of a number is the same as that of its negative. In symbols, $|x| = |-x|$ for all real numbers x.

PROOF As in the proof of Property 1, we consider three possible cases:

Case   I We have $x > 0$ so that $-x < 0$. Then by definition $|x| = x$ and $|-x| = -(-x) = x$. Hence $|x| = |-x|$.

Case II We have $x < 0$ so that $-x > 0$. In this case $|x| = -x$ and $|-x| = -x$ also. Again, $|x| = |-x|$.

Case III Since $0 = -0$, it automatically follows that $|0| = |-0|$.   ▶

Since it is obvious that all properties of absolute value will hold when $x = 0$, we shall henceforth omit the consideration of case (III) from our proofs. It will, of course, be understood that the given property holds for $x = 0$.

Geometrically, absolute value may be interpreted as *distance*. If x and y are two points on a line and $x < y$ (see Figure 7), the distance between x and y is $y - x$. In terms of absolute value, this distance may be written *either* as $|y - x|$ *or* as $|x - y|$. As long as we use absolute value, we can subtract either way and it is not necessary to know in advance which number is larger (an application of Property 2).

FIGURE 7

$$\begin{array}{ccc} 0 & x & y \end{array}$$

Property 3 The absolute value of a product is equal to the product of absolute values. In symbols, $|xy| = |x| \cdot |y|$ for all real numbers x and y.

PROOF With two numbers there are four possible cases:

Case I Let $x > 0$ and $y > 0$. In this case, $xy > 0$ also, so $|x| = x$, $|y| = y$, and $|xy| = xy$. Hence $|xy| = |x| \cdot |y|$.

Case II Let $x > 0$ and $y < 0$. Here $xy < 0$, so $|x| = x$, $|y| = -y$, and $|xy| = -xy$. Since $-xy = x(-y)$, again we have $|xy| = |x| \cdot |y|$.

Case III Let $x < 0$ and $y > 0$. This is like case (II) with x and y interchanged.

Case IV Let $x < 0$ and $y < 0$. Here $xy > 0$, so $|x| = -x$, $|y| = -y$, and $|xy| = xy$. Since $xy = (-x)(-y)$, then $|xy| = |x| \cdot |y|$ in this case also.

Since we have considered all possible cases, the proof of Property 3 is completed. ▶

Property 4 Whenever division is defined, the absolute value of a quotient is equal to the quotient of absolute values. That is, if $y \neq 0$, then

$$\left| \frac{x}{y} \right| = \frac{|x|}{|y|}$$

PROOF The proof is similar to that of Property 3 and is left as an exercise. ▶

Property 5 For each real number x, $-|x| \leq x \leq |x|$. Equivalently, for any x we have both

$$x \leq |x| \quad \text{and} \quad -x \leq |x|$$

PROOF If $x > 0$, then $-x < 0 < x = |x|$. If $x < 0$, then $x < 0 < -x = |x|$. ▶

Property 6 *The Triangle Inequality* The absolute value of a sum is never greater than the sum of the respective absolute values. Symbolically,

$$|x + y| \leq |x| + |y|$$

for all numbers x and y.

PROOF Again there are four possible cases:

Case I Let $x > 0$ and $y > 0$. Here $x + y > 0$, so $|x + y| = x + y = |x| + |y|$. Thus equality holds.

Case II Let $x > 0$ and $y < 0$ so that $|x| = x$ and $|y| = -y$. Since $x + y$ could be either positive or negative, there are two subcases to consider:

Subcase (1) Let $x + y > 0$. Then $|x + y| = x + y \leq |x| + |y|$ by Property 5. Inequality may hold. For instance, $|3 + (-2)| = |1| = 1 < |3| + |-2| = 5$.

Subcase (2) Let $x + y < 0$. Then $|x + y| = -(x + y)$ and again using Property 5, we find

$$|x + y| = -(x + y) = -x - y \leq |x| + |y|$$

Case III Let $x < 0$ and $y > 0$. This case is similar to case (II) and is left as an exercise. Inequality can hold in this case.

Case IV Let $x < 0$ and $y < 0$. Here $x + y < 0$ so $|x + y| = -(x + y) = -x - y = |x| + |y|$. Thus equality holds. ▶

Property 7 When $a > 0$, $|x| \leq a$ if and only if $-a \leq x \leq a$. That is, the sets $\{x: |x| \leq a\}$ and $\{x: -a \leq x \leq a\}$ are identical.

PROOF The proof is left as an exercise. ▶

In words, Property 7 says that since $|x| = |x - 0|$, the number $|x|$ may be interpreted as the distance between 0 and x. This distance is less than a if and only if x lies between $-a$ and a (see Figure 8). A similar statement holds if we measure the distance from a point p rather than from zero. The point x (Figure 9) lies within distance a of p if and only if

$$|x - p| \leq a$$

or, equivalently,

$$p - a \leq x \leq p + a$$

FIGURE 8

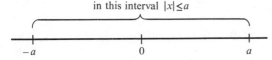

in this interval $|x| \leq a$

FIGURE 9

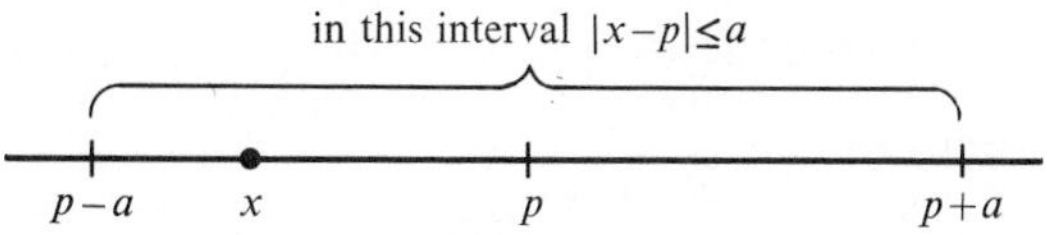

in this interval $|x-p| \le a$

$$p-a \qquad x \qquad\qquad p \qquad\qquad\qquad p+a$$

Property 8 For all x, $\sqrt{x^2} = |x|$. Note that $\sqrt{x^2}$ is equal to x *only* when x is positive or zero! If $x < 0$, then $\sqrt{x^2} = -x$.

PROOF Recall that $\sqrt{t}$ is defined for $t \ge 0$ as the non-negative number whose square is t (for example, $\sqrt{4} = 2$, $\sqrt{9} = 3$). Hence if $x > 0$, $\sqrt{x^2} = x$ since x is non-negative and its square is x^2. Similarly, if $x < 0$, then $\sqrt{x^2} = -x$ since $-x$ is non-negative and its square is x^2. In this case $\sqrt{x^2} \neq x$ since x is negative and $\sqrt{x^2}$ is, by definition, positive. But in either case, $\sqrt{x^2} = |x|$. ▶

The proofs of the four remaining properties are left as exercises.

Property 9 For all x, $|x|^2 = x^2$.

Property 10 For all x and y, $xy \le |x| \cdot |y|$.

Property 11 For all x and y, $|x - y| \le |x| + |y|$.

Property 12 For all x and y, $|x - y| \ge ||x| - |y||$. ▶

PROBLEMS

4. Verify each of the 12 properties of absolute value using $x = 4$ and $y = -3$.

5. In each of the following cases, determine whether the relation "$=$" or "$>$" holds for the statement $|a| + |b| \ge |a + b|$ (Property 6, the triangle inequality).

(a) $a = 4$, $\quad b = -2$ $\qquad\qquad$ (b) $a = -4$, $b = -2$
(c) $a = 4$, $\quad b = 2$ $\qquad\qquad\;$ (d) $a = 0$, $\quad b = -2$
(e) $a = 0$, $\quad b = 0$

6. Determine whether the relation "$=$" or "$>$" holds for the statement $|a - b| \ge ||a| - |b||$ (Property 12).

(a) $a = 0$, $\quad b = 2\theta$ for $\theta = $ some constant
(b) $a = -\theta$, $b = \sqrt{2}$, for $\theta > 0$
(c) $a = 4$, $\quad b = 0$
(d) $a = -1$, $b = -2$
(e) $a = 1$, $\quad b = -2$

7. Show that the inequality $|a - b| \ge ||a| - |b||$ is equivalent to the inequality $|ab| \ge ab$. (*Hint:* Square both sides.)

8. Show that the triangle inequality

$$|x| + |y| \geq |x + y|$$

can be restated equivalently as

$$\sqrt{x^2} + \sqrt{y^2} \geq \sqrt{(x + y)^2}$$

9. Show that if $xy \geq 0$ then xy is at least as big as the smaller of x^2 and y^2.

10. Solve the inequality $|x - 3| < 6$. Sketch the solution set as in Figure 9. (*Hint:* Use Property 7.)

11. Solve the inequality $|1 - 2x| < 1$. Sketch the solution set.

12. Solve the inequality $|2x + 3| \leq 13$. Sketch the solution set.

13. Prove that if $y > 0$, then $|x| > y$ if and only if either $x < -y$ or $x > y$. Illustrate with a sketch.

14. Interpret Problems 10–13 in terms of distance.

15. Prove Properties 4, 7, 9, 10, 11, and 12 for absolute value. In each case give an example where equality holds and, if possible, another example in which inequality holds.

4.3 BOUNDED SETS OF REAL NUMBERS

In this section we shall discuss some basic concepts concerning sets of real numbers. Some of the ideas which we want to introduce are already familiar to you. A typical case is the concept of the *maximum*, or largest element, of a set. When we say that m is the maximum of a set S, written $m = \max S$, we mean that

$$m \geq x \quad \text{for } \textit{every} \text{ element } x \in S$$

and

$$m = y \quad \text{for } \textit{some} \text{ element } y \in S.$$

For instance, for the set $S = \{7, 3, 9, -6, 2\}$, $\max S = 9$.

Example 1 Not every set of real numbers has a maximum. Consider, for instance, the set $I = \{x : 0 < x < 1\}$ of all real numbers lying between zero and one. The number 1 is not a maximum, for while it is true that $1 \geq x$ for every $x \in I$, it is not true that 1 is an element of I. On the other hand, no member of I can be a maximum. For if $m \in I$, then $0 < m < 1$ and it follows that $m < (m + 1)/2 < 1$. That is, the number $(m + 1)/2$ is an element of I which is greater than m. We conclude that I has no maximum. ▶

If B is a set of real numbers and u is a number such that $u \geq x$ for every $x \in B$, then u is called an *upper bound* for B. A set which has an upper bound is said to be *bounded above*.

Our discussion in Example 1 shows that 1 is an upper bound for *I*. Obviously, any number $u > 1$ is also an upper bound. But we have also argued that no number smaller than 1 can be an upper bound. In this sense the number 1 is the smallest or least upper bound for the set *I*.

Definition 3 If *C* is a set of real numbers and *b* is a number such that

  (i) *b* is an upper bound for *C*

  (ii) if *u* is any upper bound for *C*, then $b \leq u$

then *b* is called the *least upper bound* for *C* and is written $b = \text{lub } C$. ▶

A set can have only one least upper bound. For if *b* and *d* were both least upper bounds of a set *C*, then since *b* is a least upper bound and *d* is an upper bound we must have $b \leq d$ (Property (ii) above). For entirely similar reasons we also have $d \leq b$, and thus $b = d$.

PROBLEMS

1. For each of the following sets, what is the maximum? What is the least upper bound?

   (a) $\{1, 2, 3\}$ (b) $\{x: 0 < x \leq 1\}$

   (c) The union $A = \{x: 0 < x \leq 1\} \cup \{x: 1 < x < 2\}$

It is a fact of great importance that every nonempty set of real numbers which has an upper bound has a least upper bound. This seemingly obvious statement is, unfortunately, impossible to prove without a deeper study of the real numbers than we can indulge in here. We shall simply accept it as an axiom.*

Least Upper Bound (LUB) Axiom Every nonempty set of real numbers which is bounded above has a least upper bound. ▶

PROBLEMS

2. The concept of least upper bound includes the idea of a maximum. Prove that if *C* has a maximum then $\max C = \text{lub } C$. (It is because some bounded sets do not have maxima that we have introduced the more general concept of least upper bound.)

3. Prove that a set can have at most one maximum.

A number *b* is said to be a *lower bound* for a set *C* if $b \leq x$ for every $x \in C$. The concept of lower bound is analogous to that of upper bound. If *b* is a lower bound which is a member of *C*, then *b* is called the *minimum* of *C* and

*For a discussion of the principles involved, see Bartle, R. G., *The Elements of Real Analysis* (John Wiley & Sons, Inc., New York, 1964), Sections 4–6. More extensive discussions may be found in the references which he cites.

we write $b = \min C$. The number b is the *greatest lower bound* of C, $b = $ glb C, if

    (i) b is a lower bound for C

    (ii) if d is any lower bound for C, then $b \geq d$

A set is *bounded below* if it has a lower bound. A set which has both an upper bound and a lower bound is said to be *bounded*.

Example 2   Zero is the greatest lower bound of the set

$$I = \{x: 0 < x < 1\}$$

Firstly, it is obvious from the definition of I that $0 < x$ for every $x \in I$. Secondly, no lower bound can be greater than zero. For if $b > 0$, then $0 < b/2 < b$. That is, $b/2 \in I$ and $b/2 < b$. Hence b is not a lower bound. I has no minimum. However, it is bounded since it has an upper bound, 1, and a lower bound, 0.   ▶

PROBLEMS

    4. Rework Problems 2 and 3 with the word maximum replaced by minimum and least upper bound replaced by greatest lower bound.

One of the most important results implied by the LUB Axiom is known as the Archimedean property of the real numbers.

Theorem 1   *The Archimedean Property*   Let r be any positive real number. Then there exists a positive integer n such that $n > r$.

PROOF   An indirect proof is best. Suppose that the conclusion is false. That is, suppose that for every positive integer n we have $r \geq n$. Then the number r is an upper bound for the set $N = \{1, 2, 3, 4, \ldots\}$ of positive integers. It follows from the LUB Axiom that N has a least upper bound b. Now if n is a positive integer, then so is $n + 1$. Hence $b \geq n + 1$ for all $n \in N$ or, equivalently, $b - 1 \geq n$ for all n. But this means that $b - 1$ is an upper bound for N, an impossibility since $b - 1 < b$ and b is the least upper bound.

We arrived at this contradiction by assuming that N was bounded. This supposition must be false and hence the theorem is true.   ▶

PROBLEMS

    5. Since N, the set of positive integers, is unbounded, it follows that any set (such as the set R of real numbers) of which N is a subset, must also be unbounded. Prove this.

An alternative version of the Archimedean property which we shall find useful is the following theorem.

Theorem 1a If r is any positive number, no matter how small, there exists a positive integer n such that $1/n < r$.

PROOF This is merely a rephrasing of Theorem 1. For if $r > 0$, then $1/r > 0$ also. Choosing $n > 1/r$ yields $1/n < r$. ▶

Theorem 1a says essentially that there is no such thing as an arbitrarily small positive number. No matter how small a positive number we may choose, there is a number of the form $1/n$ which is smaller.

Example 3 Let $D = \{1, \frac{1}{2}, \frac{1}{3}, \frac{1}{4}, \frac{1}{5}, \ldots\} = \{1/n: n \text{ is a positive integer}\}$. Since $1 \in D$ and $1 \geq 1/n$ for each positive integer n, we see that $1 = \max D = $ lub D. Zero is a lower bound for D since $0 < 1/n$ for all n. But if $r > 0$, Theorem 1a shows that there exists $1/n \in D$ such that $1/n < r$. Hence r is not a lower bound and glb $D = 0$. There is no minimum element in D. ▶

PROBLEMS

6. State precisely, using proper notation, what is meant by "b is not a lower bound for the set A."

7. Show that a set cannot have two distinct upper bounds both of which belong to the set.

8. Prove that a nonempty set has at most one glb.

9. Given the set $A = \{\frac{2}{3}, \frac{4}{5}, \frac{6}{7}, \frac{8}{9}, \frac{10}{11}, \ldots, 2n/(2n+1), \ldots\}$.

   (a) Find the glb and lub of this set.

   (b) If $c = $ lub A, find a member of the set which is greater than $c - \frac{1}{100}$.

10. Let A be a nonempty set and define the set B by $B = \{x: -x \in A\}$. Prove that $c = $ lub A if and only if $-c = $ glb B.

11. Prove that if B is a bounded set of real numbers and if A is a nonempty subset of B, then

$$\text{glb } A \geq \text{glb } B \quad \text{and} \quad \text{lub } A \leq \text{lub } B$$

12. Prove that if $b \in H$ and b is an upper bound for H, then $b = $ lub H.

SUPPLEMENTARY READING

Anderson, K. W., and Hall, D. W., *Sets, Sequences and Mappings* (John Wiley & Sons, Inc., New York, 1963), Chapter 1.

Good, R. A., *Introduction to Mathematics* (Harcourt, Brace & World, Inc., New York, 1966), Chapters 9 and 10.

SEQUENCES **5**

5.1 BASIC CONCEPTS

A function whose domain is the set $N = \{1, 2, 3, \ldots\}$ of positive integers is called a *sequence*. Since a function is a collection of ordered pairs and its domain is the set of first elements in these pairs, it follows that a sequence must be a set like

$$S = \{(1, a_1), (2, a_2), (3, a_3), (4, a_4), \ldots\}$$

Such notation is quite cumbersome and since we understand that the domain of the sequence is the set N, we use the simpler notation

$$S = (a_1, a_2, a_3, \ldots) \quad \text{or simply } S = (a_n)$$

to denote the sequence. For any positive integer n, the quantity a_n is the *value* of S corresponding to the integer n. The symbol ( ), rather than { }, is used to denote a sequence in order to emphasize that we are concerned not only with the set of values of the sequence, but also with the order in which they appear.

Example 1 Suppose we begin at one end of a line that is two units long and move toward the other end in a succession of steps, each of which covers half the remaining distance. From our initial position zero, shown in Figure 1, we step half the total distance to arrive at the first new position $P_1 = 1$.

FIGURE 1

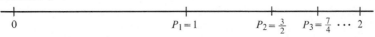

The remaining distance is now one unit, and stepping half this distance brings us to position $P_2 = \frac{3}{2}$. Continuing in this fashion brings us successively to $P_3 = \frac{7}{4}$, $P_4 = \frac{15}{8}$, $P_5 = \frac{31}{16}$, $P_6 = \frac{63}{32}$, ... In this way we obtain the sequence

$$(P_n) = (P_1, P_2, P_3, \ldots) = (1, \tfrac{3}{2}, \tfrac{7}{4}, \tfrac{15}{8}, \tfrac{31}{16}, \tfrac{63}{32}, \ldots)$$

It is easy to verify that in general the nth position will be

$$P_n = 2 - \frac{1}{2^{n-1}} \quad \text{for } n = 1, 2, 3, \ldots \tag{1}$$

For instance, $P_3 = \frac{7}{4} = 2 - (1/2^2)$ and $P_5 = \frac{31}{16} = 2 - (1/2^4)$. ▶

One common way of indicating a sequence is to list the first few values and to use dots to indicate that the sequence continues according to the pattern thereby established. For instance, the set E of positive even integers might be indicated by $E = (2, 4, 6, 8, \ldots)$. This notation has the slight disadvantage that one can never be absolutely certain what the pattern is. The notation $(2, 4, 6, 8, \ldots)$ also corresponds, for instance, to the first four values of the sequence given by the formula

$$S(n) = 2n + (n - 1)(n - 2)(n - 3)(n - 4) \tag{2}$$

whose next value is $S(5) = 10 + (5 - 1)(5 - 2)(5 - 3)(5 - 4) = 34$, rather than 10. However, as a practical matter, the use of a notation like $(2, 4, 6, 8, \ldots)$ rarely, if ever, causes confusion. Everyone "knows" that the sequence of even integers is intended. Hence we will use such notation as a matter of convenience, writing a defining relation such as (2) in less obvious cases.

Specifying the values of a sequence by a formula, as in (1) and (2), allows no ambiguity as to what the sequence is. The same clarity is obtained if the sequence is specified by a *recursive formula*. That is, we may specify the first value a_1 and give a rule for finding a_n ($n \geq 2$) once a_{n-1} is known. Or, more generally, we may specify a_1 and give a rule for obtaining a_n from $a_1, a_2, \ldots, a_{n-1}$. For example, the sequence $E = (a_1, a_2, a_3, \ldots)$ of even integers is given recursively either by

$$a_1 = 2 \quad \text{and} \quad a_n = a_{n-1} + 2, n \geq 2$$

or by

$$a_1 = 2 \quad \text{and} \quad a_n = a_{n-1} + a_1, n \geq 2$$

Of course, the simplest and most natural way to specify E is to write $a_n = 2n$ for each integer $n \in N$.

Example 2 When money is borrowed at simple interest, the amount of interest which accrues is the same for each time unit of given length (say a year), and is determined by multiplying the per time unit interest rate i by the original amount borrowed P. Hence if A_n is the total amount to be re-paid after n time periods, the sequence $(A_n) = (A_1, A_2, A_3, \ldots)$ is deter-mined by the recursive formula

$$A_1 = P + iP$$

and, for $n \geq 2$,

$$A_n = A_{n-1} + iP$$

The first four terms of the sequence are

$$A_1 = P + iP$$
$$A_2 = A_1 + iP = P + 2iP$$
$$A_3 = A_2 + iP = P + 3iP$$
$$A_4 = A_3 + iP = P + 4iP$$

It is apparent that this pattern will continue and that for any $n \in N$,

$$A_n = P + niP = P(1 + in) \qquad \blacktriangleright$$

PROBLEMS

1. If \$100 is borrowed at 5% simple interest for three years, what is the final amount due? 115.00

A sequence (a_n) is said to be *increasing* if $a_n \geq a_{n-1}$ for each positive integer n. The sequences in Examples 1 and 2, the sequence E of even integers, and the sequence $(0, 0, 1, 1, 2, 2, 3, 3, \ldots)$ are all increasing. A sequence (a_n) is *decreasing* if $a_n \leq a_{n-1}$ for each integer $n \in N$. An example is the *harmonic sequence*

$$H = \left(1, \frac{1}{2}, \frac{1}{3}, \ldots, \frac{1}{n}, \ldots\right)$$

The *constant sequence* $(3, 3, 3, 3, \ldots)$ is both increasing and decreasing.

PROBLEMS

2. Give two examples of increasing sequences and two of decreasing sequences.

3. Is the sequence (a_n) defined by

$$a_1 = 1$$
$$a_n = \tfrac{1}{2}(2 - a_{n-1}), \quad n \geq 2$$

increasing? decreasing?

A real number u is called an *upper bound* for the sequence (a_n) if $a_n \leq u$ for every n. Similarly, the number c is a *lower bound* for (a_n) if $c \leq a_n$ for every n. We say that a sequence is *bounded above* if it has an upper bound,

bounded below if it has a lower bound, and *bounded* if it is bounded above
and bounded below.

The sequence $(1, \frac{3}{2}, \frac{7}{4}, \ldots)$ of Example 1 is bounded. In fact, 0 is a lower
bound and 2 an upper bound for this sequence. Similarly $H = (1, \frac{1}{2}, \frac{1}{3}, \ldots)$
is bounded below by 0 and above by 1 and so H is bounded. The sequence
$E = (2, 4, 6, \ldots)$ of even integers is bounded below by zero but is not
bounded above. The sequence $(1, -1, 2, -2, 3, -3, \ldots)$ is not bounded
either above or below.

PROBLEMS

    4. For the sequence (a_n) in Problem 3

        (a) is 2 an upper bound?

        (b) is 0 a lower bound?

        (c) is $\frac{3}{4}$ an upper bound?

If $A = (a_n)$ is a sequence and if $r_1 < r_2 < r_3 < \cdots < r_n < \cdots$ is a
strictly increasing sequence of positive integers, then the sequence

$$B = (a_{r_1}, a_{r_2}, a_{r_3}, \ldots)$$

is called a *subsequence* of A. Thus, for example, the sequences $E =
(2, 4, 6, 8, \ldots)$ and $F = (1, 3, 5, 7, \ldots)$ are both subsequences of the
sequence $N = (1, 2, 3, 4, \ldots)$ of positive integers. One possible subsequence of the sequence $(1, -1, 1, -1, 1, -1, \ldots)$ is the *constant sequence*
$(1, 1, 1, 1, \ldots)$ obtained by choosing every other term beginning with the
first. The sequence $(a_3, a_4, a_5, a_6, \ldots)$ is a subsequence of $A =
(a_1, a_2, a_3, a_4, \ldots)$. In fact any sequence obtained from A by choosing
elements a_n so that the *subscripts* are in ascending order is a subsequence
of A.

PROBLEMS

    5. One subsequence of $H = (1, \frac{1}{2}, \frac{1}{3}, \frac{1}{4}, \ldots)$ is the sequence $K = (1, \frac{1}{3}, \frac{1}{5},$
        $\frac{1}{7}, \ldots)$. Write another subsequence of H. Write a subsequence of K. Is
        every subsequence of K also a subsequence of H?

If $A = (a_n)$ and $B = (b_n)$ are two sequences, then the *sum* $S = A + B$
of A and B is the sequence whose elements are, respectively, the sums of the
corresponding elements of A and B. That is, $S = (s_n)$ is the sum of the sequences A and B if for each n, $s_n = a_n + b_n$.

Similarly, the *difference* $D = A - B$ is the sequence $D = (d_n)$ defined by
$d_n = a_n - b_n$. The *product* $P = A \cdot B$ is the sequence $P = (p_n)$ defined by
$p_n = a_n \cdot b_n$. If no element b_n is zero, the *quotient* of A and B is the sequence
$Q = (q_n) = A/B$ defined by $q_n = a_n/b_n$.

Example 3 If A and B are the sequences $A = (1, 3, 5, 7, 9, 11, \ldots)$ and $B = (1, -\frac{1}{2}, \frac{1}{3}, -\frac{1}{4}, \frac{1}{5}, -\frac{1}{6}, \ldots)$ then

$$A + B = (2, \tfrac{5}{2}, \tfrac{16}{3}, \tfrac{27}{4}, \tfrac{46}{5}, \tfrac{65}{6}, \ldots)$$

$$A - B = (0, \tfrac{7}{2}, \tfrac{14}{3}, \tfrac{29}{4}, \tfrac{44}{5}, \tfrac{67}{6}, \ldots)$$

$$A \cdot B = (1, -\tfrac{3}{2}, \tfrac{5}{3}, -\tfrac{7}{4}, \tfrac{9}{5}, -\tfrac{11}{6}, \ldots)$$

$$A/B = (1, -6, 15, -28, 45, -66, \ldots)$$

Similarly, if $C = (1, 0, -1, 1, 0, -1, 1, 0, -1, \ldots)$, we have defined $A + C$, $A - C$, and $A \cdot C$, but A/C is not defined since some of the elements of C are zero. ▶

PROBLEMS

6. Write out the first four terms of the sequences for which the nth term is

(a) $\dfrac{n}{n^2 + 2}$ (b) $(-1)^{2n} \dfrac{n}{(n+1)(n+2)}$ (c) $1 + \dfrac{1}{n^2}$

(d) $\dfrac{2^n}{2^n + 1}$ (e) $n^2 - 2n$

7. Which of the sequences in Problem 6 are bounded? Write upper and lower bounds for each bounded sequence.

8. In each case the first four terms of a sequence (a_n) are given. Write a formula for a_n which will produce the values given for $n = 1, 2, 3,$ and 4.

(a) $\dfrac{1+9}{3^2}, \dfrac{4+9}{4^2}, \dfrac{9+9}{5^2}, \dfrac{16+9}{6^2}$

(b) $\dfrac{2}{5}, \dfrac{2^2}{5^2}, \dfrac{2^3}{5^3}, \dfrac{2^4}{5^4}$

(c) $-1 + \sqrt{3}, 0 + \sqrt{4}, 1 + \sqrt{5}, 2 + \sqrt{6}$

(d) $4 - \sqrt{5}, 8 - \sqrt{8}, 12 - \sqrt{13}, 16 - \sqrt{20}$

(e) $\dfrac{4+2}{1}, \dfrac{8+2}{2 \cdot 1}, \dfrac{12+2}{3 \cdot 2 \cdot 1}, \dfrac{16+2}{4 \cdot 3 \cdot 2 \cdot 1}$

9. A trainer is attempting to condition 50 dogs to stand on their heads at the sound of a buzzer. Being of a theoretical bent, he has hypothesized that the average number of dogs who are conditioned at the completion of the kth trial is

$$N_c(k) = 50 - 50(1 - \theta)^k$$

where θ is a learning parameter, $0 < \theta < 1$.

(a) Given $\theta = 0.5$, how many dogs on the average would the trainer expect to be conditioned to the buzzer at the completion of the (i) first (ii) second, (iii) third, (iv) nth trial?

(b) Is the sequence $(N_c(k))$ bounded above (below)? If so, indicate an upper (lower) bound.

(c) If the trainer were to continue the training trials indefinitely, how many dogs would he expect never to be conditioned?

10. The Mugivit Company wishes to increase its sales revenue by 3% per quarter. If S_t is the sales revenue in quarter t and S_{t-1} is the sales revenue in the immediately preceding quarter, then the quarterly growth rate G (in percent) of the sales revenue is given by

$$G = 100 \frac{S_t - S_{t-1}}{S_{t-1}}$$

Given that $S_1 = \$100,000$ calculate the required sales revenue needed to meet the growth requirements in the (a) second, (b) third, (c) fourth, (d) nth quarters. Is the sequence bounded below (above)?

11. Prove that every subsequence of an increasing sequence is increasing.

12. Prove in general that every increasing sequence has a lower bound.

13. Prove that an increasing sequence is bounded if it has an upper bound.

14. Prove that a sequence (p_n) is bounded if there exists a non-negative number M such that $|p_n| \leq M$ for all $n \in N$.

15. Show that if (p_n) and (q_n) are bounded sequences, then the sum $(s_n) = (p_n) + (q_n)$ is also a bounded sequence.

5.2 SEQUENCES OF STATEMENTS— MATHEMATICAL INDUCTION

The forms of valid argument introduced in Chapter 2 all involve only a finite number of statements, usually two or three. In this section we present a form of argument, known as mathematical induction, which enables us to establish the truth of an infinite sequence of statements. Our procedure is based on the following property of the positive integers.

Well Ordering Principle Let S be a nonempty set of positive integers. Then S has a smallest element. ▶

For our purposes we shall accept this statement as an axiom, albeit a perfectly natural one. Of course every nonempty set of positive integers contains a minimum! How could it be otherwise? Notice, however, that this innocent looking statement would be false if the set N of positive integers were replaced by the set Q of rational numbers (fractions) or the set R of reals. In Section 4.3 we saw numerous examples of sets which, even though bounded below, did not have a minimum. An example is the set of all real numbers which are larger than 2.

Example 1 The union $\bigcup_{k=1}^{\infty} A_k$ consists of those elements which are contained in at least one of the sets of the sequence $(A_1, A_2, A_3, \ldots)$. Suppose we define a new sequence $(Z_1, Z_2, Z_3, \ldots)$ by

$$Z_1 = A_1$$

$$Z_2 = A_1' \cap A_2$$

$$Z_3 = A_1' \cap A_2' \cap A_3$$

$$\vdots$$

In general, for any positive integer $k \geq 2$,

$$Z_k = A_1' \cap A_2' \cap \cdots \cap A_{k-1}' \cap A_k$$

consists of those elements which are in A_k but are not in any A with a subscript smaller than k (see Figure 2). The Z_k are disjoint sets. For, if $j \neq k$ (say $j > k$), the set

$$Z_j = A_1' \cap \cdots \cap A_k' \cap \cdots \cap A_{j-1}' \cap A_j$$

is a subset of A_k' while Z_k is a subset of A_k. (Why?) It follows that Z_k and Z_j have no elements in common.

FIGURE 2

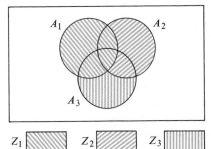

Since we always have $Z_k \subseteq A_k$, the union $\bigcup_{k=1}^{\infty} Z_k$ must be a subset of $\bigcup_{k=1}^{\infty} A_k$. Actually, these unions are equal. To see this, choose any element x of $\bigcup_{k=1}^{\infty} A_k$. Then x belongs to at least one A_k. Said another way, the set S_x of positive integers containing the subscripts of those A_k of which x is a member, is a nonempty set. By the Well Ordering Principle, S_x has a smallest element, say m. By definition of m, then, we have $x \in A_m$ but $x \notin A_k$ if $k < m$. That is, x belongs to $A_1' \cap A_2' \cap \cdots \cap A_{m-1}' \cap A_m = Z_m$, and hence is an element of $\bigcup_{k=1}^{\infty} Z_k$.

In the language of Section 3.2 we have shown that the sets $Z_1, Z_2, Z_3, \ldots$ constitute a *partition* of $\bigcup_{k=1}^{\infty} A_k$. ▶

Theorem 1 Let P be a set of positive integers such that

(a) P contains the integer 1.

(b) if n is any member of P, then $n + 1$ is also a member of P.

Then P is the set of all positive integers.

PROOF An indirect proof is best. Thus, suppose that conditions (a) and (b) hold, but that P does not contain all positive integers. Then the set V of positive integers not in P is nonempty, and by the Well Ordering Principle has a minimum m. Note that $m > 1$ since 1 is in P. Since m is the smallest element of V, $m - 1$ cannot be in V, but must be in P. If so, it follows from (b) (taking $n = m - 1$) that $(m - 1) + 1 = m$ is also in P, contradicting the fact that $m \in V$. ▶

The above discussion serves as a preliminary to the following statement, the main result of this section, which represents one of the most important of mathematical tools.

Principle of Mathematical Induction Let $S_1, S_2, S_3, \ldots$ be a sequence of statements for which

(a) S_1 is true.

(b) for every n, if S_n is true, then so is S_{n+1}.

Then all the statements $S_1, S_2, S_3, \ldots$ are true.

PROOF The proof follows immediately since the set T, containing those integers n for which the corresponding statement S_n is true, satisfies conditions (a) and (b) of Theorem 1. ▶

Intuitively, the induction principle is like the game with dominos (illustrated in Figure 3) that we all played as children. We first set the dominos on end so that if one fell it would hit and knock over the next one (Property (b) above). With this setup, the only action required to topple all the dominos is to knock over the first one (Property (a)). Obviously both properties, (a) and (b), are necessary to ensure that all dominos fall. If either is omitted, we have no guarantee whatever concerning the outcome.

FIGURE 3

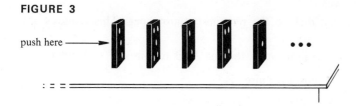

push here

Example 2 In the game shown in Figure 4, n disks, each smaller than the one below, are placed on one of three pegs. The object is to shift all the disks to another peg by moving them one at a time, always making sure that a larger disk is never placed on top of a smaller one.

FIGURE 4

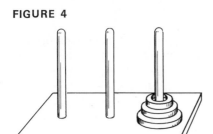

Obviously, with one disk only one move is required. With two disks we first shift the smaller one to one of the empty pegs, then move the larger one to the remaining empty peg and finally place the smaller one on top of the larger one—a total of three moves. A little experimenting shows that three disks require 7 moves, four require 15 moves, five require 31 moves, and so forth. Table 1, in which M_n denotes the number of moves required to shift n disks, summarizes these results.

TABLE 1

n	1	2	3	4	5
M_n	1	3	7	15	31

In each case (or for each n) the value of M_n is given by the formula

$$M_n = 2^n - 1$$

and it is natural to wonder whether this holds for all positive integers n. To check this conjecture, we introduce the sequence of statements $(S_1, S_2, S_3, \ldots)$ where S_n is the statement "$M_n = 2^n - 1$."

We know already that S_1 is true, as well as S_2, S_3, S_4 and S_5. Now suppose that for some value of n, S_n is true. That is, suppose that the number of moves needed to shift n disks really is $2^n - 1$. Then the moves required to shift $n + 1$ disks may be calculated as follows:

(i) Move the top n disks to another peg. This requires (by hypothesis) $2^n - 1$ moves.

(ii) Move the last disk to the remaining empty peg. This requires one move.

(iii) Move the n disks on top of the last. This requires $2^n - 1$ moves.

The total number of moves required is

$$(2^n - 1) + 1 + (2^n - 1) = 2 \cdot 2^n - 1 = 2^{n+1} - 1$$

Thus, we have shown that if $M_n = 2^n - 1$, then $M_{n+1} = 2^{n+1} - 1$. That is, if statement S_n is true, then so is S_{n+1}. By the Principle of Mathematical Induction, then, all the statements S_1, S_2, S_3, ... are true. ▶

PROBLEMS

1. Suppose we impose the restriction that a disk can be moved from one peg only to an *adjacent* peg. Prove that to move n disks from one end peg to the other now requires $3^n - 1$ moves.

Example 3 Devletoglou* used a variant of the so-called *cobweb model* as a means of studying the effect of public prediction on economic stability. We shall consider only the basic model which assumes

(i) $D_t = a + bP_t$

(ii) $S_t = c + dP_{t-1}$

(iii) $D_t = S_t$

where D_t represents the quantity demanded, S_t the quantity supplied, and P_t the price during the tth time period, and a, b, c, and d are constants with $b < 0$ and $d > 0$. In effect, the model postulates linear demand and supply curves, with demand at time t being a function of the price at time t, and supply at time t being a function of the price in the immediately preceding time period.

From (i) and (iii) we find

$$P_{t-1} = \frac{D_{t-1} - a}{b} = \frac{S_{t-1} - a}{b}$$

which, when substituted into (ii) gives

$$S_t = c + d\left[\frac{S_{t-1} - a}{b}\right] = \frac{d}{b}S_{t-1} + \frac{bc - ad}{b}$$

Writing $\alpha = d/b$ and $\beta = (bc - ad)/b$ gives the recursive formula

$$S_t = \alpha S_{t-1} + \beta \tag{3}$$

for the sequence $(S_1, S_2, S_3, \ldots)$.

Substituting a few values for t in (3) gives

$S_2 = \alpha S_1 + \beta$

$S_3 = \alpha S_2 + \beta = \alpha[\alpha S_1 + \beta] + \beta = \alpha^2 S_1 + \alpha\beta + \beta$

$S_4 = \alpha S_3 + \beta = \alpha[\alpha^2 S_1 + \alpha\beta + \beta] + \beta = \alpha^3 S_1 + \alpha^2\beta + \alpha\beta + \beta$

$S_5 = \alpha S_4 + \beta = \alpha[\alpha^3 S_1 + \alpha^2\beta + \alpha\beta + \beta] + \beta$

$\quad = \alpha^4 S_1 + \alpha^3\beta + \alpha^2\beta + \alpha\beta + \beta$

*Devletoglou, E. A., "Correct Public Prediction and the Stability of Equilibrium," *Journal of Political Economy* **69**, 142–161 (1961).

It seems reasonable to guess that, in general,

$$S_t = \alpha^{t-1}S_1 + \alpha^{t-2}\beta + \alpha^{t-3}\beta + \cdots + \alpha\beta + \beta$$

or, making use of Problem 11,

$$S_t = \alpha^{t-1}S_1 + \frac{\beta[1 - \alpha^{t-1}]}{1 - \alpha} \qquad (4)$$

In order to check on this guess, let V_t denote the statement (4). For $t = 1$, (4) becomes

$$S_1 = \alpha^0 S_1 + \frac{\beta[1 - \alpha^0]}{1 - \alpha}$$
$$= S_1$$

which is certainly true. Now suppose that for some value of t, V_t (that is, statement (4)) is actually true. Then

$$S_{t+1} = \alpha S_t + \beta \qquad \qquad \text{[by (3)]}$$

$$= \alpha\left[\alpha^{t-1}S_1 + \frac{\beta(1 - \alpha^{t-1})}{1 - \alpha}\right] + \beta \qquad \text{[inductive hypothesis (4)]}$$

A little manipulation simplifies this to

$$S_{t+1} = \alpha^t S_1 + \frac{\beta(1 - \alpha^t)}{1 - \alpha}$$

which is statement V_{t+1}.

To summarize, we have shown for the sequence of statements

$$(V_1, V_2, V_3, \ldots)$$

that

(a) V_1 is true.

(b) if V_t is true, then so is V_{t+1}.

The Principle of Mathematical Induction allows us to conclude that every statement V_t is true and that formula (4) is correct for $t = 1, 2, 3, 4, \ldots$. ▶

It should be emphasized that mathematical induction is not an inductive process at all, but rather a deductive method used primarily to establish known (or suspected) formulas. Examples 2 and 3 are typical instances. Induction enters a problem when one has observed particular values for a few cases and, from these, attempts to infer a general formula.

PROBLEMS

2. In Example 2 of Section 5.1, prove by mathematical induction that $A_n = P(1 + ni)$ for $n = 1, 2, 3, \ldots$.

3. Prove that the compound amount A_n of a principal P invested at $i\%$ interest per year is given by $A_n = P(1 + i)^n$.

4. Population growth is frequently analyzed* by making use of the relationship

$$\begin{pmatrix} \text{population size} \\ \text{at beginning of} \\ (t+1)\text{st time period} \end{pmatrix} = \begin{pmatrix} \text{population size} \\ \text{at beginning of} \\ t\text{th time period} \end{pmatrix} + \begin{pmatrix} \text{no. of births in} \\ t\text{th time period} \end{pmatrix}$$

$$- \begin{pmatrix} \text{no. of deaths in} \\ t\text{th time period} \end{pmatrix} + \begin{pmatrix} \text{no. of in-migrants} \\ \text{in } t\text{th time period} \end{pmatrix}$$

$$- \begin{pmatrix} \text{no. of out-migrants} \\ \text{in } t\text{th time period} \end{pmatrix}$$

Suppose that we make the following assumptions:

(1) Effects of out-migration and in-migration are negligible and can therefore be ignored.

(2) Birth rate α is constant over time.

(3) Death rate β is constant over time.

(a) Prove by mathematical induction that given these assumptions, the model for population size P_t at the beginning of the $(t+1)$st time period is
$$P_t = P_1(1 + \alpha - \beta)^{t-1} \quad \text{for } t = 1, 2, \ldots$$

(b) If $\alpha = 0.03$ and $\beta = 0.05$, approximately how many time periods will be required to reduce the population to 25% of its original size?

5. A model of learning proposed by Bush and Mosteller† assumes that on the nth trial of a sequence of trials, the probability p_n of obtaining a certain response is related to the probability on the immediately preceding trial by
$$p_n = p_{n-1} + a(1 - p_{n-1}) - bp_{n-1}$$
$$= a + (1 - a - b)p_{n-1}$$

Here a and b are constants which depend on the respective per trial amounts of reward and inhibition. Prove that for $n = 1, 2, 3, \ldots,$

$$p_n = p_0(1 - a - b)^n + \frac{a}{a+b}[1 - (1 - a - b)^n]$$

where p_0 is the initial probability of response. (*Hint:* See Example 3. Note that no knowledge of probability theory is required to solve this problem.)

In Problems 6–17, prove by mathematical induction that the given statements are valid for each positive integer n.

6. $1 + 2 + 3 + \cdots + n = n(n+1)/2$. (*Hint:* Let f be the function defined for positive integers by $f(n) = 1 + 2 + 3 + \cdots + n$ and let S_n be the statement $f(n) = n(n+1)/2$. Use the fact that $f(n+1) = f(n) + (n+1)$ and proceed as in the examples.)

---

*Matras, J., "Demographic Trends in Urban Areas," in *Handbook for Social Research in Urban Areas*, Hauser, P. M., ed. (Unesco, 1964), Chapter 6.

†Bush, R. R., and Mosteller, F., "A Mathematical Model for Simple Learning," *Psychological Review* **58**, 313–323 (1951).

7. $1^2 + 2^2 + \cdots + n^2 = n(n + 1)(2n + 1)/6$

8. $1 + 3 + 3^2 + \cdots + 3^{n-1} = \frac{1}{2}(3^n - 1)$

9. $1 \cdot 2 + 2 \cdot 3 + 3 \cdot 4 + \cdots + n(n + 1) = \frac{1}{3}n(n + 1)(n + 2)$

10. $2 + 2^2 + 2^3 + \cdots + 2^n = 2^{n+1} - 2$

11. $a + ar + ar^2 + \cdots + ar^{n-1} = a(1 - r^n)/(1 - r)$ for $r \neq 1$

12. $(1 + a)^n \geq 1 + na$ for all $a \geq -1$

13. $n < 2^n$ (*Hint:* Apply Problem 12.)

14. If $c > 1$, then $c^{n+1} \geq c^n \geq c$

15. If $0 < c < 1$, then $0 < c^{n+1} < c^n \leq c$

16. If a and b are positive, then $a^n < b^n$ if and only if $a < b$.

17. Let $n!$ denote the product $n(n - 1)(n - 2) \cdots (2)(1)$ of the integers 1 through n. Then $n! \geq 2^{n-1}$.

18. Apply the result of Problem 11 to Problems 8 and 10 and to Formula (4) in Example 3.

A proof using mathematical induction will fail if either S_1 is not true or if S_n does not imply S_{n+1} for all n. The next two problems illustrate these possibilities.

19. Define the function f by $f(n) = 1 + 2 + \cdots + n$ and let S_n be the statement

$$f(n) = \frac{n(n + 1)}{2} + 7$$

(a) Prove that for all n, if S_n is true, then S_{n+1} is true. (*Hint:* Argue that $f(n + 1) = f(n) + (n + 1)$.)

(b) Argue that the result of Problem 6 means that all S_n are false.

20. Referring to Example 2, let S_n be the statement $M_n = n^2 - n + 1$. Prove that S_1, S_2, and S_3 are true, but that S_n implies S_{n+1} only if $n = 1$ or $n = 2$. (*Hint:* Argue that $M_{n+1} = 2M_n + 1$.)

5.3   CONVERGENT SEQUENCES OF REAL NUMBERS

It is apparent that successive members of the sequence (P_n) in Example 1, Section 5.1, get closer and closer to the point 2. They also get closer to 3, to 4, and so forth, yet somehow we think of 2 as being special in this regard. Why is this so? The nth member of the sequence is

$$P_n = 2 - \frac{1}{2^{n-1}}$$

Suppose we choose a number $r > 0$ and mark the point $2 - r$ on a line as shown in Figure 5. We know from the Archimedean property (and Prob-

FIGURE 5

$$2-r \qquad\qquad\qquad\qquad 2$$

lem 13 of Section 5.2) that no matter how small r may be, there is an integer k such that $1/2^{k-1}$ is smaller than r. Hence for this value of k, we must have $2 - r < P_k < 2$ as shown in Figure 6. Since the sequence is increasing and

FIGURE 6

$$2-r \qquad P_k \qquad\qquad\qquad 2$$

bounded above by 2, it follows that all the remaining members of the sequence, that is, $P_{k+1}, P_{k+2}, P_{k+3}, \ldots$, must also lie between $2 - r$ and 2 as shown in Figure 7. In words we can say that the distance between the

FIGURE 7

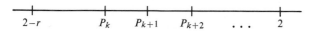

$$2-r \qquad P_k \quad P_{k+1} \quad P_{k+2} \quad \cdots \quad 2$$

fixed number 2 and a member P_k of the sequence can be made as small as we please (less than r) by choosing a member far enough along in the sequence (choosing n greater than k). The number 2 is the *only* real number about which this statement can be made and it is this property which gives 2 its special significance. In the sense of the following definition, 2 is the *limit* of the sequence (P_n).

Definition 1 Let $S = (s_n)$ be a sequence of real numbers. A number L is called the *limit* of the sequence S if for each number $r > 0$, there exists an integer k such that the distance between s_n and L is less than r for all integers n exceeding k. Symbolically,

$$|s_n - L| < r \quad \text{when } n \geq k$$

If L is the limit of S, we say that S *converges* to L and we write

$$L = \lim S \quad \text{or} \quad L = \lim_{n \to \infty} (s_n)$$

If S has no limit (that is, does not converge) it is said to *diverge*. ▶

A geometric interpretation of the limit of a sequence is given in Figure 8. The point L is the limit of the sequence (s_n) if no matter how small the interval lying between the points $L - r$ and $L + r$ may be, a point can be found in the sequence *beyond which* all members of the sequence lie inside the interval.

FIGURE 8

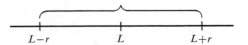

Example 1 (a) If c is a real number, the constant sequence determined by c is defined by

$$S = (s_n) = (c, c, c, c, \ldots)$$

This sequence converges to c since, if r is positive, we have

$$|a_n - c| = |c - c| = 0 < r \quad \text{for all } n \geq 1$$

(b) Let H denote the harmonic sequence $(h_n) = (1/n) = (1, \frac{1}{2}, \frac{1}{3}, \frac{1}{4}, \ldots)$ illustrated in Figure 9. The Archimedean property of the real numbers

FIGURE 9

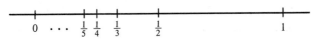

guarantees that for any $r > 0$, there exists an integer k such that $1/k < r$. Then if $n \geq k$ we have

$$0 < h_n = \frac{1}{n} \leq \frac{1}{k} < r$$

which means that the sequence H has limit zero. ▶

The definition of limit does not require that the members of the sequence all lie on one side of the limit point, as in Example 1 in Section 5.1 and Example 1 above. For instance, the sequence $(1, -1, \frac{1}{2}, -\frac{1}{2}, \frac{1}{3}, -\frac{1}{3}, \frac{1}{4}, -\frac{1}{4}, \ldots)$ is closely related to the harmonic sequence and also converges to zero. Here successive members of the sequence lie on opposite sides of the limit point zero, as shown in Figure 10.

FIGURE 10

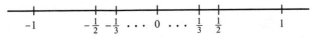

The members of a convergent sequence need not get successively closer to the limit point. The sequence

$$(S_n) = (1, 1, \tfrac{1}{2}, \tfrac{1}{4}, \tfrac{1}{3}, \tfrac{1}{9}, \tfrac{1}{4}, \tfrac{1}{16}, \tfrac{1}{5}, \tfrac{1}{25}, \ldots)$$

$$= (1, 1^2, \tfrac{1}{2}, (\tfrac{1}{2})^2, \tfrac{1}{3}, (\tfrac{1}{3})^2, \tfrac{1}{4}, (\tfrac{1}{4})^2, \ldots)$$

converges to zero as shown in Figure 11. Here the members of the harmonic sequence alternate with their squares. We never reach a point in the sequence beyond which terms become successively smaller. But for any number $r > 0$, we do reach a point beyond which all terms are less than r. And this is the criterion for convergence to zero.

FIGURE 11

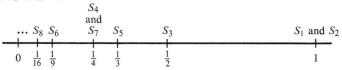

PROBLEMS

1. Prove that the sequences

$$(1, -1, \tfrac{1}{2}, -\tfrac{1}{2}, \tfrac{1}{3}, -\tfrac{1}{3}, \ldots)$$

and

$$(1, 1, \tfrac{1}{2}, (\tfrac{1}{2})^2, \tfrac{1}{3}, (\tfrac{1}{3})^2, \ldots)$$

both converge to zero.

2. (a) A sequence $S = (s_1, s_2, s_3, \ldots)$ is known to have zero as a limit. Argue that this sequence cannot also converge to 2.

(b) Give an example of a sequence which is not bounded.

(c) Give an example of a sequence which is bounded but does not converge.

We now turn to consideration of some basic properties of convergent sequences.

Property 1 A convergent sequence has a unique limit.

PROOF Suppose the convergent sequence $A = (a_n)$ has two distinct limit points L_1 and L_2 with $L_1 < L_2$ (see Figure 12). Then the number $r = \tfrac{1}{3}(L_2 - L_1)$ is positive and we have $L_1 + r < L_2 - r$.

FIGURE 12

Now if L_1 is a limit of (a_n) there exists an integer k such that when $n \geq k$, the element a_n is in the interval

$$I_1 = \{y: L_1 - r < y < L_1 + r\}$$

Similarly, there exists an integer m such that a_n is in the interval

$$I_2 = \{y\colon L_2 - r < y < L_2 + r\}$$

when $n \geq m$. But this means that when n is larger than both k and m, a_n must lie in both I_1 and I_2, an obvious impossibility. Our supposition that L_1 and L_2 were different must be false, which is what we wished to prove. ▶

Property 2 A convergent sequence is bounded.

PROOF Suppose the sequence (a_n) has limit L. Then (taking $r = 1$) there exists an integer k such that

$$L - 1 < a_n < L + 1 \quad \text{for all } n \geq k$$

If b denotes the smallest of the numbers $L - 1, a_1, a_2, \ldots, a_{k-1}$, the following statements are true:

(i) By definition $b \leq a_n$ for $n = 1, 2, 3, \ldots, k - 1$.
(ii) By definition $b \leq L - 1$.
(iii) We have $L - 1 < a_n$ for $n = k, k + 1, k + 2, \ldots$.

Taken together, statements (i)–(iii) imply that b is a lower bound for the sequence, that is, $b \leq a_n$ for $n = 1, 2, 3, \ldots$. (These statements may be geometrically intepreted as shown in Figure 13, where the elements $a_1, a_2, \ldots, a_{n-1}$ are scattered along the line. In the sketch, a_3 is the smallest of these, and since $a_3 < L - 1$, $a_3 = b$. If $n \geq k$, a_n must be to the right of the point $L - 1$.)

FIGURE 13

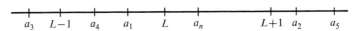

An entirely similar argument shows that the convergent sequence (a_n) also has an upper bound. Hence (a_n) is bounded. ▶

PROBLEMS

3. Complete the proof of Property 2 by showing that the largest of the numbers $L + 1, a_1, a_2, \ldots, a_{k-1}$ (for example, a_5 in Figure 13) is an upper bound for the sequence.

4. Does the sequence $(1, 2, 3, 4, \ldots)$ converge? Why or why not? (*Hint:* What does the Archimedean property say?)

Property 3 If the sequence (a_n) converges to L then every subsequence of (a_n) also converges to L.

PROOF We know that in order for (b_k) to be a subsequence of (a_n) it must be obtained by choosing members of (a_n) in such a way that subscripts on the a_n appear in ascending order. That is,

$$b_1 = a_{m_1}, \quad b_2 = a_{m_2}, \quad b_3 = a_{m_3}, \ldots$$

where $m_1 < m_2 < m_3 < \cdots$. Obviously, $m_1 \geq 1$, and if $m_k \geq k$, then $m_{k+1} \geq k + 1$ since $m_{k+1} > m_k$. This simple inductive proof allows us to conclude that $m_k \geq k$ for $k = 1, 2, 3, \ldots$.

The fact that (a_n) converges to L means that for any number $r > 0$, there is an integer N such that $|a_n - L| < r$ whenever $n \geq N$. Thus if $k \geq N$, we have $|b_k - L| < r$ since $b_k = a_{m_k}$ and $m_k \geq k \geq N$. This proves that $L = \lim (b_k)$.
▶

PROBLEMS

5. Check the validity of Property 3 by looking at a few subsequences of the sequence $(1, -1, \frac{1}{2}, -\frac{1}{2}, \frac{1}{3}, -\frac{1}{3}, \ldots)$.

Example 2 Let us define a sequence (S_n) by adding the terms in the harmonic sequence. That is, define

$$S_1 = 1, \quad S_2 = 1 + \tfrac{1}{2}, \quad S_3 = 1 + \tfrac{1}{2} + \tfrac{1}{3}, \ldots$$

and, in general,

$$S_n = 1 + \frac{1}{2} + \frac{1}{3} + \cdots + \frac{1}{n}$$

Consider the subsequence of (S_n) whose subscripts are powers of 2. Members of this subsequence are bounded below by multiples of $\frac{1}{2}$. That is,

$$S_2 = 1 + \tfrac{1}{2} > 0 + \tfrac{1}{2} = \tfrac{1}{2}$$

$$S_4 = 1 + \tfrac{1}{2} + \tfrac{1}{3} + \tfrac{1}{4} > 0 + \tfrac{1}{2} + \tfrac{1}{4} + \tfrac{1}{4} = 2(\tfrac{1}{2})$$

$$S_8 = 1 + \tfrac{1}{2} + \tfrac{1}{3} + \tfrac{1}{4} + \tfrac{1}{5} + \tfrac{1}{6} + \tfrac{1}{7} + \tfrac{1}{8} > 0 + \tfrac{1}{2} + \tfrac{1}{4} + \tfrac{1}{4} + \tfrac{1}{8} + \tfrac{1}{8}$$

$$\vdots \qquad\qquad\qquad\qquad\qquad\qquad + \tfrac{1}{8} + \tfrac{1}{8} = 3(\tfrac{1}{2})$$

In general, we find that $S_{2^n} > \frac{1}{2}n$. This means that the subsequence $(S_2, S_4, S_8, S_{16}, \ldots)$ is not bounded and, using Property 2, must diverge. But this implies (Property 3) that the original sequence $(S_1, S_2, S_3, \ldots)$ also diverges.
▶

Property 4 If $A = (a_n)$ and $B = (b_n)$ are convergent sequences with $a = \lim A$ and $b = \lim B$ then

(i) $\lim (A + B) = a + b$

(ii) $\lim (A - B) = a - b$

(iii) $\lim (A \cdot B) = a \cdot b$

If, in addition, B has no zero elements and $b \neq 0$, then

(iv) $\lim (A/B) = a/b$.
▶

Result (i) is often expressed by saying that the limit of the sum of two convergent sequences is the sum of their respective limits, or simply, the limit of a sum is the sum of the limits. Similar comments apply to the other cases. We shall prove (i) and (iii), and we leave (ii) and (iv) as exercises (Problem 12).

PROOF OF PROPERTY 4 To prove (i) we must show that for any $r > 0$, there is an integer k such that

$$|(a_n + b_n) - (a + b)| < r$$

when $n \geq k$. Since $a = \lim A$, there is an integer k_1 such that

$$|a_n - a| < \frac{r}{2} \quad \text{when } n \geq k_1$$

Similarly, since $b = \lim B$, there exists an integer k_2 such that

$$|b_n - b| < \frac{r}{2} \quad \text{when } n \geq k_2$$

Now let k be the larger of k_1 and k_2. Then for $n \geq k$ we have

$$
\begin{aligned}
|(a_n + b_n) - (a + b)| &= |(a_n - a) + (b_n - b)| \\
&\leq |a_n - a| + |b_n - b| \qquad \text{[Triangle Inequality]} \\
&< \frac{r}{2} + \frac{r}{2} = r
\end{aligned}
$$

To prove (iii), we note that $|a_n b_n - ab|$ can be written

$$
\begin{aligned}
|a_n b_n - ab| &= |a_n b_n - ab_n + ab_n - ab| \\
&= |b_n(a_n - a) + a(b_n - b)| \\
&\leq |b_n| \, |a_n - a| + |a| \, |b_n - b|
\end{aligned}
$$

Since the convergent sequence B is bounded, there exists a number $u > 0$ such that $|b_n| < u$ for all n. Arguing as in the proof of (i), for any number $r > 0$ we can find an integer k such that when $n \geq k$, we have simultaneously

$$|a_n - a| < \frac{r}{2u} \quad \text{and} \quad |b_n - b| < \frac{r}{2|a|}$$

(If $a = 0$, the latter quantity is not needed.) Hence, for $n \geq k$,

$$
\begin{aligned}
|a_n b_n - ab| &\leq u|a_n - a| + |a| \, |b_n - b| \\
&\leq u \cdot \frac{r}{2u} + |a| \cdot \frac{r}{2|a|} = r \qquad \blacktriangleright
\end{aligned}
$$

Example 3 (a) If, for some constant c, the sequences (a_n) and (b_n) are related by

$$b_n = a_n + c$$

and if (a_n) converges, then Property 4(i) gives

$$\lim (b_n) = \lim (a_n + c)$$
$$= \lim (a_n) + \lim (c)$$
$$= \lim (a_n) + c$$

(b) If for some constant c, (a_n) and (b_n) are related by

$$b_n = c \cdot a_n$$

and if (a_n) converges, then Property 4(iii) implies

$$\lim (b_n) = \lim (c \cdot a_n) = [\lim (c)][\lim (a_n)] = c \lim (a_n) \qquad \blacktriangleright$$

Example 4 When a test having reliability r $(0 < r < 1)$ is made n times longer, the reliability R_n of the new test is, according to the classical Spearman–Brown formula,* given by

$$R_n = \frac{nr}{1 + (n - 1)r}$$

For a fixed value of r, the sequence (R_n) has the limit

$$\lim_{n \to \infty} (R_n) = \lim_{n \to \infty} \left(\frac{nr}{1 + (n - 1)r} \right)$$

$$= \lim_{n \to \infty} \left(\frac{r}{(1/n) + [1 - (1/n)]r} \right) \qquad \text{[Dividing top and bottom by } n\text{]}$$

$$= \frac{\lim_{n \to \infty} (r)}{\lim_{n \to \infty} (1/n) + \left[\lim_{n \to \infty} (1) - \lim_{n \to \infty} (1/n) \right] r} \qquad \text{[Properties 4(i)–(iv)]}$$

$$= \frac{r}{0 + (1 - 0)r} \qquad \text{[Example 1(b)]}$$

$$= 1$$

Thus by simply increasing the length of a test, we can make its reliability as close to unity (perfect reliability) as we please. $\qquad \blacktriangleright$

PROBLEMS

6. Suppose the sequences (a_n), (b_n), and (c_n) are related by $a_n \le b_n \le c_n$ for $n = 1, 2, 3, \ldots$. Show that if (a_n) and (c_n) have a common limit L, then the limit of (b_n) must also be L.

7. Prove that $\lim (1/2^n) = 0$. (*Hint:* From Problem 13 of Section 5.2, we know $0 < 1/2^n < 1/n$. Apply Problem 6.)

8. Find $\lim (4 - 1/2^{n-2})$.

9. Prove that if k is a positive integer, then $\lim (1/n^k) = 0$. (*Hint:* First prove by induction that $1/n \ge 1/n^k$ for all n. Then apply Problem 6.)

*See Gulliksen, H., *Theory of Mental Tests* (John Wiley & Sons, Inc., New York, 1950), Chapter 8.

10. Use Property 4 and Problem 9 to determine limits (if they exist) of the sequences whose nth terms are

(a) $3 + \dfrac{2}{n^2}$

(b) $2n + (-1)^n$

(c) $\dfrac{n^2 - 5}{n^2 + 5}$

(d) $\dfrac{n^4}{n^2 + 2} - \dfrac{n^3}{2n^2 - 2}$

(e) $\dfrac{1 - 3n}{1 + n}$

(f) $\dfrac{(-1)^n}{6n^3}$

(g) $\dfrac{3n^2 - 3n + 4}{4n^2 - 5n - 2}$

(h) $\dfrac{n(n + 2)}{(n + 1)(n + 4)}$

11. Prove that if $\lim (a_n)$ exists and u is an upper bound for (a_n), then $\lim (a_n) \leq u$. Give an example in which equality holds and another in which inequality holds.

12. Prove Properties 4(ii) and 4(iv).

13. Prove that $\lim (2/\sqrt{n^2 + 1}) = 0$. (*Hint:* $n^2 < n^2 + 1$ and therefore $1/n > 1/\sqrt{n^2 + 1} > 0$.)

14. Prove that if $A = (a_n)$ has limit a, then

(a) $\lim A^2 = a^2$.

(b) $\lim (-a_n) = -a$.

15. Show by example that if $\lim (A + B) = L$, it does not necessarily follow that either A or B converges.

5.4 CONVERGENCE OF MONOTONE SEQUENCES

Although a convergent sequence must be bounded (Property 2), in general a bounded sequence need not be convergent. For example, the sequence $(1, -1, 1, -1, 1, -1, \ldots)$, whose values are alternately 1 and -1, is certainly bounded but is not convergent. However, there are certain types of sequences, in particular those which are *monotone* (that is, increasing or decreasing), for which boundedness is tantamount to convergence.

Property 5 *Monotone Convergence Theorem* A bounded increasing sequence converges to its least upper bound. A bounded decreasing sequence converges to its greatest lower bound.

PROOF Since the two cases are quite similar, we shall prove only the first, leaving the second as an exercise.

If the increasing sequence (a_n) is bounded, it has a least upper bound u. Then u is an upper bound, but for any $r > 0$, the number $u - r$ is not an upper bound. This means that there exists a member a_k of the sequence for which

$$u - r < a_k \leq u$$

Since the sequence is increasing, we must have

$$u - r < a_k \le a_n \le u \quad \text{for all } n \ge k$$

(This is illustrated in Figure 14. Note that when $n \ge k$, the distance between u and a_n cannot exceed r.)

FIGURE 14

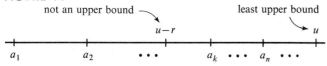

To sum up, we have shown that for any $r > 0$, there exists a number k such that $|a_n - u| < r$ for all $n \ge k$. This implies that $u = \lim (a_n)$. ▶

Example 1 Reread Example 1 of Section 5.1 and the persuant discussion at the beginning of Section 5.3. ▶

Our original definition for convergence has the distinct disadvantage that in order to be used, we must already know, or at least suspect, the correct value of the limit. We then merely verify that our suspicion is correct. Thus in Examples 1, Section 5.1 and 1(b), Section 5.3 it was readily apparent that the limits were, respectively, 2 and 0, and we were able to show that these values satisfied the requirements of the definition.

On the other hand, Property 5 allows us to assert that a limit exists even when no particular candidate presents itself. This, together with Property 3, often allows an easy solution to the problem of finding limits.

Example 2 Let (a_n) be the sequence defined recursively by

$$a_1 = 1$$

$$a_{n+1} = \frac{2a_n + 3}{4} \quad \text{for } n \ge 1$$

A simple calculation shows that $1 = a_1 < a_2 = \frac{5}{4} < 2$. Proceeding inductively, if $a_{n-1} < a_n < 2$, then

$$\frac{2a_{n-1} + 3}{4} < \frac{2a_n + 3}{4} < \frac{4 + 3}{4} < 2$$

which implies that

$$a_n < a_{n+1} < 2$$

Thus the sequence (a_n) is increasing and bounded above by 2.

It follows from Property 5 that this sequence converges to a limit L. Since $(a_{n+1}) = (a_2, a_3, a_4, \ldots)$ is a subsequence of (a_n) and hence, by Property 3,

$\lim (a_{n+1}) = \lim (a_n)$, we have

$$L = \lim (a_{n+1}) = \lim \left(\frac{2a_n + 3}{4} \right)$$

$$= \frac{\lim (2a_n) + \lim (3)}{\lim (4)}$$

$$= \frac{2L + 3}{4}$$

Solving this equation, we find $L = \frac{3}{2}$. ▶

PROBLEMS

    1. Try to find L in Example 2 in some other way.

Example 3   If a is any real number, the sequence of powers of a

$$(a, a^2, a^3, \ldots) = (a^n)$$

is called a *geometric sequence*. It is convenient for future reference to list the basic convergence properties of geometric sequences here. There are several cases to consider, depending on the value of a.

    Case I $0 \leq a < 1$. In this case, the sequence is decreasing, but bounded below by zero (Problem 15, Section 5.2). It follows from Property 5 that it converges to a limit $L \geq 0$. The subsequence $(a^2, a^4, a^6, \ldots)$ is the product of (a^n) times itself. Thus, on the one hand, using Property 3, we have

$$\lim (a^2, a^4, a^6, \ldots) = L$$

while, on the other hand, Property 4(iii) gives

$$\lim (a^2, a^4, a^6, \ldots) = [\lim (a^n)][\lim (a^n)] = L^2$$

Hence $L = L^2$ so that $L = 0$ or $L = 1$. Obviously $L = 1$ is impossible since $a < 1$ is the first member of a decreasing sequence. We conclude that if $0 \leq a < 1$, then $\lim (a^n) = 0$.

    Case II $a = 1$. In this case we obtain the constant sequence $(1, 1, 1, \ldots)$ which converges to 1.

    Case III $a > 1$. Here $1/a < 1$, so $\lim (1/a^n) = 0$ by Case I. Pick any positive number M so that $1/M > 0$ also. Then $\lim (1/a^n) = 0$ means that there is an integer k such that $1/a^n < 1/M$ when $n \geq k$. Equivalently, $a^n > M$ when $n \geq k$, so that M is not an upper bound for the sequence. It follows that the sequence is unbounded and hence (Property 2) that it diverges.

    Case IV $-1 < a < 0$. Here $-a$ lies between zero and one, so the sequence $(-a, (-a)^2, (-a)^3, \ldots)$ converges to zero (Case I). This means that for any positive number r, there is an integer k such that $|(-a)^n| < r$ whenever $n \geq k$. Since $|a^n| = |(-a)^n|$, we also have $|a^n| < r$ when $n \geq k$. Hence (a^n) converges to zero.

Case V $a = -1$. In this case we obtain the sequence $(-1, 1, -1, 1, -1, 1, \ldots)$ which diverges.

Case VI $a < -1$. Here $-a > 1$ and again $|a^n| = |(-a)^n|$. Hence the sequence is unbounded (Case III) and must diverge.

To summarize, the geometric sequence (a^n)

(i) converges to zero when $-1 < a < 1$.

(ii) converges to one when $a = 1$.

(iii) diverges in all other cases. ▶

PROBLEMS

2. Suppose $0 < a < 1$. Find the limit of $[(1 - a^n)/(1 - a)]$.

Example 4 Rainio,* in his study of opinion changes resulting from contact with other individuals, denotes by p_n the theoretical frequency with which a certain opinion is expressed after the nth contact with another individual who favors this opinion. He assumes that the sequence (p_n) is described by

$$p_{n+1} = p_n + \alpha(1 - p_n) \tag{5}$$

where α is a non-negative constant.

The sequence (q_n) defined by

$$q_n = 1 - p_n$$

converges to a limit L if and only if (p_n) converges to $1 - L$. Rewriting (5) in terms of q_n and q_{n+1} we find

$$q_{n+1} = (1 - \alpha)q_n \quad \text{for } n = 0, 1, 2, 3, \ldots$$

Thus

$$q_1 = (1 - \alpha)q_0$$
$$q_2 = (1 - \alpha)q_1 = (1 - \alpha)^2 q_0$$
$$q_3 = (1 - \alpha)q_2 = (1 - \alpha)^3 q_0$$

A simple induction shows that

$$q_n = (1 - \alpha)^n q_0 \quad \text{for } n = 0, 1, 2, 3, \ldots$$

Using Example 3 we see that (q_n) converges to zero if $0 \leq \alpha < 2$ and diverges for all other values of α. Equivalently, (p_n) converges to 1 when $0 \leq \alpha < 2$ and diverges otherwise. ▶

*Rainio, K., "A Stochastic Model of Social Interaction," *Transactions of the Westermarch Society* (Munksgaard, 1962), Vol. VIII.

PROBLEMS

3. Stephan and Mischler* have found that the rate of total participation of an individual in a small discussion group is a function of his rank order according to total number of "acts of interaction" during the discussion period and follows the form

$$p_i = ar^{i-1} \quad \text{for } i = 1, \ldots, n$$

where n is the size of the group excluding the leader, r and a are constants, and p_i is the estimated proportion of total participative acts initiated by the ith ranking group member.

The leader's interaction rate is

$$I_n = 1 - [p_1 + p_2 + \cdots + p_n]$$

since the proportions must sum to 1, that is,

$$I_n + p_1 + p_2 + \cdots + p_n = 1$$

(a) Find the limiting value, if it exists, of the sequence (I_n) representing the leader's interaction rate. (*Hint:* Use Problem 11, Section 5.2.)

(b) Does the limit exist for all real values of a and r? If not, indicate the set of values of a and r for which the limit does exist.

4. In Example 3 of Section 5.2, prove that the sequence (S_t)

(a) is convergent when $-1 < d/b < 0$.

(b) is bounded but divergent when $d/b = -1$.

(c) is unbounded and divergent when $d/b < -1$.

5. In Problem 4, Section 5.2, prove that (P_t) is convergent whenever $\beta > \alpha$ and divergent whenever $\alpha > \beta$. Interpret these results in terms of the model for population growth.

6. Suppose that (a_n) is the sequence defined recursively by

$$a_1 = 2 \quad \text{and} \quad a_{n+1} = ca_n + k \quad \text{for } n \geq 1$$

where $0 < c < 1$ and k is a real constant.

(a) Find a_n in terms of a_1, c, and k.

(b) Find $\lim_{n \to \infty} (a_n)$.

(c) Apply these results to Equation (5) of Example 4.

7. Prove that if two subsequences of a given sequence converge to different limits, then the sequence does not converge.

8. In Problem 3, Section 5.2, you have shown that the accumulated amount resulting when a principal P is invested for n years at interest rate i compounded annually is

$$y = P(1 + i)^n$$

*Stephan, F., and Mischler, E. G., "The Distribution of Participation in Small Groups: An Exponential Approximation," *American Sociological Review* **17**, 598–608 (1952).

(a) Show that if interest is compounded m times a year, the general formula is

$$y = P\left(1 + \frac{i}{m}\right)^{mn}$$

$$= P\left[\left(1 + \frac{i}{m}\right)^{m/i}\right]^{in}$$

$$= P\left[\left(1 + \frac{1}{r}\right)^{r}\right]^{in}$$

where $r = m/i$.

(b) Show that if for each positive integer r we let $s_r = [1 + (1/r)]^r$ and $t_r = [1 + (1/r)]^{r+1}$, then

(i) (s_r) is an increasing sequence. (*Hint:* Show that $s_{r+1}/s_r \geq 1$ by writing s_{r+1}/s_r in the form $(1 + 1/r)\{1 - [1/(r + 1)^2]\}^{r+1}$ and then applying Problem 12 of Section 5.2.)

(ii) (t_r) is a decreasing sequence.

(iii) $s_r < t_k$ for every k and r.

(c) Given (i), (ii), and (iii), prove that the sequence $[1 + (1/r)]^r$ has t_1 as an upper bound and therefore converges.

The mathematical constant e is defined as the limit as $r \to \infty$ of the sequence of numbers $[1 + (1/r)]^r$. Hence, if the number of compoundings within a year is extended indefinitely,

$$y = Pe^{in}$$

which is a formula frequently used in investment theory.

9. Prove that every decreasing sequence converges to its greatest lower bound.

10. Prove that if (a_n) is an increasing sequence bounded above by 1, then lim (a_n) exists and is less than or equal to 1. Give an example in which the limit is less than 1.

11. If $a_n = (-\frac{1}{3})^n$, find lim (a_n).

12. If $a_n = (-1)^{n+1}(n - 1)/2^n$, find lim (a_n).

13. In Problem 5 of Section 5.2, find $\lim_{n \to \infty} (p_n)$. What conditions must be imposed on the constants a and b in order to insure convergence?

14. As a description of the extinction process, Bush and Mosteller* propose the relation
$$p_n = p_{n-1} - bp_{n-1} = (1 - b)p_{n-1}$$

*Bush, R. R., and Mosteller, F., "A Mathematical Model for Simple Learning," *Psychological Review* **58**, 313–323 (1951).

between the probabilities p_{n-1} and p_n of response on the $(n-1)$st and nth trials, respectively. The positive constant b is called an extinction parameter. Prove that the sequence (p_n)

   (i) converges to zero if $0 < b < 2$.

  (ii) is the constant sequence $(p_0, p_0, p_0, \ldots)$ if $b = 0$.

 (iii) diverges if $b \geq 2$.

SUPPLEMENTARY READING

Anderson, K. W., and Hall, D. W., *Sets, Sequences and Mappings* (John Wiley & Sons, Inc., New York, 1963), Chapters 2 and 4.

INFINITE SERIES **6**

6.1 SUMMATION NOTATION

The capital Greek sigma $\sum$ notation is a shorthand method in mathematics for designating sums. As an example, the sum of squares of the first seven positive integers may be written

$$1^2 + 2^2 + 3^2 + 4^2 + 5^2 + 6^2 + 7^2 = \sum_{k=1}^{7} k^2$$

The symbol k^2 indicates the form of the quantities which are to be added. The symbols "$k = 1$" below and "7" above the summation sign $\sum$ indicate that 1 is the initial value taken by k and that 7 is the terminal value. It is always understood that every integer from the initial value to the terminal value, inclusive, is to be included in the sum. Conventionally, the initial value is the lesser of the two values. The symbol k is called the *variable of summation*.

Quantities to be added may take a variety of forms. If, for instance, the function f is defined for each integer k by $f(k) = 2k + 1$, then

$$\sum_{k=4}^{7} f(k)$$

stands for

$$
\begin{aligned}
f(4) + f(5) &+ f(6) + f(7) \\
&= (2 \cdot 4 + 1) + (2 \cdot 5 + 1) + (2 \cdot 6 + 1) + (2 \cdot 7 + 1) \\
&= 9 + 11 + 13 + 15 \\
&= 48
\end{aligned}
$$

Similarly,

$$\sum_{k=0}^{5} k = 0 + 1 + 2 + 3 + 4 + 5 = 15$$

Each of the above examples is an illustration of the basic definition:

$$\sum_{k=m}^{n} f(k) = f(m) + f(m+1) + \cdots + f(n) \tag{1}$$

PROBLEMS

1. Evaluate the following sums.

(a) $\sum_{r=2}^{6} (r-3)$ (b) $\sum_{i=-1}^{1} 2^i$ (c) $\sum_{k=1}^{5} (k^2 - 3k)$

The variable of summation often appears as a subscript. For instance, if we have four quantities $x_1 = 3$, $x_2 = 6$, $x_3 = -4$, and $x_4 = 2$, then

$$\sum_{k=1}^{4} x_k = x_1 + x_2 + x_3 + x_4 = 3 + 6 - 4 + 2 = 7$$

while

$$\sum_{k=1}^{4} (2x_k - 1) = (2x_1 - 1) + (2x_2 - 1) + (2x_3 - 1) + (2x_4 - 1)$$
$$= (6 - 1) + (12 - 1) + (-8 - 1) + (4 - 1) = 10$$

PROBLEMS

2. Given that $x_1 = 2$, $x_2 = 3$, $x_3 = -1$, $x_4 = 6$, and $x_5 = -4$, find

(a) $\sum_{i=1}^{5} 4x_i$ (b) $\sum_{i=1}^{5} x_i^2$

(c) $\sum_{k=1}^{5} (3x_k + 1)$ (d) $\sum_{j=1}^{5} (x_j + 1)(x_j + 2)$

3. Rewrite the following expressions in $\sum$ notation.

(a) $1 + \dfrac{t}{2} + \dfrac{t^2}{4} + \dfrac{t^3}{8} + \dfrac{t^4}{16} + \dfrac{t^5}{32}$

(b) $x_1^2 + x_2^2 + x_3^2 + x_4^2 - 4$

(c) $a + (a + d) + (a + 2d) + (a + 3d) + (a + 4d)$

(d) $1 \cdot 2 + 2 \cdot 3 + 3 \cdot 4 + 4 \cdot 5 + 5 \cdot 6$

(e) $(x_1 - \bar{x})^2 + (x_2 - \bar{x})^2 + (x_3 - \bar{x})^2$

4. Rewrite the following in ordinary notation.

(a) $\sum_{n=0}^{5} 2^n$ (b) $\sum_{j=1}^{4} x_{ij}$

(c) $\sum_{j=-3}^{0} k(j+1)$ (d) $\sum_{j=1}^{5} x_j y^j$

It is important to observe that the symbol used for the variable of summation is entirely arbitrary and for this reason is called a *dummy variable*. For instance,

$$x_1 + x_2 + \cdots + x_n = \sum_{k=1}^{n} x_k = \sum_{j=1}^{n} x_j = \sum_{t=1}^{n} x_t = \sum_{w=1}^{n} x_w$$

and

$$\log 2 + \log 3 + \log 4 = \sum_{k=2}^{4} \log k = \sum_{j=2}^{4} \log j = \sum_{r=2}^{4} \log r$$

In addition, the quantities to be added may appear in different forms and yet their sums may be identical. As examples,

$$1^2 + 2^2 + 3^2 + 4^2 = \sum_{k=1}^{4} k^2 = \sum_{k=2}^{5} (k - 1)^2$$

$$= \sum_{k=10}^{13} (k - 9)^2 = \sum_{k=-2}^{1} (k + 3)^2$$

and

$$x_1 + x_2 + \cdots + x_n = \sum_{k=1}^{n} x_k = \sum_{k=0}^{n-1} x_{k+1} = \sum_{k=2}^{n+1} x_{k-1} \qquad (2)$$

In each case the initial and terminal values of the summation variable have been altered, but the quantities being summed have had their form altered in a compensating manner so that the value of the sum is unchanged.

PROBLEMS

5. Compute

$$\sum_{k=1}^{4} k^2 + \sum_{j=-1}^{2} (3j - 1)$$

and compare with

$$\sum_{r=4}^{7} (r - 3)^2 + \sum_{t=2}^{5} [3(t - 3) - 1]$$

(You should obtain the same result in both cases.)

A convenient way of obtaining the various summation forms in (2) is by making a *change of variable*. For example, to obtain $\sum_{k=0}^{n-1} x_{k+1}$ from $\sum_{k=1}^{n} x_k$, we first replace k by $r + 1$ to obtain

$$\sum_{k=1}^{n} x_k = \sum_{r+1=1}^{n} x_{r+1} \qquad (3)$$

Now, if $r + 1$ runs from 1 to n, then r itself must take values from 0 to $n - 1$. (That is to say, if $r + 1 = 1$, then $r = 0$; while if $r + 1 = n$, then $r = n - 1$.) Thus, we have

$$\sum_{r+1=1}^{n} x_{r+1} = \sum_{r=0}^{n-1} x_{r+1} \qquad (4)$$

Replacing r by k in this last sum yields

$$\sum_{r=0}^{n-1} x_{r+1} = \sum_{k=0}^{n-1} x_{k+1} \qquad (5)$$

Combining Equations (3)–(5) produces the desired result.

PROBLEMS

6. Compute

$$\sum_{k=4}^{7} (k^2 - 3k + 2)$$

Note that this is identical with

$$\sum_{k=4}^{7} (k - 1)(k - 2)$$

Now make the change of variable $r = k - 2$ to obtain

$$\sum_{r=2}^{5} r(r + 1) = \sum_{r=2}^{5} (r^2 + r)$$

Compute this latter sum to see that its value agrees with the first computation.

The basic rules of operation for sums parallel the associative and distributive properties of numbers (Section 1.4) and are easily obtained from the definition (1). Thus, if in the sum

$$\sum_{k=1}^{n} x_k = x_1 + x_2 + x_3 + \cdots + x_n$$

each of the x_k is equal to the same constant c, then

$$\sum_{k=1}^{n} x_k = c + c + \cdots + c = nc$$

Thus we obtain our first rule for sums.

Rule 1 for Sums   If each $x_k = c$, then

$$\sum_{k=m}^{n} x_k = \sum_{k=m}^{n} c = (n - m + 1)c \qquad \blacktriangleright$$

Example 1 $\displaystyle\sum_{k=1}^{6} 4 = 6(4) = 24$ and $\displaystyle\sum_{r=-2}^{7} \tfrac{3}{2} = 10(\tfrac{3}{2}) = 15 \qquad \blacktriangleright$

Combining the two sums

$$\sum_{k=1}^{r} x_k = x_1 + x_2 + \cdots + x_r$$

and

$$\sum_{k=r+1}^{n} x_k = x_{r+1} + x_{r+2} + \cdots + x_n$$

we obtain our second rule.

Rule 2 for Sums $\quad \displaystyle\sum_{k=1}^{r} x_k + \sum_{k=r+1}^{n} x_k = \sum_{k=1}^{n} x_k$ ▶

Next, suppose that for each value of k we have $x_k = cy_k$ for some constant c. Then

$$\sum_{k=1}^{n} x_k = \sum_{k=1}^{n} (cy_k) = cy_1 + cy_2 + \cdots + cy_n$$
$$= c(y_1 + y_2 + \cdots + y_n)$$
$$= c \sum_{k=1}^{n} y_k$$

Similar considerations yield our third rule.

Rule 3 for Sums $\quad \displaystyle\sum_{k=m}^{n} (cx_k) = c \sum_{k=m}^{n} x_k$ ▶

Rule 3 says that constant factors may be written either inside or outside the summation sign. This, of course, is merely another way of writing the distributive law for numbers.

Our fourth rule follows from the commutative and associative laws for real numbers.

Rule 4 for Sums $\quad \displaystyle\sum_{k=m}^{n} (x_k + y_k) = \sum_{k=m}^{n} x_k + \sum_{k=m}^{n} y_k$ ▶

This is easily proved by noting that

$$\sum_{k=m}^{n} (x_k + y_k) = (x_m + y_m) + (x_{m+1} + y_{m+1}) + \cdots + (x_n + y_n)$$
$$= (x_m + x_{m+1} + \cdots + x_n) + (y_m + y_{m+1} + \cdots + y_n)$$
$$= \sum_{k=m}^{n} x_k + \sum_{k=m}^{n} y_k$$

PROBLEMS

7. Use the above rules for sums to simplify the form of

$$\sum_{k=1}^{7} (3k^2 + 6k + 1) + \sum_{r=2}^{8} (-3r^2 + 2r) + \sum_{k=8}^{12} 2k$$

(*Hint:* First write $r = k + 1$.)

Example 2 An investment yielding fixed periodic payments is called an *annuity*. The *present value* of an annuity is the amount which must be invested initially in order to provide the payments.

Let us suppose that you wish to receive $\$Y$ annually for the next n years from an annuity on which the remaining principal is compounded annually at $100i\%$. Since P_j dollars invested now will amount to $P_j(1 + i)^j$ dollars in j years (see Problem 3, Section 5.2), the amount to be invested now in order to obtain $\$Y$ in j years is $P_j = \$Y/(1 + i)^j$. Hence the present value of this annuity is

$$P = \sum_{j=1}^{n} P_j = \sum_{j=1}^{n} \frac{Y}{(1 + i)^j} \qquad \blacktriangleright$$

PROBLEMS

8. Use Problem 11 of Section 5.2 to show that in Example 2, the quantity

$$P = \sum_{j=1}^{n} \frac{Y}{(1 + i)^j}$$

may be rewritten in the simpler form

$$P = \frac{Y}{i}\left[1 - \left(\frac{1}{1 + i}\right)^n\right]$$

9. What amount must be invested at 5% interest in order to guarantee 20 yearly payments of $10,000?

10. Show that

(a) $\displaystyle\sum_{i=1}^{k} x_i + \sum_{i=1}^{n-k} x_{i+k} = \sum_{i=1}^{n} x_i$

(b) $\displaystyle\sum_{j=1}^{n} (kx_j + my_j) = k \sum_{j=1}^{n} x_j + m \sum_{j=1}^{n} y_j$

11. Suppose that a business enterprise is so organized that the president has n immediate subordinates, each of whom also has n subordinates, and so on. If the president is regarded as the first level, show that the total number of personnel in an organization with L levels of authority is

$$T = \frac{n^L - 1}{n - 1}$$

12. Suppose that we wished to approximate an organizational structure with six administration levels by the simple hierarchy in Problem 11.

(a) How many employees would there be required if the span of control (number of subordinates per superior) were (i) three? (ii) four?

(b) If the number of levels is doubled, how many more people are required?

13. If $\bar{x} = (1/n)\sum_{k=1}^{n} x_k$ is the *mean* (average) of the n numbers $x_1, x_2, \ldots, x_n$, prove that $\sum_{i=1}^{n} (x_i - \bar{x}) = 0$. The differences $(x_i - \bar{x})$ are

termed the *deviations* of the x_i from their average $\bar{x}$. In effect, you are being asked to prove that the sum of deviations from the mean (average) of a set of quantities is zero, a fact often used in statistical computations.

14. Show that

(a) $\displaystyle\sum_{j=1}^{k} (x_j + 1)^2 y_j = \sum_{j=1}^{k} x_j^2 y_j + 2 \sum_{j=1}^{k} x_j y_j + \sum_{j=1}^{k} y_j$

(b) $\displaystyle\sum_{j=1}^{n} (x_j - \bar{x})^2 = \sum_{j=1}^{n} x_j^2 - n\bar{x}^2$

where $\bar{x}$ is as defined in Problem 13.

15. Mr. Franklin has \$10,000 with which to purchase a 4% annuity compounded annually. What is the amount he will receive annually if he elects to receive income payments for

(a) 2 years?   (b) 5 years?   (c) n years?

(*Hint:*   Refer to Problem 8.)

16. A company with a simple hierarchal organization structure as described in Problem 11 has called in a consultant to do a time study of the communication flow from the president's office to the lower levels. Suppose that company policy requires all employees at a given organization level be contacted before the communication is transmitted to the next lower level. As a result of his analysis, the consultant reported the following element times:

t_p = the average time (in minutes) required for a subordinate to process the message

t_w = the average time required to transmit the message between subordinates at the same organizational level

t_b = the average time required to transmit the message from one organizational level to a designated contact on the next lower one

(a) If there are L levels in the organization and a span of control of n, show that the total average time required for a message originating in the president's office to be processed sequentially through the entire organization is

$$\frac{n^L - 1}{n - 1} (t_p + t_w) + L(t_b - t_w) - (t_b + t_p)$$

(b) Compute the average time for sequential circulation of a message if $n = 4$, $L = 6$, $t_p = 10$, $t_w = 30$, $t_b = 10$. What suggestions would you make?

6.2 CONVERGENCE OF INFINITE SERIES

How can a value be attached to an expression of the form

$$a_1 + a_2 + a_3 + a_4 + \cdots + a_n + \cdots \tag{6}$$

or to the form

$$\sum_{n=1}^{\infty} a_n \qquad (7)$$

which we take to be equivalent with (6)? The answer to this question of how to add a sequence of numbers $(a_1, a_2, a_3, \ldots, a_n, \ldots)$ constitutes the study of *infinite series*.

It is important to note that there is no particular value which we are able *a priori* to assign to an array of symbols which calls for an infinite number of additions. The associative and distributive properties for addition of real numbers (Section 1.4) provide rules by which two or three numbers may be added. These rules may be extended by induction to any *finite* set of numbers, but nowhere in our previous discussion is there any rule which governs the present case. We must consider new rules.

We begin with a sequence $(a_1, a_2, a_3, \ldots)$ and our problem is to "add" all these numbers. The procedure for doing this is quite simple. We consider that the first number a_1 forms a *first sum*, $s_1 = a_1$. We then add a_2 to s_1 to form a *second sum*, $s_2 = a_1 + a_2$. Similarly, we obtain a *third sum*, $s_3 = a_1 + a_2 + a_3$, a *fourth sum*, $s_4 = a_1 + a_2 + a_3 + a_4$, and, in general, for any positive integer k, a *kth sum*

$$s_k = a_1 + a_2 + \cdots + a_k = \sum_{n=1}^{k} a_n \qquad (8)$$

The sequence $S = (s_1, s_2, s_3, \ldots, s_k, \ldots)$ with s_k defined by (8) is called the *infinite series generated by the sequence* $A = (a_1, a_2, a_3, \ldots)$. If the sequence S is convergent, we call $\lim S$ the *sum* of the infinite series. If S does not converge, the series is said to *diverge* or to *have no sum*. The elements a_n are called the *terms*, and the elements s_k are called the *partial sums* of this infinite series.

It is mathematical convention to use expressions (6) and (7) both for the series generated by the sequence (a_n) and for the sum of this series. As long as it is understood that the convergence of a series must be established, the double use of these notations will not lead to any confusion. Having obtained a workable definition for the sum of a series, let us see if it gives results we would expect.

Example 1 In Example 1 of Section 5.1, we considered the problem of starting at the point zero and moving to the right in a sequence of steps each of which covered half the distance remaining to the point 2. The successive step lengths (Figure 1) form the sequence

$$A = (1, \tfrac{1}{2}, \tfrac{1}{4}, \tfrac{1}{8}, \ldots) = (1, \tfrac{1}{2}, (\tfrac{1}{2})^2, (\tfrac{1}{2})^3, \ldots)$$

while the successive positions constitute the sequence

$$S = (1, \tfrac{3}{2}, \tfrac{7}{4}, \tfrac{15}{8}, \ldots) = (2 - 1, 2 - \tfrac{1}{2}, 2 - (\tfrac{1}{2})^2, 2 - (\tfrac{1}{2})^3, \ldots)$$

FIGURE 1

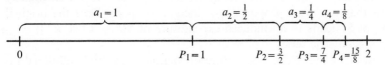

We recognize S as the sequence of partial sums of the infinite series generated by the sequence A. The kth partial sum is

$$s_k = 2 - \frac{1}{2^{k-1}}$$

Since $\lim (s_k) = 2$, we have

$$\sum_{n=0}^{\infty} (\tfrac{1}{2})^n = \lim_{k \to \infty} (s_k) = 2$$

as the sum of our infinite series of step lengths. ▶

The infinite series of Example 1 is a special kind of *geometric series* whose general form is

$$\sum_{n=0}^{\infty} x^n = 1 + x + x^2 + x^3 + \cdots$$

Simple multiplication shows that

$$(1 - x)(1 + x + x^2 + \cdots + x^{k-1}) = 1 - x^k$$

so that the kth partial sum of a geometric series is

$$s_k = 1 + x + x^2 + \cdots + x^{k-1} = \frac{1 - x^k}{1 - x} \quad (x \neq 1)$$

If $x = 1$, then $s_k = k$. For instance, in Example 1, $x = \tfrac{1}{2}$ and

$$s_k = \frac{1 - (\tfrac{1}{2})^k}{1 - \tfrac{1}{2}} = 2 - (\tfrac{1}{2})^{k-1}$$

We know already (Example 3, Section 5.4) that for a geometric sequence, $\lim (x^k) = 0$ if $-1 < x < 1$, $\lim (x^k) = 1$ if $x = 1$, and otherwise $\lim (x^k)$ does not exist. Thus when $|x| < 1$, the geometric series converges and its sum is

$$\sum_{n=0}^{\infty} x^n = \lim_{k \to \infty} \left(\frac{1 - x^k}{1 - x} \right) = \frac{1}{1 - x} \tag{9}$$

For all other values of x, this series has no sum (that is, it diverges).

Example 2 A common assumption made in economics is that if \$1 is spent in goods and services, total income will increase by \$1. If a proportion $k(0 < k < 1)$ of the increase in income is in turn spent for goods and services, total income will be further increased by $ k, and so on indefinitely assuming that the same proportion of income is spent. Thus the original spending of $1 leads to an induced expenditure of $(k + k^2 + k^3 + \cdots)$ which is a geometric series.

It follows that the initial expenditure of $1 leads to a total additional income in dollars of

$$1 + k + k^2 + \cdots = \sum_{i=0}^{\infty} k^i = \frac{1}{1 - k}$$

The sum $1/(1 - k)$ is called the *multiplier*, a term first introduced by Kohn* in 1931. ▶

PROBLEMS

1. Find the value of the economic multiplier defined in Example 2 when

   (a) $k = 0.05$ (b) $k = 0.1$ (c) $k = \frac{1}{2}$ (d) $k = 0.9$

Our results concerning geometric series show that $\sum_{n=0}^{\infty} x^n$ converges if and only if the sequence (x^n) converges to zero. In general, it is necessary (but not sufficient) that in order for the sequence (a_n) to generate a convergent series $\sum_{n=1}^{\infty} a_n$ we must have $\lim (a_n) = 0$.

Theorem 1   If $\sum_{n=1}^{\infty} a_n$ converges, then $\lim (a_n) = 0$. Equivalently, if $\lim (a_n) \neq 0$, then $\sum_{n=1}^{\infty} a_n$ diverges.

PROOF   For each $n > 1$, $a_n = s_n - s_{n-1}$, where (s_n) is the sequence of partial sums defined by (8). Convergence of this sequence means that $\lim (s_n)$ exists and Property 3 for sequences guarantees that $\lim (s_n) = \lim (s_{n-1})$. Therefore,

$$\lim (a_n) = \lim (s_n) - \lim (s_{n-1}) = 0 \qquad ▶$$

That $\lim (a_n) = 0$ is *not* a *sufficient* condition for the convergence of $\sum_{n=1}^{\infty} a_n$ is shown by the next example. Here $\lim (a_n) = 0$, but $\sum_{n=1}^{\infty} a_n$ diverges.

Example 3   The sequence $(1/n)$ generates the *harmonic series*

$$\sum_{n=1}^{\infty} \frac{1}{n} = 1 + \frac{1}{2} + \frac{1}{3} + \frac{1}{4} + \cdots$$

We have already proved (Section 5.3, Examples 1 and 2) that $\lim (1/n) = 0$ but that the sequence $(1 + \frac{1}{2} + \cdots + 1/n)$ of partial sums does not converge. ▶

*Kohn, R. F., "The Relation of Home Investment to Unemployment," *The Economic Journal* 31, 173–198 (1931).

The remainder of this section is devoted to proving theorems that are used in determining either convergence or divergence of a series. These theorems are direct consequences of previous results obtained for sequences since there is a direct relation between the limit of a sequence and the sum of a series.

Theorem 2 If the series $\sum_{n=1}^{\infty} a_n$ and $\sum_{n=1}^{\infty} b_n$ both converge, then the series $\sum_{n=1}^{\infty} (a_n + b_n)$ converges and the sums are related by

$$\sum_{n=1}^{\infty} (a_n + b_n) = \sum_{n=1}^{\infty} a_n + \sum_{n=1}^{\infty} b_n \qquad \blacktriangleright$$

Theorem 3 If the series $\sum_{n=1}^{\infty} a_n$ is convergent and if c is any real number, then the series $\sum_{n=1}^{\infty} (ca_n)$ is convergent and

$$\sum_{n=1}^{\infty} (ca_n) = c \sum_{n=1}^{\infty} a_n \qquad \blacktriangleright$$

The validity of both theorems follows directly from the definition of convergence of a series and Property 4 for sequences.

PROBLEMS

2. Show by example that convergence of $\sum_{n=1}^{\infty} (a_n + b_n)$ does not imply convergence of either $\sum_{n=1}^{\infty} a_n$ or $\sum_{n=1}^{\infty} b_n$. (*Hint:* Try $(a_n) = (1, -1, 1, -1, \ldots)$.)

Example 4 Dropping of leaflets by air is a means sometimes used in advertising to reach a large group of consumers. The effectiveness of such an advertising effort is largely dependent upon the number of people that actually come in contact with the leaflets.

Suppose that it is assumed that a constant proportion λ of the leaflets actually survive a given time period t, while the remainder are lost, destroyed, or otherwise rendered unreadable. It is further assumed that each leaflet that survives until the tth time period will reach β people, on the average, during that period. Given that N leaflets are dropped, how many people do we expect to be reached by the leaflets?

If we assume that all of the leaflets dropped actually land on the target zone, then N leaflets will survive time period $t = 0$ and will reach $N\beta$ people. Of the N leaflets surviving time period $t = 0$, only $N\lambda$ will survive period $t = 1$ and will reach $\beta N\lambda$ people. Of the $N\lambda$ leaflets surviving time $t = 1$, only $\lambda(N\lambda) = N\lambda^2$ will survive time $t = 2$ and will reach $\beta N\lambda^2$ people, and so on.

The total number of the populace reached by the air drop is, using Equation (9) and Theorem 3, given by the series

$$\beta N + \beta N\lambda + \beta N\lambda^2 + \cdots = \sum_{t=0}^{\infty} \beta N\lambda^t = \beta N \sum_{t=0}^{\infty} \lambda^t = \frac{\beta N}{1 - \lambda}$$

In this context, the quantity $\beta/(1 - \lambda)$ is analogous to the multiplier in Example 2. ▶

Theorem 4 Let (a_n) be a sequence of non-negative real numbers. Then $\sum_{n=1}^{\infty} a_n$ converges if and only if the sequence $S = (s_n)$ of partial sums is bounded. In this case

$$\sum_{n=1}^{\infty} a_n = \lim (s_n) = \text{lub } (s_n)$$

PROOF Since $a_n \geq 0$, the sequence S is increasing. The result is now a consequence of Property 5 for sequences. ▶

Example 5 The sequence $(1/n^p)$ generates the *hyperharmonic series*, sometimes called the *p series*,

$$\sum_{n=1}^{\infty} \frac{1}{n^p} = 1 + \frac{1}{2^p} + \frac{1}{3^p} + \frac{1}{4^p} + \cdots \tag{10}$$

When $p = 1$, we have the harmonic series of Example 3, the partial sums of which are not bounded (see Example 2 in Section 5.3). When $p < 1$, we have $n^p < n$ or, equivalently, $1/n^p > 1/n$. For $p < 1$, then, the kth partial sum

$$t_k = \sum_{n=1}^{k} \frac{1}{n^p}$$

of the hyperharmonic series exceeds the corresponding partial sum

$$s_k = \sum_{n=1}^{k} \frac{1}{n}$$

of the harmonic series. It follows that the sequence of partial sums (t_k) is also unbounded and (Theorem 4) that the hyperharmonic series (10) diverges when $p \leq 1$.

When $p > 1$, the hyperharmonic series converges. We prove this for $p = 2$, leaving the general case, which is quite similar, as an exercise (Problem 24). The trick is to find an upper bound for the sequence of partial sums, as required by Theorem 4. To do this we look at the subsequence of partial sums $(t_1, t_3, t_7, t_{15}, t_{31}, t_{63}, \ldots)$ whose subscripts are one less than a power of 2. We find

$$t_1 = 1$$

$$t_3 = 1 + \frac{1}{2^2} + \frac{1}{3^2} < 1 + \frac{1}{2^2} + \frac{1}{2^2} = 1 + 2\left(\frac{1}{2^2}\right)$$

$$t_7 = 1 + \frac{1}{2^2} + \frac{1}{3^2} + \frac{1}{4^2} + \frac{1}{5^2} + \frac{1}{6^2} + \frac{1}{7^2}$$

$$< 1 + \frac{1}{2^2} + \frac{1}{2^2} + \frac{1}{4^2} + \frac{1}{4^2} + \frac{1}{4^2} + \frac{1}{4^2} = 1 + 2\left(\frac{1}{2^2}\right) + 4\left(\frac{1}{4^2}\right)$$

and, in general,

$$t_{2^k-1} < 1 + 2\left(\frac{1}{2^2}\right) + 4\left(\frac{1}{4^2}\right) + \cdots + 2^{k-1}\frac{1}{(2^{k-1})^2}$$

$$= 1 + \frac{1}{2} + \frac{1}{4} + \cdots + \frac{1}{2^{k-1}}$$

$$= \frac{1 - (\frac{1}{2})^k}{1 - \frac{1}{2}} < 2 \qquad\qquad (11)$$

Hence the partial sums are bounded and the hyperharmonic series (10) converges when $p = 2$. ▶

PROBLEMS

In Problems 3–10 write out the first four terms of each of the infinite series.

3. $\displaystyle\sum_{n=1}^{\infty} \frac{n}{n^2 + 1}$ **4.** $\displaystyle\sum_{n=1}^{\infty} \frac{2}{n(n + 2)}$ **5.** $\displaystyle\sum_{n=0}^{\infty} \frac{n^2 2^n}{3^n}$

6. $\displaystyle\sum_{k=0}^{\infty} \frac{(-1)^k k}{k + 4}$ **7.** $\displaystyle\sum_{n=1}^{\infty} \frac{1}{2n^n}$ **8.** $\displaystyle\sum_{t=1}^{\infty} \frac{\log t}{t}$

9. $\displaystyle\sum_{n=0}^{\infty} (-1)^{n+1} \frac{n}{2^n}$ **10.** $\displaystyle\sum_{n=1}^{\infty} \frac{2(x - a)^{n-1}}{n(n + 2)}$

In Problems 11–14, calculate the first four partial sums of each of the given infinite series.

11. $1 + \frac{3}{2} + \frac{5}{4} + \frac{7}{8} + \cdots$ **12.** $1 - \frac{1}{2} + \frac{1}{4} - \frac{1}{8} + \cdots$

13. $1 + \frac{1}{2^2} + \frac{1}{3^2} + \frac{1}{4^2} + \cdots$ **14.** $1 - \frac{1}{3} + \frac{1}{5} - \frac{1}{7} + \cdots$

15. Find the sums of the following series.

(a) $\displaystyle\sum_{n=0}^{\infty} \frac{3^{n-1}}{4^n}$ (b) $\displaystyle\sum_{k=a}^{\infty} \frac{1}{4^k}$

(c) $1 - \frac{2}{3} + \frac{4}{9} - \frac{8}{27} + \cdots$ (d) $-4 + \frac{4}{3} - \frac{16}{9} + \frac{64}{27} - \cdots$

16. Kintsch* in a model of choice behavior states that

$$\sum_{n=2}^{\infty} (1 - b)^{n-2} b \cdot s^n = \frac{bs^2}{1 - (1 - b)s}$$

Verify his claim.

17. Why does the geometric series diverge when

(a) $x = 1$?   (b) $x = -1$?

*Kintsch, W., "A Response Time Model for Choice Behavior," *Psychometrika* **28**, 27–32 (1963).

18. In Problem 15, Section 6.1, what amount will Mr. Franklin receive if he wishes to continue to receive payments indefinitely? Check against the result obtained by using $P = \sum_{j=1}^{\infty} P_j$ in Example 2, Section 6.1.

19. Prove that

$$\lim_{n \to \infty} \sum_{r=0}^{n} x^r = \lim_{n \to \infty} \sum_{r=0}^{n-1} x^r \quad \text{when } |x| < 1$$

20. Suppose that an attitude is considered as a single dichotomous attribute. That is, during any given time interval t, an individual can be in one of two states: either he is favorably disposed towards the issue in question (state 1) or he is opposed to the issue (state 2).

Depending upon the assumptions, a change in attitude may be represented by a variety of mathematical models. One such model is as follows. Given a population of n individuals, it is assumed that during any given time period t a constant proportion α of the individuals in state 1 at the beginning of the period change to state 2, and a constant proportion β of those individuals in state 2 at the beginning of the period change to state 1. If we let n_{1t} be the number of individuals in state 1 at the beginning of the tth time period and n_{2t} be the number of individuals in state 2 at the beginning of the tth time period, then

$$n_{1t} = n_{1(t-1)} - \alpha n_{1(t-1)} + \beta n_{2(t-1)}$$
$$= n_{1(t-1)} - \alpha n_{1(t-1)} + \beta(n - n_{1(t-1)})$$
$$= n_{1(t-1)}(1 - \alpha - \beta) + \beta n$$

Show that as $t \to \infty$, the number of individuals with favorable attitudes approaches the equilibrium

$$n_1 = n \frac{\beta}{\alpha + \beta}$$

regardless of the number initially in state 1. (*Hint:* Write n_{1t} in terms of $n_{1(t-2)}$ and $n_{1(t-3)}$. Do you see a pattern? Use induction and then apply Problem 19.)

21. Let $\sum a_n$ be a given series and let $\sum b_n$ be a new series whose terms are the same as those in $\sum a_n$ except that those for which $a_n = 0$ have been omitted. Prove that $\sum a_n$ converges to a number A if and only if $\sum b_n$ converges to A.

22. Show that convergence of a series is not affected by changing or omitting (changing to zeros) a finite number of terms. (Naturally, the sum may be changed.)

23. Show that grouping the terms of a convergent series by inserting parentheses does not alter the sum. However, grouping terms in a divergent series can produce convergence. (*Hint:* After grouping, the new sequence of partial sums is a subsequence of the old.)

24. Concerning the hyperharmonic series of Example 5

(a) establish Formula (11) by induction.

(b) argue that knowing the series (10) converges for $p = 2$ automatically means that it converges when $p > 2$. (*Hint:* When $p > 2$, $n^p \geq n^2$.)

(c) prove that the hyperharmonic series converges for all $p > 1$. (*Hint:* Follow the proof for $p = 2$ step by step.)

25. Prove that for any fixed integer m,

$$\sum_{j=1}^{\infty} a_j = \sum_{j=1}^{m} a_j + \sum_{j=m+1}^{\infty} a_j$$

(*Hint:* Apply Property 4(i) for sequences. Thus $\sum_{j=1}^{\infty} a_j$ converges if and only if $\sigma_m = \sum_{j=m+1}^{\infty} a_j$ converges for every integer $m = 0, 1, 2, 3, \ldots$.)

26. Let σ_m be defined as in Problem 25. Prove that the series $\sum_{j=1}^{\infty} a_j$ converges if and only if the sequence $(\sigma_0, \sigma_1, \sigma_2, \ldots)$ converges to *zero*. Thus while convergence to zero of the sequence (a_n) of individual terms is not sufficient to guarantee convergence of $\sum_{j=1}^{\infty} a_j$, convergence to zero of the sequence (σ_m) of "tail series" is sufficient.

27. Friedman[*] has proposed that *permanent income*, defined as the income considered to be normal by a household, be estimated by

$$(1 - c) \sum_{\pi=0}^{\infty} c^{\pi} y_{\pi} \qquad (12)$$

where y_{π} is the observed income π time periods ago and c is a constant between 0 and 1. Thus, according to Friedman's concept, permanent income depends on all previous incomes with the weight given to past incomes decreasing geometrically over time.

Assume that $Y = \text{lub} (y_{\pi})$ is the largest income previously attained. Show that the sum of the series (12) cannot exceed Y.

6.3   BASES FOR THE REAL NUMBER SYSTEM

An *interval* is a set of points lying between two points on a line. If a and b are real numbers with $a \leq b$, there are four possible intervals which may be formed using points between a and b, depending on whether a or b each is or is not included in the interval. The four intervals $\{x: a \leq x \leq b\}$, $\{x: a < x \leq b\}$, $\{x: a \leq x < b\}$, and $\{x: a < x < b\}$ are, respectively, denoted by $[a, b]$, $(a, b]$, $[a, b)$, and (a, b). The bracket indicates that an endpoint is included in the interval while a parenthesis indicates that the endpoint is not included. The notation (a, b) unfortunately is the same as that used for ordered pairs, but it will always be clear from the context which meaning is intended.

[*]Friedman, M., *A Theory of the Consumption Function* (National Bureau of Economic Research, New York, 1957).

Corresponding to any positive real number x there exists an interval $[a_0, a_0 + 1)$, where a_0 is an integer, which contains x. Since the Archimedean property (Section 4.3) guarantees that *some* integer is greater than x, we simply take $a_0 + 1$ to be the smallest such integer (Figure 2).

FIGURE 2

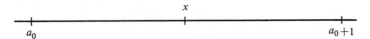

Now suppose we partition the interval $[a_0, a_0 + 1)$ into 10 smaller intervals $[a_0, a_0 + \frac{1}{10})$, $[a_0 + \frac{1}{10}, a_0 + \frac{2}{10})$, $\ldots$, $[a_0 + \frac{9}{10}, a_0 + 1)$ as shown in Figure 3. Then x lies in exactly one of these intervals, say

$$\left[a_0 + \frac{n_1}{10}, a_0 + \frac{n_1 + 1}{10}\right)$$

FIGURE 3

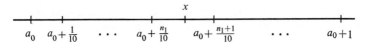

We then partition this new val into 10 equal intervals

$$\left[a_0 + \frac{n_1}{10} \quad + \frac{n_1}{10} + \frac{1}{100}\right),$$

$$\left[a_0 + \frac{n_1}{10} + \frac{1}{100}, a_0 + \frac{n_1}{10} + \frac{2}{100}\right), \ldots,$$

$$\left[a_0 + \frac{n_1}{10} + \frac{9}{100}, a_0 + \frac{n_1 + 1}{10}\right)$$

Again x lies in exactly one of these intervals, say

$$\left[a_0 + \frac{n_1}{10} + \frac{n_2}{100}, a_0 + \frac{n_1}{10} + \frac{n_2 + 1}{100}\right), \ldots$$

In this way we generate a monotone increasing sequence of points

$$(a_0, a_1, a_2, \ldots) = \left(a_0, a_0 + \frac{n_1}{10}, a_0 + \frac{n_1}{10} + \frac{n_2}{100}, \ldots\right)$$

which converges to x. (Why?) This procedure generates the *decimal expansion*

$$x = a_0 + \frac{n_1}{10} + \frac{n_2}{100} + \frac{n_3}{1000} + \cdots = a_0 . n_1 n_2 n_3 \cdots$$

Example 1 Let $x = \frac{7}{3}$. Then $a_0 = 2$, since $2 \leq \frac{7}{3} < 3$ as shown in Figure 4. Next $n_1 = 3$, since $2\frac{3}{10} \leq \frac{7}{3} < 2\frac{4}{10}$. Then $n_2 = 3$, since $2 + \frac{3}{10} +$

$\frac{3}{100} \leq \frac{7}{3} < 2 + \frac{3}{10} + \frac{4}{100}$, and $n_3 = 3$, since $2 + \frac{3}{10} + \frac{3}{100} + \frac{3}{1000} \leq \frac{7}{3} < 2 + \frac{3}{10} + \frac{3}{100} + \frac{4}{1000}$. Continuing in this way, we conclude that the decimal expansion is

$$\tfrac{7}{3} = 2.333\ldots$$

FIGURE 4

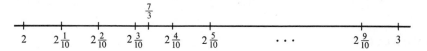

To check that our expansion is correct, note that $2.333\ldots$ may be re-written as the geometric series

$$2 + \sum_{k=1}^{\infty} \frac{3}{10^k} = 2 + \frac{3}{10} \sum_{k=1}^{\infty} \left(\frac{1}{10}\right)^{k-1}$$

Formula (9) shows that this sum is

$$2 + \frac{3}{10} \frac{1}{1 - (1/10)} = 2 + \frac{3/10}{9/10} = 2\tfrac{1}{3} = \tfrac{7}{3}$$

as required. ▶

By partitioning each interval into *two* new intervals rather than 10, we generate the *binary expansion* of any positive number.

Example 2 The binary expansion of $1\frac{2}{3}$ is obtained as follows. First, we take $a_0 = 1$, since $1 \leq 1\frac{2}{3} < 2$ as shown in Figure 5. Next we partition the interval $[1, 2)$ into two equal intervals $[1, 1 + \frac{1}{2})$ and $[1 + \frac{1}{2}, 2)$. Since $1\frac{2}{3}$ lies in the interval $[1 + \frac{1}{2}, 2)$, we take $n_1 = 1$. Now partition again to find $1\frac{2}{3}$ in the interval

$$[1 + \tfrac{1}{2}, 1 + \tfrac{1}{2} + \tfrac{1}{4}) = [1 + \tfrac{1}{2} + \tfrac{0}{4}, 1 + \tfrac{1}{2} + \tfrac{1}{4})$$

FIGURE 5

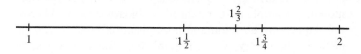

so that $n_2 = 0$. Partitioning this interval we find $1\frac{2}{3}$ in

$$[1 + \tfrac{1}{2} + \tfrac{0}{4} + \tfrac{1}{8}, 1 + \tfrac{1}{2} + \tfrac{0}{4} + \tfrac{2}{8})$$

so $n_3 = 1$, and so forth. Continuing in this way we find the binary (base 2) expansion of $1\frac{2}{3}$ is

$$1 + \tfrac{1}{2} + \tfrac{0}{4} + \tfrac{1}{8} + \tfrac{0}{16} + \tfrac{1}{32} + \cdots = 1.101010\ldots$$

To check on results, we sum the geometric series

$$1 + \sum_{k=1}^{\infty} \frac{1}{2^{2k-1}} = 1 + 2 \sum_{k=1}^{\infty} \frac{1}{2^{2k}} = 1 + \frac{2}{4} \sum_{k=1}^{\infty} \frac{1}{4^{k-1}}$$

Again applying Formula (9) gives

$$1 + \frac{2}{4} \frac{1}{1 - \frac{1}{4}} = 1\tfrac{2}{3} \qquad \blacktriangleright$$

PROBLEMS

1. Find the ternary (base 3) expansion of $\frac{3}{4}$. Check your result by summing a geometric series.

The method for writing representations of numbers in any base is essentially the same as the base 10. Thus 74.5 in base 10 means

$$(7 \times 10^1) + (4 \times 10^0) + (5 \times 10^{-1}) = (7 \times 10) + (4 \times 1) + (5 \times \tfrac{1}{10})$$

But in base 2 this becomes

$$1001010.1 = (1 \times 2^6) + (0 \times 2^5) + (0 \times 2^4) + (1 \times 2^3) + (0 \times 2^2)$$
$$+ (1 \times 2^1) + (0 \times 2^0) + (1 \times 2^{-1})$$
$$= (1 \times 64) + (0 \times 32) + (0 \times 16) + (1 \times 8) + (0 \times 4)$$
$$+ (1 \times 2) + (0 \times 1) + (1 \times \tfrac{1}{2})$$

PROBLEMS

2. Write 74.5 in base 3; in base 5.

Representation of binary numbers within any computing machine is relatively simple. A voltage may be either high or low. A specific location in a magnetic core may be polarized in one or the other direction. A switch may be open or closed. Each such dichotomous device represents one binary digit or *bit*, and a collection of bits can be used to represent a binary number as in the above example.

The requirements for the arithmetic operations of addition and multiplication are correspondingly simple to represent in the binary system. Whereas the decimal system requires an addition and multiplication table of 100 entries each, arithmetic in the binary system requires only four-element addition and multiplication tables. These are shown in Table 1. Examples of

TABLE 1. Binary Arithmetic Tables

+	0	1
0	0	1
1	1	10

×	0	1
0	0	0
1	0	1

binary arithmetic are

$$
\begin{array}{r}
100100 \\
+\quad 111110 \\
\hline
1100010
\end{array}
\qquad
\begin{array}{r}
1001011 \\
\times \quad 1101 \\
\hline
1001011 \\
1001011 \\
1001011 \\
\hline
111001111
\end{array}
$$

PROBLEMS

3. Perform the following operations of binary arithmetic. What decimal numbers are you working with?

   (a) $\quad$ 101110.1001
       $+$ 111001.0110

   (b) $\quad$ 1010110.1110
       $-$ 1110001.0010

   (c) $\quad$ 1101110.1
       $\times \quad$ 10001.1

   (d) $\quad$ 111.00011
       $\times \qquad$ 0.11

   (e) 1101.1011 ÷ 100.100 (to three binary places).

   (f) 0.0110001 ÷ 1.01101 (to five significant decimal places).

4. Add $\frac{3}{10}$ and $\frac{2}{5}$ by first writing the binary expansion of each and then summing.

Octal numbers are numbers to the base 8. For example, the decimal (base 10) number 1342.75 can be expressed as

$$(2 \cdot 8^3) + (4 \cdot 8^2) + (7 \cdot 8^1) + (6 \cdot 8^0) + (6 \cdot 8^{-1})$$

which is the octal number 2476.6.

PROBLEMS

5. Convert the following decimal numbers to octal and binary form.

   (a) 24321 $\qquad$ (b) 613428.3 $\qquad$ (c) 318.625

   (d) 4321.068359375 $\qquad$ (e) $\frac{22}{7}$

6. Construct octal addition and multiplication tables like the binary tables in Table 1.

7. Perform the following arithmetic in the octal system.

   (a) $\quad$ 240176
       $+$ 161125

   (b) $\quad$ 32176.176
       $+$ 11715.017

   (c) $\quad$ 351651
       $-$ 276316

   (d) $\quad$ 2714.076
       $-$ 3276.037

   (e) $\quad$ 210.31
       $\times \quad$ 27.07

   (f) $\quad$ 0.07761
       $\times \quad$ 0.172

   (g) 4301 ÷ 273 (to three octal places).

   (h) 22.135 ÷ 0.00243 (to five significant figures).

Many computers are now using numbers to the base 16 which are termed *hexadecimal* numbers. A hexadecimal digit is called a *byte* and is equivalent to four bits. Memory storage in a computer consists of individual words (locations) each of which will hold four bytes.

PROBLEMS

8. Convert the following decimal numbers to their hexadecimal equivalents. Use $\alpha = 10, \beta = 11, \gamma = 12, \delta = 13, \epsilon = 14,$ and $\tau = 15$.

   (a) 3480
   (b) 29146.75
   (c) $\frac{5}{3}$
   (d) $\frac{2}{5}$

9. Convert your answers in Problem 3 to hexadecimals.

10. What is the largest decimal number that can be stored in one hexadecimal word?

6.4 TECHNIQUES FOR SUMMING INFINITE SERIES

In working with a particular series, one is normally interested in finding its sum. Unfortunately, this is rarely possible, the geometric series being an exceptional case. Usually, recourse must be had to methods of approximating the sum. In any case, there is no point in trying to sum a divergent series (which, of course, has no sum) and so one begins by determining whether the series in question actually converges. Most of the techniques for checking convergence apply to series having only positive terms.

Theorem 5 *Comparison Test for Convergence* Let (a_n) and (b_n) be sequences of non-negative real numbers and suppose there is a positive integer N such that $a_n \leq b_n$ when $n \geq N$. Then convergence of $\sum_{n=1}^{\infty} b_n$ implies convergence of $\sum_{n=1}^{\infty} a_n$. Equivalently, divergence of $\sum_{n=1}^{\infty} a_n$ implies divergence of $\sum_{n=1}^{\infty} b_n$.

PROOF By Problem 22 of Section 6.2, the terms a_n and b_n for $n < N$ may be omitted without affecting the convergence of either series. We may thus assume that $N = 1$. Convergence of $\sum_{n=1}^{\infty} b_n$ means that the sequence (t_k) of partial sums

$$t_k = b_1 + b_2 + \cdots + b_k$$

is bounded. But $s_k = a_1 + a_2 + \cdots + a_k \leq t_k$ follows from $a_n \leq b_n$. Thus the partial sums of $\sum_{n=1}^{\infty} a_n$ are bounded, and by Theorem 4, $\sum_{n=1}^{\infty} a_n$ converges. ▶

Example 1 The most commonly used comparison series are the geometric and the hyperharmonic. Recall (Equation (9) and Example 5 of Section 6.2) that $\sum_{k=0}^{\infty} x^k$ converges if and only if $|x| < 1$, while $\sum_{n=1}^{\infty} 1/n^p$ converges if and only if $p > 1$.

(a) The series

$$\frac{1}{2} + \frac{\sqrt{2}}{3} + \frac{\sqrt{3}}{4} + \cdots + \frac{\sqrt{n}}{n+1} + \cdots$$

diverges since $\sqrt{n}/(n+1) \geq 1/(n+1)$ and $\sum_{n=0}^{\infty} 1/(n+1)$ is the divergent harmonic series.

(b) For every $n \geq 1$, $1/n(n+1)(n+2) < 1/n^3$. Hence the series

$$\frac{1}{1 \cdot 2 \cdot 3} + \frac{1}{2 \cdot 3 \cdot 4} + \frac{1}{3 \cdot 4 \cdot 5} + \cdots + \frac{1}{n(n+1)(n+2)} + \cdots$$

converges.

(c) The series

$$\frac{3}{1 \cdot 5} + \frac{4}{2 \cdot 6} + \frac{5}{3 \cdot 7} + \cdots + \frac{n+2}{n(n+4)} + \cdots$$

diverges since $(n+2)/[n(n+4)] > (1/2)(1/n)$ and $\sum_{n=1}^{\infty} 1/n$ diverges.

(d) If n is a positive integer, the product $1 \cdot 2 \cdot 3 \cdots n$ of the integers 1 through n is called n-*factorial* and denoted $n!$. It is easily seen that for $n \geq 4$, $n! \geq 2^n$ or $1/n! \leq 1/2^n$. Hence the series

$$\sum_{n=1}^{\infty} \frac{1}{n!} = 1 + \frac{1}{2!} + \frac{1}{3!} + \frac{1}{4!} + \cdots$$

converges since

$$\sum_{n=0}^{\infty} \frac{1}{2^n} = 1 + \frac{1}{2} + \left(\frac{1}{2}\right)^2 + \left(\frac{1}{2}\right)^3 + \cdots$$

is a convergent geometric series. ▶

Theorem 6 *Ratio Test for Convergence* Let (a_n) be a sequence of positive real numbers and suppose that there is a number $r < 1$ and an integer N such that

$$\frac{a_{n+1}}{a_n} \leq r \quad \text{when } n \geq N$$

Then $\sum_{n=1}^{\infty} a_n$ converges.

If, on the other hand, there exists a number $r \geq 1$ and an integer N such that

$$\frac{a_{n+1}}{a_n} \geq r \quad \text{when } n \geq N$$

then $\sum_{n=1}^{\infty} a_n$ is divergent.

PROOF As in the proof of Theorem 5 we may take $N = 1$. Then in the first case with $r < 1$, we have

$$a_2 \leq r a_1$$

$$a_3 \leq r a_2 \leq r^2 a_1$$

and, in general,

$$a_n \leq r^{n-1} a_1$$

The comparison test (Theorem 5) shows that

$$\sum_{n=1}^{\infty} a_n \leq \sum_{n=1}^{\infty} a_1 r^{n-1} = \frac{a_1}{1-r}$$

In case $r \geq 1$, we have

$$a_2 \geq r a_1 \geq a_1$$
$$a_3 \geq r a_2 \geq a_2 \geq a_1$$

and, in general,

$$a_n \geq a_1$$

Since $a_1 > 0$, the sequence (a_n) cannot converge to zero and hence $\sum_{n=1}^{\infty} a_n$ cannot converge (Theorem 1). ▶

Example 2   In the series

$$\frac{x}{1 \cdot 3} + \frac{x^2}{2 \cdot 4} + \frac{x^3}{3 \cdot 5} + \cdots + \frac{x^n}{n(n+2)} + \cdots$$

suppose $x > 0$ and let $a_n = x^n / n(n+2)$. Then

$$\frac{a_{n+1}}{a_n} = \frac{x^{n+1}}{x^n} \cdot \frac{n(n+2)}{(n+1)(n+3)} = x \frac{1 + (2/n)}{[1 + (1/n)][1 + (3/n)]}$$

so that

$$\lim_{n \to \infty} \frac{a_{n+1}}{a_n} = x$$

If $x < 1$, there exists a number r such that $x < r < 1$ and the definition of a limit ensures that there exists an integer N such that $a_{n+1}/a_n \leq r$ when $n \geq N$. In this case the series converges.

If $x > 1$, a similar argument ensures the existence of a number $r > 1$ and an integer M such that $a_{n+1}/a_n \geq r$ when $n \geq M$. In this case, then, the series diverges.

If $x = 1$, Theorem 6 does not apply (why?), but in this case the series converges since $1/n(n+2) < 1/n^2$ and $\sum_{n=1}^{\infty} (1/n^2)$ converges. ▶

The comparison and ratio tests may be extended to arbitrary series by introducing the concept of absolute convergence.

Definition 1   A series $\sum_{n=1}^{\infty} a_n$ is *absolutely convergent* if the series $\sum_{n=1}^{\infty} |a_n|$ converges. ▶

Thus absolute convergence of a series depends on the actual convergence of the associated series of absolute values, rather than on convergence of the series itself. However, it is not possible for a series to be absolutely convergent and not convergent.

Theorem 7 If a series converges absolutely, then it converges.

PROOF Suppose $\sum_{n=1}^{\infty} |a_n| = A$ and define a sequence (b_n) by

$$b_n = a_n + |a_n| \tag{13}$$

Then $b_n = 0$ if a_n is negative and $b_n = 2|a_n|$ if a_n is positive. In any case,

$$0 \le b_n \le 2|a_n|$$

and the partial sums of $\sum_{n=1}^{\infty} b_n$ are bounded by $2A$. Theorem 4 guarantees that $\sum_{n=1}^{\infty} b_n$ converges.

From (13), we have

$$\sum_{n=1}^{\infty} a_n = \sum_{n=1}^{\infty} b_n - \sum_{n=1}^{\infty} |a_n|$$

This being the difference of two convergent series, it follows that $\sum_{n=1}^{\infty} a_n$ also converges. This completes the proof. ▶

Example 3 (a) Since

$$\left| \frac{(-1)^{n-1}}{n^3 \sqrt{n}} \right| < \frac{1}{n^3}$$

and $\sum_{n=1}^{\infty} 1/n^3$ is a convergent hyperharmonic series, the series

$$1 - \frac{1}{2^3 \sqrt{2}} + \frac{1}{3^3 \sqrt{3}} - \frac{1}{4^3 \sqrt{4}} + \cdots$$

converges absolutely (Theorem 5) and hence must converge (Theorem 7).

(b) If in Example 2 we apply the ratio test to the series

$$\left| \frac{x}{1 \cdot 3} \right| + \left| \frac{x^2}{2 \cdot 4} \right| + \left| \frac{x^3}{3 \cdot 5} \right| + \cdots + \left| \frac{x^n}{n(n+2)} \right| + \cdots$$

we find that the assumption $x > 0$ is unnecessary and that the series converges whenever $|x| < 1$ (that is, for all x between -1 and 1). It also converges for $x = 1$ and for $x = -1$, but diverges when $|x| > 1$. ▶

Once convergence of a series has been established the problem becomes that of finding, or approximating, its sum. Since, by definition, the sequence of partial sums converges to the sum, the most obvious approach is simply to add as many terms as is practicable in terms of time and technical resources (calculators, computers, man power, etc.).

Of course, this approach has the disadvantage that it gives no indication of how closely the calculated partial sum approximates the true sum. Without some additional information, two types of errors may be committed: (1) An insufficient number of terms may be added resulting in a poor approximation, or (2) many more terms than necessary may be included, resulting in an unnecessary waste of time and resources. What is needed is an estimate of the error associated with using a partial sum to approximate the sum.

Such an estimate may be derived from the conditions of Theorem 6. For instance, suppose $N = 1$, $a_1 = 11$, and $r = \frac{1}{2}$ in that theorem. Then we have $a_n \leq r^{n-1}a_1$ for every n so that

$$\sum_{n=k+1}^{\infty} a_n \leq \sum_{n=k+1}^{\infty} r^{n-1}a_1 = \frac{a_1 r^k}{1-r} = 11\left(\frac{1}{2}\right)^{k-1}$$

Thus the error made in using the partial sum $s_{10} = a_1 + a_2 + \cdots + a_{10}$ to approximate $\sum_{n=1}^{\infty} a_n$ cannot exceed $11/512 \approx 0.02$. This error is small enough so that for most practical purposes this approximation suffices. The use of 20 terms approximates the true sum to within 0.00002.

Definition 2 A series whose terms are alternately positive and negative is called an *alternating series* and may be written in the form

$$a_1 - a_2 + a_3 - a_4 + a_5 - a_6 + \cdots$$

where each $a_n > 0$. ▶

Theorem 8 *Alternating Series Test for Convergence* Suppose $(a_1, a_2, a_3, \ldots)$ is a decreasing sequence of positive numbers which converges to zero. Then the alternating series

$$a_1 - a_2 + a_3 - a_4 + \cdots$$

must converge.

PROOF Suppose that $a_{n+1} \leq a_n$ for all n and that (a_n) converges to zero. If k is even, then

$$\sum_{n=1}^{k} a_n = a_1 - (a_2 - a_3) - (a_4 - a_5) - \cdots - (a_{k-2} - a_{k-1}) - a_k \quad \textbf{(14)}$$

and

$$\sum_{n=1}^{k} a_n = (a_1 - a_2) + (a_3 - a_4) + \cdots + (a_{k-1} - a_k) \quad \textbf{(15)}$$

Since $a_{n+1} \leq a_n$, each parenthesis in (14) and (15) is non-negative. From (14) we see that $\sum_{n=1}^{k} a_n \leq a_1$ and from (15) that the subsequence $\left(\sum_{n=1}^{k} a_n\right)$ of partial sums formed by taking even numbers of terms is increasing. Being bounded and increasing, this subsequence has a limit, say L.

If k is even, $k + 1$ is odd and we can write

$$\sum_{n=1}^{k+1} a_n = \sum_{n=1}^{k} a_n + a_{k+1}$$

Since $\lim_{k \to \infty} (a_{k+1}) = 0$, it follows that the subsequence of partial sums containing an odd number of terms also converges to L. Hence $L = \sum_{n=1}^{\infty} a_n$ and the original series is convergent. This completes the proof. ▶

A useful property of convergent alternating series is that the magnitude of the error made by using the sum of the first k terms as an approximation to the sum of the series is less than the magnitude of the $(k + 1)$st term. If k is even,

$$\sum_{n=1}^{\infty} a_n - \sum_{n=1}^{k} a_n = a_{k+1} - (a_{k+2} - a_{k+3}) - (a_{k+4} - a_{k+5}) - \cdots$$
$$= (a_{k+1} - a_{k+2}) + (a_{k+3} - a_{k+4}) + \cdots \tag{16}$$

so that this difference is positive but does not exceed a_{k+1}. If k is odd, $\sum_{n=1}^{k} a_n - \sum_{n=1}^{\infty} a_n$ has the same representations as (16) above. In either case, then,

$$\left| \sum_{n=1}^{k} a_n - \sum_{n=1}^{\infty} a_n \right| < a_{k+1}$$

Example 4 Assume that we wish to compute the value, correct to two significant figures, of the sum of the series

$$1 - \tfrac{1}{2} + \tfrac{1}{3} - \tfrac{1}{4} + \cdots$$

which is convergent by the alternating series test.

Since $1/101 \approx 0.0090$, we obtain the sum correct to two significant figures if we take the sum of the first 100 terms as an approximation to the sum. ▶

PROBLEMS

In Problems 1–9, test the given series for convergence by use of the comparison test (Theorem 5) or the alternating series test (Theorem 8).

1. $\displaystyle \sum_{n=1}^{\infty} \frac{1}{n + 7}$

2. $\displaystyle \frac{2}{\sqrt{1 \cdot 2}} + \frac{2}{\sqrt{2 \cdot 3}} + \frac{2}{\sqrt{3 \cdot 4}} + \cdots + \frac{2}{\sqrt{n(n + 1)}} + \cdots$

3. $\displaystyle \frac{1}{2} - \frac{1}{5} + \frac{1}{10} - \frac{1}{17} + \cdots + \frac{(-1)^{n+1}}{n^2 + 1} + \cdots$

4. $\displaystyle -1 + \frac{1}{\sqrt{3}} - \frac{1}{\sqrt{5}} + \frac{1}{\sqrt{7}} + \cdots + \frac{(-1)^n}{\sqrt{2n - 1}} + \cdots$

5. $\displaystyle \frac{3 + 1}{3^2 - 1} - \frac{4 + 1}{4^2 - 1} + \frac{5 + 1}{5^2 - 1} - \cdots$

6. $\displaystyle \sum_{n=2}^{\infty} \frac{\sqrt{n}}{n^2 - 1}$

7. $\displaystyle\sum_{n=2}^{\infty} \frac{(-1)^{n+1}}{(n+1)n(n-1)}$

8. $\displaystyle 1 - \frac{1}{2^2} + \frac{1}{2^4} - \cdots + \frac{(-1)^{n-1}}{2^{2(n-1)}} + \cdots$

9. $\displaystyle \left(\frac{1}{1 \cdot 2 \cdot 3}\right)^2 + \left(\frac{1}{2 \cdot 3 \cdot 4}\right)^2 + \left(\frac{1}{3 \cdot 4 \cdot 5}\right)^2 + \cdots$

10. In a model of paired-associate learning, Bower* uses the series

$$\sum_{k=1}^{\infty} \left(1 - \frac{1}{M}\right)(1-c)^{k-1}$$

Show that the sum of this series is $(1/c)[1 - (1/M)]$.

11. In Problem 10,

(a) show that

$$\sum_{k=1}^{\infty} \frac{1}{k}\left(1 - \frac{1}{M}\right)(1-c)^{k-1}$$

converges.

(b) show that

$$\sum_{k=1}^{\infty} \frac{1}{k!}\left(1 - \frac{1}{M}\right)(1-c)^{k}$$

converges.

(c) Find

$$\sum_{j=0}^{\infty} \frac{1}{M}\left(1 - \frac{1}{M}\right)^{j}(1-c)^{j}$$

12. Assume† that the relative frequency p_n of the nth most frequent reply to a word association test is

$$p_n = \frac{1}{2n-1} - \frac{1}{2n+1}$$

Show that $\sum_{n=1}^{\infty} p_n$ converges.

Test each of the series in Problems 13–17 for convergence by the ratio test (Theorem 6).

13. $\displaystyle \frac{1}{1 \cdot 2 \cdot 3} + \frac{x}{2 \cdot 3 \cdot 4} + \frac{x^2}{3 \cdot 4 \cdot 5} + \cdots$

14. $\displaystyle x + \frac{1}{2}\frac{x^3}{3} + \frac{1 \cdot 3}{2 \cdot 4}\frac{x^5}{5} + \cdots$

*Bower, G. H., "Application of a Model to Paired-Associate Learning," *Psychometrika* **26**, 255–280 (1961).

†Adapted from Haight, F. A., "Some Statistical Problems in Connection with Word Association Data," *Journal of Mathematical Psychology* **3**, 217–233 (1966).

15. $\dfrac{1}{3} + \dfrac{1 \cdot 3}{3 \cdot 6} + \dfrac{1 \cdot 3 \cdot 5}{3 \cdot 6 \cdot 9} + \dfrac{1 \cdot 3 \cdot 5 \cdot 7}{3 \cdot 6 \cdot 9 \cdot 12} + \cdots$

16. $\displaystyle\sum_{k=1}^{\infty} \dfrac{k+3}{k^3}$

17. $\displaystyle\sum_{k=1}^{\infty} \dfrac{6^{k+1}k^3(-1)^{k+1}}{8^k}$

18. Show that the series $\sum_{n=1}^{\infty} [(-1)^n / \sqrt{n}]$ converges but that

$$\sum_{n=1}^{\infty} \left| \dfrac{(-1)^n}{\sqrt{n}} \right|$$

does not.

    If a series $\sum a_n$ converges but $\sum |a_n|$ does not, then $\sum a_n$ is said to be *conditionally convergent*. Check the series in Problems 19–22 for convergence. Do any of the series converge conditionally?

19. $\frac{3}{2} - \frac{4}{3} + \frac{5}{4} - \frac{6}{5} + \cdots$

20. $\frac{1}{2} - \frac{1}{4} + \frac{1}{6} - \frac{1}{8} + \cdots$

21. $1 - \dfrac{2^3}{2 \cdot 1} + \dfrac{3^3}{3 \cdot 2 \cdot 1} - \dfrac{4^3}{4 \cdot 3 \cdot 2 \cdot 1} + \cdots$

22. $1 - \dfrac{2^{3/2}}{4} + \dfrac{3^{3/2}}{9} - \dfrac{4^{3/2}}{16} + \cdots$

23. Suppose we are interested in studying the effects of knowledge of past results on a subject's judgment regarding the weight of a stimulus object. We postulate an "ideal" subject who behaves as follows:

    (i) He alternately overestimates and underestimates the actual weight.

    (ii) Each overestimate is a constant proportion λ of the error (underestimate) on the immediately preceding trial.

    (iii) Each underestimate is a constant proportion β of the error (overestimate) on the preceding trial.

    (iv) His initial error (trial zero) is d and he overestimates on trial 1.

    (a) Find the magnitude of the estimation error for each trial.

    (b) Assuming that λ and β are both between 0 and 1, find the sum of the magnitudes determined in (a). (*Hint:* First prove that the series is convergent, then insert parentheses so as to group together successive pairs of terms.)

    (c) Find the sum of the absolute magnitudes of the errors.

24. In Problem 23, suppose that we modify the behavior of our "ideal" subject by assuming that the λ and β values vary from trial to trial. Specifically, assume that the λ value on trial $2n - 1$ is $1/(2n - 1)$ and the β value on trial $2n$ is $1/2n$, $n = 1, 2, 3, \ldots$.

    (a) Find the magnitude of the estimation error for each trial.

    (b) Prove that the alternating series is convergent.

(c) Estimate the error made by taking the sum of the first eight terms of the given alternating series as an approximation to the sum of the error magnitudes determined in (a).

25. Prove that if the series $\sum_{n=1}^{\infty} a_k$ and $\sum_{n=1}^{\infty} b_k$ are absolutely convergent, then so are the series $\sum_{n=1}^{\infty} (a_k \pm b_k)$ and $\sum_{n=1}^{\infty} c a_k$ for any constant c. Thus absolutely convergent series can be combined in the same way as ordinary finite sums.

26. Show that if a convergent series of numbers contains either a finite number of negative terms or a finite number of positive terms, then it is absolutely convergent.

27. (a) Show that the Ratio Test for Convergence (Theorem 6) implies the following result: Let (a_n) be a sequence of positive numbers and suppose that $r = \lim (a_{n+1}/a_n)$ exists. Then $\sum_{n=1}^{\infty} a_n$ is absolutely convergent if $r < 1$ and divergent if $r > 1$.

(b) Apply the result of (a) to the geometric series.

6.5 DOUBLE SUMS AND DOUBLE SERIES

In many contexts, numbers are naturally arranged in rows and columns of a table forming a rectangular pattern, thus:

$$
\begin{array}{ccccc}
x_{11} & x_{12} & x_{13} & \cdots & x_{1n} \\
x_{21} & x_{22} & x_{23} & \cdots & x_{2n} \\
\vdots & & & & \\
x_{m1} & x_{m2} & x_{m3} & \cdots & x_{mn}
\end{array}
$$

There are $m \times n$ number of x_{jk} arranged in m rows and n columns. The position occupied by a particular x quantity is indicated by the double subscript form. The variable j ranges from 1 to m and indicates the row number, while k ranges from 1 to n and designates the appropriate column. Thus $x_{4,7}$ is the element in the fourth row and seventh column, while $x_{9,2}$ lies in row 9 and column 2.

If we wish to add all $m \times n$ of the x_{jk} we could first obtain the row totals

$$
R_1 = \sum_{k=1}^{n} x_{1k}, \quad R_2 = \sum_{k=1}^{n} x_{2k}, \ldots, \quad R_m = \sum_{k=1}^{n} x_{mk}
$$

and then add these numbers to obtain the grand total

$$
T = \sum_{j=1}^{m} R_j
$$

Since $R_j = \sum_{k=1}^{n} x_{jk}$, T may be rewritten as the *double sum*

$$
T = \sum_{j=1}^{m} \left(\sum_{k=1}^{n} x_{jk} \right)
$$

If, on the other hand, we obtain column totals first, we find

$$T = \sum_{k=1}^{n} \left(\sum_{j=1}^{m} x_{jk} \right)$$

Since the total T is the same in either case, we have established the following rule.

Rule for Interchange of Order of Summation

$$\sum_{k=1}^{n} \sum_{j=1}^{m} x_{jk} = \sum_{j=1}^{m} \sum_{k=1}^{n} x_{jk} \qquad \blacktriangleright$$

Thus we see that the order of summation is immaterial when the limits of summation do not depend on the variables of summation j and k. (Sums with variable limits of summation are considered in Problem 2.)

Example 1 Suppose that x_{jk} is always equal to $k - j$. Then

$$\sum_{j=1}^{4} x_{jk} = \sum_{j=1}^{4} (k - j)$$

$$= (k - 1) + (k - 2) + (k - 3) + (k - 4) = 4k - 10$$

Hence

$$\sum_{k=1}^{5} \sum_{j=1}^{4} (k - j) = \sum_{k=1}^{5} (4k - 10) = 4 \left(\sum_{k=1}^{5} k \right) - 5(10)$$

$$= 4(15) - 5(10) = 60 - 50 = 10$$

Interchanging the order of summation gives

$$\sum_{j=1}^{4} \sum_{k=1}^{5} (k - j) = \sum_{j=1}^{4} (15 - 5j) = 4(15) - 5 \left(\sum_{j=1}^{4} j \right)$$

$$= 4(15) - 5(10) = 10 \qquad \blacktriangleright$$

PROBLEMS

1. Rewrite the following expressions without using summation notation.

(a) $\displaystyle\sum_{i=1}^{2} \sum_{j=1}^{4} x_{ij}$

(b) $\displaystyle\sum_{k=1}^{3} \sum_{j=2}^{5} \frac{k}{2^j - j}$

(c) $\displaystyle\sum_{j=1}^{4} \sum_{i=1}^{2} x_{ij}$

(d) $\displaystyle\sum_{j=2}^{5} \sum_{k=1}^{3} \frac{k}{2^j - j}$

(*Hint:* Note that interchanging the order of summation makes no difference.)

Double series constitutes the infinite counterpart to double sums. A double series is generated by a doubly infinite array of numbers (a double sequence) such as

$$
\begin{array}{ccccc}
a_{11} & a_{12} & a_{13} & \cdots & a_{1k} & \cdots \\
a_{21} & a_{22} & a_{23} & \cdots & a_{2k} & \cdots \\
a_{31} & a_{32} & a_{33} & \cdots & a_{3k} & \cdots \\
\vdots \\
a_{j1} & a_{j2} & a_{j3} & \cdots & a_{jk} & \cdots \\
\vdots
\end{array}
\tag{17}
$$

As above, the symbol a_{jk} represents the element in the jth row and kth column of the array.

Each row in the array (17) generates an infinite series. Symbolically, the sum of the elements in the jth row is

$$
R_j = \sum_{k=1}^{\infty} a_{jk}
\tag{18}
$$

This sequence of row sums itself generates a series $\sum_{j=1}^{\infty} R_j$ which, on inserting the series (18) for R_j, becomes the *double series*

$$
\sum_{j=1}^{\infty} \sum_{k=1}^{\infty} a_{jk}
\tag{19}
$$

Summing the elements in the separate columns first gives

$$
C_k = \sum_{j=1}^{\infty} a_{jk}
\tag{20}
$$

as the sum of the elements in the kth column of (17). The sequence (C_k) generates the double series

$$
\sum_{k=1}^{\infty} C_k = \sum_{k=1}^{\infty} \sum_{j=1}^{\infty} a_{jk}
\tag{21}
$$

The double series (19) and (21) are called the *iterated double series* generated by the double sequence (17). Because of the successive limiting processes involved in computing first (18) and then (19), or first (20) and then (21), it is not at all clear that these iterated series have the same value and, indeed, they may not (see Problem 5). Fortunately, however, our discussions will involve only absolutely convergent double series. To say that an iterated series $\sum_{j=1}^{\infty} \sum_{k=1}^{\infty} a_{jk}$ "converges absolutely" means that the associated double series $\sum_{j=1}^{\infty} \sum_{k=1}^{\infty} |a_{jk}|$ of absolute values converges. In this case, interchanging the order of summation still produces the same sum. The following theorem, which we state without proof, gives the details.

Theorem 9 If either of the double series (19) or (21) converges absolutely, then

(a) it converges.

(b) the other series converges absolutely.

(c) the two series have the same sum. ▶

Example 2 If for each pair of non-negative integers j and k we have $a_{jk} = (\frac{1}{2})^j(\frac{1}{3})^k$, then

$$\sum_{j=0}^{\infty} \sum_{k=0}^{\infty} a_{jk} = \sum_{j=0}^{\infty} \sum_{k=0}^{\infty} (\tfrac{1}{2})^j \, (\tfrac{1}{3})^k$$

Noting that $(\frac{1}{2})^j$ is constant relative to the inner summation (in which k is the summation variable) the double sum becomes

$$\sum_{j=0}^{\infty} \left(\frac{1}{2}\right)^j \sum_{k=0}^{\infty} \left(\frac{1}{3}\right)^k = \sum_{j=0}^{\infty} \left(\frac{1}{2}\right)^j \left(\frac{1}{1 - \frac{1}{3}}\right)$$

$$= \frac{1}{1 - \frac{1}{3}} \cdot \frac{1}{1 - \frac{1}{2}} = 3 \qquad ▶$$

Example 3 Let $a_{jk} = (\frac{1}{2})^j(\frac{1}{3})^k$ when $k \le j$, and $a_{jk} = 0$ when $k > j$. For example, $a_{42} = (\frac{1}{2})^4(\frac{1}{3})^2$, but $a_{24} = 0$. Then

$$\sum_{j=0}^{\infty} \sum_{k=0}^{\infty} a_{jk} = \sum_{j=0}^{\infty} \sum_{k=0}^{j} (\tfrac{1}{2})^j \, (\tfrac{1}{3})^k$$

Summing first on k gives

$$\sum_{j=0}^{\infty} \left(\frac{1}{2}\right)^j \frac{1 - (\frac{1}{3})^{j+1}}{1 - \frac{1}{3}}$$

which may be written in two separate sums as

$$\frac{1}{1 - \frac{1}{3}} \sum_{j=0}^{\infty} \left(\frac{1}{2}\right)^j - \frac{1}{1 - \frac{1}{3}} \sum_{j=0}^{\infty} \left(\frac{1}{2}\right)^j \left(\frac{1}{3}\right)^{j+1}$$

The first of these sums has the value 3, as in Example 2, while the second sum is

$$-\frac{1}{1 - \frac{1}{3}} \frac{1}{3} \sum_{j=0}^{\infty} \left(\frac{1}{2} \cdot \frac{1}{3}\right)^j = -\frac{\frac{1}{3}}{\frac{2}{3}} \cdot \frac{1}{1 - \frac{1}{6}} = -\frac{3}{5}$$

The total is $3 - \frac{3}{5} = \frac{12}{5}$. ▶

PROBLEMS

2. When limits of summation involve the summation variables one must be careful when interchanging the order in double sums.

(a) By writing out all the terms, show that

$$\sum_{k=1}^{3} \sum_{j=1}^{k} x_{kj} = \sum_{j=1}^{3} \sum_{k=j}^{3} x_{kj}$$

Why is

$$\sum_{k=1}^{3} \sum_{j=1}^{k} x_{kj}$$

not equal to

$$\sum_{j=1}^{k} \sum_{k=1}^{3} x_{kj}$$

(b) Show that

$$\sum_{k=1}^{4} \sum_{r=1}^{k} k(r+1) = \sum_{r=1}^{4} \sum_{k=r}^{4} k(r+1)$$

3. Find the sum

$$\sum_{k=1}^{\infty} \sum_{j=k-1}^{\infty} \frac{1}{2^j} \frac{1}{4^k}$$

4. By summing each double series separately, show that

$$\sum_{k=0}^{\infty} \sum_{j=k}^{\infty} x^{k+j} = \sum_{j=0}^{\infty} \sum_{k=0}^{j} x^{k+j}$$

(*Hint:* Let $a_{jk} = 0, j < k$ and $a_{jk} = x^{k+j}, j \geq k$.)

5. Consider the doubly infinite array of numbers

$$
\begin{array}{cccccc}
1 & -1 & 0 & 0 & 0 & 0 \ldots \\
0 & 1 & -1 & 0 & 0 & 0 \ldots \\
0 & 0 & 1 & -1 & 0 & 0 \ldots \\
0 & 0 & 0 & 1 & -1 & 0 \ldots \\
& & \vdots & & &
\end{array}
$$

For any natural number m, the mth row has 1 and -1, respectively, in the mth and $(m+1)$st positions. All other entries are zero.

(a) Show that for each m the sum of the elements in the mth row is

$$R_m = \sum_{n=1}^{\infty} a_{mn} = 0$$

Hence,

$$\sum_{m=1}^{\infty} R_m = \sum_{m=1}^{\infty} \sum_{n=1}^{\infty} a_{mn} = 0$$

(b) On the other hand, if C_n denotes the sum of the series generated by the elements in the nth column, show that

$$C_1 = 1 \quad \text{and} \quad C_n = 0 \quad \text{for all } n \geq 2$$

Thus,

$$\sum_{n=1}^{\infty} C_n = \sum_{n=1}^{\infty} \sum_{m=1}^{\infty} a_{mn} = 1$$

Interchanging the order of summation in a double series can make a difference in the value of the sum obtained! Note that neither iterated series is absolutely convergent.

SUPPLEMENTARY READING

Hirchman, I. I., *Infinite Series* (Holt, Rinehart and Winston, New York, 1962).

Rainville, E. D., *Infinite Series* (Macmillan and Company, New York, 1967).

COMBINATORIAL ANALYSIS 7

7.1 TWO BASIC PRINCIPLES OF COUNTING

The problem of determining the number of elements in a set arises so often that we shall devote this chapter to developing techniques for attacking this and other closely related problems. Our approach is based on the following two principles of counting. For convenience, the symbol $n(A)$ is used to denote the number of elements in a set A.

CP 1   If A and B are disjoint sets, then

$$n(A \cup B) = n(A) + n(B) \tag{1}$$

CP 2   If A and B are any two sets, then the product

$$n(A) \cdot n(B)$$

represents the number of ordered pairs (a, b) which can be formed by choosing the first element a from the set A and the second element b from the set B.   ▶

(Recall that the set of all such ordered pairs is called the *Cartesian product* of the sets A and B and is denoted $A \times B$. Review Problems 5–11, Section 3.1.)

PROOF OF CP 1   We note that since every element of $A \cup B$ is a member of either A or B, the sum $n(A) + n(B)$ counts each element in $A \cup B$ at least once. On the other hand, since A and B have no elements in common, no element is counted in both $n(A)$ and $n(B)$. Thus, Formula (1) must be correct since it counts each element in $A \cup B$ exactly once.   ▶

PROOF OF CP 2   Suppose A has m elements $a_1, a_2, \ldots, a_m$ and B has n elements $b_1, b_2, \ldots, b_n$. The element a_1 may be paired with each of the elements $b_1, b_2, \ldots, b_n$ in turn, making a total of n possible pairs with a_1 in first position. In the same way, we see that a_2 is the first element in n pairs, as are a_3, a_4, etc. Altogether, then, there are

$$n + n + n + \cdots + n = mn$$

pairs. ▶

The proof of CP 1 indicates the criterion which should always be applied to counting problems.

Counting Criterion   (CC)   The number of elements in a set has been correctly determined if

(a) no element has been omitted in the counting process.

(b) no element has been counted more than once. ▶

Example 1   If there are 17 alternate shipping routes from New York City to Chicago and 12 routes from Chicago to Salt Lake City, then there are $17 \times 12 = 204$ routes from NYC to SLC which go through Chicago. Here a different complete route is obtained for each different pair of routes, one from NYC to Chicago and the other from Chicago to SLC. According to CP 2, each of these is counted exactly once in our computation. ▶

Example 2   Robinson* has reported that in a randomly selected sample of 40 Democrats reelected to the Eighty-fifth Congress, 21 disapproved of, 16 moderately approved of, and 3 strongly agreed with the policies of the Department of State headed by a Republican Secretary. Hence, there were $16 + 3 = 19$ Democratic Congressmen who either moderately approved of or strongly agreed with State Department policies. The conclusion was reached by an application of CP 1. ▶

CP 2 is often rephrased in terms of a job which consists of performing a succession of tasks. If there are m ways of performing the first task and if, *no matter how the first task is performed*, there are then n ways of performing the second task, then there are $m \times n$ ways of performing the tasks in order.

Example 3   From a squadron of 50 members the commander chooses the member whom he would most like to have fly support for him and also the member whom he would least like to have. There are $50 \times 49 = 2450$ ways in which this may be done. ▶

*Robinson, J. A., "Process Satisfaction and Policy Approval in State Department–Congressional Relations," *American Journal of Sociology* **67**, 278–283 (1961).

Each of the principles CP 1 and CP 2 may be extended to any finite number of sets. The general forms of these counting principles are given below.

CP 1′ If $A_1, A_2, \ldots, A_r$ are a finite number of mutually disjoint sets, then

$$n(A_1 \cup A_2 \cup \cdots \cup A_r) = n(A_1) + n(A_2) + \cdots + n(A_r) \qquad \blacktriangleright$$

CP 2′ If there are n_1 ways to perform task T_1, if, no matter how task T_1 is performed, there are then n_2 ways of performing task T_2, if, no matter how tasks T_1 and T_2 are performed, there are then n_3 ways of performing task T_3, and so forth, then there are

$$n_1 \times n_2 \times n_3 \times \cdots \times n_r$$

ways of performing tasks $T_1, T_2, \ldots, T_r$ in the given order. $\qquad \blacktriangleright$

PROOF OF CP 1′ We have proved CP 1′ for $r = 2$. Suppose it is true for $r = k$, and consider the mutually disjoint sets $A_1, A_2, \ldots, A_k, A_{k+1}$. Then the sets

$$B = A_1 \cup A_2 \cup \cdots \cup A_k \quad \text{and} \quad C = A_{k+1}$$

must be disjoint (why?). Thus

$$n(A_1 \cup \cdots \cup A_k \cup A_{k+1}) = n(B \cup C) = n(B) + n(C)$$

which, from the inductive assumption, may be rewritten as

$$\underbrace{n(A_1) + n(A_2) + \cdots + n(A_k)}_{n(B)} + \underbrace{n(A_{k+1})}_{n(C)} \qquad \blacktriangleright$$

PROBLEMS

1. Establish CP 2′, the general form of CP 2.

Example 4 There are 3,628,800 ways in which the Big Ten football teams may be ranked in the final standings. For, any such ranking may be thought of as the outcome of performing 10 tasks; viz., choosing a team to occupy each place in the standings. The first task, choosing a team for first place, can be accomplished in 10 ways. Regardless of the team chosen, there remain nine possible choices for second, then eight for third, etc. The total number of possible rankings is $10 \times 9 \times 8 \times \cdots \times 2 \times 1 = 3,628,800$. $\qquad \blacktriangleright$

By using an argument similar to that of Example 4, it is easily seen that the number of possible ways of arranging, ordering, or ranking a collection of n different objects is the product

$$n(n - 1)(n - 2) \cdots (2)(1)$$

of the integers 1 through n. For shorthand, this number is written $n!$ and called *n-factorial*. For instance,

$$4! = 4 \cdot 3 \cdot 2 \cdot 1 = 24$$

$$7! = 7 \cdot 6 \cdot 5 \cdot 4 \cdot 3 \cdot 2 \cdot 1 = 5040$$

$$10! = 10 \cdot 9 \cdots 2 \cdot 1 = 3,628,800$$

Arrangements, or orderings, of a collection of objects are often called *permutations* of those objects. We have seen that $n!$ is the number of permutations of n objects.

PROBLEMS

2. Compute 3!, 8!, and 11!. (*Hint:* $8! = 8 \cdot 7!$, $11! = 11 \cdot 10!$, and, in general, $(n + 1)! = (n + 1)n!$.)

3. Write out the 4! ($=24$) permutations of the numbers 1, 2, 3, and 4.

4. Compute $6!/(3!)(3!)$, $12!/(8!)(4!)$, and $9!/7!$.

Example 5 Of the 3! ($=6$) ways of arranging the numbers 1, 2, and 3, only two, namely 231 and 312, satisfy the restriction that no number is in its "proper" place. We could count the number of possible arrangements by considering the job of filling three positions subject to the prescribed conditions. For the first position there are two available choices (2 or 3) since 1 cannot be used. Regardless of which is chosen, there is only one choice for second position (for example, if 2 is chosen first, then 3 must come next since it cannot be placed third) and then one choice for the remaining position. Altogether, there are $2 \cdot 1 \cdot 1 = 2$ arrangements. ▶

A convenient graphical method for analyzing the problem in Example 5 is indicated by the *tree diagram* shown in Figure 1. The two branches emanating from the topmost point correspond to the choices available for the

FIGURE 1

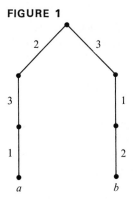

first position. If 2 is chosen first, only 3 is available next, as indicated by the single branch leading downward from the top branch labeled 2, and so forth. The two endpoints *a* and *b* correspond to the two possible arrangements. Similarly, the tree diagram of Figure 2 shows that there are nine ways of arranging the numbers 1, 2, 3, and 4 in such a way that no number appears in its proper position.

FIGURE 2

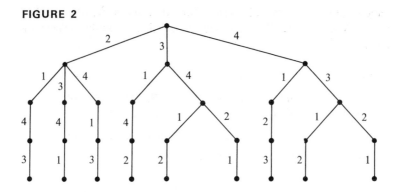

PROBLEMS

    5. A direct application of CP 2 is not possible in the tree of Figure 2 since, for example, it makes a difference, in the count for position two, whether 2 or 3 is placed in position one. Try to devise a method for counting which employs both CP 1 and CP 2. (*Hint:* First partition the set of all allowable arrangements into the union of the set *A*, of those arrangements beginning with 2, and the set *B*, of those arrangements beginning with either 3 or 4.)

Example 6 Caplow* has developed a theory of coalition formation in a triad based on the following assumptions.

    (1) Triad members may differ in strength. A stronger member can and will seek to control a weaker member.

    (2) Each triad member seeks to control all others. Control over two is preferred to control over one which in turn is preferred to control over none.

    (3) Strength is additive; that is, the strength of a coalition is the sum of the strengths of its two members.

    (4) There is a precoalition condition in every triad, in the sense that any coercion by a stronger member to force a weaker member to join a nonadvantageous coalition will provoke the formation of an advantageous coalition (if one exists) to oppose the coercion.

*Caplow, T., "A Theory of Coalitions in the Triad," *American Sociological Review* **21**, 489–493 (1956).

Let us denote the respective strengths of triad members A, B, and C by the positive real numbers a, b, and c. If, for example, $a = 5$, $b = 4$, and $c = 2$, there are three possible situations:

(i) A tries to control B. In this case B and C form an opposing coalition.

(ii) A tries to control C. Again B and C form a coalition.

(iii) B tries to control C. Here C forms a coalition with A.

The tree diagram of Figure 3 indicates the coalitions which may be formed in each of the 16 essentially different triad situations which may be obtained by knowing the relative magnitudes of a, b, and c. The three end branches in the upper left portion of the tree correspond to those cases in which the strength of one member exceeds the combined strength of the other two. In these cases, no coalitions are formed.

FIGURE 3

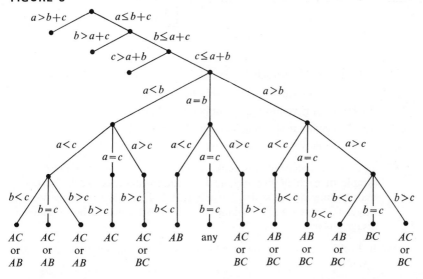

When no single member is dominant, coalitions may be formed as indicated. For instance, in the right-most branch, $a > b > c$ and the cases are enumerated as they were in the case $a = 5$, $b = 4$, and $c = 2$. ▶

PROBLEMS

6. An experimenter wishes to select a sample so that he will have at least 20 people in each possible subdivision obtained by classifying according to sex, residence (urban or rural), race (white or nonwhite), and marital status (married, divorced, or never married). What is the minimum sample size needed to meet his requirements?

7. A salesman who knows only the name and not the location of the office of the purchasing agent for the Maze Company enters the administrative

building whose shape is shown in Figure 4. If we assume that, except for the dead-ends 4 and 9, he never retraces a route, how many possible routes are there to the agent's office? (*Hint:* Construct a tree diagram using the corridor numbers shown in the diagram. Each path must end in a choice of either 5 or 6.)

FIGURE 4

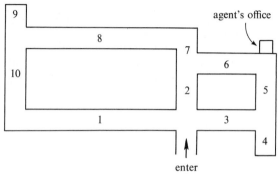

8. From the tree constructed in Problem 7, answer the following questions.

(a) How many routes require only two choices?

(b) How many routes require only three choices?

(c) How many routes require only four choices?

(d) How many routes require five or more choices?

9. Two judges independently rank in order the quality of a representative sample of each of five technical education students' work. Assuming that the ranking of the first judge is 1, 2, 3, 4, and 5, how many possible ways are there for the ranking of the second judge to be in complete disagreement?

10. In Problem 9, how many possible ways of ranking for the second judge result in

(a) exactly one disagreement?

(b) exactly two disagreements?

(c) at most two disagreements?

(d) three or more disagreements?

11. Construct a tree diagram to represent all possible outcomes for four Presidential elections if we assume that in each election, only the two major parties are represented. How many possibilities result in

(a) no party change?

(b) exactly one party change?

(c) at most three party changes?

(d) not more than two party changes?

12. Suppose that you know that in a set of 13 objects, a subject prefers one object over all others and is indifferent to the remaining objects. Assuming that in any pairing of two subsets of objects, the subject will prefer that subset containing the preferred object, how can you identify the preferred object by requiring only three subject judgments? (Compare with Problem 15, Section 3.5.)

13. A physician has a diagnostic questionnaire consisting of 10 "yes" or "no" questions. How many diagnostic categories can be identified with this instrument?

14. A model of voting behavior assumes that 10% of those voting Republican in one election will vote Democratic in the next election and 20% of those voting Democratic in one election will vote Republican in the next election. The model further assumes that the total number of voters is constant at 60 million for each election and that 40% of the population initially vote Republican.

   (a) Draw a tree to represent the possibilities for four elections.

   (b) Compute the number of Democratic and Republican voters in each of the four elections.

15. In Problem 14, what proportion of the population would eventually vote Republican if the tree were extended indefinitely? (*Hint:* Compare with Problem 20, Section 6.2.)

7.2   COUNTING NUMBERS OF SAMPLES

The process of choosing a collection of elements from a given set is called *sampling*. If r elements are chosen from a set containing n elements, we obtain an *r-sample* from an *n-set*. For concreteness, we often visualize the set as a box containing n objects distinguished either by numbers or colors. Sampling then becomes the job of filling r positions with objects from the box.

The number of distinct samples which may be obtained depends on the criteria by which samples are judged to be different—specifically, on whether the *order* in which objects are drawn is important, and on whether objects are *replaced* after being drawn.

Example 1   From the 3-set $A = \{a, b, c\}$, the following distinct 2-samples may be drawn:

   (a) If order is important and if the first sample object is replaced, we can obtain one of the nine different samples (a, a), (a, b), (a, c), (b, a), (b, b), (b, c), (c, a), (c, b), or (c, c). Our samples in this case are ordered pairs of elements of A and collectively comprise the Cartesian product $A \times A$.

   (b) If order is important but the first object is not replaced, we can obtain any one of (a, b), (a, c), (b, a), (b, c), (c, a), or (c, b). Again, we obtain ordered pairs, but pairs in which both elements are the same are not allowed.

(c) If order is not important but the first object is replaced, we can obtain one of $\{a, a\}$, $\{a, b\}$, $\{a, c\}$, $\{b, b\}$, $\{b, c\}$, or $\{c, c\}$. Here the samples are sets of objects. Disregarding order means, for instance, that $\{a, b\}$ and $\{b, a\}$ are regarded as identical.

(d) If order is not important and the first object is not replaced, we obtain one of $\{a, b\}$, $\{a, c\}$, or $\{b, c\}$. Again the samples are 2-sets. ▶

PROBLEMS

1. For each of the cases indicated in Example 1, find
   (a) the number of 2-samples from a 4-set.
   (b) the number of 3-samples from a 4-set.

2. Draw a tree diagram for cases (a) and (b) of Example 1 in each part of Problem 1.

When the order in which objects are drawn is important, the job of counting the number of r-samples from an n-set is equivalent to counting the number of ways of successively filling r positions with objects drawn from a box which initially contains n objects (Figure 5). If the objects are replaced, case (a), there are n ways of filling each position, regardless of the ways in which the other positions are filled. Applying CP 2′, it follows that there are n^r different *ordered r-samples* which may be obtained when sampling *with replacement*.

FIGURE 5

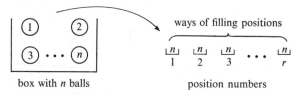

box with n balls   position numbers

If objects are not replaced, case (b), there are n ways of filling the first position after which, since one object is now unavailable, there are only $(n - 1)$ ways of filling position two, then $(n - 2)$ ways for position three, etc. Hence there are

$$n(n - 1)(n - 2) \cdots (n - r + 1) \tag{2}$$

different *ordered r-samples* which may be obtained when sampling *without replacement*. For shorthand, the product (2) is usually written $(n)_r$.

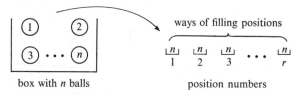

PROBLEMS

3. Compute n^r and $(n)_r$ for $n = 10$ and $r = 3$.

4. Compare the above general results with the answers obtained for cases (a) and (b) in Example 1 and Problems 1 and 2.

Example 2 (a) An inspector for the Widget Company draws a sample of 10 parts from a lot of 30 manufactured parts. If the sample is drawn with replacement, there are 30^{10}, or about 600 trillion possibilities, while if sampling is without replacement, there are $30(29)(28) \cdots (21)$ or roughly 100 trillion possible samples.

(b) Ten prizes of differing values are to be distributed among 30 people by drawing names from a hat. There are 30^{10} ways in which prizes may be awarded if names drawn are replaced and $(30)_{10}$ ways if names are not replaced. ▶

When $r = n$, the number $(n)_r$ becomes $n!$ and each ordered sample drawn without replacement corresponds to a permutation of the n objects. For this reason the problem of counting ordered r-samples is sometimes phrased as a problem of counting r-permutations.

Unordered samples drawn with replacement, case (c), are of relatively little importance for our purposes and are not considered further. The number of unordered samples which may be drawn without replacement, case (d), can be found by taking a fresh look at case (b). We obtained the number $(n)_r = n(n - 1) \cdots (n - r + 1)$ by choosing objects one at a time in order to fill the r available positions. But any particular arrangement which can be obtained in this way consists of a certain set (unordered sample) of r objects arranged in a certain order. Hence any such arrangement can be obtained by first choosing the unordered sample and then arranging the chosen objects. Using CP 2, this implies that

$$(n)_r = \begin{pmatrix} \text{number of unordered} \\ \text{samples of } r \text{ objects} \end{pmatrix} \times \begin{pmatrix} \text{number of ways} \\ \text{of arranging} \\ r \text{ chosen objects} \end{pmatrix}$$

Since the number of arrangements (permutations) of r objects is $r!$, the number of *unordered samples* of r objects drawn *without replacement* from a set of n objects is

$$\frac{(n)_r}{r!}$$

For convenience, this number will be denoted by

$$\binom{n}{r}$$

Unordered samples drawn without replacement are also called *combinations*, and problems of counting samples are called problems in *combinatorial analysis*. Hence the name of this chapter.

PROBLEMS

5. Be sure that it is clear to you why the above argument obeys the counting criterion (CC).

6. Compute

$$\binom{n}{r}$$

for $n = 10$, $r = 3$ and for $n = 5$, $r = 2$.

7. Compute

$$\binom{n}{0}, \quad \binom{n}{1}, \quad \text{and} \quad \binom{n}{n}$$

8. Compare our general results with case (d) in Example 1 and Problems 1 and 2.

9. Compare

$$\binom{n}{r} \quad \text{and} \quad \binom{n}{n-r}$$

Example 2′ The number of different unordered samples of size 10 which could be drawn without replacement from a lot of 30 parts is

$$\binom{30}{10} = \frac{(30)(29) \cdots (21)}{(10)(9) \cdots (1)} \quad \text{or} \quad \text{about 30 million} \qquad \blacktriangleright$$

The number

$$\binom{n}{r}$$

can be written entirely in terms of factorials. Since

$$n! = (n)(n-1) \cdots (n-r+1)(n-r)(n-r-1) \cdots (1) \qquad (1)$$
$$= (n)_r \cdot (n-r)!$$

we have

$$\binom{n}{r} = \frac{(n)_r}{r!} = \frac{n!}{r!(n-r)!} \qquad (3)$$

The form of (3) makes it apparent that

$$\binom{n}{r}$$

and

$$\binom{n}{n-r}$$

are identical. For a counting interpretation of this fact, we note that

$$\binom{n}{r}$$

represents the number of r-subsets which may be chosen from a given n-set. To each r-subset there corresponds a unique $(n-r)$-subset consisting of the elements *not* chosen, and thus the number of r-subsets is equal to the number of $(n - r)$-subsets.

Example 3 The number of distinguishable arrangements of r red balls and $n - r$ white balls is

$$\binom{n}{r}$$

This is because each distinct arrangement is determined by choosing from the n available positions the r positions in which the r red balls are to be placed. ▶

PROBLEMS

10. List the

$$\binom{5}{2} = 10$$

distinguishable arrangements of two letters a and three letters b.

11. Argue that the count

$$\binom{n}{r}$$

would apply to any problem of arranging n objects, r of one kind and $n - r$ of another.

Counting the number of r-subsets of an n-set is equivalent to counting the number of ways of partitioning an n-set into two subsets, the first containing r objects and the second $n - r$. By applying this argument k successive times, we may count the number of ways of partitioning an n-set into k subsets, the first containing n_1 objects, the second n_2, etc., where, of course, $n_1 + n_2 + \cdots + n_k = n$.

The n_1 objects for the first subset may be chosen in $\binom{n}{n_1}$ ways. Regardless of which objects are chosen, there remain $n - n_1$ objects from which we may choose n_2 objects for the second subset in

$$\binom{n - n_1}{n_2}$$

ways. In general, for the jth subset, there are $n - n_1 - n_2 - \cdots - n_{j-1}$ available objects from which n_j objects may be chosen in

$$\binom{n - n_1 - \cdots - n_{j-1}}{n_j}$$

ways. Hence the total number of ways of performing the required task is

$$\binom{n}{n_1} \binom{n - n_1}{n_2} \binom{n - n_1 - n_2}{n_3} \cdots \binom{n - n_1 - \cdots - n_{k-1}}{n_k}$$

Using the factorial form (3), this may be simplified to

$$\frac{n!}{n_1!n_2! \cdots n_k!}$$

which we denote by

$$\binom{n}{n_1, n_2, \ldots, n_k}$$

PROBLEMS

12. Write out the

$$\binom{4}{2, 1, 1} = 12$$

ways of partitioning a set of four objects into three subsets, the first containing two objects and the other two, one each.

13. Go through the details of the simplification indicated above.

Example 4   A bridge hand is a subset of 13 cards chosen from a standard deck of 52 cards. There are

$$\binom{52}{13}$$

or about 629 billion, such subsets.

Dealing the cards to the four players around a bridge table amounts to partitioning the set of 52 cards into four subsets of 13 cards each. There are

$$\frac{52!}{13!13!13!13!}$$

or about 437×10^{26} (437 followed by 26 zeros) different possible deals.   ▶

Example 5   In a public opinion poll, a subject is asked to express agreement or disagreement with each of 10 items. The subset A_k of response patterns in which he agrees with exactly k items contains

$$\binom{10}{k}$$

elements, one for each possible choice of k items out of the 10 presented. Hence the number of possible response patterns in which the subject agrees with at most three items is, using CP 1′,

$$n(A_0 \cup A_1 \cup A_2 \cup A_3) = n(A_0) + n(A_1) + n(A_2) + n(A_3)$$

$$= \binom{10}{0} + \binom{10}{1} + \binom{10}{2} + \binom{10}{3}$$

$$= 1 + 10 + 45 + 120 = 176 \qquad ▶$$

Example 6 Consider a committee whose members successively vote in favor of some bill. As soon as a majority of the members have voted for the bill it is declared passed. The member whose vote provided the majority is called *pivotal* and is given credit for passage of the bill.

Shapley and Shubik* have proposed the proportion of possible voting orders in which an individual is pivotal as a measure of his voting power. For instance, in a committee of three in which the chairman has two votes and the other members one each, the chairman's power is $\frac{2}{3}$ and the power of each member $\frac{1}{6}$. For, if a majority is required for passage, the chairman must be included in every winning coalition and is pivotal if he votes either second or third. Thus in the six possible voting orders

$$ABC, \ ACB, \ BAC, \ BCA, \ CAB, \text{ and } CBA$$

the chairman C is pivotal in the first four, member A is pivotal in the fifth, and B is pivotal in the last.

As a rationale for the method, we may assume that members are lined up in support of a bill in order of their intensity of feeling for the bill. In this ordering the member occupying the pivotal position is crucial to the formation of a winning coalition. If we assume that over the spectrum of bills which might be introduced, each ordering is as likely to occur as any other, the Shapley–Shubik measure provides a realistic indicator of power.

As a more complex illustration of the method, Shapley and Shubik considered a tricameral legislature in which the first house contained member A only, the second contained three members B_1, B_2, and B_3 and the third, five members C_1, C_2, C_3, C_4, and C_5. A winning coalition must contain A, at least two B_i and at least three C_j. Thus, for example, in the ordering

$$B_1 \ \dot{B}_3 \ C_4 \ C_2 \ \dot{A} \ \dot{C}_1 \ B_2 \ C_3 \ C_5$$

B_3, A, and C_1 are pivotal within their respective houses (as indicated by the dots), while C_1 (third dot) is pivotal for passage of the bill.

Member C_1 is pivotal if he is preceded in the ordering by A, exactly two other C_j and either two or three B_i. Thus the number of orderings in which he is pivotal in position six is computed by choosing two C_j (there are $\binom{4}{2} = 6$ ways), two B_i (3 ways), and A (1 way), arranging the five chosen members in the first five positions (5! ways), and the other three members in the last three positions (3! ways). There are a total of $6 \cdot 3 \cdot 1 \cdot 5! \cdot 3!$ such orderings.

Similarly, there are $6 \cdot 1 \cdot 1 \cdot 6! \cdot 2!$ ways in which C_1 is pivotal in position seven. Since there are 9! possible arrangements of the nine legislators, the power of C_1 is

$$P(C_1) = \frac{6 \cdot 3 \cdot 5! \cdot 3! + 6 \cdot 6! \cdot 2!}{9!} = \frac{5}{84} \qquad \blacktriangleright$$

*Shapley, L. S., and Shubik, M., "A Method for Evaluating the Distribution of Power in a Committee System," *American Political Science Review* **48**, 787–792 (1954).

Relatively few counting problems can be solved by a simple application of one of the formulas developed above. Aside from use of the numbers

$$\binom{n}{r}$$

we strongly recommend that all counting problems be analyzed by direct use of the basic principles CP 1 and CP 2, taking the discussion in this section as a guide.

PROBLEMS

14. An electric utility company sends teams of three men on emergency repair jobs. If the company has 10 qualified repairmen, how many different teams can be formed?

15. Five candidates for branch manager in a food store chain are ranked according to their weighted average score on a number of variables thought to be relevant to job success.

    (a) How many ways can this be done if no two candidates receive the same rank?

    (b) How many ways can the candidates be ranked if no two receive the same rank and candidate A is always ranked above candidate B?

16. A sample of five transistors is selected from a lot of 20. The sampling plan is such that the entire lot will be rejected if three or more defective transistors are found in the sample.

    (a) How many different unordered samples could be drawn?

    (b) If seven of the 20 transistors are defective, what percent of the samples would lead to rejection?

17. In Problem 16, suppose the decision strategy is as follows: If zero or one defectives are found in the sample, the lot is accepted. If three or more units are found defective, the lot is rejected. If exactly two sample units are defective, a second sample of three units is selected from the remaining 15 and the lot is accepted if at least two units in the second sample are not defective. If seven transistors in the lot are defective, what percent of the possible samples leads to rejection?

18. Analogous to Example 3, show that there are

$$\frac{n!}{n_1! n_2! \cdots n_k!}$$

distinguishable arrangements of n_1 ones, n_2 twos, ..., n_k k's. Give some other interpretations of this result.

19. In Example 5, suppose we allow the possibility that the subject may be neutral towards an item.

    (a) How many ways can he respond so that he agrees with four items, disagrees with five items, and is neutral toward the remaining item?

    (b) How many ways can he answer so that he agrees with at most one item and disagrees with at least eight items?

20. In a four-man committee, members A, B, C, and D have 3, 2, 1, and 1 votes, respectively. Simple majority wins. Compute the Shapley–Shubik power rating (see Example 6) for each member.

21. In Example 6, for reasons of symmetry, the power of each of the members C_2, C_3, C_4, and C_5 is also 5/84. Show that the power of A is 8/21 and the power of each of B_1, B_2, and B_3 is 3/28. Thus, assuming that the power of a house is the sum of the powers of its members, the three houses possess power in inverse relation to their size!

22. Shapley and Shubik (refer to Example 6) have analyzed the power of the "Big Five" in the U. N. Security Council. The council consists of 11 members, five of whom have veto power. Passage of a measure requires seven affirmative votes and no vetoes. Verify the Shapley–Shubik calculations which give combined power 76/77 to the Big Five and 1/77 to the other six members.

23. (a) In Example 3, Section 3.4, how many I scales would be obtained from a J scale having three stimuli? four stimuli? five stimuli?

(b) Develop a general formula for the number of I scales obtainable from a J scale of n stimuli.

7.3 BINOMIAL COEFFICIENTS

In addition to their use in counting, the numbers

$$\binom{n}{r}$$

also appear as coefficients in binomial expansions. For instance, $(x + y)^2 = x^2 + 2xy + y^2$ may be written as

$$(x + y)^2 = \binom{2}{0} x^2 y^0 + \binom{2}{1} x^1 y^1 + \binom{2}{2} x^0 y^2$$

and

$$(x + y)^3 = x^3 + 3x^2 y + 3xy^2 + y^3$$

as

$$(x + y)^3 = \binom{3}{0} x^3 y^0 + \binom{3}{1} x^2 y^1 + \binom{3}{2} x^1 y^2 + \binom{3}{3} x^0 y^3$$

These results are special cases of the following theorem.

Binomial Expansion Theorem   If n is a positive integer, then

$$(x + y)^n = \binom{n}{0} x^n y^0 + \binom{n}{1} x^{n-1} y + \cdots + \binom{n}{n-1} xy^{n-1} + \binom{n}{n} x^0 y^n$$

$$= \sum_{r=0}^{n} \binom{n}{r} x^{n-r} y^r \qquad (4)$$

PROOF To see why the numbers

$$\binom{n}{r}$$

appear in binomial expansions, let us first consider the case when $n = 2$. Applying the distributive law twice to $(x + y)^2 = (x + y)(x + y)$ gives

$$(x + y)(x + y) = (x + y) \cdot x + (x + y) \cdot y$$
$$= x \cdot x + y \cdot x + x \cdot y + y \cdot y$$

Each term in the sum is the product of two quantities, each of which is either x or y. Closer inspection shows, that in each product, the first number came from the first $(x + y)$ factor and the second number from the second $(x + y)$ factor. The final result consists of the sum of all possible products which may be obtained in this way.

Similarly, $(x + y)^n$ is the product of n different $(x + y)$ factors. A term in the expansion is obtained by choosing either x or y from each factor. A term of the form $x^{n-r}y^r$ is uniquely determined by the r factors from which y is selected. Since there are

$$\binom{n}{r}$$

ways in which r factors may be chosen, the term

$$\binom{n}{r} x^{n-r}y^r$$

appears in the expansion of $(x + y)^n$.   ▶

PROBLEMS

1. Write out the binomial expansions of $(1 + t)^4$ and $(s^2 - 4s)^3$.

Example 1 Putting $x = y = 1$ in (4) yields

$$2^n = (1 + 1)^n = \sum_{r=0}^{n} \binom{n}{r} = \binom{n}{0} + \binom{n}{1} + \cdots + \binom{n}{n}$$

This may be interpreted as follows. The number of subsets of size r which may be chosen from an n-set is

$$\binom{n}{r}$$

Hence, the total number of subsets of all sizes is

$$\binom{n}{0} + \binom{n}{1} + \cdots + \binom{n}{n} = 2^n$$ ▶

Example 2 If we set $x = 1$ and $y = -1$ in (4) we obtain

$$0 = (1 - 1)^n = \sum_{r=0}^{n} \binom{n}{r} (-1)^r$$
$$= \binom{n}{0} - \binom{n}{1} + \binom{n}{2} - \cdots + (-1)^n \binom{n}{n}$$

Adding this equation to that of the preceding example gives

$$2^n + 0 = \sum_{r=0}^{n} \binom{n}{r} + \sum_{r=0}^{n} \binom{n}{r} (-1)^r = \sum_{r=0}^{n} \binom{n}{r} [1 + (-1)^r]$$

Since $(-1)^r$ is 1 when r is even and -1 when r is odd, the two sums combined yield twice the sum of those numbers

$$\binom{n}{r}$$

for which r is even. Hence,

$$\sum_{r \text{ even}} \binom{n}{r} = \sum_{r \text{ odd}} \binom{n}{r} = \frac{1}{2} \sum_{\text{all } r} \binom{n}{r} = \frac{1}{2} (2^n) = 2^{n-1}$$ ▶

Example 3 We have already seen (Problem 8, Section 5.4) that if a principal P is invested at interest rate i compounded m times per year, the accumulated amount after n compounding periods is $A_n = P[1 + (i/m)]^n$. Thus the worth of \$100 two and one-half years after being invested at 4% compounded quarterly is

$$A_{10} = \$100 (1 + 0.01)^{10}$$

By writing out four terms of the binomial expansion we obtain the approximate value

$$A_{10} \approx \$100 \left[1 + \binom{10}{1} (0.01) + \binom{10}{2} (0.01)^2 + \binom{10}{3} (0.01)^3 \right]$$
$$= \$100 [1 + 0.1 + 0.0045 + 0.00012]$$
$$\approx \$110.46$$

which, as the reader may verify, is correct to the nearest penny. ▶

PROBLEMS

2. Find, to the nearest \$10, the accumulated amount of \$1000 invested for 20 years at 5% compounded annually.

Example 4 Suppose a, b, and k are positive integers with $a \geq k$ and $b \geq k$. Then

$$\binom{a+b}{k} = \binom{a}{k}\binom{b}{0} + \binom{a}{k-1}\binom{b}{1} + \cdots + \binom{a}{1}\binom{b}{k-1} + \binom{a}{0}\binom{b}{k}$$
$$= \sum_{r=0}^{k} \binom{a}{k-r}\binom{b}{r} \tag{5}$$

To see this, consider a box containing $a + b$ balls, of which a are apricot colored and b are brown. The binomial coefficient $\binom{a+b}{k}$ represents the number of different k-subsets which may be chosen from this box. A particular subset may contain any number of brown balls between zero and k, inclusive. The number of subsets containing exactly r brown balls is $\binom{a}{k-r}\binom{b}{r}$ since we may choose any r brown balls from the b available balls, and then the remaining $k - r$ balls from the a apricot balls. Since the total number of k-subsets is equal to

$$\binom{\text{number of subsets containing no brown balls}} + \binom{\text{number containing one brown ball}} + \cdots + \binom{\text{number containing} \\ k \text{ brown balls}}$$

we have established the desired formula. ▶

The definition

$$\binom{n}{r} = \frac{(n)_r}{r!}$$

for the binomial coefficients was made only for those cases when r lies between zero and n. However, it is convenient to extend this definition to other values as follows:

$$\binom{n}{r} = \frac{(n)_r}{r!}, \qquad \text{for } r \geq 0$$

$$\binom{n}{r} = 0, \qquad \text{for } r < 0$$

(6)

PROBLEMS

3. Prove that

$$(n)_r \quad \text{and} \quad \binom{n}{r}$$

are both zero if $r > n$.

This extended definition has the advantage of making Equation (5) valid even if $k > a$ or $k > b$. Obviously, there is no way of choosing a greater number of apricot balls, or brown balls, than are originally contained in the box. You should check the summation in (5) to see that the appropriate terms are zero in these cases.

PROBLEMS

4. Use the definition (6) to prove that

$$\binom{n}{r} = \binom{n-1}{r-1} + \binom{n-1}{r}$$

(7)

for all values of $n \geq 1$ and for any r. Be sure to include those cases where $r \leq 0$.

When r is a positive integer not exceeding n, Formula (7) has a counting interpretation. The

$$\binom{n}{r} \text{ } r\text{-subsets}$$

of an n-set can be classified according to whether or not they contain a specific element. The number of subsets containing the specified element is

$$\binom{n-1}{r-1}$$

since we obtain a different subset for each choice of $r - 1$ elements from the remaining $n - 1$ elements. The number of r-subsets which do not contain the specified element is

$$\binom{n-1}{r}$$

for in this case we must choose the entire subset from the other $n - 1$ elements.

Example 5 The property of binomial coefficients expressed in Problem 4 can be used to generate all these coefficients starting only from the facts that

$$\binom{n}{0} = 1 \quad \text{and} \quad \binom{n}{r} = 0$$

if $r > n$. Thus in Table 1, often called *Pascal's Triangle*, the elements in the first column must be ones while the elements above the main diagonal are all zero. In all other cases, the number in row n and column r is the sum of two numbers, one just above (row $n - 1$, column r) and the other above and to the left (row $n - 1$, column $r - 1$) of the desired number. Although only eight rows are shown, the table can, of course, be extended indefinitely.

▶

TABLE 1.   Partial table of the numbers $\binom{n}{r}$

n \ r	0	1	2	3	4	5	6	7
0	1	0	0	0	0	0	0	0
1	1	1	0	0	0	0	0	0
2	1	2	1	0	0	0	0	0
3	1	3	3	1	0	0	0	0
4	1	4	6	4	1	0	0	0
5	1	5	10	10	5	1	0	0
6	1	6	15	20	15	6	1	0
7	1	7	21	35	35	21	7	1

The method used to derive the binomial expansion theorem may be extended to the following result.

Multinomial Expansion Theorem Let n be a positive integer and x_1, $x_2, \ldots, x_k$ any real numbers. Then

$$(x_1 + x_2 + \cdots + x_k)^n = \sum \binom{n}{n_1, n_2, \ldots, n_k} x_1^{n_1} x_2^{n_2} \cdots x_k^{n_k}$$

where the indicated summation is taken over all sets of non-negative integers $(n_1, n_2, \ldots, n_k)$ for which $n_1 + n_2 + \cdots + n_k = n$.

PROOF In order to obtain the product $(x_1 + x_2 + \cdots + x_k)^n$, we must choose from each factor one of the numbers $x_1, x_2, \ldots, x_n$, multiply these choices together, and add all possible products which can be obtained in this way. Let us imagine that each factor is placed into one of k groups depending on which of the numbers $x_1, x_2, \ldots, x_k$ is chosen from that factor. The number of ways of partitioning the n factors so that exactly n_1 are placed in the x_1 group, n_2 in the x_2 group, $\ldots$, n_k in the x_k group is

$$\binom{n}{n_1, n_2, \ldots, n_k}$$

Hence this number represents the coefficient of $x_1^{n_1} x_2^{n_2} \cdots x_k^{n_k}$ in the expansion. ▶

Example 6 A collection of k non-negative integers $(n_1, n_2, \ldots, n_k)$ is called a k-*partition* of the integer n if $n_1 + n_2 + \cdots + n_k = n$. There are ten 3-partitions of the integer 3—namely, $(3, 0, 0)$, $(0, 3, 0)$, $(0, 0, 3)$, $(2, 1, 0)$, $(2, 0, 1)$, $(1, 2, 0)$, $(1, 0, 2)$, $(0, 2, 1)$, $(0, 1, 2)$, and $(1, 1, 1)$—and, correspondingly, 10 terms in the expansion of any cube of a trinomial. For instance

$$(x + y + z)^3 = \binom{3}{3, 0, 0} x^3 y^0 z^0 + \binom{3}{0, 3, 0} x^0 y^3 z^0 + \binom{3}{0, 0, 3} x^0 y^0 z^3$$

$$+ \binom{3}{2, 1, 0} x^2 y z^0 + \binom{3}{2, 0, 1} x^2 y^0 z + \binom{3}{1, 2, 0} x y^2 z^0$$

$$+ \binom{3}{1, 0, 2} x y^0 z^2 + \binom{3}{0, 2, 1} x^0 y^2 z + \binom{3}{0, 1, 2} x^0 y z^2$$

$$+ \binom{3}{1, 1, 1} xyz$$

$$= x^3 + y^3 + z^3 + 3x^2 y + 3x^2 z + 3xy^2 + 3xz^2 + 3y^2 z$$

$$+ 3yz^2 + 6xyz \qquad\qquad ▶$$

Example 7 Zannetos* uses the following argument to support the contention that the total number of channels of communication in an organization increases at least proportionately with the square of its size.

*Zannetos, Z. S., "On the Theory of Divisional Structures: Some Aspects of Centralization and Decentralization of Control and Decision Making," *Management Science* **12**, B49–B68 (1965).

In a group of n members, the number of pairwise communication channels is given by the binomial coefficient

$$\binom{n}{2} = \frac{n(n-1)}{2} \qquad (8)$$

Suppose that the group is increased c-fold, where c is an integer $(c > 1)$. Then the total number of communication channels becomes

$$\binom{cn}{2} = \frac{cn(cn-1)}{2} \qquad (9)$$

Dividing (9) by (8) we obtain

$$\frac{\binom{cn}{2}}{\binom{n}{2}} = \frac{c^2(n^2 - n/c)}{(n^2 - n)}$$

But $n/c < n$ since $c > 1$ and hence,

$$\frac{(n^2 - n/c)}{(n^2 - n)} > 1$$

Therefore,

$$\frac{\binom{cn}{2}}{\binom{n}{2}} > c^2 \qquad \blacktriangleright$$

Example 8 Davies and Davies* have suggested a scoring scheme for the Elithorn Perceptual Maze Test based on certain analytical properties of the maze. A typical test item is shown in Figure 6. The subject's task is to trace a path from the origin $(0, 0)$ to the upper diagonal (whose endpoints are both labeled 7). He is allowed to pass from one point only to the next point immediately above or to the right. His goal is to choose a path which passes through the greatest possible number of large dots. In Figure 6, three is the greatest number of large dots lying on a single path.

In an arbitrary maze the kth diagonal is the set of those points (x, y) for which $x + y = k$. A particular point (x, y) is reached from $(0, 0)$ by taking x steps to the right (R) and y steps upward (U). Since the order in which the steps are taken is immaterial, the total number of paths to (x, y) is just the number of ways of arranging x R's and y U's; that is,

$$\binom{x + y}{x}$$

The total number of paths which terminate on the kth diagonal is 2^k since the determination of any such path is equivalent to filling each of k positions

*Davies, M. G., and Davies, Ann D. M., "The Difficulty and Graded Scoring of Elithorn's Perceptual Maze Test," *British Journal of Psychology* **56**, 295–302 (1965), and "Some Analytical Properties of Elithorn's Perceptual Maze," *Journal of Mathematical Psychology* **2**, 371–380 (1965).

FIGURE 6

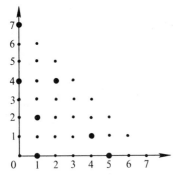

with one of the two symbols, R or U. Symbolically,

$$\sum_{\substack{x=0 \\ x+y=k}}^{k} \binom{x+y}{x} = \sum_{x=0}^{k} \binom{k}{x} = 2^k,$$

as in Example 1.

Let us denote by $u_m(x, y)$ the number of paths terminating at (x, y) which pass through exactly m large dots. Since any such path contains $(x + y)$ steps, it can pass through no more than $x + y$ points. Hence

$$u_0(x, y) + u_1(x, y) + \cdots + u_{x+y}(x, y) = \sum_{m=0}^{x+y} u_m(x, y)$$

is equal to the total number of paths

$$\binom{x+y}{x}$$

Similarly, if we denote by $U_m(k)$ the number of paths terminating at the kth diagonal which pass through exactly m large dots, we see that

$$U_m(k) = \sum_{\substack{x=0 \\ x+y=k}}^{k} u_m(x, y)$$

while

$$U_0(k) + U_1(k) + \cdots + U_k(k) = \sum_{m=0}^{k} U_m(k)$$

is equal to 2^k, the total number of paths to the kth diagonal. Thus, for instance, on the third diagonal in Figure 6, we have the results summarized in Table 2.

The authors argue that a subject should receive a score which reflects both the difficulty of the maze and the completeness of his solution. The scoring formula suggested is

$$\log_{10} \frac{U_{\tilde{m}}(D)}{U_m(D)}$$

where m is the number of dots on the path selected, the last diagonal is the Dth, and $U_{\widetilde{m}}(D)$ is the largest of the numbers $U_m(D)$. For Figure 4, where $D = 7$, we find $U_0(7) = 14$, $U_1(7) = 59$, $U_2(7) = 49$ and $U_3(7) = 6$. Thus $U_{\widetilde{m}}(7) = 59$ and a subject who selected a path traversing two dots would receive a larger score ($\log_{10}(59/49) = 0.08$) than a subject whose selected path traversed only one dot ($\log_{10}(59/59) = 0$). ▶

TABLE 2. Total number of paths to the third diagonal in Figure 6

	u_0	u_1	u_2	u_3	Total
(3, 0)	0	1	0	0	$1 = \binom{3+0}{3} = \binom{3}{3}$
(2, 1)	1	2	0	0	$3 = \binom{2+1}{2} = \binom{3}{2}$
(1, 2)	0	2	1	0	$3 = \binom{1+2}{1} = \binom{3}{1}$
(0, 3)	1	0	0	0	$1 = \binom{0+3}{0} = \binom{3}{0}$
Total	$U_0 = 2$	$U_1 = 5$	$U_2 = 1$	$U_3 = 0$	Grand total $8 = 2^3$

PROBLEMS

5. Use the binomial theorem for the following computations.

(a) Fourth term of $(a + y)^{10}$

(b) Tenth term of $[a^2 - (1/2x)]^{14}$

(c) Seventh term of $[x^3 + (y^4/x)]^{10}$

(d) Term containing b^9 of $(a + b)^{20}$

(e) Term containing x^{12} of $[x^4 + (y^2/2)]^4$

(f) $(1.01)^8$ to four decimal places

(g) $(0.98)^{12}$ to four decimal places

(h) $[(x + y) - 1]^6$, complete expansion

(i) $[2y - (1/y)]^8$, complete expansion

(j) $[(x/y^2) - (y/x^2)]^{62}$, term containing x^{11}

6. Extend the Pascal Triangle (Table 1) to $n = 12$.

7. Using a form similar to that in Problem 4, argue that

$$\binom{n+1}{n_1, n_2, n_3}$$
$$= \binom{n}{n_1 - 1, n_2, n_3} + \binom{n}{n_1, n_2 - 1, n_3} + \binom{n}{n_1, n_2, n_3 - 1}$$

8. Prove the Binomial Expansion Theorem by mathematical induction. (*Hint:* Use Formula (7).)

9. Suppose that in Example 8, the maze contains only a single dot located at (a, b). Show that the number of paths which pass through the dot and terminate at the Dth diagonal is

$$\binom{a+b}{a} 2^{D-(a+b)}$$

10. Under the conditions of the preceding problem, develop general formulas for $u_1(x, y)$ and $u_0(x, y)$.

11. Prove that

$$\sum_{k=0}^{n} \binom{n}{k}^2 = \binom{2n}{n}$$

12. Prove that for any positive integer n,

$$\sum_{k=1}^{n} k \binom{n}{k} = n2^{n-1}$$

(*Hint:* Divide out the common factor k, factor n out of the summation, and recognize what is left as the binomial expansion of $(1 + 1)^{n-1}$.)

13. Prove that if n is a positive integer,

$$\binom{n}{1} - 2 \binom{n}{2} + 3 \binom{n}{3} - \cdots + (-1)^{n-1} n \binom{n}{n} = 0$$

14. Prove that if n is a positive integer, then

$$\sum_{k=1}^{n} k^2 \binom{n}{k} = n(n+1)2^{n-2}$$

(*Hint:* Write $k^2 = k(k-1) + k$ and proceed as in Problem 12.)

15. Find the coefficient of the term involving $x^3 y^4 z^3$ in the expansion of $(x + y + z)^{10}$.

16. Write the complete expansion of $[2x^2 - x + (1/x)]^3$.

SUPPLEMENTARY READING

Goldberg, S., *Probability, An Introduction* (Prentice-Hall, Inc., Englewood Cliffs, N. J., 1960), Chapter 3.

Good, R. A., *Introduction to Mathematics* (Harcourt, Brace & World, Inc., New York, 1966), Chapter 6.

Parzen, E., *Modern Probability Theory and its Applications* (John Wiley & Sons, Inc., New York, 1960), Chapter 2.

PART II

Matrices and Linear Algebra

ADDITION AND
MULTIPLICATION OF MATRICES **8**

8.1 ADDITION OF MATRICES

A *matrix* is a rectangular array of numbers. The arrays

$$\begin{pmatrix} 2 & 7 & 0 \\ 1 & 4 & 6 \end{pmatrix} \quad \begin{pmatrix} -1 \\ 3 \end{pmatrix} \quad (2 \quad 1 \quad 9)$$

$$\begin{pmatrix} 12 & -2 & 7 & 1 & 4 \\ 0 & 0 & 2 & -16 & 1 \\ 5 & 4 & 3 & 29 & -6 \end{pmatrix}$$

(1)

are examples of matrices.

In general, a matrix A having r rows and c columns is called an $r \times c$ matrix (read "r by c") and is denoted

$$A = \begin{pmatrix} a_{11} & a_{12} & a_{13} & \cdots & a_{1c} \\ a_{21} & a_{22} & a_{23} & \cdots & a_{2c} \\ \vdots & \cdots & \cdots & a_{ij} & \cdots \\ a_{r1} & a_{r2} & a_{r3} & \cdots & a_{rc} \end{pmatrix}$$

(2)

The element in the (i, j) position of A, that is, in the ith row and jth column, is denoted by a_{ij} or, sometimes, by $(A)_{ij}$. We often denote a matrix such as (2) by the shorthand notation $A_{r \times c} = (a_{ij})$, meaning that A has r rows, c columns, and elements called a_{ij} where i takes the values $1, 2, \ldots, r$ and j the values $1, 2, \ldots, c$. The numbers r and c are called the *dimensions* of A. If

$r = c$, then A is an *r-square matrix*. In this case the elements a_{11}, a_{22}, a_{33}, ..., a_{rr} are said to constitute the *main diagonal* of A.

PROBLEMS

1. What are the dimensions of the matrices in (1)?

Example 1   Let us define a relation P on the set of objects $\{O_1, O_2, O_3\}$ by $O_i P O_j$ if and only if object O_i is *preferred* to object O_j. We shall require that for each pair of different objects one is preferred to the other and that no object is preferred to itself.

Any particular preference structure can be indicated by writing a 3×3 preference matrix $R = (r_{ij})$ in which the (i, j) entry r_{ij} is 1 if object O_i is preferred to object O_j and is 0 if no such preference exists. For example, the matrix

$$\begin{array}{c} O_1 \;\; O_2 \;\; O_3 \\ R = \begin{array}{c} O_1 \\ O_2 \\ O_3 \end{array} \begin{pmatrix} 0 & 1 & 0 \\ 0 & 0 & 1 \\ 1 & 0 & 0 \end{pmatrix} \end{array}$$

indicates that O_1 is preferred to O_2, O_2 is preferred to O_3, and O_3 is preferred to O_1. ▶

PROBLEMS

2. For a set of n objects $\{O_1, O_2, \ldots, O_n\}$, how many ones would appear in any preference matrix?

3. Could a preference matrix contain a row or column composed of nothing but zeros? of no zeros?

Two matrices A and B are said to be *equal* $(A = B)$ if they have the same dimensions and have the same elements in corresponding positions. That is, $A = (a_{ij})$ and $B = (b_{ij})$ are equal if $a_{ij} = b_{ij}$ for all choices of i and j. Thus the matrices

$$A = \begin{pmatrix} 1 & 0 & 2 \\ 1 & 1 & 3 \end{pmatrix} \quad \text{and} \quad B = \begin{pmatrix} 1 & 0 & 2 \\ 1 & 1 & 3 \end{pmatrix}$$

are equal. But

$$C = \begin{pmatrix} 1 & 0 & 2 \\ 1 & 0 & 3 \end{pmatrix}$$

is not equal to A since $a_{22} = 1 \neq 0 = c_{22}$ and

$$D = \begin{pmatrix} 1 & 0 & 2 \\ 1 & 1 & 3 \\ 4 & -1 & 6 \end{pmatrix}$$

is not equal to A since the dimensions are different.

When two matrices A and B have the same dimensions, the *sum*

$$C = A + B$$

is defined to be the matrix obtained by adding corresponding elements in A and B. That is,

$$c_{ij} = a_{ij} + b_{ij}$$

for each (i, j) pair. If A and B do not have the same dimensions, no sum is defined.

Example 2   If

$$A = \begin{pmatrix} -1 & 1 & 3 \\ 0 & 2 & 4 \end{pmatrix} \quad \text{and} \quad B = \begin{pmatrix} 3 & 0 & 2 \\ -1 & -1 & 6 \end{pmatrix}$$

then

$$A + B = \begin{pmatrix} -1+3 & 1+0 & 3+2 \\ 0-1 & 2-1 & 4+6 \end{pmatrix} = \begin{pmatrix} 2 & 1 & 5 \\ -1 & 1 & 10 \end{pmatrix}$$

Another matrix addition is

$$\begin{pmatrix} -1 & 2 & 0 \\ 1 & 6 & 7 \\ 4 & -3 & 2 \end{pmatrix} + \begin{pmatrix} 1 & 9 & -6 \\ 1 & 0 & 1 \\ -2 & -3 & -1 \end{pmatrix} = \begin{pmatrix} 0 & 11 & -6 \\ 2 & 6 & 8 \\ 2 & -6 & 1 \end{pmatrix} \qquad \blacktriangleright$$

PROBLEMS

4. Find the sums

(a) $\begin{pmatrix} -1 & 2 \\ 0 & 3 \end{pmatrix} + \begin{pmatrix} 4 & 4 \\ 4 & 4 \end{pmatrix}$ (b) $\begin{pmatrix} 1 & 1 & 2 \\ -6 & 6 & -3 \end{pmatrix} + \begin{pmatrix} 0 & 1 & 7 \\ 7 & -1 & -2 \end{pmatrix}$

5. If $A + B = C$, where A, B, and C are matrices and A is 10×5, what must be the dimensions of B and C?

The matrix $O_{r \times c}$, each of whose entries is the number 0, serves as the *identity element for addition* of $r \times c$ matrices. That is,

$$A + O = O + A = A$$

For instance,

$$\begin{pmatrix} 1 & -7 & 3 \\ 2 & 0 & 4 \end{pmatrix} + \begin{pmatrix} 0 & 0 & 0 \\ 0 & 0 & 0 \end{pmatrix} = \begin{pmatrix} 1+0 & -7+0 & 3+0 \\ 2+0 & 0+0 & 4+0 \end{pmatrix}$$

$$= \begin{pmatrix} 1 & -7 & 3 \\ 2 & 0 & 4 \end{pmatrix}$$

The matrix $-A$, each of whose elements is the negative of the corresponding element of A, is the *additive inverse* of A. That is,

$$A + (-A) = -A + A = O$$

Thus

$$\begin{pmatrix} 1 & -7 & 3 \\ 2 & 0 & 4 \end{pmatrix} + \begin{pmatrix} -1 & 7 & -3 \\ -2 & 0 & -4 \end{pmatrix} = \begin{pmatrix} 1-1 & -7+7 & 3-3 \\ 2-2 & 0-0 & 4-4 \end{pmatrix}$$

$$= \begin{pmatrix} 0 & 0 & 0 \\ 0 & 0 & 0 \end{pmatrix} = O_{2 \times 3}$$

In general, the sum of any number of matrices having the same dimensions is obtained by adding corresponding entries.

Example 3

$$\begin{pmatrix} 0 & 1 \\ 2 & -7 \end{pmatrix} + \begin{pmatrix} 6 & 6 \\ 4 & -4 \end{pmatrix} + \begin{pmatrix} 1 & -1 \\ 0 & 1 \end{pmatrix} + \begin{pmatrix} 14 & -3 \\ 0 & 2 \end{pmatrix}$$

$$= \begin{pmatrix} 0+6+1+14 & 1+6-1-3 \\ 2+4+0+0 & -7-4+1+2 \end{pmatrix} = \begin{pmatrix} 21 & 3 \\ 6 & -8 \end{pmatrix} \quad \blacktriangleright$$

Example 4 Suppose in Example 1 that each of N judges lists his preferences among the three objects. Let $R_i (i = 1, 2, \ldots, N)$ represent the preference matrix of the ith judge. Then the matrix S whose (i, j) element $(i = 1, 2, 3; j = 1, 2, 3)$ represents the number of judges who prefer object O_i to object O_j is given by

$$S = R_1 + R_2 + \cdots + R_N = \sum_{i=1}^{N} R_i \quad \blacktriangleright$$

Matrix addition is a commutative operation, since matrix addition is defined in terms of the addition of the individual real number entries. That is, since

$$a_{ij} + b_{ij} = b_{ij} + a_{ij}$$

it follows that

$$A + B = B + A$$

Example 5 An example of the commutative property of matrix addition is

$$\begin{pmatrix} 1 & 3 \\ -1 & 0 \end{pmatrix} + \begin{pmatrix} 2 & 6 \\ 3 & 2 \end{pmatrix} = \begin{pmatrix} 1+2 & 3+6 \\ -1+3 & 0+2 \end{pmatrix}$$

$$= \begin{pmatrix} 2+1 & 6+3 \\ 3-1 & 2+0 \end{pmatrix} = \begin{pmatrix} 2 & 6 \\ 3 & 2 \end{pmatrix} + \begin{pmatrix} 1 & 3 \\ -1 & 0 \end{pmatrix} \qquad \blacktriangleright$$

Matrix addition is also an associative operation. That is, if A, B, and C are three matrices having the same dimensions, then $(A + B) + C = A + (B + C)$. This follows directly from the associative property for addition of numbers. The proof is left as an exercise (Problem 12).

PROBLEMS

6. Prove that if for three matrices A, B and C we have $A + B = A + C$, then $B = C$. (*Hint:* Add $-A$ to both sides.)

The matrix $-A$ is obtained from A by replacing each element by its negative or, equivalently, by multiplying each element of A by -1. We shall describe this operation by saying that the matrix A itself is multiplied by -1. More generally, the matrix obtained from A by multiplying each element of A by the same real number c is called the product of A by the *scalar c* and is written cA. We define Ac to mean the same thing as cA.

Example 6 If

$$A = \begin{pmatrix} 1 & 1 & 6 \\ -1 & 2 & 4 \end{pmatrix}$$

then

$$2A = \begin{pmatrix} 2 & 2 & 12 \\ -2 & 4 & 8 \end{pmatrix} \quad \text{and} \quad -3A = \begin{pmatrix} -3 & -3 & -18 \\ 3 & -6 & -12 \end{pmatrix} \qquad \blacktriangleright$$

Example 7 The relative frequency of judges who prefer object i to object j in Example 4 is $\bar{s}_{ij} = s_{ij}/N$. Thus

$$\bar{S} = \frac{1}{N} S = \frac{1}{N} \sum_{i=1}^{N} R_i$$

denotes a different kind of preference matrix whose individual elements $\bar{s}_{ij}$ represent the proportion of judges who prefer object i to object j. $\qquad \blacktriangleright$

PROBLEMS

7. In Example 7, show that $\bar{s}_{ij} = 1 - \bar{s}_{ji}$ when $i \neq j$. Of course, $\bar{s}_{ii} = 0$ for all i. (Why?)

The *transpose* A' of a matrix A is the matrix obtained from A by interchanging rows and columns. More precisely, if A is $r \times c$, the transpose A' is a $c \times r$ matrix whose (i, j) element $(A')_{i,j}$ is the element in the (j, i) position of A.*

Example 8 The transpose of

$$A = \begin{pmatrix} -1 & 3 & 2 \\ 0 & 7 & 4 \end{pmatrix}$$

is

$$A' = \begin{pmatrix} -1 & 0 \\ 3 & 7 \\ 2 & 4 \end{pmatrix}$$

while that of

$$B = \begin{pmatrix} -4 & 0 & 2 \\ 4 & 1 & 7 \\ 6 & 3 & -1 \end{pmatrix}$$

is

$$B' = \begin{pmatrix} -4 & 4 & 6 \\ 0 & 1 & 3 \\ 2 & 7 & -1 \end{pmatrix}$$

▶

It is easily seen that the transpose of the sum of two matrices is the sum of the individual transposes. For if A and B are both $r \times c$, the element in the (i, j) position of the transpose of $A + B$ is

$$[(A + B)']_{ij} = (A + B)_{ji} = A_{ji} + B_{ji}$$
$$= (A')_{ij} + (B')_{ij} = (A' + B')_{ij}$$

which is the element in the (i, j) position of $A' + B'$. Since $(A + B)'$ and $A' + B'$ have the same elements in the same positions, they must be equal.

---

*The same notation is used for the transpose of a matrix A as for the complement of a set A. No confusion should arise as it will be clear from the context which usage is intended.

Example 9 If

$$A = \begin{pmatrix} 1 & 2 & 3 \\ 4 & 5 & 6 \end{pmatrix} \quad \text{and} \quad B = \begin{pmatrix} 0 & -1 & 7 \\ -6 & -3 & 2 \end{pmatrix}$$

then

$$A + B = \begin{pmatrix} 1 & 1 & 10 \\ -2 & 2 & 8 \end{pmatrix} \quad \text{and} \quad (A + B)' = \begin{pmatrix} 1 & -2 \\ 1 & 2 \\ 10 & 8 \end{pmatrix}$$

while

$$A' = \begin{pmatrix} 1 & 4 \\ 2 & 5 \\ 3 & 6 \end{pmatrix} \quad B' = \begin{pmatrix} 0 & -6 \\ -1 & -3 \\ 7 & 2 \end{pmatrix} \quad A' + B' = \begin{pmatrix} 1 & -2 \\ 1 & 2 \\ 10 & 8 \end{pmatrix} \quad \blacktriangleright$$

PROBLEMS

8. If

$$A = \begin{pmatrix} 2 & 0 \\ 1 & 3 \end{pmatrix} \quad B = \begin{pmatrix} -1 & -1 \\ 4 & -1 \end{pmatrix} \quad C = \begin{pmatrix} 1 & 7 \\ 2 & -9 \end{pmatrix}$$

find $A + B$, $B + A$, $(A + B) + C$, $A + (B + C)$, $3A - 4B$, $(A + B)'$, $A' + B'$, $4A'$ and $(4A)'$.

9. In Example 7, prove that $\bar{S} + \bar{S}'$ is a matrix U such that $u_{ii} = 0$, while $u_{ij} = 1$ when $i \neq j$. (*Hint:* See Problem 7.)

10. The transpose of the transpose of a matrix A is equal to A itself. That is, $(A')' = A$. Prove this and give an example.

11. Prove that the transpose of the sum of any number of matrices is the sum of the individual transposes.

12. Prove that matrix addition is associative. That is, prove that if A, B, and C have the same dimensions, then

$$(A + B) + C = A + (B + C)$$

13. Scalar multiplication satisfies associative, commutative, and distributive properties. Specifically, if a and b are real numbers and if A and B are matrices having the same dimensions, then

(a) $aA = Aa$

(b) $(a + b)A = aA + bA = Aa + Ab = A(a + b)$

(c) $a(A + B) = aA + aB = Aa + Ba = (A + B)a$

(d) $(ab)A = a(bA) = b(aA) = A(ab)$

(e) $1 \cdot A = A$ and $0 \cdot A = O$ (the zero matrix)

Prove these results and then look at computational details when $a = 4$, $b = -3$,

$$A = \begin{pmatrix} 2 & 0 & 3 \\ -7 & -2 & 1 \end{pmatrix} \quad \text{and} \quad B = \begin{pmatrix} -1 & -1 & 6 \\ 0 & 4 & 5 \end{pmatrix}$$

14. (a) Show that if b is a scalar and A a matrix, and if $bA = O$ (the zero matrix), then either $b = 0$ or $A = O$.

(b) Show that if $bA = bC$ and if $b \neq 0$, then $A = C$, and that if $A = C$, then $bA = bC$ for all numbers b.

15. Prove that

$$(A - B)' = A' - B'$$

The *trace* of a square matrix A is the sum of the entries on the main diagonal and is denoted as tr A.

For example,

$$\text{tr} \begin{pmatrix} 1 & \frac{1}{2} & 0 \\ 3 & 4 & 2 \\ 6 & \frac{1}{3} & -1 \end{pmatrix} = 1 + 4 - 1 = 4$$

Prove the following:

16. If A is a square matrix, then tr $A = $ tr A'.

17. If A and B are both n-square matrices, then tr $(A + B) = $ tr $A + $ tr B.

18. If A is a square matrix and c is a scalar, then tr $(cA) = c$ tr A.

8.2 MULTIPLICATION OF MATRICES

A matrix consisting of a single row is called a *row vector*, while a one-column matrix is called a *column vector*. The entries in a vector are called its *elements*, or *components*. The number of elements is the *dimension* of the vector. For instance $V_1 = (2, 0, -3, 7, 6)$ and $V_2 = (-1, 2)$ are, respectively, five- and two-dimensional row vectors, while

$$V_3 = \begin{pmatrix} 4 \\ 2 \\ 1 \\ 0 \end{pmatrix}$$

is a four-dimensional column vector.

The concept of dimension for vectors stems from our practice of using pairs, or triples, of numbers to designate points on a plane, or in space. Thus, in Figure 1(a), the point (x, y) on the plane is found x units to the right and y units above the origin $(0, 0)$. The total distance from the origin to (x, y) is $\sqrt{x^2 + y^2}$, called the *length* of the vector (x, y). Geometrically, it is convenient to draw an arrow from the origin to (x, y) and to think of the distance $\sqrt{x^2 + y^2}$ as the length of this arrow. Similarly, in Figure 1(b), the point (x, y, z) is x units toward you, y units to the right, and z units up from the origin in three-dimensional space. The length of the vector (x, y, z) is $\sqrt{x^2 + y^2 + z^2}$.

FIGURE 1

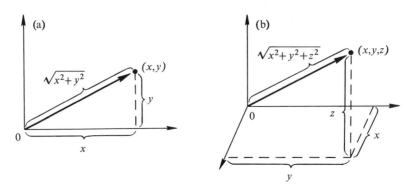

The generalization to four-, five-, and higher-dimensional vectors is made by analogy to the strictly computational aspects of two- and three-dimensional vectors, since no geometric interpretation is available. In particular, the *length* of an n-dimensional vector $V = (x_1, x_2, \ldots, x_n)$ is defined to be

$$|V| = \sqrt{x_1^2 + x_2^2 + \cdots + x_n^2}$$

Geometrically, we make no distinction between row and column vectors.

PROBLEMS

1. Find the lengths of the vectors V_1, V_2, and V_3 above.

Since vectors are matrices, the definitions of addition and multiplication by scalars apply equally well to them. The practice of drawing arrows to represent vectors enables us to give geometric interpretations to these matrix operations.

Example 1 If we multiply the vector $(1, 3)$ (shown in position (a) of Figure 2) by 2, we obtain the vector $(2, 6)$ (position (b)) which lies in the same direction from the origin as $(1, 3)$ but is twice as far away. But if we multiply by -2, we obtain the vector $(-2, -6)$ (position (c)), twice as long as $(1, 3)$ and in the opposite direction from the origin. ▶

FIGURE 2

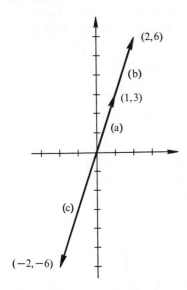

Example 2 The addition of two vectors V_1 and V_2 is accomplished geometrically as shown in Figure 3(a) by placing the tail of one vector at the head of the other. In Figure 3(b), the sum of the vectors (1, 3) and (2, −2) is the vector (3, 1) drawn from the origin to the head of the vector (2, −2) placed so as to start at the point (1, 3). ▶

FIGURE 3

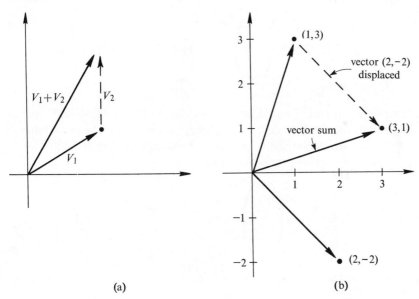

(a)   (b)

Since "the shortest distance between two points is a straight line," it is apparent from Figure 3(a) that the length of the vector $V_1 + V_2$ cannot exceed the sum of the lengths of V_1 and V_2. Thus we have the *Triangle Inequality* for vectors

$$|V_1 + V_2| \leq |V_1| + |V_2|$$

(Compare with Property 6 for absolute value in Section 4.2.)

In order to save space, we often denote a column vector V with components $a_1, a_2, \ldots, a_n$ by $V = \text{col}\,(a_1, a_2, \ldots, a_n)$ or, in transpose notation, by $(a_1, a_2, \ldots, a_n)'$. For instance, the vector V_3 above would be written either $V_3 = \text{col}\,(4, 2, 1, 0)$ or $V_3 = (4, 2, 1, 0)'$.

If $R = (r_1, r_2, \ldots, r_n)$ and $C = \text{col}\,(c_1, c_2, \ldots, c_n)$ are row and column vectors having the same dimension, the *product*

$$RC = (r_1, r_2, \ldots, r_n) \begin{pmatrix} c_1 \\ c_2 \\ \vdots \\ c_n \end{pmatrix}$$

of R and C is obtained by multiplying the corresponding elements of the two vectors and summing. Specifically,

$$RC = r_1 c_1 + r_2 c_2 + \cdots + r_n c_n = \sum_{i=1}^{n} r_i c_i$$

For example, the product of $R = (2, 0, -3)$ and $C = \text{col}\,(-1, 6, 2)$ is

$$RC = (2, 0, -3) \begin{pmatrix} -1 \\ 6 \\ 2 \end{pmatrix} = 2(-1) + 0(6) - 3(2) = -8$$

Example 3 In their linear learning models, Bush and Mosteller* operate with column vectors $P = \text{col}\,(p_1, p_2, \ldots, p_n)$, where p_i is the probability of response R_i. The requirement that these probabilities sum to unity can be expressed as the vector product equation

$$(1, 1, \ldots, 1) \begin{pmatrix} p_1 \\ p_2 \\ \vdots \\ p_n \end{pmatrix} = p_1 + p_2 + \cdots + p_n = 1$$

▶

*Bush, R. R., and Mosteller, F., *Stochastic Models for Learning* (John Wiley & Sons, Inc., New York, 1955).

PROBLEMS

2. Find the following vector products.

(a) $(2, 1)\begin{pmatrix} 4 \\ -3 \end{pmatrix}$ (b) $(0, 2, 6, 9)\begin{pmatrix} 1 \\ 6 \\ -1 \\ 4 \end{pmatrix}$ (c) $(4, 2, 3)\begin{pmatrix} -1 \\ -1 \\ 2 \end{pmatrix}$

3. A monopolist produces commodities A, B, and D having respective unit production costs of 5¢, 10¢, and 1¢. Suppose that he sells 25,000 units of commodity A, 40,000 units of B, and 62,000 units of D. Write a three-component row vector R whose elements are the respective production totals and a three-dimensional column vector C whose entries are the unit production costs. Then find the total production cost by computing the vector product RC.

4. (a) Two vectors, such as those in Problem 2(c), whose product is zero, are called *orthogonal*, or *perpendicular*. Find two different vectors each of which is orthogonal to $(1, 2, 1)$.

   (b) Find two vectors which are orthogonal both to $(1, 2, 1)$ and to each other.

Multiplication of matrices is defined, not by multiplying corresponding elements, but by multiplying corresponding rows and columns. An example will best serve to establish the procedure.

Example 4   The product

$$AB = \begin{pmatrix} 2 & 0 & -3 \\ 1 & -1 & 4 \end{pmatrix}\begin{pmatrix} -1 & 0 & 7 \\ 6 & 0 & -2 \\ 2 & 3 & 1 \end{pmatrix}$$

of the matrices A and B is the matrix

$$C = AB = \begin{pmatrix} -8 & -9 & 11 \\ 1 & 12 & 13 \end{pmatrix}$$

whose entries are obtained as follows. The element $c_{11} = -8$ in the first row and first column of C is found by multiplying the *first row of A* by the *first column of B* to obtain the product

$$(2 \quad 0 \quad -3)\begin{pmatrix} -1 \\ 6 \\ 2 \end{pmatrix} = -8$$

Similarly, element c_{12} in the first row and second column of C is found by multiplying the *first row of A* by the *second column of B* to obtain

$$c_{12} = (2 \quad 0 \;-3) \begin{pmatrix} 0 \\ 0 \\ 3 \end{pmatrix} 6- \; =$$

and so on, for the remaining four elements. ▶

In general, the product $C = AB$ of matrices A and B is defined only if the number of *columns* of A is the same as the number of *rows* of B. To find the element c_{ij} in the ith row and jth column of C, we multiply the ith *row of A* by the jth *column of B*. In this way we obtain a new matrix having the same number of rows as A and the same number of columns as B. Thus, if A is $r \times p$ and B is $q \times c$, then $C = AB$ is defined only if $p = q$, in which case C is an $r \times c$ matrix.

Example 5 If

$$A = \begin{pmatrix} -1 & 2 \\ 0 & 3 \end{pmatrix} \qquad B = \begin{pmatrix} 0 & 3 \\ 7 & 2 \end{pmatrix} \qquad C = \begin{pmatrix} 4 & -6 & 0 \\ 1 & 1 & -1 \end{pmatrix}$$

then

$$AB = \begin{pmatrix} 14 & 1 \\ 21 & 6 \end{pmatrix} \qquad AC = \begin{pmatrix} -2 & 8 & -2 \\ 3 & 3 & -3 \end{pmatrix}$$

$$BA = \begin{pmatrix} 0 & 9 \\ -7 & 20 \end{pmatrix} \qquad BC = \begin{pmatrix} 3 & 3 & -3 \\ 30 & -40 & -2 \end{pmatrix}$$

while CA and CB are not defined. ▶

It is apparent from Example 5 that *matrix multiplication is not commutative*. In that example, the product CA is not defined, even though AC is. Moreover, even though AB and BA are both defined they are not equal. However, some particular pairs of matrices do commute.

PROBLEMS

   5. Show that if

$$A = \begin{pmatrix} -1 & -2 & -2 \\ 1 & 2 & 1 \\ -1 & -1 & 0 \end{pmatrix} \quad B = \begin{pmatrix} -3 & -6 & 2 \\ 2 & 4 & -1 \\ 2 & 3 & 0 \end{pmatrix} \quad C = \begin{pmatrix} -5 & -8 & 0 \\ 3 & 5 & 0 \\ 1 & 2 & -1 \end{pmatrix}$$

then $AB = BA = C$, $BC = CB = A$, and $AC = CA = B$.

6. (a) Show that the matrices

$$I = \begin{pmatrix} 1 & 0 \\ 0 & 1 \end{pmatrix} \quad \text{and} \quad O = \begin{pmatrix} 0 & 0 \\ 0 & 0 \end{pmatrix}$$

commute with

$$A = \begin{pmatrix} 1 & 4 \\ 2 & -3 \end{pmatrix}$$

(b) Show that I and O commute with every 2×2 matrix A.

7. Argue that only square matrices can commute.

The *zero matrices* $O_{r \times c}$, whose elements are all zero, play the same role in matrix multiplication as the number zero plays in the multiplication of numbers. If A is any $r \times c$ matrix and $O_{p \times q}$ denotes a $p \times q$ matrix of zeros, then

$$O_{p \times r} A_{r \times c} = O_{p \times c} \quad \text{and} \quad A_{r \times c} O_{c \times q} = O_{r \times q} \tag{3}$$

In multiplication of real numbers, the number 1 serves as the identity element in the sense that $1 \cdot a = a \cdot 1 = a$, for every real number a. For matrices, the role of the identity is played by the *identity matrices*

$$I_{n \times n} = \begin{pmatrix} 1 & 0 & \cdots & 0 & \cdots & 0 \\ 0 & 1 & \cdots & 0 & \cdots & 0 \\ & \vdots & & & & \\ 0 & 0 & \cdots & 1 & \cdots & 0 \\ & \vdots & & & & \\ 0 & 0 & \cdots & 0 & \cdots & 1 \end{pmatrix}$$

which have main diagonal entries of 1 and elsewhere 0. In particular, it is easily verified that if A is an $r \times c$ matrix, I_r is the $r \times r$ identity matrix, and I_c is the $c \times c$ identity, then

$$I_r A = A = A I_c \tag{4}$$

The matrix I_r commutes with every $r \times r$ matrix A.

PROBLEMS

8. Go through the details of matrix multiplication to obtain Equations (3) and (4).

Example 6 Suppose that each of a group of n subjects is given t tests. In *factor analysis** it is assumed that the score s_{ji} for subject i on test j depends

*Thurstone, L. L., *Multiple-Factor Analysis* (University of Chicago Press, Chicago, 1947).

both on the degree to which subject i possesses certain underlying *ability factors* and the extent to which the test measures these factors. Specifically,

$$s_{ji} = f_{j1}p_{1i} + f_{j2}p_{2i} + \cdots + f_{jq}p_{qi} \tag{5}$$

where $f_{jm}(m = 1, 2, \ldots, q)$ denotes the degree to which test j measures factor m, and p_{mi} is the *standard score* of individual i on factor m. It is convenient to scale the observed scores and the standard scores so that

$$\sum_{i=1}^{n} s_{ji} = 0 \qquad \frac{1}{n} \sum_{i=1}^{n} s_{ji}^2 = 1$$

$$\sum_{i=1}^{n} p_{mi} = 0 \qquad \frac{1}{n} \sum_{i=1}^{n} p_{mi}^2 = 1 \tag{6}$$

In matrix form, Equation (5) may be written

$$S = FP \tag{7}$$

where $S = (s_{ji})$ is a $t \times n$ matrix of observed scores, $F = (f_{jm})$ is a $t \times q$ matrix of *factor loadings*, and $P = (p_{mi})$ is a $q \times n$ matrix of standard scores. In view of (6), the elements in each row of S and of P add to zero and the sum of squares of the elements in any row is n.

The *correlation r_{jk} between tests j and k* is defined by

$$r_{jk} = \frac{1}{n} \sum_{i=1}^{n} s_{ji}s_{ki}$$

and the *correlation z_{mv} between factors m and v* by

$$z_{mv} = \frac{1}{n} \sum_{i=1}^{n} p_{mi}p_{vi}$$

In matrix form,

$$R = \frac{1}{n} SS' \quad \text{and} \quad Z = \frac{1}{n} PP' \tag{8}$$

where R is the $t \times t$ matrix of intertest correlations and Z is the $q \times q$ matrix of interfactor correlations.

Substituting (7) into (8) gives

$$R = \frac{1}{n} SS' = \frac{1}{n} FPP'F'$$

$$= F\left(\frac{1}{n} PP'\right) F' = FZF'$$

The ideal situation occurs when correlations between different factors are zero; the factors are then said to be *uncorrelated*. In this case $Z = I$ and we have

$$R = FF'$$

as the *basic equation of factor analysis*. ▶

PROBLEMS

9. If c is a number, a matrix of the form cI, having the number c down the main diagonal and zeros elsewhere, is called a *scalar matrix*. Argue that an $r \times r$ scalar matrix commutes with all $r \times r$ matrices.

The only matrix M which has the property that $MA = A$ for all matrices A having r rows is $M = I_r$. For I_r is itself one such matrix and, supposing that M has such a property, we have

$$MI_r = I_r$$

On the other hand, we have already seen that

$$MI_r = M$$

whence $M = I_r$. A similar statement holds for multiplication on the right.

PROBLEMS

10. Suppose a matrix E has the property that for all matrices A having r rows, $EA = O$, a zero matrix. Prove that E must itself be a zero matrix. (*Hint:* What if $A = I_r$?)

11. If

$$A = \begin{pmatrix} 0 & 1 \\ 1 & 0 \end{pmatrix} \qquad B = \begin{pmatrix} 1 & 0 \\ 0 & -1 \end{pmatrix} \qquad C = \begin{pmatrix} 0 & 1 \\ -1 & 0 \end{pmatrix}$$

find

(a) A^2 (b) B^2 (c) C^2

(d) AB (e) BA (f) AC

(g) CA (h) BC (i) CB

12. If $R_n = (1, 2, \ldots, n)$, show that

$$R_n \cdot R_n' = \tfrac{1}{6}n(n + 1)(2n + 1)$$

13. If $R = (1, 3, 2)$, verify that

$$R' \cdot R = \begin{pmatrix} 1 & 3 & 2 \\ 3 & 9 & 6 \\ 2 & 6 & 4 \end{pmatrix}$$

14. Express the following economic model as a matrix equation:

$$Y = C + I$$
$$C = a + bY$$
$$I = u + vY$$

where C and I denote the respective monetary values of aggregate consumption and investment, and a, b, u, and v are constants with $b + v \neq 1$.

15. Katz* defines an irreflexive choice relation C and an irreflexive perceived choice relation P by

*Katz, L., "Identification in Sociometric Groups," Preprint of the University of Michigan.

$$p_i C p_j \Leftrightarrow \text{person } i \text{ chooses person } j$$

$$p_k P p_i \Leftrightarrow \text{person } k \text{ thinks he has been chosen by person } i$$

He contends that the tendency of person k to identify with person j is reflected in the number n_{kj} of persons by whom person k thinks he has been chosen, but who actually choose person j.

Define a choice matrix $C = (c_{ij})$ by

$$c_{ij} = \begin{cases} 1 & \text{if } p_i C p_j \\ 0 & \text{otherwise} \end{cases}$$

and a perceived choice matrix $P = (p_{ki})$ by

$$p_{ki} = \begin{cases} 1 & \text{if } p_k P p_i \\ 0 & \text{otherwise} \end{cases}$$

Show that the numbers n_{kj} are the elements of the matrix product $N = PC$.

16. Let $U = (u_1, u_2, \ldots, u_n)$ and $V = (v_1, v_2, \ldots, v_n)$ be two n-dimensional row vectors and let $X = U - V$. The *distance* between U and V is defined as the length of X. That is, the distance between U and V is $|X| = |U - V|$.

(a) Prove that

$$|U - V| = \sqrt{(u_1 - v_1)^2 + (u_2 - v_2)^2 + \cdots + (u_n - v_n)^2}$$

(b) Show that the distance between U and V can be written in the alternative form

$$|U - V| = \sqrt{(U - V)(U' - V')}$$

(Compare with Section 4.2, especially Property 8 and the interpretation of absolute value in terms of distance.)

17. Define a distance function d by $d(U, V) = |U - V|$. Prove that

(a) $d(U, V) \geq 0$, for all vectors U and V.

(b) $d(U, V) = 0$, if and only if $U = V$.

(c) the Triangle Inequality, $d(U, W) \leq d(U, V) + d(V, W)$, for all vectors U, V, and W.

18. Five stimulus objects were rated on three factors x_1, x_2, and x_3 as follows:

Object	1	2	3	4	5
Factor x_1	1	2	4	10	8
Factor x_2	1	5	0	6	4
Factor x_3	3	4	2	3	2

Compute the distance matrix $D = (d_{ij})$ where d_{ij} is the distance between objects i and j.

19. Prove that if A and B are both n-square, then tr $(AB) =$ tr (BA). (See Problems 16–18 in Section 8.1.)

20. Given the following factor matrix F:

	Reference abilities	
Tests	A	B
1	0.600	0.800
2	0.700	0.714
3	0.916	0.400
4	0.100	0.995
5	0.954	0.300

Compute the intercorrelation matrix R.

8.3 BASIC PROPERTIES OF MULTIPLICATION; BLOCK MULTIPLICATION

Although matrix multiplication is not commutative, it is associative, and multiplication distributes over addition. This means that, except for the fact that multiplication does not commute, operations involving matrix multiplication and addition are the same as similar operations involving numbers.

Theorem 1 *Associative Law for Matrix Multiplication* Let A, B, and C have dimensions $r \times q$, $q \times p$, and $p \times c$, respectively. Then

$$(AB)C = A(BC)$$

PROOF From the given dimensions for A, B, and C, we see that AB is $r \times p$, while BC is $q \times c$. It follows that $(AB)C$ and $A(BC)$ have the same dimensions, $r \times c$.

As required by the definition of equality, we must compare corresponding elements. The (i, j) element of $(AB)C$ is

$$[(AB)C]_{ij} = \sum_{k=1}^{p} (AB)_{ik} c_{kj} = \sum_{k=1}^{p} \left(\sum_{t=1}^{q} a_{it} b_{tk} \right) c_{kj} \qquad (9)$$

The (i, j) element of $A(BC)$ is

$$[A(BC)]_{ij} = \sum_{t=1}^{q} a_{it}(BC)_{tj} = \sum_{t=1}^{q} a_{it} \left(\sum_{k=1}^{p} b_{tk} c_{kj} \right) \qquad (10)$$

Interchanging the order of summation in (10) shows that (9) and (10) are identical, which completes the proof. ▶

When A is a square matrix, it is possible to multiply A by itself to form the matrix product

$$A^2 = A \cdot A$$

called, naturally, the *square of A*. Then

$$A^2 \cdot A = (A \cdot A)A = A(A \cdot A) = A \cdot A^2$$

follows from associativity. This means that the matrices A and A^2 commute and that the matrix product $A \cdot A \cdot A$ produces the same result regardless of which pair is multiplied first. A similar argument may be made in general and we define the *n*th *power A^n of A* as the product of n matrices, each of which is A. For convenience, we also define A^0 to be the identity matrix I.

Example 1 If

$$A = \begin{pmatrix} 1 & 2 \\ -1 & 1 \end{pmatrix}$$

then

$$A^0 = I = \begin{pmatrix} 1 & 0 \\ 0 & 1 \end{pmatrix} \quad A^2 = A \cdot A = \begin{pmatrix} -1 & 4 \\ -2 & -1 \end{pmatrix} \quad A^3 = \begin{pmatrix} -5 & 2 \\ -1 & -5 \end{pmatrix} \cdots$$

▶

Associativity means that the usual rules of exponents apply to matrix multiplication. That is,

$$A^n A^m = A^m A^n = A^{n+m}$$

holds for all non-negative integers m and n.

PROBLEMS

1. If

$$A = \begin{pmatrix} 1 & -1 \\ -2 & 1 \end{pmatrix}$$

find A^2, A^3, and A^4. Verify that $A \cdot A^4 = A^3 \cdot A^2$.

Theorem 2 *Distributive Law for Matrices* Let A be an $r \times q$ matrix and B and C both have dimension $q \times c$. Then

$$A(B + C) = AB + AC$$

PROOF It is easily checked that both sides have dimensions $r \times c$. Using the definitions of addition and multiplication, we have

$$[A(B + C)]_{ij} = \sum_{k=1}^{q} a_{ik}(B + C)_{kj} = \sum_{k=1}^{q} a_{ik}(b_{kj} + c_{kj})$$

$$= \sum_{k=1}^{q} (a_{ik}b_{kj} + a_{ik}c_{kj}) = \sum_{k=1}^{q} a_{ik}b_{kj} + \sum_{k=1}^{q} a_{ik}c_{kj}$$

$$= (AB)_{ij} + (AC)_{ij} = (AB + AC)_{ij}$$

This completes the proof. ▶

Example 2 A manufacturer produces three products α, β, and γ which he sells in two markets. Annual sales volumes are indicated by the matrix

$$Q = \begin{array}{c} \\ \text{Market 1} \\ \text{Market 2} \end{array} \begin{array}{ccc} \alpha & \beta & \gamma \\ \begin{pmatrix} 10{,}000 & 2{,}000 & 18{,}000 \\ 6{,}000 & 20{,}000 & 8{,}000 \end{pmatrix} \end{array}$$

If unit sales prices are given by the vector

$$\begin{array}{ccc} \alpha & \beta & \gamma \end{array}$$
$$P = (\$2.50 \quad \$1.25 \quad \$1.50)$$

then the total revenue in each market is obtained from the matrix product

$$PQ' = (2.50 \quad 1.25 \quad 1.50) \begin{pmatrix} 10{,}000 & 6{,}000 \\ 2{,}000 & 20{,}000 \\ 18{,}000 & 8{,}000 \end{pmatrix} = (\$54{,}500 \quad \$52{,}000)$$

Similarly, if the vector

$$\begin{array}{ccc} \alpha & \beta & \gamma \end{array}$$
$$C = (\$1.80 \quad \$1.20 \quad \$0.80)$$

lists unit costs for the three commodities, then the total costs (corresponding to the individual markets) are given by the matrix product

$$CQ' = (1.80 \quad 1.20 \quad 0.80) \begin{pmatrix} 10{,}000 & 6{,}000 \\ 2{,}000 & 20{,}000 \\ 18{,}000 & 8{,}000 \end{pmatrix} = (\$34{,}800 \quad \$41{,}200)$$

Making use of the distributive property of matrices, we can express the vector of gross profits as

$$PQ' - CQ' = (\$19{,}700 \quad \$10{,}800) = (P - C)Q' \qquad \blacktriangleright$$

PROBLEMS

2. Let

$$A = \begin{pmatrix} -1 & 1 \\ 2 & 4 \end{pmatrix} \qquad B = \begin{pmatrix} 1 & 2 & 0 \\ 3 & -1 & 2 \end{pmatrix} \qquad C = \begin{pmatrix} -6 & 2 & 9 \\ 1 & 1 & 3 \end{pmatrix}$$

Verify that $AB + AC = A(B + C)$.

Theorem 3 *Transpose of a Product* The transpose of a product is the product of the transposes *in reverse order*. Symbolically, if A is $r \times q$ and B is $q \times c$ then

$$(AB)' = B'A' \tag{11}$$

PROOF Since AB is $r \times c$, B' is $c \times q$, and A' is $q \times r$, it follows that both sides of (11) have dimension $c \times r$. Then

$$((AB)')_{ij} = (AB)_{ji} = \sum_{k=1}^{q} a_{jk} b_{ki}$$

while

$$(B'A')_{ij} = \sum_{k=1}^{q} (B')_{ik}(A')_{kj} = \sum_{k=1}^{q} b_{ki} a_{jk}$$

The two sums are obviously equal. ▶

Example 3 In Example 2

$$(P - C)Q' = (\$0.70 \quad \$0.05 \quad \$0.70)\begin{pmatrix} 10,000 & 6,000 \\ 2,000 & 20,000 \\ 18,000 & 8,000 \end{pmatrix}$$

$$= (\$19,700 \quad \$10,800)$$

Applying Theorem 3, we obtain

$$[(P - C)Q']' = Q \cdot (P - C)'$$

$$= \begin{pmatrix} 10,000 & 2,000 & 18,000 \\ 6,000 & 20,000 & 8,000 \end{pmatrix}\begin{pmatrix} \$0.70 \\ \$0.05 \\ \$0.70 \end{pmatrix} = \begin{pmatrix} \$19,700 \\ \$10,800 \end{pmatrix}$$

▶

PROBLEMS

3. Using the matrices A and B in Problem 2, compute $B'A'$ and verify that this is the same as $(AB)'$.

In multiplying matrices it often helps to *partition* the matrices into *submatrices*, or *blocks*. An example will best indicate the procedure.

Example 4 In Example 4 of Section 8.2, we found the product of the matrices

$$A = \begin{pmatrix} 2 & 0 & -3 \\ 1 & -1 & 4 \end{pmatrix} \quad \text{and} \quad B = \begin{pmatrix} -1 & 0 & 7 \\ 6 & 0 & -2 \\ 2 & 3 & 1 \end{pmatrix} \tag{12}$$

to be

$$AB = \begin{pmatrix} -8 & -9 & 11 \\ 1 & 12 & 13 \end{pmatrix}$$

Suppose we partition A into two submatrices $A = (A_1 \quad A_2)$ where

$$A_1 = \begin{pmatrix} 2 & 0 \\ 1 & -1 \end{pmatrix} \quad \text{and} \quad A_2 = \begin{pmatrix} -3 \\ 4 \end{pmatrix}$$

and partition B as

$$\begin{pmatrix} B_1 & B_2 \\ B_3 & B_4 \end{pmatrix}$$

where

$$B_1 = \begin{pmatrix} -1 \\ 6 \end{pmatrix} \quad B_2 = \begin{pmatrix} 0 & 7 \\ 0 & -2 \end{pmatrix} \quad B_3 = (2) \quad B_4 = (3 \quad 1)$$

These partitionings are indicated by the dotted lines in (12). If we now pretend that A is a matrix having only two elements and B a matrix with four, we might form the product

$$(A_1 \quad A_2) \begin{pmatrix} B_1 & B_2 \\ B_3 & B_4 \end{pmatrix} = (A_1 B_1 + A_2 B_3 \quad A_1 B_2 + A_2 B_4) \qquad \textbf{(13)}$$

Note that since we partitioned the columns of A in the same way as the rows of B, all the products $A_1 B_1$, $A_2 B_3$, $A_1 B_2$, and $A_2 B_4$ make sense as matrix multiplications. Moreover, $A_1 B_1$ and $A_2 B_3$ are both 2×1, while $A_1 B_2$ and $A_2 B_4$ are both 2×2. Thus the sums also make sense as matrix addition.

Simple calculation shows that

$$A_1 B_1 = \begin{pmatrix} -2 \\ -7 \end{pmatrix} \quad A_2 B_3 = \begin{pmatrix} -6 \\ 8 \end{pmatrix} \quad A_1 B_2 = \begin{pmatrix} 0 & 14 \\ 0 & 9 \end{pmatrix}$$

$$A_2 B_4 = \begin{pmatrix} -9 & -3 \\ 12 & 4 \end{pmatrix}$$

Hence

$$A_1 B_1 + A_2 B_3 = \begin{pmatrix} -8 \\ 1 \end{pmatrix} \quad \text{and} \quad A_1 B_2 + A_2 B_4 = \begin{pmatrix} -9 & 11 \\ 12 & 13 \end{pmatrix}$$

and the matrix product becomes

$$\begin{pmatrix} -8 & \vdots & -9 & 11 \\ 1 & \vdots & 12 & 13 \end{pmatrix}$$

which is the same as AB. ▶

Details of the proof that the technique illustrated in Example 4 always works are straightforward but rather lengthy and so are omitted. However, we will state the general theorem.

Block Multiplication Theorem If the $r \times q$ matrix A is partitioned as

$$A = \begin{pmatrix} A_{11} & A_{12} & \cdots & A_{1t} \\ A_{21} & A_{22} & \cdots & A_{2t} \\ \vdots & & & \\ A_{s1} & A_{s2} & \cdots & A_{st} \end{pmatrix}$$

and the $q \times c$ matrix B is partitioned as

$$B = \begin{pmatrix} B_{11} & B_{12} & \cdots & B_{1v} \\ B_{21} & B_{22} & \cdots & B_{2v} \\ \vdots & & & \\ B_{t1} & B_{t2} & \cdots & B_{tv} \end{pmatrix}$$

and if the columns of A are partitioned in the same way as the rows of B (so that all matrix multiplications make sense), then multiplication of the partitioned A and B using the blocks as individual elements produces the same result as the basic definition of matrix multiplication of the nonpartitioned A and B. ▶

Example 5 Block multiplication is especially useful when special patterns appear in the matrices. Consider the problem, which arises in probability theory, of finding powers of a matrix like

$$P = \begin{pmatrix} 1 & 0 & 0 & \vdots & 0 & 0 \\ 0 & 1 & 0 & \vdots & 0 & 0 \\ 0 & 0 & 1 & \vdots & 0 & 0 \\ \hdashline 0 & \frac{1}{6} & \frac{1}{6} & \vdots & \frac{1}{3} & \frac{1}{3} \\ \frac{1}{8} & \frac{1}{4} & \frac{1}{8} & \vdots & \frac{1}{4} & \frac{1}{8} \end{pmatrix}$$

Partitioning as indicated, we write

$$P = \begin{pmatrix} I & O \\ A & T \end{pmatrix}$$

where I is the 3×3 identity matrix, O is a 3×2 block of zeros, A is the 2×3 matrix

$$\begin{pmatrix} 0 & \frac{1}{6} & \frac{1}{6} \\ \frac{1}{8} & \frac{1}{4} & \frac{1}{8} \end{pmatrix}$$

and

$$T = \begin{pmatrix} \frac{1}{3} & \frac{1}{3} \\ \frac{1}{4} & \frac{1}{8} \end{pmatrix}$$

is 2 × 2. Then

$$P^2 = \begin{pmatrix} I & O \\ A & T \end{pmatrix}\begin{pmatrix} I & O \\ A & T \end{pmatrix} = \begin{pmatrix} I & O \\ A + TA & T^2 \end{pmatrix}$$

$$P^3 = P^2 \cdot P = \begin{pmatrix} I & O \\ A + TA & T^2 \end{pmatrix}\begin{pmatrix} I & O \\ A & T \end{pmatrix} = \begin{pmatrix} I & O \\ A + TA + T^2A & T^3 \end{pmatrix}$$

and a simple induction shows that

$$P^n = \begin{pmatrix} I & O \\ A + TA + \cdots + T^{n-1}A & T^n \end{pmatrix} = \begin{pmatrix} I & O \\ \sum_{k=0}^{n-1} T^kA & T^n \end{pmatrix}$$

where $T^0 = I$. This approach is obviously much more efficient than a direct application of the basic definition of multiplication. ▶

Example 6 Need for block multiplication also arises in problems of large-scale computation. For instance, the multiplication of two 50 × 50 matrices, each of which contains 2500 entries, requires 125,000 multiplications and almost as many additions. One method of performing such calculations on a computer whose storage capacity is limited is to partition each of the matrices into smaller matrices. If the blocks are suitably small, the machine can successively compute products of these blocks, store the results on tape to clear the internal storage for the next block of calculations, and finally call back the individual blocks one at a time for summation. ▶

PROBLEMS

4. If

$$A = \left(\begin{array}{cc|cc} 1 & 0 & a & b \\ 0 & 1 & c & d \\ \hline 0 & 0 & -1 & 0 \\ 0 & 0 & 0 & -1 \end{array}\right) = \begin{pmatrix} I & B \\ O & -I \end{pmatrix}$$

show that regardless of the values of a, b, c, and d, we have $A^2 = I_{4\times4}$.

5. If

$$V = \begin{pmatrix} 1 & -1 \\ 1 & 3 \end{pmatrix} \qquad W = \begin{pmatrix} -3 & -1 \\ 1 & -1 \end{pmatrix} \qquad A = \begin{pmatrix} I_{2\times2} & O \\ O & -I_{2\times2} \end{pmatrix}$$

$$B = \begin{pmatrix} -I & V \\ O & I \end{pmatrix} \quad \text{and} \quad C = \begin{pmatrix} -I & O \\ W & I \end{pmatrix}$$

show that $A^2 = B^2 = C^2 = I_{4\times4}$; $AB + BA = AC + CA = BC + CB = -2I_{4\times4}$.

6. A square matrix of the form

$$D = \begin{pmatrix} d_1 & 0 & \cdots & 0 \\ 0 & d_2 & \cdots & 0 \\ & \vdots & & \\ 0 & 0 & \cdots & d_n \end{pmatrix} \tag{14}$$

in which all elements off the main diagonal are zero, is called a *diagonal matrix*. The identity matrix and all square zero matrices are diagonal. We abbreviate (14) by writing $D = \text{diag}\,(d_1, d_2, \ldots, d_n)$.

(a) Show that any two diagonal matrices of the same order commute.

(b) Show that if $D = \text{diag}\,(d_1, d_2, \ldots, d_n)$, then

$$D^p = \text{diag}\,(d_1^p, d_2^p, \ldots, d_n^p).$$

That is, the pth power of a diagonal matrix D is a diagonal matrix in which the diagonal elements are the pth powers of the elements of D.

7. Verify that if A is the matrix

$$\begin{pmatrix} -1 & -2 & 1 \\ -2 & 2 & -2 \\ 1 & -2 & -1 \end{pmatrix}$$

and R is the matrix

$$\begin{pmatrix} 1/\sqrt{6} & 1/\sqrt{3} & 1/\sqrt{2} \\ -2/\sqrt{6} & 1/\sqrt{3} & 0 \\ 1/\sqrt{6} & 1/\sqrt{3} & -1/\sqrt{2} \end{pmatrix}$$

then $R'AR = \text{diag}\,(4, -2, -2)$.

8. Let $D = \text{diag}\,(d_1, d_2, d_3)$. Write a matrix equation for $\sum_{i=1}^{3} d_i$.

9. Simplify $(A'B' + 2C)'$.

10. If

$$A = \begin{pmatrix} 0 & -1 & -1 \\ 1 & 2 & 1 \\ -2 & -2 & -1 \end{pmatrix} \qquad B = \begin{pmatrix} -2 & -3 & -3 \\ -1 & 0 & -1 \\ 2 & 2 & 3 \end{pmatrix}$$

and c is any number, show that

$$[cA + (1 - c)B]^2 = I$$

11. If

$$A = \begin{pmatrix} 0 & 2 & -2 \\ 6 & -4 & 6 \\ 4 & -4 & 6 \end{pmatrix} \quad \text{and} \quad B = \begin{pmatrix} 8 & -6 & 6 \\ 4 & -2 & 4 \\ -6 & 6 & -4 \end{pmatrix}$$

show that $A^2 = B^2 = [\tfrac{1}{2}(A + B)]^2 = 4I$, while $(A - B)^2 = 0$.

12. If

$$P = \begin{pmatrix} 1 & 0 \\ \frac{1}{4} & \frac{1}{3} \end{pmatrix}$$

find the following:

(a) P^2 (b) P^3 (c) P^4 (d) P^n

(e) Find $\lim_{n \to \infty} P^n$, where by the limit of a sequence of matrices we mean the matrix whose (i, j) element is, for each pair (i, j), the limit of the sequence of numbers appearing in the (i, j) position of P, P^2, P^3, ...

13. If A and B are both $n \times n$ matrices, prove that

$$(A + B)(A - B) \neq A^2 - B^2 \quad \text{unless} \quad AB = BA$$

14. Prove that if A is a square matrix, then

$$(I + A)^n = I + nA + \frac{n(n-1)}{2} A^2 + \cdots + A^n$$

$$= \sum_{k=0}^{n} \binom{n}{k} A^k$$

(Here $A^0 = I$.)

8.4 APPLICATIONS TO DIRECTED GRAPHS AND TO INPUT–OUTPUT ANALYSIS

The theory of directed graphs has found extensive application in the study of small groups and of formal organizations. Sociometric patterns, communication networks, cliques and subgroup formation, rumor flow, patterns of influence, and interdepartmental interaction are all amenable to study by these methods. Here we are primarily concerned with showing how matrix manipulations can be used in the analysis of directed graphs.

Let R be an irreflexive relation defined on a set of objects

$$V = \{v_1, v_2, \ldots, v_n\}$$

We have seen (Section 3.5) that such a relation may be depicted graphically by a directed graph (digraph) which consists of points corresponding to the objects $v_1, v_2, \ldots, v_n$ together with directed lines (arrows) joining pairs of points. For instance, the relation R depicted by the digraph in Figure 4(a) is such that $v_1 R v_2$, $v_2 R v_3$, and $v_1 R v_3$.

The *adjacency matrix* A_R of a given relation R relating objects $v_1, v_2, \ldots, v_n$ is the $n \times n$ matrix with elements a_{ij} defined by

$$a_{ij} = \begin{cases} 1 & \text{if } v_i R v_j \\ 0 & \text{otherwise} \end{cases}$$

FIGURE 4

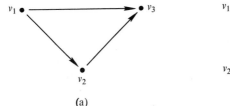 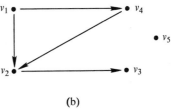

(a)   (b)

The matrix R in Example 1 of Section 8.1 is an adjacency matrix for a preference relation. The adjacency matrix for the relation whose digraph appears in Figure 4(b) is

$$
A = \begin{array}{c} \\ v_1 \\ v_2 \\ v_3 \\ v_4 \\ v_5 \end{array}
\begin{array}{c} \begin{array}{ccccc} v_1 & v_2 & v_3 & v_4 & v_5 \end{array} \\
\begin{pmatrix}
0 & 1 & 0 & 1 & 0 \\
0 & 0 & 1 & 0 & 0 \\
0 & 0 & 0 & 0 & 0 \\
0 & 1 & 0 & 0 & 0 \\
0 & 0 & 0 & 0 & 0
\end{pmatrix}
\end{array}
\qquad (15)
$$

A *sequence* of length n from an initial point v_1 to a terminal point v_n is an alternating sequence of $n + 1$ points and n directed lines v_1, $v_1 \rightarrow v_2$, v_2, $v_2 \rightarrow v_3, \ldots, v_{n-1} \rightarrow v_n, v_n$. There is no restriction on the number of times a point or a directed line may be repeated in the sequence. A *path* from v_1 to v_n is a sequence in which no point may occur more than once. A sequence or path with the same initial and terminal point is said to be *closed*.

Let A be a given adjacency matrix and consider the element

$$f_{ij} = a_{i1}a_{1j} + a_{i2}a_{2j} + \cdots + a_{in}a_{nj}$$

in the ij position of $F = A^2$. Obviously, a product $a_{ik}a_{kj}$ is nonzero only if a_{ik} and a_{kj} are both equal to 1; that is, only if there is a sequence of length two leading from v_i to v_j. It follows that f_{ij} is the *number* of sequences of length two from v_i to v_j.

In a similar way, we see that the element

$$g_{ij} = f_{i1}a_{1j} + f_{i2}a_{2j} + \cdots + f_{in}a_{nj}$$

in the ij position of $G = A^3 = F \cdot A$ indicates the number of sequences of length three which lead from v_i to v_j. A simple induction establishes the following general result.

Theorem 4 Let A_R be the adjacency matrix of a relation R. Then the ij element of A_R^p is the number of sequences of length p from v_i to v_j. ▶

Example 1 The square of the adjacency matrix A in (15) is

$$
B = A^2 = \begin{array}{c} \\ v_1 \\ v_2 \\ v_3 \\ v_4 \\ v_5 \end{array}
\begin{array}{c} \begin{array}{ccccc} v_1 & v_2 & v_3 & v_4 & v_5 \end{array} \\
\left(\begin{array}{ccccc}
0 & 1 & 1 & 0 & 0 \\
0 & 0 & 0 & 0 & 0 \\
0 & 0 & 0 & 0 & 0 \\
0 & 0 & 1 & 0 & 0 \\
0 & 0 & 0 & 0 & 0
\end{array} \right) \end{array}
$$

The fact that $b_{12} = 1$ indicates that there is only one sequence of length two from v_1 to v_2, namely v_1, $v_1 \rightarrow v_4$, v_4, $v_4 \rightarrow v_2$, v_2. Similarly, the values $b_{13} = 1$ and $b_{43} = 1$ indicate that there is exactly one sequence of length two from v_1 to v_3 and one from v_4 to v_3. In each of the above instances, the sequence is also a path since no point is repeated. ▶

PROBLEMS

1. By computing A^3 and A^4, show that in Figure 4(b) there is exactly one three-step sequence (from v_1 to v_3) and no sequence of length greater than three. Verify these calculations by tracing the paths through the digraph.

*Boolean arithmetic** on the integers 0 and 1 is the same as ordinary arithmetic except that $1 + 1 = 1$. The symbol $\oplus$ is used to denote the operation of Boolean addition. In the theory of directed graphs, Boolean addition arises as follows. Let R_1 and R_2 be two irreflexive relations and let R be the *union* of R_1 and R_2. That is, $v_i R v_j \Leftrightarrow v_i R_1 v_j \lor v_i R_2 v_j$. Then the adjacency matrix A_R for the relation R is the Boolean sum of the adjacency matrices A_{R_1} and A_{R_2} for the relations R_1 and R_2.

Similarly, if the relation S is the *intersection* of R_1 and R_2—that is, if $v_i S v_j$ means that $v_i R_1 v_j \land v_i R_2 v_j$—then the adjacency matrix A_S of S is obtained by multiplying the matrices A_{R_1} and A_{R_2} *element by element*.

Example 2 The relations R_1 and R_2 whose digraphs are shown in Figure 5 have adjacency matrices

$$
A_{R_1} = \begin{array}{c} \\ v_1 \\ v_2 \\ v_3 \\ v_4 \end{array}
\begin{array}{c} \begin{array}{cccc} v_1 & v_2 & v_3 & v_4 \end{array} \\
\left(\begin{array}{cccc}
0 & 1 & 1 & 1 \\
0 & 0 & 1 & 1 \\
0 & 0 & 0 & 1 \\
1 & 0 & 0 & 0
\end{array} \right) \end{array}
\quad \text{and} \quad
A_{R_2} = \left(\begin{array}{cccc}
0 & 1 & 0 & 1 \\
0 & 0 & 1 & 0 \\
0 & 0 & 0 & 0 \\
0 & 1 & 0 & 0
\end{array} \right)
$$

*After George Boole, nineteenth-century English mathematician.

FIGURE 5 (a) Digraph of R₁; (b) Digraph of R₂

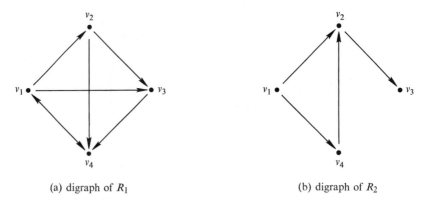

(a) digraph of R_1 (b) digraph of R_2

The Boolean sum $A_{R_1} \oplus A_{R_2}$ and element-by-element product $A_{R_1} \otimes A_{R_2}$ of these two matrices are

$$A_{R_1} \oplus A_{R_2} = \begin{pmatrix} 0 & 1 & 1 & 1 \\ 0 & 0 & 1 & 1 \\ 0 & 0 & 0 & 1 \\ 1 & 1 & 0 & 0 \end{pmatrix} \text{ and } A_{R_1} \otimes A_{R_2} = \begin{pmatrix} 0 & 1 & 0 & 1 \\ 0 & 0 & 1 & 0 \\ 0 & 0 & 0 & 0 \\ 0 & 0 & 0 & 0 \end{pmatrix}$$

These correspond to the digraphs shown in Figure 6, which should be compared with those in Figure 5.   ▶

FIGURE 6 (a) Digraph of R₁ ∨ R₂; (b) Digraph of R₁ ∧ R₂

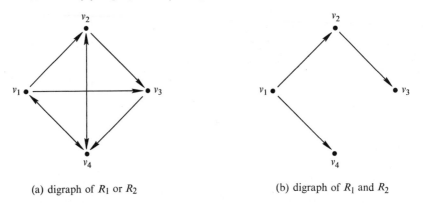

(a) digraph of R_1 or R_2 (b) digraph of R_1 and R_2

Now let us examine input–output analysis. Consider an economy composed of N industries each of which produces a single commodity. Each commodity may reenter the system as an input to any or all of the industries. Specifically, we assume that unit production (output) of the jth commodity requires a fixed quantity a_{ij} of the ith commodity as input. These *input co-efficients* a_{ij} may be arranged in an $N \times N$ matrix as

$$
A = \begin{array}{c} \\ \\ \text{input commodity} \\ \\ \end{array}
\begin{array}{cc}
& \begin{array}{cccc} \text{output commodity} \\ 1 & 2 & \ldots & N \end{array} \\
\begin{array}{c} 1 \\ 2 \\ \vdots \\ N \end{array} &
\begin{pmatrix}
a_{11} & a_{12} & \ldots & a_{1N} \\
a_{21} & a_{22} & \ldots & a_{2N} \\
\vdots & & & \\
a_{N1} & a_{N2} & \ldots & a_{NN}
\end{pmatrix}
\end{array}
\tag{16}
$$

in which entries in the jth column $(j = 1, 2, \ldots, N)$ specify the input re-quirements for production of one unit of the jth commodity. This input–output model is the Leontief model.*

Example 3   The first column of the input matrix

$$
A = \begin{array}{c} \\ \text{input} \\ \\ \end{array}
\begin{array}{c} \text{output} \\
\begin{pmatrix}
\frac{1}{2} & \frac{1}{4} & \frac{1}{3} \\
\frac{1}{3} & \frac{1}{10} & 0 \\
0 & \frac{1}{5} & \frac{1}{2}
\end{pmatrix}
\end{array}
$$

indicates that the production of one unit of commodity 1 requires $\frac{1}{2}$ unit of commodity 1, $\frac{1}{3}$ unit of commodity 2, and none of commodity 3. Similar re-quirements for commodities 2 and 3 may be read from the other columns. ▶

The simplest Leontief model arises when the output of each industry serves only as input to the various industries within the economic system and no other inputs are required. In this case we refer to a *closed* Leontief system. Let x_j $(j = 1, 2, \ldots, N)$ denote the output of the jth industry. Since pro-duction of x_j requires $a_{ij}x_j$ units of the ith output, the total quantity re-quired of output i is, for $i = 1, 2, \ldots, N$,

$$
\begin{aligned}
y_i &= a_{i1}x_1 + a_{i2}x_2 + \cdots + a_{iN}x_N \\
&= \sum_{j=1}^{N} a_{ij}x_j
\end{aligned}
\tag{17}
$$

The N equations (17) may be written in matrix form as $Y = AX$, where A is the input–output matrix (16), $Y = \text{col}(y_1, y_2, \ldots, y_N)$ is a column vector of input requirements, and $X = \text{col}(x_1, x_2, \ldots, x_N)$ is the vector of outputs.

In a closed system, there are no input sources outside the system and it is apparent that the system cannot function when an input requirement y_i exceeds output x_i. A closed system which can function, that is, for which

*After Wassily Leontief, who pioneered this approach to the analysis of economic systems.

there is a vector $X \geq O$ such that

$$Y = AX \leq X$$

is called *viable*. Here the symbol O denotes a vector of zeros. The matrix inequality $B \geq C$ means that for each i and j the ij elements b_{ij} and c_{ij} of B and C satisfy the inequality $b_{ij} \geq c_{ij}$. Thus $X \geq O$ means that no element of X is negative, reflecting the fact that negative output is impossible. A vector $X \geq O$ such that

$$AX = X \qquad \qquad (18)$$

is called an *interior equilibrium* of the system. Methods for determining whether a system is viable and for finding interior equilibria will be discussed in Chapter 9.

An *open Leontief system* consists of a *productive sector* of N industries as above, and a *nonproductive*, or *consumer sector*, which creates a demand for commodities over and above their use as inputs for production. The vector $Y = AX$ again represents the inputs required by the productive sector. The difference

$$X - AX = (I - A)X$$

thus represents the net output available to the nonproductive (open) segment of the economy. The fundamental problem is whether the economy can satisfy the demands of the open segment. That is, if D is the demand vector of the consumer sector, the problem is to determine whether there exists a vector $X \geq O$ such that

$$(I - A)X = D \qquad \qquad (19)$$

This problem, too, will be solved in the next chapter.

PROBLEMS

2. Consider an irreflexive relation, say the relation "chooses," and let A be the adjacency matrix of this relation. Prove that the number of persons chosen by both v_i and v_j is the element in the ij position of the matrix product AA'. Interpret the diagonal entries.

3. Argue that in the situation presented in Problem 2, the number of persons who choose both v_i and v_j is given by the ij element in the product $A'A$. Interpret the diagonal entries in this product.

4. Show that if there is a vector $X \geq O$ such that $(A - I)X = O$, then X is an interior equilibrium of a closed Leontief model.

5. Assume that in a three-industry open economy, the input-coefficient matrix is

$$A = \begin{matrix} 0.2 & 0.3 & 0.2 \\ 0.4 & 0.1 & 0.2 \\ 0.1 & 0.3 & 0.4 \end{matrix}$$

If the output vector is col $(10, 6, 5)$, find the net output available for the open sector of the economy.

6. Verify that the vectors col $(1, 1, 1)$ and col $(2, 2, 2)$ are both interior equilibria of the closed Leontief system with input-coefficient matrix

$$A = \begin{pmatrix} 0.2 & 0.3 & 0.5 \\ 0.4 & 0.4 & 0.2 \\ 0.1 & 0.5 & 0.4 \end{pmatrix}$$

In fact, show that every vector $X = $ col (x_1, x_2, x_3) with positive entries, and $x_1 = x_2 = x_3$ is an interior equilibrium.

7. In a digraph for a relation R defined on a set of points $v_1, v_2, \ldots, v_n$, a point v_j is said to be *reachable* from a point $v_i (v_i T v_j)$ if there is a sequence from v_i to v_j. We agree to say that any point is reachable from itself by a path of zero length, so that T is reflexive.

   (a) Argue that T is also a transitive relation.

   (b) Give examples to show that T is neither symmetric nor asymmetric.

8. In Problem 7

   (a) Argue that if v_j is reachable from v_i, it can be reached in $n - 1$ steps or less, where n is the number of points in the graph.

   (b) Let A be the adjacency matrix for the relation R and denote the ij element in A^k by $a_{ij}^{(k)}$. Show that (a) implies that v_j is reachable from v_i if and only if at least one of the elements $a_{ij}^{(0)}, a_{ij}^{(1)}, \ldots, a_{ij}^{(n-1)}$ is positive.

   (c) Let B denote the adjacency (reachability) matrix for the relation T defined by

   $$b_{ij} = \begin{cases} 1 & \text{if } v_i T v_j \\ 0 & \text{otherwise} \end{cases}$$

   Show that $B = I \oplus A \oplus A^2 \oplus \cdots \oplus A^{n-1}$ (Boolean addition).

9. In Problem 8, let two points v_i and v_j *communicate* $(v_i C v_j)$ if each is reachable from the other.

   (a) Argue that C is an equivalence relation.

   (b) Show that the adjacency matrix for C is $C = B \otimes B'$, where B is the reachability matrix and $\otimes$ denotes the product obtained by multiplying corresponding elements. Hence the sum of the elements in the ith row of C is the number of points which communicate with v_i.

   (c) Develop a matrix representation for the number of points which communicate through paths of only one step.

10. From the digraph in Figure 7 compute

   (a) the adjacency matrix.

   (b) the reachability matrix.

   (c) the communication matrix.

FIGURE 7

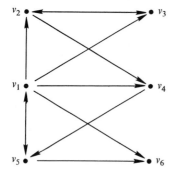

11. A social behavior model due to Rashevsky* postulates that the satisfaction accruing to an individual increases with the effort of everyone in the society, but decreases with the amount of effort personally expended. Symbolically, the satisfaction S_i of the ith individual in an N-member society is given by

$$S_i = a_i \sum_{j=1}^{N} x_j p_{ji} - b_i x_i^2$$

where x_j denotes the amount of effort exerted by the jth individual, p_{ji} is a measure of the effect on individual i of effort by individual j, and a_i and b_i are constants associated with the ith individual.

(a) Let $S = (S_1, S_2, \ldots, S_N)$ be the row vector of individual satisfactions, $P = (p_{ji})$ the matrix of the coefficients p_{ji}, A be the diagonal matrix (see Problem 6, Section 8.3) $A = \text{diag}(a_1, a_2, \ldots, a_N)$, $X = (x_1, x_2, \ldots, x_N)$ the row vector of efforts, X_D the diagonal matrix $X_D = \text{diag}(x_1, x_2, \ldots, x_N)$, and $B = (b_1, b_2, \ldots, b_N)$. Show that

$$S = XPA - BX_D^2$$

(b) Rashevsky defines a higher satisfaction $\overline{S}_i$ for individual i as the sum

$$\overline{S}_i = \sum_{j=1}^{N} k_{ij} S_j$$

where k_{ij} reflects the degree to which individual i is concerned with the satisfaction of individual j. Show that

$$\overline{S} = K[AP'X' - X_D^2 B']$$

where $\overline{S}$ is the vector of higher satisfactions and $K = (k_{ij})$ is the matrix of concern coefficients.

*See Coleman, J. S., "An Expository Analysis of Some of Rashevsky's Social Behavior Models," in *Mathematical Thinking in the Social Sciences*, Lazarsfeld, P. F., Ed. (Free Press, Glencoe, Ill., 1954), pp. 105–165.

SUPPLEMENTARY READING

Harary, F., Norman, R. Z., and Cartwright, D., *Structural Models: An Introduction to the Theory of Directed Graphs* (John Wiley & Sons, Inc., New York, 1966); Harary, F., "Graph Theory and Group Structure," in *Readings in Mathematical Psychology*, Luce, R. D., Bush, R. R., and Galanter, E., Eds. (John Wiley & Sons, Inc., New York, 1965), Vol. 2, 225–241.

Horst, P., *Matrix Algebra for Social Scientists* (Holt, Rinehart and Winston, New York, 1963).

Labovitz, S., "Application of Matrix Algebra to Social Units," *Sociology and Social Research* **51**, 220–234 (1967).

Zelinsky, D., *A First Course in Linear Algebra* (Academic Press Inc., New York, 1968).

LINEAR EQUATIONS AND LINEAR PROGRAMMING **9**

9.1 LINEAR EQUATIONS

The system of equations

$$a_{11}x_1 + a_{12}x_2 + \cdots + a_{1n}x_n = c_1$$
$$a_{21}x_1 + a_{22}x_2 + \cdots + a_{2n}x_n = c_2 \qquad (1)$$
$$\vdots \qquad\qquad\qquad \vdots$$
$$a_{m1}x_1 + a_{m2}x_2 + \cdots + a_{mn}x_n = c_m$$

is a system of m linear equations in n variables. Such a system may be rewritten in the matrix form

$$AX = C$$

where

$$A = \begin{pmatrix} a_{11} & a_{12} & \cdots & a_{1n} \\ a_{21} & a_{22} & \cdots & a_{2n} \\ \vdots & & & \vdots \\ a_{m1} & a_{m2} & \cdots & a_{mn} \end{pmatrix} \qquad X = \begin{pmatrix} x_1 \\ x_2 \\ \vdots \\ x_n \end{pmatrix} \quad \text{and} \quad C = \begin{pmatrix} c_1 \\ c_2 \\ \vdots \\ c_m \end{pmatrix} \qquad (2)$$

are, respectively, the *matrix of coefficients*, the *vector of unknowns*, and the *vector of constants* associated with the equations. Any n-dimensional vector Y for which $AY = C$ is called a *solution* of the system. The problem is to find all solutions.

The most important technique for solving a system of linear equations involves *successive elimination of variables*.

Example 1 To solve the system

$$x_1 - 2x_2 + 3x_3 = 4$$
$$-x_1 + x_2 + x_3 = 0$$
$$2x_1 + 2x_2 - x_3 = 1$$

we begin by eliminating x_1 from each equation except the first. To accomplish this, we add the first equation to the second and then add -2 times the first equation to the third equation. These operations produce the new system

$$x_1 - 2x_2 + 3x_3 = 4$$
$$- x_2 + 4x_3 = 4$$
$$6x_2 - 7x_3 = -7$$

In this system we first multiply the second equation through by -1 to obtain

$$x_2 - 4x_3 = -4 \qquad (3)$$

Then we eliminate x_2 from the first and third equations by adding twice Equation (3) to the first equation and -6 times Equation (3) to the third. The result is

$$x_1 - 5x_3 = -4$$
$$x_2 - 4x_3 = -4$$
$$17x_3 = 17$$

Dividing the last equation by 17 to obtain $x_3 = 1$, adding five times this equation to the first equation and four times this equation to the second equation gives

$$x_1 = 1$$
$$x_2 = 0$$
$$x_3 = 1$$

as the only solution to the original set of equations. ▶

Example 2 The same procedure may be followed in solving the system

$$x_1 - 2x_2 + 3x_3 = 4$$
$$-x_1 + x_2 + x_3 = 0 \qquad (4)$$
$$2x_1 + 2x_2 - 18x_3 = -16$$

which, except for the third equation, is the same as the system in Example 1. As before, we start by eliminating x_1 from the second and third equations:

$$x_1 - 2x_2 + 3x_3 = 4$$
$$- x_2 + 4x_3 = 4 \qquad (5)$$
$$6x_2 - 24x_3 = -24$$

Then, we multiply the second equation by -1 and eliminate x_2 from the first and third equations:

$$
\begin{aligned}
x_1 \qquad\quad - 5x_3 &= -4 \\
x_2 - 4x_3 &= -4 \\
0x_2 + 0x_3 &= 0
\end{aligned}
\tag{6}
$$

The third equation in (6) consists of nothing but zeros because in the preceding system (5), the third equation was proportional to $(-6$ times) the second equation. Since this equation puts no restriction on the unknowns, it may be eliminated, leaving

$$
\begin{aligned}
x_1 \qquad\quad - 5x_3 &= -4 \\
x_2 - 4x_3 &= -4
\end{aligned}
$$

We now have but two restrictions on three unknowns. Obviously one of the variables, say x_3, may be assigned any value whatever, after which corresponding values of x_1 and x_2 may be determined. For instance, if x_3 is assigned the value 1, then

$$
x_1 = -4 + 5(1) = 1 \quad \text{and} \quad x_2 = -4 + 4(1) = 0
$$

In general, if $x_3 = b$, then

$$
\begin{aligned}
x_1 &= -4 + 5b \\
x_2 &= -4 + 4b \\
x_3 &= b
\end{aligned}
\tag{7}
$$

The system (7) represents the form which all solutions of Equations (4) must have. *Particular solutions* are obtained by assigning specific values to b. ▶

Example 3 Consider the system

$$
\begin{aligned}
x_1 - 2x_2 + 3x_3 &= 4 \\
-x_1 + x_2 + x_3 &= 0 \\
2x_1 + 2x_2 - 18x_3 &= 0
\end{aligned}
\tag{8}
$$

which is identical to that in Example 2 except for the right side of the third equation. Proceeding as before, we obtain first

$$
\begin{aligned}
x_1 - 2x_2 + 3x_3 &= 4 \\
- x_2 + 4x_3 &= 4 \\
6x_2 - 24x_3 &= -8
\end{aligned}
$$

and then

$$x_1 \qquad -\quad 5x_3 = -4$$

$$x_2 - \quad 4x_3 = -4$$

$$0x_2 + \quad 0x_3 = \quad 16$$

There are no possible choices for x_1, x_2, and x_3 which will make $0x_2 + 0x_3 = 16$. It follows that there are no solutions and that the system (8) is an *inconsistent system of equations*. ▶

PROBLEMS

1. Find all solutions of the systems

(a)
$$x - y + z = 4$$
$$2x - 4y + 3z = -1$$
$$-2y + 3z = 0$$

(b)
$$x - y + z = 4$$
$$2x - 4y + 3z = -1$$
$$-x + 3y - 2z = 0$$

(c)
$$x - y + z = 4$$
$$2x - 4y + 3z = -1$$
$$-x + 3y - 2z = 5$$

In solving the systems of linear equations in Examples 1–3 above, we have performed only two basic types of operations—either multiplication of an equation by a nonzero constant or addition of a multiple of one equation to another equation. Each of these operations may be reversed by performing another operation of the *same* type. Specifically, if the ith equation is multiplied by $c \neq 0$, the original equation may be recovered by multiplying the new one by $1/c$. If c times the ith equation is added to the jth equation, the original jth equation is obtained by adding $-c$ times the ith equation to the new jth equation. Since, for each operation the original system of equations may be obtained from the new system, and vice versa, it follows that *all systems of equations obtained in this way are equivalent and have the same solutions*.

In performing a succession of operations on a system of linear equations we either reach a point at which all coefficients of the variables in one equation are zero (Examples 2 and 3) or we do not (Example 1). In the former case we may eliminate that equation if the right side is also zero (Example 2) or decide that the system of equations has no solution if the right side is not zero (Example 3). In the latter case we either obtain a unique solution (Example 1) or many solutions (Example 2). It is apparent, then, that successive elimination will always work, giving all the solutions if there are any and leading to an impossible equation if no solutions exist. The method thus has the threefold advantage of being logically simple, computationally straightforward, and foolproof.

PROBLEMS

In each of Problems 2–9, find all solutions of the given system of equations.

2. $x_1 + 3x_2 = 6$
   $2x_1 + 4x_2 = 10$

3. $x - 3y = 4$
   $2x + 6y = 12$

4. $x_1 + 6x_2 = 0$
   $4x_1 - 12x_2 = 0$

5. $x_1 + 2x_2 - 4x_3 = 0$
   $2x_1 + 4x_2 - 2x_3 = 0$

6. $x_1 + x_2 + 2x_3 = x_1$
   $-x_1 + 2x_2 + 3x_3 = 2x_2$
   $-2x_1 - 3x_2 + x_3 = x_3$

7. $x + 2y - 3z = -2$
   $4x - y + 2z = 8$
   $13x - y + 3z = 22$

8. $x + 3y + 2z + 7t = -5$
   $3x + 7y + 2z + 11t = 1$
   $2x + 5y + 3z + 12t = -7$
   $-x + z + t = -6$

9. $x + 2y - 3z = 0$
   $3x - y + z = 0$
   $x + y + z = 0$

10. A plant uses two machines M_1 and M_2 in the production of two products P and Q. Due to other production scheduling, machine M_1 is available 12 hours per day, while M_2 is available only 9 hours per day. To produce one unit of product P requires $\frac{1}{2}$ hour of M_1 time and 1 hour of M_2 time. Each unit of Q requires $\frac{3}{4}$ hour of M_1 time and $\frac{1}{4}$ hour of M_2 time. How many units of each product should be produced per day to keep each machine fully utilized?

11. Suppose that Commercial Airlines has available three types of air cargo transport capable of carrying three types of equipment according to the following load chart:

| | Transport | | |
	I	*II*	*III*
Equipment			
A	1	2	1
B	0	1	2
C	2	1	1

How many planes of each type should be dispatched to transport exactly 16 type A, 10 type B, and 12 type C machines? (Assume that each plane is fully loaded.)

12. Given the following input-coefficient matrix for a three-industry closed Leontief model

$$\begin{pmatrix} 0.3 & 0.1 & 0.4 \\ 0.2 & 0.8 & 0.6 \\ 0.5 & 0.1 & 0 \end{pmatrix}$$

show that an interior equilibrium exists. Is there a *unique* output mix that will result in an interior equilibrium condition? (*Hint:* Use Equation (18) in Section 8.4.)

13. A three-sector open economy has the following matrix of input coefficients:

| | | Consumer | |
Producer	Services	Manufacturing	Agriculture
Services	0.1	0.4	0.3
Manufacturing	0.4	0.05	0.5
Agriculture	0.2	0.3	0.1

If the demands for the three sectors from the nonproductive sector of the economy are 10, 25, and 20 billion dollars, respectively, find the outputs needed to satisfy these demands. (*Hint:* Use Equation (19) in Section 8.4.)

14. The economic model

$$Y = C + I$$
$$C = a + bY$$
$$I = u + vY$$

was introduced in Problem 14 of Section 8.2.

(a) Express this system in matrix form.

(b) Solve for Y, C, and I in terms of the constants.

9.2 ELEMENTARY OPERATIONS ON MATRICES

By now you may have noticed that in performing operations on systems of linear equations such as (1), there is no need to continue writing the "unknowns" $x_1, \ldots, x_n$. All computation is done with the coefficients a_{ij} and the scalars c_i. In Example 1, for instance, we could just as well have operated on the array

$$\begin{pmatrix} 1 & -2 & 3 & | & 4 \\ -1 & 1 & 1 & | & 0 \\ 2 & 2 & -1 & | & 1 \end{pmatrix}$$

consisting of the matrix of coefficients augmented by the column of constants to obtain the successive arrays (follow the arrows)

$$\begin{pmatrix} 1 & -2 & 3 & | & 4 \\ -1 & 1 & 1 & | & 0 \\ 2 & 2 & -1 & | & 1 \end{pmatrix} \longrightarrow \begin{pmatrix} 1 & -2 & 3 & | & 4 \\ 0 & -1 & 4 & | & 4 \\ 0 & 6 & -7 & | & -7 \end{pmatrix}$$

$$\begin{pmatrix} 1 & 0 & -5 & | & -4 \\ 0 & 1 & -4 & | & -4 \\ 0 & 0 & 17 & | & 17 \end{pmatrix} \longrightarrow \begin{pmatrix} 1 & 0 & 0 & | & 1 \\ 0 & 1 & 0 & | & 0 \\ 0 & 0 & 1 & | & 1 \end{pmatrix} \tag{9}$$

from which the solution $x_1 = 1$, $x_2 = 0$, and $x_3 = 1$ is easily read.

At each stage we have performed one of the following *elementary row operations* on the matrix in question. In fact, operation R 3 below was not performed, but might be used, for instance, if the coefficient of x_1 in the first equation were zero.

(R 1) Multiplication of a row (vector) by a nonzero real number.

(R 2) Replacement of one row by the sum of that row and a constant times another row.

(R 3) Interchange of two rows.

Example 1 Mathematical models of behavioral systems* frequently take the form of a set of linear equations. The problem facing the experimenter who wishes to use such a model is to estimate the elements of the matrix of coefficients of the equations on the basis of observed empirical results.

To illustrate the difficulties which may arise, consider a simple economic model with demand function

$$D = a + bP + cY$$

and supply function

$$S = d + eP$$

Here a, b, c, d, and e are constants, P denotes price, and Y denotes income. If we multiply the demand function by a constant α (row operation R 1) to obtain

$$\alpha D = \alpha a + \alpha bP + \alpha cY \tag{10}$$

or add β times the supply function (operation R 2) to (10) to obtain a *linear combination* of D and S of the form

$$\alpha D + \beta S = (\alpha a + \beta d) + (\alpha b + \beta e)P + \alpha cY$$

we obtain expressions having the *same general form* as D itself. Thus the experimenter, to whom the coefficients a, b, and c in D are not known but must be estimated, is unable to distinguish the true demand function D from any linear combination of D and S which he might observe. The function D is said to be *not identifiable*.

The supply function, on the other hand, is *identifiable* since there is no way the two functions can be combined using elementary operation R 2 without thereby obtaining a nonzero coefficient for Y and hence a function which is obviously not the supply function S.

The concept of identifiability can be formulated more precisely as follows. Two linear functions

$$a_1x_1 + a_2x_2 + \cdots + a_nx_n$$

$$b_1x_1 + b_2x_2 + \cdots + b_nx_n$$

*Adapted from Blalock, H. M., and Blalock, A. B., *Methodology in Social Research* (McGraw-Hill Book Company, Inc., New York, 1968).

are said to have the *same form* if for each $i = 1, 2, \ldots, n$ the corresponding coefficients a_i and b_i are either both zero or both nonzero. Thus the functions

$$2x + 6y \quad \text{and} \quad 3x - 2y$$

have the same form, whereas

$$3x + y \quad \text{and} \quad 6x + 3y - z$$

do not. A linear function L is *identifiable* if it has a unique form. That is, it must not be possible to apply row operations R 1–R 3 to the linear functions *other than* L so as to produce a linear function having the same form as L.

As a specific illustration, consider the system of four linear functions having coefficient matrix

$$\begin{pmatrix} 1 & -2 & 3 & 6 & 0 & 0 \\ 0 & 6 & 0 & 3 & 2 & 1 \\ 8 & 0 & 0 & 0 & 3 & 4 \\ 0 & 0 & 10 & 0 & 1 & 5 \end{pmatrix} \tag{11}$$

We perform three row operations:

(i) Add -4 times the second row to the third row.

(ii) Add -5 times the second row to the fourth row.

(iii) Add $-\frac{9}{5}$ times the new third row to the new fourth row.

The resulting matrix is

$$\begin{pmatrix} 1 & -2 & 3 & 6 & 0 & 0 \\ 0 & 6 & 0 & 3 & 2 & 1 \\ 8 & -24 & 0 & -12 & -5 & 0 \\ -\frac{72}{5} & \frac{66}{5} & 10 & \frac{33}{5} & 0 & 0 \end{pmatrix} \tag{12}$$

in which the first and fourth rows have the same form. Hence the first linear form is not identifiable. ▶

A matrix which is obtained from an identity matrix by performing *one* of the row operations R 1–R 3 is called an *elementary row matrix*.

Example 2 (a) The elementary row matrix

$$E = \begin{pmatrix} 1 & 0 & 0 \\ 0 & 1 & 0 \\ 2 & 0 & 1 \end{pmatrix}$$

is obtained from the 3×3 identity matrix by adding twice the first row to the third row. If B is any 3×2 matrix, the product

$$EB = \begin{pmatrix} 1 & 0 & 0 \\ 0 & 1 & 0 \\ 2 & 0 & 1 \end{pmatrix} \begin{pmatrix} b_{11} & b_{12} \\ b_{21} & b_{22} \\ b_{31} & b_{32} \end{pmatrix} = \begin{pmatrix} b_{11} & b_{12} \\ b_{21} & b_{22} \\ 2b_{11} + b_{31} & 2b_{12} + b_{32} \end{pmatrix}$$

may be obtained from B by adding twice the first row to the third row.

(b) If E is the elementary row operation matrix obtained from the 3×3 identity by interchanging the first and third rows, and if B is any 3×3 matrix, then

$$EB = \begin{pmatrix} 0 & 0 & 1 \\ 0 & 1 & 0 \\ 1 & 0 & 0 \end{pmatrix} \begin{pmatrix} b_{11} & b_{12} & b_{13} \\ b_{21} & b_{22} & b_{23} \\ b_{31} & b_{32} & b_{33} \end{pmatrix} = \begin{pmatrix} b_{31} & b_{32} & b_{33} \\ b_{21} & b_{22} & b_{23} \\ b_{11} & b_{12} & b_{13} \end{pmatrix}$$

The product EB is just B with first and third rows interchanged. ▶

PROBLEMS

1. Work out the products in Example 2.

2. Let $E = \text{diag}\,(3, 1, 1, 1)$ be the matrix which you obtain from the 4×4 identity by multiplying the first row by 3. Compare the product EB, where B is 4×5, with B itself.

The results of Example 2 and Problem 2 suggest the following general result.

Theorem 1 If $E_{n \times n}$ is an elementary row matrix obtained from the identity by performing one elementary row operation, and if A is any $n \times m$ matrix, then EA is the matrix obtained by performing that same row operation on the matrix A.

PROOF We shall give the proof only for row operation R 2, leaving the other two cases as exercises. Consider the matrix product EA, where $E_{n \times n}$ is the matrix

$$E_{n \times n} = \begin{pmatrix} 1 & 0 & \cdots & 0 & \cdots & 0 & \cdots & 0 \\ 0 & 1 & \cdots & 0 & \cdots & 0 & \cdots & 0 \\ & \vdots & & & & & & \\ 0 & 0 & \cdots & 1 & \cdots & 0 & \cdots & 0 \\ & \vdots & & & & & & \\ 0 & 0 & \cdots & c & \cdots & 1 & \cdots & 0 \\ & \vdots & & & & & & \\ 0 & 0 & \cdots & 0 & \cdots & 0 & \cdots & 1 \end{pmatrix}$$

with labels *k*th column and *r*th column over the respective columns, and *k*th row and *r*th row marking the respective rows.

obtained by adding c times the kth row of I to the rth row and A is any $n \times m$ matrix. The ij element in EA is

$$(EA)_{ij} = \sum_{k=1}^{n} e_{ik}a_{kj}$$

obtained by multiplying the ith row of E by the jth column of A. If i is not equal to r, the only nonzero element in the ith row of E is $e_{ii} = 1$ in the ith position. The element e_{ii} multiplies a_{ij}, so that in this case we have

$$(EA)_{ij} = 1 \cdot a_{ij} = a_{ij}$$

In all rows but the rth, then, the elements of EA are equal to the corresponding elements of A.

In the rth row of E, we find c in the kth position, 1 in the rth position, and zeros everywhere else. Thus

$$(EA)_{rj} = e_{rk}a_{kj} + e_{rr}a_{rj} = c \cdot a_{kj} + 1 \cdot a_{rj}$$
$$= ca_{kj} + a_{rj}$$

The elements in the rth row of EA are formed by starting with elements in the rth row of A and adding to those c times the corresponding elements from the kth row of A. This completes the proof. ▶

For each statement about rows of a matrix, there is an analogous statement about columns. Thus, there are three elementary column operations

(C 1) Multiplication of a column by a nonzero real number.

(C 2) Replacement of a column by the sum of that column and a constant times another column.

(C 3) Interchange of two columns.

An *elementary column matrix* is obtained from an identity matrix by performing a single column operation. The proof of the following theorem is left as an exercise.

Theorem 2 If $E_{n \times n}$ is an elementary column matrix obtained from the identity by performing an elementary column operation, and if A is any $m \times n$ matrix, then AE is the matrix obtained by performing that same column operation on A. ▶

Note that row operations are performed by multiplying on the *left* by an elementary row matrix, while column operations are performed by multiplying on the *right* by an elementary column matrix.

PROBLEMS

3. Find the matrix product AE, where A is any 4×3 matrix and E is the elementary column matrix

(a) $\begin{pmatrix} 1 & 2 & 0 \\ 0 & 1 & 0 \\ 0 & 0 & 1 \end{pmatrix}$
(b) $\begin{pmatrix} 0 & 0 & 1 \\ 0 & 1 & 0 \\ 1 & 0 & 0 \end{pmatrix}$

4. Write elementary row matrices which will produce the successive arrays in (9). Verify your answer by multiplying these by the initial array. (*Hint:* Each new array requires two or three row operations.)

5. In Example 1, find the elementary row matrices required to transform the initial coefficient matrix (11) to the final coefficient matrix (12).

6. Prove that the first row of the matrix (11) is not identifiable by using column instead of row operations. (*Hint:* Operate on the transpose of the coefficient matrix.)

In Problems 7–10, determine whether the first row is identifiable (see Example 1).

7. $\begin{pmatrix} 6 & 1 & 3 & 0 & 0 \\ 1 & -2 & 3 & 2 & 4 \\ 4 & 1 & 6 & 3 & 6 \end{pmatrix}$

8. $\begin{pmatrix} 1 & 6 & 5 & 4 & 0 & 0 \\ 2 & -2 & -6 & 3 & 8 & 1 \\ -1 & 4 & -3 & 2 & 3 & -1 \end{pmatrix}$

9. $\begin{pmatrix} 1 & 2 & 3 & 0 & 0 & 0 \\ -6 & 3 & -4 & 4 & 1 & 8 \\ 3 & -10 & -2 & 0 & 2 & -2 \\ 1 & 8 & 3 & 0 & 0 & 6 \end{pmatrix}$

10. $\begin{pmatrix} -8 & 1 & 6 & 2 & 0 & 0 & 0 \\ 1 & 2 & 0 & -6 & 2 & 2 & 4 \\ 3 & 0 & 4 & 0 & 0 & 6 & 2 \\ 0 & -6 & 3 & 1 & -2 & 1 & -3 \end{pmatrix}$

11. Prove Theorem 1 for row operations R 1 and R 3.

12. Prove Theorem 2.

9.3 THE INVERSE OF A MATRIX

If $a \neq 0$ is a real number, the number

$$\frac{1}{a} = a^{-1}$$

is called the *reciprocal* or the *multiplicative inverse* of a. It has the property that $a^{-1}a = aa^{-1} = 1$. By analogy, an $n \times n$ matrix B is called an *inverse* of the $n \times n$ matrix A if

$$BA = AB = I \qquad (13)$$

A matrix which has an inverse is said to be *nonsingular*. Other square matrices are called *singular*.*

A matrix, if it has an inverse at all, can have only one. For if B and C are both inverses of A, that is, if $BA = AB = I$ and $CA = AC = I$, then

$$C = CI = C(AB) = (CA)B = IB = B$$

Thus, we speak of *the* inverse of A, which we denote by A^{-1}.

Note that the definition (13) is symmetric in A and B. That is, if B is the inverse of A, then A is also the inverse of B. Symbolically,

$$(A^{-1})^{-1} = A$$

Example 1   To find the inverse of

$$A = \begin{pmatrix} 1 & 2 \\ 3 & 4 \end{pmatrix}$$

we look for a matrix

$$B = \begin{pmatrix} a & b \\ c & d \end{pmatrix}$$

with the property that $AB = I$. Writing out the matrix product gives

$$AB = \begin{pmatrix} 1 & 2 \\ 3 & 4 \end{pmatrix} \begin{pmatrix} a & b \\ c & d \end{pmatrix} = \begin{pmatrix} a + 2c & b + 2d \\ 3a + 4c & 3b + 4d \end{pmatrix} = \begin{pmatrix} 1 & 0 \\ 0 & 1 \end{pmatrix}$$

Equating corresponding coefficients, we have four linear equations

$$a + 2c = 1 \qquad b + 2d = 0$$
$$3a + 4c = 0 \qquad 3b + 4d = 1$$

whose solution is $a = -2$, $b = 1$, $c = \frac{3}{2}$, and $d = -\frac{1}{2}$. Hence

$$B = A^{-1} = \begin{pmatrix} -2 & 1 \\ \frac{3}{2} & -\frac{1}{2} \end{pmatrix}$$

▶

PROBLEMS

1. Verify in Example 1 that $AA^{-1} = A^{-1}A = I$.

Example 2   Since $I \cdot I = I$, the identity matrix is its own inverse.   ▶

*Although it is possible to invent the concept of inverse for nonsquare matrices, we do not do so in this text.

2. Find all 2 × 2 matrices A which are their own inverses. (*Hint:* Solve the equation $A \cdot A = I$.)

Example 3 The matrix

$$A = \begin{pmatrix} 1 & 2 \\ -3 & -6 \end{pmatrix}$$

has no inverse. For if

$$B = \begin{pmatrix} a & b \\ c & d \end{pmatrix}$$

is to be the inverse of A, we must have

$$\begin{pmatrix} 1 & 2 \\ -3 & -6 \end{pmatrix}\begin{pmatrix} a & b \\ c & d \end{pmatrix} = \begin{pmatrix} a + 2c & b + 2d \\ -3a - 6c & -3b - 6d \end{pmatrix} = \begin{pmatrix} 1 & 0 \\ 0 & 1 \end{pmatrix}$$

Equating corresponding elements gives the four equations

$$\begin{matrix} a + 2c = 1 & b + 2d = 0 \\ -3a - 6c = 0 & -3b - 6d = 1 \end{matrix} \tag{14}$$

The third equation $-3a - 6c = -3(a + 2c) = 0$ requires $a + 2c = 0$, while the first requires $a + 2c = 1$. Obviously, no choices for a and c can do both, so Equations (14) have no solution and A has no inverse. ▶

The above example shows that, unlike real numbers, there are nonzero matrices for which no inverse exists.

PROBLEMS

3. Prove that a zero matrix cannot have an inverse.

Example 4 We saw in Example 1, Section 9.1, that the equations

$$\begin{matrix} x_1 - 2x_2 + 3x_3 = 4 \\ -x_1 + x_2 + x_3 = 0 \\ 2x_1 + 2x_2 - x_3 = 1 \end{matrix} \tag{15}$$

have the solution $x_1 = 1$, $x_2 = 0$, and $x_3 = 1$. These equations may be written in the matrix form $AX = C$, where A is the nonsingular coefficient matrix

$$A = \begin{pmatrix} 1 & -2 & 3 \\ -1 & 1 & 1 \\ 2 & 2 & -1 \end{pmatrix}$$

whose inverse is

$$A^{-1} = \begin{pmatrix} \frac{1}{3} & -\frac{4}{3} & -\frac{1}{3} \\ \frac{1}{15} & -\frac{7}{15} & -\frac{4}{15} \\ \frac{4}{15} & \frac{2}{15} & -\frac{1}{15} \end{pmatrix}$$

With A^{-1} known, it is a simple matter to solve the linear equations (15). We multiply both sides of $AX = C$ by A^{-1} to obtain

$$A^{-1}(AX) = A^{-1}C$$

Since $A^{-1}A = I$ and $IX = X$, this reduces to

$$X = A^{-1}C$$

Performing the indicated matrix multiplication yields

$$X = \begin{pmatrix} x_1 \\ x_2 \\ x_3 \end{pmatrix} = A^{-1}C = \begin{pmatrix} \frac{1}{3} & -\frac{4}{3} & -\frac{1}{3} \\ \frac{1}{15} & -\frac{7}{15} & -\frac{4}{15} \\ \frac{4}{15} & \frac{2}{15} & -\frac{1}{15} \end{pmatrix} \begin{pmatrix} 4 \\ 0 \\ 1 \end{pmatrix} = \begin{pmatrix} 1 \\ 0 \\ 1 \end{pmatrix}$$

from which we read $x_1 = 1$, $x_2 = 0$, and $x_3 = 1$, verifying our previous computations. ▶

PROBLEMS

4. Verify in Example 4 that $A^{-1}A = I$.

At first glance it would seem that the method of Example 4, in which the matrix equation $AX = C$ is solved to give $X = A^{-1}C$, represents a considerable saving in time and effort over the method of Example 1 of Section 9.1. However, the gain is more apparent than real since, unfortunately, there remains the problem of finding the inverse A^{-1}. Moreover, the method of Example 4 obviously fails if A does not have an inverse. Thus we shall continue to solve linear equations by successive elimination of variables (row operations).

The problem of finding the inverse of a matrix is itself best approached through elementary row operations. To see how to proceed, suppose that by performing row operations we are able to transform the $n \times n$ matrix A into the identity matrix I. This means that there are elementary row matrices $R_1, R_2, \ldots, R_t$ such that

$$R_t R_{t-1} \cdots R_2 R_1 A = I \tag{16}$$

By definition, then, the product

$$R = R_t R_{t-1} \cdots R_2 R_1 \tag{17}$$

is the inverse of A since $RA = I$. But

$$A^{-1} = R = RI = R_t R_{t-1} \cdots R_1 I \tag{18}$$

and this, together with (16), provides the following scheme.

Theorem 3 In order to find the inverse of a nonsingular matrix A, we perform row operations on A and also on the identity matrix I. The *same* operations which change A into I (Equation (16)) will transform I into A^{-1} (Equation (18)). ▶

TABLE 1. Steps in Obtaining the Inverse Matrix

Row	A			I		
R_{10}	3	2	1	1	0	0
R_{20}	2	3	2	0	1	0
R_{30}	4	1	1	0	0	1
$R_{11} = \frac{1}{3}R_{10}$	1	$\frac{2}{3}$	$\frac{1}{3}$	$\frac{1}{3}$	0	0
$R_{21} = R_{20} - \frac{2}{3}R_{10}$	0	$\frac{5}{3}$	$\frac{4}{3}$	$-\frac{2}{3}$	1	0
$R_{31} = R_{30} - \frac{4}{3}R_{10}$	0	$-\frac{5}{3}$	$-\frac{1}{3}$	$-\frac{4}{3}$	0	1
$R_{12} = R_{11} - \frac{2}{5}R_{21}$	1	0	$-\frac{1}{5}$	$\frac{3}{5}$	$-\frac{2}{5}$	0
$R_{22} = \frac{3}{5}R_{21}$	0	1	$\frac{4}{5}$	$-\frac{2}{5}$	$\frac{3}{5}$	0
$R_{32} = R_{21} + R_{31}$	0	0	1	-2	1	1
$R_{13} = R_{12} + \frac{1}{5}R_{32}$	1	0	0	$\frac{1}{5}$	$-\frac{1}{5}$	$\frac{1}{5}$
$R_{23} = R_{22} - \frac{4}{5}R_{32}$	0	1	0	$\frac{6}{5}$	$-\frac{1}{5}$	$-\frac{4}{5}$
$R_{33} = R_{32}$	0	0	1	$-\frac{10}{5}$	$\frac{5}{5}$	$\frac{5}{5}$

Example 5 Table 1 shows steps in the computation of the inverse of

$$A = \begin{pmatrix} 3 & 2 & 1 \\ 2 & 3 & 2 \\ 4 & 1 & 1 \end{pmatrix}$$

We begin by appending to A the identity matrix I. The row operations which change A into I and I into A^{-1} are indicated in the margin. The symbol R_{ij} denotes the ith row at the jth stage. Note that the strategy here, as in solving linear equations, is to select operations which will successively transform the columns of A into columns of the identity matrix.

From Table 1 we read

$$A^{-1} = \begin{pmatrix} \frac{1}{5} & -\frac{1}{5} & \frac{1}{5} \\ \frac{6}{5} & -\frac{1}{5} & -\frac{4}{5} \\ -2 & 1 & 1 \end{pmatrix}$$

▶

PROBLEMS

5. In Example 5, verify that $AA^{-1} = A^{-1}A = I$.

If the reduction (16) is possible, then A has the inverse indicated in (17). Equivalently, if A has no inverse (is singular), then it will not be possible to change A into an identity matrix. We state without proof that this is always indicated by the appearance of a zero row at some stage in the procedure.

Example 6   If we attempt to find the inverse of

$$A = \begin{pmatrix} 3 & 2 & 1 \\ 2 & 3 & 2 \\ 7 & 8 & 5 \end{pmatrix}$$

we are led to row operations indicated in Table 2. Since it is not possible to transform the third column of A into a vector having one in the third row and zeros elsewhere without altering the desired values in the first and second columns, we conclude that A has no inverse.   ▶

TABLE 2.   Steps Showing No Inverse Exists

Row	A			I		
R_{10}	3	2	1	1	0	0
R_{20}	2	3	2	0	1	0
R_{30}	7	8	5	0	0	1
$R_{11} = \frac{1}{3}R_{10}$	1	$\frac{2}{3}$	$\frac{1}{3}$	$\frac{1}{3}$	0	0
$R_{21} = R_{20} - \frac{2}{3}R_{10}$	0	$\frac{5}{3}$	$\frac{4}{3}$	$-\frac{2}{3}$	1	0
$R_{31} = R_{30} - \frac{7}{3}R_{10}$	0	$\frac{10}{3}$	$\frac{8}{3}$	$-\frac{7}{3}$	0	1
$R_{12} = R_{11} - \frac{2}{5}R_{21}$	1	0	$-\frac{1}{5}$	$\frac{3}{5}$	$-\frac{2}{5}$	0
$R_{22} = \frac{3}{5}R_{21}$	0	1	$\frac{4}{5}$	$-\frac{2}{5}$	$\frac{3}{5}$	0
$R_{32} = R_{31} - \frac{10}{5}R_{21}$	0	0	0	-1	$-\frac{10}{5}$	1

PROBLEMS

In Problems 6–11, first try to find the inverse of the matrix of coefficients. If an inverse exists, use it to obtain a solution to the system of equations. If no inverse exists, try to solve the equations in the usual way by successive elimination of variables. Note that while existence of an inverse guarantees a solution, failure to have an inverse is entirely inconclusive.

6. $x_1 + 3x_2 = 4$
   $2x_1 - 3x_2 = 6$

7. $x - y = 4$
   $2x + 3y = 10$

8. $3x_1 + 2x_2 - 6x_3 = 14$
   $-x_1 + 4x_2 + 5x_3 = -3$
   $6x_1 - 10x_2 - 21x_3 = 10$

9. $x + 2y - 3z = 10$
   $3x - y + z = 20$
   $x + y + z = 5$

10. $-4x - 4y + 14z = 10$
   $x + 2y + 3z = 1$
   $-4x - 6y + z = 3$

11. $x + 2y + 2z - w = 19$
   $2x + 4y - z + 8w = 8$
   $x + 2y - 5z + 13w = -23$
   $3x + 6y + 4z + w = 45$

12. (a) Solve Problem 13 of Section 9.1 by first computing the inverse of the matrix $I - A$.

(b) Explain why the ij element of $(I - A)^{-1}$ can be interpreted as the output of industry i which is required to provide one unit of industry j product to the nonproductive sector.

13. Suppose the demand vector $D = (10, 25, 20)$ in Problem 13, Section 9.1 can be written as the sum

$$D = C + V + G$$

where C, V, and G denote vectors of personal consumption, private investment, and government demand, respectively, for the three sectors.

(a) Let F be the matrix (C, V, G) whose columns are the vectors C, V, and G. Explain why the $(2, 3)$ element in $(I - A)^{-1}F$ can be interpreted as the manufacturing output which satisfies government demand.

(b) Why is the sum of the elements in the ith row of $(I - A)^{-1}F$ equal to x_i, the output of the ith sector?

14. If A and B are nonsingular matrices, show that $(AB)^{-1} = B^{-1}A^{-1}$. (*Hint:* Compute $(AB)(B^{-1}A^{-1})$ and $(B^{-1}A^{-1})AB$.)

15. Prove that if $A_1, A_2, \ldots, A_k$ are nonsingular matrices, then the product $A_1A_2 \cdots A_k$ is nonsingular and has the inverse $A_k^{-1}A_{k-1}^{-1} \cdots A_1^{-1}$. The inverse of a product is the product of the individual inverses *in reverse order*.

16. What is the effect on A^{-1} if we

(a) interchange two columns of A?

(b) multiply one row by a constant?

(c) multiply one column by a constant?

9.4 LINEAR INEQUALITIES AND CONVEX SETS

In this section we propose to study *linear inequalities*. That is, we shall be looking for solutions of systems such as

$$a_{11}x_1 + a_{12}x_2 + \cdots + a_{1n}x_n \leq c_1$$

$$a_{21}x_1 + a_{22}x_2 + \cdots + a_{2n}x_n \leq c_2 \qquad \textbf{(19)}$$

$$\vdots$$

$$a_{m1}x_1 + a_{m2}x_2 + \cdots + a_{mn}x_n \leq c_m$$

This system is identical in form to the system (1), except that inequalities have replaced equalities.

In matrix notation, (19) may be written as

$$AX \leq C$$

where, as usual, an inequality between matrices means that the inequality applies to each pair of corresponding elements in the two matrices. It is assumed that the numbers a_{ij} and c_i are known. Our problem is to find all solutions, that is, all vectors X such that $AX \leq C$.

Example 1 Consider the system

$$2x + 3y \leq 6$$

$$x - y \leq 0$$

$$2x - y \geq -1$$

The last inequality is opposite in sense to those in (19), but this is easily remedied by multiplying through by -1 to obtain the system

$$2x + 3y \leq 6$$

$$x - y \leq 0 \qquad \textbf{(20)}$$

$$-2x + y \leq 1$$

Graphically, the set of points (x, y) for which $2x + 3y = 6$ is the straight line shown in Figure 1. Those points which satisfy $2x + 3y < 6$ lie in the shaded region below this line. Thus the points which satisfy the first inequality

$$2x + 3y \leq 6$$

lie on or below the line. Such a region is called a *closed half-plane*.

Similar comments apply to the other two inequalities, and we find that the set of solutions (x, y) of the inequalities (20) is the *intersection* of three closed half-planes. In Figure 2 this intersection is the shaded triangle, including the boundaries.

FIGURE 1

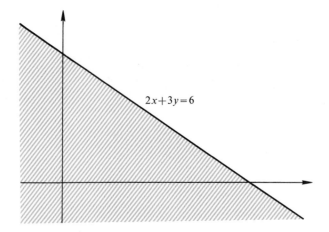

The points $(3/8, 7/4)$, $(6/5, 6/5)$, and $(-1, -1)$ are called *corner points*, or *extreme points*, of the triangular solution set. These corner points are intersections of pairs of boundary lines. Thus they may be found by changing the inequalities in (20) to equalities and solving these equations in pairs. ▶

A set C is called *convex* if whenever points X and Y lie in C, all points Z between X and Y also lie in C (see Figure 3). Typical examples of convex sets in the plane are circles, triangles, rectangles, and lines. Spheres, cubes, lines,

FIGURE 2

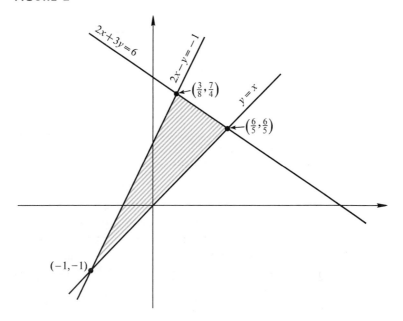

FIGURE 3   Typical two-dimensional convex sets

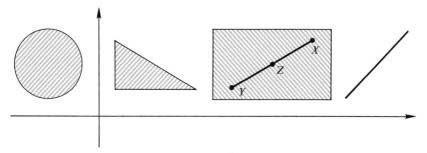

and planes in space are convex. The kidney-shaped set in Figure 4 is not convex since the point Z between X and Y is not in the set.

Our reason for introducing the concept of convexity here is that systems of linear inequalities always produce convex solution sets. (For instance, the solution set in Figure 2 is obviously convex.) A proof begins with the fact that if P and Q are two points in the plane, then any point Z lying on the line which passes through P and Q can be written in the form

$$Z = (1 - t)P + tQ \tag{21}$$

where t is a real number. If $t = 0$, for instance, Equation (21) becomes $Z = P$, while if $t = 1$, $Z = Q$ (see Figure 5). Values of t greater than one give points beyond Q, while $t < 0$ produces a point beyond P. Values between 0 and 1 yield points between P and Q. Of particular interest is the point

$$A = \frac{P + Q}{2} = \tfrac{1}{2}P + \tfrac{1}{2}Q$$

lying halfway between P and Q. This point is called the *average* of P and Q and is obtained by putting $t = \tfrac{1}{2}$ in Equation (21).

FIGURE 4   A set which is not convex

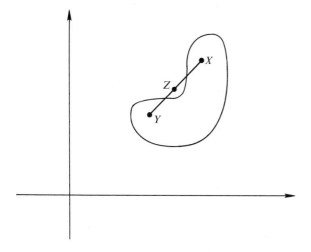

FIGURE 5

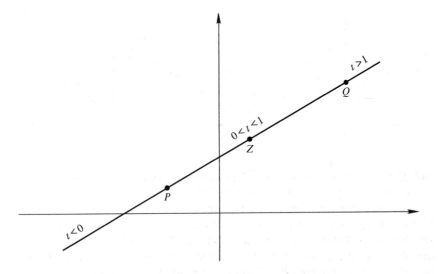

These ideas are carried over by analogy to higher dimensions. In general, if X and Y are two n-dimensional vectors, a vector Z lies *between* X and Y if and only if

$$Z = (1 - t)X + tY \qquad (22)$$

where t is a number lying between 0 and 1.

Suppose now that X and Y are solutions of the system of linear inequalities $AX \leq C$ and let Z be any point between X and Y. That is, suppose that $AX \leq C$, $AY \leq C$, and that Z satisfies Equation (22). Then

$$AZ = A[(1 - t)X + tY] = (1 - t)AX + tAY$$
$$\leq (1 - t)C + tC = C$$

and it follows that Z is also a solution. In short, any vector Z lying between two solutions X and Y is also a solution. By definition, then, the solution set is convex.

Example 2 An electronics plant has 54 employees—36 assemblers and 18 packers. The plant produces two component packages A and B, each of which requires the labor of both assemblers and packers. The labor requirements per unit are presented in Table 3.

TABLE 3

Labor classification	Man hours required per unit	
	Component A	Component B
Assembly	1	2
Packing	1	$\frac{1}{2}$

If each employee is assumed to work 8 hours per day, there are $36 \cdot 8 = 288$ assembler man-hours and $18 \cdot 8 = 144$ packer man-hours available per day. Let a and b denote the respective daily productions of components A and B. Obviously,

$$a \geq 0 \quad \text{and} \quad b \geq 0 \tag{23}$$

since we cannot have negative production. In addition, the capacity restrictions for assemblers and packers require that

$$a + 2b \leq 288 \quad \text{and} \quad a + \tfrac{1}{2}b \leq 144 \tag{24}$$

The convex set of points (a, b) that satisfy the four inequalities (23) and (24) is shaded in Figure 6. Note that of the

$$\binom{4}{2} = 6$$

points obtained by changing the four inequalities to equalities and solving in pairs, only four are actually corner points of the solution set. The question which naturally arises is this: Subject to the given restrictions, how should a and b be chosen so as to maximize profit? We shall answer this question in the next section. ▶

FIGURE 6

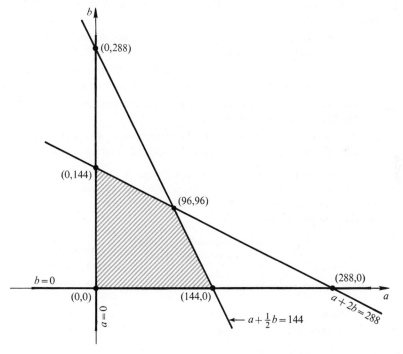

PROBLEMS

In Problems 1–6, draw the convex solution set for the given system of linear inequalities.

1. $x \geq 0$
   $y \geq 0$
   $x + 3y \leq 9$

2. $x \leq 1$
   $y \leq 2$
   $3x + 2y \geq 0$

3. $y \geq 0$
   $x \geq 0$
   $x \leq 4$

4. $-5x + y \leq 0$
   $-2x + y \geq 0$

5. $x + 3y \geq 0$
   $x - 3y \leq 0$
   $y \leq 1$

6. $2x + 4y \geq 8$
   $4x + 2y \geq 8$
   $x \geq 0$
   $y \geq 0$

In Problems 7–10, draw the solution set and find the extreme points.

7. $3x + y \geq 6$
   $x \leq 0$
   $y \leq 0$

8. $-4x + y \leq 1$
   $-x + y \geq 2$
   $x + y \leq 4$

9. $2x + y \geq -1$
   $-x + y \leq 4$
   $x \leq 2$

10. $-3x + y \leq 3$
    $x + y \leq 0$
    $y \geq 0$

11. A political candidate has decided to purchase x minutes of local radio time and y minutes of local TV time. The rates are \$50 per minute for radio and \$200 per minute for TV time.

   (a) Negative amounts of radio and TV time cannot be bought. Express these restrictions as inequalities and draw the solution set.

   (b) It is decided that not less than one minute of TV time and $\frac{1}{2}$ minute of radio time will be bought. Modify the solution set of (a).

   (c) The amount spent for radio and TV advertising cannot exceed \$1000. Modify the solution set in (b).

   (d) It is decided that at least twice as much should be spent on radio advertising as on TV advertising. Modify the solution set of (c).

12. A machine shop produces two casting types, A and B. Each casting requires an operation on each of three different machines. The times required on each machine and the available free time for each machine are given in the following table.

	Time required to produce one casting of type		Available monthly time
	A	B	
Machine I	4	6	110
Machine II	$2\frac{1}{2}$	3	100
Machine III	4	9	150

Let x denote the number of A-type castings and y the number of B-type castings to be produced. Draw the convex set of possible (x, y) pairs.

13. Let the input matrix for a simple two-sector Leontief model be

$$A = \begin{pmatrix} 0 & 2 \\ \frac{1}{3} & 0 \end{pmatrix}$$

(a) Graph the set of viable solutions to the model.

(b) Does a non-zero interior equilibrium exist?

14. Suppose that the input matrix in Problem 13 is

$$A = \begin{pmatrix} 0 & 1 \\ 2 & 0 \end{pmatrix}$$

What is the set of viable solutions?

15. Given an input matrix

$$A = \begin{pmatrix} 0 & a_{12} \\ a_{21} & 0 \end{pmatrix}$$

what are the restrictions on a_{12} and a_{21} in order to ensure a non-zero viable solution?

16. Show that the intersection of convex sets is a convex set.

17. Draw a picture to show that the union of two convex sets need not be a convex set.

9.5   LINEAR PROGRAMMING

A fundamental problem in economics is to allocate limited resources so as to best meet desired goals. One of the most important tools for attacking such problems is *linear programming*.

Example 1   Suppose in Example 2 of the previous section that components A and B yield per unit profits of $\$1.50$ and $\$1.00$, respectively. The plant management is faced with the problem of allocating available labor so as to achieve the objective of maximum daily profit.

In the notation of Example 2, this means that, subject to the restrictions (23) and (24), a and b are to be chosen so as to maximize the profit function

$$P = (1.50)a + (1.00)b = \tfrac{3}{2}a + b \tag{25}$$

Let us look at the situation geometrically as in Figure 7. If the value of P is fixed at, say, $P = 75$, the set of points (a, b) which satisfy Equation (25) is a straight line (see Figure 7(a)). Choosing other values for P will yield other lines, all of which are parallel (the lines for $P = 30$ and $P = 150$ are shown in Figure 7(a)). It is apparent that when profit is decreased, the line representing the graph of the profit function will move closer to the origin. On

FIGURE 7

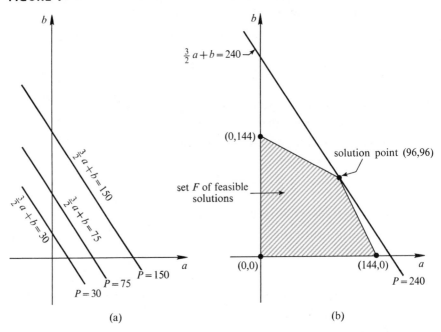

(a)   (b)

the other hand, when profit is increased, the line will move further away. Maximum profit is obtained by moving the line as far to the right as possible while at the same time intersecting the solution set F for the inequalities (23) and (24). Figure 7(b) shows that these twin objectives are realized by the line

$$\tfrac{3}{2}a + b = 240$$

which intersects the set F only at the corner point (96, 96). Thus, maximum daily profit is \$240, achieved by producing 96 units each of components A and B. ▶

PROBLEMS

1. Use the information in Table 3 to show that the optimum solution in Example 1 utilizes all available labor resources, allocating 96 assembler man-hours and 96 packer man-hours to component A, and 192 assembler man-hours and 48 packer man-hours to component B.

2. Compute the profit which would be obtained at each of the other corner points (0, 144), (144, 0), and (0, 0) of the set F.

Example 1 contains all the essential ingredients of a linear programming problem. In the general problem there are n variables $x_1, x_2, \ldots, x_n$ together with m restrictions on these variables in the form of linear inequalities:

$$a_{11}x_1 + a_{12}x_2 + \cdots + a_{1n}x_n \leq c_1$$
$$a_{21}x_1 + a_{22}x_2 + \cdots + a_{2n}x_n \leq c_2$$
$$\vdots \qquad\qquad (26)$$
$$a_{m1}x_1 + a_{m2}x_2 + \cdots + a_{mn}x_n \leq c_m$$

(In Example 1, there are two variables a and b on which restrictions are imposed by Equations (24).) In addition, there is a condition (as in Equation (23)) that each variable be non-negative:

$$x_1 \geq 0$$
$$x_2 \geq 0$$
$$\vdots \qquad\qquad (27)$$
$$x_n \geq 0$$

As we saw in Section 9.4, the inequalities (26) and (27) determine a convex set F, called the set of *feasible vectors*. The problem is to determine, from among all feasible solutions $X = \text{col}(x_1, x_2, \ldots, x_n)$, those which maximize a linear function

$$g(X) = b_1x_1 + b_2x_2 + \cdots + b_nx_n \qquad (28)$$

(like the profit function (25)).

The problem may be stated more succinctly using matrix notation: Given an $m \times n$ matrix A, an $m \times 1$ vector C, and an $n \times 1$ vector B, find those $n \times 1$ vectors X which maximize the function

$$g(X) = B'X \qquad (28')$$

subject to the restrictions

$$AX \leq C \qquad (26')$$
$$X \geq 0 \qquad (27')$$

The key to solving this problem is the following theorem.

Theorem 4 Let F be the set of feasible solutions of the linear programming problem determined by Equations (26')–(28') and suppose F is bounded. Then, among all points in F, there is a corner point of F at which g is minimized and another corner point at which g is maximized.

PROOF For simplicity, we indicate the proof only for $n = 2$. The general case is similar. Let W and Y be any two feasible solutions (points in F) and let

$$X = (1 - t)W + tY \qquad (0 < t < 1)$$

be a point in F which lies between W and Y (see Figure 8(a)). Suppose that $a = B'W$ is less than $b = B'Y$. Then

$$B'X = B'[(1 - t)W + tY] = (1 - t)B'W + tB'Y$$
$$= (1 - t)a + tb = a + t(b - a)$$

FIGURE 8

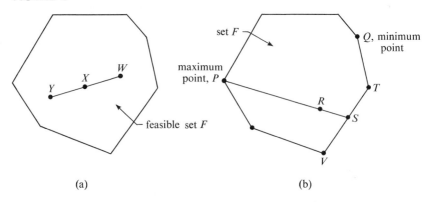

(a)　　　　　　　　　　　　(b)

Since $b - a$ is positive and t is between 0 and 1, it follows that

$$a \leq a + t(b - a) \leq b$$

or that

$$B'W \leq B'X \leq B'Y$$

The value $g(X) = B'X$ of the function g at a point X between W and Y lies between the values of g at W and Y.

Now suppose that, *among the corner points* of F, the maximum value of g occurs at P and the minimum value occurs at Q (see Figure 8(b)). Let R be any point in the feasible set F. Draw a line from P through R until it cuts the boundary of F at a point S lying between corner points V and T. We know that (1) the values of g at T and V lie between the values at Q and P, by definition of Q and P; (2) the value of g at S lies between the values at T and V, and hence between the values at Q and P; and (3) the value of g at R lies between the values at P and at S and hence between the values at Q and at P. Since this holds for any point R in F, we see that among points in the entire set F, g takes its largest value at P and its smallest value at Q. ▶

Example 2　A college interviewer is hiring for job classifications A and B. He has been instructed by his company to hire between two and four people for job A and at least four people for job B, but no more than 10 altogether.

From past experience, the interviewer knows that 80% of those hired for job A are successful and 20% unsuccessful, while for job B 90% are successful and 10% unsuccessful. A successful employee in either job is valued at \$10,000, while an unsuccessful employee costs the company \$4000 in job A and \$9000 in job B. How many people should the interviewer hire for each job in order to maximize the total average return to the company?

To solve this problem, let x denote the number to be hired for job A and y the number for job B. We expect on the average that $(0.80)x$ and $(0.90)y$ will be successful while $(0.20)x$ and $(0.10)y$ will be unsuccessful. The total average return is

$$R(x, y) = (10,000)(0.8x) - (4000)(0.2x) + (10,000)(0.9y) - (9000)(0.1y)$$
$$= 7200x + 8100y$$

The problem is to maximize R subject to the restrictions

$$x \geq 2 \quad x \quad \leq 4$$
$$y \geq 4 \quad x + y \leq 10$$

The feasible region is shown in Figure 9. The respective values of R at the corner points are

$$R(2, 8) = 7200(2) + 8100(8) = \$79{,}200$$
$$R(4, 6) = 7200(4) + 8100(6) = \$77{,}400$$
$$R(2, 4) = 7200(2) + 8100(4) = \$46{,}800$$
$$R(4, 4) = 7200(4) + 8100(4) = \$61{,}200$$

The optimal solution is to hire two persons for job A and eight for job B, yielding a total average return of \$79,200. ▶

FIGURE 9

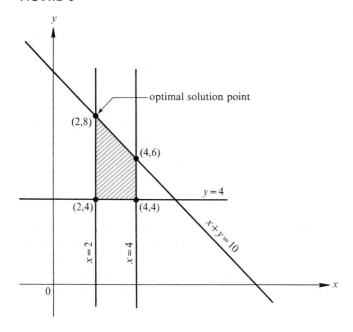

Example 3 In Example 2, suppose that a successful employee in job B is worth \$9000. Then the total average return is

$$R(x, y) = 10,000(0.8x) - 4000(0.2x) + 9000(0.9y) - 9000(0.1y)$$

$$= 7200x + 7200y = 7200(x + y)$$

The graphic solution with the same restrictions on x and y is shown in Figure 10. Instead of a solution occurring at a single corner point as in Example 2, corner points $(2, 8)$ and $(4, 6)$ both produce the maximum average return

$$R_{max}(x, y) = 7200(10)$$

$$= \$72,000$$

It follows that every boundary point of F lying on the line between $(2, 8)$ and $(4, 6)$ also yields maximum return. ▶

FIGURE 10

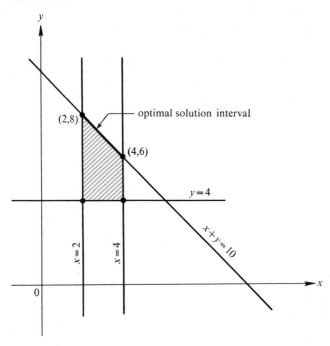

Example 4 Production of a unit output in a certain job requires the completion of three separate tasks A, B, and C. Three workers are available for assignment to the job. The efficiency of each worker in terms of the number of units produced per hour for each task is indicated in Table 4. Given that worker I costs \$2.50 per hour, worker II \$3.00 per hour, and worker III

$5.00 per hour, what is the best utilization of each worker in order to produce at least 1000 units with minimum labor cost?

TABLE 4. Number of units produced per hour for each worker

	I	II	III
Task A	40	120	60
Task B	30	50	150
Task C	60	80	100

We solve this problem by letting x, y, and z represent the number of hours production time allocated to workers I, II, and III, respectively. The linear programming problem is to minimize the cost function

$$C = (\$2.50)x + (\$3.00)y + (\$5.00)z$$

subject to the linear constraints

$$0.4x + 1.20y + 0.6z \geq 10 \qquad x \geq 0$$
$$0.3x + 0.5y + 1.5z \geq 10 \qquad y \geq 0$$
$$0.6x + 0.8y + 1.0z \geq 10 \qquad z \geq 0$$

The loci of the solution in three dimensional space is presented in Figure 11. In the case of linear inequalities in three unknowns, each linear

FIGURE 11

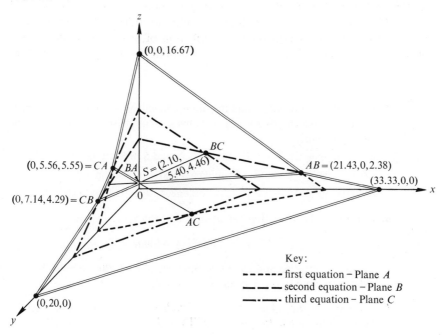

Key:
- - - - - - first equation – Plane A
- - - - second equation – Plane B
- · - · - third equation – Plane C

equality appears as a plane in space. The lines of intersection of the planes in the positive octant are:

first plane A with second plane B: $\overline{AB, BA}$

first plane A with third plane C: $\overline{AC, CA}$

second plane B with third plane C: $\overline{BC, CB}$

where the endpoints of each line are the intersections of the traces of the respective planes. The three lines $\overline{AB, BA}$, $\overline{AC, CA}$, and $\overline{BC, CB}$ intersect at the point $S = (2.10, 5.40, 4.46)$ which satisfies the set of three equalities.

The boundary planes of the unbounded convex set of feasible solutions defined by the three inequality constraints are outlined by bold lines in Figure 11. The extreme points of the convex set are:

$$E_1 = (0, 20, 0) \qquad E_5 = (0, 5.56, 5.55)$$

$$E_2 = (33.33, 0, 0) \qquad E_6 = (0, 7.14, 4.29)$$

$$E_3 = (21.43, 0, 2.38) \qquad E_7 = (2.10, 5.40, 4.46)$$

$$E_4 = (0, 0, 16.67)$$

Computing the values of the cost function $C = (\$2.50)x + (\$3.00)y + (\$5.00)z$ at each extreme point, we obtain the following costs: $60, $83.32, $65.48, $83.35, $44.43, $42.87, and $43.75, respectively. Thus, the minimum labor cost is achieved by not employing worker I, by employing worker II for 7.14 hours, and worker III for 4.29 hours. Allocation of labor according to this schedule will result in completion of approximately 1114 task A units, 1000 task B units, and 1000 task C units. In other words, in order to achieve minimum labor cost for the completion of at least 1000 items, it is necessary to complete 114 more task A units than absolutely required. ▶

PROBLEMS

3. Find both the maximum and the minimum values which the stated linear function takes in the respective convex solution sets described in the given problems.

(a) $G(x, y) = 2x + 6y$; Problems 1, 2, and 5 of Section 9.4.

(b) $F(x, y) = -5x + 2y$; Problems 3 and 4 of Section 9.4.

(c) $H(x, y) = 4x + 8y - 12$; Problems 8, 9, and 10 of Section 9.4.

4. In Problem 11 of Section 9.4, assume that the value criterion is

$$V = \$1000x + \$1000y$$

where x is TV time and y is radio time. How much time should the candidate purchase in order to maximize his value?

5. Suppose that the value criterion in Problem 4 is $V = \$500x + \$2500y$. How will this affect the amount of TV and radio time purchased?

6. In Problem 12, Section 9.4 assume that the manufacturer makes a profit of \$8 per unit for type-A castings and \$12 per unit for type-B castings.

(a) How many of each type should be produced in order to maximize his profit?

(b) Which inequality is *superfluous* in that the feasible set is the same with and without the inequality restriction?

7. Suppose that the profit on type-A castings in Problem 6 is increased to \$12 per unit. How many units of each casting should now be produced in order to maximize profits?

8. An oil refinery produces* 2000 gallons of pitch per day. A certain amount, P gallons, of this is to be blended with flux stock, of which there is an unlimited supply available, to make commercial fuel oil. The rest, $2000 - P$ gallons, is sent to a visbreaker unit which converts each gallon of pitch into 0.8 gallon of tar (see Figure 12). The tar may also be blended with flux stock to make fuel oil.

Blending requirements impose the constraints

$$5P + 11T + 37F \geq 21(P + T + F)$$

$$8P + 7T + 24F \geq 12(P + T + F)$$

where T is the amount of tar produced and F is the amount of flux stock used. Of course, $T = 0.8(2000 - P)$. Assuming that fuel oil brings a net return of 8¢ per gallon, and flux stock has a cracking value of 10¢ per gallon, find the blend of flux stock, pitch, and tar which will maximize the profit function

$$R = 8(P + T + F) - 10F$$

FIGURE 12

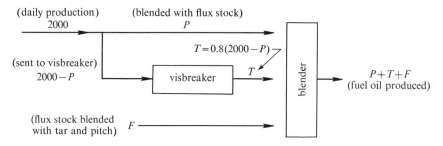

The following problems involve three variables. In each case, draw a three-dimensional picture showing the feasible region and the corner points.

9. A cotton farmer has been persuaded by his county agent to consider diversification as a means of increasing income. The local banker is will-

*Adapted from Symonds, G. H., *Linear Programming: The Solution of Refinery Problems* (Esso Standard Oil Company, New York, 1955).

ing to lend the farmer up to $6000 for the construction of facilities to handle livestock. The farmer has 80 acres of land and 3000 man-hours available per year to devote to farming.

The resource requirements for the various alternatives are as follows:

	Cotton (1 acre)	Dairy (1 cow)	Chickens (100 hens)
Labor (man-hours)	100	100	125
Land (acres)	1	2	0
Construction ($)	0	300	400

If the net unit returns are $400 per cow, $250 per 100 hens, and $200 per acre of cotton, how should the farmer allocate his resources in order to maximize his net return?

10. Rework Problem 9 with the net returns equal to $300 per cow, $400 per 100 hens, and $150 per acre of cotton.

11. A company has on hand 500 lb of peanuts, 150 lb of cashew nuts, and 75 lb of Brazil nuts. It packages and sells in eight-ounce cans three basic mixes of nuts. Type *A*, consisting of 50% peanuts, 40% cashews, and 10% Brazil nuts, sells for 39¢ per can; type *B* consists of cashew nuts only and sells for 80¢ per can; and type *C* consists of 40% cashews and 60% Brazil nuts and sells for 65¢ per can. What production schedule will provide maximum returns to the company?

SUPPLEMENTARY READING

Freund, J. E., *College Mathematics with Business Applications* (Prentice-Hall, Inc., Englewood Cliffs, N. J., 1969), Chapters 7, 8, and 10.

Gass, S. I., *Linear Programming: Methods and Applications* (McGraw-Hill Book Company, Inc., New York, 1964), 2nd ed.

Hohn, F. E., *Elementary Matrix Algebra* (Macmillan Company, New York, 1964).

CHARACTERISTIC EQUATIONS— QUADRATIC FORMS **10**

10.1 DETERMINANTS

If A is a square matrix, the *determinant* of A, written det A, is a number computed by combining the elements of A in a certain way. (Another common notation for the determinant of A is $|A|$.) For a 2×2 matrix

$$A = \begin{pmatrix} a & b \\ c & d \end{pmatrix}$$

the determinant is defined by

$$\det A = \det \begin{pmatrix} a & b \\ c & d \end{pmatrix} = ad - bc \tag{1}$$

For instance,

$$\det \begin{pmatrix} 1 & 2 \\ -1 & 3 \end{pmatrix} = (1)(3) - (2)(-1) = 3 + 2 = 5$$

The definition (1) is designed to mirror certain computations which arise in the solution of linear equations. Consider the pair of equations

$$ax + by = k$$
$$cx + dy = m$$

If we multiply both sides of the first equation by c and both sides of the second by a, we obtain

$$(ac)x + (bc)y = kc$$

$$(ac)x + (ad)y = am$$

Subtracting the first of these equations from the second in order to eliminate x, and solving the resulting equation for y gives

$$y = \frac{am - kc}{ad - bc} \tag{2}$$

Similar computations yield

$$x = \frac{kd - bm}{ad - bc} \tag{3}$$

A glance at the definition (1) shows that these solutions for x and y may be written as ratios of determinants

$$x = \frac{\det \begin{pmatrix} k & b \\ m & d \end{pmatrix}}{\det \begin{pmatrix} a & b \\ c & d \end{pmatrix}} \quad \text{and} \quad y = \frac{\det \begin{pmatrix} a & k \\ c & m \end{pmatrix}}{\det \begin{pmatrix} a & b \\ c & d \end{pmatrix}} \tag{4}$$

Equations (4) are called *Cramer's Rule** for solving linear equations.

We now wish to extend the definition of determinant from 2×2 matrices to larger square matrices (determinant is a function whose domain is the set of all square matrices). Rather than continuing to work with individual elements, as in (1), we adopt the strategy of describing the effect which elementary row operations on matrices have on the corresponding determinants. The properties of the determinant function *det* are given below.

Property 1   If matrix B is obtained from matrix A by multiplying one row of A by the constant m, then

$$\det B = m \cdot \det A$$

Multiplying a row by a constant multiplies the determinant by that same constant.   ▶

Property 2   If B is obtained from A by adding a multiple of the ith row of A to the jth row, then

$$\det B = \det A$$

Performing row operation R 2 on a matrix does not alter the value of the determinant.   ▶

*After the Swiss mathematician Gabriel Cramer (1704–1752).

Property 3 If B is obtained from A by interchanging two rows, then

$$\det B = -\det A$$

Interchanging rows changes the sign of the determinant. ▶

Property 4 The determinant of an identity matrix is 1.

$$\det I = 1$$ ▶

Computation of determinants is accomplished by performing row operations on the matrices in question.

Example 1 If A is the matrix

$$A = \begin{pmatrix} 2 & 1 & 3 \\ 1 & 0 & 1 \\ 1 & 2 & 5 \end{pmatrix}$$

then

$$\det A = \det \begin{pmatrix} 2 & 1 & 3 \\ 1 & 0 & 1 \\ 1 & 2 & 5 \end{pmatrix} = -\det \begin{pmatrix} 1 & 0 & 1 \\ 2 & 1 & 3 \\ 1 & 2 & 5 \end{pmatrix} \qquad \text{[Property 3]}$$

$$= -\det \begin{pmatrix} 1 & 0 & 1 \\ 0 & 1 & 1 \\ 0 & 2 & 4 \end{pmatrix} = -\det \begin{pmatrix} 1 & 0 & 1 \\ 0 & 1 & 1 \\ 0 & 0 & 2 \end{pmatrix} \qquad \begin{array}{l}\text{[Property 2,} \\ \text{applied twice]}\end{array}$$

$$= -2 \det \begin{pmatrix} 1 & 0 & 1 \\ 0 & 1 & 1 \\ 0 & 0 & 1 \end{pmatrix} = -2 \det \begin{pmatrix} 1 & 0 & 0 \\ 0 & 1 & 0 \\ 0 & 0 & 1 \end{pmatrix} \qquad \begin{array}{l}\text{[Property 1,} \\ \text{then} \\ \text{Property 2]}\end{array}$$

$$= -2 \det I = -2 \qquad \text{[Property 4]} \qquad ▶$$

Theorem 1 If one row of a square matrix A consists entirely of zeros, then $\det A = 0$.

PROOF Suppose the ith row of A contains only zeros and let B be the matrix obtained from A by multiplying this ith row by the constant 0. Of course, $B = A$ so $\det B = \det A$. On the other hand, Property 1 says that

$$\det B = 0 \cdot \det A = 0$$

It follows that $\det A = 0$. ▶

Example 2

$$\det \begin{pmatrix} 2 & 1 & 3 \\ 1 & 0 & 1 \\ 1 & 2 & 3 \end{pmatrix} = -\det \begin{pmatrix} 1 & 0 & 1 \\ 2 & 1 & 3 \\ 1 & 2 & 3 \end{pmatrix} \quad \text{[Property 3]}$$

$$= -\det \begin{pmatrix} 1 & 0 & 1 \\ 0 & 1 & 1 \\ 0 & 2 & 2 \end{pmatrix} \quad \text{[Property 2]}$$

$$= -\det \begin{pmatrix} 1 & 0 & 1 \\ 0 & 1 & 1 \\ 0 & 0 & 0 \end{pmatrix} \quad \text{[Property 2]}$$

$$= 0 \quad \text{[Theorem 1]} \quad \blacktriangleright$$

Since a square matrix can always be reduced by row operations either to an identity matrix or to a matrix containing a zero row, it is clear that Properties 1–4 do indeed determine a function which assigns to each square matrix a unique number. This number is nonzero if and only if the matrix can be reduced to an identity matrix. Coupling this with Theorem 3 in Section 9.3, we obtain the following important result.

Theorem 2 A square matrix is nonsingular (that is, has an inverse) if and only if its determinant is nonzero. Equivalently, a square matrix is singular if and only if its determinant is zero. $\blacktriangleright$

Example 3 (a) The matrix A in Example 1 is nonsingular and has the inverse

$$A^{-1} = \tfrac{1}{2} \begin{pmatrix} 2 & -1 & -1 \\ 4 & -7 & -1 \\ -2 & 3 & 1 \end{pmatrix}$$

(b) The matrix in Example 2 has no inverse since its determinant is zero. $\blacktriangleright$

A system of linear equations

$$AX = O \tag{5}$$

always has at least one solution, namely $X = O$. If A is square and nonsingular, then this is the only solution, as may be seen by multiplying both sides of (5) by A^{-1}.

On the other hand, if A is singular, then row operations performed on A to solve the equations will produce a zero row. This row may then be eliminated (as in Example 2 of Section 9.1) leaving a set of equations in which one unknown, say x_n, may be chosen arbitrarily and the others written as functions of x_n. In this case there are nonzero solutions to Equation (5). Together with Theorem 2 these considerations yield the following theorem.

Theorem 3   If A is a square matrix, the set of linear equations

$$AX = O$$

has a nonzero solution if and only if the determinant of A is zero.   ▶

Example 4   In Section 8.4 we saw that a vector X was an interior equilibrium for a closed Leontief system with input matrix A if

$$AX = X$$

Rewriting this equation as

$$X - AX = O \quad \text{or} \quad (I - A)X = O$$

we see that an interior equilibrium exists if and only if $\det (I - A) = 0$, that is, if and only if $I - A$ is singular.   ▶

We shall state and use without proof the following result.

Theorem 4   All statements concerning rows in Properties 1–4 of the determinant function and Theorem 1 hold for columns as well. In addition, if A is any square matrix, then

$$\det A = \det A'$$ ▶

PROBLEMS

Evaluate the following determinants:

1. $\det \begin{pmatrix} 10 & -1 \\ 6 & 2 \end{pmatrix}$

2. $\det \begin{pmatrix} -1 & 3 \\ 4 & 1 \end{pmatrix}$

3. $\det \begin{pmatrix} 4 & 2 & 3 \\ 6 & 1 & 0 \\ 0 & 2 & 1 \end{pmatrix}$

4. $\det \begin{pmatrix} 1 & 6 & 1 \\ 4 & 4 & 0 \\ 2 & 3 & 4 \end{pmatrix}$

5. $\det \begin{pmatrix} 0 & 1 & 2 & 3 \\ 2 & 1 & 5 & 2 \\ 1 & 2 & 0 & 2 \\ 2 & 2 & 3 & 0 \end{pmatrix}$

6. $\det \begin{pmatrix} a_{11} & 0 & 0 \\ a_{21} & a_{22} & 0 \\ a_{31} & a_{32} & a_{33} \end{pmatrix}$

7. Use Cramer's Rule to solve the simple national income model

$$Y = C + I_0 + G_0$$
$$C = a + bY$$

where I_0, G_0, a, and b are constants.

8. Let $U_i(i = 1, 2, 3)$ denote the utility of outcome i. Solve for u_1, u_2, and u_3 in the linear model

$$\begin{pmatrix} -\frac{2}{3} & \frac{6}{8} & -\frac{2}{5} \\ \frac{4}{3} & -\frac{1}{8} & \frac{3}{5} \\ \frac{1}{3} & \frac{3}{8} & \frac{4}{5} \end{pmatrix} \begin{pmatrix} u_1 \\ u_2 \\ u_3 \end{pmatrix} = \begin{pmatrix} u_1 \\ u_2 \\ u_3 \end{pmatrix}$$

9. Determine two different solutions for the linear equations

$$2x + 4y - 6z = 0$$
$$3x - 3y + 5z = 0$$
$$4x - y + 2z = 0$$

10. Use the results of Example 4 to rework Problem 12 of Section 9.1.

11. Use elementary row and/or column operations to establish the following equalities:

(a)
$$\det \begin{pmatrix} 1 & 1 & 1 \\ r & s & t \\ s+t & r+t & r+s \end{pmatrix} = 0$$

(b)
$$\det \begin{pmatrix} x-y & x-y & x^2-y^2 \\ 1 & 1 & x+y \\ y & 1 & x \end{pmatrix} = 0$$

(c)
$$\det \begin{pmatrix} x_1+y_1 & y_1+z_1 & z_1+x_1 \\ x_2+y_2 & y_2+z_2 & z_2+x_2 \\ x_3+y_3 & y_3+z_3 & z_3+x_3 \end{pmatrix} = 2 \det \begin{pmatrix} x_1 & x_2 & x_3 \\ y_1 & y_2 & y_3 \\ z_1 & z_2 & z_3 \end{pmatrix}$$

12. Prove that if two rows (columns) of matrix A are proportional, then $\det A = 0$.

13. (a) Show that for the 2×2 matrix

$$A = \begin{pmatrix} a & b \\ c & d \end{pmatrix}$$

Properties 1–4 of the determinant function give $\det A = ad - bc$, in agreement with Equation (1).

(b) Prove Theorem 4 for 2×2 matrices.

10.2 CHARACTERISTIC ROOTS AND CHARACTERISTIC VECTORS

The operation of multiplying a vector (row or column) by a matrix to produce a new vector has a variety of interpretations important for both theory and applications.

Example 1 The point $X = \text{col } (x_1, x_2)$ lies at a distance $r = \sqrt{x_1^2 + x_2^2}$ from the origin. Let α be the angle between the horizontal axis and the line from the origin to X (see Figure 1). From elementary trigonometry we know that

$$\cos \alpha = \frac{x_1}{r} \quad \text{and} \quad \sin \alpha = \frac{x_2}{r}$$

FIGURE 1

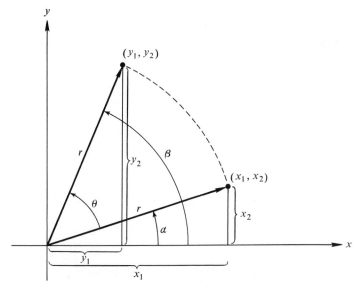

Now suppose this line is rotated through an angle θ so that the point (x_1, x_2) moves to a new position (y_1, y_2). The line from the origin through (y_1, y_2) makes an angle $\beta = \alpha + \theta$ with the x axis. Using standard trigonometric formulas we find

$$\frac{y_1}{r} = \cos \beta = \cos (\alpha + \theta) = \cos \alpha \cos \theta - \sin \alpha \sin \theta$$

$$= \frac{x_1}{r} \cos \theta - \frac{x_2}{r} \sin \theta$$

and

$$\frac{y_2}{r} = \sin \beta = \sin (\alpha + \theta) = \cos \alpha \sin \theta + \sin \alpha \cos \theta$$

$$= \frac{x_1}{r} \sin \theta + \frac{x_2}{r} \cos \theta$$

The two equations

$$y_1 = x_1 \cos \theta - x_2 \sin \theta$$

$$y_2 = x_1 \sin \theta + x_2 \cos \theta$$

may be written as the single matrix equation

$$Y = \begin{pmatrix} y_1 \\ y_2 \end{pmatrix} = \begin{pmatrix} \cos \theta & -\sin \theta \\ \sin \theta & \cos \theta \end{pmatrix} \begin{pmatrix} x_1 \\ x_2 \end{pmatrix} = AX \qquad (6)$$

▶

In Example 1, the matrix multiplication (6) is interpreted as producing a rotation of the vector X through an angle θ. Geometrically, the same result is obtained if the axes are rotated through an angle $-\theta$. This point of view is illustrated in the next example.

Example 2 In factor analysis (see Section 8.2), the elements in the jth row of the factor matrix F indicate the "loadings" of the jth test on the various factors (that is, the degrees to which the jth test measures the various factors). For instance, in the matrix

$$F = \begin{pmatrix} 0.1 & 0.2 \\ 0.3 & 0.3 \\ 0.5 & 0.4 \\ 0.6 & 0.7 \\ 0.8 & 0.6 \end{pmatrix} \qquad (7)$$

each of the five tests has approximately equal loadings on each of the two factors.

We may think of the rows of the factor matrix (7) as points in the plane. These points are plotted in Figure 2. Multiplication of F by the transpose of

$$R = \begin{pmatrix} \cos(-45°) & -\sin(-45°) \\ \sin(-45°) & \cos(-45°) \end{pmatrix} = \begin{pmatrix} 0.7071 & 0.7071 \\ -0.7071 & 0.7071 \end{pmatrix}$$

rotates the axes 45° and produces the new factor matrix

$$F^* = FR' = \begin{pmatrix} 0.2121 & 0.0707 \\ 0.4243 & 0 \\ 0.6364 & -0.0707 \\ 0.9192 & 0.0707 \\ 0.9899 & -0.1414 \end{pmatrix}$$

The result is to produce two new factors, which are combinations of the original ones, so as to have relatively large loadings on the first of these

FIGURE 2

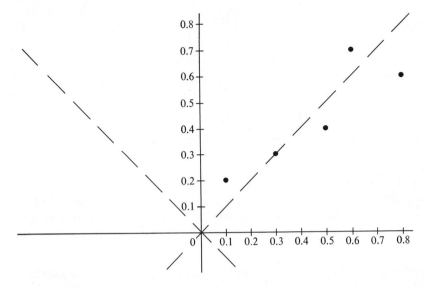

factors and small loadings on the second. Each test may thus be regarded as essentially measuring a single common factor as opposed to the more complex bifactor interpretation suggested by the original F matrix. ▶

Of particular interest relative to a square matrix A are those nonzero vectors X for which there is a constant λ such that

$$AX = \lambda X \tag{8}$$

The number λ in (8) is called a *characteristic root* of the matrix A, and X is called a *characteristic vector* of A. (Characteristic roots are also called "characteristic values," "eigenvalues," or "latent roots." Characteristic vectors are also called "eigenvectors," "latent vectors," or "invariant vectors.") Equation (8) says that multiplication of X by A produces a new vector which is just a multiple of X itself. Geometrically, multiplication of X by A produces no rotation at all but moves X along the same straight line through the origin, either lengthening it ($|\lambda| > 1$) or shrinking it ($|\lambda| < 1$), and perhaps reversing its direction ($\lambda < 0$). (See Example 1 of Section 8.2.)

Example 3   Relative to Equation (6), it is obvious geometrically that unless θ is zero or a multiple of π (a number of half-revolutions), no vector except $(0, 0)$ can be transformed into a multiple of itself. ▶

Example 4   Recall (Section 8.4) that a closed Leontief system with input matrix A has an interior equilibrium X if and only if $X \geq O$ and

$$AX = X \tag{9}$$

This corresponds to Equation (8) with $\lambda = 1$. In economic terms, (9) represents the situation where production equals input demands for each commodity.

▶

The problem of finding characteristic roots and the corresponding characteristic vectors is a problem of solving Equation (8) for λ and for X, subject to the restriction that X be nonzero. If we rewrite (8) as

$$(A - \lambda I)X = O$$

and apply Theorem 3, we see that λ must be chosen so that

$$\det (A - \lambda I) = 0 \qquad\qquad (10)$$

Equation (10) is called the *characteristic equation* for the matrix A and the determinant $\det (A - \lambda I)$ is called the *characteristic polynomial* for A. If A is an $n \times n$ matrix, this is an nth degree polynomial in λ. It follows that an $n \times n$ matrix has n (not necessarily different) characteristic roots.

Example 5 The characteristic equation $\det (A - \lambda I) = 0$ for the matrix

$$A = \begin{pmatrix} 1 & 2 \\ 2 & 1 \end{pmatrix}$$

is

$$\det \begin{pmatrix} 1 - \lambda & 2 \\ 2 & 1 - \lambda \end{pmatrix} = \lambda^2 - 2\lambda - 3 = 0 \qquad\qquad (11)$$

The solutions of (11) provide the two characteristic roots

$$\lambda_1 = -1 \quad \text{and} \quad \lambda_2 = 3$$

Solving the equation

$$(A - \lambda_1 I)X = O \quad \text{or} \quad \begin{pmatrix} 2 & 2 \\ 2 & 2 \end{pmatrix} \begin{pmatrix} x_1 \\ x_2 \end{pmatrix} = \begin{pmatrix} 0 \\ 0 \end{pmatrix}$$

shows that all characteristic vectors of A corresponding to $\lambda_1 = -1$ are multiples of the vector $\text{col} (-1, 1)$. Similarly, characteristic vectors of A which correspond to $\lambda_2 = 3$ have the form $X = \text{col} (c, c)$, where c is some constant. These are found by solving

$$(A - 3I)X = O \qquad\qquad ▶$$

PROBLEMS

    1. Verify the computations in Example 5.

A matrix obtained from a matrix C by eliminating various rows and columns of C is called a *minor matrix* of C. The determinant of a square minor matrix is called a *minor determinant* of C. If C is square, the determinant of a square minor matrix formed by choosing elements in the same numbered rows and columns is a *principal minor determinant* of C. The following theorem, which we do not prove, tells how to write the characteristic equation of a matrix in terms of its principal minor determinants.

Theorem 5 Let p_k denote the sum of all $k \times k$ principal minor determinants of an $n \times n$ matrix A. Then

$$
\det (A - \lambda I)
$$
$$
= (-\lambda)^n + p_1(-\lambda)^{n-1} + p_2(-\lambda)^{n-2} + \cdots + p_{n-1}(-\lambda) + p_n \quad \blacktriangleright
$$

Example 6 For the matrix

$$
A = \begin{pmatrix} 1 & 2 \\ 2 & 1 \end{pmatrix}
$$

in Example 5, the 1×1 principal minor determinants are formed by choosing each of the elements on the main diagonal in turn. There are two such determinants, each with value 1. The only 2×2 principal minor determinant is $\det A$. According to Theorem 5, the characteristic polynomial of A is

$$
\det (A - \lambda I) = (-\lambda)^2 + (1 + 1)(-\lambda) + \det A
$$
$$
= \lambda^2 - 2\lambda - 3
$$

as before. $\qquad\qquad\qquad\qquad\qquad\qquad\qquad\qquad\qquad\qquad\blacktriangleright$

Example 7 The matrix

$$
M = \begin{pmatrix} 3 & 4 & 10 \\ -1 & 2 & 0 \\ 2 & 1 & 5 \end{pmatrix}
$$

has first-order principal minor determinants

$$
\det (3) = 3 \quad \det (2) = 2 \quad \text{and} \quad \det (5) = 5
$$

second-order principal minors

$$
\det \begin{pmatrix} 3 & 4 \\ -1 & 2 \end{pmatrix} = 10 \quad \det \begin{pmatrix} 3 & 10 \\ 2 & 5 \end{pmatrix} = -5 \quad \text{and} \quad \det \begin{pmatrix} 2 & 0 \\ 1 & 5 \end{pmatrix} = 10
$$

and the third-order principal minor $\det M = 0$. The characteristic polynomial is

$$
\det (M - \lambda I) = (-\lambda)^3 + (3 + 2 + 5)(-\lambda)^2 + (10 - 5 + 10)(-\lambda) + 0
$$

The characteristic roots, found by solving $\det (M - \lambda I) = 0$, are $\lambda_1 = 0$, $\lambda_2 = 5 + \sqrt{10}$, and $\lambda_3 = 5 - \sqrt{10}$. $\qquad\qquad\qquad\blacktriangleright$

In Example 4 we saw that in order for a closed Leontief system to have an interior equilibrium, the number 1 must be a characteristic root of the corresponding input matrix A and there must be a non-negative characteristic vector corresponding to this root. This is not always the case. A general result which is useful in this context is the following theorem, which we do not prove.

Theorem 6 Suppose A is a square matrix with non-negative entries. Then
  (a) there is a unique positive characteristic root r which is larger in absolute value than any other root.
  (b) the root r in (a) lies between the largest and smallest of the row (or column) sums of A.
  (c) there are positive characteristic vectors corresponding to r. ▶

From this theorem, it follows that if the row (or column) sums of A are all unity, a closed Leontief system does indeed have an interior equilibrium. For instance, the input matrix

$$A = \begin{pmatrix} 0.3 & 0.2 & 0.1 \\ 0.6 & 0.3 & 0.1 \\ 0.1 & 0.5 & 0.8 \end{pmatrix}$$

has unit column sums and characteristic roots

$$1, \quad 0.2 + \sqrt{0.05} \approx 0.4236, \text{ and } 0.2 - \sqrt{0.05} \approx -0.0236$$

Any characteristic vector X associated with the root 1 has the form

$$X = \text{col}\left(\tfrac{9}{13}c, \, c, \, \tfrac{37}{13}c\right)$$

which is strictly positive whenever $c > 0$.

PROBLEMS

In each of Problems 2–9, determine the characteristic roots and associated characteristic vectors of the given matrix.

2. $\begin{pmatrix} -1 & 4 \\ 1 & 2 \end{pmatrix}$ **3.** $\begin{pmatrix} 4 & -4 \\ -4 & 4 \end{pmatrix}$ **4.** $\begin{pmatrix} -1 & 5 \\ 1 & 3 \end{pmatrix}$ **5.** $\begin{pmatrix} 3 & 6 \\ 12 & 9 \end{pmatrix}$

6. $\begin{pmatrix} 1 & 0 & 1 \\ 0 & 1 & 1 \\ 1 & 1 & 2 \end{pmatrix}$ **7.** $\begin{pmatrix} 2 & -3 & 4 \\ 0 & 2 & 1 \\ 0 & 1 & 2 \end{pmatrix}$ **8.** $\begin{pmatrix} 1 & -1 & 1 \\ 0 & 1 & 0 \\ 1 & -1 & 1 \end{pmatrix}$

9. $\begin{pmatrix} 3 & -3 & 6 & 6 \\ 6 & -3 & 12 & 6 \\ -3 & 3 & -3 & -3 \\ 6 & -3 & 6 & 3 \end{pmatrix}$

10. Find the characteristic roots of

(a) $\begin{pmatrix} 1 & 0 \\ 0 & 2 \end{pmatrix}$
(b) $\begin{pmatrix} a & 0 \\ 0 & b \end{pmatrix}$

11. What are the characteristic roots of $D = \text{diag}\,(\alpha_1, \alpha_2, \ldots, \alpha_n)$?

12. Find three different 2×2 matrices whose characteristic roots are 1 and 4.

13. Calculate the characteristic roots for the matrix

$$\begin{pmatrix} a & b \\ c & d \end{pmatrix}$$

Under what conditions are the roots

(a) real and unequal?   (b) real and equal?   (c) complex?

14. In Problem 2, the characteristic equation was found to be $\lambda^2 - \lambda - 6 = 0$. Verify that

$$\begin{pmatrix} -1 & 4 \\ 1 & 2 \end{pmatrix}^2 - \begin{pmatrix} -1 & 4 \\ 1 & 2 \end{pmatrix} - 6 \begin{pmatrix} 1 & 0 \\ 0 & 1 \end{pmatrix} = \begin{pmatrix} 0 & 0 \\ 0 & 0 \end{pmatrix}$$

In a sense, this matrix satisfies its own characteristic equation.

15. In Example 2, rotate the points by $60°$ and compute the new $F*$ matrix.

16. The *rank* of a matrix A is the *order* of the largest-order nonzero minor determinant of A. The rank of a zero matrix is zero. For instance, the matrix M in Example 7 has rank 2 since there are 2×2 minor determinants which are not zero, but the only 3×3 minor determinant, det M, is zero.

(a) Determine the ranks of the matrices in Problems 2, 6, 7 and 9.

(b) Determine the ranks of

$$\begin{pmatrix} 0 & 0 & 0 \\ 0 & 0 & 0 \end{pmatrix} \quad \text{and} \quad \begin{pmatrix} 1 & 0 & 0 & 0 \\ 0 & 0 & 0 & 1 \\ 0 & 1 & 0 & 0 \end{pmatrix}$$

17. Argue that performing elementary operations on a matrix cannot change the rank of that matrix.

18. Suppose that a system of four linear equations in seven variables has the coefficient matrix partitioned as shown (the a_{ij} are all nonzero).

$$A = \begin{pmatrix} a_{11} & a_{12} & a_{13} & a_{14} & 0 & 0 & 0 \\ a_{21} & 0 & a_{23} & 0 & a_{25} & a_{26} & 0 \\ 0 & a_{32} & a_{33} & 0 & a_{35} & a_{36} & 0 \\ a_{41} & a_{42} & 0 & a_{44} & a_{45} & 0 & a_{47} \end{pmatrix} = \begin{pmatrix} B & O \\ C & D \end{pmatrix}$$

Use the result of Problem 17 to argue that the first equation is identifiable (see Section 9.2) if and only if the 3×3 matrix D has rank 3.

10.3 SYMMETRIC MATRICES AND QUADRATIC FORMS

A matrix $A = (a_{ij})$ is *symmetric* if it is equal to its own transpose, that is, if

$$A' = A \tag{12}$$

For example, the matrices

$$\begin{pmatrix} 0 & 2 \\ 2 & 7 \end{pmatrix} \qquad \begin{pmatrix} 1 & 3 & 4 \\ 3 & 0 & 2 \\ 4 & 2 & -1 \end{pmatrix} \qquad \begin{pmatrix} 1 & 0.3 & 0.2 & 0 \\ 0.3 & 1 & 0.6 & -0.1 \\ 0.2 & 0.6 & 1 & 0.2 \\ 0 & -0.1 & 0.2 & 1 \end{pmatrix}$$

are symmetric.

In terms of rows and columns, Equation (12) says that the first row and first column of a symmetric matrix must contain the same entries in the same order; the same is true of the second row and second column, etc. In terms of elements, (12) implies that the element $(A')_{ij} = a_{ji}$ in the ij position of A' must equal the corresponding element a_{ij} of A. Thus if A is symmetric, we have

$$a_{ji} = a_{ij}$$

for all i and j. The elements which are symmetric relative to the main diagonal must be equal, as in the above examples.

Example 1 In Example 6, Section 8.2 we defined a correlation matrix $R = (r_{jk})$ in which an entry r_{jk} denoted the correlation between tests j and k. Since by definition

$$r_{jk} = \frac{1}{n} \sum_{i=1}^{n} s_{ji} s_{ki} = \frac{1}{n} \sum_{i=1}^{n} s_{ki} s_{ji} = r_{kj}$$

the matrix R is symmetric. ▶

Example 2 Define a relation R on a set of research projects by iRj if and only if project i is related to project j. Since R is a symmetric relation (see Section 3.2), its adjacency matrix $C = (c_{ij})$ defined by

$$c_{ij} = \begin{cases} 1 & \text{if } iRj \\ 0 & \text{otherwise} \end{cases}$$

must be a symmetric matrix. ▶

If $A = (a_{ij})$ is an $n \times n$ symmetric matrix and $X = \text{col}\,(x_1, x_2, \ldots, x_n)$ is an n-dimensional vector, the expression

$$Q(X) = X'AX \tag{13}$$

is called a *quadratic form* in the variables $x_1, x_2, \ldots, x_n$. Performing the multiplication in (13) we find

$$X'AX = (x_1, x_2, \ldots, x_n) \begin{pmatrix} a_{11} & a_{12} & \cdots & a_{1n} \\ a_{21} & a_{22} & \cdots & a_{2n} \\ \vdots & & & \\ a_{n1} & a_{n2} & \cdots & a_{nn} \end{pmatrix} \begin{pmatrix} x_1 \\ x_2 \\ \vdots \\ x_n \end{pmatrix}$$

$$= (x_1, x_2, \ldots, x_n) \begin{pmatrix} \sum_{j=1}^{n} a_{1j}x_j \\ \sum_{j=1}^{n} a_{2j}x_j \\ \vdots \\ \sum_{j=1}^{n} a_{nj}x_j \end{pmatrix}$$

$$= \sum_{i=1}^{n} x_i \left(\sum_{j=1}^{n} a_{ij}x_j \right) = \sum_{i=1}^{n} \sum_{j=1}^{n} a_{ij}x_i x_j$$

For each ij pair, the coefficient of $x_i x_j$ is the element in the ij position of A.

Example 3 If

$$A = \begin{pmatrix} 1 & 2 \\ 2 & 3 \end{pmatrix}$$

then the quadratic form determined by A is

$$Q(X) = X'AX$$

$$= (x_1, x_2) \begin{pmatrix} 1 & 2 \\ 2 & 3 \end{pmatrix} \begin{pmatrix} x_1 \\ x_2 \end{pmatrix} \tag{14}$$

$$= (x_1, x_2) \begin{pmatrix} x_1 + 2x_2 \\ 2x_1 + 3x_2 \end{pmatrix}$$

$$= x_1(x_1 + 2x_2) + x_2(2x_1 + 3x_2)$$

$$= x_1^2 + 2x_1 x_2 + 2x_2 x_1 + 3x_2^2 \tag{15}$$

$$= x_1^2 + 4x_1 x_2 + 3x_2^2 \tag{16}$$

Note carefully how the elements of the matrix A enter the sum (15). ▶

Example 4 Suppose, in the factor analytic setting of Example 6, Section 8.2 we wish to weight the tests differently. Let w_j be the weight assigned to test j and let $W = \text{col} (w_1, w_2, \ldots, w_t)$ be the vector of these weights. Then the *composite score* y_i for individual i is

$$y_i = w_1 s_{1i} + w_2 s_{2i} + \cdots + w_t s_{ti}$$

computed by multiplying the observed test scores for individual i by the test weights, and summing. In matrix notation

$$Y = W'S$$

where $Y = (y_1, y_2, \ldots, y_n)$ is the row vector of composite scores and $S = (s_{ji})$ is the matrix of observed scores.

The *average squared composite score* (in statistical terms, the "sample variance" of the composite scores) is

$$\sigma_Y^2 = \frac{1}{n}\sum_{i=1}^{n} y_i^2 = \frac{1}{n}YY' = \frac{1}{n}W'SS'W$$

Since $SS' = nR$, where R is the symmetric correlation matrix, we have

$$\sigma_Y^2 = \frac{1}{n}W'nRW = W'RW$$

a quadratic form in the weights. ▶

If we began with Equation (16) in Example 3 and wished to put it into matrix form, we could simply reverse the steps taken in going from (14) to (16). The reason for going first to (15) is to make sure that the resulting form involves a symmetric matrix. For instance, the matrix product

$$(x_1, x_2)\begin{pmatrix} 1 & 4 \\ 0 & 3 \end{pmatrix}\begin{pmatrix} x_1 \\ x_2 \end{pmatrix}$$

is exactly the same as (14), (15), and (16), but the coefficient matrix in the product is not symmetric.

One major reason for using only symmetric matrices in representing quadratic forms is that if we make a *change of variable*, or *linear transformation*, of the form

$$X = CY$$

where C is a nonsingular $n \times n$ matrix, so chosen that Y may be expressed in terms of X as $Y = C^{-1}X$, the quadratic form (13) becomes

$$X'AX = (CY)'ACY = Y'(C'AC)Y$$

This is a new quadratic form involving the vector $Y = \text{col}(y_1, y_2, \ldots, y_n)$. If A is symmetric, that is, if $A' = A$, then

$$(C'AC)' = C'A'(C')' = C'AC$$

so that the matrix $C'AC$ of the new form is symmetric also.

Example 5 If A is the symmetric matrix

$$A = \begin{pmatrix} 1 & 2 \\ 2 & 1 \end{pmatrix} \quad \text{and} \quad C = \begin{pmatrix} 2 & -1 \\ 1 & 2 \end{pmatrix}$$

then

$$C'AC = \begin{pmatrix} 2 & 1 \\ -1 & 2 \end{pmatrix}\begin{pmatrix} 1 & 2 \\ 2 & 1 \end{pmatrix}\begin{pmatrix} 2 & -1 \\ 1 & 2 \end{pmatrix} = \begin{pmatrix} 13 & 6 \\ 6 & -3 \end{pmatrix}$$

is also symmetric. ▶

The simplest quadratic form is one involving only squared terms. The matrix of such a form is diagonal. It is an important fact that by making an appropriate change of variable, any quadratic form can be changed into a sum of squares. In order to tell how, we need some special terminology. Two *n*-dimensional *vectors*

$$X = \text{col}(x_1, x_2, \ldots, x_n) \quad \text{and} \quad Y = \text{col}(y_1, y_2, \ldots, y_n)$$

are called *orthogonal* if

$$X'Y = x_1 y_1 + x_2 y_2 + \cdots + x_n y_n = 0 \tag{17}$$

That is, X and Y are orthogonal if multiplying corresponding elements of the two vectors, and adding, gives zero. For instance, $X = \text{col}(2, 1, -1)$ and $Y = \text{col}(1, 2, 4)$ are orthogonal since

$$X'Y = (2)(1) + (1)(2) - (1)(4) = 0$$

A *matrix* is called *orthogonal* if each row (or column) is orthogonal to every other row (or column) and if in each row (or column) the sum of squares of the elements is unity. That is, the rows (or columns) are orthogonal to each other and are each of unit length. Note that *one* matrix is orthogonal, but *two* vectors are orthogonal. For instance, the rotation matrix

$$R = \begin{pmatrix} \cos\theta & -\sin\theta \\ \sin\theta & \cos\theta \end{pmatrix}$$

in Example 1 of Section 10.2, is an orthogonal matrix. The most important property of orthogonal matrices is the following theorem.

Theorem 7 The inverse of an orthogonal matrix is the same as its transpose. Symbolically, if Q is orthogonal, then

$$Q' = Q^{-1}$$

PROOF Since the *j*th column of Q' is the *j*th row of Q, the *ij* element in the product

$$QQ'$$

is the sum of the products of corresponding elements in the *i*th and *j*th rows of Q. The definition of orthogonal matrix requires that this sum of products is zero if $i \neq j$ and one if $i = j$. Hence $QQ' = I$ and $Q' = Q^{-1}$. ▶

Theorem 8 If A is a symmetric matrix, then there is an orthogonal matrix Q such that $D = Q'AQ$ is a diagonal matrix. The diagonal entries in D are the characteristic roots of A. The columns of Q are characteristic vectors of A corresponding to the respective characteristic roots in D. ▶

We omit the proof of Theorem 8, contenting ourselves with some examples of its use.

Example 6   The characteristic roots of the symmetric matrix

$$A = \begin{pmatrix} 1 & 2 \\ 2 & 1 \end{pmatrix}$$

are $\lambda_1 = 3$ and $\lambda_2 = -1$. Characteristic vectors corresponding to these roots have the respective forms

$$V_1 = a \begin{pmatrix} 1 \\ 1 \end{pmatrix} \quad \text{and} \quad V_2 = b \begin{pmatrix} -1 \\ 1 \end{pmatrix}$$

where a and b are any constants. Note that the vectors are orthogonal, regardless of the values of a and b. The sum of squares of the elements in each vector will be unity if we choose $a = b = 1/\sqrt{2}$. Then, according to Theorem 8, the orthogonal matrix

$$Q = \begin{pmatrix} 1/\sqrt{2} & -1/\sqrt{2} \\ 1/\sqrt{2} & 1/\sqrt{2} \end{pmatrix}$$

is such that

$$Q'AQ = D = \text{diag}\,(3, -1)$$

Equivalently, if in the quadratic form

$$X'AX$$

we make the change of variable $X = QY$, we obtain the new quadratic form

$$Y'(Q'AQ)Y = Y'DY = 3y_1^2 - y_2^2$$

a sum of squares. ▶

PROBLEMS

1. Work out the computational details to obtain the results in Example 6.

Example 7   Factor analysis is an important tool in the behavioral sciences and is based on matrix concepts which we have introduced. We consider various facets of factor analysis here.

(a) Theorem 8 guarantees that, given any correlation matrix R, there is a matrix F of factor loadings such that

$$R = FF' \tag{18}$$

To determine F, we first find an orthogonal matrix C such that

$$C'RC = D = \text{diag}(\lambda_1, \lambda_2, \ldots, \lambda_n) \tag{19}$$

where λ_i are the characteristic roots of R. Writing (19) in the equivalent form

$$CC'RCC' = CDC'$$

and invoking Theorem 7 gives

$$R = CDC'$$

We now define the *square root* $D^{1/2}$ *of the matrix* D by

$$D^{1/2} = \text{diag}(\lambda_1^{1/2}, \lambda_2^{1/2}, \ldots, \lambda_n^{1/2})$$

Then

$$R = CD^{1/2}(D^{1/2})'C'$$

and we take

$$F = CD^{1/2}$$

(b) Choosing $F = CD^{1/2}$ as in (a) gives $R = FF'$. From Example 6 in Section 8.2 we know that the matrices S of observed scores and P of standard scores are related by

$$S = FP \tag{20}$$

and that, by definition,

$$R = \frac{1}{n} SS' \tag{21}$$

If the matrix F is square (that is, the number of tests equals the number of factors) and nonsingular, then the matrix Z of interfactor correlations is

$$Z = \frac{1}{n} PP' = \frac{1}{n}[F^{-1}S][S'(F')^{-1}] \qquad \text{[by (20)]}$$

$$= F^{-1}\left(\frac{1}{n} SS'\right)(F^{-1})'$$

$$= F^{-1}R(F^{-1})' \qquad \text{[by (21)]}$$

$$= I \qquad \text{[by (18)]}$$

In this case, the factors are uncorrelated.

(c) As a numerical illustration of the factorization of a correlation matrix, consider

$$R = \begin{pmatrix} 1 & 0.68 & 0.45 \\ 0.68 & 1 & 0.36 \\ 0.45 & 0.36 & 1 \end{pmatrix}$$

The characteristic equation of R is

$$\det (R - \lambda I) = (-\lambda)^3 + (1 + 1 + 1)(-\lambda)^2$$
$$+ (0.5376 + 0.7975 + 0.8704)(-\lambda) + 0.4258$$
$$= -\lambda^3 + 3\lambda^2 - 2.2055\lambda + 0.4258$$
$$= 0$$

Solving this equation we find (to four decimal places) the characteristic roots

$$\lambda_1 = 2.0067 \quad \lambda_2 = 0.6823 \quad \lambda_3 = 0.3110$$

Corresponding characteristic vectors having unit length are

$$X_1 = \begin{pmatrix} 0.627 \\ 0.600 \\ 0.496 \end{pmatrix} \quad X_2 = \begin{pmatrix} 0.243 \\ 0.454 \\ -0.857 \end{pmatrix} \quad X_3 = \begin{pmatrix} -0.740 \\ 0.658 \\ 0.140 \end{pmatrix}$$

If we use these vectors as columns of an orthogonal matrix C and write $D = \text{diag}\,(\lambda_1, \lambda_2, \lambda_3)$, we find

$$F = CD^{1/2} = C \cdot \text{diag}\,(\lambda_1^{1/2}, \lambda_2^{1/2}, \lambda_3^{1/2})$$

$$= \begin{pmatrix} 0.627 & 0.243 & -0.740 \\ 0.600 & 0.454 & 0.658 \\ 0.496 & -0.857 & 0.140 \end{pmatrix} \begin{pmatrix} 1.417 & 0 & 0 \\ 0 & 0.826 & 0 \\ 0 & 0 & 0.558 \end{pmatrix}$$

$$= \begin{pmatrix} 0.888 & 0.201 & -0.413 \\ 0.850 & 0.357 & 0.367 \\ 0.703 & 0.707 & 0.078 \end{pmatrix} \qquad \blacktriangleright$$

If A is symmetric and $Q'AQ = D$ is a diagonal matrix containing the characteristic roots of A, then the linear transformation $X = QY$ changes the quadratic form $X'AX$ into a sum of squares. For

$$X'AX = (Y'Q')A(QY)$$
$$= Y'(Q'AQ)Y$$
$$= Y'DY$$

In summation form,

$$X'AX = Y'DY = \lambda_1 y_1^2 + \lambda_2 y_2^2 + \cdots + \lambda_n y_n^2 = \sum_{i=1}^{n} \lambda_i y_i^2 \qquad (22)$$

where $\lambda_1, \ldots, \lambda_n$ are the characteristic roots of A.

If all the λ_i are positive in (22), then no matter what values for X are introduced, the quadratic form cannot be negative. Such a quadratic form is called *positive definite*. Similarly, a quadratic form is called *negative definite* if all λ_i are negative, and *indefinite* if some λ_i are positive and some negative. In the former case, the form has negative values regardless of the choice of X, while in the latter, some choices of X make the form positive and some make it negative.

Example 8 The correlation matrix R in Example 7 is positive definite since all characteristic roots are positive. The matrix A in Example 6 is indefinite. ▶

PROBLEMS

In Problems 2–5, express the given quadratic form as a sum.

2.
$$(x, y) \begin{pmatrix} 1 & 4 \\ 4 & 2 \end{pmatrix} \begin{pmatrix} x \\ y \end{pmatrix}$$

3.
$$(u, v) \begin{pmatrix} -2 & 3 \\ 3 & -10 \end{pmatrix} \begin{pmatrix} u \\ v \end{pmatrix}$$

4.
$$(x_1, x_2, x_3) \begin{pmatrix} 4 & 4 & 2 \\ 4 & 4 & 2 \\ 2 & 2 & 1 \end{pmatrix} \begin{pmatrix} x_1 \\ x_2 \\ x_3 \end{pmatrix}$$

5.
$$(y_1, y_2, y_3, y_4) \begin{pmatrix} 0 & 1 & 2 & 3 \\ 1 & 0 & 1 & 2 \\ 2 & 1 & 0 & 1 \\ 3 & 2 & 1 & 0 \end{pmatrix} \begin{pmatrix} y_1 \\ y_2 \\ y_3 \\ y_4 \end{pmatrix}$$

In Problems 6–9, express the given quadratic form as a matrix product.

6. $2u^2 - 8uv + 6v^2$

7. $3x^2 + 16xy - 2y^2$

8. $12xy - 4x^2 - y^2$

9. $x^2 + 2y^2 + z^2 + 2xy + 2xz$

10. Verify that if A is the symmetric matrix

$$A = \begin{pmatrix} -1 & -2 & 1 \\ -2 & 2 & -2 \\ 1 & -2 & -1 \end{pmatrix}$$

and R is the matrix

$$\begin{pmatrix} 1/\sqrt{6} & 1/\sqrt{3} & 1/\sqrt{2} \\ -2/\sqrt{6} & 1/\sqrt{3} & 0 \\ 1/\sqrt{6} & 1/\sqrt{3} & -1/\sqrt{2} \end{pmatrix}$$

then $R'AR = \text{diag}(4, -2, -2)$.

In Problems 11–14, find an orthogonal matrix Q such that $Q'AQ = D$, where D is diagonal.

11.
$$A = \begin{pmatrix} 1 & -1 \\ -1 & 1 \end{pmatrix}$$

12.
$$A = \begin{pmatrix} 2 & 1 \\ 1 & 2 \end{pmatrix}$$

13.
$$A = \begin{pmatrix} 4 & 2 & -2 \\ 2 & 4 & -2 \\ -2 & -2 & -1 \end{pmatrix}$$

14.
$$A = \begin{pmatrix} 5 & 2 & 1 \\ 2 & 1 & 0 \\ 1 & 0 & 1 \end{pmatrix}$$

15. Diagonalize the matrix

$$\begin{pmatrix} 4 & 4 & -2 \\ 4 & 4 & -2 \\ -2 & -2 & 1 \end{pmatrix}$$

(*Hint:* The number zero is a repeated characteristic root. Make sure the two characteristic vectors you choose are orthogonal.)

16. Verify that the C matrix of Example 7 is indeed orthogonal.

17. As in Example 7, factor the following correlation matrices finding (1) the characteristic roots; (2) the orthogonal matrix C; and (3) the factor matrix F. Verify that $FF' = R$. (*Hint:* In (a), 2 is a root.)

(a)
$$R = \begin{pmatrix} 1 & -0.4 & -0.68 \\ -0.4 & 1 & 0.4 \\ -0.68 & 0.4 & 1 \end{pmatrix}$$

(b)
$$R = \begin{pmatrix} 1 & 0.6 & 0.8 \\ 0.6 & 1 & 0 \\ 0.8 & 0 & 1 \end{pmatrix}$$

18. Determine which of the quadratic forms in Problems 2–9 are

(a) positive definite   (b) negative definite   (c) neither

19. A matrix B is said to be *skew-symmetric* if it is equal to the negative of its transpose, that is, if $B' = -B$. The matrices

$$\begin{pmatrix} 0 & -1 \\ 1 & 0 \end{pmatrix} \quad \text{and} \quad \begin{pmatrix} 0 & 4 & 3 \\ -4 & 0 & -2 \\ -3 & 2 & 0 \end{pmatrix}$$

are skew-symmetric. On the other hand the matrix

$$\begin{pmatrix} 1 & -1 \\ 1 & 1 \end{pmatrix}$$

is not skew-symmetric. Why?

20. If the $r \times r$ matrix A is symmetric (or skew-symmetric), and P is any $r \times c$ matrix, prove that $P'AP$ is symmetric (or skew-symmetric).

21. Prove that if P is any $r \times c$ matrix, then $P'P$ is symmetric.

22. Must all symmetric and skew-symmetric matrices necessarily be square? Why or why not?

23. Prove that if A is symmetric (or skew-symmetric) and if c is any scalar, then cA is symmetric (or skew-symmetric).

24. Prove that if A is any square matrix, then $S = A + A'$ is symmetric and $T = A - A'$ is skew-symmetric.

25. It is sometimes useful to diagonalize matrices which are not symmetric. In this case the matrix of characteristic vectors is ordinarily not orthogonal. For instance, suppose we have a closed Leontief system with input matrix

$$A = \begin{pmatrix} 1 - \alpha & \beta \\ \alpha & 1 - \beta \end{pmatrix}$$

where α and β lie between zero and one. Suppose further that for $t = 0$, $1, 2, \ldots$ the vector $X(t + 1)$ of system outputs in the $(t + 1)$st time period is related to the vector $X(t)$ by

$$X(t + 1) = AX(t)$$

(a) Prove that for all t,

$$X(t) = A^t X(0)$$

(b) Find the matrix $D = \mathrm{diag}\,(\lambda_1, \lambda_2)$, where λ_1 and λ_2 are the characteristic roots of A.

(c) Find a matrix $C = (V_1, V_2)$ where V_1 and V_2 are characteristic vectors corresponding to λ_1 and λ_2.

(d) Verify that $C^{-1}AC = D$ and hence that $A = CDC^{-1}$.

(e) Prove that $A^t = CD^tC^{-1}$.

(f) Prove that

$$\lim_{t \to \infty} X(t) = C \begin{pmatrix} 1 & 0 \\ 0 & 0 \end{pmatrix} C^{-1} X(0)$$

and hence that this limit is

$$\frac{1}{\alpha + \beta} \begin{pmatrix} \beta & \beta \\ \alpha & \alpha \end{pmatrix} X(0)$$

SUPPLEMENTARY READING

Hohn, F. E., *Elementary Matrix Algebra* (Macmillan Company, New York, 1964).

Thurstone, L. L., *Multiple-Factor Analysis* (University of Chicago Press, Chicago, 1947).

Zelinsky, D., *A First Course in Linear Algebra* (Academic Press, Inc., New York, 1968).

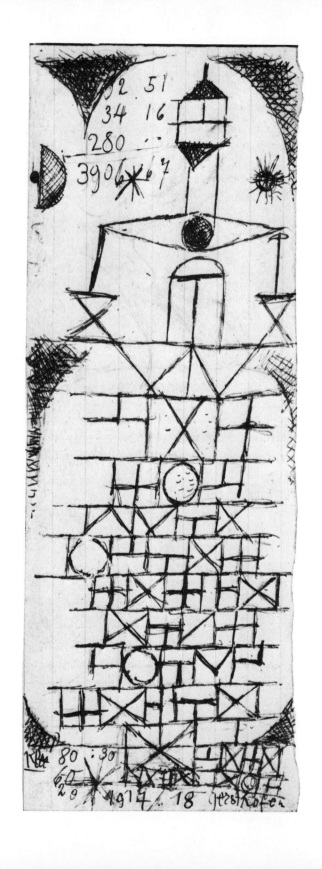

PART III

Calculus

FUNCTIONS, LIMITS, AND CONTINUITY **11**

11.1   REAL–VALUED FUNCTIONS

Recall (Section 4.1) that a function associates with each element of a certain set, called the domain of the function, an element of another set, called the range of the function. If the function f associates the element b in the range of f with the element a in the domain of f, then b is called the value of f at a and we write $b = f(a)$. A function whose range is a set of real numbers is called a *real-valued function*. Our study of calculus will be concerned with real-valued functions.

Figures 1 and 2 are graphs of two common real-valued functions, the *squaring function* $g(x) = x^2$ and the *square root function* $f(x) = \sqrt{x}$. Here we use the standard shorthand method of indicating a real-valued function by writing a formula for computing its values. Strictly speaking, the squaring function g is the set

$$g = \{(x, y): x \text{ is a real number and } y = x^2\}$$

For simplicity of expression, it is common practice instead to speak of "the function $g(x) = x^2$." As long as we keep firmly in mind the distinction between a function and its values (see Section 4.1), no harm should come from this simplification.

We also adopt the convention that the domain of any real-valued function described by a formula is the *largest* set of real numbers for which the formula makes sense (that is, produces another real number). Thus the domain of the function $f(x) = \sqrt{x}$ in Figure 2 is the set of non-negative real numbers, since negative numbers do not have square roots which are real numbers.

It is important to note that the symbol used to identify a function, whether f, g, h, or some other symbol, is an arbitrary one. The important factor is

FIGURE 1   Squaring function

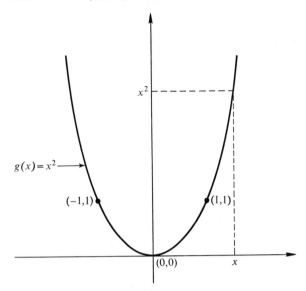

FIGURE 2   Square root function

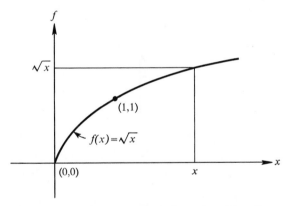

the formula itself. The symbol used to identify a member of the domain of the function is also arbitrary. Thus, for example, the notations $f(x) = \sqrt{x}$, $g(x) = \sqrt{x}$, $h(y) = \sqrt{y}$, and $m(t) = \sqrt{t}$ all describe the same function, namely the square root function. In each case the formula assigns to any number its square root. The set of ordered pairs which comprise the square root function can thus be described by any of the above notations and in this sense they are identical.

The graphs of several real-valued functions are shown in Figure 3. Many of these graphs should be familiar from previous mathematics studies. To begin a study of calculus a substantial portion of the subject matter is devoted to a thorough investigation of these and closely related functions. These functions are sometimes referred to as the *elementary functions.*

FIGURE 3 Graphs of some familiar functions:

(a) straight line function;   (b) sine function;

(c) cosine function;   (d) exponential function;

(e) natural logarithm function;   and   (f) tangent function

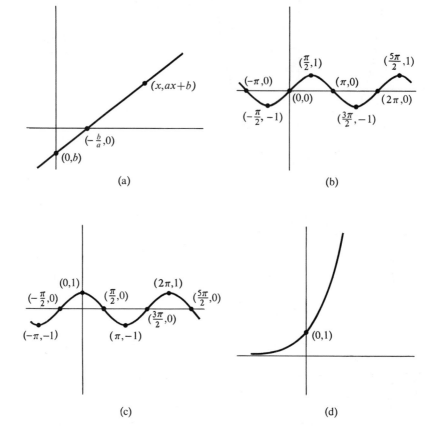

(a)

(b)

(c)

(d)

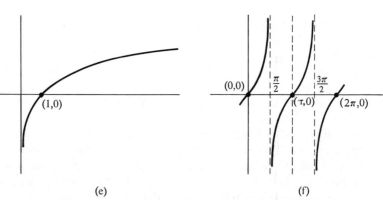

(e)

(f)

1. Draw graphs of the following functions.
   (a) *constant function* $h(x) = 2$
   (b) constant function $F(x) = c$, where c is a real number
   (c)
   $$g(x) = \begin{cases} 2 & \text{for } x < 0 \\ x & \text{for } x \geq 0 \end{cases}$$

Example 1   This example is designed to emphasize the use of functional notation. Suppose f is a function which assigns to any real number x $(\neq 0)$ the value

$$f(x) = x^2 - 1 + \frac{1}{x} \tag{1}$$

Then f assigns to the number y the value

$$f(y) = y^2 - 1 + \frac{1}{y}$$

and f assigns to 3 the value

$$f(3) = 3^2 - 1 + \tfrac{1}{3} = \tfrac{25}{3}$$

Similarly,

$$f(x - 1) = (x - 1)^2 - 1 + \frac{1}{x - 1}$$

$$f(x^2) = (x^2)^2 - 1 + \frac{1}{x^2} = x^4 - 1 + \frac{1}{x^2}$$

Thus, we see that in order to compute the value which f assigns to any quantity Q, we simply replace x by Q in Formula (1). The quantity Q may or may not be an expression involving x. ▶

PROBLEMS

2. The domain of each of the following functions is the set R of real numbers. Sketch the graph of each function by plotting a few points and drawing a smooth curve through the points.
   (a) $f(x) = 4x$ (b) $g(t) = 3 - 2t$
   (c) $h(w) = w^2 - 1$ (d) $F(z) = z^2 - 2z + 3$
   (e) $G(y) = y^3 - 6y + 2$ (f) $H(x) = 2x - x^3$
   (g) $\alpha(s) = s^2 - 4$

3. Indicate the range of each of the functions in Problem 2.

4. Write the equation of the straight line passing through the points $(2, 8)$ and $(4, 12)$. (*Hint:* Every straight line has the functional form $f(x) = ax + b$, where a and b are constants.)

5. Let $f(x) = 2x + 6$ and $g(x) = 4x - 12$.

    (a) For what number x is $f(x) = g(x)$?

    (b) For what numbers x is $f(x) < g(x)$?

    (c) Draw a graph to illustrate (a) and (b).

6. (a) Find the values of x for which

$$f(x) = x^2 + 3x + 2 = 0$$

    Draw a graph of f on which these values are displayed.

    (b) Same problem for the function $g(x) = x^2 - 6x + 8$.

7. The Spearman–Brown formula (see Example 4, Section 5.3) for reliability of a test having initial reliability r, which is then lengthened by a factor of n, is

$$f(n) = \frac{nr}{1 + (n - 1)r}$$

    (a) Compute $f(n^2)$.

    (b) Compute $f[f(n)]$.

    (c) Show that the function g defined by

$$g(n) = n\left[\frac{1}{f(n)} - 1\right]$$

    is a constant function; that is, $g(n)$ has the same value regardless of the value of n.

11.2 LIMITS

We now wish to introduce a concept of limit for arbitrary real-valued functions which depends on the idea of limit for sequences. In order to simplify our terminology, we shall call a sequence which converges to a number p, but whose elements are all different from p, a *p-deleted sequence*. That is, $(x_1, x_2, x_3, \ldots)$ is a p-deleted sequence if

$$\lim_{n \to \infty} (x_n) = p \quad \text{and} \quad \text{for all } n, \, x_n \neq p$$

Example 1 (a) The sequence $(1, \frac{1}{2}, \frac{1}{3}, \frac{1}{4}, \ldots)$ is a zero-deleted sequence since zero is not a member of the sequence and the sequence converges to zero.

    (b) The sequence

$$\left(1, -1, \frac{1}{2^2}, -\frac{1}{2^2}, \frac{1}{3^2}, -\frac{1}{3^2}, \ldots\right)$$

is also a zero-deleted sequence.

    (c) The sequence $(2, 1\frac{1}{2}, 1\frac{1}{4}, 1\frac{1}{8}, 1\frac{1}{16}, \ldots)$ is not a zero-deleted sequence since the limit is 1, not 0. It is, however, a one-deleted sequence.

(d) Although its limit is 1, the sequence $(1, 1, 1, \ldots)$ is not one-deleted, since 1 is a member of the sequence. ▶

PROBLEMS

1. Which of the following are 2-deleted sequences?

(a) $(2, 2, 2, 2, \ldots)$

(b) $(3, 2, 1, 1\frac{1}{2}, 1\frac{3}{4}, 1\frac{7}{8}, \ldots)$

(c) $(1, 1\frac{1}{2}, 1\frac{3}{4}, 1\frac{7}{8}, 1\frac{15}{16}, \ldots)$

(d) $(1, \frac{1}{2}, \frac{1}{4}, \frac{1}{8}, \frac{1}{16}, \ldots)$

The procedure to be followed in defining limit is best introduced by means of examples.

Example 2 The graph of the function f defined by

$$f(x) = \begin{cases} 2 & \text{when } x < 0 \\ 1 & \text{when } x \geq 0 \end{cases}$$

is shown in Figure 4. Any zero-deleted sequence $(x_1, x_2, x_3, \ldots)$ of *negative* numbers generates a corresponding sequence of functional values $(f(x_1), f(x_2), f(x_3), \ldots) = (2, 2, 2, \ldots)$, all of which are equal to two. The number 2 is the limit of all such sequences of functional values. Similarly, if $(y_1, y_2, y_3, \ldots)$ is a zero-deleted sequence of *positive* numbers, then $(f(y_1), f(y_2), f(y_3), \ldots)$ is the functional value sequence $(1, 1, 1, \ldots)$, whose limit is 1.

FIGURE 4

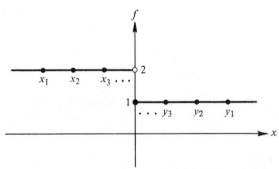

Not all zero-deleted sequences of points in the domain of f generate convergent sequences of functional values. For instance, the sequence

$$(1, -1, \tfrac{1}{2}, -\tfrac{1}{2}, \tfrac{1}{3}, -\tfrac{1}{3}, \ldots)$$

generates the sequence of functional values

$$(1, 2, 1, 2, 1, 2, \ldots),$$

which does not converge. ▶

Example 3 If $g(x) = 2x - 3$ and $(x_1, x_2, x_3, \ldots)$ is *any* 4-deleted sequence, then the limit of the corresponding functional value sequence is

$$\lim_{n \to \infty} g(x_n) = \lim_{n \to \infty} (2x_n - 3) = 2 \cdot 4 - 3 = 5$$

Every sequence (x_n) of points in the domain of f which converges to 4 generates a corresponding functional value sequence $(g(x_n))$ in the range of g which converges to 5 (see Figure 5). Note that in this case the common limit 5 is the value of g at the point 4; that is, $g(4) = 2 \cdot 4 - 3 = 5$. ▶

FIGURE 5

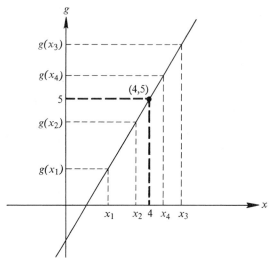

Example 4 Suppose

$$h(x) = \begin{cases} 1 & \text{when } x \neq 0 \\ 2 & \text{when } x = 0 \end{cases}$$

Then every zero-deleted sequence in the domain of h generates the sequence of functional values $(1, 1, 1, \ldots)$. Here the common limit 1 of the functional value sequences is *not* the functional value $h(0) = 2$ (see Figure 6). ▶

Definition 1 Let f be a real-valued function. If for every p-deleted sequence $(x_1, x_2, x_3, \ldots)$ of points in the domain of f, the corresponding functional value sequence $(f(x_1), f(x_2), f(x_3), \ldots)$ converges to the number L, then L is called the *limit of f at the point p*. This limit is written as

$$\lim_{x \to p} f(x) = L$$

[or in shorthand notation as]

$$\lim_{p} f = L \qquad \blacktriangleright$$

FIGURE 6

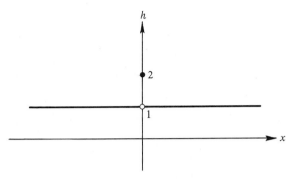

Example 5 (a) The discussion in Example 2 implies that $\lim_{x \to 0} f(x)$ does not exist since there is no common limit for all sequences of functional values generated by zero-deleted sequences of points in the domain of f.

(b) In Example 3, $\lim_{n \to \infty} g(x_n) = 5$ whenever (x_n) is a 4-deleted sequence. Hence $\lim_{x \to 4} g(x) = 5$.

(c) In Example 4, $\lim_{x \to 0} h(x) = 1$ since every zero-deleted sequence of points in the domain of h generates a sequence of functional values converging to 1. ▶

Example 6 If h is a constant function defined for all numbers x by

$$h(x) = c$$

where c is any real number, then every p-deleted sequence of points $(x_1, x_2, x_3, \ldots)$ generates a corresponding sequence of functional values $(c, c, c, \ldots)$ all equal to c. Since every such generated sequence converges to c, we see that

$$\lim_{x \to p} h(x) = c$$

regardless of the value of p. ▶

Example 7 Let I denote the identity function

$$I(x) = x$$

Then, any p-deleted sequence of numbers $(x_1, x_2, x_3, \ldots)$ produces the corresponding functional value sequence

$$(I(x_1), I(x_2), I(x_3), \ldots) = (x_1, x_2, x_3, \ldots)$$

of functional values, which also converges to p. That is,

$$\lim_{x \to p} I(x) = \lim_{x \to p} x = p$$ ▶

2. Define the function g by

$$g(x) = \begin{cases} 2x - 3 & \text{when } x \neq 4 \\ 29 & \text{when } x = 4 \end{cases}$$

Find $\lim_{x \to 4} g(x)$.

We wish to be able to find the limits of sums, differences, products, and quotients of functions. We diverge from this problem to look at the definitions of these algebraic combinations of arbitrary functions. We will see that these algebraic combinations of functions are defined in the same way as for sequences. Thus, if f and g are two real-valued functions we define for all points x common to the domains of both f and g:

(a) the *sum* $f + g$ of f and g by

$$(f + g)(x) = f(x) + g(x)$$

(b) the *difference* $f - g$ of f and g by

$$(f - g)(x) = f(x) - g(x)$$

(c) the *product* $f \cdot g$ of f and g by

$$(f \cdot g)(x) = f(x) \cdot g(x)$$

(d) the *quotient* of f and g for $g(x) \neq 0$ by

$$\left(\frac{f}{g}\right)(x) = \frac{f(x)}{g(x)}$$

Example 8 If we have the real-valued functions $f(x) = 2x^2$ and $g(x) = \sqrt{1 - x^2}$, then f is defined for all x while g is defined only for $-1 \leq x \leq 1$. Hence the sum, difference, product, and quotient of these two functions are defined only for $-1 \leq x \leq 1$. For a point x in this common domain, we have

$$(f + g)(x) = 2x^2 + \sqrt{1 - x^2}$$
$$(f - g)(x) = 2x^2 - \sqrt{1 - x^2}$$
$$(g - f)(x) = \sqrt{1 - x^2} - 2x^2$$
$$(f \cdot g)(x) = (2x^2)\sqrt{1 - x^2}$$
$$\left(\frac{f}{g}\right)(x) = \frac{2x^2}{\sqrt{1 - x^2}} \qquad \text{if } x \neq \pm 1$$
$$\left(\frac{g}{f}\right)(x) = \frac{\sqrt{1 - x^2}}{2x^2} \qquad \text{if } x \neq 0 \qquad \blacktriangleright$$

Since the concept of limit for arbitrary real-valued functions has been phrased in terms of limits of functional value sequences generated from p-deleted sequences of points in the domain of the function, it is natural to expect this new concept of limit to have the same properties as the concept of limit for sequences. (See Property 4 for sequences in Section 5.3.)

Theorem 1 Let p be a point, common to the domains of functions f and g, for which

$$\lim_{x \to p} f(x) = L \quad \text{and} \quad \lim_{x \to p} g(x) = M$$

Then

$$\lim_{x \to p} (f + g)(x) = L + M$$

In words, this says that the limit of a sum is the sum of the limits.

PROOF Our assumptions imply that if $(x_1, x_2, x_3, \ldots)$ is any p-deleted sequence, then the functional value sequences $(f(x_1), f(x_2), f(x_3), \ldots)$ and $(g(x_1), g(x_2), g(x_3), \ldots)$ converge, respectively, to L and to M. Hence, by Property 4 for sequences (see Section 5.3) and by the definition of $f + g$, the sequence

$$((f + g)(x_1), (f + g)(x_2), (f + g)(x_3), \ldots)$$

converges to $L + M$. ▶

Other properties of limits of functions which are algebraic combinations of other functions follow in similar fashion from the analogous properties for sequences. Therefore the proof of the following theorem is left as an exercise.

Theorem 2 Let p be a point, common to the domains of functions f and g, for which

$$\lim_{x \to p} f(x) = L \quad \text{and} \quad \lim_{x \to p} g(x) = M$$

Then

(i) $\lim_{x \to p} (f - g)(x) = L - M$

(ii) $\lim_{x \to p} (f \cdot g)(x) = L \cdot M$

(iii) If $M \neq 0$, $\lim_{x \to p} \left(\dfrac{f}{g}\right)(x) = \dfrac{L}{M}$

Thus, the limit of a difference, product, or quotient of two functions is the corresponding difference, product, or quotient of their respective limits. ▶

Example 9 The squaring function $g(x) = x^2$ is the product of the identity function $I(x) = x$ times itself. It follows from Theorem 2 and Example 7 that

$$\lim_{x \to p} x^2 = \left[\lim_{x \to p} I(x)\right]\left[\lim_{x \to p} I(x)\right]$$

$$= p \cdot p = p^2$$
▶

Theorem 1 and Theorem 2(ii) may be extended by induction to the sum and product of any number of functions. The proof of the following theorem is left as an exercise.

Theorem 3 Let p be a point common to the domain of the functions $f_1, f_2, \ldots, f_n$, and suppose that for $k = 1, 2, 3, \ldots, n$,

$$\lim_{x \to p} f_k(x) = L_k$$

Then

(i) $\displaystyle\lim_{x \to p} [f_1(x) + f_2(x) + \cdots + f_n(x)] = L_1 + L_2 + \cdots + L_n$

(ii) $\displaystyle\lim_{x \to p} [f_1(x)f_2(x) \cdots f_n(x)] = L_1 L_2 \cdots L_n$ ▶

Example 10 A *polynomial* is a function of the form

$$f(x) = c_0 + c_1 x + c_2 x^2 + \cdots + c_n x^n$$

which is produced by adding the products of constant functions (c_i) and identity function products (x^i). According to Theorem 3, we have

$$\lim_{x \to p} f(x) = \lim_{x \to p} c_0 + \left[\lim_{x \to p} c_1 \right]\left[\lim_{x \to p} x \right] + \left[\lim_{x \to p} c_2 \right]\left[\lim_{x \to p} x \right]^2$$

$$+ \cdots + \left[\lim_{x \to p} c_n \right]\left[\lim_{x \to p} x \right]^n$$

$$= c_0 + c_1 p + c_2 p^2 + \cdots + c_n p^n$$

$$= f(p)$$

The limit of a polynomial f at p is just the functional value $f(p)$. ▶

Example 11 An important concept in economic theory is that of *marginal revenue*, or the rate at which revenue increases relative to sales. Consider a firm whose total revenue R is related to total sales S by a function f, called the revenue function (see Figure 7). For a given total sales s_0 and correspond-

FIGURE 7

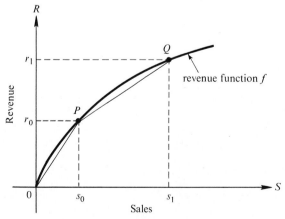

ing total revenue $r_0 = f(s_0)$, the *average revenue* per sales unit is

$$A(s_0) = \frac{f(s_0)}{s_0}$$

represented in Figure 7 by the slope of the line $0P$. As total sales increase from s_0 to s_1, so that revenue increases from $r_0 = f(s_0)$ to $r_1 = f(s_1)$, the average rate of increase is

$$\frac{r_1 - r_0}{s_1 - s_0} = \frac{f(s_1) - f(s_0)}{s_1 - s_0} \qquad \text{(2)}$$

represented by the slope of the line PQ.

If we now think of keeping point P fixed and moving Q toward P along the curve, the average rate (2) can in the limit be interpreted as the "instantaneous" rate of increase at s_0. It is this limit which defines marginal revenue. (Geometrically, this limit defines the slope of a line tangent to the graph of f at s_0.)

To take a specific case, suppose the total revenue (in millions) is given in terms of total sales (in thousands) by

$$R = f(S) = \begin{cases} 8S & \text{when } 0 \le S \le 1 \\ 12 - (4/S^2) & \text{when } S > 1 \end{cases}$$

(see Figure 8). Then the marginal revenue at the point $(2, 11)$ is

FIGURE 8

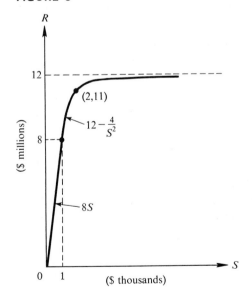

$$\lim_{S \to 2} \frac{f(S) - f(2)}{S - 2} = \lim_{S \to 2} \frac{[12 - (4/S^2)] - [12 - (4/2^2)]}{S - 2}$$

$$= \lim_{S \to 2} \frac{S^2 - 4}{S^2(S - 2)}$$

$$= \lim_{S \to 2} \frac{S + 2}{S^2}$$

$$= 1$$

Thus, if it were possible to increase sales by one unit under the conditions which prevail when $S = 2$, revenue would also increase by one unit. (Geometrically, at $S = 2$ the line tangent to the revenue function inclines 45° to the horizontal.) ▶

PROBLEMS

In Problems 3–10 assume that the domain of the function is the entire real line excluding points where the denominator is zero. Compute the indicated limits.

3. $\lim_{x \to 1} \left(x + \dfrac{1}{x} \right)$

4. $\lim_{x \to 1} (x^2 + 2x + 3)$

5. $\lim_{x \to 0} \dfrac{x - 3}{x + 3}$

6. $\lim_{x \to 2} \dfrac{x^2 - 4}{x + 2}$

7. $\lim_{x \to -2} \dfrac{x^2 - 4}{x + 2}$ (*Hint:* $x^2 - 4 = (x - 2)(x + 2)$.)

8. $\lim_{h \to 0} \dfrac{(1 + h)^2 - 1}{h}$ (*Hint:* There is a factor of h in the numerator.)

9. $\lim_{x \to 2} \dfrac{x^2 + 3x - 10}{x^2 - 6x + 8}$

10. $\lim_{h \to 0} \dfrac{(2 + h)^3 - 2^3}{h}$

11. Let

$$f(x) = \begin{cases} x + 2 & \text{when } x \le 0 \\ x + 1 & \text{when } x > 0 \end{cases}$$

Which of the following limit statements are correct?

(a) $\lim_{x \to 0} f(x) = 2$

(b) $\lim_{x \to 0} f(x) = 1$

(c) $\lim_{x \to 1} f(x) = 3$

(d) $\lim_{x \to 1} f(x) = 2$

(e) $\lim_{x \to -2} f(x) = 0$

12. Suppose that the cost function for a commodity is

$$C = 10 + 2x$$

where x is the amount produced. Show that the marginal cost is

$$MC = 2$$

for all x.

13. Suppose that the demand function for a commodity is

$$p = g(x)$$

where p denotes price and x is the quantity demanded. Define the total revenue function by

$$R(x) = x \cdot g(x)$$

If $\lim_{x \to 0} g(x) = 7$, find the marginal revenue at $x = 0$. Interpret your results.

14. Let f and g be defined by $f(x) = 2x - 6$ and $g(x) = 4x + 2$. Define the following functions and indicate the domain and range of each.

(a) $f + g$ (b) $f - g$ (c) $f \cdot g$

(d) f/g (e) g/f

15. In Problem 14 let $f(x) = x + 3$ and $g(x) = x^2 + 5x + 6$.

16. Prove Theorem 2.

17. Prove Theorem 3.

18. Prove that if $\lim_{x \to a} f(x) = \alpha$, then for any constant k,

$$\lim_{x \to a} kf(x) = k \cdot \alpha$$

19. Prove that if $\lim_{x \to p} f(x) = A$ and $\lim_{x \to p} f(x) = B$, then $A = B$.

20. Prove that if $\lim_{x \to p} g(x) = L$, then $\lim_{x \to p} [g(x) - L] = 0$

21. Find two functions f and g such that neither has a limit at $x = 0$ but whose sum $f + g$ has a limit at that point.

22. Prove that if $\lim_{x \to p} g(x) = L$, then $\lim_{x \to 0} g(p + x) = L$.

11.3 MORE ABOUT LIMITS

A sequence $(x_1, x_2, x_3, \ldots)$ is said to *diverge to infinity* if for every positive number m, no matter how large, there is an integer k such that $x_n > m$ when $n > k$. That is, a sequence diverges to infinity if, by going out far enough in the sequence (past the kth term) we can find terms as large as we wish (larger than m). Similarly $(x_1, x_2, x_3, \ldots)$ *diverges to minus infinity* if for every negative number m, no matter how large, there is an integer k such that $x_n < m$ when $n > k$. For instance, the sequences $(1, 2, 3, 4, \ldots)$ and $(1, 4, 9, 16, \ldots)$ diverge to infinity, while $(-1, -2, -3, -4, \ldots)$ diverges to minus infinity (see Figure 9).

FIGURE 9

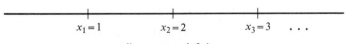

divergence to infinity

If for every sequence of numbers $(x_1, x_2, x_3, \ldots)$ which diverges to infinity the corresponding sequence of functional values

$$(f(x_1), f(x_2), f(x_3), \ldots)$$

converges and has limit L, then L is called the *limit of $f(x)$ as x approaches infinity* and we write

$$\lim_{x \to \infty} f(x) = L$$

The symbol $x \to \infty$ is read "x approaches infinity." It is intended to mean *only* that x increases without bound.

Similarly, if the sequence $(f(x_1), f(x_2), f(x_3), \ldots)$ converges to L whenever $(x_1, x_2, x_3, \ldots)$ diverges to minus infinity, then L is the *limit of $f(x)$ as x approaches minus infinity*, and we write

$$\lim_{x \to -\infty} f(x) = L$$

Example 1 If $f(x) = 1/x$, then $\lim_{x \to \infty} f(x) = 0$. To see this, let $(x_1, x_2, x_3, \ldots)$ be any sequence diverging to infinity and let r be a positive number. Then $1/r$ is positive and there is an integer k such that

$$x_n > \frac{1}{r} > 0 \qquad \text{when } r > k$$

Equivalently,

$$0 < \frac{1}{x_n} < r \qquad \text{when } r > k$$

so that the sequence $(1/x_n)$ of functional values must converge to zero. ▶

Example 2 For the revenue function in Example 11 of Section 11.2 we have

$$\lim_{S \to \infty} f(S) = \lim_{S \to \infty} [12 - (4/S^2)]$$
$$= 12 - \lim_{S \to \infty} (4/S^2)$$
$$= 12 - 0 = 12 \qquad \qquad ▶$$

If for every p-deleted sequence $(x_1, x_2, x_3, \ldots)$, the corresponding sequence of functional values $(f(x_1), f(x_2), f(x_3), \ldots)$ diverges to infinity, we write

$$\lim_{x \to p} f(x) = \infty$$

Similarly, if every p-deleted sequence in the domain of f generates a sequence of functional values which diverges to minus infinity, we write

$$\lim_{x \to p} f(x) = -\infty$$

The symbols ∞ and $-\infty$ are used simply as part of a convenient shorthand for the verbal definition and are not to be considered as numbers.

Example 3 A simple model of world population growth is

$$N(t) = N_0 e^{\theta(t - t_0)}$$

N_0 is the population size at some arbitrary initial point t_0 in time, $N(t)$ denotes population size at time $t \geq t_0$, θ is a constant growth rate, and e is a mathematical constant approximately equal to 2.718. Obviously, for any time sequence (t_n) diverging to infinity, the corresponding sequence $(N(t_n))$ also diverges to infinity. Thus

$$\lim_{t \to \infty} N(t) = \infty \qquad \blacktriangleright$$

PROBLEMS

1. Find

$$\lim_{x \to 3} \frac{1}{(x - 3)^2}$$

2. Prove that

$$\lim_{x \to 0} \frac{1}{x^2} = \infty$$

3. Find the following limits.

(a) $\displaystyle\lim_{x \to 2} \frac{1}{(x - 2)^2}$

(b) $\displaystyle\lim_{x \to \infty} \frac{x - 8}{x^2 - 4}$

(c) $\displaystyle\lim_{x \to \infty} \frac{3}{(x - 2)^2}$

(d) $\displaystyle\lim_{x \to 0} \frac{|x|}{x}$

(e) $\displaystyle\lim_{x \to -\infty} \frac{6}{(x - 4)^3}$

(f) $\displaystyle\lim_{x \to \infty} \frac{x^2 + 4}{x - 1}$

4. In Problem 7, Section 11.1 assume that n can be any positive number. Find $\lim_{n \to \infty} f(n)$.

5. Suppose, in a population of N individuals, that the average number of individuals who have completed a particular task by time t is

$$A(t) = N(1 - 2^{-\alpha t})$$

where α is a positive constant. Find $\lim_{t \to \infty} A(t)$.

6. Given certain assumptions about the nature of the items added, the *correlation** between two tests X and Y of altered length is

$$R(a, b) = \frac{\sqrt{ab}\, R_{xy}}{\sqrt{1 + (a - 1)R_x}\, \sqrt{1 + (b - 1)R_y}}$$

where R_{xy}, R_x, and R_y are, respectively, the correlation between the un-lengthened tests, the reliability of test X, and the reliability of test Y. The quantities a and b represent the lengthening factors of the two tests.

(a) For each fixed b, find $F(b) = \lim_{a \to \infty} R(a, b)$.

(b) Find $\lim_{b \to \infty} F(b)$.

7. Work out details for Examples 2 and 3.

8. Find $\lim_{x \to \infty} [(x^2 + 1)^{1/2} - x]$.

9. Prove that $\lim_{x \to \infty} [\sqrt{x(x + a)} - x] = a/2$.

*Adapted from Gulliksen, H. A., *Theory of Mental Tests* (John Wiley & Sons, Inc., New York, 1950).

11.4 CONTINUOUS FUNCTIONS

A function f is said to be *continuous at a point p* in its domain if the limit of $f(x)$ as x approaches p is the functional value $f(p)$, that is, if

$$\lim_{x \to p} f(x) = f(p)$$

If f is continuous at each point in a set S, it is said to be *continuous on S*.

Example 1 Our calculations in Example 3 of Section 11.2 show that

$$g(x) = 2x - 3$$

is continuous at $x = 4$. Actually this function is continuous at every point. For if p is any real number and $(x_1, x_2, x_3, \ldots)$ is a p-deleted sequence, then

$$\lim_{n \to \infty} (g(x_n)) = \lim_{n \to \infty} (2x_n - 3)$$
$$= 2 \lim_{n \to \infty} (x_n) - 3$$
$$= 2p - 3$$

The limit is equal to the functional value at p. ▶

Example 2 In Example 10 of Section 11.2, we saw that for any polynomial f and any point p, the limit statement

$$\lim_{x \to p} f(x) = f(p)$$

holds. Thus polynomials are continuous at all points. ▶

Example 3 (a) The function f considered in Example 2 of Section 11.2 is *not* continuous at zero, since $\lim_{x \to 0} f(x)$ does not exist. However, f is continuous at every other point.

(b) The function h in Example 4 of Section 11.2 is not continuous at zero because the limit $\lim_{x \to 0} h(x) = 1$ is not equal to the functional value $h(0) = 2$. However, h is continuous at all other points. ▶

General properties of continuous functions follow directly from corresponding properties for limits stated in Theorems 1 and 2.

Theorem 4 Suppose the functions f and g are continuous at the point p. Then the functions $f + g$, $f - g$, and $f \cdot g$ are all continuous at p. The function f/g is continuous if $g(p) \neq 0$.

PROOF We prove only the last statement, leaving the other cases for the exercises. Since f and g are continuous at p, we have $\lim_{x \to p} f(x) = f(p)$ and

$\lim_{x \to p} g(x) = g(p)$. Hence

$$\lim_{x \to p} \left(\frac{f}{g}\right)(x) = \lim_{x \to p} \frac{f(x)}{g(x)}$$

$$= \frac{\lim_{x \to p} f(x)}{\lim_{x \to p} g(x)}$$

$$= \frac{f(p)}{g(p)} \qquad\qquad \text{[Theorem 2(iii)]}$$

$$= \left(\frac{f}{g}\right)(p) \qquad\qquad \blacktriangleright$$

Intuitively, if p is a point of continuity of a function f, then f cannot take a jump at the point p. For if every sequence of functional values $(f(x_1),$ $f(x_2), f(x_3), \ldots)$ corresponding to a p-deleted sequence $(x_1, x_2, x_3, \ldots)$ must converge to $f(p)$, then for values of x near p, $f(x)$ must be near $f(p)$. For example, each of the functions in Example 3 is discontinuous at zero where the graph takes a jump (see Figures 4 and 6). A good rule of thumb for intuitively determining if a function is continuous is to see if its graph can be traced without lifting pencil from paper. Such considerations indicate that the following theorems must be true. The proofs will be omitted.

Theorem 5 *Maximum and Minimum Value Theorem* Suppose a function f is continuous throughout the interval $[a, b]$. (See Section 6.3 for interval notation.) Then f has a minimum value m and a maximum value M in the interval. That is, there are numbers α and β between a and b such that $m = f(\alpha)$ and $M = f(\beta)$, and such that for all x in the interval, the condition

$$m \leq f(x) \leq M$$

is satisfied (see Figure 10). $\qquad\qquad\qquad\qquad\qquad\qquad\qquad\blacktriangleright$

Theorem 6 *Intermediate Value Theorem* Suppose a function f is continuous throughout the interval $[a, b]$. If N is any number between $f(a)$ and

FIGURE 10

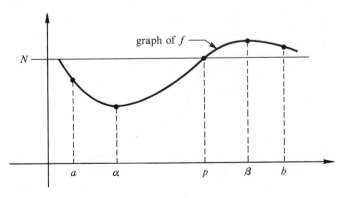

$f(b)$, there is at least one point p between a and b at which

$$f(p) = N$$

(see Figure 10). ▶

Example 4 The estimated sales volume S of a certain firm is related to the per unit price p by the demand equation

$$S = 4000 - 30p$$

The total cost C incurred in the production of S units is

$$C = 25,000 + 20S$$
$$= 105,000 - 600p \tag{3}$$

It follows that the total revenue is

$$R = pS = 4000p - 30p^2 \tag{4}$$

and that the net profit is

$$P = R - C = 4600p - 30p^2 - 105,000 \tag{5}$$

The cost and revenue functions (3) and (4) are graphed in Figure 11. As the price per unit increases, total cost steadily decreases while total revenue first increases, reaching a maximum of 133,333 at $p = 66\frac{2}{3}$, and then drops off.

The height of the shaded region above a particular per unit price p indicates the net profit corresponding to that price. Since the profit function (5) is continuous, Theorem 5 guarantees the existence of an optimum price, that

FIGURE 11

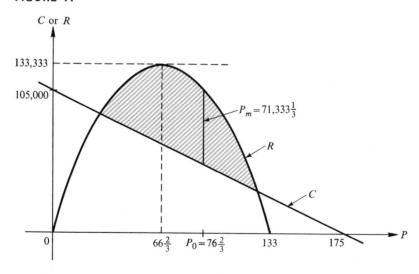

is, a price which results in the maximization of profits. To find this optimum price, we complete the square in (5) to obtain

$$P = \frac{214{,}000}{3} - 30\left(p - \frac{230}{3}\right)^2$$

Thus the maximum profit is $P_m = 214{,}000/3 = 71{,}333\frac{1}{3}$ at the optimum price $p_0 = 230/3 = 76\frac{2}{3}$. ▶

PROBLEMS

1. Find the following limits and determine in which cases the function is continuous at the point at which the limit is taken.

   (a) $\lim\limits_{x \to 3} (x^2 - 2x + 4)$

   (b) $\lim\limits_{x \to 0} (x^2 - 2x + 4)$

   (c) $\lim\limits_{x \to a} (x^2 - 2x + 4)$, where a is any real number

   (d) $\lim\limits_{t \to 0} |t|$

   (e) $\lim\limits_{z \to 4} \dfrac{z^2 - 16}{z^2 - z - 12}$

   (f) $\lim\limits_{x \to -1} f(x)$, where

$$f(x) = \begin{cases} \dfrac{x^2 + 5x + 4}{x + 1} & \text{if } x \neq -1 \\ 3 & \text{if } x = -1 \end{cases}$$

2. Each of the following functions fails to be continuous at the point p specified. Why?

   (a)
   $$f(x) = \begin{cases} \dfrac{|x|}{x} & \text{when } x \neq 0 \\ 0 & \text{when } x = 0 \end{cases} \quad \text{at } p = 0$$

   (b)
   $$g(t) = \begin{cases} \dfrac{t^2 + t - 6}{t - 2} & \text{when } t \neq 2 \\ 0 & \text{when } t = 2 \end{cases} \quad \text{at } p = 2$$

   (c)
   $$h(z) = \begin{cases} z^2 & \text{when } z \leq 1 \\ 4 - z^2 & \text{when } z > 1 \end{cases} \quad \text{at } p = 1$$

3. For each of the following functions, 0 is not in the domain. If possible, assign a value $f(0)$ so that f is continuous at 0. If this is not possible, indicate why not.

   (a) $f(x) = \dfrac{1}{x}$

   (b) $f(x) = \dfrac{1}{x^2}$

   (c) $f(x) = \sin\dfrac{1}{x}$

   (d) $f(x) = x \sin\dfrac{1}{x}$

   (e) $f(x) = \dfrac{1}{2 - 2^{1/x}}$

4. Which of the following functions have points of discontinuity, that is, points at which they are not continuous?

(a) $f(x) = \dfrac{1}{x^2}$ $x > 0$

(b)
$$f(x) = \begin{cases} x^2 \cos \dfrac{1}{x^2} & \text{if } x \neq 0 \\ 0 & \text{if } x = 0 \end{cases}$$

(c)
$$g(x) = \begin{cases} \dfrac{1}{x} & \text{if } x > 0 \\ 0 & \text{if } x \leq 0 \end{cases}$$

(d)
$$h(t) = \begin{cases} t + 1 & \text{if } t > 0 \\ -t - 1 & \text{if } t < 0 \\ 0 & \text{if } t = 0 \end{cases}$$

(e)
$$f(x) = \begin{cases} \dfrac{x^2 - 4}{x - 2} & x \neq 2 \\ 0 & x = 2 \end{cases}$$

5. Prove that any constant function is continuous.

6. Prove that if f and g are both continuous at the point p, then their sum $f + g$ is also continuous at p.

7. Prove that the difference $f - g$ between two continuous functions f and g is itself a continuous function.

8. Prove that the product of two continuous functions is a continuous function.

9. Use induction to extend to any finite number of functions the statements of Theorem 4 concerning sums and products.

10. Prove that if f is continuous on the interval from a to b, and if $f(a)$ and $f(b)$ have opposite signs, there is a point c between a and b at which $f(c) = 0$.

11. Prove that

(a) $\lim\limits_{h \to 0} [f(a + h) - f(a)] = 0$

(b) $\lim\limits_{h \to 0} f(a + h) = f(a)$

(c) $\lim\limits_{x \to a} f(x) = f(a)$

are all equivalent to saying that f is continuous at the point a.

12. Prove that a quotient of two polynomials is continuous except where the denominator is zero.

13. Given that $f(x)$ and $g(x)$ have the same domain and that $f(x) > g(x)$ for all x in the common domain, show that if $f(x)$ and $g(x)$ are continuous at $x = p$, then

$$T(x) = \frac{1}{f(x) - g(x)}$$

is also continuous at p.

14. A function f is called *additive* if

$$f(x + y) = f(x) + f(y)$$

for all real x and y.
   (a) Give an example of an additive function. (*Hint:* Try a first degree polynomial.)
   (b) Give an example of a nonadditive function. (*Hint:* Try polynomials.)
   (c) Prove that if f is additive, then $f(0) = 0$. (*Hint:* $0 = 0 + 0$.)
   (d) Prove that if an additive function is continuous at $x = 0$, then it is continuous at all points on the real line. (Use Problem 11.)

15. A real-valued function g is called *multiplicative* if

$$g(x + y) = g(x) \cdot g(y)$$

for all real x and y.
   (a) Give an example of a multiplicative function. (*Hint:* Recall the rules for exponents.)
   (b) Give an example of a nonmultiplicative function.
   (c) Show that if g is multiplicative, then either $g(0) = 1$ or $g(0) = 0$.
   (d) Show that if $g(0) = 0$, then $g(x) = 0$ for all x.
   (e) Show that if g is continuous at $x = 0$, then g is continuous at all points.

16. Give an example of a function F for which
   (a) $F(x \cdot y) = F(x) \cdot F(y)$ for all x and y;
   (b) $F(x \cdot y) = F(x) + F(y)$ for all x and y.

17. If F is a function whose domain includes the number zero and is such that

$$F(xy) = F(x) + F(y),$$

prove that F is the constant function $F(x) = 0$.

18. (a) A function f is said to be an *even function* if whenever x and $-x$ are in its domain, $f(-x) = f(x)$. Give an example of an even function.
   (b) A function f is said to be an *odd function* if whenever $-x$ and x are in its domain, $f(-x) = -f(x)$. Give an example of an odd function.

SUPPLEMENTARY READING

Anderson, K. W., and Hall, D. W., *Sets, Sequences and Mappings* (John Wiley & Sons, Inc., New York, 1963), Chapter 5.

Good, R. A., *Introduction to Mathematics* (Harcourt, Brace and World, New York, 1966), Chapter 9.

DIFFERENTIAL CALCULUS **12**

12.1 THE DERIVATIVE

As you drive your car, the speed at which you are traveling is indicated at all times on your speedometer. But Patrolman Smith, sad to say, cannot see your speedometer and must resort to indirect means of checking your speed. His solution is simple—he stretches two cables across the road a fixed distance apart and electronically measures the time required for you to cover the distance between them. In this way he determines the *average speed*, which equals "distance" divided by "time," and, assuming that your speed between the cables is constant, he uses this figure as his best estimate of your actual speed.

Let us analyze the method. The cables are stretched across at points p_1 and p_2, a distance $p_2 - p_1$ apart (see Figure 1). If these points are widely separated, say a mile apart, your speed may vary considerably and even at times exceed the speed limit and yet your average speed may be well within allowable limits. Similar considerations apply, but to a lesser extent, as the distance is reduced. Putting aside practical considerations involving the difficulty of precise measurements of time and distance, the best estimate of your "instantaneous" speed, the speed showing on your speedometer as you pass point p_1, is obtained by placing the point v_2 as close as possible to p_1.

To translate our problem into mathematical language, let us suppose that g is a function whose value $g(t)$ at time t represents your position on the road,

FIGURE 1

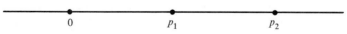

FIGURE 2

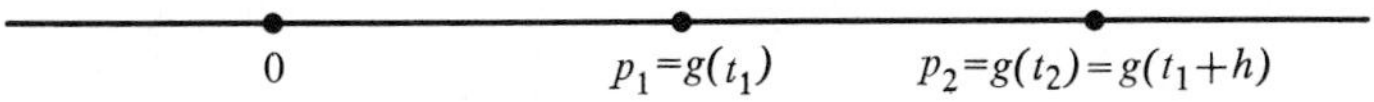

$$0 \qquad p_1 = g(t_1) \qquad p_2 = g(t_2) = g(t_1 + h)$$

measured from some arbitrary point called zero. If you pass point p_1 (Figure 2) at time t_1 and point p_2 at time $t_2 = t_1 + h$, your *average speed* in the time interval from t_1 to t_2 is

$$\frac{p_2 - p_1}{t_2 - t_1} = \frac{g(t_2) - g(t_1)}{t_2 - t_1} = \frac{g(t_1 + h) - g(t_1)}{h} \tag{1}$$

This average speed should, as h grows smaller, become a better and better estimate of the actual speed at p_1. In fact, it makes a great deal of sense to use the limit

$$\lim_{h \to 0} \frac{g(t_1 + h) - g(t_1)}{h} \tag{2}$$

as the *definition* of actual speed at p_1.

To take a specific case, suppose $g(t) = t^2$ for all $t \geq 0$. Then the average speed in the time interval from t_1 to $t_1 + h$ is

$$\frac{g(t_1 + h) - g(t_1)}{h} = \frac{(t_1 + h)^2 - t_1^2}{h} = 2t_1 + h$$

Hence the instantaneous speed at $t = t_1$ is

$$\lim_{h \to 0} (2t_1 + h) = 2t_1 \tag{3}$$

Table 1 lists average speeds in some time intervals beginning at $t_1 = 2$ and shows how the average speed gets closer to $2t_1 = 4$ as the length of the time interval shrinks.

TABLE 1. The convergence of average speed to instantaneous speed at time $t_1 = 2$ tabulated for $g(t) = t^2$.

Initial time $t_1 = 2$	Final time $t_2 = t_1 + h$ $= 2 + h$	Length of time h	Position at time t_2 $g(t_2) = (2 + h)^2$	Change in position $g(t_2) - g(t_1)$ $= 4h + h^2$	Average speed $4 + h$
2	10	8	100	96	12
2	7	5	49	45	9
2	4	2	16	12	6
2	3	1	9	5	5
2	2.5	0.5	6.25	2.25	4.5
2	2.3	0.3	5.29	1.29	4.3
2	2.1	0.1	4.41	0.41	4.1
2	2.01	0.01	4.0401	0.0401	4.01
2	2.001	0.001	4.004001	0.004001	4.001

1. According to Equation (3), the instantaneous speed at time $t_1 = 3$ is $2(3) = 6$. Make a table like Table 1 showing how the average speed over a time interval beginning at $t = 3$ and ending at $t = 3 + h$ converges to 6 as $h \to 0$.

Now let us look at another problem. In the sketch of the function f shown in Figure 3, the *slope* of the line joining the points P, with coordinates $(a, f(a))$, and Q, with coordinates $(b, f(b))$, is computed by dividing the difference $f(b) - f(a)$ in the heights of the curve at P and Q by the horizontal distance $b - a$ between these two points. Writing $b - a = h$, we have

$$\text{slope } \overline{PQ} = \frac{f(b) - f(a)}{b - a} = \frac{f(a + h) - f(a)}{h} \tag{4}$$

FIGURE 3

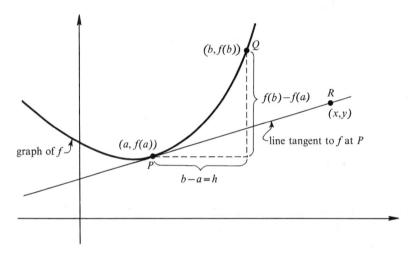

If we now think of point P as being fixed and of moving Q along the curve toward P, it seems reasonable to expect the slope of $\overline{PQ}$ to get closer and closer to the slope of the line tangent to the graph of f at point P. In fact, we shall define the slope of this tangent line (or, equivalently, the slope of f at P) to be the limit

$$\text{slope of } f \text{ at } P = \lim_{h \to 0} \frac{f(a + h) - f(a)}{h} \tag{5}$$

For example, if $f(x) = x^2$ for all x, then the slope of f at $x = a$ is

$$\lim_{h \to 0} \frac{f(a + h) - f(a)}{h} = \lim_{h \to 0} \frac{(a + h)^2 - a^2}{h}$$

$$= \lim_{h \to 0} (2a + h) = 2a \tag{6}$$

2. According to Equation (6), the slope of $f(x) = x^2$ is $2(-1) = -2$ when $x = -1$. Draw a sketch like Figure 3 and make a table like Table 1, using the same h values, to show how the slope of the line $\overline{PQ}$ converges to -2 as $h \to 0$, where P is the point $(-1, 1)$ on the curve.

If we compare the slope of the tangent line (Equations (4), (5), and (6)) with Patrolman Smith's problem (Equations (1), (2), and (3)), we find some striking resemblances. In each case we have begun with a function whose values represent the amount of a certain quantity (distance or height, as in the examples above). We have in Equations (1) and (4) computed an *average rate of change* of this quantity, and we have then, by taking limits in Equations (2) and (5), arrived at an *instantaneous rate of change*.

Definition 1 The process of moving from a function f to an instantaneous rate of change is called *differentiation*. The quantity

$$\lim_{h \to 0} \frac{f(x + h) - f(x)}{h} \tag{7}$$

is called the *derivative of the function f at the point x* in its domain. ▶

Differentiation forms the basis of one of the two major branches of calculus—the differential calculus. The derivative (7) is denoted by $f'(x)$ or by $Df(x)$. A commonly used alternative notation is to write $y = f(x)$, in which case the derivative is denoted by

$$\frac{dy}{dx}$$

This is called the *Leibniz* notation* for the derivative. Each of the three notations has its own advantages in different contexts, and the three forms are used interchangeably.

Example 1 Consider a commodity, such as gasoline, which may be produced in any non-negative quantity x. Let C be a cost function whose value $C(x)$ is the total cost associated with output level x. Then the *marginal cost* at level x is defined to be the derivative $C'(x)$.

To take a specific case, suppose $C(x) = ax - bx^2$, where a and b are constants. Then the marginal cost at level x is

$$
\begin{aligned}
C'(x) &= \lim_{h \to 0} \frac{C(x + h) - C(x)}{h} \\
&= \lim_{h \to 0} \frac{a(x + h) - b(x + h)^2 - (ax - bx^2)}{h} \\
&= \lim_{h \to 0} \frac{ax + ah - bx^2 - 2bxh - bh^2 - ax + bx^2}{h} \\
&= \lim_{h \to 0} (a - 2bx - bh) = a - 2bx
\end{aligned}
$$
▶

*After Gottfried Wilhelm Leibniz (1646–1716). Calculus was invented in the concurrent works of Leibniz and Isaac Newton.

Example 2 Other uses of the derivative in economics are the following:

(a) If f is a production function which relates the output level x of some commodity with the amount z of input of labor (or raw material, or capital), that is, if $x = f(z)$, then the derivative $dx/dz = f'(z)$ is called the *marginal product*.

(b) If p is a demand function whose value $p(x)$ at output level x specifies the price which will lead to the sale of x units of output, then $r(x) = x \cdot p(x)$ is the revenue which the firm obtains from the sale of x units. The derivative $r'(x)$ is called *marginal revenue*.

(c) If u is a utility function which attaches utility $u(x)$ to output level x, then $u'(x)$ is the *marginal utility* at the point x. ▶

Example 3 Richardson* adapts models for the spread of epidemics to the study of the diffusion of "war fever" among a population. Let $p(t)$ denote the proportion of the population favorable to war at time t so that $1 - p(t)$ is the proportion not favorable at that time. In the interval from time t to time $t + h$, the proportion of the population converted to a prowar attitude is $p(t + h) - p(t)$. The average rate of conversion is $[p(t + h) - p(t)]/h$, while the derivative

$$p'(t) = \lim_{h \to 0} \frac{p(t + h) - p(t)}{h}$$

represents the instantaneous rate of conversion.

In the simplest model, Richardson assumes that the instantaneous spread of war fever is proportional to both the number of people having prowar sentiments and to the number who are antiwar. That is,

$$p'(t) = kp(t)[1 - p(t)] \tag{8}$$

where k is a positive constant. This implies that the rate of change $p'(t)$ is near 0 when $p(t)$ is close to 0 or to 1, and is greatest when $p(t)$ is near $\frac{1}{2}$. (Equations such as (8) which involve a function and its derivative are called *differential equations*. Methods of solving such equations are discussed in Chapter 15.) ▶

Example 4 Rashevsky's† model of mass behavior applies to a population divided into two groups X and Y, for example, by allegiance to two political parties. Within each group there are a fixed number of "actives" x_0 and y_0 who are immune to attitude change. The remaining members of the groups (call their respective numbers x and y) are "passives" and subject to influence by both actives and passives.

*Richardson, L. F., "War Moods," *Psychometrika* **13**, 147–174 (1948).

†Rashevsky, N., "Studies in the Mathematical Theory of Human Relations," *Psychometrika* **4**, 221–239 (1939).

Rashevsky assumes that the numbers of passives change over time according to the equations

$$\frac{dx}{dt} = ax - by + c$$

$$\frac{dy}{dt} = \beta y - \alpha x + \gamma$$

The positive constants a and β represent the influence of passives having the same attitude, the constants b and α indicate the influence of passives holding the opposing view, while c and γ mirror the constant influence of the actives in the respective groups. ▶

Example 5 Homans* expounds a theory of group behavior based on four quantities: I, the intensity of interaction among group members; F, the level of friendliness among group members; A, the amount of activity by group members; and E, the amount of activity imposed on group members by external environmental forces. Each of these quantities represents an average over the group members and each may change with time.

The verbal treatment of Homans was translated by Simon† into the mathematical relations

$$I(t) = a_1 F(t) + a_2 A(t)$$

$$F'(t) = b[I(t) - \beta F(t)]$$

$$A'(t) = c_1[F(t) - \gamma A(t)] + c_2[E(t) - A(t)]$$

where all constants are assumed positive. The first relation indicates that interaction increases with the level of friendliness and/or amount of activity carried on within a group. The second states that the rate at which the amount of friendliness changes depends upon the disparity between the present levels of interaction and friendliness, this rate being positive when interaction is high and friendliness low, and negative in the reverse case. The third relation indicates that activity will increase when the levels of friendliness and external stimulation are high relative to the present level of activity and will decrease in the opposite situation.

PROBLEMS

In Problems 3–8, translate the verbal statements into equivalent mathematical equations.

3. The time–rate of increase of a population is proportional to the population size $N(t)$.

4. The rate of adoption of a technological innovation is proportional to the number of people $N_a(t)$ who have adopted the innovation at time t times the number of people $N_n(t)$ who have not adopted the innovation.

*Homans, G. C., *The Human Group* (Harcourt, Brace & World, Inc., New York, 1950).

†Simon, H. A., *Models of Man* (John Wiley & Sons, Inc., New York, 1957), Chapter 6.

5. Marginal revenue per unit varies with the square of the production level x.

6. The level of difficulty $D(t)$ of learning at time t decreases at a rate proportional to the product of the level of difficulty at time t and the total amount $p(t)$ of practice since the beginning of formal practice.

7. The rate at which an organization expends energy over time is proportional to the difference between the progress $p(t)$ of the organization at time t and the anticipated progress $\bar{p}(t)$.

8. The time rate of change of the number N of excited neural circuits is proportional to the average frequency of the impulses less a proportion of the number of excited neural circuits.

In Problems 9–12, find the increment $f(x + h) - f(x)$ corresponding to $h = 0.02$ and $x = 4$. Also find the average rate of change of f in the interval from $x = 4$ to $x = 4.02$.

9. $f(x) = x + 2$ **10.** $f(x) = 2x^2 - 4x + 3$

11. $f(x) = x^3 - 2x^2 + 3x - 6$ **12.** $f(x) = 2/(x + 2)$

In each of Problems 13–17, use the definition (7) to find the derivative of the indicated function.

13. $f(x) = 2x - x^2$ **14.** $g(x) = \frac{1}{3}x^3 - \frac{1}{2}x^2 + x + \sqrt{2}$

15. $u(t) = 1 + 3t + 4t^2$ **16.** $h(x) = 4x(2 - x)$

17. $v(t) = (t - 1)(t + 2)$

18. Find the slope of each of the following curves:
  (a) $y = x^2 - 2x$ at $x = 3$
  (b) $y = \frac{1}{2}x^2 + 5x + 4$ at $x = 1$

19. Suppose that total cost for a firm is given by the function

$$C(x) = 6 + 4x + 3x^2$$

where x is the amount produced. Show that the marginal cost function is

$$MC(x) = 4 + 6x$$

20. The distance d (in feet) required to stop an automobile under normal conditions is proportional to the square of the velocity (speed). If $d = 18$ when $v = 20$, find the rate of change of d when $v = 30$ and also when $v = 60$.

21. Thurstone* has developed the learning curve

$$f(x) = \frac{Lx + Lc}{x + c + a}$$

where L is the limit of practice, x is amount of formal practice, c is equivalent previous practice, and a is the rate of learning.
  (a) Find the slope of the line tangent to f at $x = 0$.

*Thurstone, L. L., "The Learning Curve Equation," reprinted in G. A. Miller, *Mathematics and Psychology* (John Wiley & Sons, Inc., New York, 1964), pp. 128–132.

(b) If $L = 80$, $a = 20$, and $c = 5$, what would be the value of $f(x)$ at $x = 25$ *if the initial rate were maintained?*

(c) How does this compare with the predicted learning performance $f(25)$?

12.2 DERIVATIVES OF SUMS, PRODUCTS, AND QUOTIENTS

The simplest results about derivatives are intuitively obvious from the interpretation of the derivative as a slope. For instance, any constant function has zero slope and the derivative reflects this. If $f(x) = c$ for all x (see Figure 4), then $f(x + h) = f(x) = c$, and we have

$$Df(x) = \lim_{h \to 0} \frac{f(x + h) - f(x)}{h} = \lim_{h \to 0} \frac{c - c}{h} = \lim_{h \to 0} 0 = 0$$

FIGURE 4 **(a) A constant function has zero slope; (b) the identity function has slope 1.**

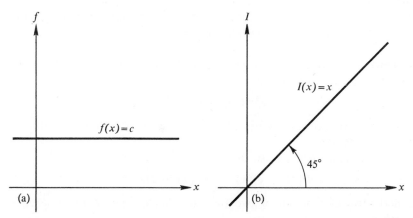

Similarly, we know from geometrical considerations that the *identity function* $I(x) = x$, whose graph is a straight line through the origin inclined $45°$ from the horizontal, must have slope 1 (see Figure 4). Using Equation (7) to obtain the derivative of I gives

$$DI(x) = \lim_{h \to 0} \frac{I(x + h) - I(x)}{h} = \lim_{h \to 0} \frac{(x + h) - x}{h} = 1$$

for every real number x. These results are both special cases of the following theorem.

Theorem 1 Let n be a non-negative integer and define the function g by $g(x) = x^n$. Then $Dg(x) = nx^{n-1}$.

PROOF   We have already proved the theorem for constant functions ($n = 0$) and the identity function ($n = 1$). For $n \geq 2$, we use the binomial expansion

$$(x + h)^n = x^n + nx^{n-1}h + \tfrac{1}{2}n(n - 1)x^{n-2}h^2 + \cdots + h^n$$

to obtain

$$Dg(x) = Dx^n = \lim_{h \to 0} \frac{(x + h)^n - x^n}{h}$$

$$= \lim_{h \to 0} \frac{[x^n + nx^{n-1}h + \tfrac{1}{2}n(n - 1)x^{n-2}h^2 + \cdots + h^n] - x^n}{h}$$

Canceling the terms involving x^n and then dividing by h gives

$$Dx^n = \lim_{h \to 0} (nx^{n-1} + \tfrac{1}{2}n(n - 1)x^{n-2}h + \cdots + h^{n-1}) = nx^{n-1} \quad \blacktriangleright$$

If n is a positive integer, the derivative of x^n is nx^{n-1}. For instance, the derivatives of x^4, x^{27} and x^2 are, respectively, $4x^3$, $27x^{26}$, and $2x$.

PROBLEMS

1. Find the derivatives of x^6, x^{11}, and x^3.

Since differentiation involves a limiting operation, it is natural to expect that properties of derivatives should parallel those of limits in general. Theorem 2 states the most important results.

Theorem 2   (a) If the derivative $Df(x)$ of the function f exists at the point x and if c is any constant, then

$$D[cf(x)] = cDf(x)$$

The derivative of a constant times a function is the constant times the derivative of the function.
   (b) If $Df(x)$ and $Dg(x)$ exist, then

$$D[(f + g)(x)] = Df(x) + Dg(x)$$

The derivative of the sum of two functions is the sum of the respective derivatives.
   (c) The results in (a) and (b) may be extended by induction to

$$D[a_1 f_1(x) + a_2 f_2(x) + \cdots + a_n f_n(x)]$$
$$= a_1 Df_1(x) + a_2 Df_2(x) + \cdots + a_n Df_n(x)$$

whenever the derivatives $Df_1(x)$, $Df_2(x)$, $\ldots$, $Df_n(x)$ exist and $a_1, a_2, \ldots, a_n$ are constants.

PROOF   For (a), we have

$$D[cf(x)] = \lim_{h \to 0} \frac{cf(x + h) - cf(x)}{h} = \lim_{h \to 0} c \frac{f(x + h) - f(x)}{h}$$

$$= c \lim_{h \to 0} \frac{f(x + h) - f(x)}{h} = cDf(x)$$

For (b), we have

$$D[f(x) + g(x)] = \lim_{h \to 0} \frac{f(x + h) + g(x + h) - [f(x) + g(x)]}{h}$$

$$= \lim_{h \to 0} \frac{f(x + h) - f(x)}{h} + \lim_{h \to 0} \frac{g(x + h) - g(x)}{h}$$

$$= Df(x) + Dg(x)$$

Finally, (c) follows by induction from (a) and (b). ▶

PROBLEMS

2. Use mathematical induction to establish Part (c) of Theorem 2.

Theorems 1 and 2 together enable us to compute the derivative of any polynomial, as in the following example.

Example 1 (a) If $f(x) = 7x^{12} - 29x^2 + 3x - 6$, then

$$\begin{aligned}
Df(x) &= D[7x^{12} - 29x^2 + 3x - 6] \\
&= D(7x^{12}) + D(-29x^2) + D(3x) + D(-6) \quad &\text{[Theorem 2(c)]} \\
&= 7Dx^{12} - 29Dx^2 + 3Dx + D(-6) \quad &\text{[Theorem 2(a)]} \\
&= 7 \cdot 12x^{11} - 29 \cdot 2x + 3 \cdot 1 + 0 \quad &\text{[Theorem 1]} \\
&= 84x^{11} - 58x + 3
\end{aligned}$$

(b) The derivative of $g(t) = -7t^3 + 21t - 2$ is

$$\begin{aligned}
g'(t) &= D[-7t^3 + 21t - 2] \\
&= D(-7t^3) + D(21t) + D(-2) \\
&= -7Dt^3 + 21Dt + 0 \\
&= -7(3t^2) + 21(1) \\
&= -21t^2 + 21
\end{aligned}$$

When $t = 4$, the derivative is $g'(4) = -21(4^2) + 21 = -315$, while at $t = 1$, $g'(1) = -21(1^2) + 21 = 0$. ▶

Example 2 Neifeld and Poffenberger* found, for $t \geq 0$, that the empirical equation

$$y(t) = a_0 + a_1 t + a_2 t^2 + a_3 t^3$$

where a_0, a_1, a_2, and a_3 are constants, described the relationship between amount $y(t)$ of work output and elapsed time t. The rate of change of work output at time t is therefore

$$y'(t) = a_1 + 2a_2 t + 3a_3 t^2$$ ▶

*Neifeld, M. R., and Poffenberger, A. T., "A Mathematical Analysis of Work Curves," *Journal of General Psychology* **1**, 448–456 (1928).

3. Calculate derivatives of

(a) $f(x) = 4x^2 - 2$

(b) $g(t) = -3t^6 + 4t^3 - 2t + 7$

(c) $h(v) = v^8 - v$

Although the derivative of a sum is the sum of the derivatives, it is not true that the derivative of a product is the product of the individual derivatives. For example, x^3 can be written as the product of x^2 and x, but the derivative $Dx^3 = 3x^2$ is not the product of $Dx^2 = 2x$ and $Dx = 1$.

Theorem 3 If the functions f and g both have derivatives at a point x, then the derivative of the product $f \cdot g$ at x is

$$D[f(x) \cdot g(x)] = f(x) Dg(x) + g(x) Df(x)$$

Thus, the derivative of the product of two functions is a sum, each term of which is the product of one of the functions times the derivative of the other.

PROOF By definition

$$D[f(x) \cdot g(x)] = \lim_{h \to 0} \frac{f(x + h)g(x + h) - f(x)g(x)}{h}$$

Rewriting the difference $f(x + h)g(x + h) - f(x)g(x)$ as

$$f(x + h)[g(x + h) - g(x)] + g(x)[f(x + h) - f(x)]$$

gives

$$D[f(x) \cdot g(x)] = \left[\lim_{h \to 0} f(x + h)\right]\left[\lim_{h \to 0} \frac{g(x + h) - g(x)}{h}\right]$$
$$+ g(x) \lim_{h \to 0} \frac{f(x + h) - f(x)}{h}$$

The second and third of the indicated limits are, of course, the respective derivatives $Dg(x)$ and $Df(x)$. As for the first, we have

$$\lim_{h \to 0} f(x + h) = \lim_{h \to 0} \left[f(x) + h\left(\frac{f(x + h) - f(x)}{h}\right)\right]$$
$$= f(x) + \left[\lim_{h \to 0} h\right]\left[\lim_{h \to 0} \frac{f(x + h) - f(x)}{h}\right]$$
$$= f(x) + 0 \cdot Df(x) = f(x) \qquad (9)$$

This completes the proof. ▶

For future reference we note that Equation (9), together with Problem 11 of Section 11.4, constitutes a proof of the following useful result.

Theorem 4   If a function has a derivative at a point x in its domain, then it must be continuous at x. ▶

Example 3   (a) $Dx^3 = x^2 Dx + xDx^2 = x^2 \cdot 1 + x \cdot 2x = 3x^2$.

(b) Since $D(3x^2 - 4x) = 6x - 4$ and $D(-6x^5 - 2x^2 + 1) = -30x^4 - 4x$, we have

$$D[(3x^2 - 4x)(-6x^5 - 2x^2 + 1)]$$
$$= (3x^2 - 4x)(-30x^4 - 4x) + (-6x^5 - 2x^2 + 1)(6x - 4)$$
$$= -126x^6 + 144x^5 - 24x^3 + 24x^2 + 6x - 4 \qquad ▶$$

Theorem 5   If the functions f and g both have derivatives at a point x and if $g(x) \neq 0$, then the derivative of the quotient f/g at x is

$$D\left[\frac{f(x)}{g(x)}\right] = \frac{g(x)Df(x) - f(x)Dg(x)}{[g(x)]^2} \qquad (10)$$

Thus, the derivative of the quotient of two functions is found by multiplying the denominator by the derivative of the numerator, subtracting the product of the numerator and the derivative of the denominator, and then dividing by the square of the denominator.

PROOF   By definition

$$D\left[\frac{f(x)}{g(x)}\right] = \lim_{h \to 0} \frac{[f(x + h)/g(x + h)] - [f(x)/g(x)]}{h}$$

The numerator $[f(x + h)/g(x + h)] - [f(x)/g(x)]$ may be rewritten as

$$\frac{f(x + h)g(x) - f(x)g(x + h)}{g(x)g(x + h)}$$

$$= \frac{g(x)[f(x + h) - f(x)] - f(x)[g(x + h) - g(x)]}{g(x)g(x + h)}$$

Dividing by h and taking limits as $h \to 0$ yields the result (10). (Remember, g is continuous so $g(x + h) \to g(x)$.) ▶

Example 4

$$D\left[\frac{3x^2 - 2x + 1}{6x^4 - 2}\right]$$

$$= \frac{(6x^4 - 2)D(3x^2 - 2x + 1) - (3x^2 - 2x + 1)D(6x^4 - 2)}{(6x^4 - 2)^2}$$

$$= \frac{(6x^4 - 2)(6x - 2) - (3x^2 - 2x + 1)(24x^3)}{(6x^4 - 2)^2}$$

$$= \frac{-36x^5 + 36x^4 - 24x^3 - 12x + 4}{(6x^4 - 2)^2} \qquad ▶$$

Using Theorem 5 we find that Theorem 1 is also valid when n is a negative integer. For if n is negative, then $-n$ is positive and

$$D(x^n) = D\left(\frac{1}{x^{-n}}\right) = \frac{x^{-n}D(1) - 1 \cdot D(x^{-n})}{(x^{-n})^2}$$

$$= \frac{x^{-n}(0) - (-n)x^{-n-1}}{x^{-2n}} \qquad \text{[since } -n \text{ is positive]}$$

$$= \frac{nx^{-n-1}}{x^{-2n}} = nx^{-n-1}x^{2n} = nx^{n-1}.$$

Example 5 (a) $D(1/x^7) = D(x^{-7}) = -7x^{-8}$.

(b) $D[6x^4 - 2x^2 + 3 - 2x^{-2} + x^{-3}] = 24x^3 - 4x + 4x^{-3} - 3x^{-4}$.

▶

Example 6 Carzo and Yanouzas* postulate that for a person whose pre-determined goal level is $g \geq 0$ and whose perceived justice per unit of reward is $a > 0$, satisfaction S is related to total reward R by

$$S(R) = \frac{aR}{g - R}$$

The rate of change of satisfaction relative to reward is

$$S'(R) = \frac{(g - R) \cdot a - aR(-1)}{(g - R)^2} = \frac{ag}{(g - R)^2}$$

Thus, the rate of change of satisfaction is inversely proportional to the square of the difference between the personal goal of the individual and the amount of reward received. ▶

Example 7 In a model of intercity migration, Galle and Taeuber† postulate that the number Y of migrants from city A to city B during a fixed time interval is given by $Y(C) = K/C^\alpha$, where C is the number of migrants competing for opportunities in city B and K and α are constants. If for illustrative purposes we assume $\alpha = 10$, then the rate of change of Y relative to C is $DY(C) = -10K/C^{11}$. ▶

PROBLEMS

In Problems 4–15, find the derivative of the given function.

4. $s(t) = 3t^4 + 2t^3 - 6t^2 + 10$ **5.** $w(z) = 4z + \frac{1}{4}z^4 + \sqrt{2}$

6. $Y(x) = (x - 1)(x^2 + 2)$ **7.** $u(r) = (r + 3)(r^2 + 2)$

*Carzo, R., and Yanouzas, J. N., *Formal Organization: A Systems Approach* (Irwin, Homewood, Ill., 1967), p. 489.

†Galle, O. R., and Taeuber, K. E., "Metropolitan Migration and Intervening Opportunities, *American Sociological Review* **31**, 5–13 (1966).

8. $s(t) = 2t^2(t^3 + 1)$

9. $w(z) = z^2(z + 2)(z - 1)$

10. $y(x) = \dfrac{x - 1}{x + 1}$

11. $z(x) = \dfrac{3x}{x + 2}$

12. $y(t) = \dfrac{(t - 1)(t^2 + t)}{t^2 + 1}$

13. $f(x) = \dfrac{x^2 + 10}{x + 1}$

14. $q(p) = \dfrac{p^3 - 1}{p^2}$

15. $u(r) = 4r + \dfrac{1}{r^2} + \dfrac{r}{(r + 2)^3}$

Find the value of the derivative at the point indicated for each of the following functions.

16. $y = (2x + 1)(x^2 + 4)$ at $x = 2$

17. $y = \dfrac{x - 2}{2 - x}$ at $x = 1$

18. $u(t) = \dfrac{t^2}{t^2 + a^2}$ at $t = a$

19. $y(x) = (x^3 + 1)(x + 2)$ at $x = 1$

20. For each of the following functions, find the value of x for which $f'(x) = 0$:

(a) $f(x) = 2x^2 - 3x + 1$ (b) $f(x) = \dfrac{1}{x^2 + 1}$

21. Prove that

$$\frac{D[f(x) \cdot g(x)]}{f(x) \cdot g(x)} = \frac{Df(x)}{f(x)} + \frac{Dg(x)}{g(x)}$$

12.3 MEAN VALUE THEOREM

For both theory and application, the most important result concerning derivatives is the Mean Value Theorem. The simplest version of this theorem is called Rolle's Theorem.

Rolle's Theorem   Suppose the function f has a derivative at each point in an interval $[a, b]$* and that $f(a) = f(b) = 0$ (see Figure 5). Then there is at least one point c between a and b at which the derivative of f is zero.

PROOF   Geometrically the statement is obvious: What goes up and comes down must level off at some point. A formal proof is not difficult. There are three cases to consider:

(i) The values of f are all zero. In this case $f'(x) = 0$ for all x between a and b and we arbitrarily take $c = (a + b)/2$.

(ii) There are points at which f is positive. By Theorem 5 of Chapter 11, there is a point c between a and b at which f takes its largest value.

*Recall (Section 6.3) that if $a \leq b$, the notation $[a, b]$ indicates the set $\{x: a \leq x \leq b\}$ of all points on the number line lying between a and b, including a and b themselves.

FIGURE 5

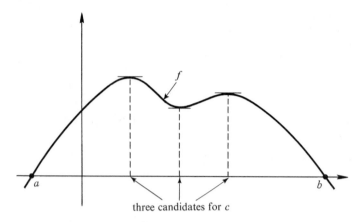

three candidates for c

That is, $f(x) \le f(c)$ for all x in the interval. Now $f'(c)$ is defined by

$$f'(c) = \lim_{h \to 0} \frac{f(c + h) - f(c)}{h}$$

Since $f(c + h) \le f(c)$, these ratios are nonpositive when $h > 0$ and non-negative when $h < 0$. At the same time, the fact that the limit exists means that for small values of h, either positive or negative, all ratios must be close to that limit. Obviously, zero is the only possible value for the limit.

(iii) The function f is never positive but there are points at which it is negative. In this case there is a point c at which f takes its smallest value. An argument similar to that in (ii) shows that here also $f'(c) = 0$. ▶

In the statement of Rolle's Theorem, the assumption that the function f possesses a derivative at each point in the interval is critical. In Figure 6, for

FIGURE 6

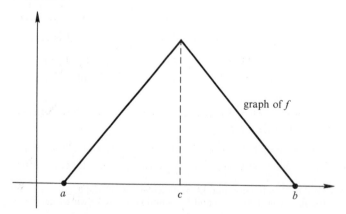

graph of f

instance, the function f takes the value zero at a and at b, but there is no point between a and b at which the derivative is zero. This is because the function has no derivative at the point c where f takes its maximum value.

Example 1 The function $f(x) = x^2 + x - 2$ is zero at $x = 1$ and at $x = -2$ (see Figure 7). Its derivative

$$f'(x) = 2x + 1$$

is zero at $x = -\frac{1}{2}$.

FIGURE 7

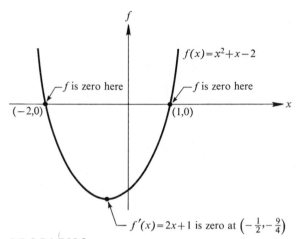

$f(x) = x^2 + x - 2$

$-f$ is zero here

$-f$ is zero here

$(-2,0)$

$(1,0)$

$f'(x) = 2x + 1$ is zero at $\left(-\frac{1}{2}, -\frac{9}{4}\right)$

PROBLEMS

1. The function $f(x) = x^3 - 4x$ is zero at $x = 0$, $x = 2$, and $x = -2$. At what point(s) is its derivative zero?

Mean Value Theorem Suppose the real-valued function f has a derivative at every point in the interval $[a, b]$. Then there is at least one point c between a and b at which

$$f'(c) = \frac{f(b) - f(a)}{b - a} \tag{11}$$

PROOF A geometric interpretation is shown in Figure 8. The ratio

$$\frac{f(b) - f(a)}{b - a}$$

is the slope of the line between the points $(a, f(a))$ and $(b, f(b))$ on the graph of f. The slope of f at a point c is $f'(c)$ and the theorem simply asserts that there is at least one point c between a and b at which the slope of the curve is the same as the slope of the line.

The proof is an easy consequence of Rolle's Theorem. The function f need not satisfy the assumptions of that theorem (since $f(b)$ and $f(a)$ may

FIGURE 8

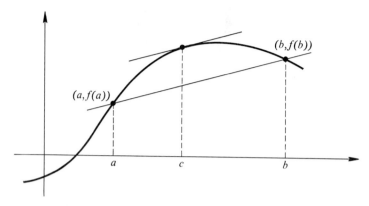

not be zero), but the function

$$g(x) = f(x) - f(a) - \frac{f(b) - f(a)}{b - a}(x - a)$$

obtained by subtracting points on the line from corresponding points of f, does. Specifically,

$$g(a) = f(a) - f(a) - \frac{f(b) - f(a)}{b - a}(a - a) = 0$$

$$g(b) = f(b) - f(a) - \frac{f(b) - f(a)}{b - a}(b - a) = 0$$

and the function g has a derivative if f does. In fact,

$$g'(x) = f'(x) - \frac{f(b) - f(a)}{b - a} \qquad (12)$$

Applying Rolle's Theorem to Equation (12) completes the proof. ▶

Example 2 If $r(x)$ is the revenue obtained by a firm from the sale of x units of output, the average revenue is $\bar{r}(x) = r(x)/x$. Assuming that zero revenue is obtained from zero output (that is, $r(0) = 0$), this may be re-written as $\bar{r}(x) = [r(x) - r(0)]/(x - 0)$. According to the Mean Value Theorem, there is a production level $x^* < x$ at which the marginal revenue $r'(x^*)$ is equal to the average revenue $\bar{r}(x)$. ▶

PROBLEMS

2. Draw a graph like Figure 8 to illustrate Example 2.

3. For the function $f(x) = x^3$, find the point c between $a = 0$ and $b = 1$ at which $f'(c) = [f(1) - f(0)]/(1 - 0)$.

A function f is said to be *increasing* in an interval if $f(x) \leq f(y)$ whenever x and y are points in the interval and $x < y$. Similarly, f is *decreasing* in an interval if, whenever x and y are in the interval and $x < y$, then $f(x) \ge f(y)$ (see Figure 9). The Mean Value Theorem enables us to determine where a function is increasing or decreasing by looking at its derivative. Specifically, if the derivative is positive throughout an interval, the function must be increasing in that interval, while if the derivative is negative, the function must be decreasing.

FIGURE 9 (a) An increasing function on $[a, b]$; (b) a decreasing function on $[a, b]$.

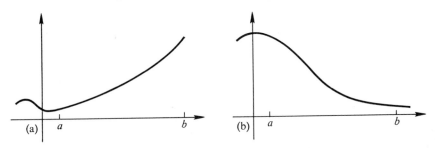

To see this, take $a \le z < y \le b$ and suppose that $f'(x) > 0$ for all x between a and b (Figure 10). From the Mean Value Theorem we know that there is a point c between z and y for which

$$\frac{f(y) - f(z)}{y - z} = f'(c)$$

or, equivalently,

$$f(y) - f(z) = f'(c)(y - z)$$

Since $f'(c) > 0$ and $y > z$, it follows that $f(y)$ exceeds $f(z)$. But z and y were any two points between a and b, so we conclude that f is increasing throughout the interval.

FIGURE 10

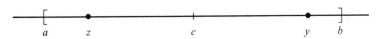

A similar argument may be used to establish the fact that a function decreases in any interval in which its derivative is negative. Together, these results provide a simple method for sketching the graph of a function.

Example 3 The derivative of

$$f(x) = x^3 - 3x^2 + 4$$

is

$$f'(x) = 3x^2 - 6x = 3x(x - 2)$$

When $x < 0$, both $3x$ and $x - 2$ are negative so that $f'(x)$, being the product of two negative numbers, is positive. For x between 0 and 2, $3x$ is positive but $x - 2$ is negative, so $f'(x)$ is negative. Finally, for $x > 2$, $f'(x)$ is positive (see Figure 11). It follows that $f(x)$ is increasing when $x < 0$ and when $x > 2$, but decreasing when $0 < x < 2$. At $x = 0$ and $x = 2$, the slope of f is zero.

FIGURE 11

Putting this information together with the fact that $f(0) = 0^3 - 3(0)^2 + 4 = 4$ and $f(2) = 2^3 - 3(2^2) + 4 = 0$ yields Figure 12 as the graph of the function. Note carefully how positive, negative, and zero values of f' are reflected in the graph.   ▶

FIGURE 12

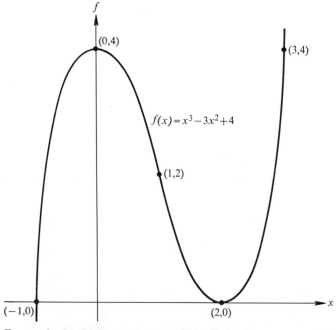

Example 4 Suppose that a political candidate assumes that the number of votes he will receive is related to the amount of money that he spends on his campaign. He postulates that when x thousands of dollars are expended, the plurality $P(x)$ of votes that he will receive is given for $x \geq 0$ by

$$P(x) = \tfrac{1}{3}x^3 - 4x^2 + 12x$$

The derivative

$$P'(x) = x^2 - 8x + 12 = (x - 2)(x - 6)$$

is positive when $0 < x < 2$ and when $x > 6$, and negative when $2 < x < 6$. A graph of P, increasing for $x < 2$ and $x > 6$, and decreasing between 2 and 6, is shown in Figure 13. ▶

FIGURE 13

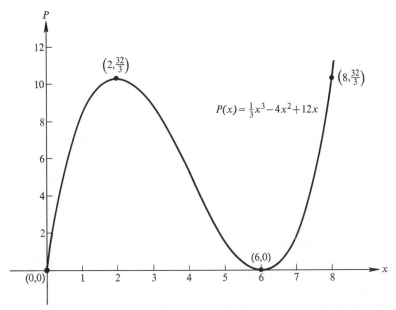

$$(2, \tfrac{32}{3})$$

$$(8, \tfrac{32}{3})$$

$$P(x) = \tfrac{1}{3}x^3 - 4x^2 + 12x$$

$$(6, 0)$$

$$(0, 0)$$

Example 5* Let $C(x)$ denote the cost of producing x units of some commodity, so that $\overline{C}(x) = C(x)/x$ is the average cost per unit. Writing

$$C(x) = x \cdot \overline{C}(x)$$

and differentiating gives

$$C'(x) = x\overline{C}'(x) + \overline{C}(x) \quad \text{or} \quad x\overline{C}'(x) = C'(x) - \overline{C}(x)$$

Recalling (Example 1 of Section 12.1) that $C'(x)$ is the marginal cost at production level x, we see that in order for average cost to rise (that is, for $\overline{C}'(x) > 0$), the marginal cost $C'(x)$ must exceed the average cost $\overline{C}(x)$. Similarly, if marginal and average costs are equal, then average cost remains constant, while marginal cost must be less than average cost in order for average cost to decline. ▶

PROBLEMS

4. For each of the following functions, find where the graph increases, where it decreases, and draw a sketch.
   (a) $f(x) = 3x^3 - 2x^2$
   (b) $g(t) = \tfrac{1}{2}t^2 + 2t + 4$
   (c) $h(z) = 2z^3 - 3z^2 + 6z - 6$

*Adapted from Baumol, W. J., *Economic Theory and Operations Analysis* (Prentice-Hall, Inc., Englewood Cliffs, N. J., 1965), 2nd ed.

5. Use the Mean Value Theorem to prove that a function must be decreasing in any interval in which its derivative is negative.

6. In what intervals are the following functions increasing? decreasing? Sketch the graph of each function.

(a) $f(x) = 2x - 3$

(b) $g(t) = t^2 - 2t + 6$ (*Hint:*   Complete the square.)

(c) $h(z) = \begin{cases} 2z - 3 & \text{if } z \le -4 \\ z^2 - 2z + 6 & \text{if } z > -4 \end{cases}$

(d) $f(y) = [y]$, where $[y]$ denotes the largest integer less than or equal to y (this is referred to as the bracket function)

7. Check that the hypotheses of Rolle's Theorem are satisfied by each of the following functions on the designated interval and find a point c such that $f'(c) = 0$.

(a) $f(x) = x^3 - x$ $[-1, 1]$ (b) $g(x) = 4x - 2x^2$ $[0, 2]$

8. Show by example that the point c in Rolle's Theorem need not be unique.

9. Prove that if the function f has a derivative at each point in the interval $[a, b]$ and if $f(a) = f(b)$, then there exists a point $c \in [a, b]$ such that $f'(c) = 0$.

10. Check that the hypotheses of Problem 9 are satisfied and find a point c in the specified interval such that $f'(c) = 0$.

(a) $f(x) = x^2$ $[-1, 1]$ (b) $f(x) = x^3 - x$ $[-1, 1]$

(c) $f(x) = |2x - 1|$ $[-1, 2]$

11. Prove that if $f'(x) = 0$ for all $x \in [a, b]$, then f is a constant function over $[a, b]$. (*Hint:*   Apply the Mean Value Theorem.)

12. Prove that if f and g are functions such that $f'(x) = g'(x)$ for every $x \in [a, b]$, then there is a constant c such that $f(x) = g(x) + c$. (*Hint:* Use Problem 11.)

12.4   COMPOSITE FUNCTIONS AND THE CHAIN RULE

In our previous discussion (Chapter 4) we have visualized a function, such as $h(x) = (2x - 3)^2$, as a system or machine which accepts an input x and produces a corresponding output $h(x) = (2x - 3)^2$ as in Figure 14. For some purposes, however, it is convenient to think of h as a two-stage system in which the input x is first transformed, by means of a function f, into the quantity $f(x) = 2x - 3$. This quantity becomes the input of a function g and is then squared (see Figure 15). Symbolically, if $f(x) = 2x - 3$ and $g(y) = y^2$ then

$$h(x) = g[f(x)] = g(2x - 3) = (2x - 3)^2$$

In general, if f has domain A and range B, and g has a domain which contains B, the function h defined for each $a \in A$ by

$$h(a) = g[f(a)]$$

FIGURE 14

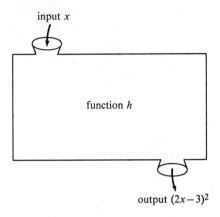

input x

function h

output $(2x-3)^2$

FIGURE 15

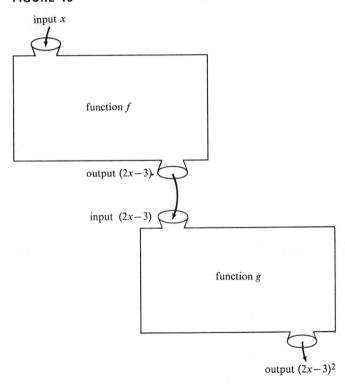

input x

function f

output $(2x-3)$

input $(2x-3)$

function g

output $(2x-3)^2$

has domain A and range C contained in the range of g (see Figure 16). We shall call h the *composite mapping of g with f* and write $h = g \circ f$. The function h is sometimes referred to as the "composition of g with f." The order of composition is important. For instance, in our first example where

FIGURE 16

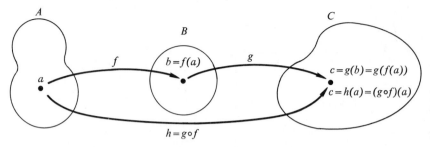

$f(x) = 2x - 3$ and $g(y) = y^2$, we have

$$(g \circ f)(x) = g[f(x)] = g(2x - 3)$$
$$= (2x - 3)^2 = 4x^2 - 12x + 9$$

while

$$(f \circ g)(x) = f[g(x)] = f(x^2)$$
$$= 2x^2 - 3$$

Example 1 Suppose that total cost is given by the function

$$f(x) = 4x^2 + 4x + 1 = (2x + 1)^2$$

where x is the amount produced. If we let

$$h(x) = 2x + 1 \quad \text{and} \quad g(y) = y^2$$

then

$$f(x) = (g \circ h)(x) \qquad \blacktriangleright$$

The theorem which tells how to differentiate a composite function is called the Chain Rule.

Chain Rule Suppose the real-valued functions f and g have continuous derivatives throughout their domains. Then the derivative of the composite function $h(x) = (g \circ f)(x)$ at the point a is given by the product

$$h'(a) = g'[f(a)] \cdot f'(a) \tag{13}$$

In words, the derivative of h at the point a is the product of the derivative of g at the point $f(a)$ times the derivative of f at the point a.

PROOF By definition

$$h'(a) = \lim_{t \to 0} \frac{h(a + t) - h(a)}{t} = \lim_{t \to 0} \frac{g[f(a + t)] - g[f(a)]}{t} \tag{14}$$

According to the Mean Value Theorem, there exists a point u between $f(a)$ and $f(a + t)$ at which

$$g[f(a + t)] - g[f(a)] = g'(u)[f(a + t) - f(a)] \tag{15}$$

Substituting (15) into (14) we obtain

$$h'(a) = \left[\lim_{t \to 0} g'(u)\right]\left[\lim_{t \to 0} \frac{f(a+t) - f(a)}{t}\right] \tag{16}$$

The second of these limits is, by definition, $f'(a)$. To evaluate $\lim_{t \to 0} g'(u)$, we note first that $\lim_{t \to 0} f(a + t) = f(a)$ since f is continuous (see Theorem 4). Since u lies between $f(a)$ and $f(a + t)$, as in Figure 17, it follows that $\lim_{t \to 0} u = f(a)$, also. Hence, using the assumed continuity of the derivative g', we obtain

$$\lim_{t \to 0} g'(u) = \lim_{u \to f(a)} g'(u) = g'(f(a)) \tag{17}$$

Inserting (17) into (16) yields the desired result (13). ▶

FIGURE 17

$f(a)$ u $f(a+t)$

It is important to remember that the three derivatives in (13) are not all evaluated at the same point. The derivatives of h and of f are evaluated at a, while the derivative of g is evaluated at $f(a)$.

The Leibniz notation (see Section 12.1) provides an easily remembered format for the Chain Rule. Let $y = f(x)$ and $z = g(y) = g(f(x))$. Then the Chain Rule (13) takes the form

$$\frac{dz}{dx} = \frac{dz}{dy} \cdot \frac{dy}{dx} \tag{18}$$

The formula works *as though* the derivatives were ratios of quantities dx, dy, and dz which could be treated just as ordinary numbers.

Example 2 (a) To differentiate $(x^4 - 3x + 2)^{14}$, let $f(x) = x^4 - 3x + 2$ and $g(y) = y^{14}$. Then $g'(y) = 14y^{13}$ and $f'(x) = 4x^3 - 3$. Taking $y = f(x)$ we have

$$D(x^4 - 3x + 2)^{14} = 14(x^4 - 3x + 2)^{13} \cdot (4x^3 - 3)$$

(b) To differentiate $[x^2 + (1/x^3)]^{-3}$, let $y = f(x) = x^2 + (1/x^3)$ and $z = g(y) = y^{-3} = [x^2 + (1/x^3)]^{-3}$. Then

$$\frac{dz}{dx} = \frac{dz}{dy} \cdot \frac{dy}{dx} = -3y^{-4}(2x - 3x^{-4}) = -3\left(x^2 + \frac{1}{x^3}\right)^{-4}\left(2x - \frac{3}{x^4}\right) \text{▶}$$

Both parts of Example 2 are special cases of the following important result.

Theorem 6 Suppose n is an integer and let $h(x) = [f(x)]^n$, where f is a differentiable function. Then h has a derivative given by

$$h'(x) = n[f(x)]^{n-1}f'(x)$$

PROOF In the Chain Rule (13), let $g(y) = y^n$. Then $g'(y) = ny^{n-1}$ and we have

$$h'(x) = g'(f(x))f'(x) = n[f(x)]^{n-1}f'(x)$$

as asserted. ▶

Example 3 (a) The derivative of

$$h(x) = (3x^4 - x^2 + 1)^{-2} \text{ is } -2(3x^4 - x^2 + 1)^{-3}(12x^3 - 2x)$$

(b) The derivative of $[(x^2 - 1)/(2x + 6)]^7$ is

$$7\left[\frac{x^2 - 1}{2x + 6}\right]^6 \left[\frac{(2x + 6)2x - (x^2 - 1)2}{(2x + 6)^2}\right] = \frac{7(x^2 - 1)^6(2x^2 + 12x + 2)}{(2x + 6)^8}$$

(c) The derivative of $(3x - 4)^2(-x^3 + 3x + 2)^4$ is

$$(3x - 4)^2[4(-x^3 + 3x + 2)^3(-3x^2 + 3)]$$
$$+ (-x^3 + 3x + 2)^4[2(3x - 4)^1 \cdot 3]$$
$$= 6(3x - 4)(-x^3 + 3x + 2)^3[2(3x - 4)(-x^2 + 1) + (-x^3 + 3x + 2)]$$
$$= 6(3x - 4)(-x^3 + 3x + 2)^3[-7x^3 + 8x^2 + 9x - 6] \quad ▶$$

We have established that the formula $Dx^n = nx^{n-1}$ is valid for all integers n whether positive, negative, or zero. Using Theorem 6 we can show that this formula is also valid if n is a rational number. To see this, let $f(x) = x^{p/q}$, where p and q are integers. Then $[f(x)]^q = x^p$ and differentiating both sides of this equation we find

$$q[f(x)]^{q-1}f'(x) = px^{p-1}$$

Solving for $f'(x)$ yields

$$f'(x) = \frac{p}{q}\frac{x^{p-1}}{[f(x)]^{q-1}} = \frac{p}{q}\frac{x^{p-1}}{[x^{p/q}]^{q-1}} = \frac{p}{q}x^{(p/q)-1}$$

Example 4 (a) $D\sqrt{x} = Dx^{1/2} = \frac{1}{2}x^{-1/2}$

(b) $D\dfrac{1}{\sqrt{x^3}} = Dx^{-3/2} = -\frac{3}{2}x^{-5/2}$

(c) $D\left(\dfrac{x^2 - 1}{x^3 + 2x + 1}\right)^{3/4} = \dfrac{3}{4}\left(\dfrac{x^2 - 1}{x^3 + 2x + 1}\right)^{-1/4}$

$$\times \left[\frac{(x^3 + 2x + 1)2x - (x^2 - 1)(3x^2 + 2)}{(x^3 + 2x + 1)^2}\right] \quad ▶$$

PROBLEMS

Find derivatives of each of the following functions. The letters a, b, and c denote constants.

1. $f(x) = (x^2 - 2x + 3)^2$

2. $g(t) = \sqrt{1 - t}$

3. $s(u) = \dfrac{u}{\sqrt{a^2 + u^2}}$

4. $h(x) = \sqrt[3]{(1 + x^2)(1 - 4x)}$

5. $u(z) = (1 - 2z)^{3/2}$

6. $f(x) = x(a^2 - x^2)^{-1/2}$

7. $s(t) = \dfrac{1}{(t^2 - 5)^2}$

8. $y(x) = ax^3 + bx^2 + c$

9. $g(x) = \sqrt{\dfrac{7 - x}{12 + x}}$

10. $f(w) = (w - 2)^5 + \dfrac{1}{(w - 2)^5}$

11. $s(t) = \dfrac{4t^2 - 3t + 1}{\sqrt{6 + 5t^2}}$

12. $h(x) = \sqrt{1 - (1 + x)^{1/2}}$

13. If Q is a demand function relating quantity $Q(p)$ and price p, the elasticity of demand is defined as

$$E(p) = \dfrac{-\dfrac{DQ(p)}{Q(p)}}{\dfrac{Dp}{p}} = -p\,\dfrac{Q'(p)}{Q(p)} \quad {}^*$$

Prove that elasticity of demand $E(p)$ is unity if $Q(p) = c/p$ for some constant c.

14. Prove that if demand is inelastic (that is, $E(p) < 0$), then a rise in price will increase consumer demand while if demand is elastic ($E(p) > 0$), then a fall in price will increase demand.

15. Sometimes the order of composition of two functions is immaterial. If $f(x) = x$ and $g(y) = y^2$ show that for all x, $(f \circ g)(x) = (g \circ f)(x)$.

16. Let $f(x) = x^2$ and $g(y) = \sqrt{y}$. Argue that $(f \circ g)(x)$ is defined only for $x \geq 0$, while $(g \circ f)(x)$ is defined for all $x \in R$. Hence, as sets of ordered pairs, $g \circ f$ and $f \circ g$ are different functions, although, when both are defined, they take the same values.

17. Prove that the operation of composing functions is associative. That is, prove that $h \circ (g \circ f) = (h \circ g) \circ f$.

12.5 DERIVATIVES OF TRIGONOMETRIC FUNCTIONS

In this section we continue our discussion of differentiation by considering the trigonometric functions, of which the sine and the cosine are the most important. The derivatives of the sine and cosine functions are obtained in Theorem 8 below. In order to prove Theorem 8, though, we need first to

---

*The minus sign is a convenience to make elasticity non-negative since demand $Q(p)$ normally decreases as p increases.

know a bit more about limits. The proof of our first theorem is quite simple and is left as an exercise (compare with Problem 6 of Section 5.3).

Theorem 7 (a) Suppose that $\lim_{x \to p} f(x)$ and $\lim_{x \to p} g(x)$ both exist and that for some positive number r, the inequality

$$f(x) \leq g(x)$$

holds throughout the interval $[p - r, p + r]$. Then

$$\lim_{x \to p} f(x) \leq \lim_{x \to p} g(x)$$

(b) Suppose that $\lim_{x \to p} f(x)$ and $\lim_{x \to p} g(x)$ exist and are equal and that for some positive number r, the inequality

$$f(x) \leq h(x) \leq g(x)$$

holds throughout the interval $[p - r, p + r]$. Then $\lim_{x \to p} h(x)$ exists and

$$\lim_{x \to p} f(x) = \lim_{x \to p} h(x) = \lim_{x \to p} g(x) \qquad \blacktriangleright$$

From elementary trigonometry we know that if the line emanating from the origin and passing through the point (x, y) makes an angle θ with the positive x axis, then

$$\cos \theta = \frac{x}{r}$$

$$\sin \theta = \frac{y}{r}$$

$$\tan \theta = \frac{y}{x}$$

where $r = \sqrt{x^2 + y^2}$ is the distance between $(0, 0)$ and (x, y) (see Figure 18(a)).

Thus in Figure 18(b) the lengths of lines PQ and TR are, respectively, $\sin \theta$ and $\tan \theta$. Since the radius of the circle is unity, the number θ represents the length of the arc PR as well as the area of the sector OPR. Comparing lengths

FIGURE 18

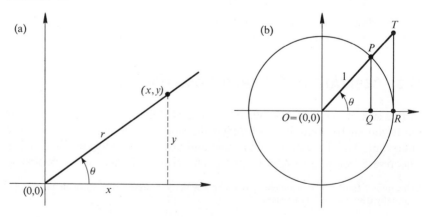

of *PQ* and *PR* shows we must have

$$\sin \theta < \theta \tag{19}$$

at least for values of θ which lie between 0 and $\pi/2$. Moreover, using the fact that $\sin(-\theta) = -\sin\theta$, we find

$$-\theta < \sin\theta < \theta \tag{20}$$

whenever $-\pi/2 < \theta < \pi/2$.

Since $\lim_{\theta \to 0} \theta = 0$, so that $\lim_{\theta \to 0}(-\theta) = -\lim_{\theta \to 0}\theta = 0$ also, it follows from Theorem 7(b) and Equation (20) that

$$\lim_{\theta \to 0} \sin\theta = 0 \tag{21}$$

Applying Equation (21) to the identity $\sin^2\theta + \cos^2\theta = 1$, we obtain

$$\lim_{\theta \to 0} \cos^2\theta = 1 - \lim_{\theta \to 0}\sin^2\theta = 1$$

Thus

$$\lim_{\theta \to 0} \cos\theta = 1 \tag{22}$$

since the cosine function has only positive values near $\theta = 0$. Since $\sin 0 = 0$ and $\cos 0 = 1$, Equations (21) and (22) imply that the sine and cosine functions are continuous at $\theta = 0$. Comparing areas of *OPR* and *OTR* and using (19), we find for $0 < \theta < \pi/2$ that

$$\sin\theta < \theta < \tan\theta$$

Dividing by $\sin\theta$ gives

$$1 < \frac{\theta}{\sin\theta} < \frac{1}{\cos\theta} \tag{23}$$

Since $\tan(-\theta) = -\tan\theta$ we also have, for $-\pi/2 < \theta < 0$,

$$\sin\theta > \theta > \tan\theta$$

If we divide by $\sin\theta$ (which is negative for these values of θ) we again obtain

$$1 < \frac{\theta}{\sin\theta} < \frac{1}{\cos\theta} \tag{24}$$

Since $\lim_{\theta \to 0} \cos\theta = 1$, we find, combining (23) and (24), that

$$\lim_{\theta \to 0} \frac{\theta}{\sin\theta} = 1 \tag{25}$$

Writing

$$\frac{\cos h - 1}{h} = \frac{\cos^2 h - 1}{h(\cos h + 1)} = \frac{-\sin^2 h}{h(\cos h + 1)} = \frac{\sin h}{h}\left(\frac{-\sin h}{\cos h + 1}\right)$$

and using Equations (21), (22), and (25), we see that

$$\lim_{h \to 0} \frac{\cos h - 1}{h} = \left(\lim_{h \to 0} \frac{\sin h}{h} \right) \left(\lim_{h \to 0} \frac{-\sin h}{\cos h + 1} \right)$$

$$= 1 \left(\frac{0}{1 + 1} \right) = 0 \tag{26}$$

A standard trigonometric formula, which holds for all values of a and θ, is

$$\sin (a + \theta) = \sin a \cos \theta + \cos a \sin \theta \tag{27}$$

If we let $x = a + \theta$ then $x \to a$ as $\theta \to 0$. Thus, taking limits in (27) we find

$$\lim_{x \to a} \sin x = \lim_{\theta \to 0} \sin (a + \theta)$$
$$= (\sin a) \lim_{\theta \to 0} \cos \theta + (\cos a) \lim_{\theta \to 0} \sin \theta$$
$$= (\sin a) \cdot 1 + (\cos a) \cdot 0$$
$$= \sin a$$

In the language of Section 11.4, this shows that *the sine function is continuous at every point.*

PROBLEMS

1. Use the trigonometric formulas

$$\cos (\theta + \alpha) = \cos \theta \cos \alpha - \sin \theta \sin \alpha$$

$$\tan \theta = \frac{\sin \theta}{\cos \theta}$$

to show that the cosine function is continuous at all points, while $\tan \theta$ is continuous except where $\cos \theta = 0$.

Formulas (25) and (26) enable us to find derivatives of the basic trigonometric functions.

Theorem 8 The derivative of the sine function is the cosine function. The derivative of the cosine function is the negative of the sine function. Symbolically,

$$D \sin x = \cos x$$

$$D \cos x = -\sin x$$

PROOF By definition,

$$D \sin x = \lim_{\theta \to 0} \frac{\sin (x + \theta) - \sin x}{\theta}$$

Since $\sin (x + \theta) = \sin x \cos \theta + \cos x \sin \theta$, this may be rewritten as

$$D \sin x = \lim_{\theta \to 0} \left[\sin x \frac{\cos \theta - 1}{\theta} + \cos x \frac{\sin \theta}{\theta} \right]$$

Using (25) and (26) gives the desired result, namely $D \sin x = \cos x$.

For the derivative of the cosine function, we obtain

$$D \cos x = \lim_{\theta \to 0} \frac{\cos (x + \theta) - \cos x}{\theta}$$

$$= \lim_{\theta \to 0} \frac{\cos x \cos \theta - \sin x \sin \theta - \cos x}{\theta}$$

$$= (\cos x) \lim_{\theta \to 0} \frac{\cos \theta - 1}{\theta} - (\sin x) \lim_{\theta \to 0} \frac{\sin \theta}{\theta}$$

$$= -\sin x \qquad \blacktriangleright$$

Example 1 The interplay between the sine and cosine functions is readily apparent from the graphs in Figure 19. The sine function is increasing when its derivative (the cosine) is positive, and decreasing when the cosine is negative. The cosine function increases when the sine function is negative (that is, when its derivative, which is the negative of the sine function, is positive), and decreases when the sine function is positive. $\qquad \blacktriangleright$

FIGURE 19

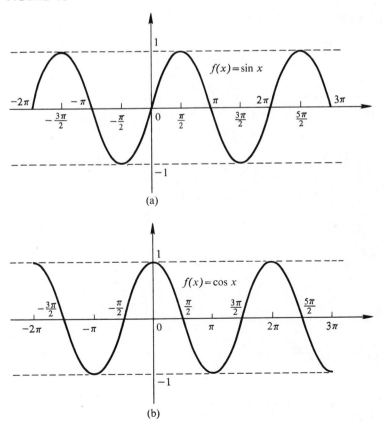

(a)

(b)

Since all other trigonometric functions can be expressed as quotients involving the sine and cosine, their derivatives can be obtained by using Theorem 8 in conjunction with Theorem 5. For instance,

$$D \tan x = D \frac{\sin x}{\cos x} = \frac{(\cos x)(\cos x) - (\sin x)(-\sin x)}{\cos^2 x}$$

$$= \frac{1}{\cos^2 x} = \sec^2 x$$

In a similar manner we find $D \cot x = -\csc^2 x$, $D \sec x = \sec x \tan x$, and $D \csc x = -\csc x \cot x$. The proofs will be left as an exercise. Table 2, Section 12.7, includes these trigonometric differentiation formulas along with other differentiation formulas.

Example 2 (a) To differentiate $h(x) = \cos(2x + 3)$, let $f(x) = 2x + 3$ and $g(y) = \cos y$ so that $h(x) = g(f(x))$. Applying the Chain Rule gives

$$D \cos(2x + 3) = h'(x) = g'(f(x))f'(x)$$

$$= -\sin(2x + 3) \cdot 2 = -2 \sin(2x + 3)$$

(b) To differentiate $h(x) = \sin(x^2 - 1)^{3/2}$, let $y = f(x) = (x^2 - 1)^{3/2}$ and $z = g(y) = \sin y = \sin(x^2 - 1)^{3/2}$. Then

$$Dh(x) = \frac{dz}{dx} = \frac{dz}{dy} \cdot \frac{dy}{dx} = (\cos y) \cdot \tfrac{3}{2}(x^2 - 1)^{1/2} \cdot 2x$$

$$= 3x(x^2 - 1)^{1/2} \cos(x^2 - 1)^{3/2}$$

(c)

$$D[\cos(2x^{1/2} - 1)]^{5/2}$$

$$= \tfrac{5}{2}[\cos(2x^{1/2} - 1)]^{3/2}[-\sin(2x^{1/2} - 1)][2 \cdot \tfrac{1}{2}x^{-1/2}]$$

$$= -\tfrac{5}{2}x^{-1/2}[\cos(2x^{1/2} - 1)]^{3/2} \sin(2x^{1/2} - 1)$$

(d)

$$D \sin\left(\frac{x^2 + x}{x^3 - 3}\right) = \cos\left(\frac{x^2 + x}{x^3 - 3}\right)\left[\frac{(x^3 - 3)(2x + 1) - (x^2 + x)3x^2}{(x^3 - 3)^2}\right]$$

$$= -\frac{x^4 + 2x^3 + 6x + 3}{(x^3 - 3)^2} \cos\left(\frac{x^2 + x}{x^3 - 3}\right)$$

(e)

$$D\left[\frac{\sin(x^2 - 1)}{x + \tan 3x}\right]$$

$$= \frac{[x + \tan 3x][\cos(x^2 - 1)]2x - [\sin(x^2 - 1)][1 + 3 \sec^2 3x]}{[x + \tan 3x]^2} \quad \blacktriangleright$$

PROBLEMS

2. Find derivatives of

   (a) $\cos (t - 1)^2$ (b) $\tan (-x^3 + 2)$

   (c) $[\sin (2x)][\cos (x - 1)^{3/2}]$

3. Use Theorem 8 and the relations $\cot x = \cos x/\sin x$, $\sec x = 1/\cos x$, and $\csc x = 1/\sin x$ to derive formulas for derivatives of the cotangent, secant, and cosecant functions.

Example 3 In a study of the beetle *pyrophorus*, Kropp* found that the average angle of orientation α at which the beetle climbed an inclined plane was related to θ, the angle of incline of the plane, according to the function

$$\alpha(\theta) = K \sin \theta + C \quad\quad K > 0$$

where C and K are constants. The rate of change of α with respect to θ is thus

$$\alpha'(\theta) = K \cos \theta \quad\quad\quad\quad \blacktriangleright$$

PROBLEMS

In Problems 4–17, find the derivative of the given function.

4. $f(x) = \tan 2x$ 5. $h(y) = y \tan y^2$

6. $g(t) = (\sin 2t)(\cos^2 3t)$ 7. $g(x) = \tan^2 (2x^2)$

8. $f(z) = z/\sin 4z$ 9. $f(x) = \cos (2x - 4)$

10. $f(t) = t^2 \cos 3t$ 11. $f(w) = \sqrt{\sin w}$

12. $f(w) = w + \cot w$ 13. $f(x) = x \tan^2 x$

14. $f(u) = \sec 2u \tan u$ 15. $f(x) = x^2/\cos^2 x$

16. $f(t) = \csc (4t - 2)$ 17. $f(x) = \tan (x + 1)^{1/2}$

18. Sketch the graphs of

   (a) $\sin (x/2)$ (b) $\cos (1/x)$ (c) $\cos (2x + \pi)$

19. Prove that

   (a) $D \sin ax = a \sin (ax + \pi/2)$

   (b) $D \cos ax = a \cos (ax + \pi/2)$

   (*Hint:* $\cos (\alpha + \beta) = \cos \alpha \cos \beta - \sin \alpha \sin \beta$)

12.6 INVERSE FUNCTIONS—THE INVERSE TRIGONOMETRIC FUNCTIONS

A function f is said to be one–one, or 1–1, on a set A if no two points in A have the same image; that is, if $f(x) \neq f(y)$ whenever $x \in A$, $y \in A$, and $x \neq y$. For instance, if $\alpha \neq 0$, the function $g(x) = \alpha x + \beta$ is 1–1 on the

---

*Kropp, B., "Geotropic Orientation in Arthropods: IV, The Beetle *pyrophorus*," *Journal of General Psychology* **2**, 484–488 (1929).

entire real line. In this case, if $x \neq y$, then $\alpha x + \beta \neq \alpha y + \beta$ (see Figure 20 (a) and (b)). On the other hand, if $\alpha = 0$, then $g(x)$ is not 1–1 (see Figure 20 (c)). In this case, $\alpha x + \beta = \alpha y + \beta = \beta$, regardless of the values of x and y.

FIGURE 20

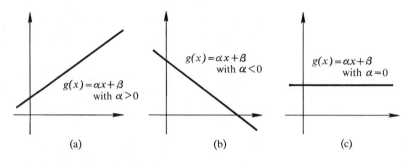

(a)   (b)   (c)

The sine function is 1–1 between 0 and $\pi/2$ and also between $\pi/2$ and $3\pi/2$, but not between 0 and π since, for instance, $\sin 0 = \sin \pi = 0$. (See Figure 19, Section 12.5.) The functions x^2 and $|x|$ are 1–1 in an interval from a to b if a and b are both positive or both negative. However, neither of these functions is 1–1 in the interval $[-2, 2]$ since $(-1)^2 = 1^2$ and $|-1| = |1|$. (See Figure 21.)

FIGURE 21

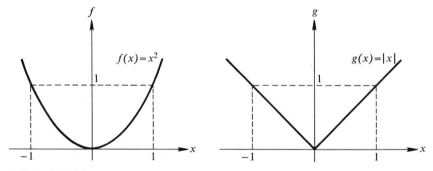

PROBLEMS

1. List some other intervals in which the sine function is one–one.

2. A function f is said to be strictly increasing if $x < y$ implies $f(x) < f(y)$ and strictly decreasing if $x < y$ means $f(x) > f(y)$ (see Section 12.3). Prove that in either case, f must be one–one.

3. Must a one–one function be either strictly increasing or strictly decreasing?

Example 1 One of the fundamental problems of measurement is to show that a given empirical domain exhibits the same structure as some arithmetical structure of numbers. If a common structure can be identified, the arithmetic system is said to be *isomorphic* to the empirical domain. Once an isomorphism has been established, questions about the empirical domain can be transferred to the arithmetic system, computations performed there, and the results transferred back and interpreted.

Suppes and Zinnes* have formulated a precise definition of isomorphism using the concept of a *relational system*. According to their usage, a relational system is a finite sequence $(S; R_1, R_2, \ldots, R_n)$ in which S is a nonempty set of elements called the *domain* of the system and $R_1, R_2, \ldots, R_n$ are relations on S. Two relational systems $(S; R)$ and $(T; Q)$ are called isomorphic if there is a one–one function f mapping S to T such that for all x and y in S,

$$xRy \Leftrightarrow f(x)Qf(y)$$

As a simple example, let $S = \{2, 4, 6, 8\}$, $T = \{1, 6, 7, 10\}$, R be $\geq$, Q be $\leq$, and define f by

$$f(2) = 10, f(4) = 7, f(6) = 6, \text{ and } f(8) = 1$$

Then it is apparent that $x \geq y$ if and only if $f(x) \leq f(y)$; that is, xRy if and only if $f(x)Qf(y)$, so that the systems $(S; R)$ and $(T; Q)$ are isomorphic.

If the function f above is not necessarily one–one, the systems are said to be *homomorphic* or, more precisely, $(T; Q)$ is a homomorphic image of $(S; R)$. Homomorphic relational systems are used by Suppes and Zinnes as the basis for a formal definition and classification of measurement scales, as follows.

Let $U = (S; R)$ be an empirical relational system and let f map U homomorphically into a system $V = (T; Q)$ in which T is some set of real numbers. Then the ordered triplet (U, V, f) is a *scale*.

Various types of scales are obtained by forming compositions $g = \phi \circ f$ in such a way that (U, V, g) is also a scale. Specifically, (U, V, g) is

(i) a *ratio scale*, if $\phi(x) = \alpha x$, where $\alpha > 0$.

(ii) an *absolute scale*, if $\phi(x) = x$.

(iii) an *interval scale*, if $\phi(x) = \alpha + \beta x$, where $\beta > 0$.

(iv) an *ordinal scale*, if ϕ is monotone.

(v) a *nominal scale*, if ϕ is one–one. ▶

Since each point y in the range of a one–one function f is the image of exactly one point x in the domain of f, we may define a new function g, called the *inverse of f*, which associates with each y in the range of f the point x in the domain of f which maps into y (Figure 22). Symbolically,

$$g(y) = x \Leftrightarrow f(x) = y \tag{28}$$

*Suppes, P., and Zinnes, J. L., "Basic Measurement Theory," in *Handbook of Mathematical Psychology*, Luce, R. D., Bush, R. R., and Galanter, E., Eds. (John Wiley & Sons, Inc., New York, 1963), Vol. 1.

FIGURE 22

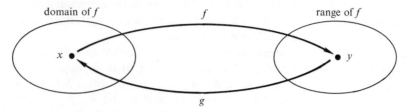

Example 2 If $f(x) = 2x + 1$, then the function $g(y) = \frac{1}{2}(y - 1)$ is inverse to f. For, if

$$y = f(x) = 2x + 1$$

then

$$x = \tfrac{1}{2}(y - 1) = g(y)$$

and conversely. (See Figure 23.) ▶

FIGURE 23 (a) The function *f*; (b) the inverse function *g* of *f*.

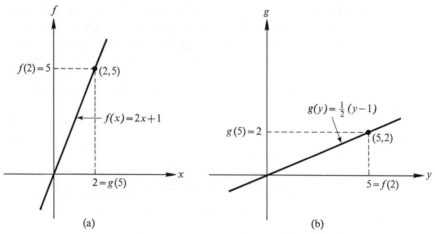

(a) (b)

The inverse of a 1–1 function f is usually denoted by f^{-1}. In general, if $f(x) = y$ and f is 1–1, then

$$f^{-1}(f(x)) = f^{-1}(y) = x$$
$$f(f^{-1}(y)) = f(x) = y \tag{29}$$

follow immediately from the defining equation (28).

Example 3 The nonsingular linear transformations introduced in Section 10.3 provide good examples of functions which have inverses, but whose range and domain are not necessarily sets of real numbers. For instance, the matrix

$$A = \begin{pmatrix} 2 & 0 \\ 1 & 3 \end{pmatrix}$$

has the inverse

$$A^{-1} = \begin{pmatrix} \frac{1}{2} & 0 \\ -\frac{1}{6} & \frac{1}{3} \end{pmatrix}$$

Thus the function f defined by

$$f(X) = AX$$

where

$$X = \begin{pmatrix} x_1 \\ x_2 \end{pmatrix}$$

has the inverse

$$f^{-1}(Y) = A^{-1}Y$$

That is,

$$f^{-1}[f(X)] = f^{-1}(AX) = A^{-1}(AX) = (A^{-1}A)X = X$$
$$f[f^{-1}(Y)] = f(A^{-1}Y) = A(A^{-1}Y) = (AA^{-1})Y = Y \qquad \blacktriangleright$$

PROBLEMS

4. If $f(x) = ax + b$ and $a \neq 0$, show that the function $g(y) = (1/a)(y - b)$ is inverse to f. Graph the functions f and g for various values of a and b.

5. The function $f(x) = ax + b$ has no inverse if $a = 0$. Why? (*Warning:* It is *not* because division by a is not possible in $(1/a)(y - b)$.)

6. Why does the function $f(x) = x^2$ have no inverse?

7. It should be obvious that if g is the inverse of f, f is also the inverse of g. But, just to make sure this is obvious, write out a proof.

The trigonometric functions sine, cosine, and tangent are defined on the entire real line, as we have previously seen, but are not one–one there (see Figure 19, Section 12.5). Hence, these functions have no inverses. However, if the domain is suitably restricted, a *local inverse* may be defined. Let the function g be defined on the interval $[-\pi/2, \pi/2]$ by $g(x) = \sin x$. Then g is 1–1 and an inverse function defined by $g^{-1}(y) = \sin^{-1} y = x$ if and only if $y = \sin x$ can be determined. Thus, $\sin^{-1}(1) = \pi/2$, $\sin^{-1}(0) = 0$, $\sin^{-1}(-1) = -\pi/2$, $\sin^{-1}(\sqrt{2}/2) = \pi/4$, and so forth (see Figure 24). Actually, if the domain of the sine is restricted to any interval of length π beginning at an odd multiple of $\pi/2$, a local inverse function can be found. However, the restriction indicated above is usually called *the inverse sine function* or sometimes, *the arcsine function*.

In similar fashion, inverses for the cosine and tangent functions are defined by restricting these functions to the domains $[0, \pi]$ and $(-\pi/2, \pi/2)$,

FIGURE 24   Graphs of the inverse trigonometric functions:

(a) inverse sine function;   (b) inverse cosine function;

(c) inverse tangent function.

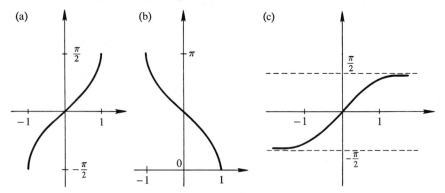

respectively (see Figure 24). The domain of the restricted tangent function does not contain the endpoints since $\tan(\pi/2)$ and $\tan(-\pi/2)$ do not exist.

Equation (29) shows that the composition of a function with its inverse always produces the identity function. That is, if f is one–one, then

$$(f \circ f^{-1})(y) = f(f^{-1}(y)) = y$$
$$(f^{-1} \circ f)(x) = f^{-1}(f(x)) = x$$

for all values x in the domain of f and y in the range of f. This, together with the Mean Value Theorem, enables us to obtain a formula for the derivative of the inverse of a function in terms of the derivative of the function itself. In so doing we accept without proof the following theorem.

Theorem 9   *Inverse Function Theorem*   Suppose f is a 1–1 function having an inverse f^{-1}. Then if f is continuous, so is f^{-1}.   ▶

Theorem 10   If f is a continuous one–one function having a continuous derivative f', then the inverse function $g = f^{-1}$ has a derivative which is

$$g'(x) = \frac{1}{f'(g(x))}$$

(Note that x is a point in the domain of g, not of f.)

PROOF   Since $f(g(x)) = x$ for all x, we can write $h = (x + h) - x = f(g(x + h)) - f(g(x))$. Applying the Mean Value Theorem to f in this expression gives

$$h = [g(x + h) - g(x)]f'(t) \quad \text{or} \quad \frac{g(x + h) - g(x)}{h} = \frac{1}{f'(t)}$$

where t is some point between $g(x)$ and $g(x + h)$. Now g is continuous (Theorem 9) so that $g(x + h) \to g(x)$ as $h \to 0$. With t trapped between

$g(x)$ and $g(x + h)$ we must also have $t \to g(x)$ as $h \to 0$. Applying the continuity of f' gives $f'(t) \to f'(g(x))$ as $h \to 0$, or

$$g'(x) = \lim_{h \to 0} \frac{g(x + h) - g(x)}{h} = \lim_{h \to 0} \frac{1}{f'(t)} = \frac{1}{f'(g(x))}$$

as asserted. ▶

To illustrate the use of Theorem 10, suppose $g(x) = \sin^{-1} x$. Then

$$Dg(x) = D \sin^{-1} x = \frac{1}{D \sin (\sin^{-1} x)} = \frac{1}{\cos (\sin^{-1} x)}$$

Now $\cos^2 \theta + \sin^2 \theta = 1$ for all θ. Hence if $\theta = \sin^{-1} x$, so that $\sin \theta = \sin (\sin^{-1} x) = x$, we have

$$\cos^2 \theta = 1 - \sin^2 \theta = 1 - x^2$$

Since $\cos \theta \geq 0$ for all θ in the range of the inverse sine function, it follows that

$$D \sin^{-1} x = \frac{1}{\sqrt{1 - x^2}} \qquad \text{for } -1 < x < 1$$

Similar considerations yield

$$D \cos^{-1} x = \frac{-1}{\sqrt{1 - x^2}} \quad \text{and} \quad D \tan^{-1} x = \frac{1}{1 + x^2}$$

These proofs are left as exercises.

Example 4   To find the derivative of $h(x) = x^2 \cos^{-1} 2x$, we first use the product rule to obtain

$$Dh(x) = x^2 D \cos^{-1} 2x + (\cos^{-1} 2x)Dx^2$$

Applying the Chain Rule gives

$$D \cos^{-1} 2x = \frac{-1}{\sqrt{1 - (2x)^2}} \, D(2x) = \frac{-2}{\sqrt{1 - 4x^2}}$$

so that

$$Dx^2 \cos^{-1} 2x = \frac{-2x^2}{\sqrt{1 - 4x^2}} + 2x \cos^{-1} 2x \qquad ▶$$

PROBLEMS

In Problems 8–13 find the derivative of the given function.

8. $m(x) = \sin^{-1} 2x$

9. $f(y) = \tan^{-1} \left(\dfrac{1}{y} \right)$

10. $g(x) = x \cot^{-1} \left(\tfrac{1}{2} x \right)$

11. $u(t) = \csc^{-1} t^2$

12. $f(x) = a^2 \cos^{-1} \left(\dfrac{x}{a} \right) + x\sqrt{a^2 - x^2}$, where a is a constant

13. $h(x) = x \tan^{-1}\left(\dfrac{x}{2}\right)$

14. Prove that

(a) $D \cos^{-1} x = \dfrac{-1}{\sqrt{1 - x^2}}$ (b) $D \tan^{-1} x = \dfrac{1}{1 + x^2}$

15. The graph of $f(x) = 2x + 1$ may be obtained from the graph of its inverse $g(y) = \frac{1}{2}(y - 1)$ by rotating the plane around the line $f(x) = x$. (See Figure 23.) Prove that the same is true of any function and its inverse.

16. Prove that if a function is strictly increasing (decreasing) then its inverse is also strictly increasing (decreasing).

17. If, in Example 1, we let

$$S = \{2, 4, 6, 8\}, \ T = \{1, 6, 7, 10\}, \ R \text{ be } \geq \ \text{ and } \ Q \text{ be } <$$

show that $(S; R)$ and $(T; Q)$ are not isomorphic.

18. (a) If $S = \{4, 8\}$, $T = \{12, 16, 20\}$, R is $\geq$, and Q is $<$, prove that $(S; R)$ and $(T; Q)$ are not isomorphic.

(b) Are they homomorphic? That is, can you find a function f, not necessarily one–one, so that xRy if and only if $f(x)Qf(y)$? (*Hint:* Try making $(S; R)$ a homomorphic image of $(T; Q)$.)

19. If A is a subset of the domain of the function f, the *image* of A under the function f is defined as

$$f(A) = \{f(x): x \in A\}$$

For any two subsets A and B of the domain of f, prove that

(a) $f(A \cup B) = f(A) \cup f(B)$

(b) $f(A \cap B) \subseteq f(A) \cap f(B)$

20. For any two subsets A and B of the range of f, prove that

(a) $f^{-1}(A \cup B) = f^{-1}(A) \cup f^{-1}(B)$. Note that for any set C in the range of f we define $f^{-1}(C)$ to be the set of all points x in the domain of f for which $f(x) \in C$. This definition is valid, and the notation $f^{-1}(C)$ will be used, whether or not f has an inverse.

(b) $f^{-1}(A \cap B) = f^{-1}(A) \cap f^{-1}(B)$

21. Prove that if $h = g \circ f$ denotes the composition of the function $f: X \rightarrow Y$ and $g: Y \rightarrow Z$, then

(a) $h(A) = g[f(A)]$ for every $A \subseteq X$

(b) $h^{-1}(C) = f^{-1}[g^{-1}(C)]$ for every $C \subseteq Z$

22. Let A be a subset of the domain and B a subset of the range of the function f. Prove

(a) $f^{-1}[f(A)] \supseteq A$

(b) $f[f^{-1}(B)] = B$

(c) $f[A \cap f^{-1}(B)] \subseteq f(A) \cap B$

12.7 LOGARITHMIC AND EXPONENTIAL FUNCTIONS

An *exponential function h* is any function of the form

$$h(x) = b^x$$

where b is some positive constant. For example, $f(x) = 2^x$, $g(x) = 10^x$, and $m(x) = 5^x$ are all exponential functions. The function referred to as *the* exponential function is

$$E(x) = e^x$$

where e is approximately 2.718. The constant e was introduced in Problem 8 of Section 5.4 as the limit: $\lim_{n \to \infty} [1 + (1/n)]^n = e$. There we saw that an amount P of money invested at i percent and compounded continuously (that is, an infinite number of times per year) would grow in n years to the quantity $P_n = Pe^{in}$.

The *natural logarithm function* ln is the inverse of the exponential function E. That is,

$$\ln y = x \Leftrightarrow y = e^x$$

This corresponds to the usual definition of the natural logarithm of a number y as the power to which e must be raised in order to obtain y. From our knowledge of inverse functions this means that

$$\ln e^x = x \quad \text{and} \quad e^{\ln x} = x \tag{30}$$

for all real numbers x. In particular, $\ln e = 1$, and e is called the *base* of the natural logarithms.

For any positive number b, the inverse of the exponential function

$$h(x) = b^x$$

is called the *logarithm function to the base b* and denoted $\log_b$. Thus

$$\log_b y = x \Leftrightarrow y = b^x \tag{31}$$

The logarithm to the base b of y is the power to which b must be raised in order to obtain y. The most commonly used logarithms are logarithms to the base 10. Such logarithms are called the *common logarithms*.

The following properties of exponential and logarithm functions, familiar from algebra, are summarized here for reference purposes:

(i) By definition,

$$b^{-x} = \frac{1}{b^x}$$

(ii) For any real numbers x and y,

$$b^{x+y} = b^x \cdot b^y$$

(iii) For any positive numbers a and c,

$$\log_b ac = \log_b a + \log_b c$$

(iv) If a is positive, then

$$\log_b \frac{1}{a} = -\log_b a$$

(v) Combining (iii) and (iv) yields

$$\log_b \frac{a}{c} = \log_b a - \log_b c$$

(vi) For any exponent x,

$$\log_b a^x = x \log_b a$$

We have seen that the derivative formula $Dx^{n+1} = (n+1)x^n$ is valid for all exponents n. This formula thus determines a function whose derivative is x^n for every n *except* $n = -1$. For in that case we obtain $Dx^0 = 0$, rather than a derivative in a form involving x^{-1}. This naturally raises the question, "Is there a function whose derivative is $1/x$?" The answer is "yes" and, as we now show, that function is the natural logarithm function.

Theorem 11 The derivative of the natural logarithm function is

$$D \ln x = \frac{1}{x}$$

PROOF To obtain the derivative of $\ln x$, we write

$$D \ln x = \lim_{h \to 0} \frac{\ln(x+h) - \ln(x)}{h} = \lim_{h \to 0} \frac{1}{h} \ln\left(\frac{x+h}{x}\right)$$

$$= \lim_{h \to 0} \frac{1}{h} \ln\left(1 + \frac{h}{x}\right) = \lim_{h \to 0} \ln\left(1 + \frac{h}{x}\right)^{1/h}$$

We now make the substitution $z = x/h$, thus changing the variable in the limit statement (note that $z \to \infty$ as $h \to 0$) to obtain

$$D \ln x = \lim_{z \to \infty} \ln\left(1 + \frac{1}{z}\right)^{z/x} = \frac{1}{x} \lim_{z \to \infty} \ln\left(1 + \frac{1}{z}\right)^{z} \tag{32}$$

Since $\lim_{z \to \infty} [1 + (1/z)]^z = e$ and since $\ln e = 1$, it follows* that

$$D \ln x = \frac{1}{x}$$

as claimed. ▶

If $y = \ln x$ so that $x = e^y$, then $\log_b x = y \log_b x = (\ln x) \log_b e$. It follows that

$$D \log_b x = (\log_b e) D \ln x = (\log_b e) \frac{1}{x}$$

Putting this result together with the Chain Rule gives the following general theorem.

---

*According to Section 11.4, a function f is continuous at a point a if $\lim_{x \to a} f(x) = f(a) = f(\lim_{x \to a} x)$. Thus the interchange of limit and ln in evaluating (32) requires continuity of the logarithm function, a fact which we accept without proof.

Theorem 12   If the function ϕ has a derivative ϕ', then

(a) $D \ln \phi(x) = \dfrac{D\phi(x)}{\phi(x)}$

(b) $D \log_b \phi(x) = (\log_b e) \dfrac{D\phi(x)}{\phi(x)}$

PROOF   We shall prove (a) only, leaving (b) as an exercise. Define the composite function $h(x)$ by $h(x) = L(\phi(x))$ where $L(y) = \ln y$. Then from the Chain Rule, we obtain

$$D \ln \phi(x) = h'(x) = L'(\phi(x))\phi'(x) = \frac{1}{\phi(x)} \phi'(x) = \frac{D\phi(x)}{\phi(x)} \qquad \blacktriangleright$$

Example 1   (a) To differentiate $h(x) = \ln (2x^2 - 1)$, we use Theorem 12(a) with $\phi(x) = 2x^2 - 1$ to obtain

$$D \ln (2x^2 - 1) = \frac{D(2x^2 - 1)}{2x^2 - 1} = \frac{4x}{2x^2 - 1}$$

(b) $D \ln (x - 2)^5 = \dfrac{D(x - 2)^5}{(x - 2)^5} = \dfrac{5(x - 2)^4}{(x - 2)^5} = \dfrac{5}{x - 2}$

Alternatively, since $\ln (x - 2)^5 = 5 \ln (x - 2)$,

$$D \ln (x - 2)^5 = 5D \ln (x - 2) = \frac{5}{x - 2}$$

(c) To differentiate $f(x) = x^x$ we first take logarithms to obtain $\ln f(x) = x \ln x$. Then

$$\frac{f'(x)}{f(x)} = D \ln f(x) = Dx \ln x = x\left(\frac{1}{x}\right) + \ln x$$

Solving for $f'(x)$ we have $f'(x) = x^x[1 + \ln x]$.

(d) $D \log_{10} (x^2 - x + 1) = (\log_{10} e) \dfrac{2x - 1}{x^2 - x + 1} \qquad \blacktriangleright$

An interesting property of the exponential function $E(x) = e^x$ is that it is its own derivative. To see this, we differentiate both sides of the equation $\ln e^x = x$ to obtain

$$1 = Dx = D \ln e^x = \frac{1}{e^x} De^x$$

from which it immediately follows that $De^x = e^x$. This result, too, may be generalized by using the Chain Rule.

Theorem 13   If the function ϕ has a derivative, then

(a) $De^{\phi(x)} = e^{\phi(x)} D\phi(x)$

(b) $Db^{\phi(x)} = (\ln b)b^{\phi(x)} D\phi(x)$

In particular, $Db^x = (\ln b)b^x$.

PROOF (a) Define the composite function h by $h(x) = E(\phi(x))$ where $E(y) = e^y$. Then using the Chain Rule,

$$De^{\phi(x)} = h'(x) = E'(\phi(x))\phi'(x) = e^{\phi(x)}\phi'(x) = e^{\phi(x)}D\phi(x)$$

(b) Since $b = e^{\ln b}$, we have $b^{\phi(x)} = e^{(\ln b)\phi(x)}$. Thus

$$Db^{\phi(x)} = De^{(\ln b)\phi(x)} = (\ln b)\phi'(x)e^{(\ln b)\phi(x)}$$

$$= (\ln b)b^{\phi(x)}D\phi(x) \qquad \blacktriangleright$$

Example 2 (a) To find the derivative of $h(x) = e^{2x^2-1}$, we use Theorem 13(a) with $\phi(x) = 2x^2 - 1$ to obtain

$$De^{2x^2-1} = [D(2x^2 - 1)]e^{2x^2-1} = 4xe^{2x^2-1}$$

(b) $De^{(x-2)5} = [D(x - 2)^5]e^{(x-2)5} = 5(x - 2)^4e^{(x-2)5}$

(c) $D[e^{x^2}\ln(2x - 1)] = e^{x^2}D\ln(2x - 1) + [\ln(2x - 1)]De^{x^2}$

$$= \frac{2e^{x^2}}{2x - 1} + 2xe^{x^2}\ln(2x - 1)$$

(d) $D(10^{x^2-x+1}) = (\ln 10)(2x - 1)10^{x^2-x+1} \qquad \blacktriangleright$

Example 3 In his classic theory of learning, Hull* assumed that habit strength is given in terms of the number N of repetitions by

$$H(N) = 100(1 - e^{-iN})$$

where i is a positive constant. Assuming for purposes of illustration that N can be any positive number, the instantaneous rate of change of habit strength is

$$H'(N) = -100e^{-iN}D(-iN) = 100ie^{-iN}$$

Here we see that the constant i indicates the rate at which habit strength is acquired. Moreover, since $H'(N)$ is always positive, habit strength is a monotone increasing function of the number of repetitions. $\qquad \blacktriangleright$

Since $De^x = e^x$ is always positive, it follows that the exponential function is increasing for all x. Moreover, since the derivative e^x increases as x increases, the function must increase at an ever increasing rate. These properties are evident in Figure 25(a). In a similar way, we see that the slope $D\ln x = 1/x$ of the natural logarithm function is positive but decreasing, which leads to the graph of Figure 25(b).

Not only does the exponential function turn out to be its own derivative, but it is essentially the only function having this property. To see this, suppose that g is another function for which $g'(x) = g(x)$ and define a function

*Hull, C. L., *Principles of Behavior* (Appleton-Century-Crofts, New York, 1943).

FIGURE 25

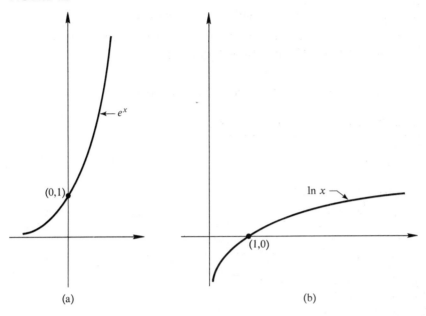

(a) (b)

h by $h(x) = g(x)/e^x$. Differentiating h gives

$$h'(x) = \frac{e^x g'(x) - g(x)De^x}{e^{2x}} = 0$$

since $g'(x) = g(x)$ and $De^x = e^x$. It follows that h is a constant function and that $g(x) = ce^x$ for some constant c.

The above discussion can be generalized to the case where the derivative g' is proportional to g; that is, where $g'(x) = kg(x)$ for some constant k. An argument similar to that used above shows that in this case g must have the form $g(x) = ce^{kx}$, where c is constant. The proof is left as an exercise.

Example 4 A typical example of natural phenomena in which the rate of change of a certain quantity is proportional to the size of that quantity is afforded by the growth of a bacterial culture. If we make the reasonable assumption that the number of reproductions taking place at any instant of time is proportional to the number of bacteria present at that time, then $N(t)$, the number of bacteria present at time t, satisfies the equation

$$N'(t) = kN(t).$$

The number of bacteria in the culture can then be expected to grow exponentially according to the law $N(t) = ce^{kt}$.

For example, if we begin (at $t = 0$) with 100 bacteria which are observed to grow in $t = 1$ day to 1000 bacteria we find

$$N(0) = c = 100 \quad \text{and} \quad N(1) = ce^k = 1000$$

Hence $c = 100$ and $k = \ln(1000/c) = \ln 10$, so that

$$N(t) = 100(e^{\ln 10})^t = 100(10)^t$$

If the growth continues at the same rate, there will be $N(6) = 100(10)^6$, or 100 million bacteria present at the end of 6 days. ▶

All our results concerning derivatives are summarized in Table 2.

TABLE 2.   Short Table of Derivatives

1. $Dc = 0$
2. $Dx = 1$
3. $Dx^n = nx^{n-1}$
4. $D[f(x) + g(x)] = Df(x) + Dg(x)$
5. $D[cf(x)] = cDf(x)$
6. $D[c_1 f_1(x) + c_2 f_2(x) + \cdots + c_n f_n(x)] = c_1 Df_1(x) + c_2 Df_2(x) + \cdots + c_n Df_n(x)$
7. $D[f(x) \cdot g(x)] = f(x)Dg(x) + g(x)Df(x)$
8. $D\left[\dfrac{f(x)}{g(x)}\right] = \dfrac{g(x)Df(x) - f(x)Dg(x)}{[g(x)]^2}$ if $g(x) \neq 0$
9. $D[g(f(x))] = Dg[f(x)] \cdot Df(x)$
10. $D[(f(x))^n] = n(f(x))^{n-1} Df(x)$
11. $D \sin x = \cos x$
12. $D \cos x = -\sin x$
13. $D \tan x = \sec^2 x$ if $\cos x \neq 0$
14. $D \cot x = -\csc^2 x$ if $\sin x \neq 0$
15. $D \sec x = \sec x \tan x$ if $\cos x \neq 0$
16. $D \csc x = -\csc x \cot x$ if $\sin x \neq 0$
17. $D \sin^{-1} x = \dfrac{1}{\sqrt{1 - x^2}}$ for $-1 < x < 1$
18. $D \cos^{-1} x = \dfrac{-1}{\sqrt{1 - x^2}}$ for $-1 < x < 1$
19. $D \tan^{-1} x = \dfrac{1}{1 + x^2}$
20. $D \ln x = \dfrac{1}{x}$ for $x > 0$
21. $D \log_b x = \dfrac{(\log_b e)}{x}$ for $x > 0$
22. $De^x = e^x$
23. $Db^x = (\ln b)b^x$

PROBLEMS

Find derivatives of each of the following functions.

1. $f(x) = \ln(x + 2)$

2. $g(z) = \ln\left(\dfrac{\sqrt{3 + 2z}}{z^2}\right)$

3. $h(t) = 2^t$

4. $f(y) = [\ln(y + 2)^2]^{5/3}$

5. $f(x) = 2x^x$

6. $f(t) = e^{(1/2)t}$

338 / DIFFERENTIAL CALCULUS

7. $g(w) = e^{w^2}$

8. $h(x) = e^{\tan x}$

9. $h(x) = x^2 e^{-x}$

10. $g(u) = e^u \ln (\sin u)$

11. $f(w) = e^{-w} \cos w$

12. $f(x) = \sin^{-1} e^x$

13. Prove part (b) of Theorem 12.

14. Derive properties (iii)–(vi) for logarithmic functions from Equation (31) and properties (i) and (ii) for exponential functions.

15. Prove that if $g'(x) = kg(x)$ where k is constant, then there is a constant c such that $g(x) = ce^{kx}$ for all x.

16. The world contains approximately 10 billion acres of arable land. Assuming that a minimum of one-quarter of an acre is required to provide sufficient food for one person, it would seem that world population is limited to at most 40 billion. The 1965 population was about three billion. If world population continues to grow at the present rate of 1.8% per year, when will the 40 billion mark be reached?

12.8 APPLICATIONS OF DIFFERENTIATION— OPTIMIZATION AND STABILITY

The classical use of differentiation is to find the optimum value or values of a function. This problem arises in economics, for example, when one wishes to maximize his profit or to minimize his loss. A function f is said to have a *relative maximum* at a point a if there exists a number $\delta > 0$ such that $f(x) \le f(a)$ for all points x in the domain of f which lie within distance δ of a. Similarly, we say that f has a *relative minimum* at a point b if there is a $\delta > 0$ for which $f(x) \ge f(b)$ whenever $x \in D(f)$ and $|x - b| < \delta$. (See Figure 26.)

A point c is called an *interior point* of a set S if there is a number $\delta > 0$ such that all points x lying within δ distance of c are members of S. (Note that here, as above, the actual size of δ is not important except that δ must be greater than zero. It may be very small or it may be large. The existence of

FIGURE 26

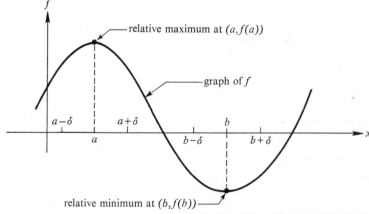

even one value of δ which works is sufficient. The same considerations were, of course, implicit in the definition, in Section 11.4, of continuity at a point.)

Example 1 (a) In the interval [0, 1], all points except the endpoints 0 and 1 are interior points. For if $0 < c < 1$, we simply take δ to be the smaller of the distances from c to the endpoints; that is, the smaller of c and $1 - c$. Then if $|x - c| < δ$, we must have x between 0 and 1 also (Figure 27).

FIGURE 27

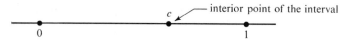

interior point of the interval

c

0   1

(b) The absolute value function has a relative minimum at zero since $|0| = 0$, and for all other x, $|x| \geq 0$ (Figure 28).

FIGURE 28

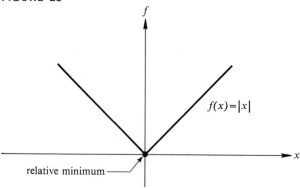

$f(x) = |x|$

relative minimum

(c) The function g defined by $g(x) = 2x - 3$ when $0 \leq x \leq 4$ has a relative minimum when $x = 0$ and a relative maximum at $x = 4$ (Figure 29). ▶

FIGURE 29

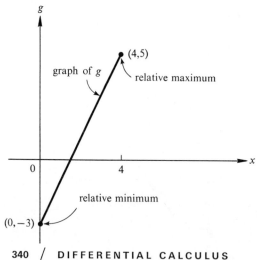

graph of g

(4,5)

relative maximum

0   4

relative minimum

(0, −3)

The basic tools for problems involving maxima and minima are the following two theorems.

Theorem 14 *Interior Maximum and Minimum Theorem* Suppose a function f has either a relative maximum or a relative minimum at a point c interior to the domain of f. Then if $f'(c)$ exists, it must be equal to zero.

PROOF Suppose f has a relative maximum at the interior point c. Then for some $\delta > 0$, $f(c + h) \leq f(c)$ for all h such that $-\delta < h < \delta$. It follows that the ratio $[f(c + h) - f(c)]/h$ is negative when $h > 0$ and positive when $h < 0$, so that zero is the only possible value for

$$f'(c) = \lim_{h \to 0} \frac{f(c + h) - f(c)}{h}$$

The proof for the case when f has a relative minimum at c is similar and is left as an exercise. ▶

Theorem 15 *Criteria for relative maxima and minima* Suppose that the function f is continuous and has a derivative throughout an interval $[a, b]$. (This interval may comprise part or all of the domain of f.)

(a) If c is an interior point of $[a, b]$ and if for some number $\delta > 0$, we have $f'(x) \geq 0$ when $c - \delta < x < c$ and $f'(x) \leq 0$ when $c < x < c + \delta$, then f has a relative maximum at c.

(b) If c is an interior point of $[a, b]$ and if for some $\delta > 0$ we have $f'(x) \leq 0$ when $c - \delta < x < c$ and $f'(x) \geq 0$ when $c < x < c + \delta$, then f has a relative minimum at c.

(c) If there is a number δ such that $f'(x) \geq 0$ (or $f'(x) \leq 0$) when $b - \delta < x \leq b$, then f has a relative maximum (or relative minimum) at b. (The maximum or minimum is, of course, relative to points in $[a, b]$ which lie near b.)

(d) If there is a number δ such that $f'(x) \geq 0$ (or $f'(x) \leq 0$) when $a \leq x < a + \delta$, then f has a relative minimum (or relative maximum) at a.

PROOF The truth of the theorem is intuitively obvious from the sketches in Figure 30. We shall give a formal proof only for part (a), the other cases being similar. We first choose points x and y, as in Figure 30(a), in such a way that $c - \delta < x < c < y < c + \delta$. Then the Mean Value Theorem guarantees the existence of points α, between x and c, and β, between c and y, for which

$$f(c) - f(x) = (c - x)f'(\alpha) \quad \text{and} \quad f(y) - f(c) = (y - c)f'(\beta)$$

The assumptions for part (a) guarantee that $f'(\alpha)$ is positive while $f'(\beta)$ is negative. Since $c - x > 0$ and $y - c > 0$, it follows that $f(c) \geq f(x)$ and $f(c) \geq f(y)$. This being true for arbitrarily chosen x and y points near c, we see that f has a relative maximum at c. ▶

FIGURE 30

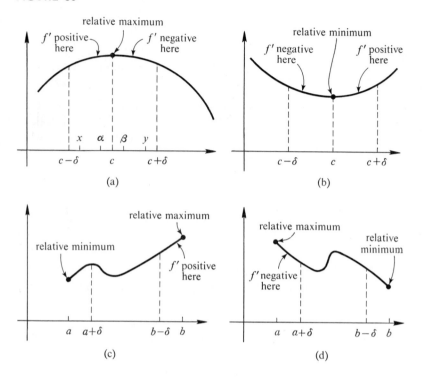

(a)

(b)

(c)

(d)

The procedure for finding maximum and minimum values (or extreme values) of a function is now clear from Theorems 14 and 15. These values may occur at the endpoints of the domain of the function. (Typical cases are linear programming problems like Examples 1–3 of Section 9.5.) Parts (c) and (d) of Theorem 15 also apply here. When extreme values occur at interior points, the derivative may not exist (as in Example 1(b)), but if it does, then according to Theorem 14, it must be zero. In summary, then, to find extreme values of a function:

(i) Check the endpoints of the domain by applying Theorem 15(c) or (d).

(ii) Check points where the derivative fails to exist, applying Theorem 15(a) and (b).

(iii) Check those points where the derivative is zero, applying Theorem 14 and Theorem 15(a) and (b).

Example 2 In Example 3 of Section 12.7, the instantaneous rate of change of habit strength was found to be $H'(N) = 100ie^{-iN}$, where i is a positive constant. Since N must be non-negative and $H'(N)$ is always positive, Theorem 15(d) implies that H has a minimum at $N = 0$. (See Figure 31.) ▶

FIGURE 31

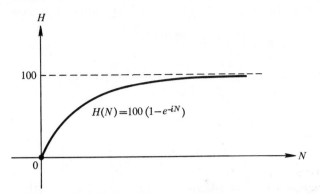

$$H(N)=100\,(1-e^{-iN})$$

Example 3   The derivative of $f(t) = t^3 - 5t^2 + 3t + 12$ is $f'(t) = 3t^2 - 10t + 3 = (3t - 1)(t - 3)$. A little checking shows that this derivative is positive when $t < \frac{1}{3}$, zero at $t = \frac{1}{3}$, negative between $t = \frac{1}{3}$ and $t = 3$, zero again at $t = 3$, and positive when $t > 3$. It follows from Theorem 15(a) and (b) that f has a relative maximum when $t = \frac{1}{3}$ and a minimum at $t = 3$. (See Figure 32.)   ▶

FIGURE 32

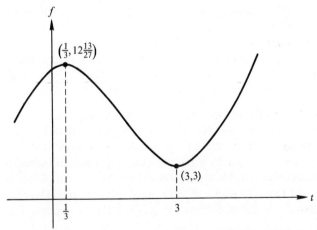

Example 4*   Suppose that the total number n of repairs over the useful life t of a consumer durable is given by the function $n(t) = t^\alpha/\beta$, where α and β are positive constants. Since an original unit is replaced after time t, the average number of replacements per unit of time is $1/t$ and the average replacement cost per unit of time is $c(t) = p/t$, where p is the price of a new unit.

From the definition of n, the number of repairs per unit of time is $n(t)/t$,

*Adapted from Brems, H., *Quantitative Economic Theory* (John Wiley & Sons, Inc., New York, 1968), pp. 38–39.

and if the average cost of a repair is r, the average repair cost per unit of time is $r[n(t)/t]$. Ignoring capital costs, rental costs RC per unit of time may thus be defined as

$$RC(t) = \frac{p}{t} + r\frac{n(t)}{t}$$

Substituting t^α/β for $n(t)$ gives

$$RC(t) = \frac{p}{t} + \frac{r}{\beta}t^{\alpha-1}$$

the derivative of which is

$$RC'(t) = -\frac{p}{t^2} + \frac{r(\alpha-1)}{\beta}t^{\alpha-2} = \frac{p}{t^2}\left[\frac{r(\alpha-1)}{\beta p}t^\alpha - 1\right]$$

This derivative is zero when $t^\alpha = \beta p/r(\alpha-1)$ or $t = [\beta p/r(\alpha-1)]^{1/\alpha}$, negative to the left of this point and positive to the right. The point

$$t = \left[\frac{\beta p}{r(\alpha-1)}\right]^{1/\alpha}$$

yields the minimum value

$$\frac{p\alpha}{\alpha-1}\left[\frac{r(\alpha-1)}{\beta p}\right]^{-1/\alpha}$$

for the average rental cost per unit time.

A closer analysis of $RC(t)$ shows that for small t, the per-unit-time replacement cost p/t dominates. As t increases, this term decreases to zero while the per-unit-time repair cost $(r/\beta)t^{\alpha-1}$ increases. In terms of average rental costs, $[\beta p/r(\alpha-1)]^{1/\alpha}$ represents the optimal time for replacement of the item in question. ▶

Example 5 *Equilibrium and Stability* In his analysis of the forces which activate an international arms race, Richardson* postulates that the rate of defense expenditure by a nation tends to be increased by such items as expenditures by a rival nation and by grievances held by a nation independently of existing defense budgets. He also assumes that the burden of maintaining a defense budget tends to decrease the rate of expenditure. Thus for two nations, Richardson's model of defense expenditure can be expressed as

$$E'(t) = aR(t) - \alpha E(t) + g$$
$$R'(t) = bE(t) - \beta R(t) + h \tag{33}$$

where E and R denote expenditures of the respective nations, a and b are positive constants reflecting the degree of fear and insecurity regarding the intent of the other nation, α and β (>0) reflect the burden of maintaining a defense budget, and g and h indicate the respective grievances.

The system (33) is said to be in equilibrium when there is no tendency for E and R to change. That is, equilibrium occurs when $E'(t)$ and $R'(t)$ are both

*Richardson, L. F., *Arms and Insecurity* (Boxwood Press, Pittsburgh, Pa., 1960).

zero. The solution to the simultaneous equations

$$E' = aR - \alpha E + g = 0$$
$$R' = bE - \beta R + h = 0 \tag{34}$$

is called the equilibrium point, or the *point of balance of power* of the system.

If $ab = \alpha\beta$, the two lines in (34) are parallel and no equilibrium point exists. The system is obviously unstable in this case. When $ab \neq \alpha\beta$, the point (E_0, R_0) of balance of power is given by

$$E_0 = \frac{ah + \beta g}{\alpha\beta - ab} \qquad R_0 = \frac{bg + \alpha h}{\alpha\beta - ab}$$

To obtain a better picture of the behavior of the system, let us indicate by horizontal shading the region in the E-R plane in which $E' > 0$ and by vertical shading the region in which $R' > 0$. Two examples are shown in Figure 33. Double shading indicates the region in which both E and R tend to increase and hence in which there is a drift toward a mutual arms race, while in the unshaded region there is a tendency toward mutual armament reductions and greater cooperation.

FIGURE 33   Two examples of an arms' race. Heavy arrows indicate the direction of movement of the system.

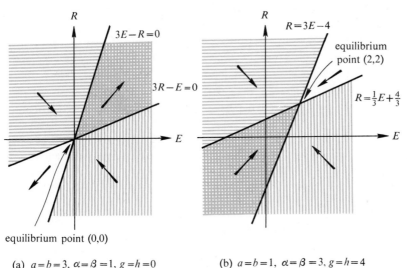

(a) $a=b=3$, $\alpha=\beta=1$, $g=h=0$

(b) $a=b=1$, $\alpha=\beta=3$, $g=h=4$

Stability of the system depends upon whether the movement over time is toward or away from the point of balance of power. In Figure 33(a), for instance, the doubly shaded region lies above and to the right of the equilibrium point $(0, 0)$. In this region, then, the tendency is toward ever-increasing expenditure by both nations and away from the point of equilibrium. An unstable situation obtains.

On the other hand, in Figure 33(b), the drift in both the shaded regions (increasing expenditure) and the unshaded region (decreasing expenditure) is toward the equilibrium point (2, 2). The system has a built-in stability in that any perturbance away from the point of balance of power is met by a gradual return to that point.

▶

PROBLEMS

In Problems 1–13, find the relative maxima and/or minima of indicated functions.

1. $f(x) = 2x^3 - 3x$

2. $g(x) = \frac{1}{3}x^3 - 2x^2 + 3x + 7$

3. $h(t) = \dfrac{2t}{1 + t^2}$

4. $f(t) = \dfrac{1}{t} + \dfrac{1}{1 - t}$

5. $f(w) = \dfrac{\sqrt{w}}{4 + w}$

6. $g(z) = (z - 1)^{2/3}(z - 4)^2$

7. $h(x) = \dfrac{1 - 2x + x^2}{1 + x + 2x^2}$

8. $f(x) = \dfrac{(4 - x)^3}{3 - 2x}$

9. $k(y) = y\sqrt{1 - 4y^2}$

10. $h(x) = x^{\ln x}$

11. $g(t) = \ln(\sin t)$

12. $g(w) = e^{-w^2}$

13. $r(\theta) = e^{-a\theta} \sin b\theta$

14. What is the largest product which can be obtained using two positive numbers whose sum is 2?

15. The product of two numbers is 56. What is the smallest possible sum of the squares of the two numbers?

16. What number exceeds its square by the greatest amount?

17. A model of class size postulates an educational return of 10 units per student for a class size not exceeding 25. Return per student is postulated to decrease by 2% for each student in excess of 25. What class size gives maximum educational return?

18. As in Example 2 of Section 12.1, let x denote quantity demanded, and let $p(x)$, $r(x) = x \cdot p(x)$, $C(x)$ and $\theta(x) = r(x) - C(x)$ be, respectively, the demand, revenue, cost, and profit functions, all of which are assumed to be differentiable for $x > 0$.

(a) Argue that maximum or minimum profit can occur (aside from $x = 0$) only when marginal revenue equals marginal cost.

(b) Show that if we assume $p(x) = \alpha - \beta x$ and $C(x) = bx^2 + ax + k$, where all constants are positive, then profit is maximized at output level $x_m = (\alpha - a)/2(\beta + b)$.

19. Prove that marginal revenue curve intersects the average revenue $AR(x) = r(x)/x$ at the point where average revenue is a maximum.

20. Assume the total cost function for a firm is $TC(x) = 75 + 2x - 2x^2 + \frac{1}{2}x^3$ where x denotes output level.

(a) Find the output level for which total cost is a minimum.

(b) Determine the output level which minimizes average cost. Verify that at this level, marginal cost is equal to average cost.

21. Referring to Example 5, draw a diagram depicting the behavior of the system when $a = b = 2$, $\alpha = \beta = 1$, and $g = h = 3$.

22. Prove Theorem 14 for the case of a relative minimum.

23. Prove parts (b), (c), and (d) of Theorem 15.

12.9 HIGHER–ORDER DERIVATIVES; TAYLOR'S THEOREM

We have defined the derivative f' of a function f to be a new function which assigns to each point x in the domain of f the number

$$f'(x) = \lim_{h \to 0} \frac{f(x + h) - f(x)}{h}.$$

The derivative of f', that is, the derivative of the derivative of f, is called the *second derivative* of f, denoted by f'' or by $D^2 f$. In symbols,

$$f''(x) = \lim_{h \to 0} \frac{f'(x + h) - f'(x)}{h} \tag{35}$$

When the Leibniz notations

$$y = f(x) \quad \text{and} \quad f'(x) = \frac{dy}{dx}$$

are used, the second derivative $f''(x)$ is usually denoted by

$$\frac{d^2 y}{dx^2}$$

Example 1   (a) If $y = f(x) = 4x^3 + \frac{1}{2}x^2 + x - 1$, then

$$f'(x) = \frac{dy}{dx} = 12x^2 + x + 1$$

$$f''(x) = \frac{d^2 y}{dx^2} = 24x + 1$$

(b) The learning function

$$f(x) = \frac{Lx + Lc}{x + c + a}$$

introduced in Problem 21 of Section 12.1 has the derivative

$$f'(x) = \frac{La}{(x + c + a)^2}$$

The second derivative is

$$f''(x) = D \frac{La}{(x + c + a)^2} = \frac{-2La}{(x + c + a)^3}$$

(c) In Example 6 of Section 12.2, the satisfaction function $S(R) = aR/(g - R)$ was found to have the derivative $S'(R) = ag/(g - R)^2$. Its second derivative is

$$D^2\left[\frac{aR}{g - R}\right] = D\left[\frac{ag}{(g - R)^2}\right] = \frac{2ag}{(g - R)^3} \quad \blacktriangleright$$

Some insight may be gained into the meaning of the second derivative if we recall our original interpretation of a derivative as a rate of change. The second derivative indicates the rate of change of the first derivative, which in turn represents the slope of the function being differentiated. If the second derivative is positive, the slope must increase, while if the second derivative is negative the slope decreases. The two cases are shown in Figure 34. A function whose second derivative is positive is called *convex*, or, sometimes, concave upward; while one whose second derivative is negative is *concave*, or concave downward.

FIGURE 34

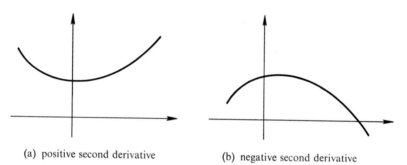

(a) positive second derivative   (b) negative second derivative

Example 2   The first derivative of $f(x) = x^3 - 3x^2 + 4$ (see Figure 35) is

$$f'(x) = 3x^2 - 6x = 3x(x - 2)$$

and the second derivative is

$$f''(x) = 6x - 6 = 6(x - 1)$$

(See Example 3 of Section 12.3.) When x is less than 1, the second derivative is negative, and hence the function is concave. The second derivative is positive and the curve convex when x exceeds 1. $\quad \blacktriangleright$

The point $(1, 2)$ in the above example is the point at which the graph changes from concave to convex. It is useful to distinguish such points from other points on a given graph by the name *point of inflection*.

Third- and higher-order derivatives are obtained by successive differentiation, that is, the third derivative is the derivative of the second derivative, the fourth derivative is the derivative of the third derivative and so forth.

FIGURE 35

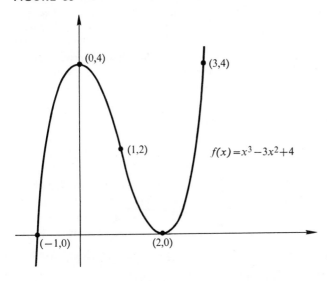

(0,4)

(3,4)

(1,2)

$f(x) = x^3 - 3x^2 + 4$

(−1,0)

(2,0)

The third derivative of a function f is denoted by f''' or by D^3f and, in general, the nth derivative is denoted by $f^{(n)}$ or by $D^n f$. (Following the Leibniz notation, the notations d^3y/dx^3 and $d^n y/dx^n$ are also in common use.)

Example 3 (a) If $g(x) = x^3 - x^2 + 9x + 2$, then $g'(x) = 3x^2 - 2x + 9$, $g''(x) = 6x - 2$, $g'''(x) = 6$, and $g^{(4)}(x) = 0$. In fact, $g^{(n)}(x) = 0$ for all x if $n \geq 4$.

(b) If $h(t) = \sin t$, then $Dh(t) = \cos t$, $D^2h(t) = -\sin t$, $D^3h(t) = -\cos t$, $D^4h(t) = \sin t$, etc. The cycle is repeated at every fourth derivative.

(c) The third derivative of the learning function (see Example 1(b)) is $f'''(x) = 6La/(x + c + a)^4$. ▶

PROBLEMS

1. Show by induction that the nth derivative of the learning function

$$f(x) = \frac{Lx + Lc}{x + c + a}$$

is given by

$$f^{(n)}(x) = \frac{(-1)^{n-1} n! La}{(x + c + a)^{n+1}}$$

2. Show that for $n = 1, 2, 3, \ldots$, the nth derivative of $\ln x$ is

$$\frac{(-1)^{n-1}(n - 1)!}{x^n}$$

3. Find formulas for the nth derivatives of e^x and $\cos x$.

The most important applications of second and higher derivatives arise in connection with an extension of the Mean Value Theorem, called Taylor's Theorem.*

Taylor's Theorem Suppose the function f and its first n derivatives f', f'', ..., $f^{(n-1)}$, $f^{(n)}$ are defined and continuous in an interval $[a, b]$. Then there is at least one point c between a and b for which

$$f(b) = f(a) + f'(a)(b - a) + \frac{f''(a)}{2!} (b - a)^2 + \frac{f'''(a)}{3!} (b - a)^3 + \cdots$$
$$+ \frac{f^{(n-1)}(a)}{(n - 1)!} (b - a)^{n-1} + \frac{f^{(n)}(c)}{n!} (b - a)^n \tag{36}$$

▶

Before proving Taylor's Theorem, let us see how it relates to the Mean Value Theorem. When $n = 1$, Taylor's Theorem states that, subject to the assumed existence of the first derivative, there is at least one point c (where $a < c < b$) for which $f(b) = f(a) + f'(c)(b - a)$. This, of course, is the statement of the Mean Value Theorem.

PROOF OF TAYLOR'S THEOREM We begin, as in the proof of the Mean Value Theorem, by inventing a function which serves our purpose and to which we can apply Rolle's Theorem. It is easily verified that

$$\phi(x) = f(b) - f(x) - f'(x)(b - x) - \frac{f''(x)}{2!} (b - x)^2 - \cdots$$
$$- \frac{f^{(n-1)}(x)}{(n - 1)!} (b - x)^{n-1}$$
$$- \left[f(b) - f(a) - f'(a)(b - a) - \frac{f''(a)}{2!} (b - a)^2 - \cdots \right.$$
$$\left. - \frac{f^{(n-1)}(a)}{(n - 1)!} (b - a)^{n-1} \right] \frac{(b - x)^n}{(b - a)^n}$$
$$= f(b) - f(x) - \sum_{k=1}^{n-1} \frac{f^{(k)}(x)}{k!} (b - x)^k$$
$$- \left[f(b) - f(a) - \sum_{k=1}^{n-1} \frac{f^{(k)}(a)}{k!} (b - a)^k \right] \frac{(b - x)^n}{(b - a)^n}$$

is such a function. That is, $\phi(a) = \phi(b) = 0$ and ϕ' exists since f has n derivatives. Hence Rolle's Theorem guarantees the existence of a point c between a and b at which $\phi'(c) = 0$.

Differentiating ϕ gives

$$\phi'(x) = -f'(x) - \sum_{k=1}^{n-1} \frac{f^{(k)}(x)}{k!} k(b - x)^{k-1}(-1) - \sum_{k=1}^{n-1} \frac{f^{(k+1)}(x)}{k!} (b - x)^k$$
$$- \left[f(b) - f(a) - \sum_{k=1}^{n-1} \frac{f^k(a)}{k!} (b - a)^k \right] \frac{n(b - x)^{n-1}(-1)}{(b - a)^n}$$

*After Brook Taylor (1685–1731), an English mathematician.

Combining the first term $-f'(x)$ with the first sum and then replacing k by $k - 1$ in this sum shows that all terms in the first two sums cancel, except the term $[-f^{(n)}(x)/(n - 1)!] (b - x)^{n-1}$. Inserting $x = c$, where $\phi'(c) = 0$, gives

$$0 = \frac{-f^{(n)}(c)}{(n - 1)!} (b - c)^{n-1}$$
$$+ \left[f(b) - f(a) - \sum_{k=1}^{n-1} \frac{f^{(k)}(a)}{k!} (b - a)^k \right] \frac{n(b - c)^{n-1}}{(b - a)^n}$$

Cancelling the (nonzero) common factor $(b - c)^{n-1}$, multiplying by $(b - a)^n/n$, and solving for $f(b)$ yields the desired equation (36). ▶

One typical use of Taylor's Theorem is in conjunction with the Interior Maximum and Minimum Theorem. According to that theorem, at any maximum or minimum point of a function f which is interior to the domain of f, the derivative—if it exists—must be equal to zero. Suppose, then, that $f'(x) = 0$ at $x = a$ and that the second derivative of f exists in an interval which contains a as an interior point. Taylor's Theorem with $n = 2$ becomes

$$f(b) - f(a) = \frac{f''(c)}{2} (b - a)^2$$

where c lies between a and b. If we assume that f'' is continuous at a, so that $f''(c)$ will have the same sign as $f''(a)$ if we take b (and hence c, also) sufficiently close to a, the following facts may be observed (a glance at Figure 34 will make the statements geometrically obvious):

(i) If $f''(a) > 0$, then $\frac{1}{2}f''(c)(b - a)^2 > 0$ and $f(b) > f(a)$ for all b near a. Hence a is a point of relative minimum.

(ii) If $f''(a) < 0$, then $\frac{1}{2}f''(c)(b - a)^2 < 0$ and $f(b) < f(a)$ for points b near a. In this case, a is a point of relative maximum of f.

(iii) If $f''(a) = 0$, no definite statement can be made since f'' may be both negative and positive at different points near a.

Example 4 Let $R(x)$ and $C(x)$ denote, respectively, the revenue and cost associated with the production of x units of a certain commodity. Then the profit function $P = R - C$ is maximized at $x = a$ if $P'(a) = 0$ and $P''(a) < 0$.

The requirement $P'(a) = R'(a) - C'(a) = 0$ states that marginal revenue must equal marginal cost. The condition $P''(a) = R''(a) - C''(a) < 0$ or, equivalently, $R''(a) < C''(a)$, indicates that at the point of greatest profit, marginal cost must be increasing more rapidly than marginal revenue. ▶

If $f''(a)$ is zero when $f'(a) = 0$, our test for a relative maximum or minimum yields no results. However, this may be remedied as follows. Suppose the derivatives $f'(a), f''(a), \ldots,$ and $f^{(n-1)}(a)$ are all zero and that $f^{(n)}(a) \neq 0$ is the first nonzero derivative of f at a. Then Taylor's Theorem becomes

$$f(b) - f(a) = \frac{f^{(n)}(c)}{n!} (b - a)^n \qquad (a < c < b)$$

There are two cases to consider:

(i) If n is even, the argument is the same as for $n = 2$. In this case $(b - a)^n/n!$ is positive so that everything depends on $f^{(n)}(c)$. If $f^{(n)}(a) > 0$, so that $f^{(n)}(c) > 0$ also, then a is a point of relative minimum. If $f^{(n)}(a) < 0$, then a is a relative maximum.

(ii) If n is odd, then $(b - a)^n$ is positive when $b > a$, but negative when $b < a$. Hence, whether $f^{(n)}(a)$ is positive or negative, there are points near a at which $f(b) > f(a)$ and other points at which $f(b) < f(a)$. The point a is neither maximum nor minimum.

FIGURE 36

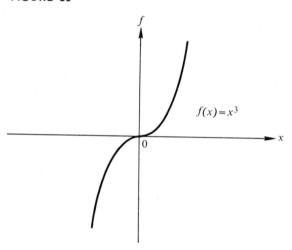

$f(x) = x^3$

FIGURE 37

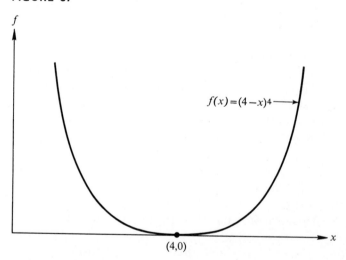

$f(x) = (4-x)^4$

$(4,0)$

Example 5 (a) The graph of $f(x) = x^3$ is shown in Figure 36. At $x = 0$, both the first derivative $f'(x) = 3x^2$ and the second derivative $f''(x) = 6x$ are zero. The third derivative $f'''(x) = 6$ being the first nonzero derivative, it follows that $x = 0$ is neither a maximum nor a minimum point for f.

(b) The function $f(x) = (4 - x)^4$ has first, second, and third derivatives $f'(x) = -4(4 - x)^3$, $f''(x) = 12(4 - x)^2$, and $f'''(x) = -24(4 - x)$ equal to zero when $x = 4$. The fourth derivative $f^{(4)}(x) = 24$ being positive, we conclude that $f(4) = 0$ is a relative minimum value for f. (See Figure 37.)

PROBLEMS

In Problems 4–13, find the first and second derivatives of the given function.

4. $f(x) = x^3 + 3x^2 - 2x + 6$ **5.** $g(u) = u^5 - 2u^3 + \frac{7}{2}u$

6. $s(z) = \sqrt[3]{z^4} - \sqrt[3]{2z + 8}$ **7.** $w(x) = \sqrt{x} - \dfrac{6}{\sqrt{x}}$

8. $h(y) = \sqrt{y^2 + 1}$ **9.** $f(y) = \dfrac{2 - \sqrt{y}}{2 + \sqrt{y}}$

10. $y(x) = x \sin 2x$ **11.** $f(x) = \dfrac{2x}{e^x}$

12. $h(t) = t \sin^{-1} t$ **13.** $g(x) = \ln (x - \sqrt{x^2 + a^2})$

14. Find the first four derivatives of

    (a) $x^3 - 2x^2 + x - 8$ (b) $x^3 - \ln x$

    (c) $e^{-x} + \cos x$ (d) $(\sin x)^4$

15. In each case, the behavior of a function is described, for all points x in an interval $[a, b]$, in terms of its first and second derivatives. Sketch a possible graph of the function.

    (a) $f'(x) > 0$ and $f''(x) > 0$ (b) $f'(x) > 0$ and $f''(x) < 0$

    (c) $f'(x) > 0$ and $f''(x) = 0$ (d) $f'(x) < 0$ and $f''(x) > 0$

    (e) $f'(x) < 0$ and $f''(x) < 0$

16. In each case the behavior of a function at a *fixed* point c is described in terms of its derivatives. Sketch a possible graph of the function.

    (a) $f'(c) = 0$ and $f''(c) > 0$

    (b) $f'(c) = 0$ and $f''(c) < 0$

    (c) $f'(c) = 0, f''(c) = 0$ and $f'''(c) \neq 0$

    (d) $f'(c) > 0$ and $f''(c) = 0$

    (e) $f'(c) < 0$ and $f''(c) = 0$

17. Use the second derivative test to verify that the value

$$\frac{p\alpha}{\alpha - 1}\left[\frac{r(\alpha - 1)}{\beta p}\right]^{-1/\alpha}$$

obtained in Example 4 of Section 12.8 is actually a minimum value.

18. Suppose we have the cost function $C(x) = ax^3 + bx^2 + cx + d$, where, in order to make economic sense, we assume $d > 0$ and require that C be monotonically increasing for $x \geq 0$. What restrictions do these conditions impose on the coefficients?

19. A monopolist has respective revenue and cost functions $R(x) = -\alpha x^2 + \beta x$ and $C(x) = ax^2 + bx + c$, where all constants are assumed positive. The government imposes a tax at the rate of t dollars per unit output. Find the tax rate t which will maximize the tax revenue function $T(x) = tx_t$, where x_t is the output which maximizes profit *after* imposition of the tax at rate t.

20. Prove that the function $f(x) = x^3 + ax^2 + bx + c$ has neither a relative maximum nor a relative minimum if and only if $a^2 \leq 3b$.

SUPPLEMENTARY READING

Stein, S. K., *Calculus in the First Three Dimensions* (McGraw-Hill Book Company, New York, 1967), Chapters 2–5, 10, 11, and 21–23.

Thomas, G. B., *Calculus and Analytic Geometry*, 4th ed. (Addison-Wesley Publishing Company, Reading, Massachusetts, 1968), Chapters 1–4 and 7.

DIFFERENCE EQUATIONS **13**

13.1 CONCEPT OF A DIFFERENCE

The data with which a behavioral or management scientist works is ordinarily *discrete* in nature. That is, it takes the form of a sequence of values, usually obtained at different points in time. Economic data, for instance, is typically available through periodic reports, daily, monthly, yearly, or in the case of the census every ten years. In psychology and sociology, experiments often proceed through a sequence of trials, the experimental data consisting of various measures associated with trial outcomes.

Two different points of view can be adopted for the analysis of such data. In the first point of view, it may be assumed that the underlying process operates continuously in time and that the experimental data constitute a *sampling* of the process at isolated time points. In this case, a typical model describes the process by continuous functions, the derivatives of which mirror various changes associated with the process. Thus calculus techniques predominate in the analysis of the data. Examples 3–5 of Section 12.1, Example 4 of Section 12.7, and Examples 4 and 5 of Section 12.8 provide typical illustrations.

The second point of view is that the observed data constitute the entire process. In this case, the data appear at isolated time points rather than throughout an interval so that techniques of calculus are inappropriate and one works directly with the data sequence. A useful method for doing this is the *method of differences* and of *difference equations*, the subjects of this chapter.

Let us assume that there is a function f whose value $f(n)$ at time (or trial) n provides a numerical measure of the experimental outcome for that time period. The change in the process between time n and time $n + 1$ is then represented by the *first difference*

$$\Delta f(n) = f(n + 1) - f(n)$$

which is positive or negative depending on whether there was an increase or a decrease in the experimental results.

Example 1 A typical economic example is the cumulative growth of a sum of money at compound interest. An amount A_0 invested at $100r\%$ compounded annually grows to an amount A_n after n years. The amount A_{n+1} after $n + 1$ years is related to A_n by

$$A_{n+1} = (1 + r)A_n$$

The difference

$$\Delta A_n = A_{n+1} - A_n = rA_n \tag{1}$$

represents the accumulated interest on the amount A_n during the period from time n to time $n + 1$.

Because it involves the difference $\Delta A_n = A_{n+1} - A_n$, Equation (1) is called a *difference equation*. The solution of this equation, as we saw in Problem 3 of Section 5.2 is

$$A_n = (1 + r)^n A_0 \tag{2}$$

Methods for solving difference equations will be investigated in detail in Sections 13.4–13.6.   ▶

Example 2 If $g(n) = 2n^2 - 3n + 2$, the difference between the functional values at $n = 0$ and $n = 1$ is

$$\Delta g(0) = g(1) - g(0) = 1 - 2 = -1$$

Similarly, the change between $n = 1$ and $n = 2$ is

$$\Delta g(1) = g(2) - g(1) = 4 - 1 = 3$$

In general, for any value of n, the first difference function Δg evaluated at n is

$$\Delta g(n) = g(n + 1) - g(n)$$
$$= 2(n + 1)^2 - 3(n + 1) + 2 - [2n^2 - 3n + 2]$$
$$= 4n - 1 \qquad\qquad ▶$$

Higher differences of functions are defined successively, as were derivatives. Thus the *second difference* $\Delta^2 f$ of f is the difference of the first difference:

$$\Delta^2 f(n) = \Delta[\Delta f(n)] = \Delta f(n + 1) - \Delta f(n)$$

the third difference is the difference of the second difference:

$$\Delta^3 f(n) = \Delta[\Delta^2 f(n)] = \Delta^2 f(n+1) - \Delta^2 f(n)$$

and, in general, the kth difference

$$\Delta^k f(n) = \Delta[\Delta^{k-1} f(n)] \tag{3}$$

is the difference of the $(k-1)$st difference.

Example 3 For the function g in Example 2, we find

$$\begin{aligned}
\Delta^2 g(n) &= \Delta[\Delta g(n)] \\
&= \Delta[4n - 1] \\
&= [4(n+1) - 1] - [4n - 1] \\
&= 4
\end{aligned}$$

and

$$\begin{aligned}
\Delta^3 g(n) &= \Delta[\Delta^2 g(n)] \\
&= \Delta^2 g(n+1) - \Delta^2 g(n) \\
&= 4 - 4 = 0
\end{aligned}$$

for every value of n. Then $\Delta^4 g(n) = 0$, $\Delta^5 g(n) = 0$, etc. In fact, for this example $\Delta^k g(n)$ is 0 for every value of n whenever k is greater than 2.

A part of our results are exhibited in Table 1, called a *difference table*. Of course, the values of n may be extended indefinitely in each direction. ▶

TABLE 1. Difference table for $g(n) = 2n^2 - 3n + 2$

n	$g(n)$	$\Delta g(n)$	$\Delta^2 g(n)$	$\Delta^3 g(n)$	$\Delta^4 g(n)$
-2	16				
		-9			
-1	7		4		
		-5		0	
0	2		4		0
		-1		0	
1	1		4		0
		3		0	
2	4		4		0
		7		0	
3	11		4		
		11			
4	22				

Repeated use of Equation (3) allows us to write any difference in terms of values of f alone. For example,

$$\Delta^2 f(n) = \Delta(\Delta f(n)) = \Delta f(n+1) - \Delta f(n)$$
$$= [f(n+2) - f(n+1)] - [f(n+1) - f(n)]$$
$$= f(n+2) - 2f(n+1) + f(n) \qquad (4)$$

and

$$\Delta^3 f(n) = \Delta(\Delta^2 f(n)) = \Delta^2 f(n+1) - \Delta^2 f(n)$$
$$= [f(n+3) - 2f(n+2) + f(n+1)]$$
$$- [f(n+2) - 2f(n+1) + f(n)]$$
$$= f(n+3) - 3f(n+2) + 3f(n+1) - f(n) \qquad (5)$$

Other formulas are derived in the exercises but will not be required in the text.

Example 4 Differences may be given physical interpretations. If, for example, $f(t)$ denotes the distance traveled by an automobile in time t, then $\Delta f(t) = f(t+1) - f(t)$ represents the change in distance in the unit of time from t to $t+1$ and as such gives the average velocity over this time interval (compare with Section 12.1).

When Patrolman Smith checks your speed by stretching two cables across the road, he computes such an average velocity. However, in this case the elapsed time might be any amount h rather than 1 unit. The computation for average speed becomes

$$\frac{\Delta_h f(t)}{h} = \frac{f(t+h) - f(t)}{h}$$

the quantity $\Delta_h f(t)$ being the first difference of f over the interval from t to $t+h$.
▶

Example 5 In Harrod's* classic one-country model of economic growth, he postulates that

(i) A country's net savings are proportional to net income.

(ii) A country's desired investment is proportional to its increase in income.

Thus, denoting national income, net savings, and desired investment during time period t by $Y(t)$, $S(t)$, and $I(t)$, respectively, he writes

(i) $S(t) = cY(t)$

(ii) $I(t) = g[Y(t) - Y(t-1)] = g\Delta Y(t-1)$

*Discussed in Baumol, W. J., *Economic Dynamics* (Macmillan, Inc., New York, 1959), pp. 37–44.

where c and g are positive constants representing the propensities to save and to invest.

Harrod assumes that net investment is realized only if $S(t) = I(t)$, in which case (i) and (ii) yield the result that

$$\Delta Y(t-1) = Y(t) - Y(t-1) = \frac{c}{g} Y(t)$$

The amount of growth being proportional to total income, it follows that national income must grow by ever increasing amounts. ▶

PROBLEMS

In Problems 1–9, find (a) $\Delta y(0)$, $\Delta y(1)$, $\Delta y(2)$, $\Delta y(3)$, $\Delta y(n)$; and (b) $\Delta^2 y(0)$, $\Delta^2 y(1)$, $\Delta^2 y(2)$, $\Delta^2 y(3)$, $\Delta^2 y(n)$.

1. $y(n) = 2$ 2. $y(n) = n + 1$ 3. $y(n) = 5n + 2$

4. $y(n) = n^2$ 5. $y(n) = n(n-1)$ 6. $y(n) = n^2 - 2n + 1$

7. $y(n) = \dfrac{1}{n}$ 8. $y(n) = 2^n$ 9. $y(n) = \dfrac{1}{3^n}$

10. Construct a difference table through the fourth difference for $g(n) = n^3$ using $n = 1, 2, \ldots, 6$.

11. Find general formulas for $\Delta^4 f(n)$ and $\Delta^5 f(n)$. (See Equations (4) and (5).)

12. By induction, prove that the general difference formula is

$$\Delta^r f(n) = \sum_{k=0}^{r} \binom{r}{k} (-1)^{r-k} f(n+k)$$

[Hint: Use the recursive definition $\Delta^k f(n) = \Delta(\Delta^{k-1} f(n))$.]

13. Verify that

(a) $\Delta^2 n^2 = 2$ (b) $\Delta^3 n^3 = 3 \cdot 2 \cdot 1$ (c) $\Delta^4 n^4 = 4 \cdot 3 \cdot 2 \cdot 1$

14. Recall from Section 7.2 that $(n)_r = n(n-1)(n-2) \cdots (n-r+1)$. Prove that

$$\Delta(n)_r = r(n)_{r-1}$$

15. Find $\Delta^k(n)_5$ for $k = 2, 3, 4, 5$, and 6.

16. Show that

(a) $\Delta a^n = a^n(a-1)$

(b) $\Delta 2^n = 2^n$

(c) $\Delta \sin an = 2 \sin \dfrac{a}{2} \cos a(n + \tfrac{1}{2})$

(d) $\Delta \cos an = -2 \sin \dfrac{a}{2} \sin a(n + \tfrac{1}{2})$

(e) $\Delta \ln n = \ln\left(1 + \dfrac{1}{n}\right)$

17. If $f(n)^{(r)}$ is defined as $f(n)^{(r)} = f(n) \cdot f(n-1) \cdots f(n-r+1)$, verify that $\Delta(a + bn)^{(r)} = br(a + bn)^{(r-1)}$.

18. Find the following higher-order differences.

(a) $\Delta^2 a^n$ (b) $\Delta^r 2^n$

13.2 BASIC PROPERTIES OF DIFFERENCES

In this section we discuss certain properties of differences which will be of importance in the solution of difference equations.

Property 1 Operations of taking differences commute with one another. Symbolically, for all non-negative integers k and r,

$$\Delta^k[\Delta^r f(n)] = \Delta^r[\Delta^k f(n)] = \Delta^{k+r} f(n)$$

PROOF The truth of this assertion is apparent from the recursive nature of Equation (3). In order to obtain the $(k + r)$th difference of f we may first compute r differences and then k more, or first k and then r more. ▶

Property 2 If c is any constant and f a function of n, then

$$\Delta^k[cf(n)] = c\Delta^k f(n) \tag{6}$$

That is, the operations of differencing and multiplying by a constant commute.

PROOF The result holds for $k = 1$ since

$$\Delta^1[cf(n)] = \Delta[cf(n)] = cf(n + 1) - cf(n)$$
$$= c[f(n + 1) - f(n)] = c\Delta f(n) \tag{7}$$

If (6) is true for $k = m$, then

$$\Delta^{m+1}[cf(n)] = \Delta[\Delta^m cf(n)]$$
$$= \Delta[c\Delta^m f(n)] \quad \text{[by the inductive assumption]}$$
$$= c\Delta[\Delta^m f(n)] \quad \text{[using (7)]}$$
$$= c\Delta^{m+1} f(n) \quad \text{[from (3)]} \qquad\qquad ▶$$

Property 2 does not hold if c is not constant. For instance,

$$\Delta[nf(n)] = (n + 1)f(n + 1) + nf(n)$$
$$= n\Delta f(n) + f(n + 1)$$

Property 3 The operation of taking differences distributes over sums of functions. In symbols, if

$$g(n) = f_1(n) + f_2(n) + \cdots + f_r(n)$$

then

$$\Delta^k g(n) = \Delta^k f_1(n) + \Delta^k f_2(n) + \cdots + \Delta^k f_r(n) \tag{8}$$

for every positive integer k. (Loosely speaking, this says that the difference of a sum equals the sum of the differences. Compare this statement with Theorem 2(c) in Section 12.2.)

PROOF   It is convenient to use summation notation. Since,

$$g(n) = \sum_{j=1}^{r} f_j(n)$$

we have

$$\Delta g(n) = \Delta \left[\sum_{j=1}^{r} f_j(n) \right] = \sum_{j=1}^{r} f_j(n + 1) - \sum_{j=1}^{r} f_j(n)$$

$$= \sum_{j=1}^{r} [f_j(n + 1) - f_j(n)]$$

$$= \sum_{j=1}^{r} \Delta f_j(n) \tag{9}$$

proving that (8) is true for first differences.

As an inductive assumption suppose that (8) is true for $k = m$. Then

$$\Delta^{m+1} g(n) = \Delta^{m+1} \left[\sum_{j=1}^{r} f_j(n) \right]$$

$$= \Delta \left\{ \Delta^m \left[\sum_{j=1}^{r} f_j(n) \right] \right\}$$

$$= \Delta \left\{ \sum_{j=1}^{r} \Delta^m f_j(n) \right\} \qquad \text{[inductive assumption]}$$

$$= \sum_{j=1}^{r} \Delta [\Delta^m f_j(n)] \qquad \text{[using (9)]}$$

$$= \sum_{j=1}^{r} \Delta^{m+1} f_j(n)$$

completing the proof.   ▶

Example 1   In their studies of the sense of time in humans, McGrath and O'Hanlon* found that a person's subjective estimate $E(t)$ of time is not, in general, equal to the elapsed time t but is related by a function of the form

$$E(t) = a + bt$$

*McGrath, J. J., and O'Hanlon, J. F., Jr., "Methods for Measuring the Rate of Subjective Time," *Perceptual and Motor Skills* **24** (3, Part 2), 1235–1240 (1967). The authors actually use the derivative of $E(t)$ rather than a ratio of differences to define *RST*.

where a and b are constants. The *rate of change RST* of subjective time is then defined as

$$RST = \frac{\Delta E(t)}{\Delta t} = \frac{\Delta(a + bt)}{\Delta t}$$

$$= \frac{\Delta a + \Delta(bt)}{\Delta t} \qquad \text{[Property 3]}$$

$$= \frac{\Delta a + b\Delta t}{\Delta t} \qquad \text{[Property 2]}$$

Since $\Delta a = a - a = 0$ and $\Delta t = t + 1 - t = 1$, we find $RST = b$. ▶

Example 2 Suppose that when x units of a certain commodity are produced, the unit production cost is

$$U(x) = a - bx$$

where a and b are constants. Then the total production cost is

$$C(x) = x \cdot U(x) = ax - bx^2$$

In terms of differences, marginal cost $M(x)$ is the change in total production costs resulting from a unit increase in production from x to $x + 1$ units. Here

$$M(x) = \Delta C(x) = \Delta(ax - bx^2)$$

$$= a\Delta x - b\Delta x^2 \qquad \text{[Properties 2 and 3]}$$

Since $\Delta x = (x + 1) - x = 1$ and $\Delta x^2 = (x + 1)^2 - x^2 = 2x + 1$, we have

$$M(x) = a - b(2x + 1) = a - b - 2bx$$

(Compare this example with Example 1 of Section 12.1.) ▶

We complete our list of properties of differences by introducing a mathematical shorthand definition which allows us to distribute a function over a "sum" of differences.

Property 4 In an expression of the form

$$a_k\Delta^k f(n) + a_{k-1}\Delta^{k-1} f(n) + \cdots + a_1\Delta f(n) + a_0 f(n) \qquad (10)$$

where $a_0, a_1, a_2, \ldots, a_n$ are constants, we treat $f(n)$ as though it could be factored out and we rewrite (10) as

$$[a_k\Delta^k + a_{k-1}\Delta^{k-1} + \cdots + a_1\Delta + a_0]f(n) \qquad (11) \quad ▶$$

The expression

$$\phi_k(\Delta) = a_k\Delta^k + a_{k-1}\Delta^{k-1} + \cdots + a_1\Delta + a_0$$

appearing in (11) is called a *polynomial difference operator of degree k*. The definition, or meaning, of $\phi_k(\Delta)$ is specified through its operation on a function in the manner indicated by the equivalence of (10) and (11). Specifically, $\phi_k(\Delta)$ operates in such a way as to change a function $f(n)$ into another function $\phi_k(\Delta)f(n)$, which is a linear combination of differences of $f(n)$. From the general point of view of Section 4.1, $\phi_k(\Delta)$ is itself a function whose domain is a set of functions. To avoid confusion, the word *operator* is used in mathematics for such functions.

In Theorem 1 of Section 1.4, we discovered rules for operating with sets which were quite similar to certain algebraic operations with numbers, namely, the commutative, associative, and distributive laws. In the same way, Properties 1–4 above constitute rules for operating with differences which have direct analogies to ordinary algebraic operations with numbers. Specifically, Property 1 shows that repeated operations of taking differences combine as though the symbol Δ were an algebraic quantity being raised to powers. (Of course, this is *not* what is happening, but the symbols combine *as though* it is.) Property 2 states a commutative law for Δ and its "powers," while Properties 3 and 4 provide distributive laws. The net result is that differences combine with each other and with constants according to precisely the same laws that govern ordinary numbers. Numbers and differences are said to exhibit the same *mathematical structure*.

Example 3 The repeated operation of taking differences indicated in

$$(\Delta + a)[(\Delta + a)f(n)]$$

may be rewritten without calculations as

$$[\Delta^2 + 2a\Delta + a^2]f(n)$$

by simply "multiplying" the indicated "product" $(\Delta + a)(\Delta + a)$. A formal proof that this is the case may be obtained from Properties 1–4 as follows:

$$
\begin{aligned}
(\Delta + a)[(\Delta + a)f(n)] &= (\Delta + a)[\Delta f(n) + af(n)] &&\text{[Property 4]} \\
&= \Delta[\Delta f(n) + af(n)] + a[\Delta f(n) + af(n)] &&\text{[Property 4]} \\
&= \Delta^2 f(n) + \Delta[af(n)] + a\Delta f(n) + a^2 f(n) &&\text{[Property 3]} \\
&= \Delta^2 f(n) + 2a\Delta f(n) + a^2 f(n) &&\text{[Property 2]} \\
&= (\Delta^2 + 2a\Delta + a^2)f(n) &&\text{[Property 4]}
\end{aligned}
$$

We shall take advantage of this correspondence between the properties of differences and numbers by writing $(\Delta + a)^2 f(n)$ in place of

$$(\Delta + a)[(\Delta + a)f(n)]$$

Similarly,

$$(\Delta + a)\{(\Delta + b)[(\Delta + a)f(n)]\}$$

may be rewritten as

$$(\Delta + b)(\Delta + a)^2 f(n)$$

Other examples will appear in the exercises and in succeeding sections. ▶

We conclude this discussion by summarizing the most important results concerning polynomial difference operators. No proofs are necessary beyond citing the correspondence already noted between differences and real numbers and observing that each of these results would be valid if Δ were a number and the operation were multiplication rather than differencing.

Property 5 Let $\phi_k(\Delta)$ and $\psi_m(\Delta)$ be any two polynomial difference operators and let c be a constant. Then

   (i) $\phi_k(\Delta)[cf(n)] = c\phi_k(\Delta)f(n)$
   (ii) $\phi_k(\Delta)[f(n) + g(n)] = \phi_k(\Delta)f(n) + \phi_k(\Delta)g(n)$
   (iii) $[\phi_k(\Delta) + \psi_m(\Delta)]f(n) = \phi_k(\Delta)f(n) + \psi_m(\Delta)f(n)$
   (iv) $\phi_k(\Delta)[\psi_m(\Delta)f(n)] = \psi_m(\Delta)[\phi_k(\Delta)f(n)] = [\phi_k(\Delta) \times \psi_m(\Delta)]f(n)$

The symbol $\times$ denotes ordinary multiplication of polynomials. ▶

Example 4 Since

$$(\Delta - 1)2^n = \Delta 2^n - 2^n = (2^{n+1} - 2^n) - 2^n = 0$$
$$(\Delta - 2)3^n = \Delta 3^n - 2 \cdot 3^n = (3^{n+1} - 3^n) - 2 \cdot 3^n = 0$$

it follows that for any constants a and b,

$$(\Delta^2 - 3\Delta + 2)[a \cdot 2^n + b \cdot 3^n] = 0$$

For,

$(\Delta^2 - 3\Delta + 2)[a \cdot 2^n + b \cdot 3^n]$

$= (\Delta - 1)(\Delta - 2)[a \cdot 2^n + b \cdot 3^n]$

$= (\Delta - 1)(\Delta - 2)[a \cdot 2^n] + (\Delta - 1)(\Delta - 2)[b \cdot 3^n]$ [Property 5(ii)]

$= a(\Delta - 2)[(\Delta - 1)2^n] + b(\Delta - 1)[(\Delta - 2)3^n]$ [Properties 5(i) and (iv)]

$= a(\Delta - 2)[0] + b(\Delta - 1)[0]$

$= 0 + 0 = 0$ ▶

PROBLEMS

1. In Example 5, Section 13.1, modify Harrod's approach by assuming that

$$I(t) = g[Y(t) - Y(t - 1)] + kY(t) + L$$

where k and L are constants. Find $\Delta I(t)$, $\Delta^2 I(t)$, and, in general, $\Delta^r I(t)$.

2. Suppose that population size at time t is given by

$$N(t) = N_0(1 + \alpha - \beta)^t$$

where N_0 is the initial population size and α and β are constant birth and death rates, respectively. Find $\Delta N(t)$, $\Delta^2 N(t)$, and, in general, $\Delta^r N(t)$.

3. Verify for any function $Y(n)$ that

(a) $Y(n + 1) = Y(n) + \Delta Y(n)$

(b) $Y(n + 2) = Y(n) + 2\Delta Y(n) + \Delta^2 Y(n)$

(c) $Y(n + 3) = Y(n) + 3\Delta Y(n) + 3\Delta^2 Y(n) + \Delta^3 Y(n)$

4. Prove by induction that for the general function $Y(n + r)$ with r any positive integer:

$$Y(n + r) = (1 + \Delta)^r Y(n)$$

Hence, prove that

$$Y(n + r) = \sum_{k=0}^{r} \binom{r}{k} \Delta^k Y(n)$$

5. Verify the following formulas for differencing products and quotients. (Compare with Theorems 3 and 5 in Section 12.2.)

(a) $\Delta[f(n) \cdot g(n)] = g(n)\Delta f(n) + f(n + 1)\Delta g(n)$

(b) $\Delta\left[\dfrac{f(n)}{g(n)}\right] = \dfrac{g(n)\Delta f(n) - f(n)\Delta g(n)}{g(n) \cdot g(n + 1)}$ if $g(n)$ and $g(n + 1)$ are

not zero.

6. If $(n)_r = n(n - 1) \cdots (n - r + 1)$, prove that

(a) $n^2 = (n)_2 + (n)_1$

(b) $n^3 = (n)_3 + 3(n)_2 + (n)_1$

(c) $n^4 = (n)_4 + 6(n)_3 + 7(n)_2 + (n)_1$

7. Extending Problem 6 show by induction that for any integer r, n^r can be written as a linear combination of the form

$$n^r = a_r(n)_r + a_{r-1}(n)_{r-1} + \cdots + a_1(n)_1 \tag{12}$$

where $a_1, \ldots, a_r$ are constants.

[Hints: $n^{r+1} = n \cdot n^r$, $n = (n - k) + k$, and $(n - k)(n)_k = (n)_{k+1}$.]

8. If $Y(n) = n^4$, find $\Delta Y(n)$:

(a) from the definition of Δ.

(b) by first expressing n^4 as in Problem 6(c) and then using the result of Problem 14 in Section 13.1.

(c) by writing $n^4 = n^3 \cdot n$ and using Problem 5(a).

9. Find the marginal cost function if when n units are produced, the unit cost is $a - b^{-cn}$, where a, b, and c are constants. (See Example 2 of this section.)

10. *Elasticity of demand* is defined as the ratio of the percent change in quantity demanded to the percent change in price. Let $P(t)$ denote the unit price at time t, and let $Q(t)$ be the number of units demanded at time t. If $P(t) = a - bQ(t)$, show that the elasticity of demand is

$$\frac{100 \times \Delta Q(t)/Q(t)}{100 \times \Delta P(t)/P(t)} = -\frac{1}{b}\frac{P(t)}{Q(t)}$$

11. In Problem 10, find the first difference of elasticity of demand.

12. If $P(t) \cdot Q(t) = k$, where k is a constant, show that the elasticity of demand is $-Q(t+1)/Q(t)$.

13. Prove the results concerning polynomial difference operators stated in Property 5.

13.3 APPLICATIONS

Let us recall from Section 11.2 that a polynomial is a function f of the form

$$f(n) = a_0 + a_1 n + a_2 n^2 + \cdots + a_{k-1} n^{k-1} + a_k n^k \qquad \text{(13)}$$

where $a_0, a_1, a_2, \ldots, a_k$ are constants and k is a non-negative integer. If $a_k \neq 0$, f is a polynomial of degree k. Some polynomials are given special names. Thus, the function f whose values are given by $f(n) = 3n^4 + 7n - 2$ is a polynomial of degree four or quartic, while $g(n) = 6 - n^3$ is a third degree, or cubic, polynomial. Polynomials of degrees zero, one, and two are called constant, linear, and quadratic functions, respectively.

The first difference of a constant function is zero, since if $f(n) = c$ for each value of n, then $\Delta f(n) = f(n+1) - f(n) = c - c = 0$. If $f(n) = cn + b$, a linear function, then $\Delta f(n) = [c(n+1) + b] - [cn + b] = c$, a constant function. Similarly, the first difference of the quadratic function $f(n) = an^2 + bn + c$ is

$$\Delta f(n) = f(n+1) - f(n) = a(n+1)^2 + b(n+1) + c - [an^2 + bn + c]$$
$$= 2an + (a+b)$$

which is a linear function. These results are generalized in the following theorem (which should be compared with Theorem 1, Section 12.2).

Theorem 1 The first difference of a polynomial of degree $k \geq 1$ is a polynomial of degree $k - 1$.

PROOF It suffices to prove the theorem for the special case $f(n) = n^k$; that is, to prove that differencing reduces by one the degree of each term in a polynomial. In this case we have

$$\Delta f(n) = f(n+1) - f(n) = (n+1)^k - n^k$$

Using the binomial expansion (Section 7.3),

$$\Delta f(n) = \left[n^k + k n^{k-1} + \frac{k(k-1)}{2} n^{k-2} + \cdots + kn + 1 \right] - n^k$$
$$= k n^{k-1} + \frac{k(k-1)}{2} n^{k-2} + \cdots + kn + 1$$

which establishes the result. ▶

Since each operation of taking differences drops the degree by one, it follows that the rth difference of a polynomial of degree $k \geq r$ is a polynomial of degree $k - r$. If $k < r$, the rth difference is zero.

Example 1 If the output of a production system for time period t is given by the output function

$$P(t) = 5t^3 - 2t^2 + 10t + 2$$

then $\Delta P(t) = 15t^2 + 11t + 13$, $\Delta^2 P(t) = 30t + 26$, and $\Delta^3 P(t) = 30$. Obviously, $\Delta^k P(t) = 0$ for all $k > 3$. ▶

In most applications of differences it is very important to be able to reverse the operation of taking differences in order to be able to find functions whose differences are given. If, for example, we are asked to find a function f for which $\Delta f(n) = 0$, we know that $f(n) = c$ is such a function. Again, if $\Delta f(n) = 4$, then any linear function of the form $f(n) = 4n + c$ is an answer to the problem. The next theorem generalizes these results to polynomials of any degree; the proof is omitted (but see Problem 8).

Theorem 2 Every polynomial is the first difference of some other polynomial. Specifically, if $g(n)$ is a polynomial of degree r, then there is a polynomial $f(n)$ of degree $r + 1$ such that $\Delta f(n) = g(n)$. ▶

One application of this theorem is the following method of evaluating sums. Let g be any function and consider the sum $S = \sum_{k=m}^{n} \Delta g(k)$ where $m \leq n$. Since $\Delta g(k) = g(k + 1) - g(k)$, S may be rewritten as

$$S = [g(m + 1) - g(m)] + [g(m + 2) - g(m + 1)] + \cdots$$
$$+ [g(n) - g(n - 1)] + [g(n + 1) - g(n)]$$

All terms cancel except two, leaving

$$S = g(n + 1) - g(m)$$

Once the function g is known the sum S may easily be computed.

We use this result as follows. Suppose we wish to evaluate the sum $\sum_{k=m}^{n} f(k)$, where f is a given function. If we can find a function g whose first difference Δg is the same as f, our task is greatly simplified. For then,

$$\sum_{k=m}^{n} f(k) = \sum_{k=m}^{n} \Delta g(k) = g(n + 1) - g(m) \qquad (14)$$

as above.

Example 2 If we wish to find the sum $S = \sum_{k=1}^{n} k$ of the first n positive integers, our problem is solved if we can find a function g such that

$$\Delta g(k) = k \qquad (15)$$

Such a function exists according to Theorem 2 and, in fact, is a polynomial of the form

$$g(k) = ak^2 + bk + c \qquad (16)$$

Taking differences in (16) and imposing condition (15) leads to

$$\Delta g(k) = 2ak + a + b = k$$

In order for this to hold for all k, we choose $a = \frac{1}{2}$ and $b = -\frac{1}{2}$ so that

$$g(k) = \tfrac{1}{2}k^2 - \tfrac{1}{2}k$$

(The value of c is irrelevant, so choose $c = 0$.) Inserting this into Equation (14) gives

$$
\begin{aligned}
S = \sum_{k=1}^{n} k &= g(n + 1) - g(1) \\
&= \tfrac{1}{2}(n + 1)^2 - \tfrac{1}{2}(n + 1) - [\tfrac{1}{2} \cdot 1^2 - \tfrac{1}{2} \cdot 1] \\
&= \tfrac{1}{2}n(n + 1)
\end{aligned}
$$

For instance, $1 + 2 = 3 = \tfrac{1}{2}(2)(3)$, $1 + 2 + 3 = 6 = \tfrac{1}{2}(3)(4)$, $1 + 2 + 3 + \cdots + 100 = \tfrac{1}{2}(100)(101) = 5050$, and so forth. (Compare with Problem 6 in Section 5.2.) ▶

Example 3 Similar techniques may be applied to find sums of values of functions other than polynomials. For instance, to find the geometric sum $h(x, n) = \sum_{k=0}^{n} x^k$, we note that if $g(k) = x^k/(x - 1)$, then

$$\Delta g(k) = \frac{x^{k+1}}{x - 1} - \frac{x^k}{x - 1} = \frac{x^k(x - 1)}{x - 1} = x^k$$

Thus

$$
\begin{aligned}
h(x, n) &= g(n + 1) - g(0) \\
&= \frac{x^{n+1}}{x - 1} - \frac{x^0}{x - 1} = \frac{x^{n+1} - 1}{x - 1}
\end{aligned}
$$

a result familiar from our discussion in Section 6.2 concerning partial sums of the geometric series. ▶

We can also apply a difference method to the problem of determining the maximum and minimum values of a function in a given interval, which we have previously treated by the methods of calculus (Section 12.8). When, for a function f, $\Delta f(n) = f(n + 1) - f(n)$ is positive, it means that $f(n + 1)$ exceeds $f(n)$, or that the value of f has increased as we move from n to $n + 1$. Similarly, if $\Delta f(n) < 0$, then $f(n + 1) < f(n)$, indicating a decrease in the value of f.

Thus, if we wished to find the point at which f attains its maximum, or largest value (assuming such a point exists), we would look for a value n for which

$$\Delta f(n - 1) > 0 \quad \text{and} \quad \Delta f(n) < 0 \tag{17}$$

For if (17) holds, we have both

$$f(n) > f(n-1) \quad \text{and} \quad f(n) > f(n+1)$$

so that, at least relative to points on either side, f attains its largest value at n. That is, n is a point of relative maximum.

Example 4 In a monopolistic market, the monopolist may adjust his output and price to obtain maximum profits. Let us suppose that the per-unit price p and the output n are related by $p = 10 - (1/1000)n$, so that the total revenue is

$$R(n) = n \cdot p = 10n - \tfrac{1}{1000}n^2$$

Then if the total cost function is

$$C(n) = 8 + 4n + \tfrac{1}{500}n^2$$

the profit obtained from the production of n items is

$$P(n) = R(n) - C(n) = 6n - \tfrac{3}{1000}n^2 - 8$$

In order to find the output which results in maximum profit, we look for that value m of n for which $\Delta P(m-1) > 0$ and $\Delta P(m) < 0$. Now

$$\Delta P(n) = 6 - \tfrac{3}{1000}(2n + 1)$$

so $\Delta P(n) > 0$ when

$$6 > \tfrac{3}{1000}(2n + 1)$$

and $\Delta P(n) < 0$ when

$$6 < \tfrac{3}{1000}(2n + 1)$$

These two cases yield $m = 1000$ as the number of items which should be produced. The maximum profit is

$$P(1000) = 6(1000) - \tfrac{3}{1000}(1000)^2 - 8$$
$$= 2992 \qquad\qquad \blacktriangleright$$

PROBLEMS

1. Extend the result of Example 2 by showing that

(a) $\displaystyle\sum_{k=1}^{n} k^2 = \frac{n(n+1)(2n+1)}{6}$

(b) $\displaystyle\sum_{k=1}^{n} k^3 = \frac{n^2(n+1)^2}{4}$

(c) $\displaystyle\sum_{k=1}^{n} (2k-1) = n^2$

2. Use the results of Problem 1 to find

$$\sum_{k=1}^{10} (k^3 - 6k^2 + 7)$$

3. The total cost to the Nu-U Cosmetic Company for the production of n units is

$$T(n) = 10{,}000 + 400n - 3n^2 + \tfrac{1}{5}n^3$$

(a) Find the marginal cost (see Example 2, Section 13.2) for the tenth unit.

(b) When is the marginal cost a minimum and what is the minimal marginal cost?

(c) What is the minimum total cost? Is total cost minimized at the same point as marginal cost?

4. If $Y(n) = n^5$, find $\Delta Y(n)$

(a) from the definition $\Delta Y(n) = Y(n + 1) - Y(n)$.

(b) by writing $n^5 = n^3 \cdot n^2$ and using the result of Problem 5(a), Section 13.2.

5. Find a function $Y(t)$ for which

(a) $\Delta Y(t) = Y(t + 1) - Y(t) = t^2 + 3t + 2$

(b) $\Delta Y(t) = 2t + 3$

(c) $\Delta Y(t) = 2 \cdot 2^t + 1$

6. Prove that if $Y(n)$ is the rth-degree polynomial

$$Y(n) = a_r n^r + a_{r-1} n^{r-1} + \cdots + a_1 n + a_0$$

then $\Delta^r Y(n) = a_r \cdot r!$ and all succeeding differences are zero.

7. If when n batches of an item are produced, the per unit cost is

$$C(n) = n^2 - 6n + 10$$

show that the unit costs are minimized when three batches are produced.

8. Combine the results of Problem 14, Section 13.1, and Problem 7, Section 13.2, to prove Theorem 2.

13.4 THE SIMPLEST DIFFERENCE EQUATIONS

Equations involving differences of functions arise in certain mathematical models employed in the behavioral sciences. Such equations naturally are called *difference equations*. To *solve* a difference equation is to find a function which, when substituted into the equation, yields an identity—an expression valid for all values of the variable. As a simple example, the equation $\Delta f(n) = 2n + 1$ has the solution $f(n) = n^2 + 3$, since

$$\Delta(n^2 + 3) = (n + 1)^2 + 3 - [n^2 + 3] = 2n + 1$$

holds regardless of the value of n.

We shall limit our general remarks to *linear difference equations*, these being the simplest and the type most often encountered in practice. A linear difference equation has the form

$$\phi(\Delta)f(n) = g(n) \tag{18}$$

where $\phi(\Delta)$ is a polynomial difference operator of the type introduced in Equations (10) and (11) in Section 13.2, g is a known function, and f is the function for which a solution is sought. Equation (18) is called linear because the function f and its differences appear to the first power. As examples, the equation $\Delta^2 f(n) - 3\Delta f(n) = 2n^3$ is linear, while $\Delta f(n) + 2[f(n)]^2 = 0$ is not linear.

The *order* of the linear difference equation (18) is defined to be the degree of the polynomial difference operator $\phi(\Delta)$. For instance, $\Delta f(n) = 3n^2$ is a first-order equation, while $\Delta^4 f(n) + 6\Delta^2 f(n) - f(n) = 2$ is of fourth order.

Solutions to the simplest linear difference equations may be easily obtained from our prior knowledge of differences. We know, for example, that the first difference of a constant function is zero (Section 13.3). Hence the equation $\Delta f(n) = 0$ has the solution $f(n) = c$, where c may be any constant. For this equation there are no other solutions since

$$\Delta f(n) = f(n + 1) - f(n) = 0$$

means that $f(n + 1) = f(n)$ for all values of n. It follows that all functional values are the same and we have proved the following theorem.

Theorem 3 A function f is a solution of $\Delta f(n) = 0$ if and only if f is a constant function. ▶

An immediate consequence of Theorem 3, and a key to solving difference equations, is the next theorem.

Theorem 4 Two functions which have the same difference function can differ at most by a constant amount. That is, if $\Delta f(n) = \Delta g(n)$, then there is a constant c (which may be zero) such that for all values of n, $f(n) = g(n) + c$.

PROOF Assume $\Delta f(n) = \Delta g(n)$ for all values of n and define a new function h by $h(n) = f(n) - g(n)$. Then $\Delta h(n) = \Delta f(n) - \Delta g(n) = 0$ and it follows from Theorem 3 that h must be a constant function. This completes the proof. ▶

A major concern in solving any difference equation is to make sure, as we were able to do in Theorem 3, that *all* solutions have been found. For instance, any constant function $f(n) = c$ is a solution of the equation

$$\Delta^2 f(n) = 0 \tag{19}$$

But there are other solutions, such as $f(n) = 6n$ and $f(n) = 3 - 2n$, as may easily be verified. To find all possible solutions of the difference equation (19), we first substitute $g(n) = \Delta f(n)$, obtaining the equivalent form $\Delta g(n) = 0$. According to Theorem 3, the only solutions of this latter equation are constants, so that the task of solving (19) may be replaced by that of solving

$$g(n) = \Delta f(n) = c \tag{20}$$

One solution of (20) is $h(n) = cn$, since $\Delta(cn) = c(n+1) - cn = c$. If $f(n)$ is any other solution, that is, if $\Delta f(n) = c$ also, then it follows from Theorem 4 that there is a constant b such that $f(n) = cn + b$ for all values of n. In short, we have proved the following theorem.

Theorem 5 Every solution of the equation $\Delta^2 f(n) = 0$ has the form $f(n) = cn + b$, where b and c are constants. ▶

Example 1 Suppose* that reaction time y is related to the number n of units of information in a choice situation having 2^n alternatives by the equation

$$y(n+2) = 2y(n+1) - y(n)$$

Then

$$y(n+2) - 2y(n+1) + y(n) = \Delta^2 y(n) = 0$$

and it follows from Theorem 5 that $y(n) = a + bn$. Hyman's results show that y is approximately equal to 0.2 when $n = 0$ and is about 0.8 when $n = 4$. Hence

$$y(0) = 0.2 = a + b(0)$$
$$y(4) = 0.8 = a + b(4)$$

from which we obtain $a = 0.2$, $b = 0.15$, and $y(n) = 0.2 + 0.15n$. ▶

In order to evaluate the constants a and b in Example 1, it was necessary to know the values of y corresponding to two different values of n. Functional values, such as $y(0) = 0.2$ and $y(4) = 0.8$ in Example 1, which are specified in order to obtain a unique solution to a difference equation, are said to constitute the *initial conditions* on the problem.

In the proof of Theorem 5 the solution of the second-order equation (19) was facilitated by replacing it by the first-order equation (20). A similar *reduction of order* enables us to solve higher-order equations and leads to the following theorem.

Theorem 6 Let k be any positive integer. A function f is a solution of the equation $\Delta^k f(n) = 0$ if and only if f is a polynomial of degree $k - 1$ or less; that is, if and only if f has the form

$$f(n) = c_0 + c_1 n + c_2 n^2 + \cdots + c_{k-1} n^{k-1}$$

where $c_0, c_1, \ldots, c_{k-1}$ are constants.

*Adapted from Hyman, R., "Stimulus Information as a Determinant of Reaction Time," *Journal of Experimental Psychology* **45**, 188–196 (1953).

PROOF Theorems 3 and 5 have established this result for $k = 1$ and $k = 2$. Suppose, as an inductive assumption, that the theorem is true for $k = r$. That is, suppose that all solutions of $\Delta^r f(n) = 0$ have the form $f(n) = b_0 + b_1 n + \cdots + b_{r-1} n^{r-1}$, where $b_0, b_1, \ldots, b_{r-1}$ are constants. To solve the $(r + 1)$st order equation $\Delta^{r+1} f(n) = 0$, we first substitute $g(n) = \Delta f(n)$, reducing the order and obtaining $\Delta^r g(n) = 0$. The inductive assumption then states that $g(n)$ must have the form

$$g(n) = \Delta f(n) = b_0 + b_1 n + \cdots + b_{r-1} n^{r-1} \tag{21}$$

Adapted to our present needs, Theorem 2 guarantees that there is a polynomial $h(n) = c_0 + c_1 n + c_2 n^2 + \cdots + c_r n^r$ with the property that $\Delta h(n) = b_0 + b_1 n + \cdots + b_{r-1} n^{r-1}$. That is, h is a solution of (21). But any other solution $f(n)$ is such that $\Delta f(n) = \Delta h(n)$. It follows from Theorem 4 that there is a constant c such that $f(n) = h(n) + c$, whence f itself must be an rth degree polynomial. This completes the proof. ▶

Example 2 Let t denote time, in years, after the year 1779 and suppose that $S(t)$, the number of scientists in the United States at time t, is governed by the difference equation $\Delta^3 S(t) = 0$. Suppose further that the numbers of U.S. citizens classified as scientists in the years 1779, 1780, and 1781 were, respectively, $S(0) = 300$, $S(1) = 310$, and $S(2) = 345$.

From Theorem 5, we know that S has the form $S(t) = a + bt + ct^2$, where a, b, and c are constants to be determined from the specified values of S. Substituting $t = 0, 1$, and 2 we obtain

$$S(0) = a + b(0) + c(0^2) = a = 300$$
$$S(1) = a + b(1) + c(1^2) = a + b + c = 310$$
$$S(2) = a + b(2) + c(2^2) = a + 2b + 4c = 345$$

Solving these equations for a, b, and c yields $a = 300$, $b = -25$, and $c = 12.5$ so that
$$S(t) = 12.5t^2 - 25t + 300$$

According to this equation the number of U.S. scientists in the year 1979 is predicted to be

$$S(200) = 12.5(200)^2 - 25(200) + 300 = 499{,}800 \qquad ▶$$

PROBLEMS

1. Which of the following are linear difference equations? What is the order of those equations which are linear? (The letter c denotes a constant.)

   (a) $\Delta y(n) + 2y(n) = 0$
   (b) $\Delta y(n) + c^2 y(n) = 2$
   (c) $y(n)\Delta^2 y(n) = 1$
   (d) $\Delta^3 y(n) - \Delta^2 y(n) + c = 6n$
   (e) $[\Delta y(n)]^{1/2} + y(n) = n^2$

2. In each case, verify that the given function is a solution of the difference equation. (The letters a, b, and c denote constants.)

(a) $y(n + 1) - y(n) = 0$; $y(n) = 1$

(b) $y(n + 1) - y(n) = 1$; $y(n) = n + c$

(c) $f(n + 1) - f(n) = n$; $f(n) = n(n - 1)/2$

(d) $\Delta f(n) = 3n(n + 1)/2$; $f(n) = [n(n^2 - 1)/2] + c$

(e) $h(n + 1) - 2h(n) = 0$; $h(n) = 2^n$

(f) $h(n + 1) - 2h(n) = 0$; $h(n) = c \cdot 2^n$

(g) $h(n + 1) - 2h(n) = 0$; $h(n) = c \cdot 2^{n+b}$

(h) $y(n + 2) - 3y(n + 1) + 2y(n) = 0$; $y(n) = a + b \cdot 2^n$

3. The difference equation $[\Delta y(n)]^2 + [y(n)]^2 = -1$ has no solution. Why?

4. Verify in each case that the given function is a solution to the difference equation together with the indicated initial conditions.

(a) $Y(t) = 4Y(t - 1) + 3$, $Y(0) = 5$
(Solution: $Y(t) = 6 \cdot 4^t - 1$)

(b) $Y(t + 2) = Y(t)$, $Y(0) = 0$, $Y(1) = 1$
(Solution: $Y(t) = \frac{1}{2}[1 - (-1)^t]$)

5. (a) Show that the function $f(n) = (c_1 + c_2 n)2^n$ is a solution of the equation $f(n + 2) - 4f(n + 1) + 4f(n) = 0$, regardless of the values assigned to the constants c_1 and c_2.

(b) Find the particular solution (that is, the particular choice of c_1 and c_2) which satisfies the initial conditions $f(0) = 2$ and $f(1) = 12$.

6. Find the function $g(t)$ which satisfies the following conditions:

(a) $g(t + 2) - 2g(t + 1) + g(t) = 0$; $g(0) = 1$, $g(2) = 4$

(b) $g(t + 3) - 3g(t + 2) + 3g(t + 1) - g(t) = 0$;
$g(0) = 10$, $g(1) = 20$, $g(2) = 40$

7. Prove that the difference equation $f(n + 1) - \lambda f(n) = 0$ has the solution $f(n) = c \cdot \lambda^n$, where c is a constant, by

(a) induction.

(b) rewriting the equation in the form

$$\frac{f(n + 1)}{\lambda^{n+1}} - \frac{f(n)}{\lambda^n} = 0$$

8. Given the economic model

$$Y(t + 1) - C(t + 1) - I(t + 1) = 0$$
$$C(t + 1) - bY(t) = a \quad (a > 0 \text{ and } 0 < b < 1)$$
$$I(t + 1) - I(t) = 0$$

with known initial values $Y(0)$, $C(0)$, and $I(0)$.

(a) Show that the difference equations
(i) $Y(t + 1) - bY(t) = a + I(0)$
(ii) $C(t + 1) - bC(t) = a + bI(0)$
follow from the model.

(b) Verify that

$$Y(t) = b^t \left(Y(0) - \frac{a + I(0)}{1 - b} \right) + \frac{a + I(0)}{1 - b}$$

$$C(t) = b^t \left(C(0) - \frac{a + bI(0)}{1 - b} \right) + \frac{a + bI(0)}{1 - b}$$

satisfy the respective difference equations.

(c) Verify that $Y(t)$ and $C(t)$ satisfy the economic model.

13.5 GENERAL LINEAR DIFFERENCE EQUATIONS WITH RIGHT-HAND SIDE ZERO

Thus far we have learned only how to solve linear difference equations which involve iterations of the operator Δ itself. Theorem 6 summarizes our efforts and at the same time provides the basis for solving more complicated equations. Recall that $(\Delta - a)f(n)$ is short for $\Delta f(n) - af(n)$ and that $(\Delta - a)^k f(n)$ indicates that the operator $\Delta - a$ is to be applied k successive times. The operators $\Delta - a$ and Δ are related as follows.

Theorem 7 For all positive integers n and k,

$$\Delta^k[(1 + a)^{-n}f(n)] = (1 + a)^{-(n+k)}[(\Delta - a)^k f(n)] \qquad (22)$$

PROOF Since the expression (22) is somewhat complicated, let us look first at the situation when $k = 1$. We wish to compute the first difference of $(1 + a)^{-n}f(n)$. This is

$$\begin{aligned}
\Delta[(1 + a)^{-n}f(n)] &= (1 + a)^{-(n+1)}f(n + 1) - (1 + a)^{-n}f(n) \\
&= (1 + a)^{-(n+1)}[f(n + 1) - (1 + a)f(n)] \\
&= (1 + a)^{-(n+1)}[f(n + 1) - f(n) - af(n)] \\
&= (1 + a)^{-(n+1)}[\Delta f(n) - af(n)] \\
&= (1 + a)^{-(n+1)}[(\Delta - a)f(n)] \qquad (23)
\end{aligned}$$

For $k = 1$, then, Equation (22) is valid.

The proof may be completed by induction. If (22) is true for $k = r$, then we have the inductive assumption

$$\Delta^r[(1 + a)^{-n}f(n)] = (1 + a)^{-(n+r)}[(\Delta - a)^r f(n)] \qquad (24)$$

It follows that

$$\begin{aligned}
\Delta^{r+1}[(1 + a)^{-n}f(n)] &= \Delta\{\Delta^r[(1 + a)^{-n}f(n)]\} \\
&= \Delta[(1 + a)^{-(n+r)}(\Delta - a)^r f(n)] \qquad \text{[by (24)]} \\
&= (1 + a)^{-(n+r+1)}[(\Delta - a)^{r+1}f(n)] \qquad \text{[from (23)]}
\end{aligned}$$

and this completes the proof. ▶

The principal worth of Theorem 7 is that it enables us to find solutions to all equations of the form $(\Delta - a)^k f(n) = 0$.

Theorem 8 Let k be any positive integer and $a \neq -1$. A function f is a solution of the equation $(\Delta - a)^k f(n) = 0$ if and only if f has the form

$$f(n) = [c_0 + c_1 n + \cdots + c_{k-1} n^{k-1}](1 + a)^n$$

where $c_0, c_1, \ldots, c_{k-1}$ are constants. (Theorem 6 is the special case $a = 0$.)

PROOF Since $(1 + a)^{-(n+k)}$ is not zero, the equation $(\Delta - a)^k f(n) = 0$ is equivalent to $(1 + a)^{-(n+k)}(\Delta - a)^k f(n) = 0$. But, by Theorem 7, this latter expression is the same as $\Delta^k[(1 + a)^{-n} f(n)] = 0$. Applying Theorem 6 gives

$$(1 + a)^{-n} f(n) = c_0 + c_1 n + \cdots + c_{k-1} n^{k-1}$$

from which the conclusion of the theorem follows upon multiplying both sides by $(1 + a)^n$. ▶

The restriction $a \neq -1$ in Theorem 8 is actually no loss of information. For in this case, $(\Delta - a)f(n) = 0$ becomes

$$(\Delta + 1)f(n) = f(n + 1) - f(n) + f(n) = f(n + 1) = 0$$

and thus the solution is already known. Similar comments apply to $(\Delta + 1)^k f(n)$ when $k > 1$.

Example 1 Harrod's one-country model of economic growth (Example 5, Section 13.1) leads to the equation

$$Y(t + 1) - Y(t) = \frac{c}{g} Y(t + 1)$$

where c and g are constants and $Y(t)$ denotes national income during time period t. Rewriting this equation in the form

$$Y(t + 1) = \frac{g}{g - c} Y(t)$$

and then subtracting $Y(t)$ to form a first difference yields the equation

$$Y(t + 1) - Y(t) = \frac{g}{g - c} Y(t) - Y(t)$$

or

$$\left(\Delta - \frac{c}{g - c}\right) Y(t) = 0$$

According to Theorem 8, any solution of this equation looks like

$$Y(t) = a\left(1 + \frac{c}{g - c}\right)^t = a\left(\frac{g}{g - c}\right)^t$$

where a is a constant. Since $Y(0) = a[g/(g - c)]^0 = a$, the solution is

$$Y(t) = Y(0)\left(\frac{g}{g - c}\right)^t$$ ▶

Example 2 In developing his *multiplier–acceleration principle*, Samuelson*
assumes that national income $Y(t)$ at time t is the sum of estimated consumer
demand $C(t)$ and estimated investment demand $I(t)$. He further assumes that
consumer demand is given by

$$C(t) = kY(t - 1) \tag{25}$$

where k is the *multiplier constant*, and investment demand by

$$I(t) = A[Y(t - 1) - Y(t - 2)] \tag{26}$$

where A is the *acceleration constant*. Putting (25) and (26) into the original
equation $Y(t) = C(t) + I(t)$ yields the *backward difference equation*

$$Y(t) = (k + A)Y(t - 1) - AY(t - 2)$$

The equivalent (forward) difference equation is

$$Y(t + 2) = (k + A)Y(t + 1) - AY(t) \tag{27}$$

Assume for purposes of illustration that $k = \frac{3}{4}$ and $A = 2\frac{1}{4}$. Then
Equation (27) is

$$Y(t + 2) = 3Y(t + 1) - \tfrac{9}{4}Y(t)$$

Since this equation involves values of Y at three consecutive time points,
it is a second-order difference equation. To put it into difference notation,
we begin by adding and subtracting terms to form a second difference, as in
Equation (4) of Section 13.1. This gives

$$Y(t + 2) - 2Y(t + 1) + Y(t) = Y(t + 1) - \tfrac{5}{4}Y(t)$$

or

$$\Delta^2 Y(t) = Y(t + 1) - \tfrac{5}{4}Y(t)$$

Next we add and subtract $Y(t)$ to incorporate $Y(t + 1)$ into a first
difference. The equation now looks like

$$\Delta^2 Y(t) = Y(t + 1) - Y(t) - \tfrac{1}{4}Y(t)$$

or, in operator notation,

$$\Delta^2 Y(t) - \Delta Y(t) + \tfrac{1}{4}Y(t) = 0$$

Factoring the operator gives

$$(\Delta - \tfrac{1}{2})^2 Y(t) = 0$$

the solution of which, by Theorem 8, is

$$Y(t) = (c_0 + c_1 t)(\tfrac{3}{2})^t$$

In order to illustrate how specific values may be obtained for the constants
c_0 and c_1, let us suppose that Y values of 2 and 3.5 are observed at times

*Cited in Baumol, W. J., *Economic Dynamics* (Macmillan Company, New York, 1959),
2nd ed.

$t = 0$ and $t = 1$, respectively. Then

$$2 = Y(0) = [c_0 + c_1(0)](\tfrac{3}{2})^0 = c_0$$

$$3.5 = Y(1) = [c_0 + c_1(1)](\tfrac{3}{2})^1 = \tfrac{3}{2}(c_0 + c_1)$$

From this we find $c_0 = 2$ and $c_1 = \tfrac{1}{3}$, so that the solution of (27) is

$$Y(t) = (2 + \tfrac{1}{3}t)(\tfrac{3}{2})^t$$

Obviously, $Y(t) \to \infty$ as $t \to \infty$. ▶

With the information now at hand it is but a small step to the solution of complicated equations involving mixtures of operators. Let us look at the simplest case

$$(\Delta - a)(\Delta - b)f(n) = 0 \tag{28}$$

where $a \neq b$ and neither a nor b is equal to -1, in order to see how the general case goes.

First, substituting $g(n) = (\Delta - b)f(n)$ reduces the order and gives $(\Delta - a)g(n) = 0$ for which we already know the solution

$$g(n) = (\Delta - b)f(n) = c(1 + a)^n$$

where c is some constant. We now use Theorem 7 to replace $(\Delta - b)f(n)$ by

$$(1 + b)^{n+1}\Delta[(1 + b)^{-n}f(n)]$$

obtaining

$$(1 + b)^{n+1}\Delta[(1 + b)^{-n}f(n)] = c(1 + a)^n$$

Dividing through by $(1 + b)^{n+1}$ gives

$$\Delta[(1 + b)^{-n}f(n)] = \frac{c}{1 + b}\left[\frac{1 + a}{1 + b}\right]^n = dx^n$$

where $d = c/(1 + b)$ and $x = (1 + a)/(1 + b)$ are constants.

The results of Example 3 in Section 13.3 show that $dx^n/(x - 1)$ is a function whose first difference is dx^n. It follows (Theorem 4) that we must have

$$(1 + b)^{-n}f(n) = \frac{dx^n}{x - 1} + q$$

where q is another constant. That is, letting $p = d/(x - 1)$, we obtain

or

$$(1 + b)^{-n}f(n) = p\left[\frac{1 + a}{1 + b}\right]^n + q$$

$$f(n) = p(1 + a)^n + q(1 + b)^n \tag{29}$$

for some constants p and q. To sum up, we have shown that every solution of (28) looks like (29).

On the other hand, no matter what values of p and q may be chosen, (29) is a solution of (28). For, using the basic properties of differences, we find

$$(\Delta - a)(\Delta - b)[p(1 + a)^n + q(1 + b)^n]$$
$$= (\Delta - b)(\Delta - a)[p(1 + a)^n] + (\Delta - a)(\Delta - b)[q(1 + b)^n]$$
$$= p(\Delta - b)[(\Delta - a)(1 + a)^n] + q(\Delta - a)[(\Delta - b)(1 + b)^n]$$
$$= p(\Delta - b)[0] + q(\Delta - a)[0]$$
$$= 0 + 0 = 0$$

Now that we have gone through all the formalities, let us see how our job can be simplified. Comparing (28) and (29) we note that each factor of the operator produces a corresponding term in the solution. This is no accident, as the following reasoning shows. Suppose h is a function which makes $(\Delta - b)h(n) = 0$. Then

$$(\Delta - a)[(\Delta - b)h(n)] = (\Delta - a)(0) = 0$$

and it follows that h is a solution of the original equation (28). Similarly, if a function g is such that $(\Delta - a)g(n) = 0$, then

$$(\Delta - a)(\Delta - b)g(n) = (\Delta - b)[(\Delta - a)g(n)] = (\Delta - b)(0) = 0$$

and g, too, is a solution of (28).

We know already that $(\Delta - a)g(n) = 0$ has $(1 + a)^n$ as a solution and that $(1 + b)^n$ is a solution of $(\Delta - b)h(n) = 0$. Combining our results, we see that for any constants p and q, the function $f(n) = p(1 + a)^n + q(1 + b)^n$ is a solution of (28).

Example 3   Let us suppose in the multiplier–acceleration model of Example 2 that $A = 2\frac{2}{3}$ and $k = \frac{2}{3}$ and that the initial conditions are $Y(0) = 300$ and $Y(1) = 450$. Then the general equation (27) becomes

$$Y(t + 2) = \tfrac{10}{3} Y(t + 1) - \tfrac{8}{3} Y(t)$$

which may be rearranged to read

$$Y(t + 2) - 2Y(t + 1) + Y(t) - \tfrac{4}{3}[Y(t + 1) - Y(t)] + \tfrac{1}{3}Y(t) = 0$$

or, in operator notation,

$$(\Delta^2 - \tfrac{4}{3}\Delta + \tfrac{1}{3})Y(t) = 0$$

Multiplying by 3 and factoring yields the equation

$$(3\Delta - 1)(\Delta - 1)Y(t) = 0$$

for which the solution is

$$Y(t) = c_0(1 + \tfrac{1}{3})^t + c_1(1 + 1)^t = c_0(\tfrac{4}{3})^t + c_1(2)^t$$

The initial conditions

$$Y(0) = 300 = c_0 + c_1 \quad \text{and} \quad Y(1) = 450 = \tfrac{4}{3}c_0 + 2c_1$$

yield

$$Y(t) = 225(\tfrac{4}{3})^t + 75(2)^t \qquad\qquad \blacktriangleright$$

It should be apparent that the approach outlined above generalizes to any problem in which the operator can be factored. (A general proof using induction is possible, but lengthy, and will be omitted.) To cite just two further examples, the equation

$$(\Delta - 2)(\Delta + 3)^3 f(n) = 0$$

has the solution

$$f(n) = a(1 + 2)^n + b(1 - 3)^n + cn(1 - 3)^n + dn^2(1 - 3)^n$$

while all solutions of

$$(\Delta + 2)^2(\Delta - 7)(\Delta - 1)^4 f(n) = 0$$

look like

$$(a + bn)(1 - 2)^n + c(1 + 7)^n + (d + en + gn^2 + hn^3)(1 + 1)^n$$

(The letters a, b, c, d, e, g, and h denote constants.)

PROBLEMS

1. Let S_0, S_1, S_2, ... be a sequence of objective stimulus values in which $\Delta S_i = S_{i+1} - S_i$ represents a just noticeable change in stimulus intensity. If R_i is a subjective psychological response to S_i, the Weber–Fechner law* states that $\Delta R_i = c(\Delta S_i/S_i)$, where c is a constant. Show that if $\Delta R_i = 1$ for all i, then $S_i = (1 + \alpha)^i S_0$, where $\alpha = 1/c$.

In Problems 2–6, find the solution of the indicated difference equation which satisfies the given initial conditions.

2. $Y(t + 2) = 4Y(t + 1) - 4Y(t)$; $Y(0) = 10$, $Y(1) = 12$

3. $f(t + 2) = -2f(t + 1) - f(t)$; $f(0) = 2$, $f(1) = 8$

4. $g(n) = 4g(n - 1) - 3g(n - 2)$; $g(0) = 2$, $g(1) = 4$

5. $h(n + 1) = 5h(n) - 6h(n - 1)$; $h(0) = -9$, $h(1) = -12$

6. $Y(t + 2) = 2Y(t)$; $Y(0) = 3$, $Y(1) = 8$

*Guilford, J. P., *Psychometric Methods* (McGraw-Hill Book Company, New York, 1954), 2nd ed., pp. 37–42.

7. Show that if Harrod's model is modified so that savings $S(t)$ is assumed to be proportional to income $Y(t + 1)$ in the *next* time period, then income $Y(t)$ satisfies the difference equation

$$Y(t + 2) = (g/c)[Y(t + 1) - Y(t)].$$

Solve this equation for the case when $Y(0) = 1$, $Y(1) = 5$, and $g = 4c$.

8. In Example 1, argue that the income sequence $Y(t)$ is:

(i) monotone increasing and unbounded when $0 < c < g$.

(ii) oscillatory and unbounded when $g < c < 2g$.

(iii) oscillatory but convergent to zero when $c > 2g$.

(iv) constant when $c = 0$.

(v) oscillatory and bounded when $c = 2g$.

In Problems 9–13, solve the indicated equation and discuss the limiting behavior of the solution.

9. $Y(t + 1) = 2Y(t); \ Y(0) = 3$

10. $f(t + 1) + 3f(t) = 0; \ f(0) = 1$

11. $g(n + 1) - \frac{1}{3}g(n) = 0; \ g(0) = 4$

12. $f(n + 1) + \frac{1}{4}f(n) = 0; \ f(0) = 5$

13. $Y(n + 1) + 4Y(n) = 0; \ Y(0) = -1$

14. Solve the multiplier–acceleration equation (27) in the case $k = \frac{1}{2}$, $A = 4$, $Y(0) = 2$, and $Y(1) = 5$. What happens as $t \to \infty$?

15. Show that the linear difference equation

$$Y(n + 2) - \alpha Y(n + 1) + \beta Y(n) = 0$$

has a real solution as long as $\beta \leq \alpha^2/4$.

16. (a) Show that all solutions of the equation

$$Y(n + 2) - \alpha Y(n + 1) + \beta Y(n) = 0$$

have the form $Y(n) = c_1 r_1^n + c_2 r_2^n$, where $r_1 - 1$ and $r_2 - 1$ are the roots of the quadratic equation $x^2 + (2 - \alpha)x + (1 + \beta - \alpha) = 0$, provided these roots are unequal.

(b) Prove that if $|r_1| > |r_2|$ and $c_1 \neq 0$, then

$$\lim_{n \to \infty} Y(n) = \lim_{n \to \infty} c_1 r_1^n$$

In particular,

(i) $Y(n)$ is unbounded if $r_1 > 1$.

(ii) $Y(n)$ converges to zero if $-1 < r_1 < 1$.

(iii) $Y(n)$ oscillates without bound if $r_1 < -1$.

(iv) $Y(n)$ converges to c_1 if $r_1 = 1$.

(v) $Y(n)$ oscillates but is bounded if $r_1 = -1$.

13.6 METHOD OF COMPLEMENTARY AND PARTICULAR SOLUTIONS

Having determined, at least in principle, a method for solving difference equations of the form

$$\phi(\Delta)f(n) = 0 \qquad (30)$$

let us now consider equations like

$$\phi(\Delta)f(n) = a(n) \qquad (31)$$

where $a(n)$ is some given function of n.

Suppose that we have somehow found a solution $g(n)$ of (31) and suppose that $f(n)$ is *any* other solution. Then since $\phi(\Delta)g(n) = \phi(\Delta)f(n) = a(n)$, we have

$$\phi(\Delta)[f(n) - g(n)] = \phi(\Delta)f(n) - \phi(\Delta)g(n) = a(n) - a(n) = 0$$

This means that the difference $f(n) - g(n)$ is a solution of (30) and hence that *every* solution $f(n)$ of (31) can be written as the sum of the particular solution $g(n)$ and a solution $f(n) - g(n)$ of (30). In short, to find all solutions of (31), we need find only one solution of that equation to which we add all solutions of (30).

Standard terminology calls (30) the equation *complementary* to (31) and any solution of (30) is called a *complementary solution* of (31). The problem of finding all solutions of $\phi(\Delta)f(n) = a(n)$ has been reduced to the problem of finding any particular solution together with that of finding all complementary solutions. Since we know how to solve (30), our only problem (in theory at least) is that of finding a particular solution.

Example 1 To solve the equation

$$-\Delta^2 f(n) + 2\Delta f(n) = n \qquad (32)$$

we first solve the complementary equation $\Delta(\Delta - 2)f(n) = 0$ to obtain complementary solutions of the form $C(n) = a + b \cdot 3^n$, where a and b are constants.

To find a particular solution of (32),* we ask what function of n is such that combinations of its first two differences would yield n. Since differences of polynomials are polynomials of lower degree and since differencing has produced the first-degree polynomial n, it seems reasonable to guess that we began with a third-degree polynomial of the form

$$p(n) = c_0 + c_1 n + c_2 n^2 + c_3 n^3 \qquad (33)$$

If (33) is to be a solution of (32), we must have $-\Delta^2 p(n) + 2\Delta p(n) = n$ or

$$-\Delta^2(c_0 + c_1 n + c_2 n^2 + c_3 n^3) + 2\Delta(c_0 + c_1 n + c_2 n^2 + c_3 n^3) = n$$

---

*Note that we still have no solutions of (32) since the so-called complementary solutions are actually solutions of $-\Delta^2 f(n) + 2\Delta f(n) = 0$ and are *not* solutions of (32) itself.

Writing the indicated differences and then simplifying leads to the equation $2c_1 - 4c_3 + 4c_2 n + 6c_3 n^2 = n$. To make this an identity in n, we need $c_3 = 0$, $c_2 = \frac{1}{4}$, and $c_1 = 0$. No restriction is placed on c_0, and we arbitrarily choose it to be zero.

With these choices, our particular solution is $p(n) = \frac{1}{4}n^2$ which, together with the complementary solutions already obtained, means that all solutions of (32) look like

$$f(n) = \tfrac{1}{4}n^2 + a + b \cdot 3^n$$

where a and b are constants. ▶

In guessing at the form of a particular solution, we found that the terms c_0, $c_1 n$, and $c_3 n^3$ were superfluous. At the same time, we ran the risk of guessing incorrectly and consequently of finding no solution at all. This is an unsatisfactory state of affairs but is easily remedied, as follows.

Suppose $f(n)$ is a solution of (32), that is, suppose $-\Delta^2 f(n) + 2\Delta f(n) = n$. Since identical functions have identical differences, it follows that

$$\Delta^2[-\Delta^2 f(n) + 2\Delta f(n)] = \Delta^2 n = 0$$

and hence *every* solution of (32) is also a solution of this new equation. Factoring the operator gives

$$\Delta^3(\Delta - 2)f(n) = 0$$

so that

$$f(n) = a + bn + cn^2 + d \cdot 3^n$$

is the form which every solution of (32) must have.

The terms a and $d \cdot 3^n$ being part of the complementary solution, only $bn + cn^2$ is necessary for a particular solution. Substituting this in (32) gives $b = 0$ and $c = \frac{1}{4}$ so that the complete solution is

$$f(n) = \tfrac{1}{4}n^2 + a + d \cdot 3^n$$

as before.

The key to this approach is that in choosing the operator Δ^2 in order to annihilate (reduce to zero) the right-hand side, we arrive at an equation whose solution form is known. All guesswork concerning the form of a particular solution is thus eliminated.

Example 2 Using Equations (4) and (5) of Section 13.1 and proceeding as in Example 2 of Section 13.5 we can write the equation

$$y(n + 3) - 6y(n + 2) + 3y(n + 1) + 10y(n) = 8n^2 + 4n \qquad (34)$$

in operator form as

$$(\Delta - 4)(\Delta + 2)(\Delta - 1)y(n) = 8n^2 + 4n$$

The operator Δ^3 annihilates the right-hand side. Thus the complete solution has the form

$$y(n) = c_0 + c_1 n + c_2 n^2 + c_3 \cdot 5^n + c_4(-1)^n + c_5 \cdot 2^n$$

in which the constants c_0, c_1, and c_2 must be chosen to provide a particular solution, while the last three terms constitute the complementary solution.

Substituting the particular solution $c_0 + c_1 n + c_2 n^2$ into the original equation (34) and equating coefficients of like powers of n gives $c_0 = 3$, $c_1 = 2$, and $c_2 = 1$. The complete solution is

$$y(n) = 3 + 2n + n^2 + c_3 \cdot 5^n + c_4(-1)^n + c_5 \cdot 2^n \qquad \blacktriangleright$$

PROBLEMS

1. Complete the analysis of Example 2 by verifying that we must have $c_0 = 3$, $c_1 = 2$, and $c_2 = 1$.

Example 3 The linear difference equation $Y(n + 1) = \lambda Y(n) + k$, where $\lambda \neq 0$ and k are constants, may be written in operator form as

$$[\Delta - (\lambda - 1)]Y(n) = k$$

Since the operator Δ annihilates k, the complete solution is $Y(n) = c + b\lambda^n$, the term $b\lambda^n$ representing the complementary solution.

To determine the constant c, we require $[\Delta - (\lambda - 1)]c = k$, or

$$c = \frac{k}{1 - \lambda}$$

The solution $Y(n) = [k/(1 - \lambda)] + b\lambda^n$ contains an arbitrary constant b. If the initial value $Y(0)$ is known, then we have $Y(0) = k/(1 - \lambda) + b$ or $b = Y(0) - k/(1 - \lambda)$. The solution is

$$Y(n) = \frac{k}{1 - \lambda} + \left[Y(0) - \frac{k}{1 - \lambda} \right]\lambda^n$$

and it is apparent that limiting behavior of the sequence $(Y(n))$ depends on the term λ^n, the other quantities being constant. In particular, since (λ^n) diverges when $|\lambda| > 1$ and when $\lambda = -1$ and converges otherwise, we find that

(i) $Y(n)$ is unbounded when $\lambda > 1$.

(ii) $Y(n)$ converges to $k/(1 - \lambda)$ when $0 \leq \lambda < 1$.

(iii) $Y(n)$ oscillates, but converges to $k/(1 - \lambda)$ when $-1 < \lambda < 0$.

(iv) $Y(n)$ oscillates between $Y(0)$ and $2k/(1 - \lambda) - Y(0)$ when $\lambda = -1$.

(v) $Y(n)$ oscillates and is unbounded when $\lambda < -1$.

The limit $\lim_{n \to \infty} Y(n) = k/(1 - \lambda)$, when it exists (that is, when $|\lambda| < 1$), is called the *equilibrium state* of the solution, while the term $[Y(0) - k/(1 - \lambda)]\lambda^n$ represents the *transient part* of the solution. $\qquad \blacktriangleright$

Example 4 The cobweb model introduced in Example 3 of Section 5.2 generates the linear difference equation

$$S(t+1) - \left(\frac{d}{b}\right) S(t) = \frac{bc - ad}{b}$$

whose solution is

$$S(t) = \frac{bc - ad}{b - d} + \left[S(0) - \frac{bc - ad}{b - d} \right] \left(\frac{d}{b}\right)^t \tag{35}$$

Since by assumption, $b < 0$ and $d > 0$, the sequence $(S(t))$ is always oscillatory, being unbounded when $d > |b|$, cyclic when $d = |b|$, and convergent to the equilibrium state $(bc - ad)/(b - d)$ when $d < |b|$. (Compare with Problem 4 in Section 5.4.)

The time path for a convergent cobweb model is shown in Figure 1. The linear demand function $D(t) = a + bP(t)$ and the lagged linear supply function $S(t) = c + dP(t - 1)$ are plotted against price $P(t)$.

FIGURE 1

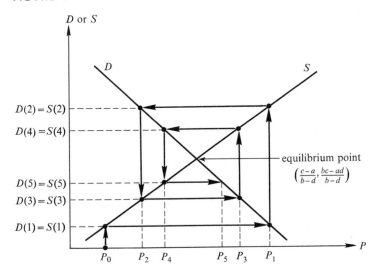

An initial price $P(0) = P_0$ leads to an initial value $S(1) = c + dP_0$ on the supply curve. (Follow the arrows in Figure 1.) The model requires that initial demand $D(1)$ equal $S(1)$, which leads to a new price $P(1) = P_1$ obtained by solving the equation $D(1) = a + bP_1$ for P_1. The price P_1 leads to a new supply $S(2) = c + dP_1$, which in turn generates a new demand $D(2) = S(2)$, a new price P_2 obtained by solving $D(2) = a + bP_2$ for P_2, etc. The oscillatory nature of the time path and the convergence of the cobweb to the point of intersection of the demand and supply curves (the point where $D = S = (bc - ad)/(b - d)$ are apparent from the figure. ▶

PROBLEMS

2. Verify that (35) is indeed the correct solution to the cobweb equation.

3. Show that the equilibrium price is $(c - a)/(b - d)$.

Example 5 According to Theorem 8, the operator $(\Delta - a)^k$ serves as an annihilator for $n^{k-1}(1 + a)^n$. Thus to solve the equation

$$Y(t + 2) - 6Y(t + 1) + 9Y(t) = 2t + 3^t \qquad (36)$$

or, in operator form, $(\Delta - 2)^2 Y(t) = 2t + 3^t$, we first apply the operator $\Delta^2(\Delta - 2)$ to both sides to obtain $\Delta^2(\Delta - 2)^3 Y(t) = 0$. The complete solution of this equation has the form

$$Y(t) = c_0 + c_1 t + (c_2 + c_3 t + c_4 t^2)3^t$$

Because of the operator $(\Delta - 2)^2$ in the original equation, the terms $(c_2 + c_3 t)3^t$ constitute the complementary solution, leaving

$$c_0 + c_1 t + c_4 t^2 3^t$$

as the form of the particular solution.

A little computation shows that $(\Delta - 2)^2[c_0 + c_1 t + c_4 t^2 3^t] = 2t + 3^t$ if and only if $c_0 = \frac{1}{2}$, $c_1 = \frac{1}{2}$, and $c_4 = \frac{1}{18}$. Hence the complete solution of (36) is

$$Y(t) = (c_2 + c_3 t)3^t + \tfrac{1}{2} + \tfrac{1}{2}t + \tfrac{1}{18}t^2 3^t \qquad \blacktriangleright$$

PROBLEMS

In Problems 4–11, find all solutions of the indicated equation.

4. $Y(t + 2) = 5Y(t + 1) - 4Y(t) + 22$

5. $f(n + 2) = 4f(n + 1) - 3f(n) + 6$

6. $g(t + 2) + 6g(t + 1) - 7g(t) + 16 = 0$

7. $Y(n + 2) = 2Y(n + 1) + 3Y(n) + 2n^2 + 6n - 3$

8. $8Y(t + 2) - 6Y(t + 1) + Y(t) = 3^t$

9. $g(n + 2) - 4g(n + 1) + 3g(n) = 2^n - 3n$

10. $h(t + 2) - 6h(t + 1) + 9h(t) = t \cdot 6^t$

11. $4Y(n + 2) + 4Y(n + 1) - 3Y(n) = 2^n(n + 1)$

Solve the equations in Problems 12–15 subject to the given initial conditions.

12. $Y(t + 2) + Y(t + 1) - 2Y(t) = 12; \; Y(0) = 8, \; Y(1) = 10$

13. $f(t + 2) - 4f(t + 1) + 4f(t) = 10; \; f(0) = 1, f(1) = 8$

14. $f(n + 2) - 2f(n + 1) - 3f(n) = 12; \; f(0) = 12, f(1) = 4$

15. $Y(t + 2) - 4Y(t + 1) + 3Y(t) = 6; \; Y(0) = 1, \; Y(1) = 2$

16. Prove that the difference equation

$$Y(t + 2) - (1 + \beta)Y(t + 1) + \beta Y(t) = k$$

has the solution

$$Y(t) = c_0 + c_1\beta^t + kt/(1 - \beta)$$

(β, k, c_0, and c_1 are constants.)

17. Anderson* proposed the difference equation model of response extinction

$$R(n + 1) = R(n) - \theta[R(n) - R(\infty)]$$

where $R(n)$ is the response on extinction trial n, $R(\infty)$ is the extinction response level at equilibrium, and $0 < \theta \leq 1$ is the extinction rate. Show that the general solution is

$$R(n) = R(\infty) - [R(\infty) - R(1)](1 - \theta)^{n-1}$$

18. Given that the demand and supply functions for a commodity are, respectively, $D(t) = 10 - 4P(t)$ and $S(t) = 3P(t - 1)$

(a) deduce a cobweb model for the price of the commodity.

(b) solve the resulting equation.

(c) compute $P(t)(t = 0, 1, 2, 3, 4, 5)$ if $P(0) = 1$.

(d) graph the cobwebs around the demand and supply functions for $t = 0, 1, 2, \ldots, 5$.

19. Modify the Harrod model (see Example 5 of Section 13.1) to the extent that total investment during period t is assumed to be

$$I(t) = g[Y(t) - Y(t - 1)] + kY(t) + L$$

where k and $L > 0$ are constants. Given no other changes,

(a) show that income $Y(t)$ satisfies the difference equation

$$Y(t + 1) = \frac{-g}{c - g - k} Y(t) + \frac{L}{c - g - k}$$

(b) solve the difference equation for $Y(t)$.

(c) verify that if g and c are both positive and if $k < c$, the ratio

$$\frac{Y(t + 1) - Y(t)}{Y(t + 1)}$$

must now increase, whereas in the original model this ratio was the constant c/g.

20. Consider the nonlinear difference equation

$$c_1 f(n + 1) = \frac{c_2 f(n)}{1 + c_2 f(n)}$$

where c_1 and c_2 are constants and $f(0) > 0$.

(a) Show that by making the substitution $y(n) = 1/f(n)$, the nonlinear equation is transformed into the linear difference equation

$$y(n + 1) = \frac{c_1}{c_2} y(n) + c_1$$

*Anderson, N. H., "Comparison of Different Populations: Resistance to Extinction and Transfer," *Psychological Review* **70**, 162–179 (1963).

(b) Use the transformed equation to find the solution of the original equation.

21. (a) Show that if b is constant,

$$\Delta \sin bn = [\cos b - 1] \sin bn + \sin b \cos bn$$

$$\Delta \cos bn = [\cos b - 1] \cos bn - \sin b \sin bn$$

(b) Use the information in part (a) to solve the equation

$$Y(n + 2) - Y(n) = \sin \tfrac{1}{2}\pi n$$

[*Hint:* Proceeding as in Example 1, guess a particular solution of the form $c_1 \sin bn + c_2 \cos bn$, where c_1 and c_2 are constants.*]

22. Suppose that a sum of money A is borrowed at 6% interest compounded annually and is retired by a payment of amount R at the end of each annual interest period. Let $P(n)$ be the outstanding principle just after the nth payment of R.

(a) Show that $P(n)$ satisfies the difference equation

$$P(n + 1) = (1 + 0.06)P(n) - R$$

(b) Determine the periodic payment R needed to retire the debt in k periods.

23. Let S be a set of N community residents. At time period t, $N(t)$ residents favor and $N - N(t)$ residents oppose a local community issue. In each time period, 100 β% of those who previously favored the issue and 100 α% of those who previously opposed the issue change their position.

(a) Show that

$$N(t) = \frac{N\alpha}{\alpha + \beta} + \left[N(0) - \frac{N\alpha}{\alpha + \beta} \right] (1 - \alpha - \beta)^t$$

(b) Discuss the limiting behavior of $N(t)$. Compare with Problem 20 of Section 6.2.

24. In the cobweb model, it was assumed that the market was cleared at every time period, that is, $D(t) = S(t)$. Suppose instead that we assume

$$D(t) \quad = a - bP(t)$$

$$S(t) \quad = -c + dP(t)$$

$$P(t + 1) = P(t) + \alpha[D(t) - S(t)]$$

where $\alpha > 0$ denotes a demand-induced price adjustment coefficient and a, b, c, and d are all positive. Show that the model has the time path

$$P(t) = \left(P(0) - \frac{a + c}{b + d} \right) [1 - \alpha(b + d)]^t + \frac{a + c}{b + d}$$

*Special methods, such as those illustrated in Problems 20 and 21, for solving equations not covered by our general discussion, may be found in Goldberg, S., *Introduction to Difference Equations: With Illustrative Examples from Economics, Psychology and Sociology* (John Wiley & Sons, Inc., New York, 1958).

25. In relation to Problem 24, discuss the behavior of the time path $P(t)$ when

(i) $0 < \alpha < \dfrac{1}{b + d}$

(ii) $\alpha = \dfrac{1}{b + d}$

(iii) $\dfrac{1}{b + d} < \alpha < \dfrac{2}{b + d}$

(iv) $\alpha = \dfrac{2}{b + d}$

(v) $\alpha > \dfrac{2}{b + d}$

26. Suppose that we modify the multiplier–acceleration model (Example 2 of Section 13.5) to assume that

$$Y(t) = C(t) + I(t) + G$$
$$C(t) = \gamma Y(t - 1) \qquad 0 < \gamma < 1$$
$$I(t) = \alpha[C(t) - C(t - 1)] \qquad \alpha > 0$$

where G is a constant. Find the time path $Y(t)$.

27. In Problem 26, discuss the stability of the equilibrium in each of the following cases:

(a) $\gamma > \dfrac{4\alpha}{(1 + \alpha)^2}$; $\qquad \alpha\gamma > 1$

(b) $\gamma > \dfrac{4\alpha}{(1 + \alpha)^2}$; $\qquad \alpha\gamma < 1$

(c) $\gamma = \dfrac{4\alpha}{(1 + \alpha)^2}$; $\qquad \alpha < 1$

(d) $\gamma = \dfrac{4\alpha}{(1 + \alpha)^2}$; $\qquad \alpha > 1$

28. Show that if $Z(t)$ is a particular solution of

$$Y(t + 2) - \alpha Y(t + 1) + \beta Y(t) = f(n)$$

and if $W(t)$ is a particular solution of

$$Y(t + 2) - \alpha Y(t + 1) + \beta Y(t) = g(n)$$

then $Z(t) + W(t)$ is a particular solution of

$$Y(t + 2) - \alpha Y(t + 1) + \beta Y(t) = f(n) + g(n)$$

SUPPLEMENTARY READING

Goldberg, S., *Introduction to Difference Equations: with Illustrative Examples from Economics, Psychology and Sociology* (John Wiley & Sons, Inc., New York, 1958).

INTEGRAL CALCULUS 14

14.1 AREA

The area A of the region in Figure 1 bounded by heavy lines is easily computed to be 3/2. But what of the area in Figure 2? Is there a simple means of computing it? The answer is "yes" and is provided by the second of the two major concepts of calculus—the integral. (A mathematically precise definition of area is given at the end of Section 14.2. In the present section, however, we shall proceed intuitively, using only the familiar definition for area of a rectangle.)

FIGURE 1

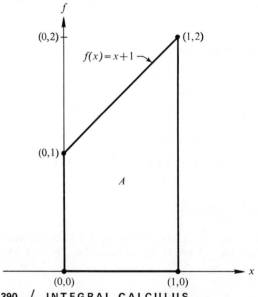

FIGURE 2

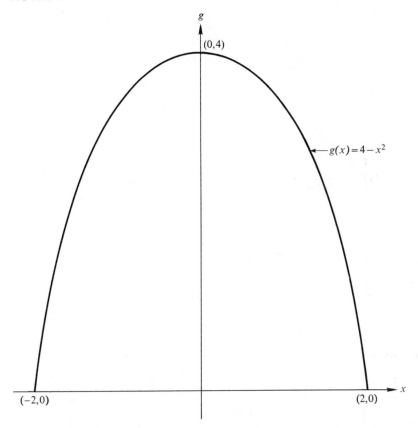

Before formally introducing the idea of an integral, let us look at an alternative and seemingly roundabout method of computing the area in Figure 1. In the interval [0, 1] the graph of $f(x) = x + 1$ lies between the graphs of the two constant functions $g(x) = 1$ and $h(x) = 2$. Thus the required area A is no greater than the area of a rectangle having height two and base one, and no less than that of a smaller rectangle with height and base both equal to unity. In short, Figure 3 shows that we have $1 \leq A \leq 2$. A better approximation using rectangles may be obtained by partitioning the interval [0, 1] into two intervals $[0, \frac{1}{2}]$ and $[\frac{1}{2}, 1]$ as shown in Figure 4. The dotted rectangles in the intervals $[0, \frac{1}{2}]$ and $[\frac{1}{2}, 1]$ have respective areas $\frac{1}{2}$ and $\frac{3}{4}$. Since both lie entirely within the area A, we must have

$$\tfrac{1}{2} + \tfrac{3}{4} = \tfrac{5}{4} \leq A$$

Similarly, if we use the two rectangles with bases on the horizontal axis and dashed tops, we find $A \leq \frac{3}{4} + 1 = \frac{7}{4}$. Our second approximation,

$$\tfrac{5}{4} \leq A \leq \tfrac{7}{4}$$

is a considerable improvement over the initial result $1 \leq A \leq 2$.

FIGURE 3

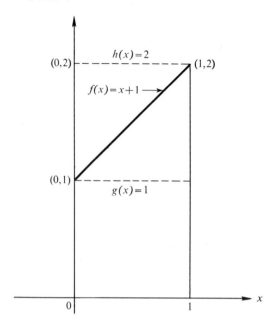

FIGURE 4

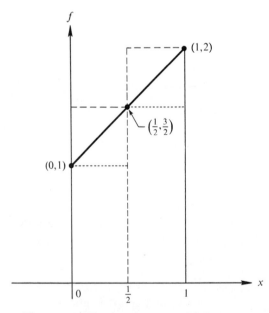

The pattern having been established, let us proceed to the general case. We partition the interval $[0, 1]$ into n smaller intervals $[0, 1/n]$, $[1/n, 2/n]$, $\ldots, [(n - 1)/n, 1]$ each of length $1/n$. (See Figure 5.) In each smaller interval

FIGURE 5

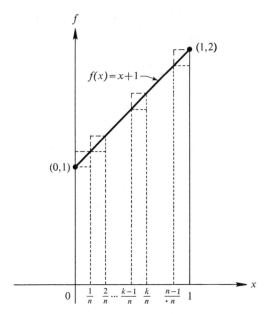

we choose, as above, one rectangle inscribed within the area A and another rectangle which circumscribes a part of A. In the kth interval $[(k-1)/n, k/n]$, the height of the smaller rectangle is $(k-1)/n + 1$, obtained by evaluating $f(x) = x + 1$ at the left endpoint $(k-1)/n$, while the height of the larger rectangle is $k/n + 1$, obtained by evaluating f at the right endpoint k/n. Each rectangle has width $1/n$. Computing each of the areas and summing, we find

$$\frac{1}{n}\left[1 + \left(1 + \frac{1}{n}\right) + \left(1 + \frac{2}{n}\right) + \cdots + \left(1 + \frac{n-1}{n}\right)\right]$$

$$\leq A \leq \frac{1}{n}\left[\left(1 + \frac{1}{n}\right) + \left(1 + \frac{2}{n}\right) + \cdots + \left(1 + \frac{n}{n}\right)\right]$$

or, in summation notation,

$$\frac{1}{n}\left[n + \frac{1}{n}\sum_{k=0}^{n-1} k\right] \leq A \leq \frac{1}{n}\left[n + \frac{1}{n}\sum_{k=1}^{n} k\right]$$

Since $\sum_{k=1}^{n} k = n(n+1)/2$ (from Problem 6 of Section 5.2), this equation may be rewritten as

$$\frac{1}{n}\left[n + \frac{1}{n}\frac{(n-1)n}{2}\right] \leq A \leq \frac{1}{n}\left[n + \frac{1}{n}\frac{n(n+1)}{2}\right]$$

or, on simplifying,

$$\frac{3}{2} - \frac{1}{2n} \leq A \leq \frac{3}{2} + \frac{1}{2n} \tag{1}$$

As $n \to \infty$, both the left and right sides of this inequality converge to $\frac{3}{2}$. It follows that $A = \frac{3}{2}$, which coincides with our original computation for the area.

Now that we are acquainted with a scheme, albeit a rather cumbersome one, for approaching the problem of area, and having gained confidence that our scheme does produce reasonable results, let us apply it to find the area outlined in Figure 2. Since the graph of $g(x) = 4 - x^2$ is symmetric with respect to the vertical axis, we may simplify the problem slightly by computing the area above the interval $[0, 2]$ and then multiplying by 2.

We first partition the interval $[0, 2]$ into n smaller intervals (Figure 6), each of width $2/n$. In the kth interval $[2(k - 1)/n,\ 2k/n]$, the maximum height of the curve is

$$g\left[\frac{2(k - 1)}{n}\right] = 4 - \left[\frac{2(k - 1)}{n}\right]^2 = 4 - \frac{4}{n^2}(k^2 - 2k + 1)$$

while the minimum height is

$$g\left(\frac{2k}{n}\right) = 4 - \left(\frac{2k}{n}\right)^2 = 4 - \frac{4k^2}{n^2}$$

FIGURE 6

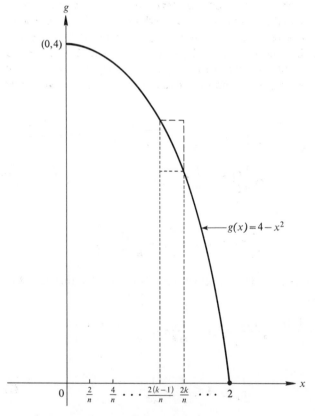

$(0,4)$

$g(x) = 4 - x^2$

$0 \quad \frac{2}{n} \quad \frac{4}{n} \ \ldots \ \frac{2(k-1)}{n} \ \frac{2k}{n} \ \ldots \ 2$

The area A_k which lies under the curve in the kth interval thus satisfies the inequality

$$\frac{2}{n}\left[4 - \frac{4k^2}{n^2}\right] \le A_k \le \frac{2}{n}\left[4 - \frac{4}{n^2}(k^2 - 2k + 1)\right]$$

so that the total area $A = \sum_{k=1}^{n} A_k$ is restricted by

$$\sum_{k=1}^{n} \frac{2}{n}\left[4 - \frac{4k^2}{n^2}\right] \le A \le \sum_{k=1}^{n} \frac{2}{n}\left[4 - \frac{4}{n^2}(k^2 - 2k + 1)\right] \tag{2}$$

The sums in (2) may be simplified using the formulas

$$\sum_{k=1}^{n} 1 = n$$

$$\sum_{k=1}^{n} k = \frac{n(n+1)}{2}$$

$$\sum_{k=1}^{n} k^2 = \frac{n(n+1)(2n+1)}{6}$$

to give

$$8 - \frac{8}{n^3}\frac{n(n+1)(2n+1)}{6} \le A$$

$$\le 8 - \frac{8}{n^3}\left[\frac{n(n+1)(2n+1)}{6} - \frac{2n(n+1)}{2} + n\right]$$

As $n \to \infty$ both the left and right sides of this inequality converge to $8 - 8/3 = 16/3$. It follows that $A = 16/3$ and that the entire area in Figure 2 is $2A = 32/3$. Thus we see again that our scheme for computing areas does produce reasonable results.

PROBLEMS

In each of the following problems, use the method outlined in this section to find the area which lies below the graph of the given function and above the indicated interval $[a, b]$ on the horizontal axis.

1. $f(x) = x$ $a = 0, b = 1$
2. $g(x) = 3x$ $a = 0, b = 1$
3. $h(x) = 2$ a and b arbitrary, except that $a < b$
4. $g(t) = t^2$ $a = -2, b = 1$
5. $h(t) = -t + 1$ $a = -2, b = 1$
6. $m(t) = t^2 - t + 1$ $a = -2, b = 1$
7. $f(z) = z + 2$ $a = 0, b = 1$
8. $f(z) = z + 2$ $a = 1, b = 2$
9. $f(z) = z + 2$ $a = 0, b = 2$

10. Let c be a positive constant and f be a function which is non-negative throughout the interval $[a, b]$. Formulate a theorem which relates the

area below f with that below $c \cdot f$, both areas lying above the interval $[a, b]$ on the horizontal axis. (*Hint:* Compare Problems 1 and 2.)

11. Let f and g be two functions each non-negative throughout the interval $[a, b]$. Formulate a theorem relating the area under $f + g$ with those under f and under g. Again each area lies above $[a, b]$. (*Hint:* Compare the results of Problems 4–6.)

12. Let f be a function non-negative throughout the interval $[a, b]$ and let c be any number between a and b. Formulate a theorem which relates the area below f and above $[a, b]$ with those below f and above $[a, c]$ and $[c, b]$, respectively. (*Hint:* Compare the results of Problems 7–9.)

13. Using the method for computing area indicated in this section, prove the statements formulated in Problems 10–12. Give a geometric interpretation of each result.

14. Let functions f and g be such that $f(x) \geq g(x)$ throughout the interval $[a, b]$. Formulate a method for computing the area in the interval $[a, b]$ which lies between f and g. Use this method to find the area in the interval from $a = 0$ to $b = 1$ which lies between the graphs of $f(x) = 3x$ and $g(x) = x^2 - x$.

14.2 THE INTEGRAL

The method for computing areas introduced in Section 14.1 is closely related to the concept of the integral of a function. However, since integrals have wide application in situations where an area interpretation is not meaningful, we shall develop a definition of the integral which does not depend on any knowledge of area. Unless stated otherwise, all functions considered in this chapter are assumed to be continuous.

The integral of a function f over an interval $[a, b]$ is denoted by the various notations

$$\int_a^b f \quad \text{or} \quad \int_a^b f(x) \quad \text{or, most often,} \quad \int_a^b f(x)\, dx$$

To give meaning to these notations we begin, as with areas, by partitioning the interval $[a, b]$ into n smaller intervals of equal width $(b - a)/n$ (see Figure 7). For $k = 1, 2, \ldots, n$, we choose in the kth interval

$$\left[a + (k - 1)\frac{b - a}{n}, \, a + k\frac{b - a}{n} \right]$$

a point x_k at which the continuous function f takes its largest value and another point y_k at which f takes its smallest value. We then form the *upper sum*

$$M_n = \sum_{k=1}^{n} \frac{b - a}{n} f(x_k) = \frac{b - a}{n} f(x_1) + \frac{b - a}{n} f(x_2) + \cdots$$

$$+ \frac{b - a}{n} f(x_n) \qquad \textbf{(3)}$$

FIGURE 7

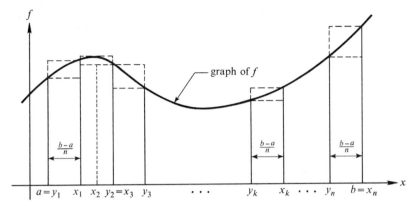

and the *lower sum*

$$m_n = \sum_{k=1}^{n} \frac{b-a}{n} f(y_k) = \frac{b-a}{n} f(y_1) + \frac{b-a}{n} f(y_2) + \cdots$$

$$+ \frac{b-a}{n} f(y_n) \qquad (4)$$

Example 1 The function h defined in the interval $[1, 3]$ by

$$h(t) = 1 + 4t - t^2 = 5 - (t - 2)^2$$

is shown in Figure 8. Cutting the interval into four smaller intervals each of width $\frac{1}{2}$, we obtain the upper sum

$$M_4 = \tfrac{1}{2}h(\tfrac{3}{2}) + \tfrac{1}{2}h(2) + \tfrac{1}{2}h(2) + \tfrac{1}{2}h(\tfrac{5}{2})$$

$$= \tfrac{1}{2}(\tfrac{19}{4} + 5 + 5 + \tfrac{19}{4}) = \tfrac{39}{4}$$

FIGURE 8

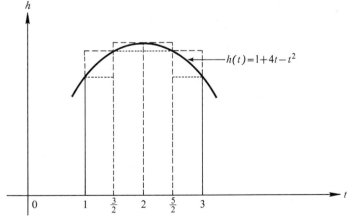

$h(t) = 1 + 4t - t^2$

and the lower sum

$$m_4 = \tfrac{1}{2}h(1) + \tfrac{1}{2}h(\tfrac{3}{2}) + \tfrac{1}{2}h(\tfrac{5}{2}) + \tfrac{1}{2}h(3)$$
$$= \tfrac{1}{2}(4 + \tfrac{19}{4} + \tfrac{19}{4} + 4) = \tfrac{35}{4} \quad \blacktriangleright$$

Example 2 When the interval $[0, 1]$ is partitioned into smaller intervals each of width $1/n$, the upper and lower sums for the function $f(x) = x + 1$ are (see Figure 1 and Equation (1) in Section 14.1)

$$M_n = \frac{3}{2} + \frac{1}{2n} \quad \text{and} \quad m_n = \frac{3}{2} - \frac{1}{2n} \quad \blacktriangleright$$

If the sequences of upper and lower sums,

$$(M_n) = (M_1, M_2, M_3, \ldots)$$

and

$$(m_n) = (m_1, m_2, m_3, \ldots)$$

determined by Equations (3) and (4), respectively, both converge and have a common limit

$$I = \lim_{n \to \infty} (M_n) = \lim_{n \to \infty} (m_n)$$

this limit is called the *integral of f on the interval* $[a, b]$ and we write

$$I = \int_a^b f(x)\, dx \tag{5}$$

Example 3 In Example 2 the quantities M_n and m_n are given for the function $f(x) = x + 1$ on the interval $[0, 1]$ by

$$M_n = \frac{3}{2} + \frac{1}{2n} \quad \text{and} \quad m_n = \frac{3}{2} - \frac{1}{2n}$$

The sequences (M_n) and (m_n) converge to the common limit $\tfrac{3}{2}$. Thus the integral for this function is

$$\int_0^1 f(x)\, dx = \int_0^1 (x + 1)\, dx = \tfrac{3}{2}$$

Of course, this is the same answer as the area A found in Figure 1. $\quad \blacktriangleright$

If f is a positive function throughout the interval $[a, b]$, the sums (3) and (4) have area interpretations. For M_n, for instance, the product

$$\frac{b - a}{n} f(x_k)$$

represents the area of a rectangle having width $(b - a)/n$ and height $f(x_k)$ (see Figure 7). Since in the kth interval, f is largest at x_k, the area of this rectangle is equal to or larger than the area under the graph of f. The sum M_n thus represents an upper bound to the area under the curve between a and b. Similar comments apply to show that m_n is a lower bound for this same area. Since, when the integral exists, M_n and m_n converge to the same limit, we see that the two concepts area and integral are equivalent when the function is positive throughout the interval.

Example 4 Suppose the function g is defined on the interval $[-1, 2]$ by $g(x) = x^3$. If we divide the interval into six smaller intervals, each of length $\frac{1}{2}$, we find (see Figure 9)

$$m_6 = \tfrac{1}{2}(-1)^3 + \tfrac{1}{2}(-\tfrac{1}{2})^3 + \tfrac{1}{2}(0)^3 + \tfrac{1}{2}(\tfrac{1}{2})^3 + \tfrac{1}{2}(1)^3 + \tfrac{1}{2}(\tfrac{3}{2})^3$$

$$= \tfrac{27}{16}$$

If an interpretation in terms of area is attempted here, we notice that some rectangles in Figure 9 lie *below* the horizontal axis. The areas of these

FIGURE 9

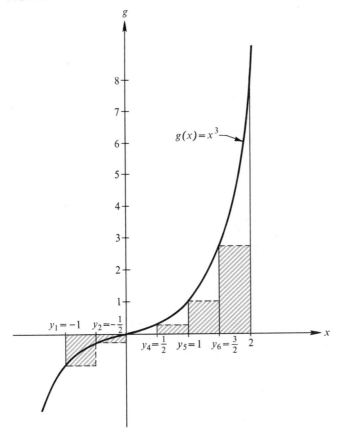

rectangles are counted negatively or *subtracted,* while the areas of the rectangles which lie above the axis are counted positively or added. ▶

PROBLEMS

1. For the function g defined in Example 4, find M_4 and m_4.

Example 5 Suppose f is the constant function

$$f(x) = c \qquad a \le x \le b$$

Then, regardless of the values of $x_1, x_2, \ldots, x_n$, we have

$$\sum_{k=1}^{n} \frac{b-a}{n} f(x_k) = \sum_{k=1}^{n} \frac{b-a}{n} \cdot c = c(b-a)$$

This means that (M_n) and (m_n) are constant sequences and that

$$\int_a^b f(x)\, dx = \int_a^b c\, dx = c(b-a) \tag{6}$$

If c is positive, this integral represents the area of the rectangle in Figure 10(a). On the other hand, if c is negative, then $c(b - a)$ is negative and the integral itself cannot be interpreted as area. However, the absolute value $|c(b - a)|$ represents the area of the region in Figure 10(b). ▶

FIGURE 10

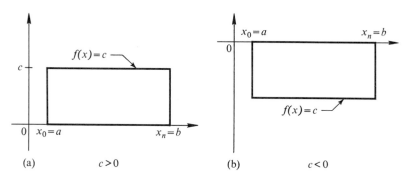

(a) $c > 0$ (b) $c < 0$

Example 6 By contrast, the function f defined on $[0, 1]$ by

$$f(x) = \begin{cases} 1 & \text{if } x \text{ is rational} \\ 0 & \text{if } x \text{ is irrational} \end{cases}$$

does not have an integral. For we know that every interval, no matter how small, contains both rational and irrational points. Hence, by choosing each x_k to be a rational number and each y_k to be irrational in Equations (3) and (4), we obtain the sums

$$M_n = \sum_{k=1}^{n} \frac{1}{n} f(x_k) = \sum_{k=1}^{n} \frac{1}{n} = 1$$

$$m_n = \sum_{k=1}^{n} \frac{1}{n} f(y_k) = \sum_{k=1}^{n} 0 = 0$$

The two sequences converge, but not to the same limit. ▶

The following definition summarizes our discussion and introduces some useful terminology.

Definition of Integral Let f be a continuous real-valued function defined throughout an interval $[a, b]$. The number I is called the *definite integral of f on the interval $[a, b]$* if I is the common limit of the sequences (M_n) and (m_n) defined in Equations (3) and (4). The function f is called the *integrand*, and if the integral exists, f is said to be *integrable* on $[a, b]$. The number I is denoted by any of the notations

$$I = \int_a^b f, \quad \int_a^b f(x), \quad \text{or} \quad \int_a^b f(x)\, dx$$

The numbers a and b are called, respectively, the *lower and upper limits of integration.* ▶

Although the computations and inequalities developed in Section 14.1 were concerned with area, careful reading will show that the concept of area itself remained intuitive and was never defined. Although area provided a prime motivation for the definition of integral, in the definition we carefully avoided any mention of area. Having made the concept of integral independent of area, it is now logically permissible to *define* area in terms of integral.

Definition of Area Let f be a function which is *non-negative* throughout the interval $[a, b]$. Then the *area A* of the region shaded in Figure 11(a) is

$$A = \int_a^b f(x)\, dx$$

If $f(x) \geq g(x)$ for all $x \in [a, b]$, then the area of the region shaded in Figure 11(b) is

$$A = \int_a^b [f(x) - g(x)]\, dx$$ ▶

FIGURE 11

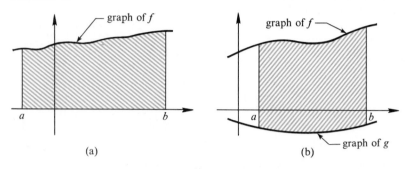

(a)

(b)

— graph of g

Example 7 If $f(x) = 0$ and $g(x) = c$ are constant functions with $c < 0$, then the area outlined in Figure 10(b) is

$$A = \int_a^b [f(x) - g(x)]\, dx = \int_a^b [0 - c] = \int_a^b (-c) = -c(b - a) \quad \blacktriangleright$$

PROBLEMS

For Problems 2–8, the given function f is defined on the interval $[0, 1]$. Find the upper and lower sums M_n and m_n which are obtained from the partition $\{0, 1/n, 2/n, \ldots, (n-1)/n, 1\}$. If possible, simplify your results by applying appropriate summation formulas.

2. $f(x) = x$ **3.** $f(x) = 2x + 3$ **4.** $f(x) = \frac{1}{2} - x$

5. $f(x) = x^{1/2}$ **6.** $f(x) = x^2$ **7.** $f(x) = 3x^2 + 2x$

8. $f(x) = 3x^3 + 2x^2 - x + 6$

9. By taking limits as $n \to \infty$, find the definite integral of f on $[0, 1]$ in Problems 2–4 and 6–8.

10. Find the area of the region between $f(x) = x$ and $g(x) = x^2$ over the interval $[0, 1]$.

11. Evaluate these definite integrals.

(a) $\int_1^3 3x\, dx$ (b) $\int_0^2 x^2\, dx$

(c) $\int_{-1}^{+1} |x|\, dx$ (d) $\int_0^4 (x + |x|)\, dx$

(e) $\int_{-1}^2 (x^2 - |x|)\, dx$ (f) $\int_1^2 (x^3 + x - 1)\, dx$

14.3 ANTIDERIVATIVES

If a function f is the derivative of another function F, that is, if $F'(x) = f(x)$ for all x in the domain of f, then F is called an *antiderivative*, or an *indefinite integral*, of f and we write

$$F(x) = \int f(x)\,dx$$

The next theorem shows that all antiderivatives of a given function are closely related. (Compare with Theorem 4 in Section 13.4)

Theorem 1   If functions F and G have the same derivative throughout an interval $[a, b]$, then there is a constant C such that for all x in $[a, b]$,

$$F(x) = G(x) + C$$

PROOF   Define the function H by $H(x) = F(x) - G(x)$. Then

$$H'(x) = F'(x) - G'(x) = 0$$

since the derivatives of F and G are assumed equal. Pick any two points y and z between a and b. Since the derivative of H exists throughout the interval $[a, b]$, the Mean Value Theorem guarantees the existence of a point c between y and z at which

$$H(z) - H(y) = H'(c)(z - y)$$

But $H'(c) = 0$ and it follows that $H(z) = H(y)$. This being true for all y and z, H must be a constant function and our proof is complete.   ▶

Example 1   (a) The fact that $D(x^2 + 2x + 7) = 2x + 2$ means that $F(x) = x^2 + 2x + 7$ is an antiderivative of $f(x) = 2x + 2$. On the other hand, if G is any other antiderivative of f, then $DG(x) = 2x + 2 = DF(x)$ and by Theorem 1, F and G can differ at most by a constant. Thus *all* antiderivatives of f are of the form $G(x) = x^2 + 2x + 7 + C$ or, since $7 + C$ is itself some constant, are of the form

$$G(x) = x^2 + 2x + B$$

where B is a constant.
   (b) The derivative of $G(x) = -\cos x + \ln x + \frac{1}{3}x^3$ is

$$G'(x) = \sin x + \frac{1}{x} + x^2$$

Hence every antiderivative of $\sin x + (1/x) + x^2$ has the form

$$F(x) = -\cos x + \ln x + \frac{x^3}{3} + C$$

where C is a constant.   ▶
   The "equals" in $F(x) = \int f(x)\,dx$ must be properly interpreted to mean "F is *an* antiderivative of f" and not to mean "F is *the* antiderivative of f." For if a function f has one antiderivative F, then it has an infinite number, these being of the form $F + C$, where C is a constant.
   On the other hand, Theorem 1 guarantees that two antiderivatives of the same function f can differ at most by a constant amount. Once one anti-

derivative is known, then, all are known. This equivalence of meaning between $F(x) = \int f(x)\,dx$ and $F'(x) = f(x)$ means that each derivative relation listed in Table 2 of Chapter 12 has a counterpart in terms of anti-derivatives. For easy reference we list the most useful cases in Table 1. A great many antiderivatives may be found by direct use of Table 1, together with the Chain Rule (Section 12.4). (Techniques for obtaining antiderivatives of more complicated functions are discussed in Sections 14.7–14.9.) Note that absolute value is indicated in lines 6 and 7. This is because if x is positive, then $D \ln x = 1/x$, while if x is negative, $D \ln (-x) = D(-x)/-x = 1/x$, also. The formula $\int (1/x)\,dx = \ln |x| + C$ takes care of both cases.

TABLE 1. A short table of antiderivatives

1. $\displaystyle\int DF(x)\,dx = F(x) + C$

2. $\displaystyle\int 0\,dx = C$

3. $\displaystyle\int 1\,dx = x + C$

4. $\displaystyle\int x^n\,dx = \frac{x^{n+1}}{n+1} + C \qquad$ if $n \neq -1$

5. $\displaystyle\int [f(x)]^n f'(x)\,dx = \frac{[f(x)]^{n+1}}{n+1} + C \qquad$ if $n \neq -1$

6. $\displaystyle\int x^{-1}\,dx = \ln |x| + C$

7. $\displaystyle\int \frac{f'(x)}{f(x)}\,dx = \ln |f(x)| + C$

8. $\displaystyle\int [f(x) + g(x)]\,dx = \int f(x)\,dx + \int g(x)\,dx$

9. $\displaystyle\int af(x)\,dx = a \int f(x)\,dx \qquad$ if a is a constant

10. $\displaystyle\int [c_1 f_1(x) + c_2 f_2(x) + \cdots + c_n f_n(x)]\,dx$

$$= c_1 \int f_1(x)\,dx + c_2 \int f_2(x)\,dx + \cdots + c_n \int f_n(x)\,dx$$

11. $\displaystyle\int e^x\,dx = e^x + C$

12. $\displaystyle\int a^x\,dx = \frac{a^x}{\ln a} + C$

13. $\displaystyle\int \cos x\,dx = \sin x + C$

14. $\displaystyle\int \sin x\,dx = -\cos x + C$

15. $\displaystyle\int \sec^2 x\,dx = \tan x + C$

Example 2 (a) Since $D \cos x = -\sin x$, all antiderivatives of

$$f(x) = -\sin x$$

look like

$$F(x) = \cos x + C$$

where C is a constant.

(b) If ϕ has a derivative then

$$\int e^{\phi(x)}\phi'(x)\,dx = e^{\phi(x)}$$

Hence all antiderivatives of $f(x) = 2xe^{x^2}$ look like

$$F(x) = e^{x^2} + C$$

where C is a constant.

(c) All antiderivatives of $g(x) = \sin 3x - e^{-2x} + x$ look like

$$G(x) = -\tfrac{1}{3}\cos 3x + \tfrac{1}{2}e^{-2x} + \frac{x^2}{2} + C$$

(d) If f has a derivative, then

$$D[f(x)]^{n+1} = (n+1)[f(x)]^n f'(x)$$

In terms of antiderivatives,

$$\int [f(x)]^n f'(x)\,dx = \frac{1}{n+1}[f(x)]^{n+1} + C$$

Thus, for example, $x(x^2 + 3)^{-1/3}$ is the derivative of

$$\frac{1}{2}\frac{(x^2 + 3)^{2/3}}{\tfrac{2}{3}} = \frac{3}{4}(x^2 + 3)^{2/3}$$

Similar reasoning applied to the equation

$$Dh(x) = 3x(x^2 + 3)^{-1/3} + (x - 1)(2x^2 - 4x + 1)^6$$

yields

$$h(x) = \frac{3}{2}\frac{(x^2 + 3)^{2/3}}{\tfrac{2}{3}} + \frac{1}{4}\frac{(2x^2 - 4x + 1)^7}{7} + C$$

$$= \tfrac{9}{4}(x^2 + 3)^{2/3} + \tfrac{1}{28}(2x^2 - 4x + 1)^7 + C$$

▶

PROBLEMS

Find the following antiderivatives.

1. $\displaystyle\int x^6 \, dx$

2. $\displaystyle\int x^{-4} \, dx$

3. $\displaystyle\int \left(x^3 - 2x^2 + 3 - \frac{4}{x^6} \right) dx$

4. $\displaystyle\int y^4 (1 - y)^2 \, dy$

5. $\displaystyle\int \frac{1}{\sqrt[3]{z}} \, dz$

6. $\displaystyle\int \frac{1 - u}{\sqrt{u}} \, du$

7. $\displaystyle\int x\sqrt{1 + x^2} \, dx$

8. $\displaystyle\int \sin x \cos x \, dx$

9. $\displaystyle\int e^x (e^x - 2)^2 \, dx$

10. $\displaystyle\int \frac{x}{x^2 + 1} \, dx$

11. $\displaystyle\int \cos^2 x \sin x \, dx$

12. $\displaystyle\int (2 + e^x) e^x \, dx$

13. $\displaystyle\int \frac{x + 1}{x + 2} \, dx$

14. $\displaystyle\int \frac{x}{a^2 - x^2} \, dx$

15. $\displaystyle\int \frac{\sin t \cos t}{2 + \sin^2 t} \, dt$

16. $\displaystyle\int \tan x \, dx$

17. $\displaystyle\int e^{\sin x} \cos x \, dx$

18. $\displaystyle\int \sin 2kx \, dx$

19. $\displaystyle\int \frac{e^{1/x}}{x^2} \, dx$

20. $\displaystyle\int \frac{x^2 - x}{x + 1} \, dx$

21. $\displaystyle\int 2 \frac{e^x}{e^x + 1} \, dx$

22. $\displaystyle\int \frac{3x^3 + 6x^2 + 3x + 1}{x^2 + 2x + 1} \, dx$

(*Hint:* In Problems 4, 9, and 12, multiply out first. In Problems 13, 20, and 22, perform the indicated division first. In Problems 7–9, 11, 12, 17, and 19, apply Formula 5 in Table 1. In Problems 10, 14, 15, and 21, apply Formula 7 in Table 1.)

14.4 FUNDAMENTAL THEOREM OF CALCULUS OR INTEGRATION MADE EASY

To evaluate an integral directly from the definition in Section 14.2 is ordinarily an arduous task. However, by making a slight restriction on the kinds of functions for which we try to find integrals, we can make the problem of evaluation relatively easy, while at the same time losing almost no practical applicability. The result which simplifies the computation of integrals also establishes the link between the two main operations of calculus—differentiation and integration.

Fundamental Theorem of Calculus If $\int_a^b f(x)\,dx$ exists and if there is a function $F(x)$ whose derivative is $f(x)$, then

$$\int_a^b f(x)\,dx = F(b) - F(a)$$ ▶

The statement of this theorem is remarkable. If f is integrable, all we need do in order to evaluate its integral is to find another function, any one at all, whose derivative is f. The rest is a matter of simple computation. Before proving the Fundamental Theorem, let us first apply it to the problem of computing the area in Figure 2.

Example 1 The function $g(x) = 4 - x^2$ is integrable, as we have already seen. The area in Figure 2 is thus equal to

$$\int_{-2}^2 g(x)\,dx = \int_{-2}^2 (4 - x^2)\,dx$$

Since $G(x) = 4x - \frac{1}{3}x^3$ is one antiderivative of $4 - x^2$, the desired area is, according to the Fundamental Theorem,

$$G(2) - G(-2) = 4(2) - \frac{2^3}{3} - \left[4(-2) - \frac{(-2)^3}{3}\right] = \frac{32}{3}$$

verifying the computations in Section 14.1. ▶

PROOF OF THE FUNDAMENTAL THEOREM Let $r = (b - a)/n$ and define points $t_0, t_1, t_2, \ldots, t_n$ (Figure 12) by

$$t_0 = a, \; t_1 = a + r, \; t_2 = a + 2r, \; \ldots, \; t_n = b$$

By successively adding and subtracting the numbers $F(t_1), F(t_2), \ldots, F(t_{n-1})$, the difference

$$F(b) - F(a) = F(t_n) - F(t_0)$$

FIGURE 12

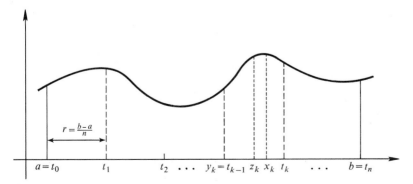

may be written in the form

$$F(b) - F(a) = [F(t_n) - F(t_{n-1})] + [F(t_{n-1}) - F(t_{n-2})] + \cdots$$
$$+ [F(t_1) - F(t_0)]$$
$$= \sum_{k=1}^{n} [F(t_k) - F(t_{k-1})] \tag{7}$$

Since f is the derivative of F, the Mean Value Theorem guarantees that within each interval $[t_{k-1}, t_k]$ there is a point z_k at which

$$F(t_k) - F(t_{k-1}) = f(z_k)[t_k - t_{k-1}]$$

Substituting this into (7) gives

$$F(b) - F(a) = \sum_{k=1}^{n} f(z_k)[t_k - t_{k-1}] = \sum_{k=1}^{n} f(z_k) \frac{b-a}{n}$$

If, as in Equations (3) and (4), we choose x_k and y_k to be points in the interval $[t_{k-1}, t_k]$ at which f takes its largest and smallest values, respectively (Figure 12), then

$$f(y_k) \leq f(z_k) \leq f(x_k)$$

which means that

$$\sum_{k=1}^{n} f(y_k) \frac{b-a}{n} \leq \sum_{k=1}^{n} f(z_k) \frac{b-a}{n} \leq \sum_{k=1}^{n} f(x_k) \frac{b-a}{n}$$

or, equivalently,

$$m_n \leq F(b) - F(a) \leq M_n$$

for all values of n. Since the number $F(b) - F(a)$ is trapped between corresponding members of the sequences (M_n) and (m_n), and since both these sequences converge to the value of the integral, it follows that

$$F(b) - F(a) = \int_a^b f(x)\,dx$$

This completes the proof. ▶

Example 2 Let $F(x) = \frac{1}{2}x^2 + x$ and $f(x) = x + 1$. Since $F'(x) = f(x)$ it follows from the Fundamental Theorem that

$$\int_0^1 f(x)\,dx = \int_0^1 (x+1)\,dx$$

$$= F(1) - F(0) = (\tfrac{1}{2} + 1) - (0) = \tfrac{3}{2}$$

This represents the area in Figure 1 and verifies our previous calculations. ▶

The result of Theorem 1 in Section 14.3 indicates clearly why, in the statement of the Fundamental Theorem, no particular antiderivative F of f was required in the computation

$$\int_a^b f(x)\,dx = F(b) - F(a)$$

For if G is any other antiderivative of f, then there is a constant C such that $G(x) = F(x) + C$ for all x. This being the case, we find

$$G(b) - G(a) = [F(b) + C] - [F(a) + C] = F(b) - F(a)$$

The constant C is simply added in and subtracted out, the actual value chosen being of no consequence.

It is convenient to use the notation

$$F(x)\Big]_a^b$$

to denote the difference $F(b) - F(a)$ which appears in the Fundamental Theorem. That is, $F(x)]_a^b$ indicates that the function F is to be evaluated first at b, then at a, and the difference computed.

Example 3 We know that $D \sin x = \cos x$. Thus

$$\int_0^{\pi/2} \cos x\,dx = \sin x\Big]_0^{\pi/2}$$

$$= \sin \frac{\pi}{2} - \sin 0 = 1 - 0 = 1$$

The area under the cosine curve between zero and $\pi/2$ is unity as shown in Figure 13. ▶

FIGURE 13

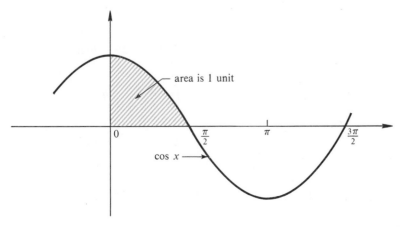

area is 1 unit

$\cos x$ ⟶

Example 4   The area lying between the graphs of the parabola $f(x) = x^2$ and the line $g(x) = x + 2$ is shaded in Figure 14. By definition, the required area is given by the integral

$$\int_{-1}^{2} (g(x) - f(x))\, dx = \int_{-1}^{2} (x + 2 - x^2)\, dx$$

Since $D(\frac{1}{2}x^2 + 2x - \frac{1}{3}x^3) = x + 2 - x^2$, the value of this integral is

$$\frac{x^2}{2} + 2x - \frac{x^3}{3}\Big|_{-1}^{2} = \frac{2^2}{2} + 2(2) - \frac{2^3}{3} - \left[\frac{(-1)^2}{2} + 2(-1) - \frac{(-1)^3}{3}\right] = \frac{9}{2}$$

▶

FIGURE 14

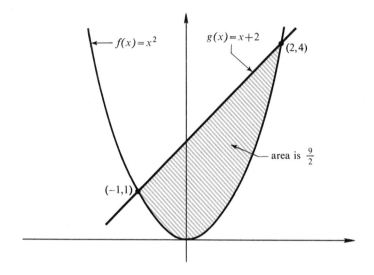

PROBLEMS

Evaluate the integrals in Problems 1–14.

1. $\int_{0}^{2} 2x^2\, dx$

2. $\int_{1}^{3} (y^3 + 2y - 6)\, dy$

3. $\int_{2}^{4} x(x^2 - 2)^2\, dx$

4. $\int_{1}^{4} \frac{1}{\sqrt{2t + 1}}\, dt$

5. $\int_{0}^{2} \frac{x}{\sqrt{x^2 + 3}}\, dx$

6. $\int_{2}^{4} \frac{2x^2 + 3}{x}\, dx$

7. $\int_{0}^{\pi} \sin^2 x \cos x\, dx$

8. $\int_{0}^{1} (2 + e^x)^2 e^x\, dx$

9. $\displaystyle\int_0^1 \frac{x^2 - x}{x - 2}\, dx$

10. $\displaystyle\int_0^{2\pi} 3\,\frac{\sin 2x}{1 + \sin^2 x}\, dx$

11. $\displaystyle\int_1^2 \frac{e^{\sqrt{x}}}{\sqrt{x}}\, dx$

12. $\displaystyle\int_0^{\pi} \sin 2x\, dx$

13. $\displaystyle\int_{-2}^1 |x|\, dx$

14. $\displaystyle\int_1^3 \frac{x^2 - 1}{x + 1}\, dx$

15. Find the area bounded by $f(x) = 3x^2$ and $g(x) = 16 - x^2$.

14.5 BASIC PROPERTIES OF INTEGRALS

In order to apply the Fundamental Theorem in the examples of the preceding section, we tacitly assumed that the functions encountered were integrable. Since Example 6 of Section 14.2 contains a function which cannot be integrated, we naturally ask the question: "What functions are integrable?" A partial answer, but an extremely important one, is contained in the following theorem, which we state without proof.*

Theorem 2 (a) All continuous functions are integrable. Specifically, if the function f is continuous throughout the interval $[a, b]$, then $\int_a^b f(x)\, dx$ exists.

(b) If the function f is continuous in $[a, b]$ except for a finite number of points of discontinuity, then $\int_a^b f(x)\, dx$ exists. ▶

Integrals which are not covered by Theorem 2 together with the Fundamental Theorem rarely arise in practice. By relying on the Fundamental Theorem we shall lose little of essence for applications while gaining a great deal in simplicity. Thus, although our theorems are actually valid in a broader sense, *from now on in this book we shall consider only those cases to which the Fundamental Theorem, in conjunction with Theorem 2, may be applied.* One of the immediate benefits of this approach is that properties of integrals may be derived directly from already familiar properties of derivatives.

Theorem 3 (a) If $f(x)$ is integrable and c is any constant, then $cf(x)$ is integrable and

$$\int_a^b cf(x)\, dx = c \int_a^b f(x)\, dx$$

*The proof of this result is beyond the scope of the present text. A good source is Bartle, R. G., *The Elements of Real Analysis* (John Wiley & Sons, Inc., New York, 1964), pp. 283–284. Roughly, the argument rests on the fact that points close together in the domain of a continuous function generate functional values which are close together. Hence, by taking n large enough, the quantities M_n and m_n will have values which are nearly the same.

In words, the integral of a constant times a function is equal to the constant times the integral of the function.

(b) If $f(x)$ and $g(x)$ are both integrable, then $f(x) + g(x)$ is integrable and

$$\int_a^b [f(x) + g(x)] \, dx = \int_a^b f(x) \, dx + \int_a^b g(x) \, dx$$

That is, the integral of a sum of two functions is equal to the sum of the integrals of the individual functions.

PROOF (a) Suppose

$$\int_a^b f(x) \, dx = F(b) - F(a)$$

where $F(x)$ is a function whose derivative is $f(x)$. Then the derivative of $cF(x)$ is $cf(x)$ and it follows that

$$\int_a^b cf(x) \, dx = cF(b) - cF(a) = c \int_a^b f(x) \, dx$$

(b) Let

$$\int_a^b f(x) \, dx = F(b) - F(a) \quad \text{and} \quad \int_a^b g(x) \, dx = G(b) - G(a)$$

where $F'(x) = f(x)$ and $G'(x) = g(x)$. Then, if $H(x) = F(x) + G(x)$, we have $H'(x) = f(x) + g(x)$ so that

$$\int_a^b [f(x) + g(x)] \, dx = H(b) - H(a) = F(b) + G(b) - [F(a) + G(a)]$$

$$= \int_a^b f(x) \, dx + \int_a^b g(x) \, dx \qquad \blacktriangleright$$

Theorem 3 can be extended by induction to linear combinations of any finite number of functions. The proof of the next theorem is left as an exercise.

Theorem 4   If $f_1(x)$, $f_2(x)$, $\ldots$, $f_n(x)$ are n functions integrable on $[a, b]$ and if $c_1, c_2, \ldots, c_n$ are constants, then

$$\int_a^b [c_1 f_1(x) + c_2 f_2(x) + \cdots + c_n f_n(x)] \, dx$$

$$= c_1 \int_a^b f_1(x) \, dx + c_2 \int_a^b f_2(x) \, dx + \cdots + c_n \int_a^b f_n(x) \, dx \qquad \blacktriangleright$$

Example 1

(a) $\int_1^2 (t^7 - 3t^2 + 11)\,dt = \int_1^2 t^7\,dt - \int_1^2 3t^2\,dt + \int_1^2 11\,dt$

$$= \frac{t^8}{8}\Big|_1^2 - t^3\Big|_1^2 + 11t\Big|_1^2$$

$$= \frac{2^8 - 1^8}{8} - (2^3 - 1^3) + 11 \cdot 2 - 11 \cdot 1$$

$$= 32 - \tfrac{1}{8} - 7 + 11 = 35\tfrac{7}{8}$$

(b) $\int_0^1 (2x + 1)(x^2 - 1)^3\,dx = \int_0^1 2x(x^2 - 1)^3\,dx + \int_0^1 (x^2 - 1)^3\,dx$

In the first integral on the right, we recognize $2x$ as the derivative of $x^2 - 1$. Recalling that an antiderivative of $[f(x)]^n f'(x)$ is $[f(x)]^{n+1}/(n + 1)$, we have

$$\int_0^1 2x(x^2 - 1)^3\,dx = \frac{(x^2 - 1)^4}{4}\Big|_0^1 = \frac{(0)^4}{4} - \frac{(-1)^4}{4} = -\frac{1}{4}$$

In the second integral we simply multiply out to obtain

$$\int_0^1 (x^2 - 1)^3\,dx = \int_0^1 (x^6 - 3x^4 + 3x^2 - 1)\,dx$$

$$= \frac{x^7}{7} - \frac{3x^5}{5} + x^3 - x\Big|_0^1$$

$$= \tfrac{1}{7} - \tfrac{3}{5} + 1 - 1 = -\tfrac{16}{35}$$

The original integral is equal to

$$-\tfrac{1}{4} - \tfrac{16}{35} = -\tfrac{99}{140} \qquad \blacktriangleright$$

Theorem 5 If $a < c < b$ and if $f(x)$ is integrable in each of the intervals $[a, b]$, $[a, c]$, and $[c, b]$, then

$$\int_a^b f(x)\,dx = \int_a^c f(x)\,dx + \int_c^b f(x)\,dx \qquad (8)$$

(In terms of area, if $f(x) \geq 0$ for all x in $[a, b]$, Equation (8) says that the total area of the region shaded in Figure 15 is equal to the area shaded to the left of the point c plus the area shaded to the right of c.)

PROOF If F is an antiderivative of f, then

$$\int_a^b f(x)\,dx = F(b) - F(a) = [F(b) - F(c)] + [F(c) - F(a)]$$

$$= \int_c^b f(x)\,dx + \int_a^c f(x)\,dx \qquad \blacktriangleright$$

FIGURE 15

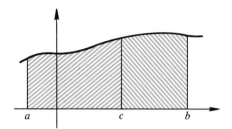

Thus far we have defined the integral $\int_a^b f(x)\,dx$ only when $a < b$. It is convenient to extend this definition to allow any relation between a and b. We do this by requiring that Theorem 5 (Equation (8)) hold for all possible choices of a, b, and c. In particular, if $a = b = c$, we have

$$\int_a^a f(x)\,dx = \int_a^a f(x)\,dx + \int_a^a f(x)\,dx = 2\int_a^a f(x)\,dx$$

so that $\int_a^a f$ must be zero for any choice of a. Choosing $a = b$ and $a < c$ gives

$$\int_a^a f(x)\,dx = \int_a^c f(x)\,dx + \int_c^a f(x)\,dx$$

Since $\int_a^a f(x)\,dx = 0$, this means that

$$\int_c^a f(x)\,dx = -\int_a^c f(x)\,dx$$

These results are summarized in the following definition.

Definition 1

(a) $\displaystyle\int_a^a f(x)\,dx = 0$ for all a

(b) If $\displaystyle\int_a^b f(x)\,dx$ exists, then

$$\int_b^a f(x)\,dx = -\int_a^b f(x)\,dx \qquad\qquad \blacktriangleright$$

Example 2   Suppose that $y(z)$, the proportion of a population scoring less than z on a behavioral measurement, is given by

$$y(z) = \int_0^z 2xe^{-x^2}\,dx$$

Then the proportion scoring between a and b is the proportion scoring below b less the proportion scoring below a. That is, the proportion between a and

b is

$$y(b) - y(a) = \int_0^b 2xe^{-x^2}\, dx - \int_0^a 2xe^{-x^2}\, dx$$

$$= \int_a^b 2xe^{-x^2}\, dx$$

Since $D(-e^{-x^2}) = 2xe^{-x^2}$, this latter proportion is

$$-e^{-x^2}\Big]_a^b = e^{-a^2} - e^{-b^2} \qquad \blacktriangleright$$

PROBLEMS

1. Find the area bounded by the vertical axis and the functions f, g, and h defined by $[f(x)]^2 = 2x$, $g(x) = 1$, and $h(x) = 4$.

2. Prove that if $f(-x) = f(x)$ for all x, then

$$\int_{-a}^a f(x)\, dx = 2\int_0^a f(x)\, dx$$

3. Show that if $f(a - x) = f(x)$ for all x, then

$$\int_0^a f(x)\, dx = 2\int_0^{a/2} f(x)\, dx$$

4. Prove Theorem 4.

14.6   APPLICATIONS

The definite integral

$$\int_a^b f(x)\, dx \qquad (9)$$

is defined (Section 14.2) as the limit of a sequence of sums of the form

$$\sum_{k=1}^n f(z_k)\frac{b - a}{n}$$

On the other hand, the Fundamental Theorem (Section 14.4) relates the integral (9) to an antiderivative F of f by

$$\int_a^b f(x)\, dx = F(b) - F(a)$$

An important point of view for applications of the integral is obtained from this result if we define a function G by

$$G(x) = \int_a^x f(t)\, dt$$

G is simply a function of the upper limit of an integral. Thus we have $G(x) = F(x) - F(a)$ and $G(a) = 0$. Since F is an antiderivative of f, we have $DG(x) = DF(x) = f(x)$ for all x, from which we learn that G is also an antiderivative of f. Put another way, f represents the rate of change of G.

The equivalent integral notation

$$D \int_a^x f(t)\, dt = f(x)$$

produces a still different sounding result. The derivative of an integral with variable upper limit is just the integrand evaluated at that upper limit. Thus we have a number of different ways of looking at integrals. As the following examples show, each way leads to important applications.

Example 1 We know, of course, that the speed at which an automobile moves represents the rate of change of the distance traveled. It follows that distance is the integral of speed. Specifically, if $v(t)$ represents the velocity of an automobile at time t, then the distance covered between times $t = a$ and $t = b$ is

$$D(a, b) = \int_a^b v(t)\, dt$$

To take a particular case, if $v(t) = t + 1$, the distance covered between $t = 0$ and $t = 1$ is

$$D(0, 1) = \int_0^1 (t + 1)\, dt = \left. \frac{t^2}{2} + t \right]_0^1 = \frac{3}{2}$$

(Compare with Example 2 of Section 14.4.) ▶

Example 2 Suppose that the marginal cost function for a certain commodity is $f(x) = ax^{1/2}$, where x denotes quantity produced and $a > 0$ is a constant. Since marginal cost is by definition the derivative of the total cost function C, the total cost to produce x units is

$$C(x) = C_0 + \int_0^x f(t)\, dt = C_0 + \int_0^x at^{1/2}\, dt$$

$$= \left. \tfrac{2}{3}at^{3/2} \right]_0^x + C_0 = \tfrac{2}{3}ax^{3/2} + C_0$$

where C_0 is the cost associated with zero production. ▶

Example 3 The amount of capital stock C is related to the rate of net investment I by the equation $DC(t) = I(t)$. Thus, if the rate of net investment at time t is given (in thousands of dollars per year) by $I(t) = 2t^{1/2} + 3$, the capital stock formation in the nth year is

$$\int_n^{n+1} (2t^{1/2} + 3) \, dt = \tfrac{4}{3}t^{3/2} + 3t \Big]_n^{n+1} = \tfrac{4}{3}[(n+1)^{3/2} - n^{3/2}] + 3 \qquad \blacktriangleright$$

Example 4   Suppose that the number $N(t)$ of animals which remain in a conditioning box t seconds after the onset of shock is given, for $t \geq 0$, by

$$N(t) = N_0 e^{-\lambda t}$$

Here, $\lambda > 0$ is a constant and N_0 denotes the number initially in the box. We wish to find the total number $A(T)$ of animal-seconds spent in the box by time T.

To approximate $A(T)$, we partition the time interval $[0, T]$ into n smaller intervals, each of length T/n (see Figure 16). If n is large, the number of animal-seconds spent in the box in the kth short interval $[(k-1)T/n, kT/n]$ is given approximately by the product $(T/n)N(kT/n)$ of the number $N(kT/n)$ of animals in the box at time kT/n times the number T/n of seconds which elapse in the interval. (This product is also the area of the dotted rectangle in Figure 16.) The total number of animal-seconds for the entire interval $[0, T]$ is given approximately by the sum

$$\sum_{k=1}^{n} \frac{T}{n} N\left(\frac{kT}{n}\right) \tag{10}$$

obtained by adding up the approximate numbers for the n small intervals.

FIGURE 16

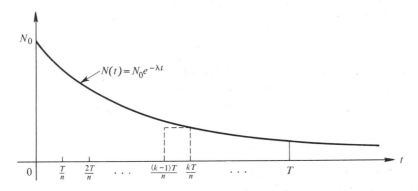

As $n \to \infty$, the sum (10) should converge to the actual number of animal-seconds spent in the box. But the limit

$$\lim_{n \to \infty} \sum_{k=1}^{n} \frac{T}{n} N\left(\frac{kT}{n}\right)$$

is the integral

$$\int_0^T N(t)\, dt$$

(that is, the area under $N(t)$ from $t = 0$ to $t = T$). Hence

$$A(T) = \int_0^T N_0 e^{-\lambda t}\, dt = -\left.\frac{N_0}{\lambda} e^{-\lambda t}\right]_0^T$$

$$= \frac{N_0}{\lambda}[1 - e^{-\lambda T}]$$

The eventual number of animal-seconds spent in the box is

$$\lim_{T \to \infty} A(T) = \frac{N_0}{\lambda} \qquad \blacktriangleright$$

Example 5 An amount A_0 of money invested at r percent compounded annually becomes $A(t) = (1 + r)^t A_0$ in t years. If interest is compounded n times per year, this formula is replaced by

$$A(t) = \left(1 + \frac{r}{n}\right)^{nt} A_0$$

Since $\lim_{n \to \infty} (1 + r/n)^n = e^r$ (see Problem 8 in Section 5.4), we see that as $n \to \infty$ (that is, if interest is compounded continuously), $A(t)$ becomes

$$A(t) = e^{rt} A_0 \qquad (11)$$

The present value of an amount $A(t)$ to be received t years in the future is defined to be the amount which, invested now, would yield $A(t)$ at time t. From (11) we see that if interest is compounded continuously at r percent per year, the present value of an amount $A(t)$ due in t years is

$$A_0 = e^{-rt} A(t)$$

The present value of a stream of incomes $A(1)$, $A(2)$, ..., $A(T)$ due, respectively, in 1, 2, ..., T years is thus given by

$$P(T) = e^{-r} A(1) + e^{-2r} A(2) + \cdots + e^{-Tr} A(T)$$

$$= \sum_{t=1}^T e^{-tr} A(t) \qquad (12)$$

If we think of having partitioned the interval $[0, T]$ into T smaller intervals each of unit length, the sum (12) may be interpreted as a sum corresponding to the integral

$$\int_0^T e^{-rt} A(t)\, dt \qquad (13)$$

The integral (13) may itself be interpreted as the present value of a stream of income to be received continuously over the time period $[0, T]$, where $A(t)$

represents the rate of change of income at time t. That is, if income were to remain constant for one year at the rate $A(t)$, then $A(t)$ dollars would be received in that year. ▶

Example 6 Net savings resulting from a proposed manufacturing facility are assumed to be a continuous flow at a constant annual rate of A dollars. If the market interest rate r is assumed constant, the present value of future savings to be realized over T years is, using (13),

$$P = A \int_0^T e^{-rt} \, dt = \frac{A}{r} (1 - e^{-rT})$$

Suppose that construction cost $C(t)$ increases with the durability of the materials used according to the function $C(t) = kt^{1/2}$, where k is a constant and t denotes durability in years. Then the optimal life T^* of the proposed facility is that value of T which maximizes the ratio of present value of future savings to cost. That is, T^* is the value of T which maximizes

$$R(T) = \frac{A(1 - e^{-rT})}{rkT^{1/2}}$$

Applying the Interior Maximum and Minimum Theorems (Theorems 14 and 15 of Section 12.8) we find that $R(T)$ is maximized when $e^{rT} = 1 + 2rT$. The optimal life of the new facility is given approximately by

$$T^* = 1.25/r \qquad\qquad ▶$$

PROBLEMS

1. Verify the result $T^* = 1.25/r$ quoted in Example 6.

2. Suppose that the marginal rate of information transmission between the input and output of a complex system is given by $A\mu^{-x}$ where A and μ are constants and x is a measure of system complexity. If $F(x)$ denotes the amount of information transmitted by a system of complexity x, and if $F(0) = 0$, show that

$$F(x) = \frac{A}{\ln \mu} [1 - \mu^{-x}]$$

3. Assume that total world population N can be described at any time t by

$$N(t) = N_0 e^{\lambda t}$$

where λ is a positive constant and N_0 denotes population size at an arbitrary reference point called $t = 0$. Assume further that Man's history to date may be characterized by a life expectancy of 25 years. Show that the total number $G(T)$ of people who have lived on the earth during the time period $[0, T]$ is approximately $(N_0/25\lambda)[e^{\lambda T} - 1]$. (*Hint:* Let $T = 25n$, where n is an integer. Argue that $G(T)$ is given approximately by $\sum_{k=1}^n N_0 e^{25k\lambda}$ and that this sum is in turn approximated by $\frac{1}{25} \int_0^T N_0 e^{\lambda t} \, dt$.)

4. Suppose that when an intercity passenger bus has been driven m hundred thousand miles, average repair costs in dollars per mile are given by

$$R(m) = Am^{1/2}$$

where A is a constant. Find the average total repair cost for a bus which has been driven 400,000 miles.

5. Upon the successful completion of a research project, each of N staff members reports, to the nearest day, the total time which he spent on the project. If the proportion $p(t)$ of the staff who report exactly t days spent on the project is given by

$$p(t) = \tfrac{1}{3}(\tfrac{2}{3})^t \qquad \text{for } t = 1, 2, 3, \ldots, 50$$

and

$$p(0) = 1 - [p(1) + p(2) + \cdots + p(50)]$$

find

(a) the exact labor cost (in summation form), assuming labor is billed at \$100 per man-day.

(b) an integral approximation to the exact cost.

6. The marginal direct labor function for a certain chemical plant is

$$\frac{dY}{dx} = Kx^{-r}$$

where Y is direct man-hours, x is quantity produced (100 tons), and $K > 0$ and $r (0 < r < 1)$ are constants. Find

(a) the total man-hours required to produce quantity x_0.

(b) the average direct labor per 100 tons required to produce quantity x_0.

7. Suppose in Problem 6 $K = 1000$ and $r = 0.322$. Find the total man-hours and the average direct labor per 100 tons required to produce

(a) the first 1000 tons.

(b) the second 1000 tons.

(c) the third 1000 tons.

8. If a constant force f is applied through a distance d, the work done is, by definition,

$$W = f \cdot d \tag{14}$$

If the force changes continuously over an interval from point a to point b, with the force at $x \in [a, b]$ being $f(x)$, the work done is

$$W = \int_a^b f(x)\,dx \tag{15}$$

(a) Show that if $f(x)$ is constant over the interval $[a, b]$ then Formula (15) becomes Formula (14).

(b) By partitioning the interval $[a, b]$ into n smaller intervals, using (14) to approximate the work done in each small interval and taking limits as $n \to \infty$, derive Formula (15).

(c) A laborer, in pushing a cart containing a leaking container a distance of 80 feet, loses $\frac{1}{4}$ of the contents on the way. The cart weighs 50 pounds, the container 20 pounds, and the contents originally weighed 160 pounds. Assume that the force required to push the cart is equal to $\frac{1}{3}$ of the total weight and that the amount which has leaked out is proportional to the distance traveled. Find the work done in moving the container.

9. In a model of extinction, Bush and Mosteller* derive the equation

$$\frac{dp}{dt} = -wbp^2$$

where w and b are constants. Given the initial condition $p(0) = p_0$, integrate to obtain the theoretical extinction curve $p(t)$.

10. As a further elaboration of their model (Problem 9), Bush and Mosteller show that the response rate is

$$R'(t) = \frac{dR}{dt} = \frac{wp_0}{1 + wbp_0 t}$$

where $R(t)$ denotes the number of responses emitted by time t.

(a) Find $R(t)$.

(b) Find the number of responses emitted during the time interval from $t = 10$ to $t = 50$.

11. In a waiting-line model,† Y, the time spent waiting in line for service, is approximated by

$$Y = \int_0^{-(k+1)/[(1/A)-(1/S)]} \left[k + 1 + t\left(\frac{1}{A} - \frac{1}{S}\right) \right] dt$$

and W, the service time, is approximated by

$$W = \int_{-(k+1)/[(1/A)-(1/S)]}^{P} \left(\frac{S}{A}\right) dt$$

where P, k, A, and S are constants. Find $Y + W$, the cumulative waiting time for the entire cycle.

14.7 INTEGRATION BY PARTS

The discussion in this and the following two sections will be concerned with techniques for finding antiderivatives, and hence for evaluating integrals, of certain types of functions which often arise in applications of the integral concept.

*Bush, R. R., and Mosteller, F., "A Mathematical Model for Simple Learning," *Psychological Review* **58,** 313–323 (1951).

†Adapted from Fabrycky, W. J., and Torgersen, P. E., *Operations Economy: Industrial Applications of Operations Research* (Prentice-Hall, Inc., Englewood Cliffs, N. J., 1966), Chapter 12.

The first of these techniques, *integration by parts*, is merely a restatement of the formula for differentiating a product. If f and g are two differentiable functions, we know (Section 12.2) that

$$D[f(x) \cdot g(x)] = f(x)g'(x) + g(x)f'(x)$$

In terms of antiderivatives, this is equivalent to saying that

$$\int f(x)g'(x)\,dx + \int g(x)f'(x)\,dx = f(x) \cdot g(x) + C$$

or, rewritten in the form that is usually called integration by parts:

$$\int f(x)g'(x)\,dx = f(x) \cdot g(x) - \int g(x)f'(x)\,dx \qquad \textbf{(16)}$$

In terms of integrals, this latter formula reads

$$\int_a^b f(x)g'(x)\,dx = f(x) \cdot g(x) \Big]_a^b - \int_a^b g(x)f'(x)\,dx$$

An important special case of (16) is

$$\int f(x)\,dx = xf(x) - \int xf'(x)\,dx$$

obtained by taking g to be the identity function $g(x) = x$.

Example 1 Suppose in Example 6 of Section 14.6 that future savings decline according to the linear function $A(t) = a - bt$ where a and b are positive constants. Then the present value of savings recouped over T years is

$$P = \int_0^T (a - bt)e^{-rt}\,dt = \int_0^T ae^{-rt}\,dt - b \int_0^T te^{-rt}\,dt$$

The first of these integrals is easily evaluated as

$$\int_0^T ae^{-rt}\,dt = -\frac{a}{r}e^{-rt}\Big]_0^T = \frac{a}{r}(1 - e^{-rT})$$

The second may be integrated by parts by taking $f(t) = t$ and $g'(t) = e^{-rt}$ in Formula (16). We obtain

$$\int_0^T te^{-rt}\,dt = -\frac{1}{r}te^{-rt}\Big]_0^T - \int_0^T \left(-\frac{1}{r}e^{-rt}\right)dt$$

$$= -\frac{1}{r}Te^{-rT} + \frac{1}{r^2}(1 - e^{-rT})$$

For the original integral we have

$$P = \int_0^T (a - bt)e^{-rt}\,dt = \frac{1}{r^2}(ar - brT - b)(1 - e^{-rT}) + \frac{bT}{r}$$

Note that the advantage of the integration-by-parts formula comes in the replacement of $f(t) = t$ by $f'(t) = 1$, resulting in an integral that is easily evaluated.

▶

PROBLEMS

    1. Verify the final calculations in Example 1.

Example 2 Let $E(t) = B(t) - G$ denote the deviation of system behavior $B(t)$ at time t from a goal G. If $E(t) = e^{-at} \sin bt$, where $a > 0$ and b are constants, then the cumulative deviation from the goal over the time interval $[0, T]$ is

$$\int_0^T E(t)\, dt = \int_0^T (e^{-at} \sin bt)\, dt$$

    This integral may be evaluated by integrating twice by parts. First, let $f(t) = \sin bt$ and $g'(t) = e^{-at}$ in Formula (16) to obtain

$$\int_0^T e^{-at} \sin bt\, dt = -\frac{1}{a} e^{-at} \sin bt \Big|_0^T - \int_0^T \left[-\frac{1}{a} e^{-at} b \cos bt \right] dt$$

$$= -\frac{1}{a} e^{-aT} \sin bT + \frac{b}{a} \int_0^T e^{-at} \cos bt\, dt \qquad (17)$$

Next, put $f(t) = \cos bt$ and $g'(t) = e^{-at}$ in (16) to obtain

$$\frac{b}{a} \int_0^T e^{-at} \cos bt\, dt = -\frac{b}{a^2} e^{-at} \cos bt \Big|_0^T - \frac{b}{a} \int_0^T \left[-\frac{1}{a} e^{-at} (-b) \sin bt \right] dt$$

$$= \frac{b}{a^2} (1 - e^{-aT} \cos bT) - \frac{b^2}{a^2} \int_0^T e^{-at} \sin bt\, dt \qquad (18)$$

Inserting (18) into (17) and solving gives

$$\int_0^T e^{-at} \sin bt\, dt = \frac{b - e^{-aT}(a \sin bT + b \cos bT)}{a^2 + b^2}$$

▶

PROBLEMS

    2. Check the calculations in Example 2.

Example 3 Many formulas listed in a table of integrals* express the integral of a function which involves the nth powers of some expression, in

---

*Extensive tables of antiderivatives (integrals) are available which include all the cases we will consider plus many more. Recourse to such a table is a must for many problems involving integrals. One of the most widely available sources is the *C.R.C. Handbook of Standard Mathematical Tables*, published by the Chemical Rubber Publishing Company, Cleveland, Ohio.

terms of the integral of a function involving $(n - 1)$st or lower powers of that expression. These recursive formulas are called *reduction formulas*. They are usually obtained by integration by parts.

A typical illustration is the formula

$$\int \cos^n x \, dx = \frac{\sin x \cos^{n-1} x}{n} + \frac{n-1}{n} \int \cos^{n-2} x \, dx$$

valid for $n \geq 2$. It may be derived by taking $f(x) = \cos^{n-1} x$ and $g'(x) = \cos x$ in (16) to obtain first

$$\int \cos^n x \, dx = \cos^{n-1} x \sin x + (n - 1) \int \sin x \cos^{n-2} x \sin x \, dx$$

Writing $\sin^2 x = 1 - \cos^2 x$ in the integral on the right side gives

$$\int \cos^n x \, dx = \cos^{n-1} x \sin x$$

$$+ (n - 1) \int \cos^{n-2} x \, dx \quad - (n - 1) \int \cos^n x \, dx$$

Solving for

$$\int \cos^n x \, dx$$

yields the stated formula. ▶

PROBLEMS

3. Find the following antiderivatives.

(a) $\int x \ln x \, dx$ (b) $\int y^n \ln y \, dy$ (c) $\int \sin^{-1} x \, dx$

(d) $\int x^2 e^{-x} \, dx$ (*Hint:* Integrate by parts twice.)

(e) $\int \frac{1}{x^3} \sin \frac{1}{x} \, dx$ $\left(Hint: \text{ Let } f(x) = \frac{1}{x} \text{ in (16).} \right)$

4. Evaluate the integrals:

(a) $\int_0^4 2x^2 e^{2x} \, dx$ (b) $\int_0^{\pi/2} x^2 \sin x \, dx$

5. In Example 2, find the cumulative deviation from G in the interval $[0, T]$ if $E(t) = e^{-at} \cos bt$.

6. Modify Example 1 by putting $A(t) = (a - bt)^2$. Find the present value of the savings generated over T years.

7. Use the method of Example 3 to obtain the reduction formula

$$\int \sin^n x \, dx = - \frac{\sin^{n-1} x \cos x}{n} + \frac{n-1}{n} \int \sin^{n-2} x \, dx$$

8. The following is known as Wallis' formula for π:*

(a) Use the reduction formula in Problem 7 to show that

$$\int_0^{\pi/2} \sin^n x \, dx = \frac{n-1}{n} \int_0^{\pi/2} \sin^{n-2} x \, dx$$

(b) Let

$$S_n = \int_0^{\pi/2} \sin^n x \, dx$$

Show that $S_0 = \pi/2$ and $S_1 = 1$.

(c) Show by induction that

$$S_{2n} = \frac{2n-1}{2n} \frac{2n-3}{2n-2} \cdots \frac{3}{4} \frac{1}{2} \frac{\pi}{2}$$

$$S_{2n+1} = \frac{2n}{2n+1} \frac{2n-2}{2n-1} \cdots \frac{4}{5} \frac{2}{3}$$

(d) Show by induction that the sequence (S_n) is monotone decreasing. (*Hint:* $0 \le \sin x \le 1$.)

(e) From (d), argue that

$$\frac{2n}{2n+1} S_{2n} < \frac{2n}{2n+1} S_{2n-1} = S_{2n+1} < S_{2n}$$

and hence that

$$\lim_{n \to \infty} \frac{S_{2n+1}}{S_{2n}} = 1$$

(f) By using (c) to actually write S_{2n+1}/S_{2n} and then applying (e), show that

$$\lim_{n \to \infty} \frac{2 \cdot 2}{1 \cdot 3} \frac{4 \cdot 4}{3 \cdot 5} \frac{6 \cdot 6}{5 \cdot 7} \cdots \frac{(2n)(2n)}{(2n-1)(2n+1)} = \frac{\pi}{2}$$

(g) Compute $2(S_9/S_8)$ and compare the result with $\pi = 3.14159$.

9. Obtain the reduction formula

$$\int \sin^r x \cos^t x \, dx = \frac{\sin^{r+1} x \cos^{t-1} x}{r+t} + \frac{t-1}{r+t} \int \sin^r x \cos^{t-2} x \, dx$$

10. Use the result of Problem 9 to show that

$$\int \sin^4 x \cos^5 x \, dx = \sin^5 x (\tfrac{1}{9} \cos^4 x + \tfrac{4}{63} \cos^2 x + \tfrac{2}{35}) + C$$

11. Rework Problem 10 by first writing $\cos^4 x = (1 - \sin^2 x)^2$, multiplying this out, and using the formula

$$\int \sin^n x \cos x \, dx = \frac{1}{n+1} \sin^{n+1} x$$

*After John Wallis (1616–1703), an English mathematician who helped lay the groundwork for the later development of calculus by Isaac Newton.

to obtain

$$\int \sin^4 x \cos^5 x \, dx = \tfrac{1}{5} \sin^5 x - \tfrac{2}{7} \sin^7 x + \tfrac{1}{9} \sin^9 x + C$$

Which method do you prefer?

14.8 INTEGRATION BY SUBSTITUTION

A special technique of integration called *integration by substitution* is derived by inverting the Chain Rule (Section 12.4)

$$D[f \circ g(x)] = f'(g(x))g'(x)$$

to obtain the equivalent form

$$\int f'(g(x))g'(x) \, dx = f(g(x)) + C \tag{19}$$

in terms of antiderivatives. A few examples will help to show how the f and g functions should be chosen in the substitution method.

Example 1 To evaluate the integral

$$\int_0^1 x(x^2 + 1)^2 \, dx$$

we substitute $g(x) = x^2 + 1$ and $f(y) = \tfrac{1}{3}y^3$. Then $g'(x) = 2x$, $f'(y) = y^2$, and $(x^2 + 1)^2 = f'(g(x))$. Applying Formula (19) gives

$$\int_0^1 x(x^2 + 1)^2 \, dx = \tfrac{1}{2}\int_0^1 2x(x^2 + 1)^2 \, dx = \tfrac{1}{2}\int_0^1 f'(g(x)) \, dx$$

$$= \tfrac{1}{2} f(g(x)) \Big]_0^1 = \tfrac{1}{2}\tfrac{1}{3}(x^2 + 1)^3 \Big]_0^1$$

$$= \tfrac{1}{6}[2^3 - 1^3] = \tfrac{7}{6}$$

This value may be verified by writing $(x^2 + 1)^2 = x^4 + 2x^2 + 1$ to obtain

$$\int_0^1 x(x^2 + 1)^2 \, dx = \int_0^1 (x^5 + 2x^3 + x) \, dx = \tfrac{1}{6}x^6 + \tfrac{1}{2}x^4 + \tfrac{1}{2}x^2 \Big]_0^1$$

$$= \tfrac{1}{6} + \tfrac{1}{2} + \tfrac{1}{2} = \tfrac{7}{6} \qquad\blacktriangleright$$

PROBLEMS

1. The method used in Example 1 is actually just another form of Line 5 in Table 1 in Section 14.3. How should the f and g functions be chosen in (19) to obtain the formula in Line 5?

2. How should the f and g functions be chosen in (19) to obtain Line 7 of Table 1?

Example 2 To find

$$\int \frac{x^3}{1 + x^4} \, dx$$

let $g(x) = 1 + x^4$ and $f(y) = \ln y$. Then $g'(x) = 4x^3$, $f'(y) = 1/y$, and $f'(g(x)) = 1/(1 + x^4)$. Hence

$$\int \frac{x^3}{1 + x^4} \, dx = \tfrac{1}{4} \int f'(g(x))g'(x) \, dx$$

$$= \tfrac{1}{4} f(g(x)) + C = \tfrac{1}{4} \ln (1 + x^4) + C \qquad \blacktriangleright$$

At the risk of introducing some rather questionable notation, we can provide a mnemonic aid to the use of the substitution method. First, recall that

$$\int f'(u) \, du = f(u) + C \tag{20}$$

If we use the notation $u = g(x)$, so that $du/dx = g'(x)$, and then treat the derivative du/dx *as though it were a ratio* of two quantities du and dx, we obtain $du = g'(x) \, dx$. A glance at (19) shows that with the substitution of u for $g(x)$ we can write

$$\int f'(g(x))g'(x) \, dx = \int f'(u) \, du$$

Using (20) this becomes $f(u) + C$ and, on replacing u by $g(x)$, yields the correct answer $f(g(x)) + C$.

Although the procedure of treating du/dx as a ratio is questionable from a rigorous point of view, it does work and eliminates the need for thinking through the Chain Rule each time. We shall use it in all our future discussions.

Example 3 To compute the antiderivative

$$\int \frac{x^2}{\sqrt{x - 1}} \, dx$$

we substitute $u = (x - 1)^{1/2}$ so that $x = u^2 + 1$. Then $dx = 2u \, du$ and $x^2 = u^4 + 2u^2 + 1$. This gives

$$\int \frac{x^2}{\sqrt{x - 1}} \, dx = \int \frac{u^4 + 2u^2 + 1}{u} \, 2u \, du = 2 \int (u^4 + 2u^2 + 1) \, du$$

$$= \tfrac{2}{5} u^5 + \tfrac{4}{3} u^3 + 2u + C$$

Replacing u by $(x - 1)^{1/2}$ gives

$$\int \frac{x^2}{\sqrt{x - 1}} \, dx = \tfrac{2}{5}(x - 1)^{5/2} + \tfrac{4}{3}(x - 1)^{3/2} + 2(x - 1)^{1/2} + C \qquad \blacktriangleright$$

Example 4 To find

$$\int \frac{x}{1 + x^4}\, dx$$

let $u = x^2$ and $du = 2x\, dx$. Then

$$\int \frac{x}{1 + x^4}\, dx = \frac{1}{2} \int \frac{2x}{1 + x^4}\, dx = \frac{1}{2} \int \frac{1}{1 + u^2}\, du$$

From a table of integrals* we read that

$$\int \frac{1}{1 + u^2}\, du = \tan^{-1} u + C$$

and, on substituting x^2 for u,

$$\int \frac{x}{1 + x^4}\, dx = \tfrac{1}{2} \tan^{-1} x^2 + C$$

This example should be compared with Example 2. ▶

Example 5 If in Example 4 we make the substitution $u = x^4$, then $du = 4x^3\, dx$ and $x\, dx = du/4x^2 = du/4\sqrt{u}$. The integral becomes

$$\int \frac{x}{1 + x^4}\, dx = \int \frac{1}{4\sqrt{u}\,(1 + u)}\, du$$

and our substitution has created a problem more complicated than the original one. Skill in finding the proper substitution comes with practice. ▶

Example 6 Hecht† assumes that the photochemical process by which a retinal substance S decomposes when exposed to light of intensity I is characterized by the equation

$$\frac{dy}{dt} = bI(a - y) - ky^2 \tag{21}$$

Here a denotes the concentration of S prior to stimulation, b and k are positive constants, and $y(t)$ is the amount of substance S remaining at time t. Rewriting (21) as

$$\frac{y'(t)}{-k[y^2 + (bI/k)y - bIa/k]} = 1$$

and completing the square in the denominator gives

$$-\frac{1}{k} \int \frac{y'(t)}{(y + bI/2k)^2 - bIa/k - b^2I^2/4k^2}\, dt = \int 1\, dt = t + C_1$$

*See footnote on page 423.

†Hecht, S., "Vision II. The Nature of the Photoreceptor Process," in *A Handbook of General Experimental Psychology*, C. Murchison, Ed. (Clark University Press, Worcester, Mass., 1934).

This may be simplified by substituting $u = y + bI/2k$, so that $u'(t) = y'(t)$, and defining a constant B by $B^2 = bIa/k + b^2I^2/4k^2$. To integrate the resulting form

$$-\frac{1}{k} \int \frac{u'(t)}{u^2 - B^2}\, dt$$

we use Figure 17 to make a *trigonometric substitution* of the form $\cos \theta = B/u$ and $\tan \theta = \sqrt{u^2 - B^2}/B$. This gives $u = B \sec \theta$, $\sqrt{u^2 - B^2} = B \tan \theta$, and $du/d\theta = B \sec \theta \tan \theta$. Hence $u'(t)\, dt = du = B \sec \theta \tan \theta\, d\theta$ and the integral becomes

$$-\frac{1}{k} \int \frac{u'(t)}{u^2 - B^2}\, dt = -\frac{1}{k} \int \frac{B \sec \theta \tan \theta}{B^2 \tan^2 \theta}\, d\theta = -\frac{1}{kB} \int \csc \theta\, d\theta$$

From a table of integrals, we read

$$\int \csc \theta\, d\theta = -\ln |\csc \theta + \cot \theta| + C_2{}^*$$

Replacing $\csc \theta$ by $u/\sqrt{u^2 - B^2}$, $\cot \theta$ by $B/\sqrt{u^2 - B^2}$, u by $y + bI/2k$, and solving for y gives the solution

$$y(t) = -\frac{bI}{2k} - \frac{B(1 + Ae^{2Bkt})}{1 - Ae^{2Bkt}}$$

where A is a constant of integration to be determined by the initial conditions and B is as defined above. ▶

FIGURE 17

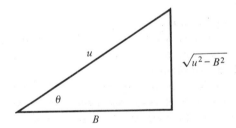

PROBLEMS

3. Complete the details of Example 6.

4. Rework Example 6 by reading

$$\int \frac{1}{u^2 - B^2}\, du$$

from a table of integrals. Which approach do you prefer?

---

*Actually, the integral

$$\int \frac{u'(t)}{u^2 - B^2}\, dt = \int \frac{1}{u^2 - B^2}\, du$$

may be read directly from an integral table. The remaining details have been included to illustrate the method of trigonometric substitution.

Using the Fundamental Theorem (Section 14.4), Equation (19) (the integration-by-substitution method) may be rewritten in terms of integrals as

$$\int_a^b f'(g(x))g'(x)\,dx = f(g(x))\Big]_a^b = f(g(b)) - f(g(a)) \qquad (22)$$

We also know that

$$\int_{g(a)}^{g(b)} f'(u)\,du = f(u)\Big]_{g(a)}^{g(b)} = f(g(b)) - f(g(a)) \qquad (23)$$

Since the right sides of (22) and (23) are identical, the left sides must be equal. That is,

$$\int_{g(a)}^{g(b)} f'(u)\,du = \int_a^b f'(g(x))g'(x)\,dx$$

In this equation, only the derivative f' appears, rather than f itself. Writing $h = f'$ to simplify the notation, we obtain the formula

$$\int_{g(a)}^{g(b)} h(u)\,du = \int_a^b h(g(x))g'(x)\,dx \qquad (24)$$

If the function g is strictly monotone (increasing or decreasing) on the interval $[a, b]$, then g has an inverse g^{-1}. Writing $c = g(a)$ and $d = g(b)$, Equation (24) becomes

$$\int_c^d h(u)\,du = \int_{g^{-1}(c)}^{g^{-1}(d)} h(g(x))g'(x)\,dx \qquad (25)$$

the form most often used in integrating by substitution.

Again, the device of writing $du = g'(x)\,dx$ when $u = g(x)$ simplifies the computational problem of actually making a *substitution*, or *change of variable* in evaluating an integral. Thus, working from the left side of (25), we "reason" as follows: Suppose we make the substitution $u = g(x)$. Then du should be replaced by $g'(x)\,dx$ and $h(u)$ by $h(g(x))$. Moreover, the limits on the integral should be changed to reflect the fact that the argument of the functions is now x rather than u. When $u = g(x) = c$, we have $x = g^{-1}(c)$, while if $u = g(x) = d$, we have $x = g^{-1}(d)$. Thus the substitution $u = g(x)$ leads to

$$\int_c^d h(u)\,du = \int_{g^{-1}(c)}^{g^{-1}(d)} h(g(x))g'(x)\,dx$$

Again, we emphasize that this is a convenient line of reasoning and it works, but it is not strictly rigorous.

Example 7 To find the value of

$$\int_0^1 \frac{1}{\sqrt{9 - x^2}}\,dx$$

make the trigonometric substitution $\frac{1}{3}x = \sin \theta$ or $x = 3 \sin \theta$ (see Figure 18). Then $\sqrt{9 - x^2} = 3 \cos \theta$, $dx/d\theta = 3 \cos \theta$, and the integral becomes

$$\int_0^1 \frac{1}{\sqrt{9 - x^2}} \, dx = \int_0^{\sin^{-1}(1/3)} \frac{3 \cos \theta}{3 \cos \theta} \, d\theta$$

$$= \int_0^{\sin^{-1}(1/3)} 1 \, d\theta = \sin^{-1}\left(\tfrac{1}{3}\right)$$

The new limits 0 and $\sin^{-1}(\frac{1}{3})$, in terms of θ, come from solving the equation $x = 3 \sin \theta$ for θ when $x = 0$ and when $x = 1$. ▶

FIGURE 18

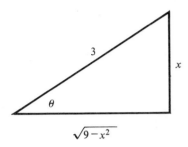

Example 8 To find

$$\int_{3/2}^{12} \frac{x}{\sqrt{2x + 1}} \, dx$$

substitute $u = \sqrt{2x + 1}$ so that $x = \frac{1}{2}(u^2 - 1)$ and $dx = u \, du$. Then

$$\int_{3/2}^{12} \frac{x}{\sqrt{2x + 1}} \, dx = \int_2^5 \frac{(u^2 - 1)u}{2u} \, du$$

$$= \tfrac{1}{2} \int_2^5 (u^2 - 1) \, du = \tfrac{1}{6}u^3 - \tfrac{1}{2}u \Big]_2^5$$

$$= \tfrac{125}{6} - \tfrac{5}{2} - [\tfrac{8}{6} - 1]$$

$$= 18$$

Note that in evaluating the integral there is no need to replace the new variable u by the original variable x. ▶

PROBLEMS

In Problems 5–22, find the indicated indefinite integrals. Consult Table 1 or a more extensive table of integrals only if absolutely necessary.

5. $\displaystyle\int \frac{t}{\sqrt{t + 2}} \, dt$ (*Hint:* Substitute $u = \sqrt{t + 2}$.)

6. $\int (y + 2)\sqrt[3]{3y + 2}\, dy$ (*Hint:* Substitute $u = \sqrt[3]{3y + 2}$.)

7. $\int \dfrac{4t + 2}{(t^2 + t)^{1/2}}\, dt$ **8.** $\int y\sqrt[3]{1 - 2y}\, dy$

9. $\int \dfrac{3\sin t \cos t}{2 + \sin^2 t}\, dt$ (*Hint:* Substitute $u = 2 + \sin^2 t$.)

10. $\int \dfrac{1 + \sin t}{\cos^2 t}\, dt$ **11.** $\int \dfrac{1}{x^2 + 25}\, dx$

12. $\int \dfrac{3\sin 2t}{2 + \sin^2 t}\, dt$ (*Hint:* $\sin 2t = 2\sin t \cos t$.)

13. $\int \dfrac{1}{\sqrt{2 - 9v^2}}\, dv$ (*Hint:* Use a trigonometric substitution.)

14. $\int (2 + \tan u)^2\, du$ **15.** $\int \dfrac{1}{\sqrt{2 - 4x^2}}\, dx$

16. $\int \dfrac{1}{y\sqrt{y^2 - 9}}\, dy$ (*Hint:* Use a trigonometric substitution.)

17. $\int \dfrac{1}{t^2\sqrt{t^2 + 9}}\, dt$ **18.** $\int \dfrac{\sqrt{v^2 + a^2}}{v^2}\, dv$

19. $\int \dfrac{y}{\sqrt{8 + 2y - y^2}}\, dy$ (*Hint:* First complete the square.)

20. $\int \dfrac{1}{(9 - s^2)^2}\, ds$ **21.** $\int \dfrac{x}{(1 + x^2)^n}\, dx$

22. $\int f'(x + a)\, dx$ (*Hint:* Substitute $y = x + a$.)

Evaluate the definite integrals in Problems 23–28.

23. $\int_{\pi/4}^{\pi/2} \dfrac{2 + \cos x}{\sin^2 x}\, dx$ **24.** $\int_0^{\pi/2} \dfrac{3\cos^2 t}{1 + \sin t}\, dt$

25. $\int_0^{\pi/2} \dfrac{2\sin\theta}{\cos^2\theta + 9}\, d\theta$ **26.** $\int_0^{\pi/4} \dfrac{\sin x}{\cos^2 x}\, dx$

27. $\int_0^5 \dfrac{1}{(y^2 + 25)^{5/2}}\, dy$ **28.** $\int_1^{\sqrt{2}} \dfrac{1}{(x + 1)(x^2 - 1)^{1/2}}\, dx$

Use a table of integrals to find the following antiderivatives and definite integrals.

29. $\int \sin^4 x\, dx$ **30.** $\int \sin^3 x \cos^4 x\, dx$

31. $\displaystyle\int_0^{\pi/3} \sin^4 \theta \cos^4 \theta \, d\theta$

32. $\displaystyle\int \sin 2y \cos 4y \, dy$

33. $\displaystyle\int \cos \theta \sin 3\theta \, d\theta$

34. $\displaystyle\int \sec^4 \theta \, d\theta$

35. $\displaystyle\int \tan^3 u \, du$ (*Hint:* Use the identity $\tan^2 u = \sec^2 u - 1$.)

36. Verify the following formulas:

(a) $\displaystyle\int \frac{1}{\sqrt{u^2 + a^2}} \, du = \ln |u + \sqrt{u^2 + a^2}| + C$

(b) $\displaystyle\int \frac{1}{\sqrt{u^2 - a^2}} \, du = \ln |u + \sqrt{u^2 - a^2}| + C$

14.9 RATIONAL FUNCTIONS AND PARTIAL FRACTION METHODS

A *rational function* is a quotient of two polynomials. Thus

$$\frac{t^2 - t + 2}{t^3 - 7t} \quad \text{and} \quad \frac{3x - 1}{x^4 + 2x^2 + 2}$$

are both rational functions. What we wish to find is a systematic way of integrating rational functions, since integrals involving rational functions often arise in practice. In our general discussion of rational functions we shall always assume that the degree of the numerator is *less than* the degree of the denominator. If this is not the case, we can make it so by long division. As an example, upon division the rational function $(x^3 - 3x + 2)/(x^2 - 1)$ becomes

$$\frac{x^3 - 3x + 2}{x^2 - 1} = x - \frac{2x - 2}{x^2 - 1}$$

the sum of a polynomial and a rational function of the desired type.

The simplest rational functions have the form

$$\frac{k}{ax + b} \tag{26}$$

where a, b, and k are constants. As is easily verified by differentiation, the antiderivative of such a function is

$$\int \frac{k}{ax + b} \, dx = \frac{k}{a} \ln |ax + b| + C \tag{27}$$

The trick in evaluating integrals of rational functions is to write the function as a sum of simple functions like (26). A few examples will best serve to indicate the method.

Example 1 (a) The function $1/(2x^2 + 3x - 2)$ may be rewritten

$$\frac{1}{2x^2 + 3x - 2} = \frac{1}{(2x - 1)(x + 2)} = \frac{2}{5}\frac{1}{2x - 1} - \frac{1}{5}\frac{1}{x + 2}$$

Hence, using (27),

$$\int \frac{1}{2x^2 + 3x - 2}\, dx = \frac{2}{5}\int \frac{1}{2x - 1}\, dx - \frac{1}{5}\int \frac{1}{x + 2}\, dx$$

$$= \tfrac{1}{5} \ln |2x - 1| - \tfrac{1}{5} \ln |x + 2| + C$$

$$= \tfrac{1}{5} \ln \left|\frac{2x - 1}{x + 2}\right| + C$$

(b) Since

$$\frac{1}{x^3 + 2x^2 - x - 2} = \frac{1}{(x - 1)(x + 1)(x + 2)}$$

$$= \frac{1}{6}\frac{1}{x - 1} - \frac{1}{2}\frac{1}{x + 1} + \frac{1}{3}\frac{1}{x + 2}$$

it follows that

$$\int \frac{1}{x^3 + 2x^2 - x - 2}\, dx = \tfrac{1}{6} \ln |x - 1| - \tfrac{1}{2} \ln |x + 1|$$

$$+ \tfrac{1}{3} \ln |x + 2| + C$$

(c) Rewriting

$$\frac{2x - 1}{2x^3 - x^2 - 4x + 3}$$

as

$$\frac{2x - 1}{2x^3 - x^2 - 4x + 3} = \frac{2x - 1}{(x - 1)^2(2x + 3)}$$

$$= \frac{8}{25}\frac{1}{x - 1} + \frac{1}{5}\frac{1}{(x - 1)^2} - \frac{16}{25}\frac{1}{2x + 3}$$

we obtain

$$\int \frac{2x - 1}{2x^3 - x^2 - 4x + 3}\, dx = \frac{8}{25} \ln |x - 1| - \frac{1}{5}\frac{1}{x - 1}$$

$$- \frac{16}{25} \ln |2x + 3| + C \qquad \blacktriangleright$$

In Example 1 the factors of the denominator of the original rational function are directly related to the simpler functions on which the integration is performed. Although we will not prove it here, this is always the case. The following summary indicates the general procedure which may be followed.

Rule to Simplify Rational Functions To find

$$\int \frac{P(x)}{Q(x)}\, dx$$

where P and Q are polynomials with the degree of P less than that of Q,

(1) factor Q. (This may be made easier by recalling that $x - a$ is a factor of $Q(x)$ if and only if $Q(x) = 0$ when $x = a$.)

(2) if $ax + b$ appears exactly n times as a factor of $Q(x)$, then form the sum

$$\frac{d_1}{ax + b} + \frac{d_2}{(ax + b)^2} + \cdots + \frac{d_n}{(ax + b)^n}$$

where $d_1, d_2, \ldots, d_n$ are constants to be determined.

(3) Solve for the d_j.

(4) Find the required antiderivatives.

Example 2 As a typical illustration, let us look again at Example 1(c). There $Q(x) = 2x^3 - x^2 - 4x + 3$ has factors $(x - 1)$, $(x - 1)$, and $(2x + 3)$. Thus we write

$$\frac{P(x)}{Q(x)} = \frac{2x - 1}{2x^3 - x^2 - 4x + 3} = \frac{2x - 1}{(x - 1)^2(2x + 3)}$$

$$= \frac{d_1}{x - 1} + \frac{d_2}{(x - 1)^2} + \frac{d_3}{2x + 3}$$

where the d_j are to be determined. (Note the inclusion of factors with both $x - 1$ and $(x - 1)^2$ in the denominator.) Multiplying through by

$$(x - 1)^2(2x + 3)$$

gives

$$2x - 1 = d_1(x - 1)(2x + 3) + d_2(2x + 3) + d_3(x - 1)^2$$

Since this expression is to hold for all values of x, we set first $x = 1$ to obtain $2 \cdot 1 - 1 = d_2(2 \cdot 1 + 3)$ or $d_2 = \frac{1}{5}$. Then, setting $x = -\frac{3}{2}$, we find $2(-\frac{3}{2}) - 1 = d_3(-\frac{3}{2} - 1)^2$ or $d_3 = -\frac{16}{25}$. To find d_1, we now set $x = 0$ on both sides to obtain

$$-1 = -3d_1 + 3d_2 + d_3$$

Inserting the known values $d_2 = \frac{1}{5}$ and $d_3 = -\frac{16}{25}$ gives $d_1 = \frac{8}{25}$, as indicated in Example 1. ▶

Example 3 Dodd* assumes that rumors or messages spread through a population at a rate proportional to the frequency of contact between those

*Dodd, S. C., "Diffusion is Predictable: Testing Probability Models for Laws of Interaction," *American Sociological Review* **20**, 392–401 (1955).

who have received the message and those who have not. He proposes the *diffusion*, or *contagion*, model

$$\frac{dy}{dt} = ky(1 - y) \tag{28}$$

where $y(t)$ and $1 - y(t)$ represent the respective proportions of the population who have and have not heard the message by time t, and k is an activity coefficient.

The time path $y(t)$ is obtained by solving the Equation (28). Rewriting this in the form

$$\frac{y'(t)}{y(1 - y)} = k$$

and using the partial fraction expansion

$$\frac{1}{y(1 - y)} = \frac{1}{y} + \frac{1}{1 - y}$$

gives

$$\int \frac{y'(t)}{y(t)} dt + \int \frac{y'(t)}{1 - y(t)} dt = \int k \, dt$$

or

$$\ln y(t) - \ln [1 - y(t)] = kt + C$$

If $y = y_0$ when $t = 0$, then $C = \ln y_0 - \ln (1 - y_0)$. Solving for $y(t)$, we find the time path

$$y(t) = \frac{y_0 e^{kt}}{1 - y_0(1 - e^{kt})}$$

which is called the *logistic curve* (Figure 19).

As $t \to \infty$, $y(t) \to 1$ so that eventually the message will diffuse throughout the population. ▶

FIGURE 19 Logistic Curve

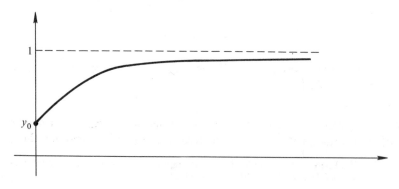

PROBLEMS

Find the following antiderivatives and definite integrals.

1. $\int \dfrac{1}{x^2 - 25} \, dx$

2. $\int \dfrac{x + 4}{x^2 - x - 2} \, dx$

3. $\int \dfrac{x^2}{x^2 - 16} \, dx$

4. $\int \dfrac{t + 1}{t(t + 2)} \, dt$

5. $\int_3^4 \dfrac{t^2 + 2t + 4}{2t^3 + 12t^2 - 32t} \, dt$

6. $\int_1^4 \dfrac{1}{y^3 + y^2} \, dy$

7. $\int_2^5 \dfrac{3w}{(1 - w^2)^2} \, dw$

8. $\int \dfrac{x - 6}{x^3 - 6x^2 + 9x} \, dx$

9. $\int_0^1 \dfrac{2x^2 + 4x + 1}{(x + 4)(x^2 + 1)} \, dx$

(*Hint:* Use the form $A/(x + 4) + (Bx + C)/(x^2 + 1)$ where A, B, and C are constants. Solve for A, B, and C.)

10. Carry out the solution for $y(t)$ indicated in Example 3.

11. In Example 3, modify the model so that

$$\frac{dy}{dt} = ky(1 - y - a)$$

where a is the fraction of the population who never hear the message. Find the time path $y(t)$.

12. Can the antiderivative

$$\int \frac{1}{x^2 + x + 4} \, dx$$

be found by partial fraction methods? If so, do it. If not, use an alternative procedure.

13. A growth model posits that the rate of growth of a certain population is governed by

$$\frac{dx}{dt} = \gamma(x - \alpha)(x - \beta)$$

where α and β denote the initial sizes of two interacting subpopulations, γ is an activity parameter, and $x(t)$ is the total population size at time t. Find the time path for $x(t)$.

14. Graham and Gagné* assume that the rate of change of response strength R varies with time t according to the equation

$$\frac{dR}{dt} = R(k_1 + k_2 - k_3 R)$$

where k_1, k_2, and k_3 are constants.

*Graham, C. H., and Gagné, R. M., "The Acquisition, Extinction and Spontaneous Recovery of a Conditioned Operant Response," *Journal of Experimental Psychology* **26**, 251–280 (1940).

(a) Integrate to obtain the function $R(t)$. (Let R_0 be the response strength at $t = 0$.)

(b) Find $\lim_{t \to \infty} R(t)$.

14.10 IMPROPER INTEGRALS

In some applications, particularly in probability and statistics, integrals arise whose limits include ∞ or $-\infty$. Examples are

$$\int_{-\infty}^{\infty} e^{-x^2}\, dx \qquad \int_{1}^{\infty} \ln x\, dx \qquad \int_{-\infty}^{2} \frac{1}{2 + x^2}\, dx$$

$$\int_{0}^{\infty} \frac{\sin x}{x}\, dx \qquad \int_{-\infty}^{\infty} xe^{-x^2}\, dx$$

Such integrals are called *improper*. However, this should be considered as merely a name attached to such objects and not as a comment on their logical validity. Mathematically, improper integrals are defined as limits of ordinary integrals, as follows.

Definition of Improper Integrals Let a be a fixed real number. Then by the symbol

$$\int_{a}^{\infty} f(x)\, dx$$

we mean

$$\lim_{b \to \infty} \int_{a}^{b} f(x)\, dx$$

provided this limit exists. Similarly,

$$\int_{-\infty}^{a} f(x)\, dx = \lim_{b \to -\infty} \int_{b}^{a} f(x)\, dx$$

$$\int_{-\infty}^{\infty} f(x)\, dx = \int_{-\infty}^{0} f(x)\, dx + \int_{0}^{\infty} f(x)\, dx$$

$$= \lim_{b \to -\infty} \int_{b}^{0} f(x)\, dx + \lim_{c \to \infty} \int_{0}^{c} f(x)\, dx$$

provided, of course, that the indicated limits exist. When an improper integral exists it is called *convergent*.* ▶

Example 1 (a) Since $D(x \ln x - x) = \ln x$ we have

$$\int_{1}^{\infty} \ln x\, dx = \lim_{b \to \infty} x \ln x - x \Big]_{1}^{b}$$

$$= \lim_{b \to \infty} b[\ln b - 1] + 1 = \infty$$

*Compare this definition with that for convergence of an infinite series (Section 6.2).

(b) The region shaded in Figure 20 has total area equal to $\frac{1}{2}$ since

$$\int_0^\infty xe^{-x^2}\,dx = \lim_{b\to\infty} -\frac{1}{2}\int_0^b D(e^{-x^2})\,dx$$

$$= \lim_{b\to\infty} -\frac{1}{2}e^{-x^2}\Big]_0^b = \lim_{b\to\infty}(-\frac{1}{2}e^{-b^2} + \frac{1}{2}) = \frac{1}{2}$$

FIGURE 20

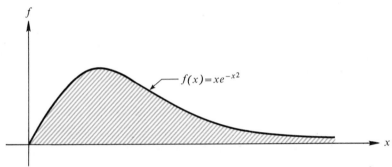

Although the shaded region is infinite in width, the height xe^{-x^2} decreases so rapidly as x increases that the total area is finite.

(c) $\displaystyle\int_0^\infty \sin x\,dx = \lim_{b\to\infty} -\cos x\Big]_0^b$

$$= \lim_{b\to\infty}(1 - \cos b)$$

But $\lim_{b\to\infty} \cos b$ does not exist so the symbol $\displaystyle\int_0^\infty \sin x\,dx$ is meaningless.

$\blacktriangleright$

Example 2 For the improper integral

$$\int_{-\infty}^\infty \frac{1}{x^2 + 2x + 5}\,dx = \int_{-\infty}^0 \frac{1}{x^2 + 2x + 5}\,dx + \int_0^\infty \frac{1}{x^2 + 2x + 5}\,dx$$

we find

$$\int_a^0 \frac{1}{(x+1)^2 + 2^2}\,dx = \frac{1}{2}\tan^{-1}\left(\frac{x+1}{2}\right)\Big]_a^0$$

$$= \frac{1}{2}\left(\tan^{-1}\frac{1}{2} - \tan^{-1}\frac{a+1}{2}\right)$$

and

$$\int_0^b \frac{1}{(x+1)^2 + 2^2}\,dx = \frac{1}{2}\left(\tan^{-1}\frac{b+1}{2} - \tan^{-1}\frac{1}{2}\right)$$

Since

$$\lim_{a\to-\infty}\tan^{-1}\frac{a+1}{2} = -\frac{\pi}{2} \quad\text{and}\quad \lim_{b\to\infty}\tan^{-1}\frac{b+1}{2} = \frac{\pi}{2}$$

the value of the integral is

$$\int_{-\infty}^{\infty} \frac{1}{x^2 + 2x + 5}\, dx = \frac{1}{2}\left[-\left(-\frac{\pi}{2}\right) + \frac{\pi}{2}\right] = \frac{\pi}{2} \qquad \blacktriangleright$$

An integral is also called improper when the integrand is not bounded within the interval of integration. In this case we first restrict the interval and then take limits.

Definition 2 (i) If $|f(x)| \to \infty$ as $x \to a$, then

$$\int_a^b f(x)\, dx = \lim_{h \to 0} \int_{a+h}^b f(x)\, dx \qquad (h > 0)$$

(ii) If $|f(x)| \to \infty$ as $x \to b$, then

$$\int_a^b f(x)\, dx = \lim_{h \to 0} \int_a^{b-h} f(x)\, dx \qquad (h > 0)$$

(iii) If $|f(x)| \to \infty$ as $x \to c$ where $a < c < b$, then

$$\int_a^b f(x)\, dx = \int_a^c f(x)\, dx + \int_c^b f(x)\, dx \qquad \blacktriangleright$$

Example 3 (a) The function $f(x) = 1/\sqrt{x-1}$ becomes infinite as $x \to 1$. Hence, to evaluate the integral

$$\int_1^2 \frac{1}{\sqrt{x-1}}\, dx$$

we write

$$\int_1^2 \frac{1}{\sqrt{x-1}}\, dx = \lim_{h \to 0} \int_{1+h}^2 \frac{1}{\sqrt{x-1}}\, dx \qquad (h > 0)$$

$$= \lim_{h \to 0} 2\sqrt{x-1}\,\Big]_{1+h}^2$$

$$= \lim_{h \to 0} 2(1 - \sqrt{h}) = 2$$

(b) The function $g(x) = 1/(x-1)^2$ becomes infinite as $x \to 1$. Thus

$$\int_1^2 \frac{1}{(x-1)^2}\, dx = \lim_{h \to 0} \int_{1+h}^2 \frac{1}{(x-1)^2}\, dx$$

$$= \lim_{h \to 0} -\frac{1}{x-1}\Big]_{1+h}^2$$

$$= \lim_{h \to 0}\left(-1 + \frac{1}{h}\right)$$

Since this limit does not exist, the integral has no value (diverges). $\qquad \blacktriangleright$

PROBLEMS

In Problems 1–18, evaluate those improper integrals which are convergent.

1. $\displaystyle\int_0^1 \frac{1}{x^3}\, dx$

2. $\displaystyle\int_0^2 \frac{1}{x}\, dx$

3. $\displaystyle\int_0^1 \frac{1}{\sqrt{x}}\, dx$

4. $\displaystyle\int_0^1 \frac{x}{\sqrt{1-x}}\, dx$

5. $\displaystyle\int_0^{\pi/2} \csc\theta\, d\theta$

6. $\displaystyle\int_0^{2a} \frac{1}{(y-a)^2}\, dy$

7. $\displaystyle\int_0^4 \frac{x}{(16-x^2)^{3/2}}\, dx$

8. $\displaystyle\int_{-a}^a \frac{1}{(a^2-y^2)^{1/2}}\, dy$

9. $\displaystyle\int_0^{\pi/4} \tan 2t\, dt$

10. $\displaystyle\int_0^\infty e^{2x}\, dx$

11. $\displaystyle\int_0^\infty e^{-2x}\, dx$

12. $\displaystyle\int_2^\infty \frac{1}{\sqrt{x}}\, dx$

13. $\displaystyle\int_0^\infty \frac{xe^{-x}}{x}\, dx$

14. $\displaystyle\int_0^\infty \frac{y}{\sqrt{y+1}}\, dy$

15. $\displaystyle\int_1^\infty \frac{x}{1+x^2}\, dx$

16. $\displaystyle\int_{-\infty}^0 \frac{1}{s^2+9}\, ds$

17. $\displaystyle\int_{-\infty}^\infty \frac{1}{x^{1/3}}\, dx$

18. $\displaystyle\int_{-\infty}^\infty \frac{1}{x^2+4x+20}\, dx$

19. For what values of k is

$$\int_0^\infty e^{kx}\, dx$$

convergent?

20. For what values of a is

$$\int_1^\infty x^a\, dx$$

convergent?

21. For what values of b is

$$\int_0^1 x^b\, dx$$

convergent?

22. Find the error:

$$\int_{-1}^{1} \frac{1}{x^2}\, dx = -\frac{1}{x}\bigg]_{-1}^{1} = -\frac{1}{1} - \left(-\frac{1}{-1}\right) = -2$$

(The integrand is positive, yet the integral is negative!)

23. Use integration by parts to show that if $a > 0$,

$$\int_{0}^{\infty} x^2 e^{-ax}\, dx = \frac{2}{a^3}$$

24. A law of income states that

$$y(x) = \int_{x}^{\infty} (\alpha - 1)u^{-\alpha}\, du \qquad (x \geq 1)$$

where x denotes income level, $\alpha > 1$ is a constant, and $y(x)$ is the proportion of persons whose incomes exceed x. Find the proportion of persons falling in the income bracket between 3 and 4.

25. In Example 2 of Section 14.7, find

$$\int_{0}^{\infty} E(t)\, dt$$

26. In Example 5 of Section 14.6, find the present value of a constant yearly income A over an infinite period (a perpetuity).

27. In Example 1 of Section 14.7, find the present value of future savings over an infinite period. Interpret this result when $b > ar$.

SUPPLEMENTARY READING

Stein, S. K., *Calculus in the First Three Dimensions* (McGraw-Hill Book Company, New York, 1967), Chapters 1, 6–8, 14, and 21–23.

Thomas, G. B., *Calculus and Analytic Geometry*, 4th ed. (Addison-Wesley Publishing Company, Reading, Massachusetts, 1968), Chapters 5, 6, and 9.

DIFFERENTIAL EQUATIONS **15**

15.1 SEPARATION OF VARIABLES

Equations involving derivatives are called *differential equations*. To *solve* a differential equation is to find all functions which, when substituted into the equation, yield an identity—that is, an expression valid for every value of the variable. The key result for determining all solutions of a differential equation is Theorem 1 of Section 14.3, which we repeat here for convenience. (Compare with Theorem 4 in Section 13.4.)

Theorem 1   If functions F and G have the same derivative, then there is a constant C such that for all x, $F(x) = G(x) + C$.   ▶

Example 1   In attempting to explain how excitatory tendencies E depend on the distance x from a goal, Hull[*] postulates the differential equation

$$DE(x) = -\frac{b}{x}$$

where b is a positive constant. Solving, we find

$$E(x) = \int -\frac{b}{x}\,dx = -b \ln x + C$$

If $E = E_0$ when $x = x_0$, then $E_0 = -b \ln x_0 + C$, or $C = E_0 + b \ln x_0$. Hence the function E is given by

$$E(x) = -b \ln x + E_0 + b \ln x_0 = E_0 + b \ln (x_0/x)$$ ▶

[*]Hull, C. L., "The Goal-Gradient Hypothesis Applied to Some Field-Force Problems in the Behavior of Young Children," *Psychological Review* **45**, 271 (1938).

Example 2 Difference and differential equations are quite closely related and, depending on the approach preferred or the data available, may both be used to describe the same behavioral situation. For instance, Harrod's model of economic growth (Example 5 of Section 13.1) asserts that

$$Y(t + 1) - Y(t) = \frac{c}{g} Y(t)$$

where c and g are positive constants and $Y(t)$ is national income at time t. The term $Y(t + 1) - Y(t)$ can be interpreted as the average change

$$\frac{Y(t + 1) - Y(t)}{1}$$

over a unit time period. Harrod's model assumes this average change to be proportional to $Y(t)$ itself.

If we assume that investment demands are sensitive to instantaneous changes in production (income), and that the instantaneous rate of change of income is proportional to income, then the average change over unit time is replaced by the limit

$$Y'(t) = \lim_{h \to 0} \frac{Y(t + h) - Y(t)}{h}$$

of the average change over an arbitrary time interval of length h, and the basic equation becomes

$$Y'(t) = \frac{c}{g} Y(t)$$

The results of Problem 15 in Section 12.7 imply that $Y(t)$ must have the form $Y(t) = ke^{(c/g)t}$ where the constant k is equal to the initial income $Y(0)$. This compares to the solution $Y(t) = Y(0)[g/(g - c)]^t$ of the original difference equation. (See Section 13.5.) ▶

A technique which makes it possible to solve many differential equations simply by finding antiderivatives is *separation of variables*. A differential equation involving a function $f(x)$ is said to have separable variables if it is possible to write the equation in the form

$$\phi(f(x))f'(x) + g(x) = 0$$

where ϕ and g are real-valued functions. The solution of such an equation is obtained by finding the antiderivative of the left side, that is, by writing

$$\int \phi(f(x))f'(x)\, dx + \int g(x)\, dx = C$$

where C is a constant. This yields an equation involving $f(x)$ and x which may be solved for $f(x)$.

Example 3 To solve the differential equation

$$f'(x) = \frac{xf(x)}{\sqrt{1 - x^2}}$$

we first separate the variables to obtain

$$\frac{f'(x)}{f(x)} = \frac{x}{\sqrt{1 - x^2}}$$

Writing antiderivatives gives

$$\int \frac{f'(x)}{f(x)} \, dx = \int \frac{x}{\sqrt{1 - x^2}} \, dx$$

or, using Lines 5 and 7 of Table 1, Section 14.3,

$$\ln f(x) = -(1 - x^2)^{1/2} + C$$

Solving for $f(x)$ yields

$$f(x) = ke^{-(1-x^2)^{1/2}}$$

where $k = e^C$ is a constant. ▶

Example 4 Let $N(t)$ denote the number of individuals in a society at time t and let $N_i(t)$ be the number of inventors in the society at time t. Rashevsky* proposes

$$r(t) = \frac{N_i(t)}{N(t)}$$

as a measure of the technological development of the society and assumes that

$$r'(t) = [A + BN(t)][kr(t) - Lr^2(t)] \tag{1}$$

where $A, B, k,$ and L are positive constants with $L \geq k$.

Let us suppose an exponential population growth

$$N(t) = N_0 e^{\lambda t}$$

where N_0 is initial population size and λ is a growth constant. If we separate the variables, we can write the differential equation (1) as

$$\frac{r'(t)}{kr(t) - Lr^2(t)} = A + BN_0 e^{\lambda t}$$

Computing the antiderivative of each side yields

$$-\frac{1}{k} \ln \frac{k - Lr(t)}{r(t)} = At + \frac{BN_0}{\lambda} e^{\lambda t} + C \tag{2}$$

*Rashevsky, N., *Mathematical Biology of Social Behavior* (University of Chicago Press, Chicago, Ill., 1951), p. 215.

where C is a constant. Thus the solution for $r(t)$ is

$$r(t) = \frac{k}{L + e^{-k\phi(t)}}$$

where $\phi(t)$ denotes the right side of (2). Letting t become large, we find

$$\lim_{t \to \infty} r(t) = k/L \qquad \blacktriangleright$$

PROBLEMS

Solve the differential equations in Problems 1–10.

1. $f'(t) = 3tf(t)$

2. $\dfrac{dy}{dx} = 2x^2 y$

3. $(xy^2 + x) + (x^2 y - y)\dfrac{dy}{dx} = 0$

4. $f'(x) = \dfrac{f^2(x) + 1}{\sqrt{1 + x^2}}$

5. $\dfrac{dy}{dx} = \dfrac{xy + y}{x + xy}$

6. $x^2 - y\dfrac{dx}{dy} - 9 = 0$

7. $2x\dfrac{dy}{dx} + y = 0$

8. $2xf'(x) + f(x) = f^2(x)$

9. $y^2 - x^4\dfrac{dy}{dx} = 0$

10. $\sin^2 y + \cos^2 x\dfrac{dy}{dx} = 0$

In Problems 11–13, find the particular solution of the differential equation which satisfies the given initial condition.

11. $2xf'(x) + f(x) = 0; \quad f(1) = 2$

12. $4y \cos x + 3 \sin x\dfrac{dy}{dx} = 0; \quad y = 2$ when $x = \tfrac{1}{2}\pi$

13. $\dfrac{ds}{d\theta} = \dfrac{\sin \theta + e^{2s} \sin \theta}{2e^s + e^s \cos 2\theta}; \quad s = 0$ when $\theta = \tfrac{1}{2}\pi$

14. Deese* proposed the differential equation $dR/dN = k(M - R)$, where R denotes response strength, N is the number of reinforcements, M is the fixed maximum value of R, and k is a constant. Show that if $R = 0$ when $N = 0$, the learning curve is $R = M(1 - e^{-kN})$.

15. Show that by making the substitution $y = vx$ the differential equation

$$\frac{dy}{dx} = G\left(\frac{y}{x}\right)$$

becomes an equation with separable variables. (*Hint:* If $y = vx$, then $y' = v + xv'$.)

16. Using the result of Problem 15, solve

(a) $\dfrac{dy}{dx} = \dfrac{y}{x} + \cos^2\dfrac{y}{x}$

(b) $(2x - y) + (4x - 2y)y' = 0$

*Deese, J., *The Psychology of Learning* (McGraw-Hill Book Company, New York, 1952).

15.2 LINEAR DIFFERENTIAL EQUATIONS

The results obtained in Chapter 12 show that the derivative operator D has properties which parallel those listed in Properties 1–5 of Section 13.2 for the difference operator Δ. Specifically, we know that:

(i) Operations of taking derivatives commute with each other and with the operation of multiplying by a constant. That is, for any positive integers k and r,

$$D^k[D^r f(x)] = D^r[D^k f(x)] = D^{k+r} f(x)$$

and, if c is a constant,

$$D^k[cf(x)] = cD^k f(x)$$

(ii) The operation of differentiation distributes over sums of functions. That is,

$$D^k[f_1(x) + f_2(x) + \cdots + f_n(x)]$$
$$= D^k f_1(x) + D^k f_2(x) + \cdots + D^k f_n(x)$$

(iii) If we define the *polynomial derivative operator*

$$\phi_n(D) = c_0 + c_1 D + c_2 D^2 + \cdots + c_n D^n$$

by

$$\phi_n(D)f(x) = c_0 f(x) + c_1 Df(x) + c_2 D^2 f(x) + \cdots + c_n D^n f(x)$$

then

$$\phi_n(D)[af(x)] = a\phi_n(D)[f(x)]$$
$$\phi_n(D)[f(x) + g(x)] = \phi_n(D)[f(x)] + \phi_n(D)[g(x)]$$

If $\psi_m(D) = a_0 + a_1 D + \cdots + a_m D^m$ is another such operator, then

$$[\phi_n(D) + \psi_m(D)]f(x) = \phi_n(D)[f(x)] + \psi_m(D)[f(x)]$$
$$\phi_n(D)[\psi_m(D)f(x)] = \psi_m(D)[\phi_n(D)f(x)]$$
$$= [\phi_n(D) \times \psi_m(D)][f(x)]$$

the symbol $\times$ denoting ordinary multiplication of polynomials.

A differential equation is called *linear* if it has the form

$$\phi(D)f(x) = g(x)$$

where $\phi(D)$ is a polynomial derivative operator, g is a known function, and f is a function for which a solution is sought. It follows from the above discussion that *all techniques employed in solving linear difference equations,* such as factoring the operator polynomial and the method of complementary and particular solutions (see Sections 13.4–13.6), *apply equally well to solving linear differential equations.* The only change lies in the specific form of the solution of the equation

$$(D - a)^k f(x) = 0$$

which arises, as in difference equations, from factoring the operator polynomial and which forms the basis for writing down by inspection the solution to any linear equation.

The following three theorems, which parallel Theorems 6, 7, and 8 for differences (Sections 13.4 and 13.5), provide the necessary tools.

Theorem 2 A function f is a solution of the differential equation

$$D^k f(x) = 0$$

if and only if f is a polynomial of degree $k - 1$ or less; that is, if and only if f has the form

$$f(x) = c_0 + c_1 x + c_2 x^2 + \cdots + c_{k-1} x^{k-1} \tag{3}$$

where $c_0, c_1, \ldots, c_{k-1}$ are constants.

PROOF Line 2 of Table 1, Section 14.3, constitutes a proof for $k = 1$. Assume now that the statement of the theorem is true when $k = r$ and consider the $(r + 1)$st-order equation $D^{r+1} f(x) = 0$. The substitution $g(x) = Df(x)$ reduces the order and gives the equation $D^r g(x) = 0$. The inductive assumption states that $g(x)$ has the form

$$g(x) = a_0 + a_1 x + a_2 x^2 + \cdots + a_{r-1} x^{r-1}$$

where $a_0, a_1, a_2, \ldots, a_{r-1}$ are constants. In terms of f, this means that

$$Df(x) = g(x) = a_0 + a_1 x + \cdots + a_{r-1} x^{r-1}$$

Computing antiderivatives (using Lines 4 and 10 of Table 1, Section 14.3) gives

$$f(x) = a_0 x + \frac{a_1}{2} x^2 + \frac{a_2}{3} x^3 + \cdots + \frac{a_{r-1}}{r} x^r + C$$

which has the form of (3) with $k = r + 1$. ▶

Example 1 Let $Y(t)$ be the number of out-migrants from a community by time t. If $D^3 Y(t) = 0$, Theorem 2 requires that $Y(t)$ have the form $Y(t) = c_0 + c_1 t + c_2 t^2$, where c_0, c_1, and c_2 are constants.

This solution is easily verified by noting that $DY(t) = c_1 + 2c_2 t$, $D^2 Y(t) = 2c_2$, and hence that $D^3 Y(t) = 0$. ▶

Theorem 3 For all positive integers n and k,

$$D^k [e^{-ax} f(x)] = e^{-ax} (D - a)^k [f(x)] \tag{4}$$

PROOF Again we proceed by induction. For $k = 1$, differentiating the product $e^{-ax}f(x)$ gives

$$D[e^{-ax}f(x)] = e^{-ax}Df(x) - af(x)e^{-ax}$$

$$= e^{-ax}(D - a)[f(x)] \qquad (5)$$

If (4) is valid for $k = r$, then

$$D^{r+1}[e^{-ax}f(x)]$$
$$= D\{D^r[e^{-ax}f(x)]\}$$
$$= D[e^{-ax}(D - a)^r f(x)] \qquad \text{[by the inductive assumption]}$$
$$= e^{-ax}(D - a)[(D - a)^r f(x)] \qquad \text{[using (5) with } (D - a)^r f(x)$$
$$\qquad\qquad\qquad\qquad\qquad\qquad \text{in place of } f(x)]$$
$$= e^{-ax}(D - a)^{r+1}[f(x)] \qquad\qquad\qquad\qquad \blacktriangleright$$

Theorem 4 All solutions of the differential equation

$$(D - a)^k f(x) = 0$$

have the form

$$f(x) = (c_0 + c_1 x + \cdots + c_{k-1}x^{k-1})e^{ax}$$

where $c_0, c_1, \ldots, c_{k-1}$ are constants.

PROOF Since e^{-ax} can never be zero, it follows that $(D - a)^k f(x) = 0$ is equivalent to $e^{-ax}(D - a)^k f(x) = 0$. In turn, this latter expression is, according to Theorem 3, the same as $D^k[e^{-ax}f(x)] = 0$. Applying Theorem 2 gives

$$e^{-ax}f(x) = c_0 + c_1 x + \cdots + c_{k-1}x^{k-1}$$

from which the result follows upon multiplying both sides by e^{ax}. $\blacktriangleright$

The various techniques employed in solving linear difference equations may now be adapted to linear differential equations, as in the following examples. The only change is the use of Theorem 4 in place of Theorem 8, Section 13.5, since we now deal with derivatives rather than differences.

Example 2 Individuals in a certain population of size N may, at any time t, be either in state A or in state B. Let $B(t)$ denote the number in state B at time t, and suppose that during a time interval $(t, t + h)$ of length h, a proportion $\alpha h + o(h)$ of these individuals shift to state A. (Here $o(h)$ denotes a quantity such that $\lim_{h \to 0} [o(h)/h] = 0$, and hence which is negligible for small values of h.) Then the number $B(t + h)$ of individuals in state B at time $(t + h)$ is given by

$$B(t + h) = B(t) - B(t)[\alpha h + o(h)]$$

The average rate of change in the interval is

$$\frac{B(t + h) - B(t)}{h} = -B(t)\left[\alpha + \frac{o(h)}{h}\right]$$

Passing to the limit we find

$$B'(t) = \lim_{h \to 0} \frac{B(t + h) - B(t)}{h} = -\alpha B(t)$$

Rewriting this equation as $(D + \alpha)B(t) = 0$ and applying Theorem 4 yields $B(t) = ce^{-\alpha t}$, where the constant c is the initial number $B(0)$ of individuals in state B. Note that $B(t) \to 0$ as $t \to \infty$. ▶

Example 3 The equation $f'''(x) + 3f''(x) = 4f(x)$ may be written in operator form as $(D^3 + 3D^2 - 4)f(x) = 0$ or, in factored form, as

$$(D - 1)(D + 2)^2 f(x) = 0 \tag{6}$$

The separate equations $(D - 1)f(x) = 0$ and $(D + 2)^2 f(x) = 0$ have solutions $c_1 e^x$ and $(c_2 + c_3 x)e^{-2x}$, respectively. By analogy to the argument used for difference equations, this means that the complete solution of (6) is

$$f(x) = c_1 e^x + (c_2 + c_3 x)e^{-2x}$$ ▶

Example 4 To solve the equation

$$(D - 2)(D - 1)f(x) = 6 + 2x \tag{7}$$

we first take two derivatives, annihilating the right side and obtaining

$$D^2(D - 2)(D - 1)f(x) = D^2[6 + 2x] = 0 \tag{8}$$

Every solution of (8) has the form $f(x) = c_0 + c_1 x + c_2 e^{2x} + c_3 e^x$, of which the terms $c_2 e^{2x} + c_3 e^x$ form the complementary solution of (7), while c_0 and c_1 must be determined in order to find the particular solution of the form $c_0 + c_1 x$.

The equation $(D - 2)(D - 1)[c_0 + c_1 x] = 6 + 2x$ requires

$$2c_0 - 3c_1 + 2c_1 x = 6 + 2x$$

Hence $c_0 = \frac{9}{2}$ and $c_1 = 1$, and the complete solution of (7) is

$$f(x) = c_2 e^{2x} + c_3 e^x + \frac{9}{2} + x$$ ▶

PROBLEMS

1. Verify the computations for c_0 and c_1 in Example 4.

Example 5 In attempting to quantify Miller's theory of conflict, Anderson[*] assumes that $dx/dt = G(x) - F(x)$ where, for a goal seeking system, $x(t)$ denotes the distance from the goal at time t, $G(x)$ is the avoidance gradient, and $F(x)$ the approach gradient. It is assumed that the avoidance and approach gradients have the forms

$$G(x) = a(x_0 - x) + b \quad \text{and} \quad F(x) = c(x_0 - x) + d \qquad (9)$$

where the positive constants c and a represent reward and punishment parameters, respectively, $x_0 = x(0)$ is the initial distance from the goal, and b and d are parameters reflecting the experimental situation.

Substituting Equations (9) into the original differential equation gives $dx/dt = (c - a)x + (a - c)x_0 + b - d$ or, in operator notation,

$$[D + (a - c)]x(t) = (a - c)x_0 + (b - d) \qquad (10)$$

Since the right side of (10) is constant, we have $D[D + (a - c)]x(t) = 0$, giving a general solution form $x(t) = c_0 + c_1 e^{-(a-c)t}$. The particular solution c_0, obtained by substituting into (10), is

$$c_0 = x_0 + \frac{b - d}{a - c}$$

This gives

$$x(t) = x_0 + \frac{b - d}{a - c} + c_1 e^{-(a-c)t}$$

from which, since $x(0) = x_0$, we find $c_1 = -(b - d)/(a - c)$. The complete solution is

$$x(t) = x_0 + \frac{b - d}{a - c}[1 - e^{-(a-c)t}]$$

If $a > c$, then $\lim_{t \to \infty} x(t) = x_0 + (b - d)/(a - c)$, while

$$\lim_{t \to \infty} |x(t)| = \infty \quad \text{if } c > a$$

We conclude that the system has an equilibrium point $x_0 + (b - d)/(a - c)$ if and only if $a > c$; that is, if and only if the punishment parameter exceeds the reward parameter.

Since the system is unchanging at equilibrium, the equilibrium state may also be obtained by setting $dx/dt = 0$. This means that $F(x) = G(x)$ or, solving for x in $a(x_0 - x) + b = c(x_0 - x) + d$, that

$$x = x_0 + \frac{b - d}{a - c}$$

as before. Although this formulation obscures the requirement $a > c$, it does make clear that at equilibrium, the positive effect of the approach gradient must exactly balance the negative effect of the avoidance gradient. ▶

[*]Anderson, N. H., "On the Quantification of Miller's Conflict Theory," *Psychological Review* **69**, 400–414 (1962).

Example 6 The operator $(D - a)^k$ serves as an annihilator for $x^{k-1}e^{ax}$ (Theorem 4). To solve the equation $f''(x) + f'(x) - 6f(x) = xe^{2x} + 2x^2$ or, in operator form,

$$(D + 3)(D - 2)f(x) = xe^{2x} + 2x^2 \tag{11}$$

we first apply the operator $(D - 2)^2 D^3$ to both sides to obtain

$$(D + 3)(D - 2)^3 D^3 f(x) = 0$$

Solutions of this latter equation have the form

$$f(x) = c_1 e^{-3x} + (c_2 + c_3 x + c_4 x^2)e^{2x} + c_5 + c_6 x + c_7 x^2$$

The terms $c_1 e^{-3x} + c_2 e^{2x}$ constitute the complementary solution leaving $(c_3 x + c_4 x^2)e^{2x} + c_5 + c_6 x + c_7 x^2$ as the form of the particular solution.

Substituting the particular solution into (11) and equating coefficients of like functions gives $c_3 = -\frac{1}{25}$, $c_4 = \frac{1}{10}$, $c_5 = -\frac{7}{54}$, $c_6 = -\frac{1}{9}$, and $c_7 = -\frac{1}{3}$. Thus the complete solution of (11) is

$$f(x) = c_1 e^{-3x} + c_2 e^{2x} + (-\tfrac{1}{25}x + \tfrac{1}{10}x^2)e^{2x} - \tfrac{7}{54} - \tfrac{1}{9}x - \tfrac{1}{3}x^2 \quad \blacktriangleright$$

PROBLEMS

2. Verify the computations for Example 6.

3. Show that the differential equation $D^2 g(t) - b^2 g(t) = f(t)$ has a complementary solution of the form $c_0 e^{-bt} + c_1 e^{bt}$.

In Problems 4–8, solve the differential equation subject to the given initial conditions.

4. $D^2 y(t) - 4Dy(t) + 4y(t) = 0$; $y(0) = 6$, $y(1) = e^2$

5. $KDx(t) + 6x(t) = t^2$; $x(0) = 1$

6. $D^2 f(t) - 6Df(t) + 5f(t) = 6$; $f(0) = 4$, $Df(0) = 10$

7. $6g''(x) - 36g'(x) + 54g(x) = 12x$; $g(0) = 1$, $Dg(0) = 8$

8. $D^2 h(t) = 3Dh(t) + 12$; $h(0) = 3$, $Dh(0) = 10$

Find particular solutions for the differential equations in Problems 9–11.

9. $D^2 f(t) - 3Df(t) + 2f(t) - t^2 e^{2t} = 0$

10. $D^2 x(t) + 2Dx(t) = 3x(t) + e^t \sin t$ (*Hint:* Assume a particular solution of the form $(c_0 \sin t + c_1 \cos t)e^t$.)

11. $D^2 g(t) + 8Dg(t) + 16g(t) = e^{4t} \cos 2t$ (*Hint:* Assume a particular solution of the form $(c_0 \cos 2t + c_1 \sin 2t)e^{4t}$.)

12. In a model of buying behavior, Longton* assumes that a customer's decision to buy a brand item depends upon the image the customer has

*Longton, P. A., "Mathematics, Models, and Marketing," *Human Relations* **18**, 289–296 (1965).

of the brand. The image is assumed to be composed of a set S of N elements which come to be associated with the brand in an all-or-none manner. (No partial associations are allowed.) An element once associated may also be disassociated from the brand.

Let λh be the proportion of disassociated elements that become associated with the brand during time h, μh be the proportion of associated elements that become disassociated from the brand during time h, and $N_1(t)$ be the number of elements associated with the brand at time t. If $p(t) = N_1(t)/N$ is the proportion of the set of total elements which are associated with the brand at time t,

(a) show that

$$p(t) = \frac{\lambda}{\lambda + \mu} [1 - e^{-(\lambda+\mu)t}] + p(0)e^{-(\lambda+\mu)t}$$

(b) find the equilibrium state and indicate whether it may be obtained by setting the derivative $p'(t)$ equal to zero.

13. In a study of quantal absorption in a visual receptor, Cornsweet† assumes that the net rate of change of regenerated molecules in the receptor is

$$\frac{dN}{dt} = K(M - N) - AQN$$

where K, M, A, and Q are constants. Find $N(t)$ assuming that $N(0) = M$.

14. The demand and supply functions for a certain product are, respectively,

$$D(t) = a + bP(t) + cDP(t) \qquad (a > 0, \ b < 0)$$
$$S(t) = \alpha + \beta P(t) \qquad (\alpha < 0, \ \beta > 0)$$

where $P(t)$ is the price at time t, and a, b, c, α, and β are constants.

(a) Solve the differential equation for $P(t)$ which results if we assume that $D(t) = S(t)$.

(b) Determine relations among the constants which lead to an equilibrium state for the system and determine the equilibrium state which obtains.

15. Modify Problem 14 by assuming that

$$D(t) = a + bP(t) + cDP(t) + dD^2P(t) \qquad (a > 0, \ b < 0)$$
$$S(t) = \alpha + \beta P(t) + \gamma DP(t) + \delta D^2P(t) \qquad (\alpha < 0, \ \beta > 0)$$

Again assuming $D(t) = S(t)$, find the time path for price (that is, solve for $P(t)$) when

(a) $\left(\dfrac{c - \gamma}{d - \delta}\right)^2 = 4\left(\dfrac{b - \beta}{d - \delta}\right)$
(b) $\left(\dfrac{c - \gamma}{d - \delta}\right)^2 > 4\left(\dfrac{b - \beta}{d - \delta}\right)$

In each case, determine necessary and sufficient conditions for a stable equilibrium.

†Cornsweet, T. N., "Changes in the Appearance of Stimuli of Very High Luminance," *Psychological Review* **69**, 257–273 (1962).

16. The Domar burden-of-debt model* assumes that

$$D'(t) = \alpha Y(t) \qquad 0 < \alpha < 1$$

where $D(t)$ denotes the total outstanding public debt and $Y(t)$ the national income at time t. Given that the interest rate i is constant,

$$T(t) = iD(t)$$

where $T(t)$ is the interest induced tax. Income is assumed to grow at a constant relative rate β so that

$$Y'(t) = \beta Y(t) \qquad 0 < \beta < 1$$

The burden of debt $B(t)$ is therefore defined as

$$B(t) = \frac{T(t)}{Y(t)} = i\frac{D(t)}{Y(t)}$$

(a) Find the time path for $B(t)$.

(b) Determine the equilibrium state for the burden of debt.

17. A system consists of two sequential stages. Let $V_1(t)$ and $V_2(t)$ denote the respective outputs at time t of stages 1 and 2. If $DV_1(t) = \alpha V_1(t)$ and $cDV_2(t) + [V_2(t)/R] = \beta V_1(t)$ where c, R, α, and β are constants, find $V_2(t)$.

SUPPLEMENTARY READING

Coddington, E. A., *An Introduction to Ordinary Differential Equations* (Prentice–Hall, Inc., Englewood Cliffs, N. J., 1961).

---

*Cited in Chiang, A. C., *Fundamental Methods of Mathematical Economics* (McGraw-Hill Book Company, New York, 1967), pp. 466–468.

16.1 TAYLOR SERIES

If $b = x$, Taylor's Theorem (Section 12.9) states that if a function f has n derivatives f', f'', f''', $\ldots$, $f^{(n)}$ throughout an interval $[a, x]$, then there exists a point c between a and x at which

$$f(x) = f(a) + f'(a)(x - a) + \frac{f''(a)}{2!}(x - a)^2 + \cdots$$

$$+ \frac{f^{(n-1)}(a)}{(n - 1)!}(x - a)^{n-1} + \frac{f^{(n)}(c)(x - a)^n}{n!} \tag{1}$$

For fixed a, the numbers $f(a)$, $f'(a)$, $\ldots$, $f^{(n-1)}(a)$ are constants, and the Taylor Formula (1) may be thought of as a means of approximating the value of $f(x)$ by calculating the following $(n - 1)$st-degree *Taylor polynomial*:

$$p_{n-1}(x) = f(a) + f'(a)(x - a) + \cdots + \frac{f^{(n-1)}(a)}{(n - 1)!}(x - a)^{n-1} \tag{2}$$

The term

$$\frac{f^{(n)}(c)}{n!}(x - a)^n$$

in (1) indicates the magnitude of the error made in this approximation.

Example 1   Since $e^0 = 1$ and all derivatives of $f(x) = e^x$ are also e^x, the successive polynomial approximations to e^x obtained by using $a = 0$ in

(2) are

$$p_0(x) = e^0 = 1$$
$$p_1(x) = e^0 + e^0 x = 1 + x$$
$$p_2(x) = e^0 + e^0 x + \frac{e^0 x^2}{2!} = 1 + x + \frac{x^2}{2}$$
$$\vdots$$
$$p_n(x) = 1 + x + \frac{x^2}{2!} + \frac{x^3}{3!} + \cdots + \frac{x^n}{n!}$$

Figure 1 shows the function e^x and the Taylor polynomial approximations for $n = 0, 1, 2,$ and 3. Each of the polynomials passes through the functional value $e^0 = 1$ when $x = 0$. For values of x near zero, the higher-degree polynomials provide the better approximations. However, although this situation holds for positive x, it does not hold for all negative x. In fact, the first approximation $p_0(x) = 1$ is the best of those pictured for large negative values of x. ▶

FIGURE 1 Successive Taylor's polynomial approximations to $f(x) = e^x$.

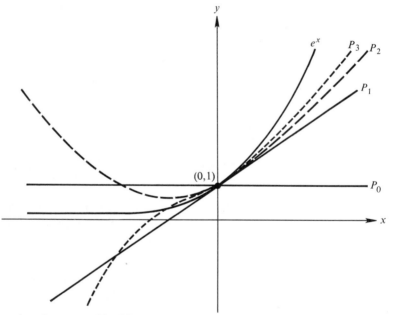

Another way of looking at the Taylor Formula (1) is to interpret the Taylor polynomial (2) as the nth-partial sum of the infinite series

$$f(a) + f'(a)(x - a) + \frac{f''(a)}{2!}(x - a)^2 + \cdots + \frac{f^{(n)}(a)}{n!}(x - a)^n + \cdots$$
$$= \sum_{n=0}^{\infty} \frac{f^{(n)}(a)}{n!}(x - a)^n \ ^* \tag{3}$$

*In this sum $f^{(0)}$ denotes the function f itself.

called the *Taylor series of the function f*, *taken about the point a*. It is apparent from (1) that this series converges to $f(x)$ if and only if the sequence of remainder, or error, terms $[f^{(n)}(c)/n!](x - a)^n$ converges to zero. Luckily, this is the case for the functions of particular interest to us.

Example 1' For the exponential function $f(x) = e^x$, the Taylor series about the point zero (that is, $a = 0$) is

$$e^x = 1 + x + \frac{1}{2!}x^2 + \frac{1}{3!}x^3 + \cdots$$

$$= \sum_{n=0}^{\infty} \frac{1}{n!}x^n \tag{4}$$

The error term for the *n*th approximation is $(1/n!)e^c x^n$, which cannot exceed $(1/n!)e^x x^n$ when x is positive, or $(1/n!)e^0|x|^n = (1/n!)|x|^n$ when x is negative. For each fixed x, the sequence of error terms converges to zero and it follows that (4) does indeed represent e^x for all values of x. ▶

Example 2 For $f(x) = \sin x$, the Taylor series of f about the point zero is determined by computing successive derivatives at zero. Since $\sin 0 = 0$, $\cos 0 = 1$, $D \sin x = \cos x$, and $D \cos x = -\sin x$, we have

$$f(0) = \sin 0 = 0$$

$$f'(0) = \cos 0 = 1$$

$$f''(0) = -\sin 0 = 0$$

$$f'''(0) = -\cos 0 = -1$$

$$\vdots$$

In general, $f^{(n)}(0) = 0$ if n is even, $f^{(n)}(0) = 1$ for $n = 1, 5, 9, 13, \ldots$, and $f^{(n)}(0) = -1$ for $n = 3, 7, 11, 15, \ldots$. Thus the Taylor series for $\sin x$ is

$$\sin x = x - \frac{x^3}{3!} + \frac{x^5}{5!} - \frac{x^7}{7!} + \frac{x^9}{9!} - \cdots$$

$$= \sum_{k=0}^{\infty} (-1)^{k+1} \frac{x^{2k+1}}{(2k+1)!} \tag{5}$$

The fact that the sine function is an odd function (that is, $\sin(-x) = -\sin x$) is reflected in the appearance of only odd powers of x in the series (5).

The error term for the *n*th approximation has the form

$$\frac{f^{(n)}(c)}{n!}x^n$$

where $f^{(n)}(c)$ is either $\pm\sin c$ or $\pm\cos c$ and c is between 0 and x. In either case, $|f^{(n)}(c)| \leq 1$ and the magnitude of the error term cannot exceed

$|x^n/n!|$. Since $(|x^n|/n!)$ converges to zero as $n \to \infty$, it follows that for every real number x, the Taylor series (5) represents $\sin x$.

Using standard trigonometric identities, values of $\sin x$ can be determined for any x if we know values for x between $-\pi/2$ and $\pi/2$. In this range, the series (5) provides a good approximation with relatively few terms. For instance, using the first two nonzero terms of (5) with $x = \pi/6$ gives $\sin \pi/6$ approximately as

$$\sin \frac{\pi}{6} \approx \frac{\pi}{6} - \frac{1}{3!}\left(\frac{\pi}{6}\right)^3$$

Since $\pi/6 \approx 3.1416/6 = 0.5236$, our approximation is

$$\sin \frac{\pi}{6} \approx (0.5236) - \frac{(0.5236)^3}{6}$$

$$= 0.5236 - 0.0239$$

$$= 0.4997$$

to four decimal places.

Since we know $\sin \pi/6 = \frac{1}{2}$, the actual error in this approximation is easily computed as $\frac{1}{2} - 0.4997 = 0.0003$. However, in a real application of the Taylor series, the true value would be unknown and it would be necessary to estimate the magnitude of the error. In the present case, since (5) is an alternating series, the next term $(1/5!)(\pi/6)^5 = 0.0003$ provides an upper bound on the possible error and shows that our estimate 0.4997 is correct to at least three decimal places. ▶

PROBLEMS

1. Use the first three terms of the series (5) to estimate $\sin 75°$. Use the fourth term to provide an upper bound on the error.

2. Same as Problem 1 for $\sin(-2\pi/5) = \sin(-72°)$.

3. Same as Problem 1 for $\sin(21\pi/5)$. (*Hint:* Compute in terms of an angle between $-\pi/2$ and $\pi/2$.)

4. Same as Problem 1 for $\sin(-12\pi/7)$.

Example 3 By an argument similar to that of Example 2, we find

$$\cos x = 1 - \frac{x^2}{2!} + \frac{x^4}{4!} - \frac{x^6}{6!} + \cdots$$

$$= \sum_{k=0}^{\infty} (-1)^k \frac{x^{2k}}{(2k)!} \tag{6}$$

We leave the verification of this formula, including the fact that the sequence of error terms converges to zero, as an exercise (Problem 29). ▶

Taylor series for the other trigonometric functions may be determined in a similar fashion. However, they are somewhat messy, and it is usually easier to proceed by computing appropriate values of the sine and cosine functions from (5) and (6) and then determining the other desired values from these values.

Example 4 A Taylor series about the point zero is called a *McLaurin series*.* We cannot write a McLaurin series for the natural logarithm function since ln 0 does not exist. In this case the most convenient point about which to write a Taylor series is $a = 1$, since ln 1 = 0.

The successive derivatives of ln x are

$$D \ln x = \frac{1}{x} = x^{-1}$$

$$D^2 \ln x = Dx^{-1} = -x^{-2}$$

$$D^3 \ln x = D(-x^{-2}) = 2x^{-3}$$

$$\vdots$$

Hence the Taylor series for the logarithm function, taken about $x = 1$, is

$$\ln x = (x - 1) - \frac{(x-1)^2}{2} + \frac{(x-1)^3}{3} - \frac{(x-1)^4}{4} + \cdots$$

$$= \sum_{k=1}^{\infty} (-1)^{k-1} \frac{(x-1)^k}{k} \qquad (7)$$

We leave it to the reader to show that this series converges only for $0 < x \leq 2$.

A more commonly used form of (7), obtained by substituting $1 + y$ for x, is

$$\ln (1 + y) = y - \frac{y^2}{2} + \frac{y^3}{3} - \frac{y^4}{4} + \cdots \qquad (8)$$

which converges for $-1 < y \leq 1$. Replacing y by $-y$ in (8) gives

$$\ln (1 - y) = -y - \frac{y^2}{2} - \frac{y^3}{3} - \frac{y^4}{4} - \cdots \qquad (9)$$

again converging for $-1 \leq y < 1$. Combining (8) and (9) yields

$$\ln \left(\frac{1+y}{1-y} \right) = \ln (1 + y) - \ln (1 - y)$$

$$= 2 \left(y + \frac{y^3}{3} + \frac{y^5}{5} + \cdots \right)$$

which converges for $-1 < y < 1$.   ▶

*After Colin McLaurin (1698–1746), a student of Sir Isaac Newton.

Example 5 *Binomial Expansions* In Section 7.3 we found that if n is a positive integer, then

$$(1 + t)^n = \sum_{k=0}^{n} \binom{n}{k} t^k \tag{10}$$

The binomial coefficient

$$\binom{n}{k}$$

is defined by

$$\binom{n}{k} = \frac{(n)_k}{k!} = \frac{n(n-1)(n-2)\cdots(n-k+1)}{k!}$$

and has a counting interpretation as the number of different subsets (samples) of size k which may be chosen from a set of n objects.

When n is not a positive integer, no such counting interpretation is possible and our original proof of Formula (10) is no longer valid. However, a McLaurin series for $f(t) = (1 + t)^n$ is easily obtained regardless of the value of n. We have

$$f(0) = (1 + t)^n \,|_{t=0} = (1 + 0)^n = 1^n = 1$$
$$f'(0) = n(1 + t)^{n-1} \,|_{t=0} = n$$
$$f''(0) = n(n-1)(1 + t)^{n-2} \,|_{t=0} = n(n-1)$$

and, by induction, for $k = 0, 1, 2, \ldots,$

$$f^{(k)}(0) = n(n-1)\cdots(n-k+1) = (n)_k$$

Note that this holds *whether or not n is a positive integer*. Thus for any value whatsoever of n, we find, using Equation (3), that

$$(1 + t)^n = 1 + nt + \frac{n(n-1)}{2!}t^2 + \frac{n(n-1)(n-2)}{3!}t^3 + \cdots$$
$$= \sum_{k=0}^{\infty} \binom{n}{k} t^k \tag{11}$$

In particular, since

$$\binom{n}{k} = 0$$

when n is an integer and $k > n$, Formula (10) is merely a special case of (11).

We shall make great use of (11) in later sections. Here we consider only one particular case. When $n = -1$ and $t = -x$ in (11), we obtain

$$\frac{1}{1 - x} = (1 - x)^{-1} = \sum_{k=0}^{\infty} \binom{-1}{k} (-x)^k$$

But

$$\binom{-1}{k} = \frac{(-1)_k}{k!} = \frac{(-1)(-2)(-3)\cdots(-k)}{k!} = \frac{(-1)^k k!}{k!} = (-1)^k$$

so

$$\binom{-1}{k}(-x)^k = x^k$$

and we obtain the familiar geometric series

$$\frac{1}{1-x} = \sum_{k=0}^{\infty} x^k \qquad \blacktriangleright$$

Example 6 Taylor's Theorem has been used by Sidman* to show that if in a group of N persons, individual learning curves have the form

$$y_i(t) = 1 - e^{-k_i t} \qquad (i = 1, 2, \ldots, N)$$

then the average

$$\bar{y}(t) = \frac{1}{N} \sum_{i=1}^{N} y_i(t)$$

cannot possibly have the same form $\bar{y}(t) = 1 - e^{-kt}$, unless all the constants k_i are equal (to k). Since the Taylor series for $y_i(t)$ is

$$y_i(t) = 1 - e^{-k_i t} = k_i t - \frac{(k_i t)^2}{2!} + \frac{(k_i t)^3}{3!} - \cdots$$

we must have

$$\bar{y}(t) = \frac{1}{N} \sum_{i=1}^{N} y_i(t) = \frac{1}{N} \sum_{i=1}^{N} k_i t - \frac{1}{2! N} \sum_{i=1}^{N} k_i^2 t^2 + \frac{1}{3! N} \sum_{i=1}^{N} k_i^3 t^3 - \cdots$$

On the other hand, if $\bar{y}(t) = 1 - e^{-kt}$, then

$$\bar{y}(t) = kt - \frac{1}{2!} k^2 t^2 + \frac{1}{3!} k^3 t^3 - \cdots$$

Equating coefficients of the first two powers of t gives

$$k = \frac{1}{N} \sum_{i=1}^{N} k_i \quad \text{and} \quad k^2 = \frac{1}{N} \sum_{i=1}^{N} k_i^2$$

Together, these equations require

$$\frac{1}{N^2} \left(\sum_{i=1}^{N} k_i \right)^2 = \frac{1}{N} \sum_{i=1}^{N} k_i^2 \qquad \textbf{(12)}$$

But it is easily verified that

$$\frac{1}{N} \sum_{i=1}^{N} (k_i - \bar{k})^2 = \frac{1}{N} \sum_{i=1}^{N} k_i^2 - \frac{1}{N^2} \left(\sum_{i=1}^{N} k_i \right)^2$$

where $\bar{k} = (1/N) \sum_{i=1}^{N} k_i$. Hence, (12) holds if and only if $k_i = \bar{k}$ for all i. (See Problem 14(b), Section 6.1.) $\qquad \blacktriangleright$

*Sidman, M., "A Note on Functional Relations Obtained from Group Data," *Psychological Bulletin* **49**, 263–269 (1952).

PROBLEMS

In Problems 5–14 find the Taylor series of the indicated function in powers of $x - a$ where a is as indicated.

5. $6 + 3x + 4x^2 + 2x^3 + x^4;\quad a = 0$

6. $6 + 3x + 4x^2 + 2x^3 + x^4;\quad a = 1$

7. $2^x;\quad a = 0$

8. $\cos x;\quad a = \pi/3$

9. $\dfrac{1}{x};\quad a = 2$

10. $\sqrt{x};\quad a = 4$

11. $\dfrac{\sin x}{x};\quad a = 0$ (Expand $\sin x$ and then divide by x.)

12. $xe^{-2x};\quad a = 0$

13. $\dfrac{e^x - e^{-x}}{2};\quad a = 0$

14. $e^{-x^2};\quad a = 0$

In Problems 15–20, write the nth-degree Taylor polynomial around $a = 0$ for the given function and value of n and indicate the form of the remainder (error) term.

15. $e^{-x};\quad n = 4$

16. $\ln(1 - x^2);\quad n = 4$

17. $\sin^{-1} x;\quad n = 4$

18. $\tan^{-1} x;\quad n = 5$

19. $\sqrt{1 + x};\quad n = 5$

20. $1/(1 + e^x);\quad n = 4$

21. In Problem 19, use the polynomial you obtained to estimate $\sqrt{1.05}$. Give an upper bound for the error in your estimate.

22. Same as Problem 21 for $\sqrt{0.90}$.

23. Same as Problem 21 for $\sin^{-1} \frac{1}{2}$. (Use Problem 17.) Compare your results with the true value $\pi/6 = \sin^{-1} \frac{1}{2}$.

In Problems 24–27, use the Taylor Formula (1) to find the indicated quantity correct to three decimal places.

24. $\ln 1.05$ (Choose $a = 1$.)

25. $e^{2.2}$ (Assume $e^2 = 7.3891$ is known. What problems do you encounter if you take $a = 0$?)

26. $\sin 92°$

27. $\cos 48°$

28. Define a function f by $f(0) = 0$ and $f(x) = e^{-1/x^2}$ when $x \neq 0$. Show that the McLaurin series for f has only zero coefficients. Hence this series does not represent the function except when $x = 0$.

29. Complete the details of Example 3.

30. Complete the details of Example 4.

16.2 DIFFERENTIATION AND INTEGRATION OF POWER SERIES

A series of the form

$$C(x) = c_0 + c_1 x + c_2 x^2 + c_3 x^3 + \cdots = \sum_{n=0}^{\infty} c_n x^n \qquad \textbf{(13)}$$

where $c_0, c_1, c_2, \ldots$ are constants, is called a *power series*. Every Taylor series about zero, whether or not it converges to the values of the function from which it was generated, is such a power series.

The series (13) converges at $x = 0$, since $C(0) = c_0$. It may also converge for values of x other than zero.

Example 1 (a) Applying the ratio test (Section 6.4) to the series

$$x + \frac{x^2}{2} + \frac{x^3}{3} + \frac{x^4}{4} + \cdots \qquad \textbf{(14)}$$

leads to

$$\lim_{n \to \infty} \left| \frac{x^{n+1}/(n+1)}{x^n/n} \right| = \lim_{n \to \infty} \left| \frac{n}{n+1} x \right| = |x|$$

Thus this series converges absolutely if $|x| < 1$ and diverges if $|x| > 1$. When $|x| = 1$, the ratio test fails and we must investigate each case separately. For $x = -1$ the series is a convergent alternating series

$$-1 + \tfrac{1}{2} - \tfrac{1}{3} + \tfrac{1}{4} - \cdots$$

while for $x = 1$, the series is

$$1 + \tfrac{1}{2} + \tfrac{1}{3} + \tfrac{1}{4} + \cdots$$

the divergent harmonic series. Thus the series (14) converges when $-1 \le x < 1$ and diverges outside this interval.

(b) The Taylor series for e^x, $\sin x$, and $\cos x$ (Examples 1–3 of Section 16.1) converge for all real x.

(c) The ratio test applied to the series

$$1 + x + 2!x^2 + 3!x^3 + \cdots$$

leads to the limit

$$\lim_{n \to \infty} \left| \frac{(n+1)! x^{n+1}}{n! x^n} \right| = \lim_{n \to \infty} (n+1)|x|$$

which exists only if $x = 0$. ▶

In each part of Example 1, the set of real numbers x for which the series converges is an *interval*, either $[-1, 1)$, the entire real line, or the set containing only the point zero. This pattern of convergence is typical of all power series.

To investigate the details of convergence, we first define the *radius of convergence* of a power series $C(x) = c_0 + c_1 x + c_2 x^2 + \cdots$ to be the supremum of the absolute values of those x for which the series converges. Thus, in Example 1 the radii of convergence for the series in parts (a), (b), and (c) are, respectively, 1, ∞, and 0. The justification for the term radius of convergence is contained in the following theorem.

Theorem 1   Let R denote the radius of convergence of the series

$$C(x) = c_0 + c_1 x + c_2 x^2 + \cdots$$

$$= \sum_{n=0}^{\infty} c_n x^n$$

Then $C(x)$ converges absolutely for all x such that $|x| < R$ and diverges if $|x| > R$ (Figure 2).

FIGURE 2

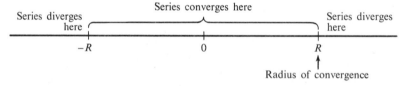

Series converges here

Series diverges here   Series diverges here

$-R$ 0 R

Radius of convergence

PROOF   The fact that $C(x)$ diverges if $|x| > R$ is part of the definition of radius of convergence. Moreover, if $R = 0$, there is nothing to prove. Thus, suppose $R > 0$ and choose a number y such that $0 < y < R$. Since by definition R is the least upper bound of the absolute values of those x for which the series converges, there exists a number $z > y$ such that either $C(z)$ or $C(-z)$ converges. Whichever of these series converges, the terms of that series form a bounded sequence. That is, there exists a positive number M such that $|c_n z^n| < M$ for $n = 0, 1, 2, 3, \ldots$.

For $|x| < y$, we have

$$|c_n x^n| \leq |c_n z^n| \left|\frac{x}{z}\right|^n \leq M \left|\frac{x}{z}\right|^n$$

Hence

$$\sum_{n=0}^{\infty} |c_n x^n| \leq M \sum_{n=0}^{\infty} \left|\frac{x}{z}\right|^n$$

Since $|x/z| < 1$, this latter series is a convergent geometric series. This completes the proof.   ▶

We see that for every power series, there is an interval inside of which the series converges and outside of which the series diverges. As suggested by Example 1, the ratio test constitutes the most useful tool for determining the interval of convergence of a power series.

The particular importance of the interval of convergence lies in the fact that within this interval the usual operations that one may perform on polynomials, such as term-by-term integration and differentiation, apply equally well to power series. The next three theorems state the key results. We shall omit the proofs.

Theorem 2 Inside its interval of convergence, a power series is continuous. ▶

Theorem 3 Inside its interval of convergence, a power series may be integrated term by term and the resulting series is the integral of the original series. Except possibly for the endpoints, the integral series has the same interval of convergence as the original series. ▶

Theorem 4 Inside its interval of convergence, a power series may be differentiated term by term and the resulting series is the derivative of the original series. Except possibly for the endpoints, the derivative series has the same interval of convergence as the original series. ▶

Example 2 By Taylor's theorem,

$$\cos x = 1 - \frac{x^2}{2!} + \frac{x^4}{4!} - \frac{x^6}{6!} + \cdots + \frac{(-1)^n x^{2n}}{(2n)!} + \cdots$$

Term-by-term differentiation of the series yields

$$-\sin x = -x + \frac{x^3}{3!} - \frac{x^5}{5!} + \cdots$$

$$= -\left(x - \frac{x^3}{3!} + \frac{x^5}{5!} - \cdots\right)$$

which agrees with the results obtained in Example 2 of Section 16.1. ▶

Example 3 The definite integral

$$\int_0^x \frac{\sin \theta}{\theta}\, d\theta$$

may be evaluated by the use of series as follows: From the McLaurin expansion of $\sin \theta$, we have

$$\frac{\sin \theta}{\theta} = 1 - \frac{\theta^2}{3!} + \frac{\theta^4}{5!} - \cdots + (-1)^{n+1} \frac{\theta^{2(n-1)}}{(2n-1)!} + \cdots$$

Integrating this series term by term, we obtain

$$\int_0^x \frac{\sin\theta}{\theta}\,d\theta = \left[\theta - \frac{\theta^3}{3\cdot 3!} + \frac{\theta^5}{5\cdot 5!} - \frac{\theta^7}{7\cdot 7!} + \cdots\right]_0^x$$

$$= x - \frac{x^3}{3\cdot 3!} + \frac{x^5}{5\cdot 5!} - \frac{x^7}{7\cdot 7!} + \cdots$$

▶

Example 4 Consider a population which changes only through birth and death.* Let $\beta(t)$ be the number of births at time t, $p(a)$ denote the proportion which survive to age a, and $N(t)$ be the population size at time t. Then the number of persons of age a at time t is $\beta(t - a)p(a)$ and it follows that

$$N(t) = \int_0^\infty \beta(t - a)p(a)\,da \tag{15}$$

We wish to be able to express $\beta(t)$ in terms of $N(t)$. To do this, suppose that

$$\beta(t) = c_0 N(t) - c_1 N'(t) + \frac{c_2}{2!}N''(t) + \cdots$$

$$+ (-1)^k \frac{c_k}{k!} N^{(k)}(t) + \cdots \tag{16}$$

where the constants $c_0, c_1, c_2, \ldots$ are to be determined.

Expanding $\beta(t - a)$ in a Taylor series about t gives

$$\beta(t - a) = \beta(t) - a\beta'(t) + \frac{a^2}{2!}\beta''(t) - \cdots + \frac{(-a)^k}{k!}\beta^{(k)}(t) + \cdots \tag{17}$$

If we differentiate the series (16) term by term and insert the results in (17) we find, after some simplification,

$$\beta(t - a) = c_0 N(t) - (c_1 + ac_0)N'(t) + \left(\frac{c_2}{2} + ac_1 + \frac{a^2 c_0}{2}\right)N''(t)$$

$$- \left(\frac{c_3}{6} + \frac{ac_2}{2} + \frac{a^2 c_1}{2} + \frac{a^3 c_0}{6}\right)N'''(t) + \cdots$$

This series may be inserted into (15) and the series integrated term by term to yield

$$N(t) = c_0 L_0 N(t) - (c_1 L_0 + c_0 L_1)N'(t)$$

$$+ (c_2 L_0 + 2c_1 L_1 + c_0 L_2)N''(t) - \cdots \tag{18}$$

where

$$L_k = \int_0^\infty a^k p(a)\,da$$

*Adapted from Keyfitz, N., *Introduction to the Mathematics of Population* (Addison–Wesley Publishing Company, Reading, Mass., 1968), Chapter 9.

To make (18) an identity, we equate coefficients of $N(t)$ and its derivatives on both sides, giving

$$c_0 = \frac{1}{L_0}$$

$$c_1 = -\frac{L_1}{L_0^2}$$

$$c_2 = -\frac{L_2}{L_0^2} + \frac{2L_1^2}{L_0^3}$$

$$\vdots$$

Substituting back into (16), we obtain

$$\beta(t) = \frac{N(t)}{L_0} + \frac{L_1}{L_0^2} N'(t) + \frac{1}{2}\left(2\frac{L_1^2}{L_0^3} - \frac{L_2}{L_0^2}\right) N''(t) + \cdots \qquad \blacktriangleright$$

PROBLEMS

Write the McLaurin expansion for each of the following functions. Indicate the radius of convergence of the series obtained.

1. $(a + x)^n$ **2.** $(1 - x)^n$ **3.** $(1 + x)^{1/2}$

4. $(1 + x)^{1/3}$ **5.** $(1 + x^2)^{1/2}$ **6.** $(1 - x)^{1/2}$

7. $(1 + x)^{3/2}$ **8.** $(1 - x^2)^{1/2}$ **9.** $(1 - x)^{1/3}$

10. $(1 - x)^{3/2}$

Use series to approximate to four decimal places the values of the following definite integrals.

11. $\displaystyle\int_0^1 \frac{e^x - 1}{x} \, dx$ **12.** $\displaystyle\int_0^1 \frac{\sin x}{\sqrt{x}} \, dx$

13. $\displaystyle\int_0^{1/2} e^{-x^2} \, dx$ **14.** $\displaystyle\int_0^{0.2} \ln\frac{1 + x}{x} \, dx$

15. $\displaystyle\int_0^1 (\cos x)^2 \, dx$ (*Hint:* Use the identity $(\cos x)^2 = \frac{1}{2}(1 - \cos 2x)$.)

Use the results of Problems 1–10 and Theorem 4 to find the series expansions of the following functions.

16. $(1 + x)^{-1/2}$ **17.** $(1 - x)^{-2}$

18. $(1 + x)^{-3/2}$ **19.** $(1 + x^2)^{-1/2}$

20. Use

$$\tan^{-1} x = \int_0^x \frac{1}{1 + t^2} \, dt$$

to derive a power series expansion for $\tan^{-1} x$.

21. Show that the series $A(x) = \sum_{n=0}^{\infty} a_n x^n$, $A'(x) = \sum_{n=0}^{\infty} n a_n x^{n-1}$, and

$$\int A(x)\, dx = \sum_{n=0}^{\infty} \frac{a_n x^{n+1}}{n+1} + C$$

all have the same radius of convergence. (*Hint:* Use the ratio test.)

22. Suppose that the function f can be represented by the power series expansion

$$f(x) = \sum_{k=0}^{\infty} a_k x^k$$

and that this series has radius of convergence $R > 0$. Prove that for $k = 0, 1, 2, \ldots$, the coefficient a_k is equal to $(1/k!) f^{(k)}(0)$. (*Hint:* Successively differentiate the series term by term, setting $x = 0$ after each differentiation.) This proves that McLaurin series representations are unique or, what is the same thing, that coefficients of the same powers of x in two different power series representations of a function must be equal.

23. In Example 4, let $t = \theta - L_1/L_0$ and expand $\beta(t)$ around the point θ.

SUPPLEMENTARY READING

Hirchman, I. I., *Infinite Series* (Holt, Rinehart and Winston, Inc., New York, 1962).

Rainville, E. D., *Infinite Series* (Macmillan Company, New York, 1967).

FUNCTIONS OF SEVERAL VARIABLES **17**

17.1 PARTIAL DERIVATIVES

In our discussion of calculus so far, we have only considered functions of a single variable, that is, functions whose domain is a set of real numbers. However, functions of n variables, which map n-dimensional vectors into numbers, are frequently encountered in applications.

Example 1 (a) The grade a student receives in a course in which he scores S_1, S_2, S_3, and S_4 on four quizzes is determined by his average score $A = \frac{1}{4}(S_1 + S_2 + S_3 + S_4)$. Here we have a mapping which transforms four-dimensional vectors (S_1, S_2, S_3, S_4) into a single real number A.

(b) The *length* of a three-dimensional vector (x, y, z) is (see Section 8.2)

$$L(x, y, z) = \sqrt{x^2 + y^2 + z^2}$$

Here L is a function of three variables. ▶

The graph of a function of one variable is two dimensional. (See, for example, Figures 12 and 13 in Section 12.3.) A three-dimensional graphical representation is necessary for a function of two variables.

Example 2 (a) The graph of the function

$$z = f(x, y) = x^2 + y^2$$

is shown three-dimensionally in Figure 1. The graph is a bullet-shaped surface whose height above a point $p = (x, y)$ in the bottom plane is the value of the function at that point. As an aid to visualizing the graph, note that at points (x, y) where f takes the constant value $c^2 > 0$ we must have

$$x^2 + y^2 = c^2 \tag{1}$$

FIGURE 1

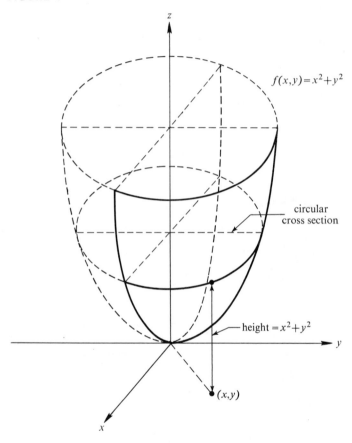

The points which satisfy (1) form a circle with radius c parallel to the bottom plane. It follows that this circle represents the cross section of the surface which lies at a height c^2 above the x, y plane. (See Figure 1.)

   (b) The graph of $f(x, y) = \frac{1}{3}\sqrt{36 - 9x^2 - 4y^2}$ is shown in Figure 2. The domain of this function is the set of all points (x, y) for which $9x^2 + 4y^2 \leq 36$. Its range is the interval $[0, 2]$. The shape is that of an inverted bowl. The height above (x, y) represents the functional value $\frac{1}{3}\sqrt{36 - 9x^2 - 4y^2}$. ▶

   A function f whose domain is a set of n-dimensional vectors and whose range is a set of real numbers is said to be a *real-valued function of n variables*.

FIGURE 2

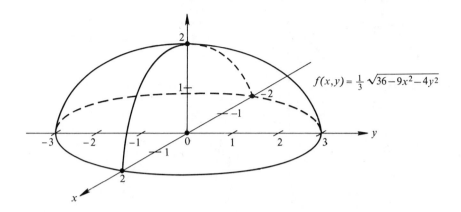

$$f(x,y) = \tfrac{1}{3}\sqrt{36 - 9x^2 - 4y^2}$$

Such a function associates one real number with a collection of n numbers. Each of these n numbers may be altered either in conjunction with, or independently of, all the others. Indeed, a classical experimental method studies the change in the value of f which occurs when one of the n variables is changed while the others are held fixed.

Example 3   That part of the graph of the function $f(x, t) = 4 - x^2 - t^2$ which falls in the first octant (that is, the region where x, t, and f are all positive) is shown in Figure 3. The value $f(x, t)$ depends on both quantities x and t. However, if we fix the value of t, say at $t = 1$, then

$$f(x, t) = 4 - x^2 - t^2$$

becomes $g(x) = f(x, 1) = 3 - x^2$, a function of x alone. It then makes sense to speak of the derivative of g, $Dg(x) = Df(x, 1) = -2x$, the second derivative $D^2g(x) = -2$, etc., and these derivatives have the same interpretations as any other derivatives. For instance, referring to Figure 3, the slope of the curve

$$t = 1 \quad \text{and} \quad g(x) = f(x, 1) = 3 - x^2$$

at the point $(x, t, f(x, t)) = (1, 1, 2)$ is

$$Dg(1) = -2(1) = -2$$

Note that this may also be interpreted as the slope *in the x direction* at the point $(1, 1, 2)$ on the surface of $f(x, t) = 4 - t^2 - x^2$. For this reason, this derivative is sometimes referred to as a *directional derivative*.   ▶

FIGURE 3

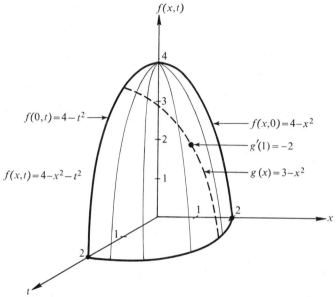

$f(x,t)$

$f(0,t)=4-t^2$

$f(x,t)=4-x^2-t^2$

$f(x,0)=4-x^2$

$g'(1)=-2$

$g(x)=3-x^2$

If f is a function mapping vectors $x = (x_1, x_2, \ldots, x_n)$ to numbers $f(x) = f(x_1, x_2, \ldots, x_n)$, then the derivative of f at x which is obtained by varying x_i ($1 \leq i \leq n$) while holding the variables $x_1, x_2, \ldots, x_{i-1}, x_{i+1}, \ldots, x_n$ fixed is called *the partial derivative of f with respect to x_i*. In symbols, if we let $e_i = (0, 0, \ldots, 0, 1, 0, \ldots, 0)$ be the vector with 1 in the *i*th position and zeros elsewhere, then the partial derivative of f with respect to x_i at the point $x = (x_1, x_2, \ldots, x_n)$ is defined to be the limit

$$\lim_{h \to 0} \frac{f(x + he_i) - f(x)}{h} \tag{2}$$

which is denoted by

$$\frac{\partial f(x)}{\partial x_i}$$

Obviously, by considering all the x_k except x_i to be held fixed (that is, to be constant), the partial derivative may be computed using the same formulas as for ordinary derivatives.

Example 4 Let $f(x) = f(x_1, x_2, x_3, x_4) = x_1^2 + 2x_2^2 x_3 + x_4^3$. Then

$$\begin{aligned}
\frac{\partial f(x)}{\partial x_2} &= \lim_{h \to 0} \frac{f(x + he_2) - f(x)}{h} \\
&= \lim_{h \to 0} \frac{f(x_1, x_2 + h, x_3, x_4) - f(x_1, x_2, x_3, x_4)}{h} \\
&= \lim_{h \to 0} \frac{2(x_2 + h)^2 x_3 - 2x_2^2 x_3}{h} \\
&= \lim_{h \to 0} \frac{4hx_2 x_3 + 2h^2 x_3}{h} = 4x_2 x_3
\end{aligned}$$

Note that the same result would be obtained from the original expression for f by considering x_1, x_3, and x_4 as constants and applying the differentiation formula $Dx_2^n = nx_2^{n-1}$. ▶

Example 5   (a) The partial derivative with respect to y of

$$f(x, y) = x^2 - 3xy + \ln(x^2 + y^2)$$

is

$$\frac{\partial f(x, y)}{\partial y} = -3x + \frac{2y}{x^2 + y^2}$$

(b) The partial derivative of $g(x, y, z, w) = (x^2 + y^2 + z^2 + w^2)^{1/2}$ with respect to w is

$$\frac{\partial g(x, y, z, w)}{\partial w} = \tfrac{1}{2}(x^2 + y^2 + z^2 + w^2)^{-1/2} \cdot 2w$$

(c) Let $x = (r, s, t, u)$ and suppose that

$$f(r, s, t, u) = re^s - \sin\frac{r}{t} + s^3 u^2$$

Then

$$\frac{\partial f(x)}{\partial r} = e^s - \frac{1}{t}\cos\frac{r}{t} \qquad \frac{\partial f(x)}{\partial s} = re^s + 3s^2 u^2$$

$$\frac{\partial f(x)}{\partial t} = \frac{r}{t^2}\cos\frac{r}{t} \qquad \frac{\partial f(x)}{\partial u} = 2s^3 u$$

▶

Example 6   The equations

$$I(t) = a_1 F(t) + a_2 A(t)$$

$$F'(t) = b[I(t) - \beta F(t)]$$

were introduced in Example 5 of Section 12.1 as representing the relations in group behavior among interaction I, friendliness F, and activity A. The assumption that an increase in either A or F will bring about an increase in I is expressed mathematically as

$$a_1 = \frac{\partial I}{\partial F} > 0 \quad \text{and} \quad a_2 = \frac{\partial I}{\partial A} > 0$$

The assumption that the rate of change of friendliness increases with an increase in I may be written

$$b = \frac{\partial F'}{\partial I} > 0$$

▶

Since partial derivatives may be computed just as ordinary derivatives, it follows that all theorems concerning derivatives apply equally well to them. In particular, there is a mean value theorem for partial derivatives. For simplicity, we state the theorem for functions of two variables. The generalization to any number of variables should be apparent.

Theorem 1 *Mean Value Theorem for Partial Derivatives* Let $f(x, y)$ be a function of two variables having partial derivatives

$$\frac{\partial f(x, y)}{\partial x} \quad \text{and} \quad \frac{\partial f(x, y)}{\partial y}$$

which exist throughout its domain. Then if the line joining points (a, y) and (b, y) lies entirely in the domain of f, there exists a number c between a and b at which

$$f(b, y) - f(a, y) = \frac{\partial f(c, y)}{\partial x}(b - a)$$

Similarly, if the line joining (x, a) and (x, b) lies in the domain of f, there exists a number d between a and b for which

$$f(x, b) - f(x, a) = \frac{\partial f(x, d)}{\partial y}(b - a)$$

Note that in the first instance y is held fixed, while in the second, x is fixed. ▶

Partial derivatives of higher order may be computed by applying standard differentiation formulas to partial derivatives already obtained.

Example 7 For the function $f(x) = f(r, s, t, u) = re^s - \sin(r/t) + s^3 u^2$ appearing in Example 5(c) we found

$$\frac{\partial f(x)}{\partial r} = e^s - \frac{1}{t} \cos \frac{r}{t}$$

as the partial derivative of f with respect to r. The *second partial of f with respect to r*, denoted by $\partial^2 f(x)/\partial r^2$, is

$$\frac{\partial^2 f(x)}{\partial r^2} = \frac{\partial}{\partial r}\left[\frac{\partial f(x)}{\partial r}\right] = \frac{\partial}{\partial r}\left(e^s - \frac{1}{t}\cos \frac{r}{t}\right)$$

$$= \frac{1}{t^2} \sin \frac{r}{t}$$

The third partial with respect to r is

$$\frac{\partial^3 f(x)}{\partial r^3} = \frac{1}{t^3} \cos \frac{r}{t}$$

The first partial derivative with respect to u is $\partial f(x)/\partial u = 2s^3 u$, the second is $\partial^2 f(x)/\partial u^2 = 2s^3$, while the third and all higher partials are zero.

Partial derivatives may also be mixed. Thus the second partial derivative of f, first with respect to r and then with respect to s, is

$$\frac{\partial^2 f(x)}{\partial s\, \partial r} = \frac{\partial}{\partial s}\left[\frac{\partial f(x)}{\partial r}\right] = \frac{\partial}{\partial s}\left[e^s - \frac{1}{t}\cos\frac{r}{t}\right] = e^s$$

The second partial, first with respect to s and then with respect to r, is

$$\frac{\partial^2 f(x)}{\partial r\, \partial s} = \frac{\partial}{\partial r}[re^s + 3s^2 u^2] = e^s$$

Note that $\partial^2 f(x)/\partial s\, \partial r$ equals $\partial^2 f(x)/\partial r\, \partial s$. Also note that the order in which the symbols r and s appear indicates the sequence in which the partial derivatives are to be computed. ▶

The fact that $\partial^2 f(x)/\partial s\, \partial r = \partial^2 f(x)/\partial r\, \partial s$ in Example 7 is not accidental. The following theorem, stated without proof, indicates that in all ordinary cases the order of computing partial derivatives is irrelevant.

Theorem 2 If the second mixed partial derivatives $\partial^2 f(x, y)/\partial x\, \partial y$ and $\partial^2 f(x, y)/\partial y\, \partial x$ are continuous, then they are equal. ▶

Example 8 Suppose* the utility function for consumption of two products is given by

$$u = kc_1^\alpha c_2^\beta$$

where c_1 and c_2 denote respective quantities consumed, and α, β, and k are constants restricted by $0 < \alpha < 1$, $0 < \beta < 1$, and $k > 0$. The partial derivative of u with respect to c_1 is

$$\frac{\partial u}{\partial c_1} = k\alpha c_1^{\alpha-1}c_2^\beta = \frac{\alpha}{c_1} kc_1^\alpha c_2^\beta = \frac{\alpha}{c_1} u$$

Similarly, we compute

$$\frac{\partial u}{\partial c_2} = \frac{\beta}{c_2}\, u, \qquad \frac{\partial^2 u}{\partial c_1^2} = \frac{\alpha(\alpha-1)}{c_1^2}\, u, \qquad \frac{\partial^2 u}{\partial c_2^2} = \frac{\beta(\beta-1)}{c_2^2}\, u$$

and

$$\frac{\partial^2 u}{\partial c_1\, \partial c_2} = \frac{\partial^2 u}{\partial c_2\, \partial c_1} = \frac{\alpha\beta}{c_1 c_2}\, u$$

Analysis of the partial derivatives shows that if we make the natural assumption that c_1 and c_2 are positive, then $\partial u/\partial c_1$, $\partial u/\partial c_2$, and $\partial^2 u/\partial c_1 \partial c_2$ are positive, while $\partial^2 u/\partial c_1^2$ and $\partial^2 u/\partial c_2^2$ are negative. Hence the marginal utility of each product is positive. The marginal utility of each product diminishes with increased consumption of that product, but increases with increased consumption of the other product. ▶

*Adapted from Brems, H., *Quantitative Economic Theory: A Synthetic Approach* (John Wiley & Sons, Inc., New York, 1968), p. 18.

PROBLEMS

In Problems 1–8, find all first partial derivatives of the indicated functions.

1. $S(x, y) = x^2 y - y^2 x$
           2. $S(x, y) = x^2 y^2 + y^3 x$

3. $z(x, y) = (x^2 + y^2)^{1/2}$
         4. $R(s, t) = e^{-s^2 t}$

5. $f(r, s) = \dfrac{r}{s} - \dfrac{s}{r}$
           6. $u = x \sin y + ay \sin x$

7. $g(x, y) = (x^2 - 6y)(x + 2)$
    8. $u = \dfrac{s^2 + 1}{st - 2}$

9. If $f(x, y) = e^x \cos y$, find

$$\frac{\partial f}{\partial x}(0, \pi/3) \quad \text{and} \quad \frac{\partial f}{\partial y}(0, \pi/3)$$

10. Find the slopes at the point $(x, y, z) = (2, -3, 14)$ of the curves cut from surface $z = 2x + y^2 + 1$ by the respective planes $x = 2$ and $y = -3$.

In Problems 11–14 find all second partial derivatives of the given functions.

11. $u(x, y) = y \sin x$
         12. $f(r, t) = a \sin^{-1}(r/t)$

13. $g(x, y, z) = e^{xyz}$
        14. $h(u, v, w) = e^u \cos v + e^v \sin w$

15. Rapoport* considered a two-person game characterized by the equations

$$S_1 = \ln(1 + px + qy) - \beta x$$
$$S_2 = \ln(1 + qx + py) - \beta y$$

where x and y are positive quantities denoting the respective efforts of the players, p, q, and β are positive constants with $p + q = 1$ and $\beta \leq p$, and S_1 and S_2 are the respective satisfaction functions of the players. Show that $\partial S_1/\partial x = \partial S_2/\partial y = 0$ when $p \neq q$ and

$$x = y = \frac{p}{\beta} - 1$$

16. If $f(x, y) = \cos(x + y) + \cos(x - y)$, show that

$$\frac{\partial^2 f}{\partial x^2} - \frac{\partial^2 f}{\partial y^2} = 0$$

17. Show that if $f(x, t) = e^{-\beta^2 t} \sin \beta x$, then $\partial^2 f/\partial x^2 = \partial f/\partial t$ regardless of the value of β.

18. Given the economic model

$$Y = C + I + G$$
$$C = \alpha + \beta(Y - T)$$
$$T = \gamma + \delta Y$$

*Rapoport, A., "Some Game-Theoretical Aspects of Parasitism and Symbiosis," *Bulletin of Mathematical Biophysics* **18**, 15–30 (1956).

where Y is national income, C is consumption, I is investment, G is government expenditure, T is tax collection, and α, β, γ, and δ are positive constants with β and δ both less than unity:

(a) Solve for Y in terms of the constants and I and G.

(b) Show that $\partial Y/\partial G > 0$ and $\partial Y/\partial \gamma < 0$, while $\partial Y/\partial \delta$ has the same sign as $\beta\gamma - I - G - \alpha$.

19. Define the function f by

$$f(x, y) = \frac{xy(x^2 - y^2)}{x^2 + y^2}$$

when $(x, y) \neq (0, 0)$ and $f(x, y) = 0$ when x and y are both zero. Show that the second partial derivatives

$$\frac{\partial^2 f(0, 0)}{\partial x\, \partial y} \quad \text{and} \quad \frac{\partial^2 f(0, 0)}{\partial y\, \partial x}$$

of f at $(0, 0)$ exist but are not equal. (This gives an example of a function for which Theorem 2 does not hold. Of course, these second partials are not continuous at $(0, 0)$.)

17.2 CHAIN RULE FOR FUNCTIONS OF SEVERAL VARIABLES

The Chain Rule for functions of several variables is a generalization of the formula

$$D[g \circ f(x)] = g'(f(x))f'(x)$$

for differentiating a composite function $h(x) = g \circ f(x)$ of a single variable (Section 12.4). To take the simplest multivariate case, suppose that H is a function of two variables and has continuous first-order partial derivatives, while functions f and g are both differentiable functions of a single variable. Then the function G defined by

$$G(t) = H(f(t), g(t)) \tag{3}$$

is a function of a single variable. Specifically, G changes a real number t into the number $G(t)$ by first computing $f(t)$ and $g(t)$ and then combining these results, by means of the function H, to obtain $H(f(t), g(t))$. (See Figure 4.)

FIGURE 4

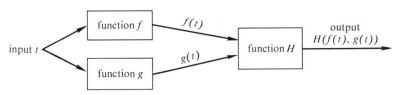

Example 1   Suppose $H(x, y) = x^2 + y^3, f(t) = t$, and $g(t) = \sin t$. Then
$$G(t) = H(f(t), g(t)) = H(t, \sin t) = t^2 + \sin^3 t \qquad \blacktriangleright$$

Being an ordinary function of a single variable, the function G in (3) has a derivative defined in the usual way by
$$G'(t) = \lim_{h \to 0} \frac{G(t + h) - G(t)}{h}$$
$$= \lim_{h \to 0} \frac{H[f(t + h), g(t + h)] - H[f(t), g(t)]}{h} \qquad (4)$$

In order to develop the Chain Rule, we first add and subtract the quantity $H[f(t + h), g(t)]$ in (4) to obtain
$$G'(t) = \lim_{h \to 0} \frac{H[f(t + h), g(t + h)] - H[f(t + h), g(t)]}{h}$$
$$+ \lim_{h \to 0} \frac{H[f(t + h), g(t)] - H[f(t), g(t)]}{h}$$

Now applying the Mean Value Theorem for Partial Derivatives (Theorem 1) gives
$$G'(t) = \lim_{h \to 0} \frac{\partial H[f(t + h), c]}{\partial y} \left[\frac{g(t + h) - g(t)}{h} \right]$$
$$+ \lim_{h \to 0} \frac{\partial H[d, g(t)]}{\partial x} \left[\frac{f(t + h) - f(t)}{h} \right]$$

where c is a number between $g(t)$ and $g(t + h)$ and d lies between $f(t)$ and $f(t + h)$.

The assumed continuity of f and g means that $c \to g(t)$ and $d \to f(t)$ as $h \to 0$. Taking limits we obtain the following important result (in its simplest version).

Chain Rule   If the function G is defined by (3), then
$$G'(t) = \frac{\partial H[f(t), g(t)]}{\partial x} \cdot f'(t) + \frac{\partial H[f(t), g(t)]}{\partial y} \cdot g'(t) \qquad (5)$$
$\blacktriangleright$

Example 2   In Example 1, we have $H(x, y) = x^2 + y^3$, $f(t) = t$, and $g(t) = \sin t$. Thus
$$\frac{\partial H}{\partial x}(x, y) = 2x, \qquad \frac{\partial H}{\partial y}(x, y) = 3y^2, \qquad f'(t) = 1, \qquad g'(t) = \cos t$$

Applying (5) gives
$$G'(t) = \frac{\partial H[f(t), g(t)]}{\partial x} f'(t) + \frac{\partial H[f(t), g(t)]}{\partial y} g'(t)$$
$$= 2f(t)f'(t) + 3g^2(t)g'(t)$$
$$= 2t + 3 \sin^2 t \cos t \qquad \blacktriangleright$$

The computation in Example 2 may easily be verified from known differentiation formulas since $DG(t) = D[t^2 + \sin^3 t] = 2t + 3\sin^2 t \cos t$. Obviously, the Chain Rule is not useful for so simple a problem, nor was it developed for this purpose. Instead, what we have in mind is the following more general problem involving partial derivatives.

Suppose that H, f, and g are all functions of two variables. Then the composite function F defined by

$$F(r, s) = H(f(r, s), g(r, s))$$

is also a function of two variables r and s. The partial derivatives of F may be found by applying the Chain Rule (5) to each variable separately. To simplify notation, let us agree to write x for $f(r, s)$ and y for $g(r, s)$ and to indicate the partial derivatives

$$\frac{\partial H}{\partial x}(x, y), \quad \frac{\partial H}{\partial y}(x, y), \quad \frac{\partial f}{\partial r}(r, s), \quad \frac{\partial f}{\partial s}(r, s), \quad \frac{\partial g}{\partial r}(r, s), \quad \text{and} \quad \frac{\partial g}{\partial s}(r, s)$$

by

$$\frac{\partial H}{\partial x}, \quad \frac{\partial H}{\partial y}, \quad \frac{\partial x}{\partial r}, \quad \frac{\partial x}{\partial s}, \quad \frac{\partial y}{\partial r}, \quad \text{and} \quad \frac{\partial y}{\partial s}$$

respectively. Then, holding s fixed so that H is a function of r alone, and applying (5), we have

$$\frac{\partial F(r, s)}{\partial r} = \frac{\partial H}{\partial x}\frac{\partial x}{\partial r} + \frac{\partial H}{\partial y}\frac{\partial y}{\partial r} \tag{6}$$

Similarly, if r is held fixed, we find

$$\frac{\partial F(r, s)}{\partial s} = \frac{\partial H}{\partial x}\frac{\partial x}{\partial s} + \frac{\partial H}{\partial y}\frac{\partial y}{\partial s} \tag{7}$$

Equations (6) and (7) constitute the *Chain Rule for functions of two variables.*

Example 3 Let $H(x, y) = e^{xy}$, $x = f(r, \theta) = r \cos \theta$, $y = g(r, \theta) = r \sin \theta$, and define F by $F(r, \theta) = H(f(r, \theta), g(r, \theta))$. Then

$$\frac{\partial F(r, \theta)}{\partial r} = \frac{\partial H}{\partial x}\frac{\partial x}{\partial r} + \frac{\partial H}{\partial y}\frac{\partial y}{\partial r}$$

$$= ye^{xy}\cos \theta + xe^{xy}\sin \theta$$

$$= r \sin \theta e^{r^2 \sin \theta \cos \theta}\cos \theta + r \cos \theta e^{r^2 \sin \theta \cos \theta}\sin \theta$$

$$= 2r \sin \theta \cos \theta e^{r^2 \sin \theta \cos \theta}$$

Similarly,

$$\frac{\partial F(r, \theta)}{\partial \theta} = \frac{\partial H}{\partial x}\frac{\partial x}{\partial \theta} + \frac{\partial H}{\partial y}\frac{\partial y}{\partial \theta}$$

$$= ye^{xy}(-r \sin \theta) + xe^{xy}(r \cos \theta)$$

$$= r^2(\cos^2 \theta - \sin^2 \theta)e^{r^2 \sin \theta \cos \theta} \qquad \blacktriangleright$$

Example 4 The real advantage of the Chain Rule for two-variable functions comes when it is impossible or impractical to solve for some of the variables in terms of the others. For instance, suppose we consider the equation

$$H(x, y) = 0$$

as implicitly specifying y as a function of x. Applying (6) we compute

$$0 = DH(x, y) = \frac{\partial H}{\partial x} + \frac{\partial H}{\partial y} y'(x)$$

from which it follows that

$$y'(x) = \frac{-\partial H/\partial x}{\partial H/\partial y}$$

To take a specific illustration, suppose

$$x^2 y + x \sin y = 1$$

Then $H(x, y) = x^2 y + x \sin y - 1 = 0$, $\partial H/\partial x = 2xy + \sin y$,

$$\frac{\partial H}{\partial y} = x^2 + x \cos y$$

and

$$y'(x) = -\frac{\partial H/\partial x}{\partial H/\partial y} = -\frac{2xy + \sin y}{x^2 + x \cos y} \qquad \blacktriangleright$$

Example 5 Consider a single-commodity market in which demand D is assumed to be a function $f(P, Y)$ of both price P and income Y, while supply S is a function $g(P)$ of price alone. The equilibrium position of the market is determined by the condition

$$g(P) = f(P, Y) \qquad (8)$$

which determines price as a function P^* of income. Differentiating the relation (8) gives

$$g'(P^*) \frac{\partial P^*}{\partial Y} = \frac{\partial f}{\partial P^*} \frac{\partial P^*}{\partial Y} + \frac{\partial f}{\partial Y}$$

from which the derivative of P^* is obtained as

$$\frac{\partial P^*}{\partial Y} = \frac{\partial f/\partial Y}{g'(P^*) - \partial f/\partial P^*}$$

Let us assume that demand increases as income rises ($\partial f/\partial Y > 0$) and decreases with a rise in price ($\partial f/\partial P < 0$) and that supply increases with price ($\partial S/\partial P > 0$). Then $\partial P^*/\partial Y$ is positive and it follows that an increase (decrease) in income will result in a corresponding increase (decrease) in equilibrium price. $\qquad \blacktriangleright$

PROBLEMS

In Problems 1–4, apply Equation (5) to find $u'(t)$. (The symbols u, x, y, and z are shorthand for functions $u(t)$, $x(t)$, $y(t)$, and $z(t)$ of t.) Check by eliminating x, y, and z before differentiating.

1. $u = x^2 + y^2; x = x(t) = (t + 1)/t, y = y(t) = t/(t - 1)$

2. $u = x^{1/2} + 2xy^2 - 2y^{1/2}; x = 1 + t^2, y = 1 - t^2$

3. $u = t^2 \sin xy; x = t^{1/2}, y = \ln t$

4. $u = x \cos y + y \sin z; x = t, y = e^t, z = 1/2t$

In Problems 5–10, w is a function of x and y (or of x, y, and z) which are, in turn, functions of r and t. Use Equations (6) and (7) to find $\partial w/\partial r$ and $\partial w/\partial t$. Check by eliminating x, y, and z first.

5. $w = x^2 - y^2; x = x(r, t) = r \sin t, y = y(r, t) = r \cos t$

6. $w = e^{-x/y}; x = 2r - t, y = r + 2t$

7. $w = \tan^{-1}(y/x); x = r + a \sin t, y = r - a \cos t$

8. $w = xe^{2y} - ye^x; x = rt, y = t/r$

9. $w = \sqrt{x^2 + y^2 + z^2}; x = rt, y = r \sin t, z = t \cos r$

10. $w = \dfrac{x^2 - y^2}{y^2 + z^2}; x = e^r, y = re^{-t}, z = \dfrac{1}{t}$

11. Suppose $u = u(x, y)$, while $x = r \cos \theta$ and $y = r \sin \theta$.
   (a) Find $\partial u/\partial r$ and $\partial u/\partial \theta$.
   (b) Show that

$$\frac{\partial u}{\partial x} = \cos \theta \, \frac{\partial u}{\partial r} - \frac{\sin \theta}{r} \frac{\partial u}{\partial \theta}$$

$$\frac{\partial u}{\partial y} = \sin \theta \, \frac{\partial u}{\partial r} + \frac{\cos \theta}{r} \frac{\partial u}{\partial \theta}$$

12. Suppose $w = f(u/v)$. Prove that

$$v \frac{\partial w}{\partial v} + u \frac{\partial w}{\partial u} = 0$$

In Problems 13–17, assume that the given relations determine y as a function of x. Find dy/dx in each case.

13. $x^3 - y^2 + 2axy = 0$

14. $xy - \sin y = x^2$

15. $e^x \sin y + e^y \sin x = 1$

16. $\sin \dfrac{x}{y} - \cos \dfrac{y}{x} = \dfrac{1}{2}$

17. $\ln y + ye^x = 10$

18. Samuelson's model of income determination* assumes that net national income Y is the sum of domestic consumption C and net private investment I (that is, $Y = C + I$), and that C itself is a function of I (that is, $C = f(I)$).

(a) Assuming the function f has an inverse f^{-1}, show how to write Y as a function g of C alone.

(b) Show how to write Y as a function h of I alone.

(c) Assuming that (a) implicitly determines C as a function m of Y alone, prove that

$$h'(I) = \frac{1}{1 - m'(Y)}$$

(Equivalently, $dY/dI = (1 - dC/dY)^{-1}$.)

19. Prove that if z is implicitly defined as a function of x and y by the equation $F(x, y, z) = 0$ and if $\partial F/\partial z \neq 0$, then

$$\frac{\partial z}{\partial x} = -\frac{\partial F/\partial x}{\partial F/\partial z} \quad \text{and} \quad \frac{\partial z}{\partial y} = -\frac{\partial F/\partial y}{\partial F/\partial z}$$

20. If the functions $r = \sqrt{x^2 + y^2}$ and $\theta = \tan^{-1}(y/x)$ implicitly define the functions $x = G(r, \theta)$ and $y = H(r, \theta)$, find

(a) $\dfrac{\partial x}{\partial r}$ (b) $\dfrac{\partial x}{\partial \theta}$

(c) $\dfrac{\partial y}{\partial r}$ (d) $\dfrac{\partial y}{\partial \theta}$

17.3 TAYLOR'S FORMULA—OPTIMIZATION

The Taylor Formula for functions of n variables is used to approximate the unknown value of a function H at a point $b = (b_1, b_2, \ldots, b_n)$ in terms of known values of the function and its derivatives at a point

$$a = (a_1, a_2, \ldots, a_n)$$

For simplicity we shall derive the formula for a function of two variables only, taking the point a to be the origin $(0, 0)$. From this, the generalization to any number of variables and any point a will be easily obtained.

Let H be a function of two variables, let $b = (b_1, b_2)$ be a fixed two-dimensional vector, and define a function G by

$$G(t) = H(tb) = H(tb_1, tb_2)$$

*Samuelson, P. A., "Simple Mathematics of Income Determination," in *Income, Employment and Public Policy: Essays in Honor of Alvin H. Hansen*, Metzler, L. A., et al., eds. (W. W. Norton, New York, 1948), pp. 133–155.

Writing $x = f(t) = tb_1$ and $y = g(t) = tb_2$ in the Chain Rule (5) gives

$$G'(t) = \frac{\partial H(tb)}{\partial x} \cdot f'(t) + \frac{\partial H(tb)}{\partial y} \cdot g'(t)$$

$$= b_1 \frac{\partial H(tb)}{\partial x} + b_2 \frac{\partial H(tb)}{\partial y} \qquad (9)$$

for the first derivative of G.

The Chain Rule may now be applied to obtain the second derivative of G. Each term in (9), being a function of both x and y, will produce two new terms. We obtain

$$G''(t) = DG'(t) = b_1 D \frac{\partial H(tb)}{\partial x} + b_2 D \frac{\partial H(tb)}{\partial y}$$

$$= b_1 \left[\frac{\partial^2 H(tb)}{\partial x^2} f'(t) + \frac{\partial^2 H(tb)}{\partial y\, \partial x} g'(t) \right]$$

$$+ b_2 \left[\frac{\partial^2 H(tb)}{\partial x\, \partial y} f'(t) + \frac{\partial^2 H(tb)}{\partial y^2} g'(t) \right]$$

$$= b_1^2 \frac{\partial^2 H(tb)}{\partial x^2} + b_2 b_1 \frac{\partial^2 H(tb)}{\partial y\, \partial x} + b_1 b_2 \frac{\partial^2 H(tb)}{\partial x\, \partial y} + b_2^2 \frac{\partial^2 H(tb)}{\partial y^2}$$

$$= b_1^2 \frac{\partial^2 H(tb)}{\partial x^2} + 2 b_1 b_2 \frac{\partial^2 H(tb)}{\partial x\, \partial y} + b_2^2 \frac{\partial^2 H(tb)}{\partial y^2}$$

The pattern for higher derivatives should now be apparent. Differentiating a third time gives $G'''(t)$ on the left side, while on the right side each term generates two new terms corresponding, respectively, to the partial derivatives with respect to x and to y. Each differentiation with respect to x produces another factor $b_1 = f'(t)$, while $b_2 = g'(t)$ appears with each differentiation with respect to y. Terms which differ only in the order of differentiation are equal and can be combined. Thus,

$$G'''(t) = b_1^3 \frac{\partial^3 H(tb)}{\partial x^3} + 3 b_1^2 b_2 \frac{\partial^3 H(tb)}{\partial x^2\, \partial y} + 3 b_1 b_2^2 \frac{\partial^3 H(tb)}{\partial x\, \partial y^2} + b_2^3 \frac{\partial^3 H(tb)}{\partial y^3}$$

$$G^{(4)}(t) = b_1^4 \frac{\partial^4 H(tb)}{\partial x^4} + 4 b_1^3 b_2 \frac{\partial^4 H(tb)}{\partial x^3\, \partial y} + 6 b_1^2 b_2^2 \frac{\partial^4 H(tb)}{\partial x^2\, \partial y^2}$$

$$+ 4 b_1 b_2^3 \frac{\partial^4 H(tb)}{\partial x\, \partial y^3} + b_2^4 \frac{\partial^4 H(tb)}{\partial y^4}$$

and, in general,

$$G^{(n)}(t) = \sum_{k=0}^{n} \binom{n}{k} b_1^k b_2^{n-k} \frac{\partial^n H(tb)}{\partial x^k\, \partial y^{n-k}}$$

may be proved by induction.

We now substitute these results and put $t = 1$ in the Taylor Formula

$$G(t) = G(0) + G'(0)t + \frac{G''(0)}{2!} t^2 + \cdots + \frac{G^{(n-1)}(0)}{(n-1)!} t^{n-1} + \frac{G^{(n)}(c)}{n!} t^n$$

for one variable (here c is between 0 and t). In terms of H, this gives

$$H(b_1, b_2) = H(b)$$

$$= H(0, 0) + \frac{\partial H(0, 0)}{\partial x} b_1 + \frac{\partial H(0, 0)}{\partial y} b_2$$

$$+ \frac{1}{2!} \left[\frac{\partial^2 H(0, 0)}{\partial x^2} b_1^2 + 2 \frac{\partial^2 H(0, 0)}{\partial x\, \partial y} b_1 b_2 + \frac{\partial^2 H(0, 0)}{\partial y^2} b_2^2 \right]$$

$$+ \cdots + \frac{1}{(n-1)!} \sum_{k=0}^{n-1} \binom{n-1}{k} \frac{\partial^{n-1} H(0, 0)}{\partial x^k\, \partial y^{n-1-k}} b_1^k b_2^{n-1-k}$$

$$+ \frac{1}{n!} \sum_{k=0}^{n} \binom{n}{k} \frac{\partial^n H(cb_1, cb_2)}{\partial x^k\, \partial y^{n-k}} b_1^k b_2^{n-k} \tag{10}$$

where c is between 0 and 1, as the *Taylor Formula for two variables.*

Obviously, we may go through the same derivation with an arbitrary point $a = (a_1, a_2)$ in place of the point $(0, 0)$. The quantities b_1 and b_2 are then replaced by $b_1 - a_1$ and $b_2 - a_2$, respectively.

Example 1 For the function $f(x, y) = e^x \sin y$, the first, second, and third partial derivatives are

$$\frac{\partial f}{\partial x} = \frac{\partial^2 f}{\partial x^2} = \frac{\partial^3 f}{\partial x^3} = e^x \sin y$$

$$\frac{\partial f}{\partial y} = \frac{\partial^2 f}{\partial x\, \partial y} = \frac{\partial^3 f}{\partial x^2\, \partial y} = e^x \cos y$$

$$\frac{\partial^2 f}{\partial y^2} = \frac{\partial^3 f}{\partial x\, \partial y^2} = -e^x \sin y$$

$$\frac{\partial^3 f}{\partial y^3} = -e^x \cos y$$

At $(x, y) = (0, 0)$, the respective values are 0, 1, 0, and -1. Together with $f(0, 0) = 0$, this gives, for the Taylor polynomial of third degree,

$$p(x, y) = 0 + 0 \cdot x + 1 \cdot y + \frac{1}{2!} \cdot 0 \cdot x^2 + \frac{2}{2!} \cdot 1 \cdot xy + \frac{1}{2!} \cdot 0 \cdot y^2$$

$$+ \frac{1}{3!} \cdot 0 \cdot x^3 + \frac{3}{3!} \cdot 1 \cdot x^2 y + \frac{3}{3!} \cdot 0 \cdot xy^2 + \frac{1}{3!} (-1) y^3$$

$$= y + xy + \tfrac{1}{2} x^2 y - \tfrac{1}{6} y^3$$

The error term involves all fourth partials. ▶

Example 2 For the utility function $u = kC_1^\alpha C_2^\beta$ in Example 8 of Section 17.1, the first and second partials were found to be

$$\frac{\partial u}{\partial c_1} = \frac{\alpha}{c_1}\, u, \qquad \frac{\partial u}{\partial c_2} = \frac{\beta}{c_2}\, u, \qquad \frac{\partial^2 u}{\partial c_1^2} = \frac{\alpha(\alpha - 1)}{c_1^2}\, u$$

$$\frac{\partial^2 u}{\partial c_1\, \partial c_2} = \frac{\alpha\beta}{c_1 c_2}\, u, \qquad \frac{\partial^2 u}{\partial c_2^2} = \frac{\beta(\beta - 1)}{c_2^2}\, u$$

Hence the Taylor polynomial of second degree around the point $a = (2, 1)$ is

$$u(c_1, c_2) = u(2, 1) + \frac{\partial u(2, 1)}{\partial c_1}(c_1 - 2) + \frac{\partial u(2, 1)}{\partial c_2}(c_2 - 1)$$

$$+ \frac{1}{2!}\left[\frac{\partial^2 u(2, 1)}{\partial c_1^2}(c_1 - 2)^2 + 2\frac{\partial^2 u(2, 1)}{\partial c_1\, \partial c_2}(c_1 - 2)(c_2 - 1)\right.$$

$$\left. + \frac{\partial^2 u(2, 1)}{\partial c_2^2}(c_2 - 1)^2\right]$$

$$= k \cdot 2^\alpha \cdot 2^\beta \left[1 + \frac{\alpha}{2}(c_1 - 2) + \beta(c_2 - 1)\right]$$

$$+ \tfrac{1}{2}k \cdot 2^\alpha \cdot 2^\beta \left[\frac{\alpha(\alpha - 1)}{4}(c_1 - 2)^2 + \frac{2\alpha\beta}{2}(c_1 - 2)(c_2 - 1)\right.$$

$$\left. + \beta(\beta - 1)(c_2 - 1)^2\right] \quad \blacktriangleright$$

The ideas concerning points of relative maximum and minimum of a function of several variables are much the same as for functions of a single variable (Section 12.8). Thus if f is a function of n variables, the point $a = (a_1, a_2, \ldots, a_n)$ is a point of *relative maximum* for f if there is a positive number δ such that $f(x) \le f(a)$ for all points x which lie within δ distance of a. Similarly, if the inequality $f(x) \ge f(a)$ holds for all x within δ distance of a, then a is a point of *relative minimum* for f.

Example 3 The point $(x, y, f(x, y)) = (1, 2, 4)$ is a relative maximum point for the function $f(x, y) = 2x + 4y - x^2 - y^2 - 1$, as may be seen by writing f in the equivalent form $f(x, y) = 4 - (x - 1)^2 - (y - 2)^2$. Consider the function

$$g(x) = f(x, 2) = 2x + 8 - x^2 - 4 - 1 = 2x - x^2 + 3$$

obtained by fixing $y = 2$. Geometrically, the graph of g is the intersection of the surface representing f and the plane whose equation is $y = 2$. (See

Figure 5.) Hence g should have a relative maximum at $x = 1$. This is easily verified since $g'(x) = 2 - 2x$ is zero and $g''(x) = -2$ is negative when $x = 1$.

FIGURE 5

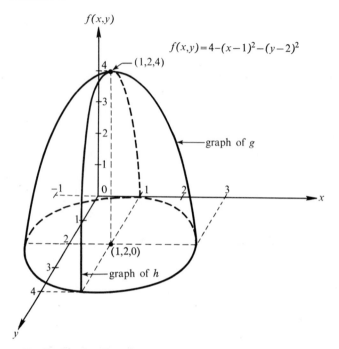

Similarly, the function

$$h(y) = f(1, y) = 2 + 4y - 1 - y^2 - 1 = 4y - y^2$$

obtained from f by fixing $x = 1$, has a relative maximum at $y = 2$. (Here, $h'(y) = 4 - 2y$ and $h''(y) = -2$.) ▶

In Example 3, the derivative of the function g at a point x is, of course, the partial derivative of f with respect to x evaluated at the point $(x, 2)$. Similarly, the derivative $h'(y)$ is the partial derivative $\partial h(1, y)/\partial y$. Our calculations show that at the point $(1, 2)$ of relative maximum for f, both these partial derivatives are zero.

It should be apparent that similar considerations apply at any point of relative maximum or minimum for any function of any number of variables. We are thus led to the following theorem, which we state without further proof.

Interior Maximum and Minimum Theorem   Let

$$f(x) = f(x_1, \ldots, x_n)$$

be a function of n variables and let a be a point of relative maximum or relative minimum which is interior to the domain of f. Then all the first partial derivatives of f, evaluated at $x = a$, are zero. That is, if the interior point a is a relative maximum or minimum, then

$$\frac{\partial f(a)}{\partial x_1} = \frac{\partial f(a)}{\partial x_2} = \cdots = \frac{\partial f(a)}{\partial x_n} = 0 \qquad \blacktriangleright$$

Example 4 The function $f(x, y) = x^2 - y^2$, shown in Figure 6, has neither a maximum nor a minimum at $(x, y) = (0, 0)$. However, both partial derivatives $\partial f/\partial x = 2x$ and $\partial f/\partial y = -2y$ are zero at $(0, 0)$. To a person walking along the surface in the plane $y = 0$, the origin looks like a minimum. To a person walking in the plane $x = 0$, the origin looks like a maximum. Such a point is called a *minimax*, or a *saddle point*, of the function.

$\qquad\qquad\qquad\qquad\qquad\qquad\qquad\qquad\qquad\qquad\qquad\qquad\qquad\qquad\blacktriangleright$

FIGURE 6

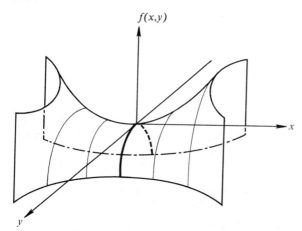

Example 4 shows that all partial derivatives being zero at a certain point a is not sufficient to guarantee that a function has a relative maximum or minimum at a. The same situation arose with functions of one variable. In that case we resorted either to examining the first derivative near a or to a test involving the second derivative. For functions of several variables the latter alternative is best. We use Taylor's Formula to develop such a test, again developing the theory only for a function of two variables.

Suppose, then, that for the function $f(x, y)$ both partial derivatives $\partial f/\partial x$ and $\partial f/\partial y$ are zero at the point a. Taking $n = 2$ and choosing a point $b = (b_1, b_2)$ near $a = (a_1, a_2)$, the Taylor Formula (10) becomes

$$f(b) = f(a) + \frac{1}{2!}\left[\frac{\partial^2 f(c)}{\partial x^2}(b_1 - a_1)^2 + \frac{2\,\partial^2 f(c)}{\partial x\,\partial y}(b_1 - a_1)(b_2 - a_2)\right.$$
$$\left. + \frac{\partial^2 f(c)}{\partial y^2}(b_2 - a_2)^2\right]$$

where $c = (c_1, c_2)$ is a point between a and b.

The difference $f(b) - f(a)$ is thus represented as a quadratic form' (see Section 10.3) $1/2! \, X'AX$, where X' is the row vector $X' = (b_1 - a_1, b_2 - a_2)$ and A is the symmetric matrix

$$A = \begin{pmatrix} \dfrac{\partial^2 f(c)}{\partial x^2} & \dfrac{\partial^2 f(c)}{\partial x \, \partial y} \\[2ex] \dfrac{\partial^2 f(c)}{\partial x \, \partial y} & \dfrac{\partial^2 f(c)}{\partial y^2} \end{pmatrix}$$

of second partials of f evaluated at c. In this form it is obvious (assuming that the second partials are continuous) that a is a point of relative minimum for f if and only if this quadratic form is positive for all points b near a. This is the case if and only if the matrix A is *positive definite* (see Section 10.3).

Similarly, a is a point of relative maximum for f if and only if A is a negative definite matrix. If A is an indefinite form, a is neither a relative maximum nor a relative minimum, while if A is semidefinite, the test fails and no specific result may be claimed.

Example 5   (a) For the function $f(x, y) = 2x + 4y - x^2 - y^2 - 1$ considered in Example 3, we have

$$\frac{\partial f}{\partial x} = 2 - 2x, \quad \frac{\partial f}{\partial y} = 4 - 2y, \quad \frac{\partial^2 f}{\partial x^2} = -2, \quad \frac{\partial^2 f}{\partial y^2} = -2, \text{ and } \frac{\partial^2 f}{\partial x \, \partial y} = 0$$

The first partial derivatives are both zero at the point $(1, 2)$. The matrix

$$A = \begin{pmatrix} -2 & 0 \\ 0 & -2 \end{pmatrix}$$

of second partials is negative definite everywhere. Hence f has a relative maximum at $(1, 2)$.

(b) For the function $f(x, y) = x^2 - y^2$ in Example 4, $\partial f/\partial x = 2x$, $\partial f/\partial y = -2y$, $\partial^2 f/\partial x^2 = 2$, $\partial^2 f/\partial y^2 = -2$, and $\partial^2 f/\partial x \, \partial y = 0$. Both first partials are zero at the origin, but the matrix

$$A = \begin{pmatrix} 2 & 0 \\ 0 & -2 \end{pmatrix}$$

of second partials is indefinite. Hence the point $(0, 0)$ is neither a maximum nor a minimum.   ▶

Example 6   A common statistical problem is to find the straight line which best fits a set of data points plotted in the plane. What straight line best fits the seven points plotted in Figure 7 which record the results of seven experiments relating amount of advertising A and total profit P? The statistician answers as follows. Ideally, he would like the chosen line

$$P = f(A) = \alpha A + \beta$$

FIGURE 7

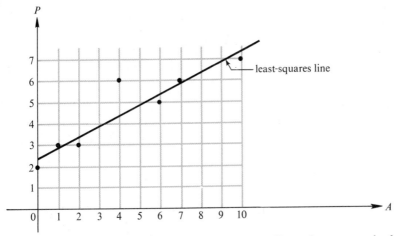

to pass through all the points. In general, this will not happen, so he looks for the line, called the *least-squares line*, which minimizes the *sum of squares* of the vertical distances between the line and the plotted points. That is, for given points $(A_1, P_1), (A_2, P_2), \ldots, (A_n, P_n)$, he seeks to minimize

$$g(\alpha, \beta) = (P_1 - \alpha A_1 - \beta)^2 + (P_2 - \alpha A_2 - \beta)^2$$
$$+ \cdots + (P_n - \alpha A_n - \beta)^2$$
$$= \sum_{i=1}^{n} (P_i - \alpha A_i - \beta)^2$$

The distances are squared in order to avoid the situation where negative and positive deviations cancel, leaving a zero sum even though the line does not pass through all the points.

The first partial derivatives of g are

$$\frac{\partial g}{\partial \alpha} = 2 \sum_{i=1}^{n} (P_i - \alpha A_i - \beta)(-A_i)$$

$$\frac{\partial g}{\partial \beta} = 2 \sum_{i=1}^{n} (P_i - \alpha A_i - \beta)(-1)$$

Setting these equal to zero leads to the equations

$$\alpha \sum_{i=1}^{n} A_i^2 + \beta \sum_{i=1}^{n} A_i = \sum_{i=1}^{n} A_i P_i \qquad (11)$$

$$\alpha \sum_{i=1}^{n} A_i + n\beta = \sum_{i=1}^{n} P_i$$

which must be solved for α and β.

The second partial derivatives of g are

$$\frac{\partial^2 g}{\partial \alpha^2} = 2 \sum_{i=1}^{n} A_i^2, \quad \frac{\partial^2 g}{\partial \alpha \, \partial \beta} = 2 \sum_{i=1}^{n} A_i, \quad \text{and} \quad \frac{\partial^2 g}{\partial \beta^2} = 2n$$

It is easily verified that the matrix

$$M = 2 \begin{pmatrix} \sum_{i=1}^{n} A_i^2 & \sum_{i=1}^{n} A_i \\ \sum_{i=1}^{n} A_i & n \end{pmatrix}$$

of second partials is positive definite. Hence the solutions of the Equations (11) yield a point of relative minimum for g.

For the points in Figure 7, we compute $n = 7$,

$$\sum_{i=1}^{7} A_i = 30, \quad \sum_{i=1}^{7} P_i = 32, \quad \sum_{i=1}^{7} A_i^2 = 206, \text{ and } \sum_{i=1}^{7} A_i P_i = 175$$

The equations

$$206\alpha + 30\beta = 175$$

$$30\alpha + 7\beta = 32$$

have the solution $\alpha = \frac{265}{542}$ and $\beta = \frac{1342}{542}$, giving $f(A) = \frac{265}{542}A + \frac{1342}{542}$ (or approximately, $f(A) = \frac{1}{2}A + \frac{5}{2}$) as the equation of the least-squares line.

▶

PROBLEMS

In Problems 1–10, find maximum and minimum points of the indicated functions. Show all tests.

1. $f(x, y) = x^2 + 3y^2 - 2x$

2. $g(s, t) = s^3 - 3t^2 + st^2$

3. $h(x, y) = 2xy$

4. $u(r, \theta) = r^2 - 2\theta^2$

5. $f(u, v) = 1/u + uv - 2/v$

6. $u(x, y) = xy(6 - x - y)$

7. $h(x_1, x_2) = (x_1 + x_2)^2 - x_1^2 - x_2^2$

8. $f(x, y, z) = x^2 + y^2 + z^2$

9. $g(x_1, x_2, x_3) = (x_1 + x_2 + x_3)^2$

10. $h(x_1, x_2, x_3) = (x_1 + x_2 + x_3)^2 - x_1^2 - x_2^2 - x_3^2$

11. Find the largest product $p_1 p_2 p_3$ which can be formed subject to the restriction $p_1 + p_2 + p_3 = 1$.

12. A monopolist produces commodities A and B having demand functions $x(a) = 18 - 2a$ and $y(b) = 10 - b$, where a and b are the respective unit prices. If the joint cost function is $C(a, b) = [x(a)]^2 + [y(b)]^2$, find the prices and corresponding outputs which maximize profit.

13. A firm producing two commodities has the revenue function

$$R = p_1 q_1 + p_2 q_2$$

where p_i is the price and q_i the quantity of the ith product ($i = 1, 2$). The cost function is

$$C = 3q_1^2 + 2q_1 q_2 + 3q_2^2$$

Assuming fixed prices p_1 and p_2, show that profit is maximized when

$$q_1 = \frac{3p_1 - p_2}{16} \quad \text{and} \quad q_2 = \frac{3p_2 - p_1}{16}$$

14. If the cost function in Problem 13 is $C = q_1^2 + 3q_1 q_2 + q_2^2$, is there an interior maximum for the profit function? Why or why not?

15. Take $\alpha = 2$ and $\beta = 3$ in the utility function of Example 8 of Section 17.1 and Example 2 of Section 17.3. Write the third-degree Taylor polynomial approximation to this function

(a) around the point $(0, 0)$.

(b) around the point $(1, 2)$.

16. Find the least-squares line of the form $y = ax + b$ for the data below. Plot the points and sketch the least-squares line.

i	1	2	3	4	5
x_i	0	2	1	6	4
y_i	1	9	4	21	15

17. Suppose that observed data consists of a set of n ordered 4-tuples (x_i, y_i, z_i, w_i), $i = 1, 2, \ldots, n$.

(a) In terms of the quantities x_i, y_i, z_i, and w_i find equations for β_1, β_2, and β_3 to give a least-squares surface of the form

$$w = \beta_1 x + \beta_2 y + \beta_3 z$$

(*Hint:* Minimize $\sum_{i=1}^{n} (w_i - \beta_1 x_i - \beta_2 y_i - \beta_3 z_i)^2$.)

(b) Apply your results to the data

i	1	2	3	4	5
x_i	0	1	4	2	1
y_i	1	−1	3	1	0
z_i	0	2	0	1	1
w_i	3	6	5	4	3

17.4 OPTIMIZATION WITH RESTRAINTS— LAGRANGE MULTIPLIERS

The tests for relative maxima and minima described in the preceding section depend on finding these points in the interior of the domain of the function concerned. However, it is not unusual in applications to find functional restrictions imposed on the domain by the nature of the problems being analyzed. Geometrically, such restrictions confine the domain of interest to some curve or surface in n-dimensional space.

To take an already familiar example, we have seen that maximum and minimum solutions of linear programming problems always occur on the boundary of the set of feasible solutions (see Section 9.5). Thus, solutions would be unchanged if the boundary of the feasible region were retained and the interior region deleted. In this case, the domain of the linear function to be optimized becomes a collection of lines or planes, the points of which lie on the boundary, rather than in the interior of the feasible region.

Since we already know how to solve linear programming problems, we shall restrict our attention here to problems involving either nonlinear functions or domains which are not simply lines or planes.

For functions of two variables, the general problem takes the following form: Maximize or minimize the function $f(x, y)$ subject to the restriction

$$h(x, y) = 0 \tag{12}$$

The relation (12) may be thought of as determining a function g such that

$$h(x, g(x)) = 0$$

for all x. If it is inconvenient or impossible to solve for $g(x)$, we may set $t = f(t) = x$ in Equation (3) and apply the Chain Rule (5) to obtain

$$\frac{\partial h}{\partial x} + \frac{\partial h}{\partial y} g'(x) = 0$$

so that

$$g'(x) = -\frac{\partial h/\partial x}{\partial h/\partial y} \tag{13}$$

To maximize f we define a function k by $k(x) = f(x, g(x))$, again use the Chain Rule to differentiate k, and equate the result to zero. This gives

$$\frac{dk}{dx} = \frac{\partial f}{\partial x} + \frac{\partial f}{\partial y} g'(x) = 0$$

or, using (13),

$$\frac{\partial f}{\partial x} + \frac{\partial f}{\partial y}\left(-\frac{\partial h/\partial x}{\partial h/\partial y}\right) = 0 \tag{14}$$

Solving this equation simultaneously with Equation (12) provides the critical points of the function.

Example 1 Consider the problem of maximizing the function

$$f(x, y) = x^2 - y^2 - y \tag{15}$$

subject to the restriction

$$h(x, y) = x^2 + y^2 - 1 = 0 \tag{16}$$

While we could solve for either x or y in (16) and substitute the result into (15), for purposes of illustration we prefer to apply the Chain Rule to differentiate the functions as they stand.

We find

$$\frac{df}{dx} = \frac{\partial f}{\partial x} + \frac{\partial f}{\partial y}\frac{dy}{dx} = 2x - (2y + 1)\frac{dy}{dx} \tag{17}$$

$$\frac{dh}{dx} = \frac{\partial h}{\partial x} + \frac{\partial h}{\partial y}\frac{dy}{dx} = 2x + 2y\frac{dy}{dx} = 0 \tag{18}$$

Substituting $dy/dx = -x/y$ from (18) into (17) and equating to zero in order to maximize gives

$$2x - (2y + 1)\left(-\frac{x}{y}\right) = 0$$

or

$$x\left[2 + 2 + \frac{1}{y}\right] = 0$$

This is zero when $x = 0$ or when $y = -\frac{1}{4}$. Using (16), the critical points are thus seen to be $(0, 1)$, $(0, -1)$, $(\sqrt{15}/4, -\frac{1}{4})$, and $(-\sqrt{15}/4, -\frac{1}{4})$.

Now $Df(x) = 2x - (2y + 1)(-x/y) = 4x + x/y$ so that

$$D^2f(x) = \frac{\partial Df(x)}{\partial x} + \frac{\partial Df(x)}{\partial y}\frac{dy}{dx}$$

$$= 4 + \frac{1}{y} - \frac{x}{y^2}\left(-\frac{x}{y}\right)$$

$$= 4 + \frac{1}{y} + \frac{x^2}{y^3}$$

This is positive at $(0, 1)$ and $(0, -1)$ and negative at $(\sqrt{15}/4, -\frac{1}{4})$ and $(-\sqrt{15}/4, -\frac{1}{4})$. Hence the latter two points are the points at which f has a relative maximum. The value of f at each of these points is

$$\left(\frac{\sqrt{15}}{4}\right)^2 - \left(-\frac{1}{4}\right)^2 - \left(-\frac{1}{4}\right) = \frac{9}{8} \qquad \blacktriangleright$$

The method of *Lagrange multipliers** arrives at the same results by different, but often simpler, means. To maximize $f(x, y)$ subject to $h(x, y) = 0$, we introduce a function $F(x, y) = f(x, y) - \lambda h(x, y)$, where λ is a constant called a Lagrange multiplier, to be determined.

*After Joseph Louis Lagrange (1736–1813), a French mathematician.

Since $h(x, y) = 0$, F should be maximized at the same points as f. We use the Chain Rules (6) and (7) to find partial derivatives of F. The two equations obtained by equating these partial derivatives to zero, together with the restriction $h(x, y) = 0$, may then be solved for x, y, and λ.

To look at details, we compute

$$\frac{\partial F}{\partial x} = \frac{\partial f}{\partial x} - \lambda \frac{\partial h}{\partial x} \quad \text{and} \quad \frac{\partial F}{\partial y} = \frac{\partial f}{\partial y} - \lambda \frac{\partial h}{\partial y}$$

Equating these to zero gives

$$\frac{\partial f}{\partial x} - \lambda \frac{\partial h}{\partial x} = 0 \tag{19}$$

$$\frac{\partial f}{\partial y} - \lambda \frac{\partial h}{\partial y} = 0 \tag{20}$$

Solving for λ in (20) and substituting into (19) gives

$$\frac{\partial f}{\partial x} - \frac{\partial f}{\partial y}\left(\frac{\partial h/\partial x}{\partial h/\partial y}\right) = 0$$

as in Equation (14).

Example 2 We use the Lagrange multiplier method to reconsider Example 1. To maximize $f(x, y) = x^2 - y^2 - y$ subject to the restriction

$$h(x, y) = x^2 + y^2 - 1 = 0$$

we define

$$F(x, y) = f(x, y) - \lambda h(x, y)$$

$$= x^2 - y^2 - y - \lambda(x^2 + y^2 - 1)$$

Then

$$\frac{\partial F}{\partial x} = 2x - 2x\lambda = 2x(1 - \lambda)$$

$$\frac{\partial F}{\partial y} = -2y - 1 - 2y\lambda = -2y(\lambda + 1) - 1$$

Solving the equations $\partial F/\partial x = 0$ and $\partial F/\partial y = 0$ gives $x = 0$ or $\lambda = 1$, and $2y(\lambda + 1) + 1 = 0$. For $x = 0$, $h(x, y) = 0$ yields $y = \pm 1$. If $\lambda = 1$, then $2y(\lambda + 1) + 1 = 0$ requires $y = -\frac{1}{4}$, which in turn gives $x = \pm\sqrt{15}/4$. The critical points are $(\sqrt{15}/4, -\frac{1}{4})$, $(-\sqrt{15}/4, -\frac{1}{4})$, $(0, 1)$, and $(0, -1)$ as before. Notice how much more easily the results are obtained here than in Example 1. ▶

The method of Lagrange multipliers applies equally well when there is more than one restriction, provided only that the number of variables is greater than the number of restrictions.

Example 3 *An Allocation Problem* The quantities Q_1 and Q_2 of output of a two-product firm depend on labor L and capital K according to the production functions

$$Q_1 = a_1 L_1^{\alpha_1} K_1^{\beta_1} \quad \text{and} \quad Q_2 = a_2 L_2^{\alpha_2} K_2^{\beta_2}$$

Here a_1 and a_2 are positive constants, while $\alpha_1, \alpha_2, \beta_1,$ and β_2 are constants between 0 and 1. There is a fixed total quantity S of labor and a fixed total quantity T of capital available to the firm. If we assume that the respective prices p_1 and p_2 are fixed, the problem becomes that of allocating labor and capital to the two commodities in order to maximize revenue. Symbolically, we seek to maximize

$$R = p_1 Q_1 + p_2 Q_2 = p_1 a_1 L_1^{\alpha_1} K_1^{\beta_1} + p_2 a_2 L_2^{\alpha_2} K_2^{\beta_2}$$

subject to the restrictions

$$L_1 + L_2 - S = 0 \quad \text{and} \quad K_1 + K_2 - T = 0 \tag{21}$$

To do this we introduce a function

$$F(L_1, L_2, K_1, K_2) = p_1 a_1 L_1^{\alpha_1} K_1^{\beta_1} + p_2 a_2 L_2^{\alpha_2} K_2^{\beta_2}$$
$$- \lambda_1 (L_1 + L_2 - S) - \lambda_2 (K_1 + K_2 - T)$$

involving two Lagrange multipliers λ_1 and λ_2. The partial derivatives are

$$\frac{\partial F}{\partial L_1} = p_1 a_1 \alpha_1 L_1^{\alpha_1 - 1} K_1^{\beta_1} - \lambda_1$$

$$\frac{\partial F}{\partial L_2} = p_2 a_2 \alpha_2 L_2^{\alpha_2 - 1} K_2^{\beta_2} - \lambda_1$$

$$\frac{\partial F}{\partial K_1} = p_1 a_1 \beta_1 L_1^{\alpha_1} K_1^{\beta_1 - 1} - \lambda_2$$

$$\frac{\partial F}{\partial K_2} = p_2 a_2 \beta_2 L_2^{\alpha_2} K_2^{\beta_2 - 1} - \lambda_2$$

Equating these to zero and solving these equations in conjunction with Equations (21) give

$$p_1 a_1 \alpha_1 L_1^{\alpha_1 - 1} K_1^{\beta_1} - p_2 a_2 \alpha_2 (S - L_1)^{\alpha_2 - 1} (T - K_1)^{\beta_2} = 0 \tag{22}$$

$$p_1 a_1 \beta_1 L_1^{\alpha_1} K_1^{\beta_1 - 1} - p_2 a_2 \beta_2 (S - L_1)^{\alpha_2} (T - K_1)^{\beta_2 - 1} = 0 \tag{23}$$

Multiplying Equation (22) by $\beta_2 (S - L_1)$ and Equation (23) by

$$\alpha_2 (T - K_1)$$

and subtracting the resulting equations yields

$$a_1 p_1 L_1^{\alpha_1 - 1} K_1^{\beta_1 - 1} [\alpha_1 \beta_2 K_1 (S - L_1) - \alpha_2 \beta_1 L_1 (T - K_1)] = 0$$

Since $\alpha_1, p_1, L_1,$ and K_1 are all positive, this simplifies to

$$\alpha_1 \beta_2 K_1 (S - L_1) = \alpha_2 \beta_1 L_1 (T - K_1)$$

or, equivalently,

$$\frac{\alpha_1/\alpha_2}{\beta_1/\beta_2} = \frac{(L_1/S) - L_1}{(K_1/T) - K_1}$$

For simplicity of interpretation, suppose that $\alpha_1 = \alpha_2$ and $\beta_1 = \beta_2$. Then the revenue function R has an extremum wherever

$$(L_1/S) - L_1 = (K_1/T) - K_1$$

that is, whenever capital is assigned to the two products in the same ratio as labor. ▶

Example 4 Symmetric matrices of the form

$$A = \begin{pmatrix} 1 & r_{12} & r_{13} & \cdots & r_{1n} \\ r_{12} & 1 & r_{23} & \cdots & r_{2n} \\ r_{13} & r_{23} & 1 & \cdots & r_{3n} \\ \vdots & & & & \\ r_{1n} & r_{2n} & r_{3n} & \cdots & 1 \end{pmatrix} \quad \text{and} \quad B = \begin{pmatrix} r_{11} & r_{12} & r_{13} & \cdots & r_{1n} \\ r_{12} & r_{22} & r_{23} & \cdots & r_{2n} \\ r_{13} & r_{23} & r_{33} & \cdots & r_{3n} \\ \vdots & & & & \\ r_{1n} & r_{2n} & r_{3n} & \cdots & r_{nn} \end{pmatrix}$$

arise in the theory of mental tests. A common problem is to find a vector of weights

$$W = (w_1, w_2, \ldots, w_n)$$

which will maximize the reliability function

$$R(W) = \frac{WBW'}{WAW'}$$

subject to the restriction $WAW' = 1$. Note that $R(W)$ is the ratio of two quadratic forms with matrices B and A, respectively.

Following Gulliksen,* we introduce a function

$$L(W) = WBW' - \lambda(WAW' - 1)$$

involving a Lagrange multiplier λ. The partial derivatives of $L(W)$ with respect to $w_1, w_2, \ldots, w_n$ may be written in matrix form as $WB - \lambda WA$. Equating these partial derivatives to zero gives

$$WB - \lambda WA = O \quad \text{or} \quad W(B - \lambda A) = O \tag{24}$$

as the matrix equation which the maximizing W vector must satisfy.

We recall (Section 10.1) that Equations (24) have a nonzero solution only if

$$\det (B - \lambda A) = 0 \tag{25}$$

However, Equation (25) may have more than one solution and we must decide which to choose. We argue as follows. If we multiply the first of

*Gulliksen, H., *Theory of Mental Tests* (John Wiley & Sons, Inc., New York, 1950), pp. 346–348.

Equations (24) by W', we obtain $WBW' - \lambda WAW' = 0$ as an equation which λ must satisfy. Solving for λ gives

$$\lambda = \frac{WBW'}{WAW'}$$

a value of the reliability function R. Since we wish to maximize R, it follows that we should first choose λ as the largest solution of (25), then substitute into (24) and solve for the corresponding W. ▶

PROBLEMS

1. Maximize $f(x, y) = x^2 - 2y^2 + 2xy + 4x$ subject to the restriction $2x - y = 0$.

2. Maximize $x(y + 6)$ subject to $x + y = 2$.

3. Maximize $g(x, y, w) = 2x + y + 2w + x^2 - 3w^2$ subject to

$$x + y + w - 1 = 0 \quad \text{and} \quad 2x - y + w = 2$$

4. Minimize $h(x, y) = x^2 + y^2$ subject to $x^2 + y^2 - 2x - 4y + 2 = 0$.

5. Maximize $R(u, v, w) = 10uvw$ subject to $u^2 + v^2 + w^2 = 4$.

6. Maximize $f(x_1, x_2, \ldots, x_n) = (x_1 x_2 \cdots x_n)^2$ subject to the restriction $x_1^2 + x_2^2 + \cdots + x_n^2 = 1$.

7. Sales S of a firm depend on material cost C and labor cost L according to

$$S(C, L) = 5CL - 3L^2$$

If the budget is such that

$$C + L = 15$$

find the maximum sales attainable.

8. Assume that the utility of quantities x and y of two commodities is given by a function $U = F(x, y)$ where $\partial F/\partial x$ and $\partial F/\partial y$ are positive. Given a fixed budget total B and fixed prices p and q for the two commodities, prove that the consumer, in order to maximize utility, must allocate his budget so as to equalize the ratio of marginal utility to price for each commodity.

9. An *indifference curve* is defined as the locus of points (x, y) such that

$$F(x, y) = \text{constant}$$

Show that in order to maximize utility in Problem 8, the consumer must allocate his budget so as to equate the slope of his budget line with the slope of some indifference curve.

10. Find the utility-maximizing demand for each of two goods if the utility function is

$$U(x, y) = 10x + 30y - x^2 - 4y^2$$

and the budgetary restriction is

$$B = 20 = px + qy = 3x + 4y$$

11. Let B be a p-component column vector and $\sum$ a $p \times p$ symmetric matrix. Given the problem of maximizing $B'\sum B$ subject to the constraint that $B'B = 1$, show that

$$B'\sum B = \lambda B'B = \lambda$$

where λ is a Lagrange multiplier. (*Hint:* See Example 4 of this section.)

17.5 ITERATED INTEGRALS

In Chapter 14, we used the concept of area to motivate the definition of the integral of a function of a single variable. In a similar way, the problem of computing the volume of a three-dimensional solid leads to the definition of an *iterated integral* of a function of two variables.

Example 1 To find the volume of the solid lying in the first octant and bounded above by the graph of $f(x, t) = 4 - x - t^2$ (see Figure 8), we begin by partitioning the interval from $x = 0$ to $x = 4$ into n smaller intervals determined by the points $0, 4/n, 8/n, \ldots, 4$. We then approximate that part of the total volume which lies in the kth interval $[4(k - 1)/n, 4k/n]$ by multiplying the width $4/n$ of the interval by the area $A(4k/n)$ of the right-hand face of the section cut by the kth interval (shaded in Figure 8). Adding

FIGURE 8

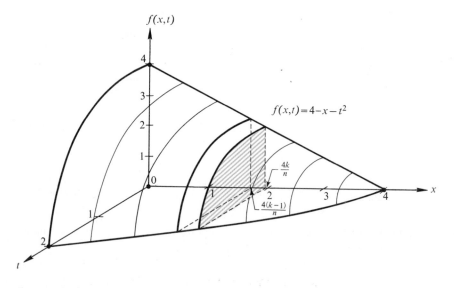

these volumes gives

$$\sum_{k=1}^{n} A\left(\frac{4k}{n}\right) \cdot \frac{4}{n}$$ (26)

as an approximation to the actual volume.

As the number n of intervals is increased, it is reasonable to expect the sum (26) to converge to the actual volume. On the other hand, the limit

$$\lim_{n \to \infty} \sum_{k=1}^{n} A\left(\frac{4k}{n}\right) \frac{4}{n}$$

is, by definition, the integral

$$\int_{0}^{4} A(x)\, dx$$ (27)

of the function A over the interval from $x = 0$ to $x = 4$. We are thus led to *define* the volume in question by the integral (27).

The area $A(x)$ may itself be computed as an integral. For instance, when $x = 1$, $f(x, t) = f(1, t) = 4 - 1 - t^2 = 3 - t^2$ and $A(1)$ is the area

$$A(1) = \int_{0}^{\sqrt{3}} (3 - t^2)\, dt$$

shaded in Figure 9. (This is the same as the area shaded in Figure 8 if $4k/n = 1$.)

FIGURE 9

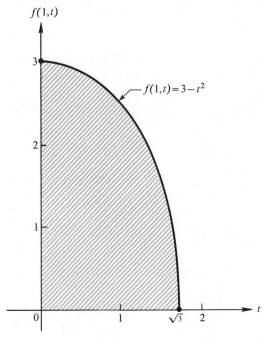

$f(1,t)$

$f(1,t) = 3 - t^2$

In general, for each *fixed* value of x between 0 and 4,

$$A(x) = \int_0^{(4-x)^{1/2}} (4 - x - t^2)\, dt$$

$$= (4 - x)t - \frac{t^3}{3} \Big]_0^{(4-x)^{1/2}}$$

$$= \tfrac{2}{3}(4 - x)^{3/2}$$

The required volume is

$$V = \int_0^4 A(x)\, dx = \int_0^4 \left[\int_0^{(4-x)^{1/2}} (4 - x - t^2)\, dt \right] dx$$

$$= \int_0^4 \tfrac{2}{3}(4 - x)^{3/2}\, dx$$

$$= -\tfrac{2}{3}\tfrac{2}{5}(4 - x)^{5/2} \Big]_0^4 = \tfrac{2}{3}\tfrac{2}{5}(4)^{5/2} = \tfrac{128}{15}$$

▶

The integral

$$\int_0^4 A(x)\, dx = \int_0^4 \left[\int_0^{(4-x)^{1/2}} (4 - x - t^2)\, dt \right] dx$$

appearing in Example 1 is called an *interated integral,* or *double integral.* Note that in computing this integral we *first hold x fixed* and integrate the function $f(x, t) = 4 - x - t^2$ as though it were a function of t alone. (Compare with our procedure for partial derivatives.) The result is a function $A(x) = \tfrac{2}{3}(4 - x)^{3/2}$ of x alone. Integrating this function then gives the desired result.

Example 2 The volume V in Example 1 may also be computed by first holding t fixed. For each value of t between 0 and 2 we obtain the area

$$B(t) = \int_0^{4-t^2} (4 - x - t^2)\, dx$$

$$= (4 - t^2)x - \frac{x^2}{2} \Big]_0^{4-t^2} = \tfrac{1}{2}(4 - t^2)^2$$

generated by cutting the volume by a plane perpendicular to the t axis. The integral of B then gives the required volume

$$V = \int_0^2 B(t)\, dt = \int_0^2 \tfrac{1}{2}(4 - t^2)^2\, dt$$

$$= \tfrac{1}{2}\int_0^2 (16 - 8t^2 + t^4)\, dt = \tfrac{1}{2}\left(16t - \tfrac{8}{3}t^3 + \tfrac{1}{5}t^5 \Big]_0^2 \right)$$

$$= \tfrac{1}{2}(32 - \tfrac{64}{3} + \tfrac{32}{5}) = \tfrac{128}{15}$$

as before.

▶

Example 3 The interated integral of the constant function $f(x, t) = 1$ over a plane region B may be interpreted as the volume of a solid of unit thickness having B as its base. Alternatively, this is simply the area of B. For instance, the area of the region B shaded in Figure 10 is given either by

$$\int_0^2 \left[\int_0^{4-x^2} 1 \, dt \right] dx = \int_0^2 \left(t \Big|_0^{4-x^2} \right) dx$$

$$= \int_0^2 (4 - x^2) \, dx$$

$$= 4x - \tfrac{1}{3}x^3 \Big|_0^2 = \tfrac{16}{3}$$

or by

$$\int_0^4 \left[\int_0^{(4-t)^{1/2}} 1 \, dx \right] dt = \int_0^4 \left(x \Big|_0^{(4-t)^{1/2}} \right) dt$$

$$= \int_0^4 (4 - t)^{1/2} \, dt$$

$$= -\tfrac{2}{3}(4 - t)^{3/2} \Big|_0^4 = \tfrac{2}{3}(4)^{3/2} = \tfrac{16}{3}$$

FIGURE 10

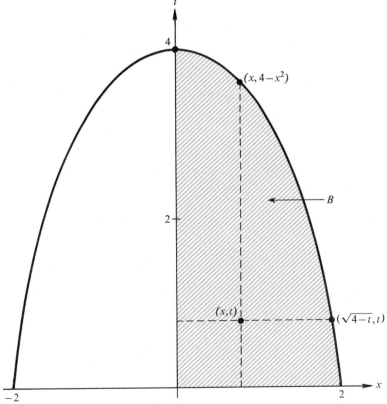

Note the procedure for obtaining the limits of integration. If we first hold x constant and vary t, then for each fixed value of x between 0 and 2, the values of t range between 0 and $4 - x^2$ (see Figure 10). On the other hand, if t is held fixed (between 0 and 4), then x ranges from 0 to $(4 - t)^{1/2}$. In each case the values of the nonfixed variable range along that portion of the line, determined by the value of the fixed variable, which intersects the region of integration.

▶

Iterated integrals of functions of more than two variables may also be computed by repeated one-dimensional integration.

Example 4 In the triple integral

$$\int_0^1 \int_0^z \int_0^y (x^2 + y^2 + z^2)\, dx\, dy\, dz \tag{28}$$

the order of integration and the variables to which the limits of integration pertain are indicated by the order of writing the symbols dx, dy, and dz. We first hold y and z fixed and integrate with x ranging from zero to y to obtain

$$\int_0^1 \int_0^z \left(\frac{x^3}{3} + y^2 x + z^2 x \Big]_0^y\right) dy\, dz = \int_0^1 \int_0^z \left(\frac{y^3}{3} + y^3 + z^2 y\right) dy\, dz$$

Next, z is held fixed and y varied from 0 to z, giving

$$\int_0^1 \left(\frac{y^4}{12} + \frac{y^4}{4} + \frac{z^2 y^2}{2}\Big]_0^z\right) dz = \int_0^1 \left(\frac{z^4}{12} + \frac{z^4}{4} + \frac{z^4}{2}\right) dz$$

Evaluating this integral yields

$$\int_0^1 \frac{5}{6} z^4\, dz = \frac{5}{6}\frac{z^5}{5}\Big]_0^1 = \frac{1}{6}$$

as the value of (28).

▶

Improper iterated integrals are computed by utilizing the definitions already stated for the one-variable case.

Example 5

$$\frac{1}{2\pi}\int_0^{2\pi} \int_0^\infty r e^{-(1/2)r^2}\, dr\, d\theta = \frac{1}{2\pi}\int_0^{2\pi} \left(\lim_{t\to\infty} \int_0^t r e^{-(1/2)r^2}\, dr\right) d\theta$$

$$= \frac{1}{2\pi}\int_0^{2\pi} \left(\lim_{t\to\infty} - e^{-(1/2)r^2}\Big]_0^t\right) d\theta$$

$$= -\frac{1}{2\pi}\int_0^{2\pi} \lim_{t\to\infty}\left(e^{-(1/2)t^2} - 1\right) d\theta$$

$$= -\frac{1}{2\pi}\int_0^{2\pi} (-1)\, d\theta = \frac{1}{2\pi}\theta\Big]_0^{2\pi} = 1 \quad ▶$$

The change of variable formula for one-dimensional integrals is

$$\int_{g(a)}^{g(b)} h(u)\, du = \int_{a}^{b} h(g(x)) g'(x)\, dx \qquad (29)$$

(See Formula (24) and the accompanying discussion in Section 14.8.) For computational purposes we think of making the substitution $u = g(x)$ so that $du = g'(x)\, dx$ and the rest is just a matter of determining the correct limits for the new integral.

A careful reading of Section 14.8 will show that the computational shortcut of writing $du = g'(x)\, dx$ is a device that *happens* to work for single-variable cases. However, in cases involving more than one variable, this shortcut *does not work and, if used, will give entirely erroneous results.* What must be done is to use in place of the function $g'(x)$ appearing in (29) a quantity which generalizes the idea of derivative to functions which map n variables into n variables. Let us look first at a particular case.

A point P in the plane may be located by specifying the horizontal distance x and the vertical distance y of P from the origin $(0, 0)$. The point P may also be located by stating the angle θ formed between the horizontal axis and a line from the origin to P, and the distance r moved along this line to reach P (see Figure 11). In terms of r and θ, we have

$$x = r \cos \theta \quad \text{and} \quad y = r \sin \theta \qquad (30)$$

The Equations (30) may be considered as defining a function, or transformation, F, which maps pairs (r, θ) of numbers into new pairs (x, y). The *derivative*, or *Jacobian matrix*, of this vector function

$$F(r, \theta) = \begin{pmatrix} r \cos \theta \\ r \sin \theta \end{pmatrix} = \begin{pmatrix} x \\ y \end{pmatrix}$$

is defined to be the matrix

$$DF(r, \theta) = \begin{pmatrix} \dfrac{\partial x}{\partial r} & \dfrac{\partial x}{\partial \theta} \\[2mm] \dfrac{\partial y}{\partial r} & \dfrac{\partial y}{\partial \theta} \end{pmatrix} = \begin{pmatrix} \cos \theta & -r \sin \theta \\ \sin \theta & r \cos \theta \end{pmatrix}$$

FIGURE 11

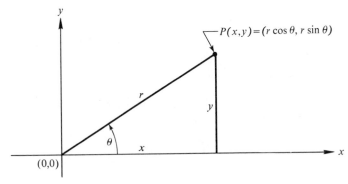

The *Jacobian* of the transformation F is defined to be the determinant

$$J = \begin{vmatrix} \cos\theta & -r\sin\theta \\ \sin\theta & r\cos\theta \end{vmatrix} = r\cos^2\theta + r\sin^2\theta = r$$

of the derivative of F.

In general, the *derivative* of a vector function

$$F = \begin{pmatrix} f_1 \\ f_2 \\ \vdots \\ f_n \end{pmatrix}$$

which maps points $(x_1, x_2, \ldots, x_n)$ to points $(y_1, y_2, \ldots, y_n)$ by means of the real-valued functions

$$y_1 = f_1(x_1, x_2, \ldots, x_n)$$
$$y_2 = f_2(x_1, x_2, \ldots, x_n)$$
$$\vdots$$
$$y_n = f_n(x_1, x_2, \ldots, x_n)$$

(31)

is defined by

$$DF(x_1, x_2, \ldots, x_n) = \begin{pmatrix} \dfrac{\partial f_1}{\partial x_1} & \dfrac{\partial f_1}{\partial x_2} & \cdots & \dfrac{\partial f_1}{\partial x_n} \\[2mm] \dfrac{\partial f_2}{\partial x_1} & \dfrac{\partial f_2}{\partial x_2} & \cdots & \dfrac{\partial f_2}{\partial x_n} \\ \vdots & & & \\ \dfrac{\partial f_n}{\partial x_1} & \dfrac{\partial f_n}{\partial x_2} & \cdots & \dfrac{\partial f_n}{\partial x_n} \end{pmatrix}$$

(32)

The *Jacobian* J of the function F is the determinant of the matrix (32). It is this Jacobian which takes the place of the derivative $g'(x)$ in (29).

Example 6 In order to evaluate the integral

$$\frac{1}{2\pi} \int_{-\infty}^{\infty} \int_{-\infty}^{\infty} e^{-(1/2)(x^2+y^2)} \, dx \, dy$$

(33)

we make the substitution $x = r\cos\theta$ and $y = r\sin\theta$. The integrand $e^{-(1/2)(x^2+y^2)}$ is replaced by

$$e^{-(1/2)(r^2\cos^2\theta + r^2\sin^2\theta)} \cdot r = re^{-(1/2)r^2}$$

the quantity r being the Jacobian of the transformation. (Compare with Equation (29).) The integral (33) is equal to

$$\frac{1}{2\pi} \int_0^{2\pi} \int_0^{\infty} re^{-(1/2)r^2} \, dr \, d\theta$$

(34)

which was shown in Example 5 to have the value 1. The limits 0 to ∞ and 0 to 2π for r and θ in (34) are chosen so as to cover the same region (the entire plane) as that covered in the integral (33). ▶

In general, if we make a substitution like (31) in an integral

$$\int\int \cdots \int_R g(y_1, y_2, \ldots, y_n) \, dy_1 \, dy_2 \ldots dy_n$$

where R is the region over which the integral is to be evaluated, the new integral is

$$\int\int \cdots \int_S g[f_1(x_1, \ldots, x_n), \ldots, f_n(x_1, \ldots, x_n)] \, |J| \, dx_1 \, dx_2 \ldots dx_n$$

Here S is the region in terms of the new variables $x_1, \ldots, x_n$ which matches the original region R, and J is the Jacobian of the transformation. The absolute value of J is used to ensure that both integrals have the same sign.

PROBLEMS

In Problems 1–5, evaluate the indicated double integrals.

1. $\displaystyle\int_0^1 \int_0^3 x \, dy \, dx$

2. $\displaystyle\int_0^2 \int_{y^2}^y x \, dx \, dy$

3. $\displaystyle\int_a^{3a} \int_y^{ay^2} (x + y) \, dx \, dy$

4. $\displaystyle\int_0^\pi \int_0^{a\cos\theta} r^3 \sin\theta \, dr \, d\theta$

5. $\displaystyle\int_1^3 \int_1^{ax^2} \frac{x}{y^2} \, dy \, dx$

6. Evaluate the improper double integral

$$\int_1^2 \int_{-\infty}^y \frac{1}{x^2 + y^2} \, dx \, dy$$

In Problems 7–9, evaluate the iterated triple integrals.

7. $\displaystyle\int_0^1 \int_1^{2x} \int_0^{x^2} x \, dz \, dy \, dx$

8. $\displaystyle\int_0^\infty \int_0^{x_3} \int_0^{2x_2} e^{-(a_1x_1 + a_2x_2 + a_3x_3)} \, dx_1 \, dx_2 \, dx_3$

9. $\displaystyle\int_0^1 \int_0^{\pi/6} \int_0^{r^2\sin^2\theta} r \, dz \, d\theta \, dr$

In Problems 10–11, use the transformation $F(r, \theta)$ as defined in Example 6.

10. Evaluate the integral

$$\int_0^t \int_0^{(t^2-x^2)^{1/2}} (x^2 + y^2)\, dy\, dx$$

11. Evaluate the integral

$$\int_0^t \int_0^{(t^2-x^2)^{1/2}} 1\, dy\, dx$$

12. Find the area of the region R bounded by the parabolas $f(x) = x^2$ and $g(x) = 9 - x^2$.

13. Find the area of the region R bounded by $y^2 = 2x^3$ and $y = 2x$.

14. Find the volume of the solid bounded by the surfaces $x^2 + y^2 = 9$, $z = x + y + 4$, and $z = 0$.

15. Find the volume in the first octant bounded by the surfaces $x^2 + y^2 = 4$, $z = x$, and $z = x^2$.

16. Find the volume of the solid in the first octant bounded by the two surfaces $z = 9 - x^2$ and $y = 9 - x^2$.

17. Find the volume bounded by the surfaces $x^2 + y^2 = z$, $x^2 + y^2 = 2x$, and $z = 0$.

SUPPLEMENTARY READING

Stein, S. K., *Calculus in the First Three Dimensions* (McGraw—Hill Book Company, New York, 1967), Chapters 9, 11, and 21–23.

Thomas, G. B., *Calculus and Analytic Geometry*, 4th ed. (Addison—Wesley Publishing Company, Reading, Massachusetts, 1968), Chapters 15 and 16.

PART IV

Probability

BASIC CONCEPTS OF PROBABILITY **18**

18.1 SAMPLE SPACES

The theory of probability is a mathematical system that provides models for experimental phenomena exhibiting some measure of unpredictability. The word "experimental" is applied to any operation which generates outcomes, whether or not an observer exercises any measure of planning or control on the operation. Many studies in the behavioral and management sciences are of this nature. The economist studying the national economy, the sociologist investigating group behavior, and the psychologist conducting a learning experiment all observe behavior whose outcome is seemingly incapable of exact prediction. By its very nature, then, probability theory has become one of the major mathematical tools used in the behavioral sciences. Underlying any use of the theory of probability is a real or conceptual experiment. We begin by considering how experiments can be described mathematically.

Suppose we observe a rat running a *T*-maze. He can turn either left or right. If we denote these outcomes by *L* and *R*, respectively, then each possible outcome of the experiment corresponds to exactly one element of the set $S = \{L, R\}$. This set of possible outcomes is called a *sample space* for the experiment. We say *a* rather than *the* sample space since there are many ways of specifying the outcomes of this experiment. If, for example, we are interested in the rat's behavior at the choice point as well as the choice made, we might use

$$T = \{LH, LW, RH, RW\}$$

as our sample space. Here *LH* indicates that the rat turned left after some hesitation, *LW* denotes a left turn without hesitation, and, similarly, for *RH* and *RW*. This example is typical of experimental situations, in that many sample spaces can be constructed to describe the experiment. Which sample space *should* be used is not specified by the theory but depends upon the goals of the experiment.

In general, a sample space Ω for an experiment is a set having the property that to each possible outcome of the experiment there corresponds *exactly one* element of Ω. Since each outcome corresponds to exactly one element of the sample space Ω, it is apparent that Ω represents a *partition* (see Section 3.2) of the collection of possible outcomes. Note that a single element of Ω may represent more than one possible outcome.

Example 1 Suppose an experiment consists of tossing a nickel and a dime. There are four possible outcomes:

(a) Both coins show heads.

(b) The nickel shows heads and the dime shows tails.

(c) The nickel shows tails and the dime heads.

(d) Both coins show tails.

A possible sample space for this experiment is

$$\Omega = \{HH, HT, TH, TT\}$$

where the pairs of letters correspond, respectively, to outcomes (a)–(d). An equally valid sample space is

$$T = \{0, 1, 2\}$$

in which the elements denote the number of heads showing on the two coins. The symbol $1 \in T$ corresponds to both outcomes (b) and (c). On the other hand, the set $V = \{HH, HT, 1\}$ cannot serve as a sample space for this experiment since the elements HT and 1 in V both correspond to outcome (b), while outcome (d) fails to be represented at all. In general, a sample space represents a classification, or partition, of the possible outcomes of an experiment and one must be careful not to mix classifications. ▶

Example 2 An experiment may have an infinite number of possible outcomes. In a study of rats trained to run down an alley to get food, Graham and Gagne* measured the time spent in a starting box. For a particular rat, any non-negative real number is a possible experimental outcome (time value) and the set of these numbers is a valid sample space. However, due to measurement limitations, time was measured only to the nearest hundredth of a second. The sample space used consisted of the elements 0, 0.01, 0.02, ... ▶

*Graham, C. H., and Gagne, R. M., "The Acquisition, Extinction and Spontaneous Recovery of a Conditioned Operant Response," *Journal of Experimental Psychology* **26**, 251–280 (1940).

It is often useful to be able to count the number of elements in a sample space, even though it may be impractical to make a complete list of the individual elements. In such cases we apply the basic principles of counting developed in Chapter 7.

Example 3 The Rich Employment Agency specializes in placing engineering applicants in one of 12 defined fields of occupational specialization. From a pool of applicants, 6 engineers are chosen and their specializations are recorded. We want to designate an appropriate sample space for this experiment. An outcome of the experiment is specified by a listing of occupations of the 6 engineers. Since each engineer is classified in one of the 12 specializations, the most natural sample space contains $12^6 = 2,985,984$ elements, each element corresponding to a possible listing. ▶

PROBLEMS

1. From a group of three men and three women, three people are assigned to a discussion group.

(a) List the 20 elements of the sample space Ω where each element corresponds to a possible discussion group.

(b) Find the subset of Ω containing those elements which correspond to discussion groups in which there are three women and no men. How many elements are there in this set?

(c) Find the subset of Ω containing those elements representing discussion groups which contain a majority of males. How many elements are there in this set?

2. Subjects are asked to place each of three foods in one of two preference categories. A week later, the experiment is repeated using the same three foods. Define a suitable sample space describing the choices made by a single subject. Write the sets of outcomes in which

(a) there are no preference reversals.

(b) there is exactly one preference reversal.

(c) there is at least one preference reversal.

3. Humphreys* has conducted a number of experiments in which a subject is asked to predict whether a light will be on or off. Suppose that a subject is run through four trials. Write an appropriate sample space Ω for this experiment. Let E be the subset of Ω whose elements denote outcomes for which the number of "on" predictions is greater than one, and F the subset whose elements denote outcomes for which the number of "on" predictions equals the number of "off" predictions. Determine the elements in the following sets.

(a) $E \cap F$ (b) $E \cap F'$

*See, for example, the analysis of L. G. Humphreys' work contained in Bush, R. R., and Mosteller, F., *Stochastic Models for Learning* (John Wiley & Sons, Inc., New York, 1955).

(c) $E \cup F$ (d) E'

(e) $E' \cap F$ (f) F'

4. A firm wishes to buy three lots of material. There are four domestic and five foreign suppliers from which to choose. Write a sample space for this experiment. How many elements denote purchases from one domestic and two foreign suppliers?

5. An experiment consists of observing two-person communication links which are established among three individuals. Once established, a communication is assumed to persist for the duration of the experiment. Write an appropriate sample space for this experiment.

6. The sample space for Problem 5 contains eight elements. How many elements would be required if there were
   (a) 4 individuals?
   (b) 5 individuals?
   (c) n individuals?

7. One method of estimating the reliability of a test is to split the test into two equal parts and obtain the correlation between the two parts. However, as Kuder and Richardson* have argued, this method does not give a unique estimate, due to the number of different ways a test can be divided.
   (a) Suppose that an experiment consists of dividing a 6-item test into two 3-item tests. Denote a sample space where each element represents a unique division.
   (b) Develop a general formula for the number of distinct ways that a test of $2n$ items can be split into two subtests of n items each.

18.2 EVENTS

A sample space serves as the *universal set* for all probability statements related to an experiment. The subsets of this universal set serve as mathematical representations of the various events which may occur in a performance of the experiment. For example, suppose an economist observes the fluctuations of the stock market on three consecutive days. Since the market can either rise R or fail to rise F on any given day, the set

$$\Omega = \{RRR, RRF, RFR, RFF, FRR, FRF, FFR, FFF\}$$

is a possible sample space for this experiment. For each outcome of the experiment we can determine whether a given event does or does not occur. If, for instance, we consider the event "the market rose at least twice in three days," we find that this event occurs if the experiment results in an outcome corresponding to one of the elements RRR, RRF, RFR, or FRR. With any

---

*Kuder, G. F., and Richardson, M. W., "The Theory of the Estimation of Test Reliability," *Psychometrika* **2**, 151–160 (1937).

other outcome the event fails to occur. We have thus associated with this event the subset

$$E = \{RRR, RRF, RFR, FRR\}$$

of Ω.

In a similar fashion we may associate with any event a collection of elements from the sample space. For reasons of simplicity and mathematical precision, we simply equate an event with its corresponding subset. That is, if Ω is a sample space corresponding to a real or conceptual experiment, then an *event* is a subset of Ω. An event E is said to *occur* if the experiment results in any outcome corresponding to an element of E.

Since an event is a subset of the sample space Ω, it follows that Ω itself and the empty set ϕ are events. The event Ω is called the *certain event*, which always occurs, while ϕ is the *impossible event*, which can never occur. A subset containing a single element of Ω is called a *simple event*, while those subsets containing more than one element are *compound events*.

Example 1 The XYZ Manufacturing Company maintains an inventory subject to certain conditions of supply and demand. Due to peculiarities in their production processes, the company can supply only 0, 100, or 200 units in a given time period. Moreover, in the same period, demand may be only 50, 100, or 200 units.

Let

$$\Omega = \{(50, 0), (50, 100), (50, 200), (100, 0), (100, 100),$$

$$(100, 200), (200, 0), (200, 100), (200, 200)\}$$

be the associated sample space, where the members of the ordered pairs represent demand and supply, respectively. The event described by "demand exceeds supply" is the subset

$$E = \{(50, 0), (100, 0), (200, 0), (200, 100)\}$$

while the event "supply exceeds demand" is

$$F = \{(50, 100), (50, 200), (100, 200)\}$$

and the event "demand equals supply" is

$$G = \{(100, 100), (200, 200)\}$$

These events may not be the most informative partition of the sample space. For example, we may be interested in the amount of discrepancy between supply and demand. In this case, to say that the event "the discrepancy between demand and supply is 100 units" occurs means that an experimental outcome has been observed which corresponds to one of the elements in the subset

$$H = \{(100, 0), (100, 200), (200, 100)\}$$

The event "demand exceeds supply *and* demand differs from supply by 100 units" is

$$E \cap H = \{(100, 0), (200, 100)\}$$

The event "demand exceeds supply and supply exceeds demand" is the impossible event ϕ since E and F have no common elements. ▶

Example 2 Returning to Example 3 of Section 18.1, let E denote the event that at least two engineers selected have the same occupational speciality. Suppose we wish to find $n(E)$, the number of elements in the subset E. This number is not easily found directly, but $n(E)$ can be calculated indirectly by using Counting Principle CP1 (Section 7.1) to obtain

$$n(E) + n(E') = n(\Omega) = 12^6$$

Here E' is the event (that is, the subset of the sample space Ω) which is the complement of E.

Now $n(E')$ is the number of ways in which the 6 engineers can be placed in 12 categories so that no 2 engineers share the same category. Invoking Counting Principle CP 2 (Section 7.1), we conclude that

$$n(E') = 12 \times 11 \times \cdots \times 7$$

and therefore that

$$n(E) = 12^6 - 12 \times 11 \times \cdots \times 7$$
$$= 2,320,704 \qquad ▶$$

PROBLEMS

1. Each of two judges ranks four subjects. Determine an appropriate sample space. Find the number of elements in the event "the ratings agree for at least one subject."

2. Let A, B, and C be events of a sample space Ω. Using only the symbols $\cup$, $\cap$, and $'$ express the following statements in set notation.

   (a) At least one of the events A, B, or C occurs.

   (b) Only A occurs.

   (c) None occurs.

   (d) At most two occur.

   (e) All three occur.

   (f) A and B occur, but not C.

   (g) Exactly two occur.

3. Observations are recorded about three members of a control group and three members of an experimental group. The six observations are ranked in order of magnitude and the sum of the ranks assigned the experimental group computed. Write the event "the sum of the ranks assigned the experimental group is 8 or less."

4. How many distinct events are contained in a sample space having
   (a) two elements?   (b) three elements?   (c) n elements?

5. We are interested in the seating arrangement selected by five male and three female subjects doing a group task at a round table. Write out the elements of the event "no two female subjects are seated adjacent to each other."

6. In a study reported by Sommer,* observers recorded the establishment of two-person communication links among eight persons seated at a rectangular cafeteria table with one person at each end and three on each side. Suppose that we wish to record only the first establishment of a two-person communication link. Define an appropriate sample space and write out the elements of the events A, B, and C described, respectively, by the following.

   A: Communication occurs between adjacent persons on the same side of the table.

   B: Communication occurs between persons seated face-to-face.

   C: Communication occurs between adjacent persons seated corner to corner.

7. In Problem 20 of Section 7.2, suppose we add a chairman E who has veto power but no vote. List the elements of the event "the chairman is pivotal."

18.3   THE AXIOMS OF PROBABILITY

Probability theory begins with the notions of "sample space" and "event." Once these are specified for a particular experimental situation, we then can proceed to assign probabilities to events in accord with the following axioms. Logically, the meaning to be assigned to the terms *probability* and *probability function* is contained in these axioms and all other statements about probability must be derived from them. The examples and theorems of this section and the next illustrate the procedure to be used and contain the basic results.

Axioms for Probability   Let Ω be a sample space for some experiment and let E_i ($i = 1, 2, 3, \ldots$) denote events (subsets of Ω). A *probability function* P is a function which assigns to each event $E_i \subseteq \Omega$ a real number $P(E_i)$, called the probability of E_i, subject to the following restrictions.

P1 $P(E_i) \geq 0$ for each event $E_i \subseteq \Omega$

P2 $P(\Omega) = 1$

P3 If $E_1, E_2, E_3, \ldots$ is a sequence of disjoint events (that is, $E_i \cap E_j = \phi$ whenever $i \neq j$) then

$$P\left(\bigcup_i E_i\right) = \sum_i P(E_i) \qquad \blacktriangleright$$

*Sommer, R., "Studies in Personal Space," *Sociometry* **22**, 247–260 (1959).

Example 1 Consider again the rat running a T-maze. Let us take

$$\Omega = \{L, R\}$$

and define a probability function P on the subsets of Ω by

$$P(\Omega) = 1, \quad P(\{L\}) = \tfrac{3}{4}, \quad P(\{R\}) = \tfrac{1}{4} \quad \text{and} \quad P(\phi) = 0$$

It is easily checked that Axioms P1–P3 are satisfied.

Note that *any* assignment of the form

$$P(\Omega) = 1, \quad P(\{L\}) = p, \quad P(\{R\}) = 1 - p \quad \text{and} \quad P(\phi) = 0$$

where $0 \le p \le 1$, represents a valid assignment of probabilities. Except for these restrictions, the axioms say nothing about which values *should* be assigned. ▶

Example 2 A favorite carnival game is played by tossing pennies on a board roughly 5-feet square marked off in 1-inch squares. If the penny falls entirely within a small square, the player wins a nickel, while if it falls on a line, the penny is lost.

Let us take as a sample space the set of points indicated in Figure 1(a) and make the rough approximation that the probability of landing in any region is proportional to the area of the region. Since a penny is $\tfrac{3}{4}$ inch in diameter, its center must land at least $\tfrac{3}{8}$ inch away from any line if the player is to win (see Figure 1(b)). Since probabilities are proportional to areas, the probability of winning is $(\tfrac{1}{4})^2/(1)^2 = \tfrac{1}{16}$. ▶

Example 3 Suppose

$$\Omega = \{a_1, a_2, \ldots, a_n\}$$

is a sample space containing n elements. For $j = 1, 2, \ldots, n$, let $S_j = \{a_j\}$ be the simple event containing the single element a_j. Writing $p_j = P(S_j)$,

FIGURE 1

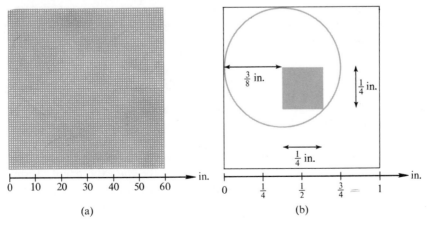

(a)　　　　　(b)

we must have $p_j \geq 0$ (Axiom P1), and

$$p_1 + p_2 + \cdots + p_n = 1$$

since $\Omega = S_1 \cup S_2 \cup \cdots \cup S_n$ is a partitioning of Ω into disjoint sets (Axioms P2 and P3).

Now let E be the compound event $\{a_1, a_2, \ldots, a_k\}$, where $k \leq n$. Then, using Axiom P3,

$$P(E) = P(S_1 \cup S_2 \cup \cdots \cup S_k) = P(S_1) + P(S_2) + \cdots + P(S_k)$$
$$= p_1 + p_2 + \cdots + p_k \qquad \blacktriangleright$$

The argument in Example 3 is easily generalized to any compound event and we obtain the following theorem.

Theorem 1 If E is an event containing a finite number of elements, then the probability of E is the sum of the probabilities of the simple events contained in E. $\qquad\blacktriangleright$

It follows that once probabilities are assigned to the simple events, probabilities of compound events may immediately be determined. For finite sample spaces, then, we shall ordinarily specify probabilities only for simple events and shall compute other probabilities by applying Theorem 1.

Example 4 Take $\Omega = \{1, 2, 3, 4, 5, 6\}$ as the sample space for the experiment of rolling a single die. If the die is "honest," we would assign probability $\frac{1}{6}$ to each simple event. The probability of the compound event $E = \{2, 4, 6\}$, described by "the outcome is an even number," is

$$P(E) = P(\{2\}) + P(\{4\}) + P(\{6\}) = \tfrac{1}{6} + \tfrac{1}{6} + \tfrac{1}{6} = \tfrac{3}{6} = \tfrac{1}{2} \qquad \blacktriangleright$$

If, as in Example 4, equal probabilities are assigned to the simple events, then computing probabilities of compound events becomes a simple task. Suppose Ω has n elements, so that each simple event carries probability $1/n$. To find the probability of a compound event E containing k elements, it is necessary to add k probabilities, each of which is $1/n$. Hence

$$P(E) = k\left(\frac{1}{n}\right) = \frac{k}{n}$$

All we need to know in order to compute the required probability is the number of elements in E. The counting techniques discussed in Chapter 7 are useful in this case.

The assignment of equal probabilities to the simple events is frequently implied by the choice of adjective used to describe the experiment. Terms such as "an honest die," "selected at random," "according to chance," and so forth, will be understood to call for equal assignments of probabilities.

Recall (Section 7.2) that the process of choosing a collection of elements from a given set is called *sampling*. The set from which the objects are chosen is called the *population*. Sampling may be done with replacement, in which case an object drawn from the population is replaced before the next object is drawn, or it may be done without replacement. Changing the method of sampling often changes appreciably the probability of obtaining a particular result.

Example 5 Edwards* describes an experiment to test a farmer's contention that he could detect the presence of hidden water by using a curved whalebone. The farmer was shown 10 covered cans and told that 5 contained water and 5 were empty. He was instructed to divide the cans into two equal groups, one group to consist of all cans containing water, the other to consist of all empty cans. We are interested in the probability that the farmer correctly places exactly k cans in the water group just by chance.

We may regard the cans the farmer places in the water group as a sample of size 5 drawn from a population of 10 items. Since the cans are selected without replacement, the sample space Ω contains

$$\binom{10}{5}$$

elements, corresponding to the possible sets of cans chosen.

Let A_k denote the event that exactly k cans are correctly placed in the water group. These k cans may be chosen in

$$\binom{5}{k}$$

ways from the 5 cans containing water. Similarly the other $5 - k$ cans can be incorrectly chosen in

$$\binom{5}{5 - k}$$

ways. Therefore

$$n(A_k) = \binom{5}{k}\binom{5}{5 - k}$$

Assuming equal assignments of probability to the simple events, we have

$$P(A_k) = \frac{\binom{5}{k}\binom{5}{5 - k}}{\binom{10}{5}}$$

▶

*Edwards, W., "Probability Preferences in Gambling," *American Journal of Psychology* **66**, 349–364 (1953).

Example 6 Suppose now that we alter the experiment of Example 5 to the extent that the farmer is no longer required to divide the cans into two equal groups. Instead, on each of five trials his task is to select a can containing water from the group of 10 cans. After each trial, the cans are rearranged and the task is repeated. This is an instance of sampling with replacement. As before, we are interested in the probability that the farmer selects exactly k cans containing water just by chance.

Since the cans are rearranged but not removed after each trial, any one of the 10 cans may be chosen on each of the five trials. Hence Ω contains 10^5 elements. We can compute the number of elements in event A_k as follows. First, we must choose k trials on which correct choices are made. This can be done in

$$\binom{5}{k}$$

ways. On each of these trials one of the 5 cans containing water must be chosen. There are 5^k ways since the cans are replaced after each trial. Similarly, there are 5^{5-k} ways of choosing an empty can on each of the remaining $5 - k$ trials. Thus

$$n(A_k) = \binom{5}{k} 5^k \cdot 5^{5-k}$$

Under the assumption that all simple events are equally likely, that is, have the same probability,

$$P(A_k) = \frac{\binom{5}{k} 5^k \cdot 5^{5-k}}{10^5}$$

The probability that the farmer correctly identifies k cans by chance under each experimental condition is shown in Table 1. ▶

TABLE 1. Probabilities of the events A_k in Examples 5 and 6

k	0	1	2	3	4	5
Sampling without replacement	0.004	0.099	0.397	0.397	0.099	0.004
Sampling with replacement	0.032	0.156	0.312	0.312	0.156	0.032

We see from the two previous examples that probability assignments differ according to the experimental design. In these examples, we saw designs of sampling with and without replacement. The difference of these two approaches is most marked for the events A_0 and A_5, where the farmer is 8 times more likely to identify all five cans correctly by chance when he is allowed to sample the cans with replacement than when he samples without replacement. Similarly, the probability of correctly identifying 4 or more cans by chance is 0.103 when choosing without replacement and 0.188 when choosing with replacement. Hence, if we wish to minimize the probability of ascribing to the farmer any power to divine water when in fact he is just guessing, the experimental design characterized by sampling without replacement is the better test of his contention.

It will be helpful in motivating new ideas and applying theoretical results to interpret probability in terms of the concept *relative frequency*. Thus when we say that "an event E has probability p," we will take this to mean that if the experiment is repeated a large number of times, then the event E is expected to occur in about $(100p)\%$ of these repetitions. For example, if the probability is 0.6 that the toss of a coin yields heads, then we would expect heads on about 60% of the tosses of this coin. Note that we do not say that heads *will* appear 60% of the time—the coin's relative frequency. For it is logically possible—not likely, but still possible—that every toss comes up heads. No guarantee about the performance of the coin is contained in our interpretation of relative frequency.

While the concept of relative frequency is not intended to serve as a substitute for the Probability Axioms, the analogy is a helpful one. Certainly all relative frequencies are non-negative real numbers. Moreover, every outcome of the experiment corresponds to an element of Ω, so that Ω, the certain event, occurs at each repetition of the experiment and hence always has a relative frequency of unity. Finally, if two events cannot occur simultaneously, the number of occurrences of one *or* the other is the sum of the individual numbers of occurrences (Counting Principle CP 1). Hence the same statement is true for the respective frequencies. Our analogy is now complete.

Example 7 Suppose the experiment of tossing a single die is repeated 100 times, giving the results in Table 2. Let $\Omega = \{1, 2, 3, 4, 5, 6\}$ be the

TABLE 2

Number k	1	2	3	4	5	6
Frequency of occurrence of k	14	20	19	13	18	16
Relative frequency of k	0.14	0.20	0.19	0.13	0.18	0.16

sample space for this experiment and let E_i (where $i = 1, 2, \ldots, 6$) denote the event that the number i is uppermost on the die. Then the relative frequency of the event $O = E_1 \cup E_3 \cup E_5$, that an odd number is obtained, is 0.51, this being the sum of the respective relative frequencies 0.14, 0.19, and 0.18 of the disjoint events E_1, E_3, and E_5 comprising O. The relative frequency of the event $O \cup E_4$ that the outcome was either odd or a 4 is $0.51 + 0.13 = 0.64$. The relative frequency of $\Omega = E_1 \cup E_2 \cup \cdots \cup E_6$ is $0.14 + 0.20 + 0.19 + 0.13 + 0.18 + 0.16 = 1$. ▶

PROBLEMS

1. The Fry Company has 10 district sales managers each supervising 10 salesmen. What is the probability that a group of 4 randomly selected salesmen contains at least 2 who have the same manager?

2. It is known that 45% of the 100 voters in a given precinct are Republican and 55% are Democrats. From the list of 100 eligible voters, two people are selected at random and their political affiliation determined. Find the probability that at least one of the persons selected is a Republican if the sample is drawn

(a) without replacement.   (b) with replacement.

3. From each lot of 100 items produced by a certain machine, a sample of size 10 is drawn. Since it is unreasonable to require that each lot be perfect, it is decided to accept a lot if the sample contains no more than 1 defective. What is the probability that a lot is accepted if it actually contains 5 defectives?

4. Three voters order 3 candidates according to degree of preference. If the candidates are randomly ordered by the voters, what is the probability that each voter selects a different preference ordering?

5. Suppose that an examinee guesses the answers on a multiple-choice test of n items, each item having m alternatives. Assuming the guesses to be randomly allocated among alternatives, what is the probability of getting exactly r correct answers?

6. From an operations research group consisting of 5 engineers, 3 economists, and 4 mathematicians, a task group of 4 is randomly selected. Find the probability that the group will consist of

(a) 3 engineers and 1 mathematician.

(b) 2 economists, 1 engineer, and 1 mathematician.

(c) 4 mathematicians.

7. Pascal and Suttell* describe an experiment in which a professional graphologist is presented with pairs of handwriting samples from 10 persons diagnosed as psychotics and 10 normal persons. The graphologist was able to identify correctly the sample handwriting of the psychotic in 6 of the 10 pairs presented.

(a) What is the probability of correctly identifying 6 or more pairs by chance?

(b) How many pairs would have to be judged correctly in order for the probability of getting at least this many by chance to be 0.05 or less?

8. Let the sample space Ω be partitioned into the disjoint events E, F, and G. Suppose that $P(F \cup G) = 0.3$ and $P(E) = 2P(F) + 0.5$. Find $P(E)$, $P(F)$, and $P(G)$.

*Pascal, G. R., and Suttell, Barbara, "Testing the Claims of a Graphologist," *Journal of Personality* **16**, 192–197 (1947).

9. Each member of a group is required to nominate another member for an elective office. We are interested in individual membership in n-cycles, an n-cycle being defined as a path from one member back to himself in n steps by means of directed choices (nominations). For example, in a group of 4, an individual can be in a

2-cycle • ⟷ • , a 3-cycle ◁ , or a 4-cycle ⬓

(One-cycles are not allowed.) What is the probability that in a group of size N where nominations are made at random, an individual is a member of a cycle of size

(a) 2?

(b) 3?

(c) k, where $2 \leq k \leq N$?

10. What is the probability in Problem 9 that an individual will not be a member of a cycle of any size?

18.4 BASIC THEOREMS

In order to facilitate consideration of applications of probability in the behavioral and social sciences, we list in this section some of the most useful elementary consequences of the Probability Axioms P1–P3. The proofs are simple, but should be studied carefully as they indicate the fundamentals of probabilistic thinking. We assume throughout this section that a sample space Ω and a probability function P have already been specified consistent with the axioms.

Theorem 2   The probability of the impossible event is $P(\phi) = 0$.

PROOF   According to Theorem 1, Section 1.4, we have both

$$\Omega \cup \phi = \Omega \quad \text{and} \quad \Omega \cap \phi = \phi$$

Axiom P3 applies, yielding

$$1 = P(\Omega) = P(\Omega \cup \phi) = P(\Omega) + P(\phi) = 1 + P(\phi)$$

The result follows by subtracting 1 from each side of the equality.   ▶

Thus the empty set must always have probability zero. However, Example 2 of Section 18.3 shows that the converse of this statement is *not* true. That is, a *set having zero probability need not be empty.* For in that example, a set consisting of a single point is not empty, but it has zero area and hence zero probability.

Theorem 3   For any event A, $P(A') = 1 - P(A)$. The probability of the complement of an event is 1 minus the probability of the event.

PROOF From the Venn diagram of Figure 2 we see that

$$A \cup A' = \Omega \quad \text{and} \quad A \cap A' = \phi$$

FIGURE 2

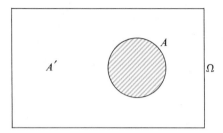

Since A and A' are disjoint, Axiom P3 applies, yielding

$$1 = P(\Omega) = P(A \cup A') = P(A) + P(A')$$

Subtracting $P(A)$ from both sides gives the result. ▶

Example 1 In Example 1 of Section 18.3 we must have

$$P(\{R\}) = 1 - P(\{L\})$$

whatever the value assigned as $P(\{L\})$. ▶

Example 2 In Example 2 of Section 18.2, if the 6 engineers are chosen at random, then the probability that at least 2 have the same occupational speciality is

$$P(E) = \frac{n(E)}{n(\Omega)} = \frac{2,320,704}{2,985,984} \approx 0.78$$

But

$$n(E) = n(\Omega) - n(E')$$

so

$$P(E) = \frac{n(\Omega) - n(E')}{n(\Omega)} = 1 - \frac{n(E')}{n(\Omega)}$$

$$= 1 - P(E') = 1 - \frac{12 \times 11 \times \cdots \times 7}{12^6}$$

$$\approx 0.78 \qquad \blacktriangleright$$

Theorem 4 If A and B are two events such that $A \subseteq B$, then $P(A) \leq P(B)$.

PROOF If $A \subseteq B$, then (see Figure 3)

$$B = A \cup (B \cap A')$$

is a union of disjoint sets. Using Axiom P3

$$P(B) = P(A) + P(B \cap A')$$

FIGURE 3

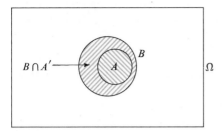

from which the result follows since $P(B \cap A')$, by Axiom P1, cannot be negative. ▶

Since every event A is a subset of the sample space Ω, it follows from Theorem 4 that $P(A) \leq P(\Omega) = 1$. Combining this with Axiom P1, we have the following theorem.

Theorem 5 For every event A, $0 \leq P(A) \leq 1$. All probabilities lie between 0 and 1. ▶

Theorem 6 Let A and B be any two events. Then

$$P(A \cup B) = P(A) + P(B) - P(A \cap B)$$

PROOF A glance at Figure 4 shows that

$$A \cup B = A \cup (B \cap A')$$

FIGURE 4

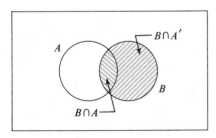

expresses $A \cup B$ as the union of disjoint sets. Hence, using Axiom P3,

$$P(A \cup B) = P(A) + P(B \cap A') \tag{1}$$

Similarly, $B = (B \cap A) \cup (B \cap A')$ is a disjoint union, from which it follows that

$$P(B) = P(B \cap A) + P(B \cap A') \tag{2}$$

Solving for $P(B \cap A')$ in (2) and substituting into (1) yields the desired result. ▶

A geometric interpretation of Theorem 6 can be obtained from Example 2 of Section 18.3. The total area covered by two overlapping regions is the sum of their individual areas minus the area of the overlap, since this was measured twice.

Example 3 Suppose that it is known that 60% of 500 male employees favor union representation and 30% oppose representation and of the 1000 female employees 30% favor union representation and 60% are opposed. The rest have no opinion. We wish to know the probability that an employee selected at random will favor union representation.

Let Ω be the sample space. Then $n(\Omega) = 1500$. Let M be the event a male employee favors a union representation and F be the event that a female employee favors union representation. Then $n(M) = 0.6 \times 500 = 300$ and $n(F) = 0.3 \times 1000 = 300$.

Since $M \cap F = \phi$, we have

$$P(M \cup F) = P(M) + P(F)$$

$$= \tfrac{300}{1500} + \tfrac{300}{1500} = \tfrac{2}{5} \qquad \blacktriangleright$$

Example 4 Complexity of cognitive structure was investigated by Scott[*] using a technique in which a subject was asked to specify a number of objects and to group the objects into as many groupings as were meaningful to him. Suppose that a given subject groups 15 objects into three groups in such a way that

12 objects are placed in group I

8 objects are placed in group II

8 objects are placed in group III

6 objects are placed in groups I and II

6 objects are placed in groups I and III

3 objects are placed in groups II and III

2 objects are placed in all three groups

Then the probability that a randomly chosen object was classified in either group II or group III is

$$P(\text{II or III}) = P(\text{II}) + P(\text{III}) - P(\text{II and III})$$

$$= \tfrac{8}{15} + \tfrac{8}{15} - \tfrac{3}{15} = \tfrac{13}{15} \qquad \blacktriangleright$$

It is useful to extend Theorem 6 to obtain a formula for the probability of the union of any number of events. For n events, $E_1, E_2, \ldots, E_n$, we shall

[*]Scott, W., "Cognitive Complexity and Cognitive Flexibility," *Sociometry* **25**, 405–414 (1962).

write s_k to denote the sum of the probabilities of all intersections obtained by taking exactly k of the events. For example, for $n = 4$,

$$s_1 = P(E_1) + P(E_2) + P(E_3) + P(E_4)$$

$$s_2 = P(E_1 \cap E_2) + P(E_1 \cap E_3) + P(E_1 \cap E_4) + P(E_2 \cap E_3)$$
$$+ P(E_2 \cap E_4) + P(E_3 \cap E_4)$$

$$s_3 = P(E_1 \cap E_2 \cap E_3) + P(E_1 \cap E_2 \cap E_4) + P(E_1 \cap E_3 \cap E_4)$$
$$+ P(E_2 \cap E_3 \cap E_4)$$

$$s_4 = P(E_1 \cap E_2 \cap E_3 \cap E_4)$$

Theorem 7 If $E_1, E_2, \ldots, E_n$ are any n events and the numbers s_k are as defined above, then

$$P(E_1 \cup E_2 \cup \cdots \cup E_n) = s_1 - s_2 + s_3 - s_4 + \cdots + (-1)^{n-1} s_n \quad \textbf{(3)}$$

PROOF We first consider the case when $n = 3$. Here Formula (3) becomes

$$P(E_1 \cup E_2 \cup E_3) = s_1 - s_2 + s_3$$
$$= P(E_1) + P(E_2) + P(E_3) - P(E_1 \cap E_2)$$
$$- P(E_1 \cap E_3) - P(E_2 \cap E_3) + P(E_1 \cap E_2 \cap E_3) \quad \textbf{(4)}$$

From Figure 5 we see that the union of three events partitions naturally into seven sections. The probability of $E_1 \cup E_2 \cup E_3$, as well as the individual events and their intersections, can be written in terms of sets I–VII marked on Figure 5. For example,

$$P(E_1 \cup E_2 \cup E_3) = P(\text{I}) + P(\text{II}) + \cdots + P(\text{VII}) \quad \textbf{(5)}$$

$$P(E_2) = P(\text{II}) + P(\text{IV}) + P(\text{V}) + P(\text{VI}) \quad \textbf{(6)}$$

$$P(E_1 \cap E_3) = P(\text{III}) + P(\text{IV}) \quad \textbf{(7)}$$

$$\vdots$$

In order to show that Formula (4) is valid, we need only show that it is equivalent to (5); that is, that the form $s_1 - s_2 + s_3$ effectively adds the probability of each of the event sections exactly once. That this is so may be

FIGURE 5

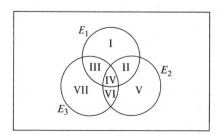

verified by substituting representations similar to (6) and (7) for each term in (4). Since section IV is contained in each of E_1, E_2, and E_3, its probability is added 3 times, subtracted 3 times, and then added once. Probabilities of sections II, III, and VI are added twice and subtracted once, while those of sections I, V, and VII are each added once.

In the general case, we wish to argue in a similar fashion that the probability of each section is added once in Formula (3). Suppose M is a section which is contained in exactly k of the n events. Then the probability of M is added exactly k times in s_1—once for each event which contains M. This probability will be subtracted out in s_2 once for each *pair* of events containing M—a total of

$$\binom{k}{2}$$

times. Similarly, $P(M)$ is added once in s_3 for each of the

$$\binom{k}{3}$$

triples of events containing M, etc. Thus $P(M)$ is added a total of

$$\binom{k}{1} - \binom{k}{2} + \binom{k}{3} - \cdots + (-1)^{k-1}\binom{k}{k}$$

times. Putting $x = 1$, $y = -1$, and $n = k$ in Formula (4) of Section 7.3, we see that this number is 1, as required. ▶

Example 5 The reliability of ratings is frequently assessed by having two judges independently order the same set of n elements according to some specified attribute. A *match* is said to occur if one element is assigned the same rank by both judges. Since reliability (consistency) appears as a departure from randomness, we are interested in the probability that matches occur by chance.

For simplicity, we consider the ordering established by one of the judges as fixed and assume that each possible ordering determined by the other has probability $1/n!$, where n is the number of elements to be ranked. Let

$$E_j \, (j = 1, 2, \ldots, n)$$

be the event that the item ranked in jth position by the first judge is also given rank j by the second judge; that is, that the two rankings match at the jth position. Then the probability of the event M_n that at least one match occurs may be computed from Theorem 7 as

$$P(M_n) = P(E_1 \cup E_2 \cup \cdots \cup E_n) = s_1 - s_2 + s_3 - \cdots + (-1)^{n-1}s_n$$

For each k, s_k denotes the sum of all probabilities corresponding to events that matches occur at k specified positions, the other positions being unrestricted. For any particular set of k positions, there is one way for the second judge to fill these positions in order to match the first judge, and

$(n - k)!$ ways of filling the other positions with the $n - k$ remaining ranks. Since there are

$$\binom{n}{k}$$

different ways to choose k positions, it follows that

$$s_k = \binom{n}{k}(n - k)! \frac{1}{n!} = \frac{1}{k!}$$

Thus

$$P(M_n) = 1 - \frac{1}{2} + \frac{1}{3!} - \frac{1}{4!} + \cdots + (-1)^{n-1} \frac{1}{n!}$$

Note that $1 - P(M_n)$ represents the first $n + 1$ terms in the expansion of e^{-1} (see Equation (4), Section 16.1). Thus as $n \to \infty$,

$$P(M_n) \to 1 - e^{-1} \approx 0.632 \qquad \blacktriangleright$$

Among the more colorful versions of the matching problem (Example 5) is the following. A group of n guests arrives at a party and each guest checks his hat. Later the hat-check girl, being bored, decides to join the party. By checkout time she is unable to distinguish one hat from another and gives them out at random. What is the probability that no guest gets his own hat?

The matching problem has potential application in any experiment where the ordering of n things established by a subject is compared with an arbitrary ordering. For example, a subject may attempt to match descriptions of objects with the objects themselves, or the preference ordering of n colors established by a subject may be compared with the ordering of the same n colors according to degree of brightness.*

The preceding discussion exhibits a recurrent pattern in the methods for attacking problems in probability. In each of the proofs of the theorems as well as in the examples, at some stage a partitioning of an event was made and Axiom P3 was applied. Since this axiom is the only one which indicates how probabilities of different events can be related, it is the key operational tool for probability theory. It is worth emphasizing as a *modus operandi* for problems in probability that the solution of a problem can often be found by a judicious partitioning of the event whose probability is desired. Indeed, in many cases the proper partitioning essentially *is* the solution, in the sense that once the partition is determined, the answer follows in a more or less routine fashion.

Thus, so far, we have begun the analysis of our problems by forming an event (set) statement, generally in the form of a partition appropriately chosen. The recognition of this approach is essential and is so often missed,

*For an application of the matching problem to personality measurement, see Vernon, P. E., "The Matching Method Applied to Investigations of Personality," *Psychological Bulletin* **33**, 149–177 (1936).

that the reader is advised to get it firmly in mind before proceeding. To sum up, the steps in solving a probability problem are to choose a sample space and a probability function, to write an event statement linking events whose probabilities are known with the event whose probability is desired, and, finally, to compute the required probability.

PROBLEMS

1. Two proposed bills are soon to be voted upon in Congress. It is estimated that 70% of the Congressmen favor bill A, 40% are opposed to bill B, and 60% are opposed to at least one of the two bills. A Congressman is selected at random. What is the probability that he will favor at least one of the bills?

2. An experimenter observing the behavior of rats in a conditioning box has established three categories of interest: E the event "rat remains relatively stationary," F the event "rat engages in bar-pressing activity," and G the event "rat engages in exploratory behavior." He assumes that $E \cap G = \phi$ and that $F \cap G = \phi$ and makes the following assignments of probability:

$$P(E \cup F) = 0.5 \qquad P(G \cup F) = 0.1$$
$$P(E \cup G) = 0.7 \qquad P(E \cup F \cup G)' = 0.4$$

Comment on the assignments of probabilities.

3. In a learning experiment, a subject is presented with a list of nonsense syllables. After a short exposure period, the list is removed and the subject is asked to recall the list. The procedure is then repeated, each repetition constituting a trial. Let A_k be the event that the subject never makes an error in recalling the list and *learns* on trial k. (All correct responses prior to learning are due to guessing.) Suppose that $P(A_k) = (g\alpha)^{k-1}g(1 - \alpha)$, where g is the probability of correctly reciting the list by chance, and $1 - \alpha$ is the per trial probability of learning.* Find the probability that the subject makes at least one error before learning the list. (Assume that the probability of eventually learning the list is 1.)

4. Let E and F be any two events. Suppose that $P(E) = 0.3$, $P(F) = 0.1$, and $P(E \cap F) = 0.1$. Find the probabilities of the events

   (a) $E' \cup F'$ (b) $E' \cup F$
   (c) $(E \cap F)'$ (d) $E \cap F'$
   (e) $E' \cap F'$ (f) $E' \cap F$

5. For any two events A and B, show that

$$P[(A \cap B') \cup (B \cap A')] = P(A) + P(B) - 2P(A \cap B)$$

Give a verbal description of this event and contrast its probability with the probability of the event $A \cup B$.

*Derived from the one-element model of paired associate learning. For a discussion see Atkinson, R. C., Bower, G. H., and Crothers, E. J., *An Introduction to Mathematical Learning Theory* (John Wiley & Sons, Inc., New York, 1965), pp. 84–108.

6. A subject is asked to match each of k facial photographs with a corresponding emotional state.

   (a) What is the probability that he correctly matches one or more photographs by chance when

   (i) $k = 3$ (ii) $k = 4$

   (iii) $k = 5$ (iv) $k = 6$

   (b) Compare the probabilities obtained in part (a) with the limiting value $1 - e^{-1}$. Argue that for $k \geq 5$, the probabilities are essentially independent of k. (See Section 16.1.) (Is this a surprising result?)

7. Use mathematical induction to prove that

$$P(E_1 \cup E_2 \cup E_3 \cup \cdots \cup E_k) = P(E_1) + P(E_2) + \cdots + P(E_k)$$

if the E_i are pairwise disjoint (that is, $E_i \cap E_j = \phi$ when $i \neq j$).

8. Prove that for any events $E_1, E_2, \ldots, E_n$,

$$P(E_1 \cup E_2 \cup \cdots \cup E_n) \leq P(E_1) + P(E_2) + \cdots + P(E_n)$$

9. Prove that $P(F \cap E') = P(F) - P(E \cap F)$.

10. Prove that for any two events E and F, $P(E) = P(E \cap F) + P(E \cap F')$.

18.5 CONDITIONAL PROBABILITY

Suppose an experiment is performed and we are interested in the probability of an event F. Suppose further that we are given additional information about the experiment, specifically, that another event E occurred. How is the probability of F affected by this knowledge? It is helpful to consider first examples in which this question can be answered on intuitive grounds. The answers thus obtained will then be used to formulate a precise mathematical definition.

Example 1   A card is drawn from an ordinary deck. Let the sample space Ω consist of 52 elements corresponding to the individual cards and assign probabilities so that each simple event carries the same probability $\frac{1}{52}$. Then the event E described by "the card drawn is a heart" has probability $P(E) = \frac{1}{4}$, as does the event F "the card drawn is a spade," while the event G "the card drawn is the seven of hearts" has probability $\frac{1}{52}$.

Now suppose we learn that the card drawn is a heart; that is, suppose we are somehow informed that event E has occurred. Obviously, event F is now impossible and should reasonably carry probability zero. Event G, on the other hand, becomes one of only 13 possibilities which, from the information available, should still be equally likely, and hence should now have probability $\frac{1}{13}$.

▶

Example 2 A coin is tossed three times. Let us take

$$\Omega = \{HHH, HHT, HTH, HTT, THH, THT, TTH, TTT\}$$

denote the simple events by $E_1 = \{HHH\}$, $E_2 = \{HHT\}$, ..., $E_8 = \{TTT\}$, and assign probabilities $\frac{1}{4}, \frac{1}{4}, \frac{1}{8}, \frac{1}{16}, \frac{1}{16}, \frac{1}{16}, \frac{1}{16}, \frac{1}{8}$, respectively, to these simple events. Let

$$E = \{HHT, HTH, HTT, THH, THT, TTH\} = E_2 \cup E_3 \cup \cdots \cup E_7$$

be the event described by "at least one of each face occurs" and

$$F = \{HHH, HHT, HTH, HTT\} = E_1 \cup E_2 \cup E_3 \cup E_4$$

the event "the first toss is heads." On the basis of our assignment of probabilities

$$P(E) = \tfrac{5}{8} \quad \text{and} \quad P(F) = \tfrac{11}{16}$$

If, now, the experiment is performed and we are told that at least one of each face turned up, what is the probability that the first toss was heads? We reason as follows. We know E has occurred, but have no other information. From the initial assignment of probabilities, it is more likely that $E_2 = \{HHT\}$ occurred than that $E_6 = \{THT\}$ occurred. Indeed, in the absence of other information and in keeping with our original assignment of probabilities, the occurrence of E_2 is 4 times as likely as that of E_6. Thus the *new* probabilities of E_2 and E_6 (and of any other pair of events contained in E) should be in the *same proportion* as the old. That is, there is a constant of proportionality c such that each new probability of an event in E is obtained by multiplying the corresponding original probability by c. Denoting the new probabilities by Q, this means that $Q(E_2) = cP(E_2) = c(\frac{1}{4})$, $Q(E_6) = cP(E_6) = c(\frac{1}{16})$, ... We may solve for c by noting that since E must occur, its new probability $Q(E)$ should be unity. Thus

$$1 = Q(E) = cP(E)$$

or

$$c = \frac{1}{P(E)}$$

We now assign new probabilities so that

(i) each simple event in E is assigned its old probability *divided* by $P(E)$;

(ii) each simple event in E' is assigned probability zero.

Returning to the original question, we have

$$Q(F) = Q(\{HHH\}) + Q(\{HHT\}) + Q(\{HTH\}) + Q(\{HTT\})$$

$$= 0 + \frac{P(\{HHT\})}{P(E)} + \frac{P(\{HTH\})}{P(E)} + \frac{P(\{HTT\})}{P(E)}$$

$$= \frac{P(\{HHT, HTH, HTT\})}{P(E)} = \frac{P(E \cap F)}{P(E)} \tag{8}$$

For our example,

$$Q(F) = \frac{P(E \cap F)}{P(E)} = \frac{\frac{7}{16}}{\frac{5}{8}} = \frac{7}{10} \qquad \blacktriangleright$$

Equation (8) provides a formula for revising original assignments of probabilities when it is known that an event E has occurred. For any event F, such a revised probability is called the *conditional probability of F given E*, and is denoted by $P(F \mid E)$. Note that since $E \cap F \subseteq E$, and thus by Theorem 4, $P(E \cap F) \leq P(E)$, the formula

$$P(F \mid E) = \frac{P(E \cap F)}{P(E)} \qquad (9)$$

always gives a number between 0 and 1. Thus it makes sense to call $P(F \mid E)$ a probability. We note that Formula (9) is meaningful only if $P(E) > 0$. If $P(E) = 0$, conditional probabilities given E are not defined. However, (9) makes sense even when E and F have no elements in common. For in this case $E \cap F = \phi$ so that $P(F \mid E) = 0$, consistent with case (ii) of Example 2.

Example 3   In a study by Hake and Hyman,* subjects were asked to predict whether a horizontal H or vertical V row of lights would be illuminated on the next trial. The experimenters controlled the sequence of the stimulus presentations so that $P(H \mid H) = P(V \mid V) = 0.8$ for one group and $P(H \mid H) = 0.9$ and $P(V \mid V) = 0.7$ for another group. In other words, the probability of the stimulus events H or V being repeated on the next trial differed for the two groups. $\blacktriangleright$

If in Equation (9) we think of keeping E fixed and letting F vary over the subsets of the sample space Ω, we obtain a function $P(\cdot \mid E)$, called the *conditional probability function corresponding to E*, which assigns numbers to events. It is an important fact that any such function is itself a probability function defined on Ω. That is to say, any conditional probability function satisfies the basic Axioms P1–P3.

Theorem 8   Let E be an event having positive probability. Then the function $P(\cdot \mid E)$ defined for each event F by

$$P(F \mid E) = \frac{P(E \cap F)}{P(E)}$$

(Equation (9)) is a probability function.

PROOF   (1) $P(F \mid E) = P(E \cap F)/P(E) \geq 0$ for all F since $P(E) > 0$ by hypothesis and $P(E \cap F) \geq 0$.

(2) $P(\Omega \mid E) = P(E \cap \Omega)/P(E) = P(E)/P(E) = 1$.

*Hake, H. W., and Hyman, R., "Perception of the Statistical Structure of a Random Series of Binary Symbols," *Journal of Experimental Psychology* **45**, 64–74 (1953).

(3) If the events $F_1, F_2, F_3, \ldots$ are disjoint, then so are the events $F_1 \cap E$, $F_2 \cap E$, $F_3 \cap E$, $\ldots$, and we have

$$P\left(\bigcup_{n=1}^{\infty} F_n \mid E\right) = \frac{P\left[E \cap \left(\bigcup_{n=1}^{\infty} F_n\right)\right]}{P(E)} = \frac{P\left[\bigcup_{n=1}^{\infty} (E \cap F_n)\right]}{P(E)}$$

$$= \frac{\sum_{n=1}^{\infty} P(E \cap F_n)}{P(E)} = \sum_{n=1}^{\infty} \frac{P(E \cap F_n)}{P(E)} = \sum_{n=1}^{\infty} P(F_n \mid E) \qquad \blacktriangleright$$

As an immediate consequence of this result, it follows that the function $P(\cdot \mid E)$ has all the properties derived in Section 18.4 for probability functions. In particular, corresponding to Theorems 2–6, we have the following theorem.

Theorem 9 The following statements are valid for conditional probabilities:

   (i) $P(\phi \mid E) = 0$

   (ii) $P(F' \mid E) = 1 - P(F \mid E)$

   (iii) If $F \subseteq G$, $P(F \mid E) \le P(G \mid E)$

   (iv) For any event F, $0 \le P(F \mid E) \le 1$

   (v) $P(F \cup G \mid E) = P(F \mid E) + P(G \mid E) - P(F \cap G \mid E)$ $\blacktriangleright$

PROBLEMS

1. A state legislature is composed of 60% Republicans and 40% Democrats. It is known that 35% of the Republicans and 60% of the Democrats favor a pending piece of legislation. What is the conditional probability that a legislator who favors the legislation is a Republican? a Democrat?

2. The Keep Cool Company specializes in the sale of air conditioners. It finds that the seasonal demand for its product varies depending upon whether the summer is abnormally cool, average, or abnormally warm. On the basis of past experience, they assign the following conditional probabilities to the various states of demand:

	demand (number of units)		
	0–100	100–300	300–500
x_1: cool summer	0.8	0.2	0
x_2: average summer	0.1	0.5	0.4
x_3: warm summer	0	0.4	0.6

Suppose that $P(x_1) = 0.2$, $P(x_2) = 0.5$, and $P(x_3) = 0.3$. The company wishes to stock only enough units to insure an adequate supply to meet the upper limit in the demand state most likely to occur. If air conditioners can be purchased only in 100-unit lots and the company wishes to order early for the purpose of taking advantage of preseason price discounts, how many lots should they order?

3. It is known that in an unincorporated community of 100 families, 60% favor incorporation and 40% are opposed. A sample of 10 families is drawn and their preferences noted.

   (a) Find the conditional probability that the fifth family sampled favors incorporation given that 7 of the 10 families in the sample favored incorporation

   (i) if the sample is drawn with replacement.

   (ii) if the sample is drawn without replacement.

   How do you interpret these results?

   (b) Show that if a sample of size n, whether drawn with or without replacement, contains k families who favor incorporation, then the probability is k/n that any particular family sampled favors incorporation.

4. Consider a family which has two children. Assume that each child is equally likely to be a boy or a girl. If we know that the conditional probability is $\frac{1}{2}$ that both children are male given that the older child is male, what is the conditional probability that both are male given that at least one of the children is a male?

5. From a three-man interview team, one man is to be selected at random for a special assignment. One of the members, having some familiarity with probability, reasons that the probability of his being selected is $\frac{1}{3}$. He asks the boss to tell him which one of his fellow members will not be chosen, claiming that since he already knows that at least one member will not be chosen, there can be no harm in divulging this information. The boss refuses on the grounds that if our probabilistically inclined man knew which member would not be chosen, then the probability of his being chosen would increase to $\frac{1}{2}$, since he would be one of two members, one of whom would be chosen for the special assignment. Show that the probability is still $\frac{1}{3}$ even if the boss answered the question, providing that in the event that our inquisitive man were to be selected the boss is just as likely to name one of the remaining two members as the other.

6. Suppose that a population* of applicants for entrance to a private college, where all applicants take a certain scholastic aptitude test (SAT), is divided into four categories: x_1, high school graduate, SAT above 70; x_2, not high school graduate, SAT above 70; x_3, high school graduate, SAT below 70; and x_4, not high school graduate, SAT below 70. Assume on the basis of past experience that for a given applicant selected, $P(x_1) = 0.60$, $P(x_2) = 0.15$, $P(x_3) = 0.05$, and $P(x_4) = 0.20$. Suppose further that each applicant is interviewed and a decision made to accept y_1 or to reject y_2 the applicant. Let the tabulations of the interviewees'

*Example adapted from Cronbach, L. J., and Gleser, G. C., *Psychological Tests and Personnel Decisions* (University of Illinois Press, Urbana, Ill., 1957).

conditional probabilities $P(y_i \mid x_j)$ be as follows:

	decision	
category	accept	reject
x_1	0.80	0.20
x_2	0.80	0.20
x_3	0.20	0.80
x_4	0.10	0.90

Find the array of conditional probabilities $P(x_i \mid y_j)$ where each $P(x_i \mid y_j)$ denotes the probability that the applicant belongs in category x_i given the result (y_j) of the interviewer's decision.

7. If B is an event with positive probability, show that for any event A
   (a) $A \subseteq B$ implies $P(A \mid B) = P(A)/P(B)$.
   (b) $B \subseteq A$ implies $P(A \mid B) = 1$.

8. Prove Theorem 9 directly from Formula (9) without using theorems established in Section 18.4.

9. Prove that if F and G are any events, then

$$P(F \cap G \mid E) \le P(F \mid E) \le P(F \cup G \mid E) \le P(F \mid E) + P(G \mid E)$$

10. Prove the following statements.
    (a) $P(E \mid E) = 1$ if $P(E) > 0$.
    (b) If $P(E) = 1$ and F is any event, then $P(F \mid E) = P(F)$.
    (c) If $P(E) > 0$ and E and F are mutually exclusive, then $P(F \mid E) = 0$.
    (d) $P(F \mid E') = [P(F) - P(E \cap F)]/[1 - P(E)]$.

11. Suppose that E and F are two events, each with positive probability. Show that only one of the following statements is, in general, true.
    (a) $P(E \mid F) + P(E' \mid F') = 1$
    (b) $P(E \mid F) + P(E' \mid F) = 1$
    (c) $P(E \mid F) + P(E \mid F') = 1$

12. Let E be an event having positive probability and suppose that events $F_1, F_2, F_3, \ldots$ partition the sample space Ω. Prove that $\sum_i P(F_i \mid E) = 1$
    (a) from basic principles.
    (b) using Theorem 8.

18.6 APPLICATIONS OF CONDITIONAL PROBABILITY

The definition (9) specifies $P(F \mid E)$ in terms of previously assigned probabilities $P(E)$ and $P(E \cap F)$. In practice, however, it is often more natural to assign conditional probabilities first and to use these to specify the probability function P. This is done by solving (9) for $P(E \cap F)$ to obtain the following theorem.

Theorem on Compound Probabilities For any events E and F such that $P(E) > 0$

$$P(E \cap F) = P(E) \cdot P(F \mid E) \tag{10}$$

The probability of E *and* F is the probability of E times the conditional probability of F, given E. ▶

Example 1 Two balls are drawn at random and without replacement from an urn* containing two red balls and three green balls. What is the probability that a red ball is drawn first followed by a green ball?

Since draws are to be made "at random," equal probabilities are assigned to each ball in the urn. The probability of obtaining a red ball on the first draw is thus $P(R_1) = \frac{2}{5}$. If a red ball is drawn first, one red and three green balls remain so that the (conditional) probability of green on the second draw would be $P(G_2 \mid R_1) = \frac{3}{4}$. These assignments together imply that

$$P(R_1 \cap G_2) = P(R_1)P(G_2 \mid R_1) = \tfrac{2}{5} \cdot \tfrac{3}{4} = \tfrac{3}{10}$$ ▶

The Theorem on Compound Probabilities may be extended to provide a rule for computing the probability of any intersection of events by multiplying conditional probabilities.

Theorem on Compound Probabilities **General Form** Let E_1, E_2, $\ldots$, E_n be any collection of events for which $P(E_1 \cap E_2 \cap \cdots \cap E_{n-1}) > 0$. Then

$$P(E_1 \cap E_2 \cap \cdots \cap E_n)$$
$$= P(E_1)P(E_2 \mid E_1)P(E_3 \mid E_1 \cap E_2) \cdots P(E_n \mid E_1 \cap E_2 \cap \cdots \cap E_{n-1}) \tag{11}$$

PROOF When $n = 2$, (11) becomes $P(E_1 \cap E_2) = P(E_1)P(E_2 \mid E_1)$, which is equivalent to (10). Assume now that (11) is true for $n = k \geq 2$. Using (10) with $E = E_1 \cap E_2 \cap \cdots \cap E_k$ and $F = E_{k+1}$, we have

$$P(E_1 \cap E_2 \cap \cdots \cap E_k \cap E_{k+1})$$
$$= P(E_1 \cap E_2 \cap \cdots \cap E_k)P(E_{k+1} \mid E_1 \cap \cdots \cap E_k)$$

The inductive assumption applied to $P(E_1 \cap \cdots \cap E_k)$ produces the required result. ▶

Example 2 The one-element learning model has been devised to characterize the learning of a simple stimulus–response association. It is assumed that on each trial the stimulus element is in one of two states—C associated

---

*Examples of the type discussed in this section were first considered in 18th and 19th century writings. The use of urns and other classical objects is derived from traditional terminology.

with the correct response or $\overline{C}$ not associated with the correct response. If the element is in state $\overline{C}$ on the ith trial, it is assumed to move to state C on the $(i + 1)$st trial with probability p. Once the element is in state C, it is assumed to remain there indefinitely. At the start of the experiment, the stimulus element is assumed to be in state $\overline{C}$. Symbolically,

$$P(\overline{C}_1) = 1 \quad P(C_{i+1} \,|\, \overline{C}_i) = p \quad \text{and} \quad P(\overline{C}_{i+1} \,|\, C_i) = 0 \qquad \text{(12)}$$

We are interested in the probability of the event C_n that the subject is in state C on trial n. Since on any trial, the stimulus element must be either in C or $\overline{C}$, $P(C_n) = 1 - P(\overline{C}_n)$ and we need only derive $P(\overline{C}_n)$.

In order for an element to be in state $\overline{C}$ on trial n, it must have been in state $\overline{C}$ on each of the preceeding $n - 1$ trials; that is,

$$\overline{C}_n = \overline{C}_1 \cap \overline{C}_2 \cap \cdots \cap \overline{C}_n \qquad \text{(13)}$$

Applying the Theorem on Compound Probabilities gives

$$P(\overline{C}_n) = P(\overline{C}_1)P(\overline{C}_2 \,|\, \overline{C}_1)P(\overline{C}_3 \,|\, \overline{C}_1 \cap \overline{C}_2) \cdots P(\overline{C}_n \,|\, \overline{C}_1 \cap \overline{C}_2 \cap \cdots \cap \overline{C}_{n-1})$$

Using (12) and (13) this reduces to

$$P(\overline{C}_n) = P(\overline{C}_1)P(\overline{C}_2 \,|\, \overline{C}_1) \cdots P(\overline{C}_n \,|\, \overline{C}_{n-1}) = (1 - p)^{n-1}$$

so that

$$P(C_n) = 1 - (1 - p)^{n-1} \qquad \blacktriangleright$$

Example 3 A member of Congress participates in successive roll-call votes on an issue. Suppose that if he votes in favor of the issue on the ith roll call, event F_i, he will vote favorably on the $(i + 1)$st roll call with probability $1 - p$. If he votes against the issue on the ith roll call, event U_i, he will vote favorably on the $(i + 1)$st roll call with probability p. If the Congressman is initially in favor of the issue, what is the probability that he will vote favorably on (a) the fourth roll-call vote? (b) the nth roll-call vote?

The solution is as follows. (a) Let $p_n = P(F_n)$ denote the probability of voting favorably on the nth roll call. For the case $n = 4$, the problem may be solved with the aid of the tree diagram in Figure 6. The entries p and $1 - p$ indicate the appropriate conditional probabilities for the various branches. According to the Theorem on Compound Probabilities, the probability for any path is found by multiplying the composite branch probabilities. The probability $P(F_4)$ of a favorable vote on the fourth trial, given a favorable position initially, is the sum of the probabilities of all paths ending in F on trial 4. Thus

$$p_4 = P(F_4) = (1 - p)^4 + p^4 + 6(1 - p)^2 p^2$$

(b) For the general case, we note first that, regardless of the particular sequence of votes cast by the Congressman, he must vote either for the issue or against it on the nth roll call. This means that

$$F_n \cup U_n = \Omega$$

FIGURE 6

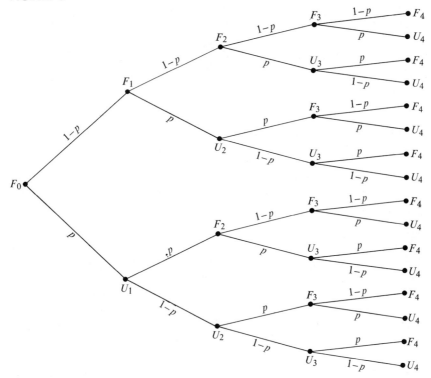

and thus

$$F_{n+1} = F_{n+1} \cap \Omega = F_{n+1} \cap (F_n \cup U_n)$$
$$= (F_{n+1} \cap F_n) \cup (F_{n+1} \cap U_n) \tag{14}$$

follows from the distributive law for sets.

Writing probabilities in (14) gives

$$p_{n+1} = P(F_{n+1}) = P(F_n \cap F_{n+1}) + P(U_n \cap F_{n+1})$$
$$= P(F_n)P(F_{n+1} \mid F_n) + P(U_n)P(F_{n+1} \mid U_n)$$
$$= P(F_n)(1-p) + P(U_n)p$$

Since $P(U_n) = 1 - P(F_n) = 1 - p_n$, this may be written in the form of a recursive relation as

$$p_{n+1} = p_n(1-p) + (1-p_n)p$$
$$= (1-2p)p_n + p$$

or as a difference equation

$$(\Delta + 2p)p_n = p$$

Solving this equation we find

$$p_n = (1 - 2p)^n p_0 + \frac{1 - (1 - 2p)^n}{2}$$

For large n, p_n is close to $\frac{1}{2}$ regardless of the initial probability p_0, a fact which may help explain the standard parliamentary device of minority delaying tactics. ▶

Example 4 *Polya Urn Model** An urn initially contains r red and g green balls. A ball is drawn at random. It is replaced and c balls of the *same* color are added to the urn (c may be zero, positive, or negative). A second ball is then drawn, replaced, and again c balls of the color drawn are added to the urn. The outcome of drawing a red ball on the ith drawing is denoted by R_i, and the outcome of drawing a green ball on the ith drawing by G_i. This process may be continued indefinitely.

The addition to the urn of balls of the same color drawn provides a rough model of phenomena where the occurrence of one case increases the probability of further cases. For example, this model could be used to describe the spread of rumors or contagious diseases. The Polya urn model becomes the statistical model of sampling with replacement from a finite population when $c = 0$, and it becomes the sampling-without-replacement model when $c = -1$. Thus, we see that the Polya urn model is a versatile model and has many applications.

We shall consider various aspects of the Polya urn model and leave others as exercises.

(a) The probability that the second ball is green G_2, given that the first ball is red R_1, is

$$P(G_2 \mid R_1) = \frac{g}{r + g + c}$$

For, if the first ball is red, the second drawing is made at random from an urn containing $r + c$ red balls and g green balls.

(b) The event "red on draw two" can occur following either red or green on draw one. Symbolically,

$$R_2 = (R_1 \cap R_2) \cup (G_1 \cap R_2)$$

(see Figure 7) so that we have

$$P(R_2) = P(R_1 \cap R_2) + P(G_1 \cap R_2)$$

$$= P(R_1)P(R_2 \mid R_1) + P(G_1)P(R_2 \mid G_1)$$

*After George Polya (1887–   ), who was born in Budapest, but who has spent most of his teaching career in this country at Stanford University.

FIGURE 7

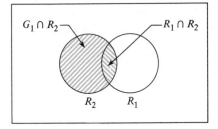

Arguing as in (a) gives

$$P(R_2) = \frac{r}{r+g} \cdot \frac{r+c}{r+g+c} + \frac{g}{r+g} \cdot \frac{r}{r+g+c}$$
$$= \frac{r}{r+g}$$

and we find that $P(R_2) = P(R_1)$.

Let us examine what this means for the case $c = -1$. Suppose that among 50 new cars there is one "lemon." Your chances of getting it are the same whether you are the first or the second customer. In Problem 12, we find that in the Polya urn model, $P(R_i) = P(R_j)$ for all i and j. For the present situation this means that no matter how many cars have already been sold, your chance of being unlucky is still $\frac{1}{50}$. This assumes, of course, that you are ignorant of which car is defective. If the defective car has actually been sold prior to your purchase, your chance is reduced to zero; while if it remains, your chance has increased over that of the initial customer. In the absence of precise knowledge, the probabilities of these two possibilities offset each other so as to yield $\frac{1}{50}$.

(c) Proceeding from the definition of conditional probability (Equation (9)), we have

$$P(R_1 \mid R_2) = \frac{P(R_1 \cap R_2)}{P(R_2)} = \frac{P(R_1)P(R_2 \mid R_1)}{P(R_2)}$$

But from (b), $P(R_1) = P(R_2)$ and thus

$$P(R_1 \mid R_2) = P(R_2 \mid R_1) = \frac{r+c}{r+g+c}$$

This appears to be a surprising result since we might be tempted to reason that "R_1 occurs before R_2 and hence cannot be affected by R_2." However, it is easy to see that this is *false* reasoning. Using (10) we may write, for any events R_1 and R_2 having positive probabilities,

$$P(R_1 \cap R_2) = P(R_1)P(R_2 \mid R_1) = P(R_2)P(R_1 \mid R_2) \qquad \textbf{(15)}$$

From this it follows that if $P(R_2 \mid R_1) \neq P(R_2)$, then $P(R_1 \mid R_2) \neq P(R_1)$. Obviously this argument applies to any two events, say E and F. Note that the conditional probability $P(F \mid E)$ is defined for *any* events E and F for which $P(E) > 0$. There is no requirement that F follow E in a time sequence, our interpretation notwithstanding. In fact, there is no temporal element in probability, an event being just a subset of the sample space.

To look at the relation between $P(R_1 \mid R_2)$ and $P(R_2 \mid R_1)$ another way, suppose we draw without replacement from an urn containing one red ball and three green balls. What is the value of $P(R_1 \mid R_2)$? Obviously the answer is zero, for if the second ball is red, the first cannot be. Here the result of the second draw affects our assignment of probabilities to the first draw. Equation (15) shows that this also happens in less obvious cases.

Finally, let us return to Example 1, since it is a special case of the Polya urn model, in which $P(R_1 \mid R_2) = P(R_2 \mid R_1) = \frac{1}{4}$. This result may be interpreted in a relative-frequency sense as follows. Consider the experiment of drawing twice from the urn and repeat this experiment many times over, say 10 million times. Since $P(R_1 \cap G_2) = P(G_1 \cap R_2) = P(G_1 \cap G_2) = \frac{3}{10}$ and $P(R_1 \cap R_2) = \frac{1}{10}$, then according to the relative-frequency interpretation (Section 18.3), there will be roughly 3 million each of outcomes $R_1 \cap G_2$, $G_1 \cap R_2$, and $G_1 \cap G_2$, while $R_1 \cap R_2$ occurs about 1 million times. Now let us consider those outcomes, of which there are approximately 4 million, in which red appears on the second draw. Of these, red also appears on draw one in about 1 million cases, or about one-quarter of the total, agreeing with our first calculation. ▶

We conclude this section with two simple theorems which are widely used in applications.

Theorem 10 Let events E_1, E_2, E_3, ... be a partition of Ω; that is, the events E_i are disjoint and $\bigcup_i E_i = \Omega$. Let A be any event. Then

$$P(A) = \sum_i P(E_i)P(A \mid E_i) \tag{16}$$

PROOF We have

$$A = A \cap \Omega = A \cap \left(\bigcup_i E_i \right) = \bigcup_i (A \cap E_i)$$

using the distributive law for sets. Since the E_i are disjoint, then so are $A \cap E_1$, $A \cap E_2$, ... and thus

$$P(A) = P\left[\bigcup_i (A \cap E_i) \right] = \sum_i P(A \cap E_i)$$

Writing $P(A \cap E_i) = P(E_i)P(A \mid E_i)$ completes the proof. ▶

Theorem 11 *Bayes' Theorem** Let E_1, E_2, E_3, ... partition Ω and let A be any event. Then, for each k,

$$P(E_k \mid A) = \frac{P(E_k)P(A \mid E_k)}{\sum_i P(E_i)P(A \mid E_i)}$$

*After the Reverend Thomas Bayes (1702–1761), an English mathematician.

PROOF The result follows immediately by substituting Formula (16) into

$$P(E_k \mid A) = \frac{P(A \cap E_k)}{P(A)} = \frac{P(E_k)P(A \mid E_k)}{P(A)} \qquad \blacktriangleright$$

Example 5 Suppose that two classes of patients enter a mental hospital, say schizophrenics and neurotics. It has been found that drug therapy is most effective in the treatment of schizophrenics, whereas psychotherapy is particularly effective in the treatment of neurotics. It is expedient that a patient be correctly classified since a misclassification can prolong confinement. In order to aid in classification, a psychological test is administered to entering patients. Past experience has shown that 20% of the neurotics and 60% of the schizophrenics pass the test, that is, equal or exceed an arbitrary cutting score. It is also known that in an undiagnosed group, roughly 40% actually are schizophrenic and 60% are neurotic.

From a large sample of incoming patients, one patient is selected at random and given the test. If the patient passes the test, what is the probability that he actually is a neurotic?

Let E_1 denote the event that the patient selected is a neurotic and E denote the event that the patient passes the test. The statement of the problem requires

$$P(E_1) = 0.6 \quad P(E_1') = 0.4 \quad P(E \mid E_1) = 0.2 \quad \text{and} \quad P(E \mid E_1') = 0.6$$

From Theorem 11, we have

$$P(E_1 \mid E) = \frac{P(E_1) \times P(E \mid E_1)}{P(E_1)P(E \mid E_1) + P(E_1')P(E \mid E_1')}$$

$$= \frac{0.6 \times 0.2}{(0.6 \times 0.2) + (0.4 \times 0.6)} = \frac{1}{3}$$

Given the information that a patient attains a passing score, the probability of his being a neurotic is reduced to 0.33, as compared with an unconditional probability of 0.60. Hence, if it is known that a patient passed the test, the odds are 2 to 1 that the patient is a schizophrenic and should therefore be given drug therapy rather than psychotherapy. $\blacktriangleright$

This example illustrates the use of Bayes' Theorem. The events E_1, E_2, E_3, ... in Theorem 11 are termed *hypotheses* and are assumed to partition Ω. The probability $P(E_k)$ is called the *a priori probability* of the hypothesis E_k. The probability $P(E_k \mid E)$ is termed the *a posteriori probability* of the hypothesis E_k. Given that an event E has occurred, the hypothesis with the highest *a posteriori* probability of occurrence is assumed to be the most plausible explanation accounting for the event.

PROBLEMS

1. A questionnaire is submitted to a group of employed people, 5% of whom are unskilled laborers, 10% semiskilled laborers, 30% skilled laborers,

30% clerical workers, and 25% managerial and professional workers. Fifty percent of the unskilled, 40% of the semiskilled, 45% of the skilled, 30% of the clerical, and 20% of the managerial and professional workers report that they are dissatisfied with their present job. What is the probability that a respondent selected at random will report that he is satisfied with his job?

2. Relative to Example 2, let E_n be the event that an incorrect response occurs on trial n and $\overline{E}_n$ be the event that a correct response occurs on trial n. If the probabilities of an incorrect response, given that the subject is in the conditioned and unconditioned state, respectively, are

$$P(E_n \mid C_n) = 0 \quad \text{and} \quad P(E_n \mid \overline{C}_n) = 1 - \frac{1}{r}$$

show that

(a) the probability of an error on trial n is

$$P(E_n) = \left(1 - \frac{1}{r}\right)(1 - p)^{n-1}$$

(b) the probability of a correct response on trials 2 and 3 is

$$P(\overline{E}_2 \cap \overline{E}_3) = p + (1 - p)\frac{p}{r} + (1 - p)^2 \left(\frac{1}{r}\right)^2$$

3. Find the general formula in Problem 2 for

(a) $P(\overline{E}_2 \cap E_3)$ (b) $P(E_2 \cap \overline{E}_3)$ (c) $P(E_2 \cap E_3)$

4. Suppose that in a certain geographic area, census data reveals that 40% of the rural population move to the city yearly, while 20% of the urban dwellers move to the country each year. Initially, 60% of the population were rural dwellers. In the long run, what proportion of the total population will be urban dwellers?

5. In an attempt to apply Bayes' Theorem to mental testing, Calandra* assumes that an examinee either knows the response to a test item, probability p, or he guesses, probability $1 - p$. Let us assume that, given an examinee knows the answer, the probability of his answering the item correctly is 1. Given that an examinee does not know the answer, the probability of his answering correctly is assumed to be $1/k$, where k is the number of multiple-choice alternatives.

Given a correct answer, show that the conditional probability that the respondent knew the item is

$$\frac{kp}{1 + (k - 1)p}$$

6. Three members of a four-member discussion group initially favor an issue, while the fourth member is opposed. Suppose that a member is selected at random and given the opportunity to defend his position. Let us further suppose that after a given member speaks, one member of the opposition (if any exists) changes his position. After the first member

*Calandra, A., "Scoring Formulas and Probability Considerations," *Psychometrika* **6**, 1–9 (1941).

speaks, a member is again randomly selected and given the opportunity to speak. The process continues indefinitely.

(a) What is the probability that the members will be in agreement at the end of the (i) third speech? (ii) fourth speech? (iii) fifth speech?

(b) What is the probability that a favorably inclined member will be selected to give the (i) second speech? (ii) third speech? (iii) fourth speech?

7. The following urn scheme is proposed as a model of the spread of rumors through a primitive tribe. An urn initially contains b black balls and w white balls. If a white ball is drawn, it is replaced and E, the event that a tribe member selected at random has not heard the rumor, occurs. If a black ball is drawn, event E' occurs, the ball is replaced, and k white balls are replaced with k black balls. The sampling process continues until the white balls are exhausted. If $b = 1$, $w = 6$, and $k = 2$:

(a) What is the probability that the second tribe member sampled will not have heard the rumor?

(b) Calculate the probability that the third member sampled has not heard the rumor. Compare with (a).

(c) What is the probability that the process terminates at the end of the third sample?

(d) What is the probability that the process terminates at the end of the fourth sample?

8. Bush and Sternberg* have proposed a model which views learning as a direct change in response probabilities from one trial to the next. The basic idea is that learning a stimulus–response association occurs over a sequence of changing error probabilities denoted by $q_1, q_2, q_3 \ldots$.
   The successive q values are assumed to be related by

$$q_{n+1} = \alpha q_n$$

where $0 < \alpha < 1$.
   If p_n denotes the probability of a correct response on trial n, so that $p_n + q_n = 1$, find the probability of a correct response (a) on trial 3 and (b) on trial n. What is the limiting probability as $n \rightarrow \infty$?

9. The Atlas Machine Company has an automatic machine process which produces screws in 100-unit lots. However, the process is not without error. Suppose for simplicity we assume that the process can be in only one of three error states E_1, E_2, and E_3 denoting respective error rates of 1, 5, and 10 defectives per lot. From past experience, the company assigns the following probabilities: $P(E_1) = 0.5$, $P(E_2) = 0.45$, $P(E_3) = 0.05$. For each error state E_i ($i = 1, 2, 3$) there is a corresponding state A_i indicating the appropriate action to be taken by the company. Let A_1 denote no corrective action, A_2 denote increasing the number of inspectors, and A_3 denote stopping the process for repair. In order to determine what action should be taken, the quality control department

*Bush, R. R., and Sternberg, S. H., "A Single-Operator Model," in *Studies in Mathematical Learning Theory*, Bush, R. R., and Estes, W. K., Eds. (Stanford University Press, Stanford, Calif., 1959), pp. 204–214.

selects a lot at random and draws a sample of 5 units from the lot. The decision rule is to select the most probable action state (there is an action state corresponding to each error state) given that the sample is found to have k defectives. What action is chosen given that the sample results in

(a) exactly 1 defective?
(b) exactly 2 defectives?
(c) 3 or more defectives?

10. The True-Tone Radio Company is negotiating a contract with a seller of electronic components which supplies components in 100-unit lots. The company wishes to protect itself against the possibility that a lot contains too many defective components. The contract therefore provides that two components will be randomly selected without replacement and tested. Unfortunately, the testing procedure is not absolutely accurate. It is estimated that on the average, 90% of all good components tested are declared acceptable while only 5% of the bad components tested are accepted.

The following alternative plans are being considered as possible guidelines in making a decision whether to accept or reject a lot:

Plan 1. If both components tested are declared to be acceptable, then accept the whole lot. Otherwise reject.

Plan 2. If both components are found to be defective, then reject the lot. Otherwise accept.

Plan 3. If both components tested are satisfactory, accept the lot, if both components tested are unsatisfactory, reject the lot. If only one component tested is satisfactory, then draw a third component at random from the remaining components and accept or reject depending on whether the component is deemed satisfactory or unsatisfactory.

(a) Denote by E the event that True-Tone Radio accepts the lot. Obtain $P(E)$ for each of the three plans if there are x components in the lot that are actually defective.

(b) Compute $P(E)$ in (a) when $x = $ 5, 10, 20, 30, 50. Draw a graph for each plan with the value of x on the horizontal axis and $P(E)$ on the vertical axis. A graph of this type is called an *operating characteristic curve* (O.C. curve) for a sampling plan.

(c) Which plan is most favorable to the buyer? to the seller?

11. Find the probability of acceptance under each plan in Problem 10 if there are x defective components and no error is made in testing the components. Draw O.C. curves for this case.

12. Relative to Example 4(b)
(a) show that $P(R_3) = r/(r + g)$.
(b) Show by induction that $P(R_n) = r/(r + g)$ for $n = $ 1, 2, 3, ...

13. (a) Use the result of Problem 12 to show that

$$P(R_n \mid R_m) = P(R_m \mid R_n) \qquad \text{for any } m \text{ and } n$$

(b) Let A and B stand for either red or green; that is, the pair (A, B) can be any of the four combinations (R, R), (R, G), (G, R), or (G, G). Then show that $P(A_m \mid B_n) = P(A_n \mid B_m)$ for all m and n.

14. Audley and Jonckheere* discuss a general urn scheme designed to model learning experiments having two subject-controlled events. Initially, the urn has r red balls and w white balls. A ball is selected at random. If it is white, event E_1 occurs, the ball is replaced, and w_1 white balls, and r_1 red balls are added to the contents of the urn. If the ball is red, event E_2 occurs, the ball is replaced, and w_2 white and r_2 red balls are added to the contents of the urn. The process is repeated indefinitely, each repetition constituting a trial. The constants w_1, r_1, w_2, and r_2 may be either negative, zero, or positive with the restriction that if any are negative, the process stops when the urn contains no balls or when the number of balls of either color is negative.

   (a) Show that in general, the results of Problem 12 do not hold for Audley and Jonckheere's urn scheme.

   (b) Use a tree diagram to show that if $r = 2$, $w = 1$, $r_1 = w_1 = 2$, $r_2 = 3$, and $w_2 = 1$, then $P(R_1) = P(R_2) = P(R_3)$.

   (c) Use a tree diagram to show that if $r = 2$, $w = 1$, $r_1 = 0$, $w_1 = -1$, $r_2 = 0$, and $w_2 = 1$, then $P(R_1) = P(R_2) \neq P(R_3)$. Argue that this implies the following result—if E is an event with probability $r/(r + w)$ where r and w are integers, it is *not* always possible to replace the experimental situation in which E arises by a single urn containing r balls of one color and w of another color.

18.7 INDEPENDENT EVENTS AND INDEPENDENT TRIALS

The probability $P(E \mid F)$ of E conditional on the occurrence of F is, in general, not the same as the unconditional probability $P(E)$. The case where these two probabilities are the same, that is, when $P(E \mid F) = P(E)$, is of special interest. If this occurs we say that event E is *independent* of the event F.

Example 1 A subject is asked on each of n trials to predict whether or not a bulb will be illuminated. Suppose that on each trial, the subject makes the same choice, say, that the bulb will be illuminated. If the bulb is illuminated on a random 60% of the trials, is the subject's choice independent of the actual state of the bulb?

Let E denote a prediction that the bulb will be illuminated and F denote actual illumination of the bulb. Then, since $P(E) = P(E \mid F) = 1$, we conclude that E is independent of F. In this case, we might instinctively expect that the events should be independent since the subject's prediction was not influenced by the actual state of the bulb.

Now suppose that the subject knows that the bulb will be illuminated 60%

*Audley, R. J., and Jonckheere, A. R., "The Statistical Analysis of the Learning Process," *British Journal of Psychology* **9**, 87–94 (1956).

of the time. Instead of making the same choice, he matches the proportion of predictions that the bulb will be illuminated; that is, 60% of the time he predicts that the bulb will be illuminated.

If we assume that the subject distributes his predictions in the same 60 : 40 ratio when the bulb is actually illuminated as when it is not illuminated, then

$$P(E) = 0.6 \quad \text{and} \quad P(E \mid F) = 0.6$$

and the events E and F are independent even though the subject's choice pattern depends upon his knowledge of $P(F)$. Events E and F will always be independent as long as the proportion of "on" predictions is the same for those trials when the bulb is on as for those when it is off. In a probability sense, this is all that is meant by independence of the events E and F. ▶

When $P(F \mid E)$ and $P(E \mid F)$ are both defined, the fact that E is independent of F implies that F is also independent of E. For in this case (from Equation (10)), we have

$$P(F \mid E) = \frac{P(E \cap F)}{P(E)} = \frac{P(F)P(E \mid F)}{P(E)} = P(F)$$

since $P(E \mid F)$ is by assumption equal to $P(E)$. In either case, we find

$$P(E \cap F) = P(E) \cdot P(F)$$

and we use this equation in our formal definition of independence.

Definition 1   Two events are said to be *independent* if and only if

$$P(E \cap F) = P(E)P(F) \tag{17}$$

▶

We shall find it convenient to refer to Equation (17) as the *multiplication rule* for events E and F.

In ordinary language, quantities which are described as independent are those which are felt to be totally unrelated and thus to have nothing in common. This intuitive notion of independence often leads to the erroneous conclusion that independent events are disjoint when, in fact, the opposite is true, as is seen in the following theorem.

Theorem 12   Independent events having positive probability cannot be disjoint.

PROOF   If E and F are disjoint events then

$$P(E \cap F) = P(\phi) = 0$$

On the other hand, if E and F are independent, then

$$P(E \cap F) = P(E)P(F)$$

This product is zero if and only if at least one of the events carries zero probability, contrary to our assumption. ▶

Example 2 If in a coin toss, the event H (heads) occurs, the event T (tails) cannot occur. Here the occurrence of one event affects the occurrence of the other and we would intuitively feel that these events, though disjoint, are not independent. The result of Theorem 12 is thus perhaps not so contrary to intuition as one might at first glance suppose. ▶

Careful study of the following examples and of Problems 2–4 will serve to clarify the common sources of confusion concerning independent events.

Example 3 Let E be any event in the sample space Ω. Then $E = E \cap \Omega$ and

$$P(E \cap \Omega) = P(E) = P(E) \cdot 1 = P(E)P(\Omega)$$

so that the certain event Ω is independent of E. In particular, setting $E = \Omega$, it follows that Ω is independent of itself. That is, if we know that Ω has occurred, the probability of Ω is not altered. ▶

Example 4 (a) Let an experiment consist of tossing an honest penny twice. Let the event E be described by "not more than one head" and F by "at least one of each face." If $\Omega = \{HH, HT, TH, TT\}$, then $E = \{HT, TH, TT\}$ and $F = \{HT, TH\}$. Assigning equal probabilities gives $P(E) = \frac{3}{4}$, $P(F) = \frac{1}{2}$, and

$$P(E \cap F) = \tfrac{1}{2} \neq P(E) \cdot P(F)$$

so that E and F are not independent.

(b) Let a second experiment consist of tossing the penny three times and let E and F be described as in (a). Again assigning equal probabilities we find $P(E) = \frac{1}{2}$, $P(F) = \frac{3}{4}$, and

$$P(E \cap F) = \tfrac{3}{8} = P(E) \cdot P(F)$$

which means that the two events are independent.

Intuitively, one might feel that the results of (a) and (b) of this example should be the same, but such intuition is not always to be trusted. One must always check to see if Formula (17) holds to verify independence. Moreover, while it is true that the *descriptions* of the events are the same, the elements of the events E of parts (a) and (b) are different, since the sample spaces differ. The same applies to the two events F and thus it is not surprising that conclusions differ in parts (a) and (b). ▶

If the probability of E is unchanged by knowledge of the occurrence of F, it seems reasonable that this probability should also be unchanged by the knowledge that F fails to occur.

Theorem 13 Two events E and F are independent if and only if the pairs (E, F'), (E', F), and (E', F') are independent pairs of events.

PROOF We shall prove the first of these statements, leaving the others as exercises. We have $E = (E \cap F) \cup (E \cap F')$, so that $P(E) = P(E \cap F) + P(E \cap F')$. Using the assumed independence of E and F gives

$$P(E \cap F') = P(E) - P(E \cap F) = P(E) - P(E)P(F)$$

$$= P(E)[1 - P(F)] = P(E)P(F') \qquad \blacktriangleright$$

In extending the concept of independence to more than two events E_1, $E_2, \ldots, E_n$, it is natural to require that Formula (17) hold for each pair E_i, $E_j (i \neq j)$. In this case we say that the events are *pairwise independent*. At the same time, one might expect that independence of events E_1, E_2, and E_3 would, for instance, imply independence of combinations such as $E_1 \cap E_2$ and E_3. However, the next example shows that pairwise independence is not sufficient for this purpose.

Example 5 Two people are contacted in a public opinion poll and queried as to whether they have a favorable F or unfavorable U reaction to the labeling of cigarettes as health hazards. An appropriate sample space is $\Omega = \{FF, FU, UF, UU\}$. Let us assign probability $\frac{1}{4}$ to each single event in Ω and consider events

$E_1 = $ "the first person was favorable"

$E_2 = $ "the second person was favorable"

$E_3 = $ "both persons gave the same reaction"

Then

$$P(E_1) = P(E_2) = P(E_3) = \tfrac{1}{2}$$

$$P(E_1 \cap E_2) = P(E_1 \cap E_3) = P(E_2 \cap E_3) = \tfrac{1}{4}$$

$$P[(E_1 \cap E_2) \cap E_3] = \tfrac{1}{4} \neq \tfrac{1}{8} = P(E_1 \cap E_2)P(E_3) = P(E_1)P(E_2)P(E_3)$$

The events are pairwise independent but E_3 is not independent of $E_1 \cap E_2$.

$\blacktriangleright$

It is clear that in addition to the pairwise independence of E_1, E_2, and E_3, we must also require

$$P(E_1 \cap E_2 \cap E_3) = P(E_1)P(E_2)P(E_3) \qquad (18)$$

to define the independence of three events. Equation (18) is referred to as the *multiplication rule* for events E_1, E_2, and E_3.

Definition 2 Three events E_1, E_2, and E_3 are said to be independent if and only if (18) holds and (17) holds for each pair; that is, if and only if the multiplication rule holds for each combination of two or more of the events.

$\blacktriangleright$

Theorem 14 If E_1, E_2, and E_3 are independent events, then so are

(a) $E_1 \cap E_3$ and E_2' (b) $E_2 \cup E_3$ and E_1

(c) E_2' and $E_3 \cup E_1'$ (d) E_1, E_2', and E_3'

PROOF We prove (a) and leave the others as exercises. It is easily verified that $E_1 \cap E_3$ may be partitioned as

$$E_1 \cap E_3 = (E_1 \cap E_3 \cap E_2) \cup (E_1 \cap E_3 \cap E_2')$$

Thus

$$
\begin{aligned}
P(E_1 \cap E_3 \cap E_2') &= P(E_1 \cap E_3) - P(E_1 \cap E_2 \cap E_3) \\
&= P(E_1)P(E_3) - P(E_1)P(E_2)P(E_3) \\
&= P(E_1)P(E_3)[1 - P(E_2)] \\
&= P(E_1 \cap E_3)P(E_2')
\end{aligned}
$$

▶

More generally, it can be proved that any event expressible solely in terms of E_1 and E_2 is independent of E_3, any event expressible in terms of E_2 and E_3 is independent of E_1, and so forth. (We do not prove this statement here, but refer the interested reader to Pfeiffer's book, listed under Supplementary Reading.)

Example 6 An established decision-making scheme requires that an individual be accepted or rejected on the basis of his scores on three tests. The decision strategy is to reject every individual who falls below a minimum cutoff score on each of the three tests.

Let E_1, E_2, and E_3 denote the events that the subject exceeds the cutoff score on the first, second, and third test, respectively. Suppose that the cutoff points were established so that

$$P(E_1) = 0.6 \qquad P(E_2) = 0.3 \qquad P(E_3) = 0.1$$

Assuming that E_1, E_2, and E_3 are independent events, what is the probability that an individual selected at random will be rejected?

Since rejection implies that an individual scores below the cutoff score on all three tests, the required probability is given by

$$P(E_1' \cap E_2' \cap E_3')$$

By the independence assumption, the probability of rejection is

$$P(E_1' \cap E_2' \cap E_3') = 0.4 \times 0.7 \times 0.9 = 0.252$$

▶

In extending the definition of independence to more than three events, we shall follow the pattern of the definition for the independence of three events.

Definition 3   Events $E_1, E_2, \ldots, E_n$ are said to be independent if and only if the multiplication rule holds for each combination of two or more of the events.   ▶

Thus a multiplication rule is required for every subcollection of events $E_1, E_2, \ldots, E_n$ except those containing none or one of the events. Since there are 2^n possible subcollections, of which one is empty and n contain a single event, there are then $2^n - n - 1$ equations which must be satisfied in order that the events be independent.

Again it is possible to prove that any two set-theoretic combinations of independent events must themselves be independent if they involve no event in common. Thus for example, if $E_1, E_2, \ldots, E_7$ are independent, then so are $(E_1 \cap E_4)' \cup E_5$ and $E_2' \cap E_6$. (Details explaining these results may be found in Pfeiffer, Chapter 2.)

Situations involving several independent events most often arise in connection with experiments composed of sequences of trials.

Example 7   Let us suppose that in the public opinion poll of Example 5, 60% of the population view cigarettes as a health hazard while 40% do not. The polling of one person constitutes a trial with possible outcomes F and U. Denote by F_n and U_n the obtaining of a favorable or an unfavorable reaction, respectively, on trial n, that is, the nth person polled.

If the trials are such that on trial n, $P(F_n) = 0.6$ and $P(U_n) = 0.4$, regardless of the outcomes of other trials, then we have a sequence of *independent trials*. The probability that, for example, the first 10 persons respond F and the next three U is then $(0.6)^{10}(0.4)^3$. Indeed every particular sequence of 13 responses, 10 of which are F and 3 of which are U would have the same probability. Since there are

$$\binom{13}{3}$$

different arrangements of 10 symbols F and 3 symbols U, the probability that of the first 13 responses, exactly 10 are F is

$$\binom{13}{3} (0.6)^{10}(0.4)^3 \qquad \blacktriangleright$$

In applications of the theory of probability, independence usually arises, as in Example 7, as an *assumption* on which the assignment of probabilities is based. Whether this is a proper criterion for making the assignment is, of course, not a part of the theory, but must be decided on the basis of the experience and judgment of the experimenter. Certainly, most of us would consider repeated tossing of a single coin as involving independent trials, while repeated presentation of the same learning situation to the same subject would not be so considered. In either case, the probabilities assigned are at the discretion of the user of probability. The theory can only be applied to deduce new probabilities from assumed ones.

On the other hand, if, having made certain assumptions, results are obtained which are at variance with one's intuition, then either assumptions or intuition must be changed. Example 4 is a case in which intuition is modified, while the next example illustrates the opposite situation.

Example 8 Numerous investigations have found that human subjects are unable to generate a random sequence of responses when so instructed. In one such study conducted by Brown,* 48 college students were each instructed to generate a sequence of heads X and tails O, which would look like the results they would expect if a balanced coin were tossed 175 times.

If the trials are independent, we would expect to find that the proportion of O on trials following an O should equal 0.5, as should the proportion of O on trials following a run of two O. In fact, the proportion of O following any pattern whatsoever should equal 0.5. Table 3 shows the results obtained.

TABLE 3

Immediately preceding response pattern	Proportion of O following	Immediately preceding response pattern	Proportion of O following
O	0.43	OOO	0.30
X	0.59	OOX	0.53
		OXO	0.42
OO	0.38	OXX	0.65
OX	0.55	XOO	0.41
XO	0.48	XOX	0.56
XX	0.66	XXO	0.53
		XXX	0.67

It is obvious that the results are not in close accord with those expected if trials are assumed to be independent. Subjects apparently tended to avoid subsequences of responses that looked "patterned." For example, subjects were more likely to generate the subsequence OX than OO. The avoidance of perceived patterning is particularly apparent in the three-deep response patterns as evidenced by the preference for subsequences $OOOX$ vs $OOOO$, $XXXO$ vs XXXX, and $OXXO$ vs $OXXX$.

On the basis of the empirical evidence, the assumption of independence does not appear warranted. More sophisticated means of evaluating whether or not response runs of X or O are excessive or may be expected from genuinely independent trials are beyond the scope of this book.† ▶

*Brown, D. L., "Non-independence in Subjectively Random Binary Sequences," Educational Testing Service Research Bulletin No. RB–64–27 (Princeton, New Jersey, 1964).

†For an introduction to the theory of runs, see Feller, W., *An Introduction to Probability Theory and its Applications* (John Wiley & Sons, Inc., New York, 1968), Vol. I, 3rd ed., Chapter 13.

Example 9 The sales manager for the MT Brush Company estimates the probability that a house call results in a sale to be $\frac{1}{5}$. If house calls are assumed to be independent trials, what is the probability that a new salesman makes his first sale on the kth house call?

Let S_i denote the event "ith house call results in a sale" and F_i the event "ith house call does not result in a sale." In order for a salesman to make his *first* sale on the kth trial, he must of necessity have been unsuccessful in making a sale on the preceding $k - 1$ trials and successful in making a sale on the kth trial. Hence, the event E_k "first sale on the kth house call" is

$$E_k = F_1 \cap F_2 \cap \cdots \cap F_{k-1} \cap S_k$$

Since the trials are assumed to be independent with constant probability of S_k,

$$P(E_k) = P(F_1)P(F_2) \cdots P(F_{k-1})P(S_k)$$
$$= \tfrac{4}{5} \times \tfrac{4}{5} \times \cdots \times \tfrac{4}{5} \times \tfrac{1}{5}$$
$$= \tfrac{1}{5}(\tfrac{4}{5})^{k-1}$$

Since in this example, k may be any positive integer, we see that independence can be applied to an infinite as well as finite sequence of trials. ▶

PROBLEMS

1. Lorge and Solomon* have developed a model of group problem-solving behavior. If it is assumed that k group members work independently and that each member has probability p_I of solving the problem, show that the probability p_G that the group of k members solves the problem is

$$p_G = 1 - (1 - p_I)^k$$

Comment on the effect of increasing group size.

2. Suppose that it is further assumed that there are s independent stages in the solution of the problem in Problem 1 and that

$$p_I = p_1 p_2 \cdots p_s$$

where p_i is the individual's probability of solution at stage i. Assuming that $p_1 = p_2 = \cdots = p_s$, and that the group solves the problem only if, at each stage, at least one member solves that stage, show that

$$p_G = [1 - (1 - p_I^{1/s})^k]^s$$

Comment on the effect of increasing the number of problem stages if we assume that $p_1 = p_I^{1/s}$ is constant.

3. The Seymore Company makes rectangular glass portholes according to contract specification for installation in space vehicles. Because of the

*Lorge, I., and Solomon, H., "Two Models of Group Behavior in the Solution of Eureka-Type Problems," *Psychometrika* **20,** 139–148 (1955).

nature of the mounting, tolerances for length and width are especially critical. It is known that 2% of the units exceed tolerance specifications for length, 3% exceed tolerance specifications for width, and 1% exceed tolerance specifications for thickness. A porthole is acceptable for delivery only if its length and width are within the specified tolerances, that is, a unit may exceed the tolerances on thickness and still be acceptable provided that the length and width are within the specified tolerances. If dimensional variability is assumed to be independent, what is the probability that at least one of six units selected at random is acceptable for delivery?

4. A test has five multiple-choice items each with four choices, only one of which is correct. Suppose a student guesses the answers to each question. Assuming that his guesses are independent, define an appropriate sample space and assign probabilities to its simple events. Find the probability that the student correctly guesses

(a) none of the items.

(b) all of the items.

(c) at least one item.

(d) at most, three items.

5. (a) An experimenter needs at least one subject in order to conduct an experiment. Suppose that he has a list of people who have indicated an interest in participating in his experiment. If each person contacted has probability of $\frac{1}{3}$ of showing and persons contacted form independent trials, how many persons should the experimenter contact to insure that the probability of at least one showing is at least 0.99?

(b) What is the probability that the fourth subject contacted by the experimenter is the first to show for the experiment?

6. The ABC Brush Company has five salesmen who are assigned separate territories. If the probability is 0.2 that a salesman will make a sale on any given day, what is the probability that at least one of the five salesmen will make a sale on a given day?

7. Suppose that you are constructing a probability model for the events A and B described below. Indicate whether the assumption of independence would be appropriate and state the reasons for your opinion.

(a) A is the event that a student is in the upper 25% of the class on the first class quiz and B is the event that the same student is in the lower 25% on the second quiz.

(b) A is the event that the national economy is in an inflationary trend and B is the event that interest rates are changed.

(c) A is the event that subject A learns a list in 10 trials and B is the event that subject B learns the list in 5 trials.

(d) A is the event that a presidential candidate of one political party is elected and B is the event that a city mayor of the same party is elected.

(e) A is the event that a student has high marks in school and B is the event that the same individual becomes president of the ABC Company.

8. Each of two groups of people contains n members. Suppose that in each group, the members have been ranked in descending order according to their scores on a psychological test. If one person is selected at random from each group and if we assume that the selections are independent, what is the probability that neither of the persons drawn falls below the mth rank order?

9. Continuing Example 6, suppose that the decision strategy is to reject those individuals who fall below a minimum cutoff score on a majority of the three tests. What is the probability of rejection under this scheme?

10. A group of 1000 voters are interviewed in May and again in August concerning their party choice in the forthcoming election. Let R, D, and U denote, respectively, the events that the respondent intends to vote Republican, Democratic, or is undecided. Probabilities of pairs of events are given in the following table:

		August			
		R	D	U	
	R	0.275	0.220	0.055	0.550
May	D	0.200	0.170	0.030	0.400
	U	0.025	0.010	0.015	0.050
		0.500	0.400	0.100	

For example, 0.275 is the probability that a respondent intends to vote Republican in May and has the same intention in August, whereas 0.550 is the probability that a respondent interviewed in May intends to vote Republican and 0.500 is the probability that a respondent interviewed in August intends to vote Republican. Given the above information, is a respondent's voting intention in May independent of his voting in August?

11. In Example 9, show that if events E_1, E_2, E_3, ... are assigned probabilities $P(E_k) = p(1 - p)^{k-1}$ where $0 \leq p < 1$, then $\sum_{k=1}^{\infty} P(E_k) = 1$; that is, this is a legitimate probability assignment.

12. If E and F are independent events and G is a subset of E, is it necessarily the case that G and F are independent?

13. Is independence a transitive relation? That is, if E and F are independent and F and G are independent, does it follow that E and G are independent?

14. Show that Ω and ϕ are each independent of every event. Further, show that the only events which are independent of themselves are events having probability zero or probability one.

15. Continuing Example 4, suppose the penny is tossed n times. Show that events E and F are independent only if $n = 3$.

16. Complete the proof of Theorem 13.

17. Complete the proof of Theorem 14.

SUPPLEMENTARY READING

Feller, W., *An Introduction to Probability Theory and its Applications* (John Wiley & Sons, Inc., New York, 1968), Vol. I, 3rd ed., Chapters 1 and 5.

Goldberg, S., *Probability, An Introduction* (Prentice-Hall, Inc., Englewood Cliffs, N. J., 1960), Chapter 2. (An excellent elementary text on probability.)

Parzen, E., *Modern Probability Theory and its Applications* (John Wiley & Sons, Inc., New York, 1960), Chapters 1 and 2.

Pfeiffer, P. E., *Concepts of Probability Theory* (McGraw-Hill Book Company, New York, 1965), Chapters 1 and 2.

DISCRETE RANDOM VARIABLES 19

19.1 RANDOM VARIABLES AND PROBABILITY FUNCTIONS

Some experimental situations directly produce numbers as outcomes, while others produce results in the form of numerical values, such as ranks or ratings, that are later assigned to the actual outcomes to aid in the description and analysis of experimental results. An assignment of numbers to outcomes of an experiment or, more precisely, to elements of the sample space associated with the experiment, produces a function whose domain is the sample space and whose range is some set of real numbers. Such a function is called a *random variable*.

Example 1 Suppose we are to toss a coin three times and that we will receive $1 for each head that appears. The set

$$\Omega = \{HHH, HHT, HTH, HTT, THH, THT, TTH, TTT\}$$

is the natural sample space for this experiment.

However, we are not interested in the exact sequence of heads and tails but rather in the number of heads each sequence produces. With any particular outcome of the experiment we are interested in associating the number of heads occurring in that outcome. This association determines a random variable N whose domain is Ω and whose range is the set $\{0, 1, 2, 3\}$. The value which N assigns to a particular element of Ω is the number of H in that element. Thus,

$$N(HHH) = 3, N(HHT) = 2, N(HTH) = 2, N(HTT) = 1, N(THH) = 2,$$

$$N(THT) = 1, N(TTH) = 1, N(TTT) = 0 \qquad \blacktriangleright$$

The term "random variable" is somewhat unfortunate in that such an object is not a variable but a function, nor is it in any sense "random." However, this terminology is standard. As long as the concept of a random variable as a function is kept firmly in mind, no confusion should result.

Random variables provide a useful shorthand notation for designating certain events. For instance, the set of elements in the sample space Ω in Example 1 to which the random variable N assigns the value 2 is denoted by $(N = 2)$. That is,

$$(N = 2) = \{HHT, HTH, THH\}$$

In general, if X is a random variable whose domain is the sample space Ω, the notation $(X = x)$ is used to denote the set of those points to which X assigns the value x. We follow the common practice of using capital letters for random variables and lower case letters for real numbers. That is,

$$(X = x) = \{\omega \in \Omega : X(\omega) = x\}$$

Similarly, $(X \leq x)$ is the shorthand notation for the set $\{\omega \in \Omega : X(\omega) \leq x\}$ of those elements in Ω which are assigned values no greater than x, and so forth.

Example 2 In Example 1, $(N = 1) = \{HTT, THT, TTH\}$, $(N = 3) = \{HHH\}$, $(N \leq 1) = \{TTT, HTT, THT, TTH\}$, and $(N > 2) = \{HHH\}$, while $(N < 0)$, $(N = \frac{3}{2})$, $(N = 7)$, $(N = -2)$, and $(N > 4)$ all denote the empty set. ▶

PROBLEMS

1. Referring to Example 1, write the events symbolized by $(N = 0)$, $(N \geq 2)$, $(N \leq 0)$, $(\frac{1}{2} < N < 2)$, and $(N \leq 3)$.

2. Suppose that in Example 1 the probabilities $\frac{1}{10}, \frac{2}{10}, \frac{3}{10}, \frac{1}{10}, \frac{1}{20}, \frac{1}{10}, \frac{1}{20}$, and $\frac{1}{10}$ are assigned to the respective simple events of Ω. For each real number x, compute $P(N = x)$.

3. In Problem 2 , compute $P(N \leq x)$ for each real number x.

Corresponding to each random variable X is a function p_X defined by

$$p_X(x) = P(X = x)$$

which assigns to each real number x the probability that X takes the value x. This function is called the *probability function of* X. The function F_X defined by

$$F_X(x) = P(X \leq x)$$

is called the *distribution function of* X.

4. For the random variable N in Example 1, draw graphs of its probability function (Problem 2) and its distribution function (Problem 3).

Example 3 In Example 9, Section 18.7, define the random variable W to be the number of the house call on which the first sale is made. Then the event $(W = k)$ is the same as the event E_k described in that example. The probability function p_W of W is given by

$$p_W(k) = P(W = k) = P(E_k) = \tfrac{1}{5}(\tfrac{4}{5})^{k-1} \qquad \text{if } k \text{ is a positive integer}$$

$$p_W(k) = 0 \qquad \text{if } k \text{ is not a positive integer}$$

A graph of the function p_W is shown in Figure 1. The functional values are zero everywhere except at the isolated positive integer points.

FIGURE 1

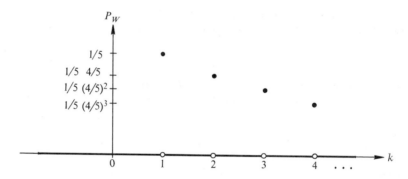

The distribution function of W is given by

$$F_W(w) = P(W \leq w) = 0 \qquad \text{for } w < 1$$

$$F_W(w) = P(W \leq w) = \sum_{k=1}^{[w]} P(W = k) = 1 - (\tfrac{4}{5})^{[w]} \qquad \text{for } w \geq 1$$

where $[w]$ denotes the largest integer less than or equal to w. For example,

$$P(W \leq 2) = P(W \leq 2.4) = P(W = 1) + P(W = 2) = 1 - (\tfrac{4}{5})^2$$

$$P(W \leq 3.29) = P(W \leq 3) = P(W = 1) + P(W = 2) + P(W = 3)$$

$$= 1 - (\tfrac{4}{5})^3$$

Figure 2 shows the graph of the distribution function F_W. Note particularly that for points between two integers the values of F_W remain the same. The only change in F_W occurs in jumps taken at the integer points $1, 2, 3, \ldots$. For each k, the magnitude of the jump at the point k equals $P(W = k)$. ▶

FIGURE 2

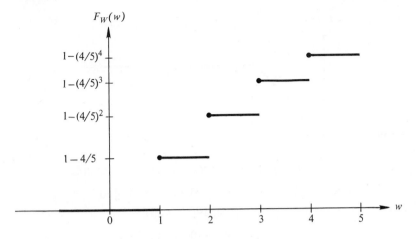

Example 4 An outcome of the experiment of flipping the spinner shown in Figure 3 is some number between 0 and 1. Let us take $\Omega = \{x : 0 \leq x < 1\}$

FIGURE 3

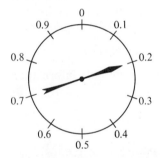

as the sample space and define the random variable X, which indicates the point at which the spinner stops, by

$$X(x) = x$$

for each $x \in \Omega$. If we assume that the probability that the spinner stops in any given interval is proportional to the length of that interval, then the distribution function of X is given by

$$F_X(x) = P(X \leq x) = \begin{cases} 0 & \text{if } x < 0 \\ x & \text{if } 0 \leq x \leq 1 \\ 1 & \text{if } x > 1 \end{cases}$$

The graph of F_X is shown in Figure 4.

FIGURE 4

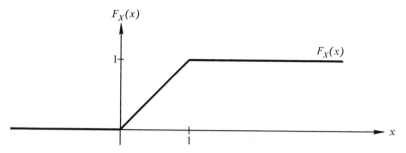

Knowing F_X, we may compute other probabilities. For example,

$$P(0.3 < X \leq 0.45) = P(X \leq 0.45) - P(X \leq 0.3)$$
$$= F_X(0.45) - F_X(0.3)$$
$$= 0.45 - 0.3 = 0.15$$

and

$$P(0.7 < X \leq 2.6) = F_X(2.6) - F_X(0.7)$$
$$= 1 - 0.7 = 0.3$$

The distribution function gives us the information we need about the random variable X. On the other hand, since the length of an interval containing a single point is zero, the probability function of X is given by

$$p_X(x) = P(X = x) = 0 \qquad \text{for all real numbers } x$$

In this case, then, the probability function gives relatively little information about X. ▶

PROBLEMS

5. In Example 4, compute $P(X > 0.6)$, $P(-0.2 \leq X \leq 0.108)$ and $P(X < 6)$.

A random variable, such as W in Example 3, whose distribution function consists only of jumps and level stretches (see Figure 2) is called *discrete*. In working with discrete random variables it is convenient to use the probability function, listing all possible values and the probabilities of taking these values. Random variables, such as X in Example 4, whose distribution functions take no jumps, and hence are continuous, are called *continuous*. For these random variables it is more convenient to use the distribution function. We shall discuss discrete random variables first, turning to the continuous case in Chapter 21.

PROBLEMS

6. The number of accidents which occur on a given day in Chapel Hill, North Carolina, is a discrete random variable A having the following probability function:

$$P(A = 0) = 0.1 \qquad P(A = 1) = 0.1c$$
$$P(A = 2) = 0.7c \qquad P(A = 3) = 2.3c$$
$$P(A = 4) = c \qquad P(A = 5) = 0.4c$$

and $P(A = k) = 0$ for all other values of k.

(a) Determine the value of the constant c.

(b) Find $P(A > 3)$, $P(1 < A \le 4)$, $P(2 < A < 5)$, and

$$P[(A < 1) \cup (A > 4)]$$

(c) Graph the distribution function of A.

7. The amount of rainfall in Seattle during the month of November is a random variable R having the distribution function

$$F_R(x) = P(R \le x) = \begin{cases} 0 & \text{if } x < 0 \\ 1 - \frac{4}{5}e^{-x} & \text{if } x \ge 0 \end{cases}$$

(a) Draw a graph of F_R.

(b) Find $P(R \le 0)$, $P(R \le 7)$, $P(2 < R \le 5)$, and $P(R > 2)$.

8. In the context of Problems 9 and 10 of Section 18.3, let the random variable L denote the length of the cycle to which the first member of a five-member group belongs. ($L = 0$ denotes absence of membership in any cycle.)

(a) Find the probability function of L.

(b) Graph its distribution function.

(c) What is the probability that this individual is a member of a cycle of length at least 3? exactly 3? at most 3?

(d) Find the smallest number t such that the probability of $(L \ge t)$ is at least $\frac{1}{2}$.

(e) What is the conditional probability that this individual is a member of a 4-cycle, given that he is a member of some cycle? That is, what is $P(L = 4 \mid L > 0)$?

9. Assume there is probability $\frac{1}{3}$ that an applicant qualifies for a position with SMART Consulting, Inc. Three applicants are independently interviewed. Let Q denote the number of applicants who qualify, N the number who do not, and define $Z = \frac{1}{2}(|Q - N| - 1)$. Define the random variable X_1 by

$$X_1 = 1 \qquad \text{if the first applicant qualifies}$$
$$X_1 = 0 \qquad \text{if the first applicant fails to qualify}$$

(a) Argue that X_1 and Z are different random variables (that is, different functions defined on the sample space), but that their probability functions are identical.

(b) Argue that there are infinitely many random variables whose probability functions are identical with that of X_1.

19.2 RANDOM VARIABLES ASSOCIATED WITH BERNOULLI TRIALS

The most important discrete random variables arise in connection with *Bernoulli* trials*. A sequence of trials is a Bernoulli sequence if

(i) The trials are independent.

(ii) Each trial results in one of two possible outcomes which are arbitrarily called success S and failure F.

(iii) The probability $p = P(S)$ of success is the same on each trial.

Repeated tossing of a coin, the public opinion poll of Example 7, Section 18.7, and drawing with replacement from an urn containing balls of two colors are typical examples of Bernoulli trials. The designation of success and failure is quite arbitrary and is used only to present a standardized terminology. In tossing a coin, heads represents success for one player but failure for the other. Each uses the model according to his own tastes.

Since there are two possible outcomes on each trial, a sample space for r trials contains $2 \times 2 \times \cdots \times 2 = 2^r$ elements, each denoting one of the possible sequences of S and F which might occur. The assumption of independence implies that the probability which is assigned to any sequence having exactly k successes S and $r - k$ failures F is $p^k(1 - p)^{r-k}$. Once p is known, the entire probability assignment is determined.

PROBLEMS

1. Suppose the model of Bernoulli trials had been applicable in Example 1 of Section 19.1 and that $p = P(H) = \frac{2}{3}$. Determine the appropriate assignment of probabilities to the simple events.

Let us denote by N_r the random variable whose values represent the number of successes in r Bernoulli trials. The possible values of N_r are $0, 1, 2, \ldots, r$. The event $(N_r = k)$ contains all those elements of the sample space which are arrangements of exactly k successes S and $r - k$ failures F. Since any such arrangement is determined by the choice of k positions in which to place S, the number of arrangements is

$$\binom{r}{k}$$

the number of ways of choosing k objects from r objects. It follows that

$$P(N_r = k) = \binom{r}{k}p^k(1 - p)^{r-k} \qquad \text{if } k = 0, 1, 2, \ldots, r \qquad (1)$$

*After James Bernoulli (1654–1705), one of the founders of the theory of probability.

Of course, if k is not one of the integers $0, 1, 2, \ldots, r$, then $P(N_r = k) = 0$. Because of the coefficient

$$\binom{r}{k}$$

appearing in (1), N_r is often called a *binomial random variable*.

Example 1 Deutsch and Madow* are concerned about the problem of distinguishing between "genuine wisdom" in bureaucratic decision makers and "pseudowisdom," which results from the accidental making of correct decisions. In their discussion, they assume that every decision of each decision maker is made independently of all others, that the organization can distinguish "correct" from "incorrect" decisions and that there is a uniform probability p of being correct which holds for all decisions and decision makers. In short, individual decisions constitute Bernoulli trials.

The probability that an individual is correct in k of n decisions is

$$\binom{n}{k} p^k (1 - p)^{n-k}$$

For $k = n$, this becomes

$$\binom{n}{n} p^n (1 - p)^0 = p^n$$

Hence, the probability that, of m individuals each making n decisions, at least one makes all n decisions correctly is

$$1 - (1 - p^n)^m$$

For a specified value v, the smallest number of individuals required so that the probability equals v that at least one person makes all n decisions correctly is approximately

$$m = \frac{\log (1 - v)}{\log (1 - p^n)}$$

If $p = 0.5$, $v = 0.8$, and $n = 8$, this number is about 410. That is, the probability is about 0.8 that an organization of 410 individuals may produce at least one person with a perfect record on eight decisions, even though the individual decision makers are no more competent than "honest" coins. ▶

PROBLEMS

2. Returning to Problem 1, let N denote the number of heads obtained. Find the probability function of N and show that it has the form of Equation (1) with $r = 3$.

*Deutsch, K. W., and Madow, W. G., "A Note on the Appearance of Wisdom in Large Bureaucratic Organizations," *Behavioral Science* **6**, 72–78 (1961).

3. By interpreting the sum

$$\sum_{k=0}^{r} \binom{r}{k} p^k (1-p)^{r-k}$$

as a binomial expansion, show that the probabilities in (1) sum to unity.

In arriving at a binomial probability function, we used as our random variable the number of successes in a fixed number of Bernoulli trials. Suppose we now take the contrary point of view and consider the number of trials required to obtain a fixed number of successes.

Consider first the random variable W representing the number of failures before the first success. Then $W + 1$ is the time (number of trials) one must wait for the first success. The first success, if it occurs at all, may come at any trial. Thus $0, 1, 2, \ldots$ all are possible values of W. The event that success never occurs, or that all trials result in failure, is denoted by $(W = \infty)$.

If $W = k$, where k is some non-negative integer, it means that the first k trials each resulted in failure, followed by success on trial $k + 1$. Conversely, if a sequence of outcomes begins with k failures followed by a success, then $W = k$. Hence,

$$(W = k) = \underbrace{F F \ldots F}_{k \text{ failures}} S$$

and we have

$$P(W = k) = (1-p)(1-p) \cdots (1-p)p$$

$$= p(1-p)^k \qquad k = 0, 1, 2, \ldots \qquad \textbf{(2)}$$

Because of the form of the probability function (2), W is called a *geometric random variable*.

If $p > 0$, the sum of the probabilities in (2) is

$$\sum_{k=0}^{\infty} P(W = k) = \sum_{k=0}^{\infty} p(1-p)^k = p \frac{1}{1 - (1-p)} = \frac{p}{p} = 1$$

In this case, $P(W = \infty)$ must be zero. On the other hand, if $p = 0$, each probability in (2) is zero and thus

$$\sum_{k=0}^{\infty} P(W = k) = 0$$

Hence

$$P(W = \infty) = 1$$

In short, if the probability of getting a success is positive, there is a probability of 1 that a success will occur eventually, while if the probability of success is zero, there is probability 1 that success never occurs.

Example 2 In psychological experiments the time interval between stimulus and response is called the latency period. Observed fluctuations in latency time are often assumed to be generated by an underlying probability process. One model of such a process is the following, due to McGill.*

Assume that a system generates a Bernoulli sequence of trials with possible outcomes R, response, and $\bar{R}$, no response. By assuming the trials are of equal duration, the number X of $\bar{R}$ outcomes occurring between two R outcomes may be considered as a latency random variable. If $p = P(R)$ denotes the probability of outcome R, the probability function of X is given by

$$P(X = k) = p(1 - p)^k \qquad \text{when } k = 0, 1, 2, \ldots$$

so X is geometric. ▶

A characteristic feature of any process which leads to a geometric random variable is its "lack of memory." To make this concept precise, let us return to the random variable W. Assuming that the first t trials result in failure, the probability of exactly k *additional* failures before the first success is

$$P(W = t + k \mid W \geq t) = \frac{P(W = t + k)}{P(W \geq t)} = \frac{p(1 - p)^{t+k}}{\sum_{j=t}^{\infty} p(1 - p)^j}$$

Since $\sum_{j=t}^{\infty} p(1 - p)^j = (1 - p)^t$, this reduces to $p(1 - p)^k$, the same as $P(W = k)$. Hence, no matter what the preceding number of failures, the probability function of the number of additional failures before the first success remains the same as it was initially.

Example 2' The "lack of memory" of geometric random variables implies that no matter how many $\bar{R}$ outcomes have occurred, the probability that R occurs in the next trial is still p, as it was for the first trial. This, of course, is a reflection of the independence of the trials. ▶

PROBLEMS

4. What is the probability that in a sequence of Bernoulli trials there are at least four failures before the first success?

5. Let $1 - p$ be the probability that your favorite baseball team wins a world series game. Assuming that games form Bernoulli trials, what is the probability that your team sweeps the series? Why does this answer coincide with that of the preceding problem?

If W_n denotes the number of failures before the nth success, then the event $(W_n = k)$ occurs if and only if the nth success is preceded by exactly k failures. That is to say, the nth success occurs on the $(k + n)$th trial, there being k failures and $n - 1$ successes on the preceding $k + n - 1$ trials.

*McGill, W. J., "Stochastic Latency Mechanisms," in *Handbook of Mathematical Psychology*, Luce, R. D., Bush, R. R., and Galanter, E. E., Eds. (John Wiley & Sons, Inc., New York, 1963).

$$k \text{ failures and } n - 1 \text{ successes in any order}$$

| $\underline{F}$ | $\underline{S}$ | $\underline{S}$ | $\cdots$ | $\underline{F}$ | $\underline{S}$ | | $\underline{S}$ |

$$k + n - 1 \text{ trials} \qquad\qquad (k + n)\text{th trial}$$

Independence of trials implies that

$$P(W_n = k) = P\binom{k \text{ failures } F \text{ and } (n - 1) \text{ successes } S}{\text{in the first } n - 1 + k \text{ trials}} \times P\binom{S \text{ on}}{\text{trial } n + k}$$

$$= \left[\binom{n - 1 + k}{k} p^{n-1}(1 - p)^k \right] \times p$$

$$= \binom{n - 1 + k}{k} p^n (1 - p)^k \qquad k = 0, 1, 2, \ldots \tag{3}$$

The first probability in the second equality of (3) is the probability of exactly $n - 1$ successes S in $n - 1 + k$ trials and is obtained directly from (1). Note that when $n = 1$, (3) reduces to (2), as it should. A short calculation shows that

$$\binom{n - 1 + k}{k}$$

is the same as

$$(-1)^k \binom{-n}{k}$$

(see Example 5 in Section 16.1), so that (3) may be rewritten in the form

$$P(W_n = k) = \binom{-n}{k} p^n (-q)^k \tag{4}$$

where $q = 1 - p = P(F)$. Because of the quantity $-n$ which appears in the binomial coefficient, W_n is called a *negative binomial random variable*. The quantity $-q$ has no probabilistic interpretation.

Example 3 Consider a legislative process in which bills are sent either to a finance committee or to a general committee. Bills are then called for action by choosing one of the committees at random and asking that committee to present one bill. If we identify "success" with the choice of the general committee, the successive choices of committees constitute Bernoulli trials with $p = P(S) = \frac{1}{2}$.

Suppose that f bills are assigned to the finance committee and g bills to the general committee. Let the random variable N be the number of bills remaining in the finance committee at the time the last bill is called from the general committee. Then for $r > 0$, $(N = r)$ occurs if and only if there are $f - r$ failures (that is, bills called from the finance committee) before the gth success. Thus, for $r > 0$,

$$P(N = r) = \binom{g - 1 + f - r}{f - r} (\tfrac{1}{2})^g (\tfrac{1}{2})^{f-r} = \binom{g + f - r - 1}{f - r} (\tfrac{1}{2})^{g+f-r}$$

For instance, if $g = f = 10$, then the probability that seven or more bills remain in the finance committee is

$$P(N \geq 7) = \sum_{r=7}^{10} P(N = r) = \sum_{r=7}^{10} \binom{19 - r}{10 - r} (\tfrac{1}{2})^{20-r} = 0.046 \qquad \blacktriangleright$$

PROBLEMS

6. Verify the calculations in Example 3.

7. In Example 3, $P(N = 0)$ is the probability

$$\sum_{k=f}^{\infty} \binom{g + k - 1}{k} (\tfrac{1}{2})^{g+k}$$

that the number of failures before the gth success is *at least f*. Why is this so?

The model of Bernoulli trials applies also to problems which concern sampling with replacement from a finite population having two types of individuals. To fix ideas, suppose we have an urn containing b black balls and g green balls. If we draw with replacement from this urn, the successive draws form Bernoulli trials with

$$p = P \text{ (black)} = \frac{b}{b + g}$$

and

$$q = 1 - p = P \text{ (green)} = \frac{g}{b + g}$$

Hence, if N_r denotes the number of black balls obtained in r draws we have

$$P(N_r = k) = \binom{r}{k} p^k q^{r-k} = \binom{r}{k} \left(\frac{b}{b + g}\right)^k \left(\frac{g}{b + g}\right)^{r-k}$$

in accordance with Equation (1).

Now let us alter our model slightly and consider draws without replacement. Then we no longer have Bernoulli trials since the trials are no longer independent. Note, however, that this is the only condition for Bernoulli trials which fails to hold. Since we are dealing with a special kind of Polya urn, the probability of black remains the same for each draw.

The result of r draws is to choose a subset of r balls from the original $b + g$ balls. We shall use a sample space Ω consisting of

$$\binom{b + g}{r}$$

elements, one for each possible r-subset and, assuming random drawing, assign the same probability

$$\frac{1}{\binom{b+g}{r}}$$

to each simple event. Again denoting by N_r the number of black balls drawn, we have

$$P(N_r = k) = \frac{\binom{b}{k}\binom{g}{r-k}}{\binom{b+g}{r}} \qquad k = 0, 1, 2, \ldots, r \qquad (5)$$

Equation (5) is derived by noting that $(N_r = k)$ is obtained by choosing any k of the b black balls and any $r - k$ balls from the g green balls. Note that our agreement about binomial coefficients gives the correct value even if $k > b$. In this case we require the probability of drawing more black balls than were originally contained in the urn. This event, being impossible, must have probability zero. But, of course, when $k > b$ the binomial coefficient

$$\binom{b}{k}$$

is zero. Similar comments apply if $r - k > g$.

A random variable whose probability function has the form of Equation (5) is called a *hypergeometric random variable*. Such a random variable appears, then, when one counts the number of "successes" in a sample drawn *without replacement* from a finite population composed of two types of individuals.

PROBLEMS

8. Use Formula (5) of Chapter 7 to show that the probabilities in (5) add to unity.

Example 4   A congressional committee consists of 6 Democrats and 5 Republicans. A subcommittee of 4 members is chosen at random. If R represents the number of Republicans in the chosen subcommittee, then

$$P(R = k) = \frac{\binom{5}{k}\binom{6}{4-k}}{\binom{11}{4}} \qquad \blacktriangleright$$

Example 5   Probabilistic considerations allow us to deduce the likely properties of samples taken from larger collections (or populations) of objects. For instance, given the proportions of Democrats and Republicans in a certain voting population, the probability that a sample of 10 contains 3 Republicans may be computed.

The problem of statistics is just the opposite. We wish to infer properties of the population as a whole from the properties of an observed sample. To take a typical case, consider Sam Yerkes, running for mayor of Anytown, U.S.A. In order to determine his popular support, an opinion poll is conducted involving 500 individuals chosen at random and without replacement from the total population of 3000 registered voters. Of those polled, 300 express support for Sam.

The proportion p of the 3000 registered voters on whose support Sam can count is still unknown. If j is the total number of voters supporting Yerkes, then $p = j/3000$ and the probability that 300 of 500 voters polled indicate support is

$$h(j) = \frac{\binom{j}{300}\binom{3000-j}{200}}{\binom{3000}{500}}$$

According to the statistical principle of *maximum likelihood*, due to Fisher,* the best estimate of p is $m/3000$, where m is the value of j which makes $h(j)$ a maximum.

A simple calculation shows that $h(j)/h(j-1)$ is greater than unity when $j < \frac{3}{5}(3001)$ and less than unity when $j > \frac{3}{5}(3001)$. Thus, as j increases $h(j)$ first increases and then decreases, reaching its maximum value when $j = 1800$. The maximum likelihood estimate of p is $1800/3000 = \frac{3}{5}$, the same as the sample proportion. ▶

PROBLEMS

9. The Vultures play the Green Sox 7 times in a given month. Assume that the Green Sox are the better team and have probability $\frac{3}{5}$ of winning and probability $\frac{2}{5}$ of losing a game.

   (a) If the games are considered as 7 independent trials, find the probability that the Green Sox win exactly

   (i) four games.

   (ii) five games.

   (iii) a majority of the games.

   (b) Do you think the assumption of independent trials is realistic?

10. (a) What is the probability in Problem 9 that the Green Sox would win if these two teams met in the World Series?

    (b) Why is the answer to (a) the same as that to Problem 9(a)(iii)?

11. (a) Show that the probabilities in Equation (1) sum to unity.

    (b) Show that the probabilities in Equation (4) sum to unity when $p > 0$. (*Hint:* Interpret the sum as a binomial expansion.)

*After R. A. Fisher (1890–1962), British statistician.

12. A plane can fly on half its engines but not on less. Assuming engines fail independently with probability q, would you rather fly in a two-engine plane or a four-engine plane?

13. Suppose that 10 people are chosen at random from a community and asked whether they favor a school bond issue. If it is assumed that 60% of the community favor the issue, what is the probability that a majority of those sampled will oppose?

14. In considering whether to develop an Atlantic coastal island, the South Sea Development Company estimates that there is probability $\frac{1}{3}$ that at least one hurricane will hit the island during any given year. The company predicts that during the first five years the development would be slightly damaged if hit during one year, in financial difficulty if hit during two years, and a complete loss if hit in three or more years. What is the probability that the company

    (a) would escape unscathed?

    (b) would be slightly damaged?

    (c) would be in financial difficulty?

    (d) would suffer complete loss?

15. In Problem 14, how many years must elapse in order for the probability of suffering at least partial damage to be at least 0.95?

16. Given that n Bernoulli trials result in exactly k successes, show that the conditional probability of a success on any particular trial is k/n.

17. (a) In Example 1 of this section, let M denote the number of decision makers who are correct in all n decisions. What is the probability that $M = j$?

    (b) Find the probability that at least one of the m decision makers is correct in at least $n - 1$ decisions.

18. Suppose that the success of any one decision maker in Problem 17 is defined in terms of a perfect record on n decisions. In order to form a "brain trust," we decide to test decision makers until we find k successful ones. Find the probability that the rth decision maker tested is the kth successful one.

19. In Example 4, suppose that the subcommittee chosen contains at least one Republican.

    (a) Find the conditional probability that all the subcommittee members are Republican.

    (b) If another 4-man subcommittee is chosen from the remaining 7 members, determine the conditional probability that all members are Democratic.

20. The No-Sneeze Company has developed a new cold pill which they hope will be effective in reducing the incidence of the common cold. In order to test their product, they have drawn a number of random samples

from the population and administered their pill. Three such tests produced the following results:

Test I Of 10 sampled, none contacted a cold.

Test II Of 17 sampled, 16 did not catch a cold and one did.

Test III Of 23 sampled, 21 did not catch a cold and 2 did.

Suppose that the normal infection rate in the population is 25%. Assuming that the subjects represent Bernoulli trials, which test gives the strongest evidence in favor of the newly developed pill? (*Hint:* Compute the probability that, assuming a 25% incidence of colds, sample results as good or better than those cited would be obtained.)

21. (a) Suppose in Problem 20 that in a test of 20 people sampled at random, the pill is declared effective if no more than 2 contact colds. What is the probability that the pill is declared effective in at least 1 of 10 such tests if in reality the pill has no effect at all?

(b) Still assuming a 25% infection rate, what is the probability that exactly 2 of 10 tests show 2 subjects with colds and 18 without?

22. A certain manufacturing process is assumed to be representable as a Bernoulli process with probability $p = \frac{2}{3}$ of a success (acceptable piece). What is the probability that the following number of pieces will have to be manufactured in order to secure one acceptable piece? Three acceptable pieces?

(a) Exactly six.   (b) Less than six.

(c) Six or less.   (d) Six or more.

(e) More than six.

23. If the process in Problem 22 produces pieces at the rate of four pieces per minute, what is the probability that the "latency time" between production of acceptable pieces will be

(a) exactly one minute?   (b) less than one minute?

(c) one minute or less?   (d) one minute or more?

(e) more than one minute?

24. In Example 3, suppose that six bills are assigned to the general committee and four bills are assigned to the finance committee. Find

(a) the probability function for N.

(b) for each real positive number r, $P(N \leq r)$.

25. Taking $f = g$ in Example 3, show that the probability that when the last bill is chosen for action from one committee, there are still r bills remaining in the other committee is, for $r > 0$,

$$\binom{2f - r - 1}{f - r} (\tfrac{1}{2})^{2f - r - 1}$$

26. In Example 5, assume that the sample is drawn with replacement. Show that the maximum likelihood estimate of p is again $\frac{3}{5}$, the proportion actually obtained in the sample.

27. Not only does a geometric random variable exhibit lack of memory, but it is the *only* discrete random variable with this property. Specifically, suppose a random variable T assumes values $0, 1, 2, \ldots$ with respective probabilities $p_0, p_1, p_2, \ldots$, and that the conditional probability that $T = k + 1$, given that $T > k$, is equal to the constant p for every k. Prove that $p_k = (1 - p)^k p$, so that T has a geometric probability function.

28. Imagine a game played over a sequence of t trials in which on each trial each of two players makes either a cooperative C or noncooperative N response.* Each player makes his response without knowledge of the response of the other. After every trial, each is informed of the other's response. One model for this experiment assumes that the two players respond independently of one another, each player's responses constituting a sequence of t Bernoulli trials and player A having probability p_A and player B probability p_B of making response N. Under these assumptions, find the probability function of X, the number of trials on which both players respond N.

29. In Problem 28, let the random variables N_A, C_A, N_B, and C_B denote, respectively, the number of N and C responses by players A and B. Show that

$$P(X = k \mid N_A = a \text{ and } N_B = b)$$

$$= \frac{t!}{k!(n-k)!(m-k)!(t-n-m+k)!} \, p_A^n p_B^m (1 - p_A)^{t-n}(1 - p_B)^{t-m}$$

whenever $t \geq n \geq k \geq 0$, $t \geq m \geq k \geq 0$, and $t \geq n + m - k$.

30. Suppose in Problem 28 that in a sequence of 10 trials player A made 6 N responses and player B made 4. This is event E.

(a) Find the conditional distribution function of X, given E. That is, compute $P(X \leq k \mid E)$ for all k.

(b) Find $P(1 < X \leq 4 \mid E)$ and $P(X > 3 \mid E)$.

(c) If, in addition to event E, we observe $X = 4$, would you feel the players were choosing independently? Present some calculations to support your conclusion.

19.3 EXPECTED VALUE OF A RANDOM VARIABLE

In our relative-frequency interpretation of probability, if the probability of heads for a given coin is $\frac{1}{2}$, then we expect that in the long run about half the tosses will result in heads and about half in tails. Put another way, we feel that the average number of heads in n tosses will be about $n/2$.

---

*For a discussion of certain games of this type, see Rapoport, A., and Chammah, A. M., *Prisoner's Dilemma: A Study in Conflict and Cooperation* (The University of Michigan Press, Ann Arbor, Mich., 1965).

Similarly, if N denotes the score obtained in tossing a die loaded so that

$$P(N = 1) = \tfrac{1}{10} \qquad P(N = 2) = \tfrac{1}{10} \qquad P(N = 3) = \tfrac{3}{10}$$

$$P(N = 4) = \tfrac{2}{10} \qquad P(N = 5) = \tfrac{2}{10} \qquad P(N = 6) = \tfrac{1}{10}$$

then out of, say, ten million tosses, we would expect roughly one million each of ones, twos, and sixes, about two million fours and fives, and about three million threes. The average score per toss, obtained by adding all the scores and dividing by ten million (10×10^6), would be about

$$\frac{1 \cdot (1 \times 10^6) + 2 \cdot (1 \times 10^6) + 3 \cdot (3 \times 10^6) + 4 \cdot (2 \times 10^6) + 5 \cdot (2 \times 10^6) + 6 \cdot (1 \times 10^6)}{10 \times 10^6}$$

$$= 1\left(\frac{1 \times 10^6}{10 \times 10^6}\right) + 2\left(\frac{1 \times 10^6}{10 \times 10^6}\right) + 3\left(\frac{3 \times 10^6}{10 \times 10^6}\right) + 4\left(\frac{2 \times 10^6}{10 \times 10^6}\right)$$

$$+ 5\left(\frac{2 \times 10^6}{10 \times 10^6}\right) + 6\left(\frac{1 \times 10^6}{10 \times 10^6}\right)$$

$$= 1(\tfrac{1}{10}) + 2(\tfrac{1}{10}) + 3(\tfrac{3}{10}) + 4(\tfrac{2}{10}) + 5(\tfrac{2}{10}) + 6(\tfrac{1}{10}) \tag{6}$$

The average score one would expect may be computed by simply multiplying each value v of the random variable N by the respective probability $P(N = v)$ and summing over all possible values.

Similar considerations apply to any discrete random variable and we formalize our discussion in the following definition.

Definition 1 Let X be a discrete random variable having possible values $x_1, x_2, x_3, \ldots$. Then the *expected value*, or the *mean value*, of X is

$$E(X) = \sum_k x_k P(X = x_k) \tag{7}$$

▶

We shall agree that the expected value of a random variable X is defined only when the series (7) is absolutely convergent. Otherwise, rearranging the terms could lead to a different sum.

Adding the terms in (6), we find $E(N) = \tfrac{36}{10}$. It is apparent, then, that the expected value of a random variable is not necessarily a value one "expects" in a single performance of the experiment. Rather, as indicated above, it should be interpreted as the *average value* to be approximated after many repetitions of an experiment.

Example 1 If S denotes the score obtained on a single toss of a fair die, then

$$E(S) = 1 \cdot P(S = 1) + 2 \cdot P(S = 2) + \cdots + 6 \cdot P(S = 6)$$

$$= 1(\tfrac{1}{6}) + 2(\tfrac{1}{6}) + 3(\tfrac{1}{6}) + 4(\tfrac{1}{6}) + 5(\tfrac{1}{6}) + 6(\tfrac{1}{6}) = \tfrac{7}{2}$$

▶

Example 2 In the "numbers game," an amount b is bet on a three-digit number, for instance, 636. A number is then chosen at random from the set $\{100, 101, \ldots, 999\}$. If it is 636, the bettor receives 600 times the amount of his bet. Otherwise, the bet is forfeited.

Letting R denote the net return when an amount b is wagered, we see that $R = 600b - b$ if the bettor wins and $R = -b$ if he loses. Hence

$$E(R) = 599bP(R = 599b) + (-b)P(R = -b)$$

$$= 599b(\tfrac{1}{900}) - b(\tfrac{899}{900}) = -\tfrac{1}{3}b$$

Regardless of the amount bet, the bettor can expect to suffer an average loss equal to one-third of that amount. ▶

Example 3 Scheff* contends that in the face of uncertainty, medical practitioners frequently adopt a decision policy that regards continuation of treatment as the less costly alternative. He suggests the use of "expected-value equations" as an alternative strategy.

For instance, the expected value of a treatment decision is

$$E_t = p \cdot c_d + (1 - p)c_n$$

where p is the probability that a patient has the disease, c_d is the treatment value for patients who have the disease, and c_n is the value for patients who do not. The expected value of the cost of a non-treatment decision is

$$E_{n_t} = pc_s + (1 - p)c_h$$

where c_s denotes the value to a person treated as though healthy but who is actually sick and c_h is the value of non-treatment for a person who is correctly diagnosed to be healthy.

As a simple illustration, suppose $c_d = 10$, $c_n = -2$, $c_s = -15$, $c_h = 5$, and $p = \tfrac{1}{5}$. Then

$$E_t = \tfrac{1}{5}(10) + \tfrac{4}{5}(-2) = 0.4$$

and

$$E_{n_t} = \tfrac{1}{5}(-15) + \tfrac{4}{5}(5) = 1$$

Nontreatment is indicated since, on the average, this decision has the greater patient value. ▶

Example 4 The analysis in Example 3 is somewhat deficient in that, assuming fixed treatment values, the decision policy for a particular disease depends entirely upon the probability p that the patient is ill. In practice, knowledge of p is augmented by diagnostic tests. For simplicity, let us assume that the patient is either sick S or healthy H and that the diagnostic procedure

*Scheff, T. J., "Decision Rules, Types of Error, and Their Consequences in Medical Diagnosis," *Behavioral Science* **8**, 97–107 (1963).

consists of two tests each of which results in either a positive $+$ or negative $-$ outcome. The possible joint outcomes are then $o_1 = (+, +)$, $o_2 = (+, -)$, $o_3 = (-, +)$, and $o_4 = (-, -)$ and we assume that the conditional probabilities of these outcomes are as shown in Table 1.

In order to utilize the test information, the diagnostician must formulate a decision strategy which tells him what diagnostic action to take for each

TABLE 1. Outcome probabilities conditional on patient's health

Patient's health	Outcome			
	o_1	o_2	o_3	o_4
S	0.75	0.10	0.10	0.05
H	0.01	0.05	0.04	0.90

possible joint outcome of the diagnostic tests. From the 16 possible strategies that might be employed, only 5 will be considered. These are shown in Table 2, T denoting treatment, and N no treatment. Note that D_1 and D_5 completely

TABLE 2

Diagnostic strategy	Outcomes			
	o_1	o_2	o_3	o_4
D_1	T	T	T	T
D_2	T	T	T	N
D_3	T	T	N	N
D_4	T	N	N	N
D_5	N	N	N	N

ignore any information provided by the diagnostic tests. Strategy D_1 represents a policy of "when in doubt, treat," a policy criticized by Scheff,[*] while strategy D_5 represents a policy of nontreatment regardless of diagnostic information.

The worth of a decision strategy is indicated by its expected value for each state of the patient's health. Using the values for c_d, c_n, c_s, and c_h given in Example 3, we find, for instance, that the expected value of strategy D_4, given that the patient is sick, is

$$0.75(10) + 0.10(-15) + 0.10(-15) + 0.05(-15) = 3.75$$

[*]Scheff, T. J., "Decision Rules, Types of Error, and Their Consequences in Medical Diagnosis," *Behavioral Science* **8**, 97–107 (1963).

Other expected values are similarly computed and are summarized in Table 3.

TABLE 3

Patient's health	Strategy				
	D_1	D_2	D_3	D_4	D_5
S	10	8.75	6.25	3.75	−15
H	−2	4.30	4.58	4.93	5
Expected value	0.4	5.19	4.91	4.69	1

Since $p = P(S) = \frac{1}{5}$, the entries in rows S and H of Table 3 may be combined to give an overall expected value for each of the decision strategies. For instance, the expected value for strategy D_2 is

$$\tfrac{1}{5}(8.75) + \tfrac{4}{5}(4.30) = 5.19$$

Since this is the largest among the five strategies considered, D_2 should be selected. Scheff's recommendations (Example 3) result in choosing strategy D_5. However, as the present example shows, the expected value can be substantially increased by the choice of a strategy in which action is contingent upon diagnostic results. ▶

PROBLEMS

1. Let T be the total score obtained when two fair dice are tossed. What is $E(T)$?

2. If X is the number of successes in a single Bernoulli trial, what is $E(X)$?

3. If Y is the number of successes in two Bernoulli trials, find $E(Y)$. Can you guess the expected number of successes in r Bernoulli trials?

The mean of the number N_r of successes in r Bernoulli trials is

$$E(N_r) = \sum_{k=0}^{r} kP(N_r = k) = \sum_{k=0}^{r} k \binom{r}{k} p^k q^{r-k}$$

Since the first term is zero, we might as well sum from $k = 1$. When $k \geq 1$ the coefficient

$$k \binom{r}{k} = \frac{k \times r!}{k!(r-k)!}$$

may be rewritten as

$$\frac{r(r-1)!}{(k-1)!(r-k)!} = r \binom{r-1}{k-1}$$

Thus, the *expectation of a binomial random variable* is

$$E(N_r) = r \sum_{k=1}^{r} \binom{r-1}{k-1} p^k q^{r-k} = rp \sum_{k=1}^{r} \binom{r-1}{k-1} p^{k-1} q^{[(r-1)-(k-1)]}$$

Making the change of variable $t = k - 1$ shows that the sum is simply the binomial expansion of $(p + q)^{r-1} = 1^{r-1} = 1$, and we obtain

$$E(N_r) = rp$$

Example 5 If from a city in which 60% of the voters are Republican, a sample of 70 voters is taken with replacement, the expected number R of Republicans in the sample is

$$E(R) = rp = 70(0.6) = 42$$

Of course, this is not to say that we will necessarily obtain a sample containing exactly 42 Republicans. For instance, there is probability 0.40 that R is 40 or less, while $P(R \geq 45) = 0.26$. ▶

Example 6 Let X be a random variable with the geometric probability function

$$P(X = k) = pq^k \qquad k = 0, 1, 2, \ldots$$

Then the expected value of X is

$$E(X) = \sum_{k=0}^{\infty} kpq^k = pq \sum_{k=1}^{\infty} kq^{k-1}$$

Since $kq^{k-1} = d(q^k)/dq$, we have, using Theorem 4 of Chapter 16,

$$\frac{d(1-q)^{-1}}{dq} = \frac{d\sum_{k=0}^{\infty} q^k}{dq} = \sum_{k=0}^{\infty} \frac{d(q^k)}{dq} = \sum_{k=1}^{\infty} kq^{k-1}$$

and it follows that

$$E(X) = pq \frac{d(1-q)^{-1}}{dq}$$

$$= \frac{q}{p}$$

This quantity q/p is the expected number of "failures" before the first success in a sequence of Bernoulli trials. For instance, in successive tosses of a fair die, the expected number of non-aces obtained before the first ace is

$$\frac{\frac{5}{6}}{\frac{1}{6}} = 5$$

In Example 5, the mean number of Republicans encountered before the first non-Republican is $0.6/0.4 = \frac{3}{2}$. ▶

Suppose X is a random variable with known probability function and that $Z = g(X)$ is a function of X. Each value of X is mapped by g into exactly one Z value, while a single Z value may be the image of many different X values. If we denote by $x_{k_1}, x_{k_2}, x_{k_3}, \ldots$ those X values for which $g(x_{k_i}) = z_k$, then the event $(Z = z_k)$ becomes

$$(Z = z_k) = \bigcup_i (X = x_{k_i})$$

Its probability is

$$P(Z = z_k) = \sum_i P(X = x_{k_i})$$

Hence,

$$E(Z) = \sum_k z_k P(Z = z_k) = \sum_k z_k \sum_i P(X = x_{k_i})$$

$$= \sum_k \sum_i g(x_{k_i}) P(X = x_{k_i}) \tag{8}$$

since, for each i, $g(x_{k_i}) = z_k$.

Noting that each X value appears exactly once in (8), we see that the expected value of $Z = g(X)$ may be computed without knowing the probability function of Z at all. For each possible value x of X, one multiplies the value $g(x)$ of the function g at x by the probability $P(X = x)$ and then sums these products. That is,

$$E(g(X)) = \sum_x g(x) P(X = x) \tag{9}$$

Example 7 If X is the number obtained in tossing an honest die once, then

$$E((X - \tfrac{7}{2})^2) = (1 - \tfrac{7}{2})^2 P(X = 1) + (2 - \tfrac{7}{2})^2 P(X = 2)$$
$$+ (3 - \tfrac{7}{2})^2 P(X = 3) + (4 - \tfrac{7}{2})^2 P(X = 4)$$
$$+ (5 - \tfrac{7}{2})^2 P(X = 5) + (6 - \tfrac{7}{2})^2 P(X = 6)$$
$$= \tfrac{25}{4}(\tfrac{1}{6}) + \tfrac{9}{4}(\tfrac{1}{6}) + \tfrac{1}{4}(\tfrac{1}{6}) + \tfrac{1}{4}(\tfrac{1}{6}) + \tfrac{9}{4}(\tfrac{1}{6}) + \tfrac{25}{4}(\tfrac{1}{6})$$
$$= \tfrac{70}{24} = \tfrac{35}{12} \qquad \blacktriangleright$$

PROBLEMS

4. Find the probability function of $Z = (X - \tfrac{7}{2})^2$ (Example 7) and then use Definition 1 to calculate $E(Z)$. Compare with the result of Example 7.

5. Let N denote the number of successes in r Bernoulli trials. Interpret the following sums as expected values of certain functions of N.

(a) $\displaystyle\sum_{k=0}^{r} 2k^2 P(N = k)$

(b) $\displaystyle\sum_{k=0}^{r} (k - rp)^2 \binom{r}{k} p^k q^{r-k}$

(c) $\displaystyle\sum_{k=0}^{r} s^k \binom{r}{k} p^k q^{r-k}$

(d) $\displaystyle\sum_{k=0}^{r} \frac{1}{k} \binom{r}{k} p^k q^{r-k}$

Theorem 1 The following list constitutes the basic properties of expected values.

(i) If c is a real number and X a random variable, then

$$E(cX) = cE(X)$$

(ii) If X is a constant random variable, that is, if $X(\omega) = c$ for all ω in the sample space, then

$$E(X) = c$$

(iii) If $g(X)$ and $h(X)$ are two functions of the random variable X, then

$$E(g(X) + h(X)) = E(g(X)) + E(h(X))$$

PROOF (i) Let $g(X) = cX$. Then from (9),

$$E(g(X)) = \sum_x cxP(X = x)$$
$$= c\sum_x xP(X = x) = cE(X)$$

(ii) Let $g(X) = c$. Then

$$E(g(X)) = \sum_x cP(X = x)$$
$$= c\sum_x P(X = x) = c \cdot 1 = c$$

(iii)

$$E(g(X) + h(X)) = \sum_x [g(x) + h(x)]P(X = x)$$
$$= \sum_x g(x)P(X = x) + \sum_x h(x)P(X = x)$$
$$= E(g(X)) + E(h(X))$$ ▶

PROBLEMS

6. Suppose $E(X) = 2$ and $E(X^2) = 6$. Find

(a) $E(4X)$ (b) $E(X^2 - 3X)$ (c) $E(X(X - 1))$

7. Prove that if a and b are constants and X is a random variable, then

$$E(aX + b) = aE(X) + b$$

If X is a random variable, the expected value of the function $g(X) = [X - E(X)]^2$ is called the *variance of the random variable X* and denoted Var (X). Thus, using μ to stand for the constant $E(X)$,

$$\text{Var }(X) = E((X - \mu)^2) = \sum_x (x - \mu)^2 P(X = x) \qquad (10)$$

The last term in (10) makes clear the rationale for the term "variance," for that sum in a sense measures the amount by which the values of X differ or vary from the expected value μ. The variance is large when there is large probability that X takes a value which differs substantially from μ, and small when the values of X cluster closely around μ. The variance of X is never negative (no negative terms appear in the sum) and is zero only when the only possible value of X is μ itself.

Example 8 Example 1 shows that the expected value of the number X obtained in tossing a fair die is

$$E(X) = \sum_{k=1}^{6} kP(X = k) = \tfrac{7}{2}$$

Hence the variance of X is $\tfrac{35}{12}$, as derived in Example 7. ▶

Theorem 2 Let a and b be constants and X be a random variable. Then

(a) Var $(aX) = a^2$ Var (X)

(b) Var $(X + b) =$ Var (X)

(c) Var $(X) = E(X^2) - [E(X)]^2$

$$= E(X(X - 1)) + E(X) - [E(X)]^2$$

PROOF

(a) Var $(aX) = E([aX - E(aX)]^2)$ [Definition of variance]

$$= E([aX - aE(X)]^2) \quad \text{[Theorem 1(a)]}$$

$$= E(a^2[X - E(X)]^2)$$

$$= a^2 E([X - E(X)]^2) \quad \text{[Theorem 1(a)]}$$

$$= a^2 \text{ Var } (X) \quad \text{[Definition of variance]}$$

(b) Var $(X + b) = E([X + b - E(X + b)]^2)$ [Definition of variance]

$$= E([X + b - E(X) - b]^2) \quad \text{[Problem 7]}$$

$$= E([X - E(X)]^2)$$

$$= \text{Var } (X) \quad \text{[Definition of variance]}$$

(c) Var $(X) = E([X - E(X)]^2)$

$$= E(X^2 - 2XE(X) + [E(X)]^2)$$

Using Theorem 1 gives

$$\text{Var } (X) = E(X^2) - 2E(X)E(X) + [E(X)]^2$$

$$= E(X^2) - [E(X)]^2$$

which is the first formula. Writing $X^2 = X(X - 1) + X$, so that

$$E(X^2) = E[X(X - 1)] + E(X) \qquad \text{[Theorem 1(c)]}$$

gives the second. ▶

Note that multiplying a random variable by a constant multiplies the variance of the random variable by the *square* of that constant. This is because variance is defined as the expected value of a squared quantity. The square root of the variance of X is called the *standard deviation* of X, denoted $SD(X)$. Theorem 2(a) implies that $SD(cX) = cSD(X)$. Adding a constant to a random variable does not change the amount of deviation of the values from the expected value and hence (Theorem 2(b)) does not change the variance at all.

Example 9 Continuing Example 4, define a random variable X_1 associated with decision rule D_1 which takes the values 10 and -2, respectively, according to whether the patient is sick or healthy. Table 3 shows that $E(X_1) = 0.4$, while

$$E(X_1^2) = \tfrac{1}{5}(10)^2 + \tfrac{4}{5}(-2)^2 = 23.2$$

Hence

$$\text{Var } (X_1) = E(X_1^2) - [E(X_1)]^2 = 23.2 - (0.4)^2 = 23.04$$

$$SD(X_1) = \sqrt{23.04} = 4.9$$

Defining random variables $X_2, \ldots, X_5$ similarly, we obtain Table 4.

TABLE 4. Standard deviations of decision
strategies

Strategy	D_1	D_2	D_3	D_4	D_5
Standard deviation	4.9	1.78	0.69	0.51	8.0

The choice of whether to maximize expected value or to minimize variation must be resolved by the practitioner. ▶

Example 10 If N_r denotes the number of successes in r Bernoulli trials, then

$$E(N_r(N_r - 1)) = \sum_{k=0}^{r} k(k - 1) \binom{r}{k} p^k q^{r-k}$$

Since the first two terms are zero, and for $k \geq 2$,

$$k(k-1)\binom{r}{k} = k(k-1)\frac{r!}{k!(r-k)!} = r(r-1)\binom{r-2}{k-2}$$

we have

$$E(N_r(N_r - 1)) = \sum_{k=2}^{n} r(r-1)\binom{r-2}{k-2}p^k q^{r-k}$$

$$= r(r-1)p^2 \sum_{k=0}^{r-2}\binom{r-2}{k}p^k q^{r-k}$$

$$= r(r-1)p^2$$

Hence

$$\text{Var }(N_r) = E(N_r(N_r - 1)) + E(N_r) - [E(N_r)]^2$$

$$= r(r-1)p^2 + rp - r^2 p^2$$

$$= rp(1-p) = rpq \qquad \blacktriangleright$$

PROBLEMS

8. A public opinion pollster finds that 40% of his mailed questionnaires are not returned. For these questionnaires, he initiates a followup which, due to extra labor, postage, handling, etc., costs $5 per questionnaire. For an initial mailing of 1000 questionnaires, what is his expected followup cost?

9. A 1967 survey of 61 male 1948 high school graduates showed the following numbers of children[*]:

Number n of children	0	1	2	3	4	5	6	7	Total
Number of graduates having n children	2	3	22	17	13	2	1	1	61

Let C be a random variable denoting number of children and suppose that the above figures indicate the probability function for C; that is, $P(C = 0) = \frac{2}{61}, \ldots, P(C = 7) = \frac{1}{61}$ and $P(C > 7) = 0$. Find the expected value and the variance of C.

10. Let X be a binary random variable with $P(X = 1) = p$ and $P(X = 0) = q = 1 - p$. Show that $E(X) = p$ and Var $(X) = pq$.

11. Show that if X is a geometric random variable with $P(X = k) = pq^k$, $k = 0, 1, \ldots$, then Var $(X) = q/p^2$. (*Hint:* Compute $E(X(X-1))$, using Example 10 as a guide.)

[*]Yoesting, D. R., Bohlen, J. M., and Beal, G. M., "Some Social Changes in Rural Iowa, 1948–1967," *Rural Sociology Reports* #69, Iowa State University, Ames, Iowa, 1967.

12. You are offered the following proposition: Toss a fair die until something other than a "one" appears uppermost. You pay 50¢ per toss and if n tosses are required, you win 2^n.

   (a) Find the expected value of your winnings (gross).

   (b) Find the expected value of your costs.

   (c) Find the expected value of your net winnings. Would you be willing to play the game?

13. In Problem 12, how much would you be willing to pay per toss if the die were replaced by a fair coin and you were required to toss until "tails" appears?

14. Let X be a random variable with $E(X) = \mu$ and Var $(X) = \sigma^2$. Define $Z = (X - \mu)/\sigma$. Show that $E(Z) = 0$ and Var $(Z) = 1$. (*Note:* Z is called the *standardized* random variable corresponding to X.)

15. If X is the number of successes in n Bernoulli trials, find the expected value and variance of X/n.

16. Show that if X is the number of "successes" in a sample of size r drawn without replacement from a population n elements, of which $S = np$ are designated as successes, then:

   (a) $E(X) = rp$

   (b) Var $(X) = rpq(n - r)/(n - 1)$. Compare with Example 10.

   (c) Find $\lim_{n \to \infty}$ Var (X). What do you conclude as to the relative effects of sampling with and without replacement?

17. A random variable U which takes values $0, 1, 2, \ldots, n$ with equal probabilities $1/(n + 1)$ is called a *uniform* random variable. Show that $E(U) = n/2$.

18. The probability function of U in Problem 17 is symmetric about the point $n/2$ in the sense that for all real numbers c, $P[U = (n/2) - c] = P[U = (n/2) + c]$. Prove that if the probability function of any random variable X is symmetric about a point m, then $E(X) = m$.

19. In a training device a malfunction is introduced at one of n possible locations and the trainee is required to test locations successively until he locates the source of the malfunction. Assuming he makes no errors in his diagnoses, find the probability function, expected value, and variance of the number T of trials required to locate the source of the malfunction if

   (a) on each trial he selects at random one of the n possible locations.

   (b) he chooses only from locations not previously tested.

20. Prove that for any real number c and a random variable X

$$E([X - c]^2) = \text{Var } (X) + [E(X) - c]^2$$

Therefore show that $E(X - c)^2$ is a minimum when $c = E(X)$.

21. A manufacturer of summer beach clothing must begin production in advance of the seasonal demand period. He estimates that if x units are produced and total demand is d units, his profit R is

$$R(x, d) = \begin{cases} \$1.50x & \text{for } 0 \leq x \leq d \\ \$1.50d - 0.50(x - d) & \text{for } x > d \end{cases}$$

Let us suppose that demand D is a random variable with probability function

d	1000	2000	3000	6000
$P(D = d)$	0.1	0.4	0.3	0.2

Assuming that all production is in lots of 1000 units, how many units should he produce if he wishes to

(a) produce for the expected demand?

(b) maximize expected profit?

(c) minimize the probability of a loss?

(d) maximize the probability of a profit of \$3000 or more?

(e) minimize the variance of profit?

19.4 JOINTLY DISTRIBUTED RANDOM VARIABLES

While our preceding discussion has dealt with random variables singly, it is often the case that there is more than one random variable of interest relative to an experimental situation. Each random variable, of course, has its own probability function, but the random variables taken together have a *joint probability function* and are said to be *jointly distributed*.

Example 1   In three tosses of a fair coin let X be the total number of heads and let Y be the number of tails on the first two tosses. The sample space, together with the corresponding values for the random variables X and Y, is

$\Omega = \{HHH,\ HHT,\ HTH,\ HTT,\ THH,\ TTH,\ THT,\ TTT\}$

X	3	2	2	1	2	1	1	0
Y	0	0	1	1	1	2	1	2

The joint probability function of X and Y is shown in Table 5. For instance, the event that $(X = 2)$ and $(Y = 1)$ simultaneously is the event $\{HTH, THH\}$ which has probability $\frac{1}{4}$, while the probability of $(X = 1) \cap (Y = 0)$ is zero. In general, the entry in row i (where $i = 0, 1, 2, 3$) and column j (where $j = 0, 1, 2$) is the probability that $(X = i)$ and $(Y = j)$.

TABLE 5

X \ Y	0	1	2	Totals
0	0	0	$\frac{1}{8}$	$\frac{1}{8}$
1	0	$\frac{1}{4}$	$\frac{1}{8}$	$\frac{3}{8}$
2	$\frac{1}{8}$	$\frac{1}{4}$	0	$\frac{3}{8}$
3	$\frac{1}{8}$	0	0	$\frac{1}{8}$
Totals	$\frac{1}{4}$	$\frac{1}{2}$	$\frac{1}{4}$	1

Probability function for X (right bracket). Probability function for Y (bottom).

The probability functions for the individual random variables appear in the margins of the table and are found by summing the elements in the appropriate row or column. For instance, the probability of the event $(X = 2)$ is obtained by adding the probabilities

$$P(X = 2, Y = 0) + P(X = 2, Y = 1) + P(X = 2, Y = 2)$$
$$= \tfrac{1}{8} + \tfrac{1}{4} + 0 = \tfrac{3}{8}$$

For simplicity, we have introduced a new notation here for the intersection of events. The intersection is denoted by commas. Thus, $(X = 2, Y = 0)$ denotes the simultaneous occurrence, or the intersection, of events $(X = 2)$ and $(Y = 0)$. ▶

If X and Y are any two random variables defined on the same sample space, the function $p_{X,Y}$ defined for all real numbers x and y by

$$p_{X,Y}(x, y) = P(X = x, Y = y)$$

is called the *joint probability function* of X and Y.

Suppose that Y has values $y_1, y_2, y_3, \ldots$. Then the event $(X = x_j)$ occurs, if it occurs at all, in conjunction with one of the events $(Y = y_1)$, $(Y = y_2)$, $(Y = y_3)$, $\ldots$. That is,

$$(X = x_j) = \bigcup_k (X = x_j, Y = y_k)$$

where the union is taken over all values y_k of the random variable Y. Since these events are disjoint, it follows that

$$P(X = x_j) = \sum_k P(X = x_j, Y = y_k) \tag{11}$$

In terms of the joint probability function, (11) says that the probability of $(X = x_j)$ may be obtained by summing the joint probability function of X

and Y over the values $y_1, y_2, \ldots$ of Y. (The number of values of Y may be either finite of infinite. Thus, no specific limits have been indicated in (11).)

A similar argument yields

$$P(Y = y_k) = \sum_j P(X = x_j, Y = y_k) \tag{12}$$

That is, the probability of $(Y = y_k)$ is obtained by summing the joint probability function over all values of X.

Example 2 Bales and his associates* obtained Table 6 of who-to-whom frequencies of verbal interaction in a six-member discussion group. Let X denote the rank order of the individual initiating a verbal communication and Y the rank order of the individual to whom the communication is directed. Then assuming that the frequencies in Table 6 represent the true probabilities of the various communications, the joint probability function for X and Y is as shown in Table 7.

TABLE 6

To individual

		1	2	3	4	5	6	Total
	1	0	1238	961	545	445	317	3506
	2	1748	0	443	310	175	102	2778
From	3	1371	415	0	305	125	69	2285
individual	4	952	310	282	0	83	49	1676
	5	662	224	144	83	0	28	1141
	6	470	126	114	65	44	0	819
	Total	5203	2313	1944	1308	872	565	12205

From Table 7, we see, for example, that the probability of a verbal communication from the first-ranked individual to the fourth-ranked individual is $P(X = 1, Y = 4) = 0.045$. Of course, any communication initiated by the first-ranked individual must be directed to one of the individuals in the group. Thus the event $(X = 1)$ occurs in conjunction with one of the events $(Y = 1), (Y = 2), \ldots, (Y = 6)$ so that

$$P(X = 1) = \sum_{j=1}^{6} P(X = 1, Y = j)$$
$$= 0 + 0.101 + 0.079 + 0.045 + 0.036 + 0.026$$
$$= 0.287 \qquad \blacktriangleright$$

*Bales, R. F., *et al.*, "Channels of Communication in Small Groups," *American Sociological Review* **16**, 461–468 (1951).

TABLE 7. Joint probability table for X and Y

		Values of Y						Probability function for X
		1	2	3	4	5	6	
	1	0	0.101	0.079	0.045	0.036	0.026	0.287
	2	0.143	0	0.036	0.026	0.015	0.008	0.228
Values of X	3	0.112	0.034	0	0.025	0.010	0.006	0.187
	4	0.078	0.025	0.023	0	0.007	0.004	0.137
	5	0.054	0.018	0.011	0.007	0	0.003	0.093
	6	0.038	0.010	0.009	0.006	0.005	0	0.068

PROBLEMS

1. Verify the calculations given in Table 7 for the remainder of the probability function for X and compute the probability function for Y.

2. Equations (11) and (12) show that individual probability functions may be obtained once the joint probability function is known. However, the converse is not true. Suppose that X and Y are random variables with $P(X = 0) = P(X = 1) = \frac{1}{2}$ and $P(Y = 0) = P(Y = 1) = P(Y = 2) = \frac{1}{3}$. Write two different possible joint probability tables for X and Y, thus showing that knowledge of individual probability functions is not sufficient to determine the joint probability function.

Consider two random variables X with values $x_1, x_2, x_3, \ldots$ and Y with values $y_1, y_2, y_3, \ldots$. For each value y_k, the *conditional probability function* of X, given $Y = y_k$, is defined by

$$P(X = x_j \mid Y = y_k) = \frac{P(X = x_j, Y = y_k)}{P(Y = y_k)} \qquad j = 1, 2, 3, \ldots$$

In this formula, k is held fixed while j varies. Note that there is a different probability function for each y_k. Similarly, the conditional probability function of Y, given $X = x_j$, is given by

$$P(Y = y_k \mid X = x_j) = \frac{P(X = x_j, Y = y_k)}{P(X = x_j)} \qquad k = 1, 2, 3, \ldots$$

Two random variables are called *independent random variables* if and only if *every* pair of events $(X = x_j)$ and $(Y = y_k)$ is a pair of independent events. Thus X and Y are independent if

$$P(X = x_j, Y = y_k) = P(X = x_j)P(Y = y_k)$$

for every pair of values (x_j, y_k). In this special case, all conditional probability functions are identical. In fact, we have

$$P(X = x_j \mid Y = y_k) = \frac{P(X = x_j, Y = y_k)}{P(Y = y_k)}$$

$$= \frac{P(X = x_j)P(Y = y_k)}{P(Y = y_k)} = P(X = x_j)$$

Example 3   From Table 7 in Example 2, we compute

$$P(X = 1 \mid Y = 4) = \frac{P(X = 1, Y = 4)}{P(Y = 4)} = \frac{0.045}{0.109} = 0.413$$

Since $P(X = 1) = 0.287$ is not equal to 0.413, we conclude that X and Y are not independent.   ▶

Example 4   Suppose X and Y are independent random variables having the same geometric probability function

$$P(X = k) = P(Y = k) = pq^k \qquad k = 0, 1, 2, \ldots$$

If Z is the sum $X + Y$, the event $(Z = n)$ occurs if X takes one of the values $0, 1, 2, \ldots, n$ and $Y = n - X$. Thus

$$(Z = n) = \bigcup_{j=0}^{n} (X = j, Y = n - j)$$

and since X and Y are independent,

$$P(Z = n) = \sum_{j=0}^{n} P(X = j)P(Y = n - j)$$

$$= \sum_{j=0}^{n} pq^j pq^{n-j} = (n + 1)p^2q^n$$

Hence the conditional probability function of X, given $Z = n$, is

$$P(X = j \mid Z = n) = \frac{P(X = j, Z = n)}{P(Z = n)} = \frac{P(X = j, Y = n - j)}{P(Z = n)}$$

$$= \frac{pq^j pq^{n-j}}{(n + 1)p^2q^n} = \frac{1}{n + 1}$$

Given that $Z = n$, X has a uniform probability function.   ▶

The following theorem is frequently used in applications of random variables. Its proof is omitted.

Theorem 3   Let X and Y be independent random variables and suppose that $Z = h(X)$ and $W = g(Y)$ are functions of X and Y, respectively. Then Z and W are independent.   ▶

Our discussion so far has been restricted to two random variables. However, most of the above considerations may be extended to any number of random variables. For instance, for three random variables X, Y, and Z we define the joint probability function $p_{X,Y,Z}$ by

$$p_{X,Y,Z}(x, y, z) = P(X = x, Y = y, Z = z)$$

Then the probability function of X alone is

$$P(X = x_j) = \sum_k \sum_l P(X = x_j, Y = y_k, Z = z_l)$$

obtained by summing the joint probability function over all values of Y and Z having positive probability, that is, summing over the values y_1, y_2, y_3, ... and z_1, z_2, z_3, ...

An important probability model involving several random variables is the generalization of Bernoulli trials to allow more than two possible outcomes on each trial. Such a probability scheme is called *multinomial trials*. In multinomial trials:

(i) Trials are independent.

(ii) Each trial results in exactly one of the outcomes O_1, O_2, ..., O_m ($m \geq 2$).

(iii) The probabilities $p_1 = P(O_1)$, $p_2 = P(O_2)$, ..., $p_m = P(O_m)$ are the same from trial to trial. (Of course, $\sum_{i=1}^{m} p_i = 1$.)

It is natural to introduce m random variables X_1, X_2, ..., X_m where X_i denotes the number of occurrences of outcome O_i in, say, r trials. For each fixed value of i ($1 \leq i \leq m$), the random variable X_i has a binomial probability function. This is because the trials are independent and, relative to outcome O_i, there are two possible outcomes—either O_i occurs or it does not—with respective probabilities p_i and $1 - p_i$. Thus the probability function of X_i is

$$P(X_i = k) = \binom{r}{k} p_i^k (1 - p_i)^{r-k} \qquad k = 0, 1, 2, \ldots, r$$

Since there are m possible outcomes on each trial, a sample space for multinomial trials contains $m \times m \times \cdots \times m = m^r$ elements, each denoting a possible sequence of r outcomes. Because of independence of trials, any particular sequence in which outcome O_1 occurs k_1 times, O_2 occurs k_2 times, ..., O_m occurs k_m times is assigned probability

$$p_1^{k_1} p_2^{k_2} \cdots p_m^{k_m}$$

We have seen (Problem 18, Section 7.2) that the number of such sequences is the multinomial coefficient

$$\binom{r}{k_1, k_2, \ldots, k_m}$$

Thus the joint probability function of the random variables $X_1, X_2, \ldots, X_m$ is

$$P(X_1 = k_1, X_2 = k_2, \ldots, X_m = k_m)$$
$$= \binom{r}{k_1, k_2, \ldots, k_m} p_1^{k_1} p_2^{k_2} \cdots p_m^{k_m} \quad \textbf{(13)}$$

for those combinations of values of $k_1, k_2, \ldots, k_m$ for which $\sum_{i=1}^{m} k_i = r$. Otherwise this probability is zero.

Example 5 In a study by Anderson* of attitude changes over time among a group of people use is made of the multinomial distribution. Anderson postulates that at any given time, a person is in one of m attitude states. To simplify our discussion, we consider only those N persons in the first attitude state at time t.† Each of these, independently of all others, is assumed to have probability p_j (where $j = 1, 2, \ldots, m$) of moving to attitude state S_j at time $t + 1$. Thus these N persons may be considered as constituting N multinomial trials. If for $j = 1, 2, \ldots, m$, we denote the number who move to state S_j by X_j, we have

$$P(X_1 = n_1, X_2 = n_2, \ldots, X_m = n_m) = \binom{N}{n_1, n_2, \ldots, n_m} p_1^{n_1} p_2^{n_2} \cdots p_m^{n_m}$$

For instance, if $N = 4$ and $p_1 = p_2 = p_3 = \frac{1}{3}$, then

$$P(X_1 = 2, X_2 = 2, X_3 = 0) = \frac{4!}{2!2!0!} \left(\frac{1}{3}\right)^2 \left(\frac{1}{3}\right)^2 \left(\frac{1}{3}\right)^0 = \frac{2}{27} \quad \blacktriangleright$$

PROBLEMS

3. Suppose that the daily closing market quotations for the Skidmore Company preferred stock form multinomial trials with outcomes O_1, an increase in market value; O_2, no change in value; and O_3, a decrease in market value, having probabilities $\frac{3}{10}$, $\frac{1}{10}$, and $\frac{6}{10}$, respectively. Let $X_1, X_2,$ and X_3 denote the respective numbers of outcomes $O_1, O_2,$ and O_3 in a three-day period. For each choice of $k_1, k_2,$ and k_3, what are the probabilities that $X_1 = k_1$, $X_2 = k_2$, and $X_3 = k_3$? (*Hint:* A partial answer is:

$$P(X_1 = 0, X_2 = 1, X_3 = 2) = 3(\tfrac{1}{10})(\tfrac{6}{10})^2 = \tfrac{108}{1000}$$
$$= P(X_1 = 1, X_2 = 1, X_3 = 1)$$
$$P(X_1 = k_1, X_2 = k_2, X_3 = k_3) = 0 \text{ unless } k_1 + k_2 + k_3 = 3.)$$

The definition of independence is extended to any number of random variables $X_1, X_2, \ldots, X_n$ by requiring that the n events $(X_1 = x_1)$, $(X_2 = x_2), \ldots, (X_n = x_n)$ be independent for *every* n-tuple of values $(x_1, x_2, \ldots, x_n)$.

*Anderson, T. W., "Probability Models for Analyzing Time Changes in Attitudes," in *Mathematical Thinking in the Social Sciences*, Lazarsfeld, P. F., Ed. (Free Press, Glencoe, Ill., 1954).

†Anderson's original model was a Markov chain.

Example 6 Let $X_1, X_2, \ldots, X_n$ be independent random variables each having the uniform probability function

$$P(X_i = k) = \frac{1}{N} \qquad \text{for } k = 1, 2, \ldots, N$$

Let us find the probability function of the largest of the X_i, a random variable we shall denote by Z.

The solution is as follows: The probability $P(Z = r)$ is difficult to obtain directly so we use an indirect approach. Note that the event $(Z \leq r)$ can be partitioned into the events $(Z = r)$ and $(Z \leq r - 1)$. Hence

$$P(Z \leq r) = P(Z = r) + P(Z \leq r - 1)$$

or

$$P(Z = r) = P(Z \leq r) - P(Z \leq r - 1)$$

But $(Z \leq r)$ if and only if all X_i are less than or equal to r. That is,

$$P(Z \leq r) = P(X_1 \leq r, X_2 \leq r, \ldots, X_n \leq r)$$

Since independence of the X_i implies independence of the events $(X_1 \leq r)$, $(X_2 \leq r), \ldots, (X_n \leq r)$, we have

$$P(Z \leq r) = P(X_1 \leq r) \cdot P(X_2 \leq r) \cdots P(X_n \leq r)$$

$$= \frac{r}{N} \times \frac{r}{N} \times \cdots \times \frac{r}{N} = \left(\frac{r}{N}\right)^n$$

Finally,

$$P(Z = r) = \left(\frac{r}{N}\right)^n - \left(\frac{r-1}{N}\right)^n \qquad \text{for } r = 1, 2, \ldots, N \qquad \blacktriangleright$$

PROBLEMS

4. Returning to Example 2,
   (a) find the conditional probability function of Y, given $X = 2$.
   (b) find the conditional probability function of X, given $Y = 1$.
   (c) find the probability function of $Z = X + Y$.
   (d) find $P(X - Y > 0)$.

5. Suppose that X and Y have the following joint probability table:

X \ Y	1	2	3	4	$P(X = x)$
0	0.07	0.13	0.05	0.10	0.35
1	0.01	0.08	0.17	0.21	0.47
2	0.06	0.00	0.10	0.02	0.18
$P(Y = y)$	0.14	0.21	0.32	0.33	1.00

(a) Find the probability function of the random variable $Z = X/Y$.

(b) Show that X and Z are dependent random variables.

6. Two subjects are observed on 10 trials of an experiment. On each trial, each subject may choose to respond cooperatively C or competitively D.

(a) Assuming that the players respond independently and that the responses of each player form Bernoulli trials with respective probabilities $p_1(C) = \frac{1}{4}$ and $p_2(C) = \frac{1}{3}$ of choosing response C, what is the probability of obtaining the following joint response frequency table?

Player II

		C	D
Player I	C	1	2
	D	1	6

(b) Find the probability of obtaining exactly k number of DD responses given that Player I makes three C and seven D responses and Player II makes two C and eight D responses.

7. Assume that a person who votes Republican in one election will in the next election vote Republican with probability $\frac{3}{4}$, vote Democratic with probability $\frac{1}{5}$, and vote Independent with probability $\frac{1}{20}$. Assuming that voting behavior is independent from one individual to another, what is the probability that of eight people who vote Republican in the Spring

(a) five vote Republican, two vote Democratic, and one votes Independent in the Fall?

(b) at least one voter changes in the Fall?

(c) the number who change equals the number who do not?

8. Assume that in Problem 7 the respective probabilities that a Democratic voter next votes Republican, Democratic, or Independent are 0.1, 0.7, and 0.2, while the corresponding probabilities for an Independent are 0.1, 0.1, and 0.8. What is the probability that of eight Republicans, seven Democrats, and four Independents:

(a) At least one voter in each party changes?

(b) Exactly one in each party changes?

(c) Writing n_{RR} for the number of Republicans who again vote Republican, n_{RD} for the number of Republicans who next vote Democratic, etc., what is the probability that

$$
\begin{array}{ccc}
n_{RR} = 5 & n_{RD} = 2 & n_{RI} = 1 \\
n_{DR} = 1 & n_{DD} = 4 & n_{DI} = 2 \\
n_{IR} = 1 & n_{ID} = 1 & n_{II} = 2
\end{array}
$$

9. Suppose in Problem 7 we are interested only in the change from "Republican" to "other" and from "other" to "Republican" at time t. Then of eight Republicans, seven Democrats, and four Independents find the probability that

(a) $n_{RR} = 5 \qquad n_{RO} = 3 \qquad n_{OR} = 2 \qquad n_{OO} = 9$

(b) $n_{RR} + n_{OR} = 7 \qquad n_{RO} + n_{OO} = 12$

10. The output of a certain filling machine of the Crunch-E Peanut Butter Company is regarded as a Bernoulli process with probability 0.9 of filling a jar within acceptable tolerances and 0.1 of not doing so. A particular jar may or may not be inspected with probability 0.1 and 0.9, respectively.

(a) If the decision to inspect any particular jar is made without prior knowledge as to fill level, what is the probability that of eight jars filled by the machine: Four are not inspected and are within acceptable tolerances, one is inspected and is within acceptable tolerances, one is inspected and is not within acceptable tolerances, and two are not inspected and are not within acceptable fill tolerances?

(b) A jar is "passed" if the jar is either within tolerances and is inspected or is not inspected. What is the probability that of seven jars passed, five are within acceptable fill tolerances?

(c) Show that if k of N jars are passed, the conditional probability that exactly n are within tolerances is

$$\binom{k}{n} p^n (1 - p)^{k-n}$$

where $p = 0.9/(0.09 + 0.81 + 0.09)$ is the probability that a jar is within acceptable tolerances given that the jar has been passed.

(d) Define the random variable X as the percent of the passed jars that are within acceptable tolerances. If in a sequence of N jars, exactly k jars are passed, find the probability that $(n/k) \times 100$ percent of the k jars are within acceptable tolerances:

(i) given the previously defined inspection plan.

(ii) if no inspection is performed.

How might you use the ratio of the two probabilities as an argument for the adoption of the inspection plan?

11. (a) Let X and Y be independent random variables each having the same geometric probability function. Find the probability function of $Z = $ larger of X and Y. (*Hint:* Proceed as in Example 6.)

(b) Obtain the probability function of Z by first deriving the joint probability function of X and Z.

(c) Find the probability function of $W = X + Y$.

12. Show that if $X_1, X_2, \ldots, X_n$ are independent random variables having the same geometric probability function, then $X_1 + X_2 + \cdots + X_n$ is a negative binomial random variable. Interpret this result in terms of waiting times.

13. Let $X_1, X_2, \ldots, X_n$ be independent random variables each having the uniform probability function

$$P(X_i = k) = \frac{1}{N} \qquad k = 1, 2, 3, \ldots, N$$

Find the probability function of U_n, the smallest of the X_i.

14. Prove that if X and Y are independent binomial random variables with respective parameters (n, p) and (m, p), then $Z = X + Y$ is a binomial random variable with parameters $n + m$ and p.

15. Suppose "at-bats" form multinomial trials with possible outcomes $H = $ hit, $W = $ walk, and $O = $ out, having respective probabilities $\frac{3}{10}$, $\frac{1}{10}$, and $\frac{6}{10}$. If X_H, X_W, and X_O denote the respective numbers of outcomes H, W, and O in three at-bats, what are the probabilities that $(X_H = a, X_W = b, X_O = c)$ for all possible choices of a, b, and c?

16. In information theory,* the basic information function H is a measure of the average amount of information or uncertainty defined over the values of a random variable. If X is a discrete random variable with r values $x_1, x_2, \ldots, x_r$ and $p_i = P(X = x_i)$, $i = 1, 2, \ldots, r$, then H is defined by

$$H(X) = -\sum_{i=1}^{r} p_i \log_2 p_i$$

Similarly, for two random variables X and Y, the joint information function is defined as

$$H(X, Y) = -\sum_{i,j} p_{ij} \log_2 p_{ij}$$

where

$$p_{ij} = P(X = x_i, Y = y_j)$$

(a) Prove that if X and Y are independent, then

$$H(X, Y) = H(X) + H(Y)$$

(b) Prove that if X and Y are dependent, then

$$H(X) + H_X(Y) = H(X, Y)$$

where

$$H_X(Y) = -\sum_{i} \sum_{j} p_{ij} \log_2 p_i(j)$$

$$p_i(j) = P(Y = y_j \mid X = x_i)$$

19.5 SUMS OF RANDOM VARIABLES

We have already seen that if X is a random variable and $g(X)$ is a function of X, then the expectation $E(g(X))$ may be found directly from the probability

---

*For a brief nontechnical discussion of information theory, see Luce, R. D., "The Theory of Selective Information and Some of Its Applications. Part I," in *Developments in Mathematical Psychology*, Luce, R. D., Ed. (Free Press, Glencoe, Ill., 1960).

function of X without knowledge of the probability function of $g(X)$. Specifically, from Equation (9) we have

$$E(g(X)) = \sum_{x_k} g(x_k)P(X = x_k) \tag{14}$$

An argument similar to that leading to (14) may be used to establish the following more general theorem, the proof of which is omitted.

Theorem 4 If $g(X_1, X_2, \ldots, X_n)$ is any function of the random variables $X_1, X_2, \ldots, X_n$, then

$$E(g(X_1, \ldots, X_n))$$
$$= \sum_{x_1,\ldots,x_n} g(x_1, x_2, \ldots, x_n)P(X_1 = x_1, X_2 = x_2, \ldots, X_n = x_n)$$

the sum being taken over all n-tuples $(x_1, x_2, \ldots, x_n)$ for which $P(X_1 = x_1, X_2 = x_2, \ldots, X_n = x_n)$ is positive. ▶

An especially important case of this theorem arises when g is the function which sums the given random variables. Thus, if $g(X, Y) = X + Y$, then

$$E(g(X, Y)) = E(X + Y) = \sum_{x_k} \sum_{y_j} (x_k + y_j)P(X = x_k, Y = y_j)$$

We break this into two separate sums

$$\sum_{x_k} \sum_{y_j} x_k P(X = x_k, Y = y_j) + \sum_{x_k} \sum_{y_j} y_j P(X = x_k, Y = y_j)$$

and sum the first of these over j first and the second over k first. Recalling that

$$\sum_{y_j} P(X = x_k, Y = y_j) = P(X = x_k)$$

and

$$\sum_{x_k} P(X = x_k, Y = y_j) = P(Y = y_j)$$

gives

$$E(X + Y) = \sum_{x_k} x_k P(X = x_k) + \sum_{y_j} y_j P(Y = y_j)$$
$$= E(X) + E(Y)$$

We have shown that the expected value of the sum of two random variables is the sum of the respective expected values.

This result can easily be extended by induction to any number of random variables (see Problem 6). This, together with Theorem 1(a) implies that for any random variables $X_1, X_2, \ldots, X_n$ and constants $c_1, c_2, \ldots, c_n$

$$E(c_1 X_1 + c_2 X_2 + \cdots + c_n X_n)$$
$$= c_1 E(X_1) + c_2 E(X_2) + \cdots + c_n E(X_n) \tag{15}$$

Example 1 The random variable N_r, denoting the number of successes in r Bernoulli trials, may be written as

$$N_r = X_1 + X_2 + \cdots + X_r$$

where X_1 is the number of successes (0 or 1) on the first trial, X_2 is the number (again 0 or 1) on the second trial, and so forth. For each $i = 1, 2, \ldots, r$ we have

$$P(X_i = 1) = p \quad \text{and} \quad P(X_i = 0) = q = 1 - p$$

Hence

$$E(X_i) = 1 \cdot p + 0 \cdot q = p$$

and

$$E(N_r) = E(X_1) + E(X_2) + \cdots + E(X_r)$$
$$= \quad p \quad + \quad p \quad + \cdots + \quad p \quad = rp \quad \blacktriangleright$$

Example 2 For the one-element learning model discussed in Example 2, Section 18.6, let us define the random variables

$$E_n = \begin{cases} 0 & \text{if a correct response occurs on trial } n \\ 1 & \text{if an error occurs on trial } n \end{cases}$$

Then if $\overline{E} = \sum_{n=1}^{\infty} E_n$, the number of errors expected during learning is

$$E(\overline{E}) = E\left(\sum_{n=1}^{\infty} E_n \right) = \sum_{n=1}^{\infty} E(E_n)$$

$$= \sum_{n=1}^{\infty} P(E_n = 1) = \sum_{n=1}^{\infty} \left(1 - \frac{1}{r} \right)(1 - p)^{n-1}$$

$$= \frac{[1 - (1/r)]}{p} \quad \blacktriangleright$$

Example 3 The control unit of an automated assembly line contains a large number N of electronic circuit boards. The engineering staff is faced with the decision of whether to

  (i) replace the circuit boards only as they fail.
  (ii) remove and replace all boards after T time periods whether they have failed or not.

The costs involved in their decision are

  C_1: cost per board of removal and replacement at some specified time period
  C_2: cost per board of replacement when failure occurs during operation

Since it is not known with certainty when each board will fail, the choice of a maintenance policy can be regarded as a decision under uncertainty.

Consider unit intervals of time as independent trials and suppose that at trial 0, all circuit boards are newly installed. For the ith board, let the random variable X_i denote the number of trials up to and including the trial on which failure occurs. Then, the probability that a new ith circuit board will fail on the kth trial is

$$P(X_i = k) = pq^{k-1}$$

where p is the probability of failure and q is the probability of survival. In addition, assume that the random variables $X_1, X_2, \ldots, X_N$ associated with the different circuit boards are mutually independent.

As each board fails, it is immediately replaced by a new board, which is in turn replaced when it fails, and so forth. Suppose that a failure occurs at trial t. This may occur if a single board fails at time $X_1 = t$; or if the first board fails and is replaced by a second board which fails such that the second failure occurs after a total time $X_1 + X_2 = t$; or, ultimately, until the tth board fails such that $X_1 + X_2 + \cdots + X_t = t$. That is, a failure which occurs on the tth trial may be either the first failure, or the second failure, or $\ldots$, or the tth failure.

Let $f_t^{(r)}$ denote the probability that the rth failure occurs on the tth trial. Then

$$f_t^{(r)} = P(X_1 + X_2 + \cdots + X_r = t)$$

$$= \binom{t-1}{t-r} p^r q^{t-r}$$

since $X_1, X_2, \ldots, X_r$ are independent random variables each with the geometric distribution $P(X = k) = pq^{k-1}$, $k = 1, 2, \ldots$ (see Problem 12, Section 19.4).

Let u_t be the probability that a failure occurs at the tth trial. Then

$$u_t = f_t^{(1)} + f_t^{(2)} + \cdots + f_t^{(r)} + \cdots + f_t^{(t)}$$

$$= \sum_{r=1}^{t} \binom{t-1}{t-r} p^r q^{t-r}$$

$$= p \sum_{r=1}^{t} \frac{(t-1)!}{(r-1)!(t-r)!} p^{r-1} q^{t-r}$$

$$= p$$

The expected number of failures or, equivalently, the expected number of replacements on the tth trial is given by

$$Np$$

The expectation principle can now be used in the choice of T. After T time periods (trials) the preventive maintenance policy calls for the replacement of all circuit boards at a cost of NC_1. For all trials $t < T$, all circuit boards that fail are replaced at an expected cost of $C_2Np(T-1)$ if it is assumed that all boards which fail at trial T are replaced at cost C_1 as part of the preventive maintenance program.

The expected maintenance cost for a given T is therefore

$$E(T) = NC_1 + C_2Np(T-1)$$

Since $E(T)$ is monotonically increasing with T, the choice of a maintenance policy is based on the expected per period cost $E(T)/T$. The decision criterion is that preventive maintenance should be undertaken at the end of any period for which the expected cost of replacing the circuit boards that fail during the period exceeds the average per period replacement cost up to and including that period. That is, preventive maintenance should be taken at the completion of trial T if

$$C_2Np \geq \frac{NC_1 + C_2Np(T-1)}{T}$$

$$\geq C_2Np - \frac{N}{T}(C_2p - C_1)$$

Hence, $C_2Np > E(T)/T$ whenever $C_2p > C_1$. Note that the difference $C_2Np - E(T)/T$ is maximized when $T = 1$.

In other words, the strategy should be to replace after every trial if the expected per unit cost of replacement of a board which fails when in use is equal to or greater than the per unit cost of preventive maintenance. Otherwise, it is more economical to replace only when a failure occurs. ▶

PROBLEMS

1. Suppose X and Y have the following joint probability table:

X \ Y	1	2	3	$P(X = x)$
1	0.1	0.3	0.2	0.6
2	0.2	0.1	0.1	0.4
$P(Y = y)$	0.3	0.4	0.3	1

Find $E(X)$, $E(Y)$, $E(X + Y)$, and $E(X \cdot Y)$ using Theorem 4. (*Hint:* Note that $E(X \cdot Y)$ is *not* equal to $E(X) \cdot E(Y)$.)

2. Suppose $X_1, X_2, \ldots, X_n$ have common mean m. Prove that the average $\bar{X} = (1/n)(X_1 + \cdots + X_n)$ also has mean m.

3. Let N_r be the number of black balls obtained in r draws without replacement from an urn containing b black balls and g green balls ($r \leq b + g$). Show that $E(N_r) = rp$ where $p = b/(b + g)$. (*Hint:* Proceed as in Example 1 and note that this is a special Polya urn.)

Unfortunately it is *not* always the case that the variance (see Equation (10)) of a sum of random variables is the sum of the individual variances. In fact, by definition,

$$\begin{aligned}
\text{Var } (X + Y) &= E([X + Y - E(X + Y)]^2) \\
&= E(\{[X - E(X)] + [Y - E(Y)]\}^2) \\
&= E([X - E(X)]^2) + 2E([X - E(X)][Y - E(Y)]) \\
&\quad + E([Y - E(Y)]^2)
\end{aligned}$$

The first and third terms represent the respective variances of X and Y but the middle term spoils the simplicity of our formula. The quantity

$$E([X - E(X)][Y - E(Y)])$$

is called the *covariance* of X and Y and is written Cov (X, Y). We have shown that

$$\text{Var } (X + Y) = \text{Var } (X) + \text{Var } (Y) + 2 \text{ Cov } (X, Y)$$

The covariance may also be written as

$$\begin{aligned}
\text{Cov } (X, Y) &= E((X - \mu_X)(Y - \mu_Y)) \\
&= \sum_j \sum_k (j - \mu_X)(k - \mu_Y)P(X = j, Y = k)
\end{aligned}$$

where $\mu_X = E(X)$ and $\mu_Y = E(Y)$. The terms in this sum are positive if $(j - \mu_X)$ and $(k - \mu_Y)$ are either both positive or both negative. Otherwise the terms are negative. Thus the covariance can be any real number. It will be positive when X and Y tend simultaneously to take values on the same side of their respective means, and negative when the opposite tendency is predominant.

Example 4   A fair coin is tossed three independent times. Let X be the number of heads on the first toss and Y the total number of heads on the three tosses. Either from their description or from their joint probability table we might expect these random variables to have positive covariance. Actual computation yields Cov $(X, Y) = \frac{1}{4}$.   ▶

PROBLEMS

4. Verify the computation in Example 4.

A more convenient form for computation of the covariance may be obtained by applying Equation (15). Since

$$(X - \mu_X)(Y - \mu_Y) = XY - \mu_X Y - \mu_Y X + \mu_X \mu_Y$$

we have

$$\text{Cov } (X, Y) = E(XY) - \mu_X E(Y) - \mu_Y E(X) + \mu_X \mu_Y$$

But $\mu_X = E(X)$ and $\mu_Y = E(Y)$ so

$$\text{Cov } (X, Y) = E(XY) - E(X) \cdot E(Y)$$

In this form it is apparent that the covariance is zero, and thus the variance of the sum $X + Y$ is the sum of the individual variances, if and only if the expectation $E(XY)$ of the product XY is equal to the product $E(X)E(Y)$ of the respective expectations. In the very important case of independent random variables, this relation holds. For if X and Y are independent, then

$$P(X = j, Y = k) = P(X = j) \cdot P(Y = k)$$

and

$$E(XY) = \sum_j \sum_k jkP(X = j, Y = k)$$
$$= \sum_j jP(X = j) \sum_k kP(X = k)$$

Summing first on k and then on j yields

$$E(XY) = E(X) \cdot E(Y)$$

In computing the variance of the sum of any number of random variables, we find

$$\text{Var } (X_1 + X_2 + \cdots + X_n) = \sum_{k=1}^{n} \text{Var } (X_k) + 2 \sum_{j<k} \text{Cov } (X_j, X_k) \quad \textbf{(16)}$$

For example,

$$\text{Var } (X_1 + X_2 + X_3 + X_4)$$
$$= \text{Var } (X_1) + \text{Var } (X_2) + \text{Var } (X_3) + \text{Var } (X_4)$$
$$+ 2 \text{ Cov } (X_1, X_2) + 2 \text{ Cov } (X_1, X_3) + 2 \text{ Cov } (X_1, X_4)$$
$$+ 2 \text{ Cov } (X_2, X_3) + 2 \text{ Cov } (X_2, X_4) + 2 \text{ Cov } (X_3, X_4)$$

If the random variables $X_1, X_2, \ldots, X_n$ are independent, then all co-variances are zero. In this case (16) becomes

$$\text{Var } (X_1 + \cdots + X_n) = \text{Var } (X_1) + \cdots + \text{Var } (X_n) \quad \textbf{(17)}$$

The variance of the sum of *independent* random variables is the sum of the individual variances.

PROBLEMS

   5. Derive Formula (16).

Example 5 Sampling without replacement from a population of size N consisting of S elements designated as "successes" and the remaining $N - S$

as "failures" may be viewed as a special kind of Polya urn. If we define random variables $X_1, X_2, \ldots, X_n$ by

$$X_k = \begin{cases} 1 & \text{if the } k\text{th draw results in "success"} \\ 0 & \text{otherwise} \end{cases} \qquad k = 1, 2, \ldots, n$$

then our previous results (see Problem 12, Section 18.6) imply

$$E(X_k) = P(X_k = 1) = \frac{S}{N}$$

$$\text{Var}(X_k) = \frac{S(N - S)}{N^2}$$

Also if $k \neq j$, then

$$E(X_k X_j) = P(X_k = 1, X_j = 1) = P(X_k = 1)P(X_j = 1 \mid X_k = 1)$$

$$= \frac{S}{N} \frac{S - 1}{N - 1}$$

so that

$$\text{Cov}(X_k, X_j) = E(X_k X_j) - E(X_k)E(X_j)$$

$$= \frac{S}{N} \frac{S - 1}{N - 1} - \left(\frac{S}{N}\right)^2 = -\frac{S(N - S)}{N^2(N - 1)}$$

Since $\text{Var}(X_k)$ and $\text{Cov}(X_j, X_k)$ are the same for all k and j, it follows from (16) that

$$\text{Var}(X_1 + X_2 + \cdots + X_n) = n\frac{S(N - S)}{N^2} - \frac{n(n - 1)S(N - S)}{N^2(N - 1)}$$

$$= npq\frac{N - n}{N - 1}$$

where $p = S/N$ and $q = (N - S)/N = 1 - p$. (Compare with Problem 16, Section 19.3.) The factor $(N - n)/(N - 1)$ serves as a correction factor for the variance when the sample is drawn without replacement. ▶

Example 6 Suppose that in a sequence of Bernoulli trials the payoff on trial k is \$$k$ for heads and \$0 for tails. If T denotes the total payoff on n trials and X_k the number of heads on trial k, we have

$$T = \sum_{k=1}^{n} k X_k$$

Hence

$$E(T) = E\left(\sum_{k=1}^{n} k X_k\right) = \sum_{k=1}^{n} k E(X_k) = \sum_{k=1}^{n} kp = p\frac{n(n + 1)}{2}$$

and

$$\text{Var}\,(T) = \sum_{k=1}^{n} \text{Var}\,(k\,X_k) = \sum_{k=1}^{n} k^2\,\text{Var}\,(X_k)$$

$$= \sum_{k=1}^{n} k^2 pq = \frac{n(n+1)(2n+1)}{6}\,pq \qquad \blacktriangleright$$

The quantity

$$\rho(X,\,Y) = \frac{\text{Cov}\,(X,\,Y)}{\sqrt{\text{Var}\,(X)\cdot\text{Var}\,(Y)}}$$

is called the *correlation* between the random variables X and Y. When X and Y are independent, $\rho(X,\,Y) = 0$ since in this case $\text{Cov}\,(X,\,Y) = 0$. It can be shown (see Problem 12) that

$$-1 \le \rho(X,\,Y) \le 1$$

and that the extreme values -1 and 1 are assumed only when one of the random variables is a *linear* function of the other. When $\rho(X,\,Y) = 0$, X and Y are said to be *uncorrelated*.

Example 7 Suppose that random variables X and Y have the following joint probability table:

Y \ X	1	$\frac{1}{2}$	0	$P(Y = y)$
1	$\frac{1}{4}$	$\frac{1}{6}$	$\frac{1}{12}$	$\frac{1}{2}$
0	$\frac{7}{24}$	$\frac{1}{12}$	$\frac{3}{24}$	$\frac{1}{2}$
$P(X = x)$	$\frac{13}{24}$	$\frac{6}{24}$	$\frac{5}{24}$	

It is obvious from inspection that X and Y are dependent since, for example, $P(X = 1,\, Y = 1) = \frac{1}{4} \ne \frac{1}{2}\cdot\frac{13}{24}$. However, it can be verified that $E(Y\cdot X) = \frac{1}{3}$, $E(Y) = \frac{1}{2}$, and $E(X) = \frac{2}{3}$ so that

$$\text{Cov}\,(X,\,Y) = E(XY) - E(X)\cdot E(Y)$$

$$= \tfrac{1}{3} - \tfrac{1}{2}\cdot\tfrac{2}{3} = 0$$

and the variables are uncorrelated. From this, we conclude that while the correlation is zero if the variables are independent, the converse is not necessarily true. $\qquad\blacktriangleright$

PROBLEMS

6. Show by induction that for any positive integer n,

$$E(X_1 + X_2 + \cdots + X_n) = E(X_1) + E(X_2) + \cdots + E(X_n)$$

7. (a) Let $Z = X - Y$. Write $E(Z)$ and Var (Z) in terms of the means, variances, and covariance of X and Y.

(b) Show that Cov $(Z, X) = $ Var $X - $ Cov (X, Y).

8. If S_r denotes the proportion of successes in r Bernoulli trials with probability p of success, show that

$$\text{Cov } (S_r, 1 - S_r) = -pq/r$$

9. Prove that the H measure of information (Problem 16, Section 19.4) has the property

$$H(XY) \le H(X) + H(Y)$$

(*Hint:* Make use of the inequality $t - 1 > \ln t$, $t > 0$, and let $t = p_i p_j / p_{ij}$.

10. As in Problem 8, Section 18.6, let the probability of an error on trial n be

$$q_n = \alpha q_{n-1} \qquad (0 < \alpha < 1)$$

Let $X_1, X_2, \ldots$ be mutually independent random variables such that X_n assumes the value 1 when an error occurs on trial n and 0 otherwise.

(a) Verify that $E(X_n) = \alpha^n q_0$.

(b) Compute the mean and variance of $T = \sum_{n=1}^{\infty} X_n$.

11. (a) Let X be a random variable with $E(X) = m$ and Var $(X) = \sigma^2$. Let Y be a binomial random variable taking the values 0 and 1 with respective probabilities $1 - p$ and p. Show that

$$\rho(X, Y) = \frac{\sqrt{p(1 - p)}}{\sigma} [E(X \mid Y = 1) - E(X \mid Y = 0)]$$

(b) Prove that $-1 \le \rho(X, Y) \le 1$.

12. Let X and Y be any two discrete random variables.

(a) Show that, for any real number c, $g(c) = E(cX + Y)^2$ is non-negative.

(b) Using the fact that $g(c)$ is a quadratic function of c (a parabola), argue that we must have

$$[E(XY)]^2 \le E(X^2) \cdot E(Y^2)$$

(This is the so-called Cauchy–Schwarz inequality.)

(c) Prove that $-1 \le \rho(X, Y) \le 1$ and that $\rho(X, Y) = \pm 1$ only if Y is a linear function of X.

13. In a sequence of r multinomial trials with possible outcomes $O_1, O_2, \ldots, O_m$, let X_i ($i = 1, 2, \ldots, m$) denote the number of occurrences of outcome O_i. Find Cov (X_i, X_j). (*Hint:* Define new random variables X_{ik} (where $i = 1, 2, \ldots, m$; $k = 1, 2, \ldots, r$) such that $X_{ik} = 1$ if trial k results in outcome E_i and $X_{ik} = 0$, otherwise.)

14. Interpret the negative binomial random variable W_n, denoting the number of failures before the nth success in a sequence of Bernoulli trials, as

the sum of n independent geometric random variables. Then use Formulas (15) and (17) to find the mean and variance of W_n.

15. Let X and Y be independent, identically distributed random variables. Define a random variable $U = X + Y$ and a random variable $V = X - Y$. Prove that the correlation between U and V is zero.

16. Prove that if X and Y are binary random variables, that is, assume only two values, then if $\rho(X, Y) = 0$, the variables are independent.

17. Faced with a pressing problem, an organization considers two options:

(a) Assign the problem to a single individual.

(b) Assign the problem to a group of k individuals.

The nature of the problem is such that it consists of three stages which must be solved in sequence. Assume that individuals work independently and that there is probability $\frac{1}{4}$ that any individual to whom the problem is assigned will, within a unit time period, achieve a satisfactory solution to the problem stage then under consideration. Assume further that a group of k individuals solves a stage of the problem if and only if at least one member arrives at a solution of that stage.

If each consultant costs the organization $50 per unit time, which option should be chosen to minimize expected cost? (*Hint:* With k individuals, let T_{ij} be the time required for the ith individual to solve the jth stage, let $S_j = \min \{T_{1j}, T_{2j}, \ldots, T_{kj}\}$ be the solution time for stage j, and let $S = S_1 + S_2 + S_3$ be the total solution time.)

18. Suppose we modify Problem 17 slightly by assuming the consultants each cost $50 for each unit of time in which the problem is *not* solved. What option should now be chosen to minimize expected cost?

19. A more realistic assumption concerning Problem 17 is that not only does each consultant cost $50 per unit time (including the time in which a solution is reached) but that, due to the possibility that a competitor will obtain the solution first, there is an additional cost for undue delay. This time, solve the problem under the assumption that if S units of time are required, then the cost is

$$C = \$50[kS + 2^S]$$

20. Prove that if $X_1, X_2, \ldots, X_N$ are independent random variables, each having mean μ and variance σ^2, and if $\overline{X} = (1/N)\sum_{i=1}^{N} X_i$, then

$$E\left(\sum_{i=1}^{N} (X_i - \overline{X})^2\right) = (N - 1)\sigma^2$$

21. The basic model of psychometric test-score theory* assumes that the test score X_i for the ith individual may be written

$$X_i = m_i + \epsilon_i$$

where m_i is that individual's true score and ϵ_i is a random error variable having expected value $E(\epsilon_i) = 0$ and Var $(\epsilon_i) = \sigma^2$. For a population

*For a discussion of the theory of test scores, see Guilford, J. P., *Psychometric Methods* (McGraw-Hill Book Company, New York, 1954), 2nd ed.

of N individuals the *reliability* of the test is defined as

$$\rho = \frac{\sum_{i=1}^{N} (m_i - \bar{m})^2}{E(\sum_{i=1}^{N} (X_i - \bar{X})^2)}$$

where

$$\bar{m} = \frac{1}{N} \sum_{i=1}^{N} m_i \quad \text{and} \quad \bar{X} = \frac{1}{N} \sum_{i=1}^{N} X_i$$

(a) Show that, assuming individual scores are independent random variables, ρ may be rewritten as

$$\rho = \frac{\sum_{i=1}^{N} (m_i - \bar{m})^2}{\sum_{i=1}^{N} (m_i - \bar{m})^2 + (N-1)\sigma^2}$$

Hence show that $0 \le \rho \le 1$, and that reliability decreases as the error variance increases.

(b) The simplifying (but unrealistic) assumption that $\sum_{i=1}^{N} \epsilon_i = 0$ is often made. Show that in this case,

$$\rho = \frac{\sum_{i=1}^{N} (m_i - \bar{m})^2}{\sum_{i=1}^{N} (m_i - \bar{m})^2 + N\sigma^2}$$

22. (a) Let T_1 and T_2 be independent geometric random variables, not necessarily having the same probability function. Let

$$Z = \min \{T_1, T_2\}$$

be the smaller of T_1 and T_2. Prove that Z itself is a geometric random variable. Interpret this result in terms of success and failure in Bernoulli trials. (*Hint:* $Z > k$ if and only if both $T_1 > k$ and $T_2 > k$.)

(b) Hence show that $E(Z) < \min \{E(T_1), E(T_2)\}$. That is, the expected value of the smaller of two geometric random variables is *less than* the smaller of the expected values.

(c) Extend the result in (b) to any two random variables. (*Hint:* We always have $Z \le T_1$ and $Z \le T_2$.)

(d) Consider a project which must be complete in a succession of stages. A number of individuals may be assigned to work independently on a particular stage, that stage being "completed" as soon as *any one* of the individuals completes his work. According to the PERT (Program Evaluation Review Technique) technique, the expected total completion time is the sum of the smallest of the expected values of individual completion times for the various stages. In view of (a)–(c) above, however, explain why PERT should tend to overestimate completion times.

19.6 GENERATING FUNCTIONS

In this section we shall briefly consider ways in which infinite series can be used to solve problems in probability. Following standard terminology, the power series $\sum_{k=0}^{\infty} a_k s^k$ will be called the *generating function* of its sequence of coefficients. Recall (Section 16.2, Problem 22) that if $f(s) = \sum_{k=0}^{\infty} a_k s^k$ converges in some interval about zero, the coefficients are *uniquely* determined by $a_k = f^{(k)}(0)/k!$. Conversely, a particular sequence $(a_0, a_1, a_2, \ldots)$ of coefficients obviously determines a unique function $f(s) = \sum_{k=0}^{\infty} a_k s^k$ within the interval of convergence of this series.

Example 1 (a) If $a_k = 1$ for each k, the generating function is

$$g(s) = \sum_{k=0}^{\infty} 1 \cdot s^k = \frac{1}{1-s}$$

(b) If N_r is the number of successes in r Bernoulli trials, the generating function of the sequence of probabilities $P(N_r = k)$ is

$$g_{N_r}(s) = \sum_{k=0}^{\infty} P(N_r = k)s^k = \sum_{k=0}^{r} \binom{r}{k} p^k q^{r-k} s^k = (q + ps)^r$$

(c) The function $e^{-\lambda(1-s)}$ generates the numbers

$$a_k = \frac{e^{-\lambda}\lambda^k}{k!} \qquad k = 0, 1, 2, \ldots \qquad \blacktriangleright$$

If X is a random variable having non-negative integer values $0, 1, 2, \ldots$, the *probability generating function* of X is the expected value

$$g_X(s) = E(s^X) = \sum_{k=0}^{\infty} P(X = k)s^k \qquad (18)$$

The closely related *moment generating function* is defined by

$$m_X(s) = E(e^{sX}) = \sum_{k=0}^{\infty} P(X = k)e^{sk} \qquad (19)$$

Obviously, the moment generating function is obtained from the probability generating function by replacing s by e^s. That is, $m_X(s) = g_X(e^s)$.

Example 2 (a) The results of Example 1(b) indicate that the probability generating function for the number of successes in r Bernoulli trials is

$$g_{N_r}(s) = E(s^{N_r}) = \sum_{k=0}^{r} \binom{r}{k} p^k q^{r-k} s^k = (q + ps)^r$$

The corresponding moment generating function is $m_{N_r}(s) = (q + pe^s)^r$.

(b) The probability generating function for a geometric random variable X with $P(X = k) = pq^k$ is

$$g_X(s) = E(s^X) = \sum_{k=0}^{\infty} pq^k s^k = \frac{p}{1 - qs}$$

The corresponding moment generating function is $m_X(s) = p/(1 - qe^s)$. $\blacktriangleright$

1. Show that for a uniform random variable Y with $P(Y = k) = 1/N$ for $k = 1, 2, \ldots, N$, the respective generating functions are

$$g_Y(s) = \frac{s}{N}\frac{1 - s^N}{1 - s} \quad \text{and} \quad m_Y(s) = \frac{e^s}{N}\frac{1 - e^{Ns}}{1 - e^s}$$

Table 8 lists generating functions of the most important discrete random variables. The primary usefulness of such a table lies in the fact that not only does any random variable have a uniquely defined generating function, but conversely, the uniqueness of coefficients in a power series means that any particular generating function corresponds to exactly one probability function.

TABLE 8

Type of random variable	Probability function	Probability generating function	Moment generating function
Binomial	$P(N_r = k) = \binom{r}{k}p^k(1 - p)^{r-k}$, $k = 0, 1, \ldots, r$	$(q + ps)^r$	$(q + pe^s)^r$
Poisson[a]	$P(X = k) = \dfrac{e^{-\lambda}\lambda^k}{k!}$, $k = 0, 1, 2, \ldots$	$e^{-\lambda(1-s)}$	$e^{-\lambda(1-e^s)}$
Geometric	$P(X = k) = pq^k$, $k = 0, 1, 2, \ldots$	$\dfrac{p}{1 - qs}$	$\dfrac{p}{1 - qe^s}$
Negative binomial	$P(X = k) = \binom{r+k-1}{k}p^rq^k$, $k = 0, 1, 2, \ldots$	$\left(\dfrac{p}{1 - qs}\right)^r$	$\left(\dfrac{p}{1 - qe^s}\right)^r$
Uniform	$P(X = k) = \dfrac{1}{n+1}$, $k = 0, 1, \ldots, n$	$\dfrac{1}{n+1}\dfrac{1 - s^{n+1}}{1 - s}$	$\dfrac{1}{n+1}\dfrac{1 - e^{s(n+1)}}{1 - e^s}$

[a] See Section 21.1.

Either the probability generating function or the moment generating function may be used to find various moments of a random variable. In particular, the first derivative of the probability generating function, evaluated at $s = 1$, gives the expected value. For,

$$g'_X(s) = \frac{d\sum_{k=0}^{\infty} P(X = k)s^k}{ds}$$

$$= \sum_{k=0}^{\infty} \frac{d[P(X = k)s^k]}{ds} \qquad \text{[Inside the interval of convergence]}$$

$$= \sum_{k=0}^{\infty} kP(X = k)s^{k-1}$$

If the interval of convergence includes the number 1, then

$$g'_X(1) = \sum_{k=0}^{\infty} kP(X = k) = E(X)$$

Similarly,

$$m'_X(s) = \frac{d}{ds} \sum_{k=0}^{\infty} P(X = k)e^{sk}$$

$$= \sum_{k=0}^{\infty} P(X = k)ke^{sk}$$

so

$$m'_X(0) = E(X)$$

The first derivative of the moment generating function, evaluated at zero, is $E(X)$.

Further differentiation leads to the following general result.

Theorem 5 The nth derivative of the moment generating function $m_X(s)$, evaluated at $s = 0$, gives the nth moment of the random variable X. That is,

$$E(X^n) = m_X^{(n)}(0)$$

The nth derivative of the probability generating function $g_X(s)$, evaluated at $s = 1$, gives the nth factorial moment of X. That is,

$$E(X(X - 1) \cdots (X - n + 1)) = g_X^{(n)}(1) \qquad \blacktriangleright$$

As a particular application of this theorem, we have

$$\text{Var}\,(X) = E(X^2) - [E(X)]^2 = m_X''(0) - [m_X'(0)]^2$$

and

$$\text{Var}\,(X) = E[X(X - 1)] + E(X) - [E(X)]^2 = g_X''(1) + g_X'(1) - [g_X'(1)]^2$$

Example 3 For the binomial random variable N_r, $g_{N_r}(s) = (q + ps)^r$. Differentiating, we find

$$g'_{N_r}(1) = rp(q + ps)^{r-1}\big|_{s=1} = rp$$

$$g''_{N_r}(1) = r(r - 1)p^2(q + ps)^{r-2}\big|_{s=1} = r(r - 1)p^2$$

Hence,

$$E(N_r) = rp$$

and

$$\text{Var } (N_r) = r(r - 1)p^2 + rp - r^2p^2$$
$$= rp - rp^2 = rp(1 - p) \qquad \blacktriangleright$$

PROBLEMS

  2. Obtain the above results using the moment generating function.

If X and Y are independent random variables, the probability generating function of the sum $Z = X + Y$ is

$$g_Z(s) = E(s^Z) = E(s^{X+Y}) = E(s^X \cdot s^Y)$$

The independence of X and Y implies independence of s^X and s^Y and we find

$$g_Z(s) = E(s^X) \cdot E(s^Y) = g_X(s) \cdot g_Y(s)$$

The probability generating function of the *sum* of two *independent* random variables is the *product* of the individual probability generating functions.

Obviously, the same considerations apply as well to moment generating functions and also to sums of more than two random variables. That is, if $X_1, X_2, \ldots, X_n$ are independent random variables, then the probability (moment) generating function of the sum

$$Z = X_1 + X_2 + \cdots + X_n$$

is the product of the probability (moment) generating functions of the individual random variables $X_1, \ldots, X_n$.

Example 4 Let X_i denote the number of successes (0 or 1) on the ith trial of a Bernoulli sequence. The X_i are independent and have the common probability generating function

$$g_{X_i}(s) = E(s^{X_i}) = s^0 P(X_i = 0) + s^1 P(X_i = 1) = q + ps$$

Hence the generating function of $N_r = X_1 + \cdots + X_r$ is

$$g_{N_r}(s) = g_{X_1}(s) \times g_{X_2}(s) \times \cdots \times g_{X_r}(s) = (q + ps)^r$$

verifying the result of Example 2. $\qquad \blacktriangleright$

Example 5 The number T_n of failures before the nth success in a sequence of Bernoulli trials can be written

$$T_n = X_1 + X_2 + \cdots + X_n$$

where $X_1, X_2, \ldots, X_n$ are independent random variables representing, respectively, the numbers of failures before the first success, between the first and second successes, ..., between the $(n - 1)$st and nth successes. Since (Example 2) the generating function of each X_i is $g_{X_i}(s) = p/(1 - qs)$, the generating function of T_n is

$$g_{T_n}(s) = \left(\frac{p}{1 - qs}\right)^n$$

verifying line 4 of Table 8. ▶

Generating functions are often useful in solving recurrence relations or difference equations. The following is a typical example in probability.

Example 6 What is the probability that the number N_r of successes in r Bernoulli trials is even, zero being considered an even number?

The solution is as follows: The event $(N_r$ even$)$ can occur in two mutually exclusive ways:

(i) N_{r-1} even and failure on trial r.

(ii) N_{r-1} odd and success on trial r.

Hence, denoting $P(N_r$ even$)$ by u_r, we have, when $r \geq 1$,

$$u_r = P(N_r \text{ even})$$
$$= P(N_{r-1} \text{ even and } F_r) + P(N_{r-1} \text{ odd and } S_r)$$
$$= P(N_{r-1} \text{ even}) P(F_r) + P(N_{r-1} \text{ odd}) P(S_r) \qquad \text{[independence of trials]}$$
$$= u_{r-1} \cdot q + (1 - u_{r-1}) \cdot p$$
$$= p + (q - p)u_{r-1}$$

Of course $u_0 = 1$.

Introducing the generating function $U(s) = \sum_{r=0}^{\infty} u_r s^r$ we find

$$U(s) = \sum_{r=0}^{\infty} u_r s^r = u_0 + \sum_{r=1}^{\infty} u_r s^r$$

$$= 1 + \sum_{r=1}^{\infty} [p + (q - p)u_{r-1}] s^r$$

$$= 1 + p \sum_{r=1}^{\infty} s^r + (q - p)s \sum_{r=1}^{\infty} u_{r-1} s^{r-1}$$

$$= 1 + \frac{ps}{1 - s} + (q - p)s U(s)$$

Solving for $U(s)$ yields

$$U(s) = \frac{1 - qs}{(1 - s)[1 - (q - p)s]}$$

or, as a partial fraction expansion,

$$U(s) = \frac{1}{2}\left[\frac{1}{1 - s} + \frac{1}{1 - (q - p)s}\right]$$

Now $U(s) = \sum_{r=0}^{\infty} u_r s^r$ and

$$\frac{1}{2}\left[\frac{1}{1-s} + \frac{1}{1-(q-p)s}\right] = \frac{1}{2}\sum_{r=0}^{\infty} s^r + \frac{1}{2}\sum_{r=0}^{\infty} (q-p)^r s^r$$

We simply equate coefficients of s^r to obtain

$$u_r = \tfrac{1}{2}[1 + (q-p)^r] \qquad \blacktriangleright$$

PROBLEMS

3. Use the probability generating function to compute expected value and variance for the random variables listed in Table 8.

4. Verify your results in Problem 3 using the moment generating function.

5. A test consists of n items. Each item has k alternatives which are assigned the weights $0, 1, 2, \ldots, k - 1$, respectively. The test score T is defined as the sum of the n weights corresponding to the alternatives selected. If the subject randomly selects an alternative for each item, show that the generating function of T is

$$g_T(s) = \left[\frac{1 - s^k}{k(1 - s)}\right]^n$$

Find $E(T)$ and Var (T).

6. For discrete random variables X and Y, define the *conditional generating function* of Y, given $X = k$, by

$$g_{Y|X}(s, k) = E[s^Y \mid X = k] = \sum_r s^r P(Y = r \mid X = k)$$

Show how the conditional moments $E(Y^n \mid X = k)$ can be obtained from this generating function.

7. Suppose that we have a population that is in the process of dying (that is, decreasing in size) as, for example, mice leaving a box at the onset of shock, dissipation of a crowd, or evacuation of people from a danger area. Let us regard each member of such a population of initial size N as a Bernoulli trial with probability $1 - e^{-t_0\lambda_1}$ of dying (leaving) in the interval $(0, t_0)$. Now suppose that at time t_0, something happens so that the probability of dying in the interval (t_0, t) changes to $1 - e^{-(t-t_0)\lambda_2}$. If $N(t)$ is the number still alive at time $t(t > t_0)$, show that $N(t)$ possesses a binomial distribution with parameters N and $p = e^{-[\lambda_1 t_0 + (t-t_0)\lambda_2]}$. (*Hint:* Use the conditional generating function.)

8. For discrete random variables X and Y, define the *joint*, or *bivariate*, *generating function* by

$$g_{X,Y}(s, t) = E[s^X t^Y] = \sum_k \sum_j s^k t^j P(X = k, Y = j)$$

(a) Show that the generating function of X alone is obtained by setting $t = 1$ in $g_{X,Y}$.

(b) Show that the generating function of $X + Y$ is obtained by setting $s = t$.

(c) Show that $E(X)$ may be obtained by taking the partial derivative of $g_{X,Y}$ with respect to s and setting $s = t = 1$.

(d) Show how the moments $E(Y)$, $E(XY)$, and $E(X^2Y)$ may be obtained from $g_{X,Y}$.

9. Let X be a random variable with generating function $g_X(s)$. Find the generating functions of $X - 1$ and of $3X$.

10. Let X_1 and X_2 be independent binomial random variables with

$$P(X_1 = k) = \binom{n_1}{k} p^k q^{n_1-k}$$

$$P(X_2 = k) = \binom{n_2}{k} p^k q^{n_2-k}$$

Use generating functions to prove that $X_1 + X_2$ has the binomial probability function

$$P(X_1 + X_2 = k) = \binom{n_1 + n_2}{k} p^k q^{n_1+n_2-k}$$

11. Given the probability generating function

$$g_X(s) = 1 - (1 - s^2)^{1/2}$$

compute $E(X)$. Interpret your results.

12. Let V_r denote the probability that the number of successes in r Bernoulli trials is divisible by three (zero is divisible by 3).

(a) Find V_0, V_1, V_2, and V_3.

(b) Give a probabilistic argument to establish the validity of the formula

$$V_r = q^r + \sum_{k=3}^{r} \binom{k}{2} p^3 q^{k-3} V_{r-k} \qquad r \geq 3$$

(c) Introduce the generating function $V(s) = \sum_{r=0}^{\infty} V_r s^r$ and use the results of (a) and (b) to show that

$$V(s) = \frac{(1 - qs)^2}{(1 - qs)^3 - (ps)^3}$$

13. Expand $V(s)$ in Problem 12 in a partial fraction expansion and argue that $\lim_{r \to \infty} V_r = \frac{1}{3}$, regardless of the value of p. Interpret this result.

SUPPLEMENTARY READING

Feller, W., *An Introduction to Probability Theory and its Applications* (John Wiley & Sons, Inc., New York, 1968), Vol. 1, 3rd ed. Chapters 6, 9, and 11.

Goldberg, S., *Probability, An Introduction* (Prentice-Hall, Englewood Cliffs, N.J., 1960), Chapters 4 and 5.

Parzen, E., *Modern Probability Theory and its Applications* (John Wiley & Sons, Inc., New York 1960), Chapters 7 and 8.

MARKOV CHAINS 20

20.1 BASIC CONCEPTS

Probability models involving independent trials are not always appropriate to behavioral processes. In this chapter we investigate a model involving dependent trials which has found wide application. Our assumptions are simple. Each trial is assumed to depend on the immediately preceding trial but, when the results of that trial are known, not on any more remote trial. A typical situation is the following case.

Example 1 *The Random Walk* A particle "walks" on a line as follows (see Figure 1). The particle starts at the origin and a coin is tossed. If it falls heads (probability p), the particle moves to the point 1. If it falls tails (probability $q = 1 - p$), the particle moves to -1. This constitutes the first trial and further trials proceed in a similar fashion. At whatever point the particle may have arrived after the first t trials, the same coin is tossed and the particle is moved either one unit to the right or to the left on the $(t + 1)$st trial. Obviously its position after $t + 1$ trials is completely determined by the outcome of trial $t + 1$ and its position after t trials. Information about preceding trials is superfluous. ▶

FIGURE 1

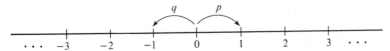

Example 1 contains the essential ingredients of a *Markov* chain* which, generally speaking, constitutes the most important and widely used probability model for the behavioral sciences. A Markov chain is described by listing the *states* of the chain, the *initial probabilities* of being in the various states and the *probabilities of transition* from one state to another. In Example 1, the states are the possible positions $0, \pm 1, \pm 2, \ldots$ of the particle. The process begins in state 0. Denoting the initial probability of being in state k by $\pi_k(0)$, we have

$$\pi_0(0) = 1 \quad \text{and} \quad \pi_k(0) = 0 \qquad \text{if } k \neq 0$$

From a given state k, it is possible to move in one step (or trial) only to state $k + 1$ or to state $k - 1$, with respective probabilities p and q. All other (one-step) transition probabilities are zero.

Example 2 In a T-maze, a rat may turn left and obtain food or turn right and receive a mild electric shock. Let us suppose that initially (on trial number zero) he is equally likely to turn left or right. Having obtained food on one trial, his probabilities of turning left and right become 0.7 and 0.3, respectively, on the following trial. Having received the shock on a given trial, his new probabilities of going left or right become 0.8 and 0.2.

We thus have a Markov chain with two states R (the rat turns right) and L (the rat turns left). The initial probabilities of R and L are, respectively, $\pi_R(0) = \frac{1}{2}$ and $\pi_L(0) = \frac{1}{2}$. The probability of transition from state R to state R (that is, the probability of turning right after having turned right on the preceding trial) is $p_{R,R} = 0.2$, the probability of transition from state R to state L is $p_{R,L} = 0.8$, and so on.

This information is conveniently summarized by writing the *initial probability vector*

$$\pi(0) = (\pi_R(0), \pi_L(0)) = (\tfrac{1}{2}, \tfrac{1}{2})$$

and the *transition matrix*

$$P = \begin{pmatrix} p_{R,R} & p_{R,L} \\ p_{L,R} & p_{L,L} \end{pmatrix} = \begin{pmatrix} 0.2 & 0.8 \\ 0.3 & 0.7 \end{pmatrix} \qquad \blacktriangleright$$

Example 3 Purchases of laundry cleaning powders over a 26-week period were studied by Styon and Smith.† Assuming that the types of laundry powders purchased constituted states of a Markov chain, they obtained the transition matrix shown in Table 1.

The transition matrix indicates the extent of brand loyalty and brand switching taking place. For example, of those households purchasing

---

*After A. A. Markov, one of the many outstanding Russian contributors to the theory of probability.

†Styon, G. P., and Smith, H., "Markov Chains Applied to Marketing," *Journal of Marketing Research* **1,** 50–55 (1964)

detergents in one week, 67% purchased only detergents during the following week—an indication of detergent loyalty. Similar evidence of brand loyalty for the other purchase behavior patterns may be noted, leading to the conclusion that purchases from week to week are not independent. ▶

TABLE 1. Transition matrix for laundry purchases

		Purchased the following week			
		Detergent	Soap powder	Both powders	No powders
	Detergent	0.67	0.09	0.02	0.22
Purchased	Soap powder	0.04	0.72	0.04	0.20
one week	Both powders	0.12	0.24	0.52	0.12
	No powder	0.15	0.26	0.02	0.57

In a Markov chain, we assume there is a collection of states, S_1, S_2, S_3, ... The event that the system is in state k at trial n is denoted by $S_k(n)$ and its probability $P(S_k(n))$ by $\pi_k(n)$. The vector

$$\pi(n) = (\pi_1(n), \pi_2(n), \pi_3(n), \ldots)$$

contains the probabilities of being in the various states at time (trial) n. Since at any trial we must be in some state and cannot be in more than one, the probabilities in the vector $\pi(n)$ must sum to unity. That is,

$$\sum_k \pi_k(n) = 1$$

The probability of transition from state S_j to state S_k (that is, the conditional probability of being in state S_k on one trial, given that the process was in state S_j on the preceding trial) is denoted by p_{jk} and the transition matrix is

To state

$$\text{From state} \quad \begin{array}{c} \\ S_1 \\ S_2 \\ S_3 \\ \vdots \end{array} \begin{pmatrix} S_1 & S_2 & S_3 & \cdots \\ p_{11} & p_{12} & p_{13} & \cdots \\ p_{21} & p_{22} & p_{23} & \cdots \\ p_{31} & p_{32} & p_{33} & \cdots \\ \vdots & \vdots & \vdots & \vdots \end{pmatrix}$$

In moving from a state we must go to *some* state, so that the elements in *each* row of the transition matrix P sum to unity. That is,

$$\sum_k p_{jk} = 1 \qquad \text{for each } j$$

A matrix with non-negative entries whose rows sum to unity is called a *stochastic matrix*. A vector whose entries are non-negative and sum to unity is a *probability vector*.

The following is the distinguishing feature of a Markov chain.

Markov Property   Suppose the system is in states $S_{j_0}, S_{j_1}, S_{j_2}, \ldots, S_{j_{n-2}}$, S_j, respectively, on trials $0, 1, 2, \ldots, n - 1$. Then the conditional probability of state S_k on trial n is the same as the conditional probability of S_k, given only that the process is in state S_j on trial $n - 1$. Symbolically,

$$P(S_k(n) \,|\, S_{j_0}(0) \cap S_{j_1}(1) \cap \cdots \cap S_{j_{n-2}}(n - 2) \cap S_j(n - 1))$$

$$= P(S_k(n) \,|\, S_j(n - 1)) = p_{jk} \quad \blacktriangleright$$

Roughly speaking, if the state of the process is known at trial $n - 1$, knowledge of previous trials adds no probabilistic information. All that matters is which state the process is in and not how it got there. This feature is sometimes called the *independence of path assumption*.

It must not be supposed, however, that two events determined by different trials are necessarily independent if the trial numbers differ by 2 or more. In the *absence* of information about intervening trials, the probability of occurrence of state k on trial n does depend on the occurrence of state j on trial m even if $n - m > 1$. The next example illustrates this point.

Example 4   In a study of the Samoan language Newman* has found that the sequence of vowels V and consonants C is adequately described as a Markov chain with

$$p_{VV} = 0.51 \qquad p_{VC} = 0.49$$

$$p_{CV} = 1 \qquad p_{CC} = 0$$

Thus, consonants never follow consonants in written Samoan, while the probabilities are approximately equal that the letter following a vowel is a vowel or a consonant.

For purposes of illustration let us arbitrarily assume that for the first letter the respective probabilities of V and C are $\frac{2}{3}$ and $\frac{1}{3}$. Then it is easily verified from Figure 2 that the probability is 0.67 that the third letter is a vowel, while the conditional probability of this event, given that the first letter is a vowel, is 0.75. Thus, the events V_1 and V_3 are not independent. Note that we have assumed nothing about the second letter. $\blacktriangleright$

PROBLEMS

1. Verify the calculations in Example 4.

*Newman, E. B., "The Pattern of Vowels and Consonants in Various Languages," *American Journal of Psychology* **64**, 369–379 (1951). Markovian properties of a number of other languages, including English, are discussed.

FIGURE 2

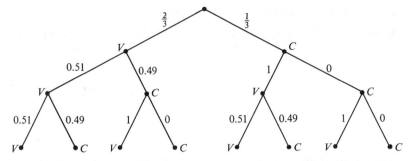

A state S_k may be reached at trial n by going through one and only one of the states $S_1, S_2, \ldots$ on trial $n - 1$. This observation allows us to partition the event $S_k(n)$ into

$$S_k(n) = \bigcup_j (S_j(n - 1) \cap S_k(n))$$

so that

$$\pi_k(n) = P(S_k(n)) = \sum_j P(S_j(n - 1) \cap S_k(n))$$

$$= \sum_j P(S_j(n - 1))P(S_k(n) \mid S_j(n - 1))$$

$$= \sum_j \pi_j(n - 1) \cdot p_{jk}$$

The term $\pi_k(n)$ is the kth element in the vector $\pi(n)$, while $\sum_j \pi_j(n - 1)p_{jk}$ is the kth element in the matrix product $\pi(n - 1) \cdot P$, of the vector $\pi(n - 1)$ times the transition matrix P. That is, for $n = 1, 2, 3, \ldots$,

$$\pi(n) = \pi(n - 1) \cdot P \tag{1}$$

It follows that probabilities of being in the various states at any time may be written in terms of the initial vector $\pi(0)$ and the transition matrix P. Using (1), we find

$$\pi(1) = \pi(0) \cdot P$$

$$\pi(2) = \pi(1) \cdot P = \pi(0) \cdot P^2$$

$$\pi(3) = \pi(2) \cdot P = \pi(0) \cdot P^3$$

and, by a simple inductive argument,

$$\pi(n) = \pi(0) \cdot P^n \tag{2}$$

We shall always write $P^0 = I$, the identity matrix, so that (2) holds for $n = 0, 1, 2, 3, \ldots$.

2. In Example 2, compute $\pi(2)$, $\pi(3)$, and $\pi(4)$.

3. Show that if the rows of a transition matrix are all equal to the same vector $R = (r_1, r_2, r_3, \ldots ,)$ then for $n = 1, 2, 3, \ldots , P^n = P$ and $\pi(n) = R$.

In terms of individual elements, Equation (2) may be written

$$\pi_k(n) = \sum_j \pi_j(0) p_{jk}(n)$$

where $p_{jk}(n)$ denotes the element in the (j, k) position of P^n. In particular, if we begin in state S_r, so that $\pi_r(0) = 1$, the probability $\pi_k(n)$ of being in state S_k at time n is

$$\pi_k(n) = 1 \cdot p_{rk}(n)$$

That is, the element $p_{rk}(n)$ in the (r, k) position of P^n may be interpreted as the probability of transition from state S_r to state S_k in n steps.

Example 5 Let us approximate the situation in Example 4 by adopting the transition matrix

$$\text{To}$$

$$\begin{array}{cc} & V \quad C \end{array}$$

$$P: \text{From} \quad \begin{array}{c} V \\ C \end{array} \begin{pmatrix} \frac{1}{2} & \frac{1}{2} \\ 1 & 0 \end{pmatrix}$$

Then

$$P^2 = \begin{pmatrix} \frac{3}{4} & \frac{1}{4} \\ \frac{1}{2} & \frac{1}{2} \end{pmatrix} \quad \text{and} \quad P^3 = \begin{pmatrix} \frac{5}{8} & \frac{3}{8} \\ \frac{3}{4} & \frac{1}{4} \end{pmatrix}$$

and, in general, a simple induction shows that

$$P^n = \begin{pmatrix} \frac{2}{3} + \frac{1}{3}(-\frac{1}{2})^n & \frac{1}{3} - \frac{1}{3}(-\frac{1}{2})^n \\ \frac{2}{3} + \frac{1}{3}(-\frac{1}{2})^{n-1} & \frac{1}{3} - \frac{1}{3}(-\frac{1}{2})^{n-1} \end{pmatrix} \tag{3}$$

From this general form we see that if we begin with a vowel, there is probability $\frac{2}{3} + \frac{1}{3}(-\frac{1}{2})^n$ that the $(n + 1)$st letter is a vowel and probability $\frac{1}{3} - \frac{1}{3}(-\frac{1}{2})^n$ that the $(n + 1)$st letter is a consonant. Similarly, the elements in the second row of P^n give the respective probabilities that the nth letter following a consonant is a vowel or a consonant.

More interesting is the fact that

$$A = \lim_{n \to \infty} P^n = \begin{pmatrix} \frac{2}{3} & \frac{1}{3} \\ \frac{2}{3} & \frac{1}{3} \end{pmatrix}$$

is a matrix with identical rows. Thus, for large n, the probability that the nth letter is a vowel is approximately $\frac{2}{3}$, regardless of whether the initial letter is a vowel or a consonant.

Let $\pi_V(n)$ and $\pi_C(n)$ denote the respective probabilities that the nth letter is a vowel or a consonant. Then if $\pi_V(1) = p$, Equation (2) shows that

$$\pi(n) = (\pi_V(n), \pi_C(n)) = (p, 1 - p) \cdot P^{n-1}$$

$$= [\tfrac{2}{3} + \tfrac{1}{3}(-\tfrac{1}{2})^{n-2} + p(-\tfrac{1}{2})^{n-1}, \tfrac{1}{3} - \tfrac{1}{3}(-\tfrac{1}{2})^{n-2} - p(-\tfrac{1}{2})^{n-1}]$$

Hence no matter what the value of p, the effect of the initial vector wears off and we have

$$\lim_{n \to \infty} \pi(n) = (\tfrac{2}{3}, \tfrac{1}{3}) \qquad \blacktriangleright$$

PROBLEMS

4. Verify the calculations of P^2 and P^3 in Example 5.

5. Go through the inductive proofs leading to Equations (2) and (3).

6. A rat is put into the maze shown in Figure 3. The rat moves through the compartments at random. That is, if there are k ways to leave a compartment, he has probability $1/k$ of choosing each of these. Write the transition matrix for this Markov chain using the compartment numbers as states of the chain.

7. Consider a system possessing an equilibrium state E_0 and which may be displaced from equilibrium in either a positive or negative direction. Examples are inflation–recession or expansion–contraction in economics

FIGURE 3

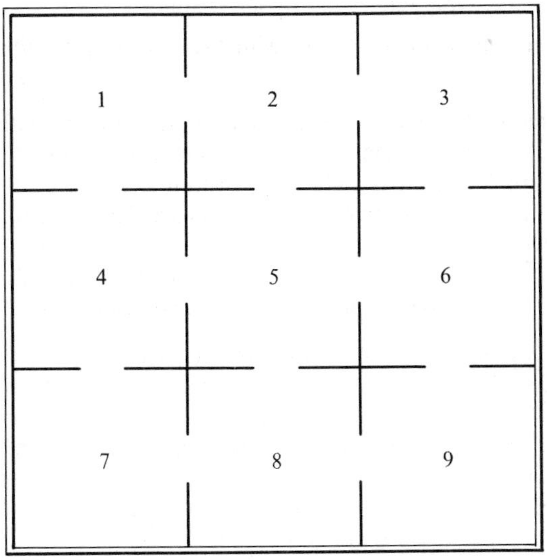

and hypoactivity–hyperactivity or elation–depression in psychology. For simplicity let us assume a system

$$\overset{\bullet}{E_{-2}} \quad \overline{\quad} \quad \overset{\bullet}{E_{-1}} \quad \overline{\quad} \quad \overset{\bullet}{E_0} \quad \overline{\quad} \quad \overset{\bullet}{E_1} \quad \overline{\quad} \quad \overset{\bullet}{E_2}$$

with five possible states in which transitions from any state are restricted to adjacent states. We thus have a finite random walk (compare with Example 1) with boundary states E_{-2} and E_2 and internal states E_{-1}, E_0, and E_1. Write the transition matrix for the system in each of the following cases:

(a) From an internal state the system may move either left or right or remain in that state, with respective probabilities p, q, and r

$$(p + q + r = 1).$$

However, once a boundary state is reached, the system is "absorbed" and remains there forever.

(b) Same problem as (a) except that when the system reaches a boundary state, it is "reflected" to the adjacent internal state. For instance, if the system enters E_{-2}, it moves to E_{-1} in the next time period.

(c) From state E_j (where $j = -2, \ldots, 2$) the system may move either to E_{j-1} or to E_{j+1} with respective probabilities $(2 + j)/4$ and $(2 - j)/4$.

8. After each successive transmission through the various levels of an organization, an order may be either in an *unaltered state U* or, due either to error or administrative reconsideration, in an *altered state A*. Assume that $p_{AU} = 1/10$ and $p_{UA} = 8/10$.

(a) Write the transition matrix P.

(b) Construct a tree diagram for the first five stages.

(c) Compute the probability vectors $\pi(1), \ldots, \pi(5)$.

(d) Compute $P^2, \ldots, P^5$. Note the trend in your results.

(e) Verify that the calculation for P^5 agrees with the results obtainable from the tree.

9. The arrival of customers at a sales station is a Bernoulli process with probability p that at least one customer arrives during a given minute. The sales clerk can complete a break only if no customers arrive in the next five minutes. Define a Markov chain whose states are defined as the number of consecutive time periods without a customer arrival. Find the transition matrix, assuming that the process terminates upon the occurrence of five successive time periods without a customer arrival.

10. Three candidates are contenders for nomination by the party convention. Suppose that on each ballot candidate A holds his delegates with probability $\frac{5}{6}$, candidate B holds his delegates with probability $\frac{3}{4}$, and candidate C holds his delegates with probability $\frac{1}{3}$. If a candidate loses his delegation on a particular ballot, he withdraws from contention. On each ballot, all delegation decisions as to whether to hold or switch are made without knowledge of the decisions of the other delegations. If the state of the system is defined as the set of candidates still in contention

for the nomination, verify that the transition matrix is

	ϕ	A	B	C	AB	AC	BC	ABC
ϕ	1	0	0	0	0	0	0	0
A	0	1	0	0	0	0	0	0
B	0	0	1	0	0	0	0	0
C	0	0	0	1	0	0	0	0
AB	$\frac{1}{24}$	$\frac{5}{24}$	$\frac{3}{24}$	0	$\frac{15}{24}$	0	0	0
AC	$\frac{2}{18}$	$\frac{10}{18}$	0	$\frac{1}{18}$	0	$\frac{5}{18}$	0	0
BC	$\frac{2}{12}$	0	$\frac{6}{12}$	$\frac{1}{12}$	0	0	$\frac{3}{12}$	0
ABC	$\frac{2}{72}$	$\frac{10}{72}$	$\frac{6}{72}$	$\frac{1}{72}$	$\frac{30}{72}$	$\frac{5}{72}$	$\frac{3}{72}$	$\frac{15}{72}$

11. (a) Show that the *Polya urn scheme* (Example 4, Section 18.6) is not a Markov chain if the state of the system is the color of the ball last drawn.

(b) However, if the state of the system lists the numbers of red and green balls in the urn, then we do have a Markov chain.

12. Diffusion of a technological innovation may be interpreted as a two-state Markov chain. At trial n, a potential user can be classified as being either familiar F or unfamiliar U with the innovation. If we assume that the transition $U \rightarrow F$ occurs with probability p and $F \rightarrow U$ with probability 0,

(a) write the transition matrix P.

(b) write the n-step transition matrix P^n directly without resort to matrix multiplication.

13. If the process in Problem 12 is assumed to start in state U, what is the probability that the system will be in state F on the

(a) first trial?   (b) second trial?   (c) nth trial?

(Compare with Example 2, Section 18.6.)

14. Suppose we have multinomial trials with possible outcomes $O_1, O_2, \ldots,$ O_r on each trial. Show that this is a Markov chain in which $P^n = P$ for all $n \geq 1$.

15. A small consulting firm employs three inexperienced trainees. In order to assure exposure to both office and field experience while at the same time avoiding partiality, the company randomly selects one of the three trainees each week and assigns him either to the office or to the field. The initial assignment is made randomly. Otherwise, if a trainee is selected, his assignment is changed from that of the previous week. If a trainee is not selected, his assignment remains unchanged. Consider as states of a Markov chain the triplets (x, y, z) where x is the number of unassigned trainees, y the number of trainees assigned to the office, and z is the number assigned to the field. Compute the transition matrix.

20.2 CLASSIFICATION OF THE STATES

It will be recalled from Chapter 3 that a relation E defined on a set Λ is called an equivalence relation if it possesses three properties:

(i) The reflexive property—for every $x \in \Lambda$, xEx.

(ii) The symmetric property—for every pair of elements x and y in Λ, if xEy, then yEx.

(iii) The transitive property—for any three elements x, y, and z in Λ, if xEy and yEz, then xEz.

The set Λ is partitioned by E into "equivalence classes" in the following sense. For each $x \in \Lambda$, let Λ_x denote the set of all elements $z \in \Lambda$ for which xEz. Then for any two elements s and t in Λ, either $\Lambda_s = \Lambda_t$ or $\Lambda_s \cap \Lambda_t = \phi$.

The equivalence relation important for Markov chains is the relation of communication. Two states S_i and S_j are said to *communicate* $(S_i C S_j)$ if there exist non-negative integers m and n such that $p_{ij}(n)$ and $p_{ji}(m)$ are both positive. Since $p_{ii}(0) = 1$, each state communicates with itself. Different states communicate when there is positive probability that each can be reached from the other in a finite number of steps.

PROBLEMS

1. The last two statements in the preceding paragraph imply that communication is reflexive and symmetric. Complete the proof that communication is an equivalence relation by establishing transitivity.

The *communication class* C_S for a state S is the set of all states which communicate with S. No communication class is empty since each state communicates with itself. And, of course, since communication is an equivalence relation, the communication classes of two states are equal if those states communicate, and disjoint if they do not.

A communication class C from which no exit is possible is said to be *closed*. In this case $p_{jk} = 0$ if S_j is in C and S_k is outside C. It follows that if we delete from the transition matrix for the chain all rows and columns not corresponding to states in C, we still have a stochastic matrix. Restricting our attention to this matrix allows us to study the closed class independently of the rest of the chain. A closed communication class which contains only one state is called *absorbing* and the same description is applied to the state itself.

A communication class which is not closed is called *open*. An open class O may be left, but never returned to. For if return were possible, a state S_j in O would communicate with a state S_k outside O. In this case, S_j and S_k would be members of the same communication class, an obvious contradiction. In leaving an open class, the process may pass either to another open class or to a closed class.

Example 1 A simple Markov chain was used by Marshall and Goldhamer[*] to model the process leading to first admission to a mental hospital. They assumed a process having the states

$$S_0 = \text{alive, sane}$$

$$S_1 = \text{alive, insane (mild), unhospitalized}$$

$$S_2 = \text{alive, insane (severe), unhospitalized}$$

$$S_3 = \text{insane, hospitalized}$$

$$S_4 = \text{dead}$$

and transition matrix

$$
\begin{array}{c c c c c c}
 & S_0 & S_1 & S_2 & S_3 & S_4 \\
S_0 & p_{00} & p_{01} & p_{02} & 0 & p_{04} \\
S_1 & 0 & p_{11} & 0 & p_{13} & p_{14} \\
S_2 & 0 & 0 & p_{22} & p_{23} & p_{24} \\
S_3 & 0 & 0 & 0 & 1 & 0 \\
S_4 & 0 & 0 & 0 & 0 & 1
\end{array}
$$

According to the model there is no return from the insane, hospitalized state nor from the state S_4. Thus, states S_3 and S_4 form individual absorbing classes for the first admission process. Each of the states S_0, S_1, and S_2 forms a separate open class. ▶

Example 2 Let us simplify Problem 15 of Section 20.1 by assuming only two sales trainees. All other conditions remain unchanged. Then the resulting Markov chain has six states with transition matrix

	(0, 1, 1)	(0, 2, 0)	(0, 0, 2)	(1, 1, 0)	(1, 0, 1)	(2, 0, 0)
S_1: (0, 1, 1)	0	$\frac{1}{2}$	$\frac{1}{2}$	0	0	0
S_2: (0, 2, 0)	1	0	0	0	0	0
S_3: (0, 0, 2)	1	0	0	0	0	0
S_4: (1, 1, 0)	$\frac{1}{4}$	$\frac{1}{4}$	0	0	$\frac{1}{2}$	0
S_5: (1, 0, 1)	$\frac{1}{4}$	0	$\frac{1}{4}$	$\frac{1}{2}$	0	0
S_6: (2, 0, 0)	0	0	0	$\frac{1}{2}$	$\frac{1}{2}$	0

[*]Marshall, A. W., and Goldhamer, H., "An Application of Markov Processes to the Study of the Epidemiology of Mental Disease," *Journal of the American Statistical Association* **50**, 99–129 (1955).

The various communication classes are easily recognized in the *flow diagram* of Figure 4. Once the process leaves state S_6, it never returns, so that S_6 forms a single open communication class. Since states S_5 and S_4 communicate with each other but with no other states, these states form a separate open class. The remaining three states form a single closed communication class.

Note that the process passes from open class $\{S_6\}$ to open class $\{S_4, S_5\}$ and then eventually to closed class $\{S_1, S_2, S_3\}$ where it is trapped. ▶

FIGURE 4

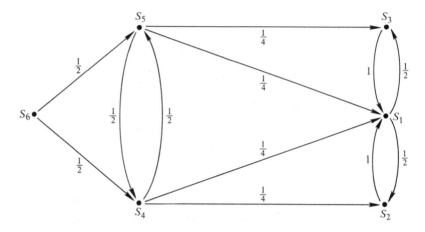

PROBLEMS

2. For the Markov chain whose flow diagram is shown in Figure 5, indicate the communication classes, determine for each class whether it is open or closed, and write the transition matrix.

If we have a Markov chain with, say, two closed classes and two open classes, it should be apparent from the above discussion that, after an appropriate relabeling of the states, the transition matrix for this chain may be written in partitioned form as

$$P = \begin{pmatrix} P_1 & O & O & O \\ O & P_2 & O & O \\ A & B & T_1 & O \\ C & D & E & T_2 \end{pmatrix} \qquad (4)$$

Here P_1 and P_2 are stochastic matrices which contain probabilities of transition within the separate closed classes, T_1 and T_2 contain transition probabilities for the respective open classes, A gives probabilities of transition

FIGURE 5

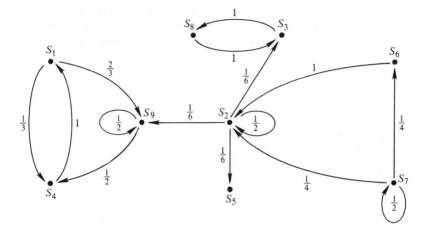

from states of the first open class to those of the first closed class, and so forth. The symbol O indicates a zero matrix.

Regardless of the number of open and closed classes, the transition matrix of a Markov chain can be partitioned in a manner similar to (4). In each case, we arrange the closed classes first, followed by the open classes, in such a way that from any open class the process may move only to an open class or a closed class listed above it. Such an arrangement will be called the *canonical form* for the transition matrix.

Example 3 The canonical form for the transition matrix corresponding to Figure 5 is

	S_5	S_3	S_8	S_1	S_4	S_9	S_2	S_6	S_7
S_5	1	0	0	0	0	0	0	0	0
S_3	0	0	1	0	0	0	0	0	0
S_8	0	1	0	0	0	0	0	0	0
S_1	0	0	0	0	$\frac{1}{3}$	$\frac{2}{3}$	0	0	0
S_4	0	0	0	1	0	0	0	0	0
S_9	0	0	0	0	$\frac{1}{2}$	$\frac{1}{2}$	0	0	0
S_2	$\frac{1}{6}$	$\frac{1}{6}$	0	0	0	$\frac{1}{6}$	$\frac{1}{2}$	0	0
S_6	0	0	0	0	0	0	1	0	0
S_7	0	0	0	0	0	0	$\frac{1}{4}$	$\frac{1}{4}$	$\frac{1}{2}$

▶

PROBLEMS

From the transition matrices indicated in Problems 3–7, draw a flow diagram, arrange the states in communication classes, and write a new transition matrix in canonical form. In Problem 7, * denotes a positive element.

3. $\begin{pmatrix} \frac{3}{4} & \frac{1}{4} & 0 \\ 0 & 1 & 0 \\ 0 & 0 & 1 \end{pmatrix}$

4. $\begin{pmatrix} 0 & \frac{1}{8} & \frac{7}{8} \\ 1 & 0 & 0 \\ 1 & 0 & 0 \end{pmatrix}$

5. The transition matrix in Problem 10 of Section 20.1.

6. The transition matrix in Problem 15 of Section 20.1.

7.

	s_1	s_2	s_3	s_4	s_5	s_6	s_7	s_8	s_9
s_1	*	0	0	0	0	0	*	0	0
s_2	0	*	0	*	0	0	0	0	0
s_3	0	0	*	0	0	0	0	0	*
s_4	0	*	0	*	0	0	0	0	0
s_5	0	0	*	0	*	0	0	0	0
s_6	0	*	0	0	0	*	*	0	*
s_7	0	0	*	0	0	0	*	*	0
s_8	0	0	0	*	*	*	0	0	0
s_9	0	0	*	0	0	0	0	0	*

8. Draw a flow diagram for the Markov chain in Problem 9, Section 20.1.

9. (a) Write a transition matrix for a five-state Markov chain with closed classes $\{s_1, s_3\}$ and $\{s_2, s_4, s_5\}$. Denote nonzero entries by *.

   (b) Partition the transition matrix according to communication classes and tell how to compute powers of this matrix.

   (c) Why can the chain be treated as two separate chains?

10. Prove that every Markov chain with a finite number of states must have at least one closed communication class.

11. For the random walk of Example 1 of Section 20.1, prove that $p_{ii}(2n + 1) = 0$ for $n = 0, 1, 2, \ldots$.

12. Prove that in a finite chain with r states, if S_j can be reached at all from S_i, it can be reached in r or fewer steps.

13. In the flow diagram in Figure 6 an arrow denotes a positive transition probability.

   (a) Argue that every state communicates with every other.

   (b) Argue that return to a given state is possible only in 4, 8, 12, ... steps; that is, in a number of steps which is a multiple of four. The process is called *periodic* (with period 4).

FIGURE 6

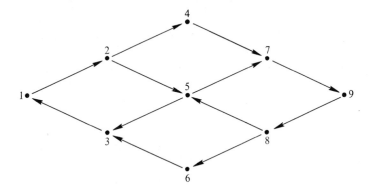

20.3 REGULAR CHAINS

A Markov chain with transition matrix P is called *regular* if some power of P contains only positive entries. It is obvious that all states communicate in a regular chain so that the chain consists of a single communication class. Moreover, it is easily seen that if $P^n > 0$, then all higher powers of P also contain only positive entries. Thus it is possible to move from any state to any other state in any number of steps greater than or equal to n.

Example 1   The chains in Examples 2, 3, and 4 of Section 20.1 are regular. Those in Examples 1, 2, and 3 of Section 20.2 are not.   ▶

PROBLEMS

    1. In Section 20.2, which of the chains in Problems 3–7 are regular and which are not?

Example 2   In Example 5 of Section 20.1

$$P = \begin{pmatrix} \frac{1}{2} & \frac{1}{2} \\ 1 & 0 \end{pmatrix}$$

and it is obvious that the states communicate. Multiplication shows that

$$P^2 = \begin{pmatrix} \frac{3}{4} & \frac{1}{4} \\ \frac{1}{2} & \frac{1}{2} \end{pmatrix}$$

has all entries positive. Indeed, we saw that for each $n \geq 2$, all entries of P^n are positive and that the powers P^n approach the limit

$$\lim_{n \to \infty} P^n = \begin{pmatrix} \frac{2}{3} & \frac{1}{3} \\ \frac{2}{3} & \frac{1}{3} \end{pmatrix}$$

▶

The behavior exhibited in Example 2 is typical of regular chains. Theorem 1 gives the details. The proof is rather lengthy and will be omitted.*

Theorem 1 If P is a regular transition matrix then the sequence (P^n) of powers of P converges to a stochastic matrix Q each of whose entries is positive and each of whose rows is the same probability vector. ▶

The next theorem provides a simple method for computing the limiting form of P^n.

Theorem 2 Let P be a regular $r \times r$ transition matrix and let $Q = \lim_{n \to \infty} P^n$ be the limiting matrix, each of whose rows is the vector $v = (v_1, v_2, \ldots, v_r)$. Then,

(a) no matter what the initial vector $\pi(0)$, the sequence $\pi(1), \pi(2), \ldots$ converges to v.

(b) v is the unique probability vector satisfying the equation

$$v = vP$$

PROOF (a) Since $\pi(n) = \pi(0) \cdot P^n$, the kth element of $\pi(n)$ is

$$\pi_k(n) = \sum_{j=1}^{r} \pi_j(0) p_{jk}(n)$$

Since, by the preceding theorem,

$$\lim_{n \to \infty} p_{jk}(n) = v_k \qquad (5)$$

holds for each j, it follows that

$$\lim_{n \to \infty} \pi_k(n) = \sum_{j=1}^{r} \pi_j(0) \cdot v_k = v_k$$

(b) Since $P^{n+1} = P^n \cdot P$, we have

$$p_{ik}(n+1) = \sum_{j=1}^{r} p_{ij}(n) \cdot p_{jk}$$

Taking limits as $n \to \infty$ and using (5) gives

$$v_k = \sum_{j=1}^{r} v_j p_{jk}$$

proving that

$$v = vP$$

If u were another probability vector such that $u = uP$, then a simple induction shows that $u = uP^n$ for all $n \geq 1$. But, since u is a probability vector, it follows from (a) that $\lim_{n \to \infty} uP^n = v$. Hence $u = v$. ▶

---

*See Kemeny, J. G., and Snell, J. L., *Finite Markov Chains* (D. Van Nostrand Company, Princeton, N.J., 1960), pp. 70–71.

Example 3 Roby* is concerned with decisions which affect income over a long period of time. He assumes:

(i) At each time (trial) t, an individual may be in one of several environmental states $S_1, S_2, \ldots, S_r$, which constitute the states of a Markov chain.

(ii) A sojourn in state S_i at trial t leads to a specific income $v_i(t)$.

Let $v(t) = \text{col } (v_1(t), v_2(t), \ldots, v_r(t))$ be the income vector at time t and $\pi(t) = (\pi_1(t), \ldots, \pi_r(t))$ be the *futurity* vector (Roby's terminology) of probabilities of being in the various states at time t. Then the decision maker's expected income at time t is

$$\pi(t) \cdot v(t) = \sum_{i=1}^{r} \pi_i(t) v_i(t) \tag{6}$$

As an example, consider a stock which goes through states of growth S_1, stability S_2, and dividend S_3 subject to the transition matrix

$$
\begin{array}{c}
\quad\quad\; S_1 \;\; S_2 \;\; S_3 \\
P: \begin{array}{c} S_1 \\ S_2 \\ S_3 \end{array}
\begin{pmatrix}
\frac{1}{2} & \frac{1}{2} & 0 \\
0 & \frac{1}{2} & \frac{1}{2} \\
\frac{2}{3} & 0 & \frac{1}{3}
\end{pmatrix}
\end{array}
$$

and an income vector $v = (1, 4, 3)$, constant over time.

According to Theorem 2, the limiting vector

$$\lim_{t \to \infty} \pi(t) = \alpha = (\alpha_1, \alpha_2, \alpha_3) \tag{7}$$

is the only probability vector satisfying the equation $\alpha = \alpha P$. Thus we want

$$
\begin{aligned}
\alpha_1 &= \tfrac{1}{2}\alpha_1 && + \tfrac{2}{3}\alpha_3 \\
\alpha_2 &= \tfrac{1}{2}\alpha_1 + \tfrac{1}{2}\alpha_2 \\
\alpha_3 &= && \tfrac{1}{2}\alpha_2 + \tfrac{1}{3}\alpha_3 \\
1 &= \alpha_1 + \alpha_2 + \alpha_3
\end{aligned}
$$

The unique solution to this set of equations is $\alpha = (\frac{4}{11}, \frac{4}{11}, \frac{3}{11})$. Hence, using (6) and (7), the limiting value of the investor's expected income per unit time is

$$\lim_{t \to \infty} \pi(t) \cdot v(t) = \alpha \cdot v = 2\tfrac{7}{11}$$

independent of the initial vector $\pi(0)$. ▶

Example 4 Markov chains have been used with some success as models of labor mobility.† However, the models require rather strong assumptions

---

*Roby, T., "Utility and Futurity," *Behavioral Science* 7, 194–210 (1962).

†See, for example, Prais, S. J., "Measuring Social Mobility," *Journal of the Royal Statistical Society* 118, 56–66 (1955).

that often are not tenable—for example, that the probability of changing industries is constant over time and does not depend on the worker's prior history or work experience, that the system is closed in that no new workers enter, or that they enter only to replace those not otherwise accounted for, etc.

In order to obtain a more realistic model, Blumen, Kogan, and McCarty* divided the working population into two classes—"stayers" and "movers." With probability 1, a stayer remains in his respective industry during the observation period (quarter). For a mover, the probability of transition from industry i to industry j is m_{ij}. (We allow the possibility that $i = j$.)

The only transitions take place among movers and for this segment of the population the transition matrix is $M = (m_{ij})$. The element $m_{ij}(n)$ in the (i, j) position of the nth power of M gives the probability that a *mover* goes from industry i to industry j in n quarters.

Let us denote by s_i the proportion of stayers in industry i, so that $1 - s_i$ is the proportion of movers. Then the probability that a person selected at random from industry i at time zero will be in industry j at time n is

$$p_{ii}(n) = s_i + (1 - s_i)m_{ii}(n)$$

$$p_{ij}(n) = \qquad (1 - s_i)m_{ij}(n) \qquad \text{for } i \neq j \qquad \text{(8)}$$

If we let S be a diagonal matrix with elements s_i, Equations (8) imply that for $n = 1, 2, \ldots$, the "transition" matrix $P(n) = (p_{ij}(n))$, for n steps, is

$$P(n) = S + (I - S)M^n$$

Note that $P(n)$ is *not* $[P(1)]^n$, so the process, strictly speaking, is not a Markov chain. Markov chain techniques are applied only to the movers.

To take a specific case, consider a simple three-industry economy in which

$$S = \begin{pmatrix} 0.9 & 0 & 0 \\ 0 & 0.2 & 0 \\ 0 & 0 & 0.4 \end{pmatrix} \quad \text{and} \quad M = \begin{pmatrix} 0.2 & 0.3 & 0.5 \\ 0.1 & 0.6 & 0.3 \\ 0.1 & 0.5 & 0.4 \end{pmatrix}$$

Then

$$\lim_{n \to \infty} P(n) = S + (I - S) \lim_{n \to \infty} M^n$$

where $\lim_{n \to \infty} M^n$ is a matrix each row of which is the probability vector $v = (\frac{1}{9}, \frac{43}{81}, \frac{29}{81})$ obtained by solving the equation $v = vM$. Thus

$$\lim_{n \to \infty} P(n) = \begin{pmatrix} 0.911 & 0.053 & 0.036 \\ 0.089 & 0.625 & 0.286 \\ 0.067 & 0.319 & 0.615 \end{pmatrix}$$

Note that the limiting probabilities are not independent of the initial industry, a reflection of the non-Markovian nature of the model and, presumably, of the relative stability of employment in the various industries. ▶

*Blumen, I., Kogan, M., and McCarty, P. J., *The Industrial Mobility of Labor as a Probability Process* (Cornell University Studies in Industrial and Labor Relations, Ithaca, N.Y., 1955), vol. 6.

PROBLEMS

2. Find the limiting transition matrix for Problem 6 of Section 20.1.

3. Find $\lim_{n \to \infty} P^n$ in Problem 8 of Section 20.1 and compare with P^5 computed in that problem.

4. In Example 3 of Section 20.1, find the eventual market share for each type of laundry powder.

5. Suppose $Q = \lim_{n \to \infty} P^n$, where P is a regular transition matrix. Consider another Markov chain with transition matrix $R = P^k$, where $k > 1$ is an integer. Prove that Q is also the limiting transition matrix for this new chain.

6. Show that the limiting vector for the general two-state chain with transition matrix

$$P = \begin{pmatrix} 1 - \alpha & \alpha \\ \beta & 1 - \beta \end{pmatrix}$$

is $v = (\beta/(\alpha + \beta), \alpha/(\alpha + \beta))$. (See Problem 25, Section 10.3.)

7. Find a transition matrix P for which $v = (\frac{1}{2}, \frac{1}{3}, \frac{1}{6})$ is the limiting probability vector.

8. Due to its favorable location in a progressive shopping center, the H. B. Nichols Department Store has been experiencing a steady increase in sales. Although the present store facilities are adequate, sustained growth may require expansion or possibly construction of new facilities. The management is attempting to predict their long-range expected share of the market. It has been determined that Nichols' present facilities cannot adequately handle annual sales in excess of $15 million. Annual sales of $15–$25 million would require expansion of present facilities, and sales in excess of $25 million would require new facilities. A market research firm retained by the company has determined that $\frac{1}{3}$ of the customers who made their last purchase at Nichols will make their next purchase from a competitor, while $\frac{1}{2}$ of those customers last patronizing a competitor will switch to Nichols for their next purchase. If annual department store sales in millions of dollars at the shopping center are forecast to increase according to the growth function

$$y = \frac{40}{1 + 9e^{-1.5t}} \qquad t = 0, 1, 2, \ldots$$

for what expansion eventuality should management begin planning? (Assume the market process is a Markov chain.)

9. Suppose that in Problem 8 some of the executives were critical of the findings of the market research firm on the grounds that consumer buying habits were not considered in sufficient detail. As a result, the market research firm proposed to study three consecutive purchases instead of two as in the initial study. For example, they proposed to estimate the proportion of customers who will make their next purchase from Nichols, given that they had made their last two consecutive purchases there. The previously unconvinced executives were satisfied and the firm was

retained to do the study. Their results are presented in the flow diagram in Figure 7. For instance, the transition $CN \rightarrow NC$ indicates a purchase at a competitor's given that the previous two purchases were at a competitor's and Nichols', respectively. Treating the new process as a Markov chain with states NN, CN, NC, and CC, find the annual sales that can ultimately be expected. Does the recommendation differ from that made on the basis of the initial study?

FIGURE 7

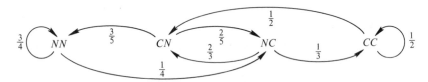

10. Consider a two-state Markov chain with transition matrix

$$P = \begin{array}{c} \\ s_1 \\ s_2 \end{array} \begin{array}{cc} s_1 & s_2 \\ \left(\begin{array}{cc} \frac{2}{3} & \frac{1}{3} \\ \frac{1}{4} & \frac{3}{4} \end{array} \right) \end{array}$$

Form a new chain whose states are pairs of states of the old chain using the rule that the new chain is in state $s_i s_j$ on trial t whenever the original chain is in states s_i and s_j, respectively, on trials $t-1$ and t. This new chain is called the *expansion* of the original chain.

(a) Compute the transition matrix for the new chain.

(b) Show that the new chain is regular.

(c) Find the fixed probability vectors for both the original and the new chains.

What is the relationship between these fixed vectors?

20.4 CHAINS WITH OPEN CLASSES

We now know that in a Markov chain consisting of a single communication class, there is a limiting behavior which is independent of the initial probabilities of being in the various states. The limiting vector v is found by solving the equation $v = vP$, where P is the transition matrix for the chain. For chains with both open and closed classes, these results apply *only after the process has entered a closed class*. Thus, in order to complete our study of Markov chains, we must consider questions concerning the behavior of the process within the open classes. The following is a key result, which we state without proof.

Theorem 3 In a finite chain, there is probability 1 that the process eventually enters a closed class. ▶

Although Theorem 3 ensures that a Markov chain will eventually pass into a closed class, to remain there forever, the behavior of the process within the open classes is itself interesting and important. In many applications of Markov chains to behavioral processes, the process is assumed to begin in an open class. Questions are then raised as to the length of time spent within the open classes and the probabilities of becoming "trapped" in one or another of the closed classes.

The simplest types of chains having open classes are those in which each closed communication class contains only a single state. Such closed classes are called *absorbing* and the same term is applied to the state within the class. For an absorbing state S_i the transition probability p_{ii} must equal unity, and the process, once it enters S_i, can never leave. (Of course, the process can never exit from any closed class, but the term absorbing is reserved for single-state classes.) The entire chain is called absorbing if all closed classes are absorbing.

Example 1 (a) In an attempt to apply the theory of Markov chains to migration into and from California, Rogers* obtained the following transition matrix:

$$
\begin{array}{c}
\\
\text{Calif.} \\
\text{U.S.} \\
\text{Death}
\end{array}
\begin{array}{ccc}
\text{Calif.} & \text{U.S.} & \text{Death} \\
\left(\begin{array}{ccc}
0.9041 & 0.0331 & 0.0628 \\
0.0068 & 0.9352 & 0.0580 \\
0 & 0 & 1
\end{array}\right)
\end{array}
$$

The notion of death as an absorbing state needs no further elaboration.

(b) Adapting and extending Rogers' work, Bartos† has proposed an absorbing Markov chain model of social mobility. He assumes that older people are more likely to stay in an occupation, region, or social class than are younger people, who are assumed to enter the mobility process via a nonabsorbing state. As a simple illustration of his model, consider the following transition matrix:

$$
\begin{array}{c}
\\
S_1 \\
S_2 \\
s_1 \\
s_2
\end{array}
\begin{array}{cccc}
S_1 & S_2 & s_1 & s_2 \\
\left(\begin{array}{cccc}
1 & 0 & 0 & 0 \\
0 & 1 & 0 & 0 \\
a & 0 & b & c \\
0 & d & e & f
\end{array}\right)
\end{array}
$$

*Rogers, A., "A Markovian Model of Interregional Migration," Center for Planning and Development Research, Berkeley, Calif., (mimeographed report), 1965. Cited in Bartos, O. J., *Simple Models of Group Behavior* (Columbia University Press, New York, 1967), Chapter 8.

†Bartos, O. J., *Simple Models of Group Behavior* (Columbia University Press, New York, 1967).

Here states refer to occupations. To say that an individual is in state s_1 means that he has occupation s_1 but has not decided to make it his life career, while, in state S_1, an individual not only has occupation s_1, but has decided to make it his lifelong career. Similar remarks apply to s_2 and S_2. From a practical point of view, of course, considering S_1 and S_2 as absorbing requires the somewhat questionable assumption that decisions regarding life career are irrevocable. ▶

In a manner similar to Example 1(b), the transition matrix for an absorbing chain having r absorbing states $A_1, A_2, \ldots, A_r$ and t states $S_1, S_2, \ldots, S_t$ in open classes may be written in the form

$$P = \begin{pmatrix} I & O \\ A & T \end{pmatrix} \tag{9}$$

where I is an $r \times r$ identity matrix listing probabilities of transition among the absorbing states, O is an $r \times t$ matrix of zeros representing probabilities of transition from the absorbing states to the other states $S_1, S_2, \ldots, S_t$, A is a $t \times r$ matrix of probabilities of transition from each of the open states to the absorbing states, and T is a $t \times t$ matrix whose entries represent probabilities of transition among the states $S_1, S_2, \ldots, S_t$.

When P has the form of Equation (9), it follows that

$$P^2 = \begin{pmatrix} I & O \\ A + TA & T^2 \end{pmatrix} \qquad P^3 = \begin{pmatrix} I & O \\ A + TA + T^2A & T^3 \end{pmatrix}$$

and simple induction shows that

$$P^n = \begin{pmatrix} I & O \\ A + TA + \cdots + T^{n-1}A & T^n \end{pmatrix}$$

$$= \begin{pmatrix} I & O \\ (I + T + \cdots + T^{n-1})A & T^n \end{pmatrix}$$

Theorem 3 implies that the sequence $(t_{ij}(n))$, of elements in the (ij) position of powers of T, must converge to zero for each pair (i, j). We indicate this by writing $\lim_{n \to \infty} (T^n) = O$, a $t \times t$ zero matrix.

As for the term $(I + T + \cdots + T^{n-1})$, it is easily seen that

$$(I - T)(I + T + \cdots + T^{n-1}) = I - T^n$$

Since $\lim_{n \to \infty} (T^n) = O$, we find

$$\lim_{n \to \infty} \{(I - T)(I + T + \cdots + T^{n-1})\} = I - \lim_{n \to \infty} (T^n) = I$$

so that

$$\lim_{n \to \infty} (I + T + \cdots + T^{n-1}) = \sum_{k=0}^{\infty} T^k = (I - T)^{-1} \tag{10}$$

In terms of the original matrix P then,

$$\lim_{n \to \infty} P^n = \begin{pmatrix} I & O \\ (I-T)^{-1}A & O \end{pmatrix} \tag{11}$$

Example 2 Suppose that in Example 1(b) we find $a = \frac{1}{5}$, $b = \frac{3}{4}$, $c = \frac{1}{20}$, $d = \frac{1}{10}$, $e = \frac{1}{10}$, and $f = \frac{8}{10}$ giving the transition matrix

$$P = \begin{pmatrix} 1 & 0 & 0 & 0 \\ 0 & 1 & 0 & 0 \\ \hline \frac{1}{5} & 0 & \frac{3}{4} & \frac{1}{20} \\ 0 & \frac{1}{10} & \frac{1}{10} & \frac{8}{10} \end{pmatrix} = \begin{pmatrix} I & O \\ A & T \end{pmatrix}$$

Then

$$(I-T) = \begin{pmatrix} \frac{1}{4} & -\frac{1}{20} \\ -\frac{1}{10} & \frac{2}{10} \end{pmatrix}, \quad (I-T)^{-1} = \frac{1}{9}\begin{pmatrix} 40 & 10 \\ 20 & 50 \end{pmatrix}$$

$$(I-T)^{-1}A = \begin{pmatrix} \frac{8}{9} & \frac{1}{9} \\ \frac{4}{9} & \frac{5}{9} \end{pmatrix}$$

giving

$$\lim_{n \to \infty} P^n = \begin{pmatrix} 1 & 0 & 0 & 0 \\ 0 & 1 & 0 & 0 \\ \frac{8}{9} & \frac{1}{9} & 0 & 0 \\ \frac{4}{9} & \frac{5}{9} & 0 & 0 \end{pmatrix}$$

Note that the elements in each row of $(I-T)^{-1}A$ sum to 1. ▶

An absorbing state S_k, once entered, can never be left. Thus the element $\lim_{n \to \infty} p_{ik}(n)$, appearing in the (i, k) position of the limiting matrix (11), represents the probability that the chain, starting in state S_i, eventually becomes "absorbed" in S_k. For instance, in Example 2, there is probability $\frac{1}{9}$ that an individual who begins in occupation s_1 decides to make occupation s_2 his life career. Similar interpretations may be made for the other elements of $(I-T)^{-1}A$.

The matrix $(I-T)^{-1} = I + T + T^2 + T^3 + \cdots$ itself has an important interpretation in terms of expected values. Recall that the element $t_{ij}(n)$ in the (i, j) position of T^n represents the probability that the process is in state S_j on the nth trial, *given it started in state S_i at trial zero* (event $S_i(0)$). If we define a random variable $X_j(n)$ to be the number of times the process is in state S_j on trial n, it is obvious that $X_j(n)$ can take only two values—1, if the process is in S_j on trial n, and 0, if the process is not in S_j on trial n. The interpretation of $t_{ij}(n)$ implies that

$$t_{ij}(n) = P(X_j(n) = 1 \mid S_i(0)) \quad \text{and} \quad 1 - t_{ij}(n) = P(X_j(n) = 0 \mid S_i(0))$$

so that

$$E(X_j(n) \mid S_i(0)) = 1 \cdot t_{ij}(n) + 0(1 - t_{ij}(n)) = t_{ij}(n)$$

The total number of trials (including trial zero) on which the process hits S_j before passing into an absorbing state is

$$X_j = X_j(0) + X_j(1) + X_j(2) + \cdots$$

Hence, when the process starts in S_i, the expected number of times it hits S_j is ,

$$E(X_j \mid S_i(0)) = \sum_{n=0}^{\infty} E(X_j(n) \mid S_i(0)) = \sum_{n=0}^{\infty} t_{ij}(n)$$

This is the element in the i, j position of $I + T + T^2 + T^3 + \cdots$ or, equivalently, the element in the i, j position of $(I - T)^{-1}$.

From this it follows that the expected number of trials on which the process remains in the open states starting from S_i is obtained by summing the elements in the ith row of $(I - T)^{-1}$.

Example 3 Let us suppose that the absorbing chain of Example 1 of Section 20.2 has the transition matrix*

State in $(i + 1)$st year

		S_4	S_3	S_2	S_1	S_0
	S_4	1	0	0	0	0
State	S_3	0	1	0	0	0
in ith	S_2	0.054	0.140	0.806	0	0
year	S_1	0.027	0.257	0	0.716	0
	S_0	0.097	0	0.004	0.003	0.896

$$= \begin{pmatrix} I & O \\ A & T \end{pmatrix}$$

Then

$$(I - T)^{-1} = \begin{pmatrix} 5.158 & 0 & 0 \\ 0 & 3.525 & 0 \\ 0.211 & 0.106 & 9.601 \end{pmatrix}$$

so that, according to the model, a sane person, on the average, spends 9.601 years in state S_0, 0.106 year in state S_1, and 0.211 year in state S_2,

*Based on estimates provided by Marshall, A. W., and Goldhamer, H., "An Application of Markov Processes to the Study of the Epidemiology of Mental Diseases," *Journal of the American Statistical Association* **50**, 99–129 (1955).

a total of 9.918 years, before being absorbed by death or a mental hospital. The prediction that a person who is severely insane but unhospitalized (S_2) will remain in that state for 5.158 years, on the average, before being absorbed, while a person who is only mildly insane but unhospitalized (S_1) can be expected to remain in that state for only 3.525 years before being absorbed appears unreasonable and tends to detract from the credibility of the model. ▶

We have not discussed chains with an infinite number of states although these occasionally arise. Results for such chains are similar to those stated above, although some additional detail is required due to the fact that the matrix $(I - T)$ may not have a unique inverse. Details may be found in the book by Feller listed under Supplementary Reading.

PROBLEMS

1. Compute $(I - T)^{-1}$ for the absorbing chain with transition matrix

$$\begin{pmatrix} 1 & 0 & 0 \\ \hline 0 & \frac{1}{3} & \frac{2}{3} \\ \frac{1}{2} & \frac{3}{8} & \frac{1}{8} \end{pmatrix} = \begin{pmatrix} I & O \\ A & T \end{pmatrix}$$

2. In Problem 10 of Section 20.1, compute $(I - T)^{-1}$, where T is the matrix of transition probabilities among states in open classes. Find the expected number of ballots required and the probability of each possible outcome of the process.

3. For the balloting process, compute the following quantities:
   (a) The expected number of ballots on which all three candidates are active contenders.
   (b) The probability that C withdraws before either A or B.
   (c) The probability that A withdraws before either B or C.
   (d) The probability that B and C withdraw on the same ballot.

4. In Problem 9 of Section 20.1, find the expected waiting time before the clerk can take a break, given that the last customer arrived two minutes ago.

5. In Example 1(a), what is the probability that a California resident will leave the state?

6. For the social mobility model of Example 1(b) show that $(I - T)^{-1}A$ is a transition matrix. (T and A are defined as in Example 2.)

7. Use Equation (10) to show that $(I - T)^{-1}T = (I - T)^{-1} - I$.

8. In Problem 8 of Section 20.1, find the expected number of times (including time zero) that the process remains in state U before passing to state A.

9. A Markov chain starts in state S_j. Let X_j denote the number of trials (including trial zero) the process remains in S_j. Prove that $E(X_j) = 1/(1 - p_{jj})$, where p_{jj} is the probability of transition from S_j to S_j.

10. For the assignment process of Example 2 of Section 20.2 compute
   (a) $(I - T)^{-1}$, where T is the matrix of transition among the states in open classes.
   (b) the expected number of times that the process remains in the open classes.
   (c) the probability that the process reaches state $(0, 0, 2)$ before
       (i) $(0, 2, 0)$ (ii) $(0, 2, 0)$ or $(0, 1, 1)$.
       (*Hint:* Make all of these states absorbing.)
   (d) the probability that the process ultimately reaches state $(0, 1, 1)$.

11. Adelman* described an application of Markov chain theory to the derivation of the equilibrium size distribution of steel firms in the United States. The range of total corporate assets was partitioned into seven asset classes which constituted the states of the growth process. The transition matrix of the chain was given as

	s_0	s_1	s_2	s_3	s_4	s_5	s_6
s_0	0.99942	0.00040	0.00016	0.00001	0.00001	0	0
s_1	0.021	0.911	0.068	0	0	0	0
s_2	0.024	0.039	0.908	0.028	0.001	0	0
s_3	0	0	0.076	0.872	0.052	0	0
s_4	0.008	0	0	0.016	0.947	0.028	0
s_5	0	0	0	0	0.037	0.926	0.037
s_6	0	0	0	0	0	0.024	0.976

$P = $ (the matrix above)

   (a) Find the average length of time (years) that a firm will spend in state s_0, state s_3, or state s_6, once that state is entered.
   (b) Find the probability that a firm starting in state s_1 moves first to state s_2 and then to one of states s_4, s_5, or s_6.
   (c) Find the conditional probability that a firm moves to state s_2 given that it leaves state s_0.

12. Bower† used the simple Markov model in Figure 8 to describe the choice behavior of an animal in a T-maze. The states s_0, s_1, and s_2 represent the animal's orientation toward stimulus sets S_0, S_1, and S_2, respectively, where S_0 denotes the stimuli available when the animal is oriented straight ahead, S_1 is the set available when the animal is oriented to the right, and S_2 is the set available when the animal is oriented to the left.

*Adelman, Irma G., "A Stochastic Analysis of the Size Distribution of Firms," *Journal of the American Statistical Association* **53**, 893–904 (1958).

†Bower, G. H., "Choice-Point Behavior," in *Studies in Mathematical Learning Theory*, Bush, R. R., and Estes, W. K., eds. (Stanford University Press, Stanford, Calif., 1959), Chapter 6.

States s_3 and s_4 represent the animal's final orientation. The experiment is so conducted that an animal always starts in state s_0.

(a) Find the probability that the process is ultimately absorbed in state s_3.

(b) Find the mean time to absorption.

(c) Find the probability that the process returns to state s_0 exactly k times before absorption.

FIGURE 8

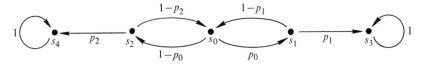

13. Prove that the expected number of *changes of state* in an absorbing chain can be calculated by forming a new matrix T^* from T by setting $p_{ii} = 0$ for all nonabsorbing states S_i and dividing each row by its new row sum. The ith row sum of $(I - T^*)^{-1}$ gives the expected number of state changes for the original process.

14. For the model of choice reaction in Problem 12, compute the expected number of changes of state.

20.5 APPLICATIONS IN SOCIOLOGY

Sociologists are frequently concerned with the processes by which the structural state of a social system changes over time. The following example illustrates the use of a Markov chain model for a social process characterized by the independence of path assumption.

Example 1 The relation "dominates" is ordinarily assumed to be irreflexive (no person dominates himself), asymmetric (if x dominates y, then y does not dominate x), and nontransitive (if x dominates y and y dominates z, one cannot necessarily conclude that x dominates z). We shall assume that the relation exists between all pairs of members of a given population. For simplicity of exposition, we shall restrict our attention to three-member groups*. The notation $x \rightarrow y$ will mean x dominates y.

From the above assumptions, it follows that in a three-member group there are only eight distinct dominance patterns (see Figure 9).

Given any dominance pattern, the *authority* of an individual i, denoted by A_i, is defined to be the number of persons whom i dominates. Thus in D_3 we have $A_x = 2$, $A_y = 1$, and $A_z = 0$.

*Adapted from Bartos, O. J., *Simple Models of Group Behavior* (Columbia University Press, New York, 1967), Chapters 5 and 6.

FIGURE 9

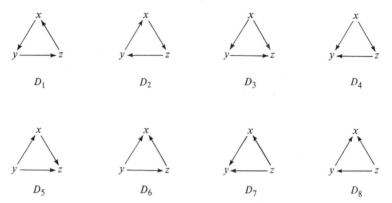

We take the point of view that the dominance pattern which exists among three individuals is a reflection of the outcomes of past *encounters* between pairs of these individuals and is subject to change through subsequent encounters. Thus, to say that D_3 is the existing pattern is to say that at the last encounter between x and y, x was dominant, that at their last encounter, x dominated z, and that y was dominant when y and z last met.

The next encounter between a pair of individuals may involve x and y, x and z, or y and z, and we assign probability $\frac{1}{3}$ to each of these events. The probability that i is dominant at a meeting between individuals i and j is *defined* to be

$$p_{ij} = \tfrac{1}{2}[1 + w(A_i - A_j)]$$

where w is a constant lying between 0 and $\frac{1}{2}$. When $w = 0$, each has probability $\frac{1}{2}$ of being dominant, while positive values of w reflect in varying degrees the difference in authority between the two individuals.

Putting all our assumptions together, we see that we have a Markov chain with states $D_1, D_2, \ldots, D_8$ and transition matrix

$$
P = \begin{array}{c|cccccccc}
 & D_1 & D_2 & D_3 & D_4 & D_5 & D_6 & D_7 & D_8 \\
\hline
D_1 & \frac{1}{2} & 0 & \frac{1}{6} & 0 & 0 & \frac{1}{6} & \frac{1}{6} & 0 \\
D_2 & 0 & \frac{1}{2} & 0 & \frac{1}{6} & \frac{1}{6} & 0 & 0 & \frac{1}{6} \\
D_3 & a & 0 & b & c & c & 0 & 0 & 0 \\
D_4 & 0 & a & c & b & 0 & 0 & c & 0 \\
D_5 & 0 & a & c & 0 & b & c & 0 & 0 \\
D_6 & a & 0 & 0 & 0 & c & b & 0 & c \\
D_7 & a & 0 & 0 & c & 0 & 0 & b & c \\
D_8 & 0 & a & 0 & 0 & 0 & c & c & b
\end{array}
$$

where $a = \frac{1}{6}(1 - 2w)$, $b = \frac{1}{6}(3 + 4w)$, and $c = \frac{1}{6}(1 - w)$.

To illustrate the computation of P, suppose the process is in state D_3. Then transition to D_2, D_6, D_7, or D_8 would require at least two dominance reversals, which is not possible in a single encounter. Transition to D_1 requires that x meet z (probability $\frac{1}{3}$) and that z dominate x (probability $\frac{1}{2}(1 - 2w)$, since in D_3 the authority of z is 0 and that of x is 2). Hence transition from D_3 to D_1 carries probability $a = \frac{1}{6}(1 - 2w)$. Other transition probabilities are computed in a similar manner.

The limiting vector is

$$\alpha = (\beta, \beta, \gamma, \gamma, \gamma, \gamma, \gamma, \gamma)$$

where $\beta = (1 - 2w)/4(2 - w)$ and $\gamma = 1/4(2 - w)$. Thus, no matter what the initial state of the process, the limiting probability of being in one of the equalitarian states D_1 or D_2 is $2\beta = (1 - 2w)/(4 - 2w)$, while the probability of being in one of the authoritarian states $D_3, \ldots, D_8$ is $6\gamma = 3/(4 - 2w)$. Since $0 \le w \le \frac{1}{2}$, it follows that the limiting probability p_A of an authoritarian structure lies between $\frac{3}{4}$ and 1, the value $\frac{3}{4}$ occurring only when no weight is given to differences in individual authority. When $w = \frac{1}{2}$, we find $p_A = 1$. In this case, $a = 0$, $\beta = 0$, and states $D_3, \ldots, D_8$ form a single closed communication class into which the process must eventually pass. ▶

PROBLEMS

1. Verify the computation of the transition matrix P.

2. Go through the calculations to obtain the limiting vector α.

3. Argue that when $w < \frac{1}{2}$, the chain consists of a single, aperiodic communication class, but that if $w = \frac{1}{2}$ there are three classes, two open and one closed, all aperiodic.

4. In the special case $w = \frac{1}{2}$, find the expected length of time the process will spend in the open classes.

5. From the point of view of distribution of authority, there are only two distinct dominance structures. In D_1 and D_2, each member dominates one of the others, while in $D_3, \ldots, D_8$ one member dominates both the other members. Let us define a new process with states $S_1 = \{D_1, D_2\}$ and $S_2 = \{D_3, \ldots, D_8\}$. The new process is in state S_1 if the original chain is in D_1 or D_2 and is in state S_2 if the original chain is in any of the states $D_3, \ldots, D_8$. Show that this new process is itself a Markov chain with transition matrix

$$\begin{array}{c} S_1 \\ S_2 \end{array} \begin{pmatrix} \frac{1}{2} & \frac{1}{2} \\ a & b + 2c \end{pmatrix}$$

where a, b, and c are defined in Example 1. That is, show that whether the original chain is in state D_1 or in state D_2, there is probability $\frac{1}{2}$ of moving to one of the states $D_3, \ldots, D_8$, while if the chain is in any one of $D_3, \ldots, D_8$, there is probability a of moving to either D_1 or D_2. (Markov chains are called *lumpable* if the states may be partitioned into classes $S_1, \ldots, S_r$ of states in such a way that for each pair of classes S_i and S_j, the probability of transition from a state U in S_i to some state in S_j is the same for each U in S_i.)

6. Show that for the lumped chain of Problem 5, the limiting vector is

$$\left(\frac{1 - 2w}{4 - 2w}, \frac{3}{4 - 2w}\right)$$

Compare with the limiting vector α in Example 1.

7. Verify that the three-state chain with transition matrix

$$\begin{array}{c} S_1 \\ S_2 \\ S_3 \end{array} \begin{pmatrix} \frac{1}{4} & \frac{1}{4} & \frac{1}{2} \\ \frac{1}{2} & 0 & \frac{1}{2} \\ \frac{1}{2} & \frac{1}{8} & \frac{3}{8} \end{pmatrix} = P$$

is lumpable relative to the partition $\{S_1\}, \{S_2, S_3\}$ but not for the partition $\{S_1, S_3\}, \{S_2\}$. Find the limiting vector for P and for the lumped chain corresponding to the first partition.

8. Leeman* has studied the dynamics of choice processes using techniques similar to those of Example 1. Consider a group of N people in which each chooses exactly one of the others. Initial choices are made at random. Choices are modified through encounters between pairs of individuals. It is assumed that all possible encounters are equally likely, that exactly one encounter occurs during each time interval (trial) and that each encounter results in one of two equally likely outcomes O_1 or O_2 for each participant in the encounter. If the encounter results in outcome O_1 for one person, then with probability 1 he chooses the other person in the encounter. If outcome O_2 results, then he chooses at random from the other $N - 1$ persons. Choices of individuals not involved in an encounter are not revised on that trial. For $N = 3$:

(a) List the states of the process (there are eight).

(b) Write the transition matrix.

(c) Find the limiting probability vector.

(d) Show that the chain is lumpable and find the limiting probabilities for the two possible choice structures

and

where $\rightarrow$ denotes the relation "chooses."

9. Consider a set of v voters who must choose among a alternatives. On each of a sequence of ballots, each voter indicates his preferred alternative. Voting is continued until some specified majority is reached.

---

*Leeman, C. P., "Patterns of Sociometric Choice in Small Groups: A Mathematical Model and Related Experimentation," *Sociometry* **15**, 220–243 (1952).

Kreweras,* in developing a model for this situation, assumes two classes of voters—those who are *resolute* in their convictions and who thus vote each time for the same alternative and those who are *floaters* in the sense that they are influenced by the results of the preceding ballot.

To take a specific case, suppose there are two alternatives, five resolute voters divided two-three in favor of the respective alternatives, and three floaters. To win, an alternative must receive at least five of the eight votes. Since the voting pattern of the resolutes is fixed, the state of the voting process can be indicated by a pair (x, y) indicating the numbers of floaters voting for the respective alternatives. Of course, $x + y$ must equal three. We shall assume that on any particular ballot, a floater votes for a particular alternative with probability proportional to the *total* number of votes cast for that alternative on the preceding ballot. For instance, if resolutes vote $(2, 3)$ and floaters (x, y) on one ballot, then the probability that a floater votes for the first alternative on the next ballot is $(x + 2)/8$. On the first ballot, alternatives are chosen at random by the floaters.

Argue that these assumptions lead to a Markov chain with 4 states, 3 of which are absorbing. Write the transition matrix and determine the probabilities that each of the respective alternatives will win.

10. Rework Problem 9 assuming three alternatives and seven voters with four resolutes divided two-one-one. Majority wins.

11. A tentative model of the spread of states in social groups is offered by Karlsson.† He assumes that the behavior of an individual group member k during time period (trial) $t + 1$ depends upon the degree of influence that each group member has upon k and upon the behavior of the group members, himself included, during the preceding time period.

A measure of the behavior of the group members is given by the matrix $B(t) = (b_{ij}(t))$ in which the ij element $b_{ij}(t)$ indicates the proportion of time period t that individual i spends in behavior state j. The intermember influence patterns are indicated by the matrix $A = (a_{ki})$ in which element a_{ki} represents the influence that individual i exerts on individual k. The coefficients are normalized so that all a_{ki} are non-negative and so that $\sum_i a_{ki} = 1$ for each k.

The basic assumption of the model is that

$$B(t + 1) = A \cdot B(t) \qquad (12)$$

which states that the proportion of time spent in behavior state j by individual k at time $(t + 1)$ is a weighted average

$$b_{kj}(t + 1) = \sum_j a_{ki} b_{ij}(t)$$

of the proportions of time spent in behavior state j by each of the group members at time t, the weights being the influence coefficients a_{ki}.

*Kreweras, G., "A Model to Weight Individual 'Authority' in a Group," in *Mathematics and Social Sciences*, Sternberg, S., Ed. (Mouton, Paris, 1960), pp. 111–118.

†Karlsson, G., "Note on the Spread of a State in Small Social Groups," *Bulletin of Mathematical Biophysics* **17**, 1–5 (1955).

(a) Use Equation (12) to show that

$$B(t) = A^t B(0)$$

where A^t denotes the tth power of A, and $B(0)$ is the matrix of initial time proportions.

(b) A set C of group members is called *closed* if for each $k \in C$ and $i \notin C$ we have $a_{ki} = 0$. A group is called *irreducible* if there are no closed subsets of group members other than the entire group.

If a group is irreducible, prove that

   (i) A^t converges to a stochastic matrix M each of the rows of which is the same probability vector.

   (ii) $\lim_{t \to \infty} B(t) = MB(0)$.

   (iii) $B(t+1) - B(t) \to O$ as $t \to \infty$.

   (iv) $\text{Max}_i\, b_{ij}(0) \geq b_{ij}(t)$ for all j and all t.

20.6 APPLICATIONS IN PSYCHOLOGY

In psychology, Markov chains have been most widely used as models of the learning process. The examples in this section illustrate the basic ideas.

Example 1 In *stimulus sampling theory*,* it is assumed that there is a set $S = \{s_1, s_2, \ldots, s_t\}$ of stimulus elements of which exactly one is sampled on each trial. The probabilities of sampling the various elements remain constant from trial to trial. Each stimulus element is assumed to be conditioned to exactly one of a set of responses $R = \{r_1, r_2, \ldots, r_w\}$. On each trial, the response given is the one to which the obtained stimulus element is conditioned.

To take the simplest case, let us consider an experiment in which a single stimulus element s is presented to the subject on each trial (see Example 2, Section 18.6) and in which the subject makes one of two responses r_1 or r_2, one of which is reinforced. For $n = 1, 2, 3, \ldots$ it is convenient to define a *response random variable*

$$R_n = \begin{cases} 1 & \text{if response } r_1 \text{ occurs on trial } n \\ 2 & \text{if response } r_2 \text{ occurs on trial } n \end{cases}$$

a *reinforcement random variable*

$$E_n = \begin{cases} 1 & \text{if response } r_1 \text{ is reinforced on trial } n \\ 2 & \text{if response } r_2 \text{ is reinforced on trial } n \end{cases}$$

*For a complete discussion see Suppes, P., and Atkinson, R. C., *Markov Learning Models for Multiperson Interactions* (Stanford University Press, Stanford, Calif., 1960).

and a *state of conditioning random variable*

$$C_n = \begin{cases} 1 & \text{if } s \text{ is conditioned to } r_1 \text{ on trial } n \\ 2 & \text{if } s \text{ is conditioned to } r_2 \text{ on trial } n \end{cases}$$

From the above assumptions we have

$$P(R_n = 1 \mid C_n = 1) = 1 \quad \text{and} \quad P(R_n = 1 \mid C_n = 2) = 0$$

We assume that $P(E_n = 1) = \pi$ and $P(E_n = 2) = 1 - \pi$ where π is a constant not dependent on trial number or on the outcomes of previous trials (independence of path). In particular, reinforcement on a trial is assumed not to be contingent on the subject's response on that trial.

If the stimulus element s is already conditioned to the response which is reinforced, it is assumed to remain so conditioned. If it is not conditioned to that response, it becomes conditioned with probability θ (the conditioning is then said to have been *effective*), while with probability $1 - \theta$ no change in conditioning occurs.

Our assumptions imply that we have a Markov chain with states S_1 and S_2 corresponding to the two responses to which s may be conditioned. Computation of transition probabilities is illustrated in the tree diagram of Figure 10. If s is conditioned to response r_1 (state S_1) the subject will make that response. With probability π, r_1 is reinforced, in which case s will again be conditioned to r_1 and state S_1 will occur again.

FIGURE 10

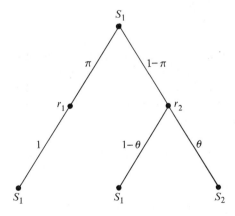

With probability $1 - \pi$, r_2 is reinforced. If the reinforcement is effective (probability θ), s becomes conditioned to r_2 and the process moves to state S_2. Otherwise (probability $1 - \theta$), state S_1 occurs again. We thus obtain the first line in the transition matrix

To state

$$
\begin{array}{c}
\text{From} \\
\text{conditioning} \\
\text{state}
\end{array}
\quad
\begin{array}{c}
S_1 \\
S_2
\end{array}
\begin{array}{cc}
\overset{\displaystyle S_1}{} & \overset{\displaystyle S_2}{} \\
\begin{pmatrix} 1 - \theta + \theta\pi & \theta - \theta\pi \\ \theta\pi & 1 - \theta\pi \end{pmatrix}
\end{array}
= P
$$

The second line may be computed similarly. ▶

PROBLEMS

1. (a) For the transition matrix P of Example 1, compute P^2, P^3, P^4, and, by induction, P^n for all $n \geq 2$. (*Hint:* See Problem 25, Section 10.3.)

   (b) Find $\lim_{n \to \infty} P^n$.

2. (a) Prove that $P(R_n = 1) = P(C_n = 1)$ for all n.

   (b) Use the result of (a) and of Problem 1 to find

   $$U_n = P(R_n = 1) \qquad n = 2, 3, \ldots$$

   in terms of $U_1 = P(R_1 = 1)$.

   (c) Find $\lim_{n \to \infty} U_n$.

3. (a) Find the mean and variance of R_n.

   (b) Find Cov (R_n, R_m) for $m \neq n$.
   (*Hint:* $E(R_n R_m) = P(R_n = 1, R_m = 1)$.)

   (c) Prove that

   $$\lim_{N \to \infty} E\left[\frac{1}{N} \sum_{n=1}^{N} R_n \right] = 2 - \pi$$

Example 2 The learning model of Bush and Mosteller,[*] while not strictly a Markov chain, is closely related. The model considers a choice situation, such as that facing a rat in a T-maze, in which there is presented a set of r alternatives $A_1, A_2, \ldots, A_r$ from which the subject must choose exactly one. This set of alternatives is repeatedly presented to the subject in a sequence of trials.

The state of the system at trial n is the alternative chosen on that trial. The probability that alternative A_k is chosen on trial n will be denoted by $v_k(n)$ and these probabilities arrayed in a probability vector

$$v(n) = (v_1(n), v_2(n), \ldots, v_r(n))$$

The difference between this model and an ordinary Markov chain lies in the assumption that, rather than a single transition matrix for the process,

---

[*]Bush, R. R., and Mosteller, F., *Stochastic Models for Learning* (John Wiley & Sons, Inc., New York, 1955).

there is a different transition matrix T_k corresponding to each alternative A_k. Thus if $v(n)$ is the probability vector at time n, and if alternative A_k $(k = 1, 2, \ldots, r)$ is chosen at trial n, the new vector of probabilities at trial $(n + 1)$ is

$$v(n + 1) = v(n)T_k$$

The inherent difficulty in the analysis of the model lies in the fact that the sequence of transition matrices used is itself determined by chance. However, some general results may be obtained and certain special cases are relatively easy to analyze.

Let us consider the simplest case of two alternatives A_1 and A_2 with corresponding transition matrices

$$T_1 = \begin{pmatrix} 1 - b_1 & b_1 \\ a_1 & 1 - a_1 \end{pmatrix} \quad \text{and} \quad T_2 = \begin{pmatrix} 1 - b_2 & b_2 \\ a_2 & 1 - a_2 \end{pmatrix}$$

Defining $c_1 = 1 - a_1 - b_1$ and $\lambda_1 = a_1/(1 - c_1)$, it is easily verified that

$$v(n)T_1 = c_1 v(n) + (1 - c_1)\Lambda_1 \tag{13}$$

where Λ_1 is the vector $(\lambda_1, 1 - \lambda_1)$. The usual models require that c_1 be non-negative.*

Equation (13) shows that the effect of applying the transformation T_1 to a vector $v(n)$ is to give a linear combination of $v(n)$ and the vector Λ_1. Since both c_1 and $1 - c_1$ lie between zero and unity, this new vector $v(n + 1) = v(n)T_1$ lies *between* $v(n)$ and Λ_1 and, in particular, is closer to Λ_1 than $v(n)$ is. In fact, if T_1 is applied to $v(n)$ k consecutive times we obtain

$$v(n + k) = v(n)T_1^k = c_1^k v(n) + (1 - c_1^k)\Lambda_1 \tag{14}$$

which, as $k \to \infty$, converges to Λ_1. Thus Λ_1 is the limiting vector for the Markov chain whose transition matrix is T_1. Our knowledge of Markov chain theory tells us that Λ_1 should be the unique solution of the equation $v = vT_1$. Putting $v(n) = \Lambda_1$ in (13) shows that this is indeed the case. ▶

PROBLEMS

4. Verify the statements made, but not proved, in Example 2. That is,
   (a) derive Equation (13).
   (b) verify that $v(n + 1) = v(n)T_1$ actually does lie between $v(n)$ and Λ_1.
   (c) prove (14) by induction.
   (d) verify that Λ_1 is the only solution of $v = vT_1$.

5. Argue that all the statements made in Example 2 concerning T_1 have direct counterparts for T_2.

*Bush, R. R., and Mosteller, F., *Stochastic Models for Learning* (John Wiley & Sons, Inc., New York, 1955).

6. We have seen that the effect of applying transformation T_1 to a vector v is to move closer to the vector Λ_1. Similarly, applying T_2 moves us closer to the vector Λ_2 satisfying $\Lambda_2 = \Lambda_2 T_2$. What happens when we apply first T_1 and then T_2? How do vT_1T_2 and vT_2T_1 compare?

7. The results of Problem 6 show that the order in which transformations are applied affects the final result and, in particular, that it is the last transformation which has the greatest single effect on the present probability vector. This difficulty would vanish if the transformations should commute. However, show that commutativity holds only if the transformations have the same limiting vectors or if one is the identity matrix.

SUPPLEMENTARY READING

Feller, W., *An Introduction to Probability Theory and its Applications* (John Wiley & Sons, Inc., New York, 1968), Vol. I, 3rd ed., Chapter XV.

Kemeny, J. G., and Snell, J. L., *Finite Markov Chains* (D. Van Nostrand Company, Inc., Princeton, N. J., 1960).

CONTINUOUS TIME PROCESSES AND CONTINUOUS RANDOM VARIABLES **21**

21.1 THE POISSON PROCESS

In the Bernoulli process we considered occurrences of a phenomenon called "success" at points on a discrete time (trial) scale 1, 2, 3, 4, ... In this chapter we turn our attention to phenomena which occur over a continuous time scale $t \geq 0$. Simple examples are the successive arrivals of voters at a polling place, successive purchases of a brand item, or the times at which automobiles pass a particular point on the highway.

We assume that observation begins at time $t = 0$ and we write X_t to denote the number of occurrences of the phenomenon in the time interval $(0, t]$. Then $X_0 = 0$ and the difference $X_u - X_s$ gives the number of occurrences in the interval $(s, u]$. (See Figure 1.)

FIGURE 1

We have a random variable X_t corresponding to each real number $t \geq 0$. By making various assumptions about the joint probability functions of these random variables we may generate various probability models for such processes. In this section we shall discuss the simplest such process, the Poisson process, which is governed by three assumptions:

(1) *Stationarity*—The number of occurrences in an interval of time depends only on the length of the interval and not on its location. Thus if

$u - s = t$, the random variable $X_u - X_s$ has the same probability function as X_t.

(2) *Independence*—The numbers of occurrences in two *disjoint* time intervals are independent. That is, if $t < s < u < v$, then the random variables $X_s - X_t$ and $X_v - X_u$ are independent. This assumption is described by saying that the process has *independent increments*.

(3) There exists a positive constant λ such that if h is small,

(a) the probability of no occurrences in $(0, h]$ is approximately $1 - \lambda h$,

(b) the probability of exactly one occurrence in $(0, h]$ is approximately λh,

(c) the probability of more than one occurrence in $(0, h]$ is negligible.

More precisely, we assume that

$$P(X_h = 0) = 1 - \lambda h + o_1(h)$$

$$P(X_h = 1) = \lambda h + o_2(h)$$

$$P(X_h \geq 2) = o_3(h)$$

where o_1, o_2, and o_3 denote functions which are negligible relative to h when h is small. That is,

$$\lim_{h \to 0} \frac{o_1(h)}{h} = \lim_{h \to 0} \frac{o_2(h)}{h} = \lim_{h \to 0} \frac{o_3(h)}{h} = 0$$

In many behavioral situations, these assumptions seem quite reasonable. For example, automobile accident rates remain fairly constant over time and those fluctuations which occur may reasonably be ascribed to chance effects. The probability function for number of accidents depends on the length of the time interval involved while, within limits, the number of accidents in one interval is unaffected by the number in another, nonoverlapping, interval.

The rationale behind assumption (3) is roughly as follows. Let $\lambda = E(X_1)$ be the expected number of occurrences in a unit time interval. Partition this unit interval into N subintervals of equal length $h = 1/N$. The probability of at least one occurrence within any one of these subintervals is

$$1 - P(X_h = 0)$$

so the expected number of subintervals containing an occurrence is

$$\frac{1}{h} [1 - P(X_h = 0)]$$

Intuitively, one feels that as $h \to 0$, this number should approach the expected number λ of occurrences in the unit interval and this is the crux of assumption (a). Implicit in this argument is the assumption that the probability of two or more occurrences in any subinterval is negligible as $h \to 0$. Obviously, such considerations may provide a model for accidents but not for numbers of cars involved in accidents. For here it is quite likely that we find multiple occurrences within a short interval, in fact, simultaneously.

To derive the probability function of X_t we write

$$X_{t+h} = X_t + (X_{t+h} - X_t) \tag{1}$$

and partition the event $(X_{t+h} = n)$ into

$$(X_{t+h} = n) = \bigcup_{k=0}^{n} [X_t = k, X_{t+h} - X_t = n - k] \tag{2}$$

Equation (1) expresses the fact that the number of occurrences up to time $t + h$ is the number up to time t plus the number which occur between times t and $t + h$. Equation (2) says that if there are n occurrences by time $t + h$, then in time t we could have zero or one or ... or n, while the remainder come in $(t, t + h]$.

Writing $p_n(t)$ for the probability $P(X_t = n)$ of n occurrences in time t, we have, using assumptions 1 and 2,

$$p_n(t + h) = \sum_{k=0}^{n} p_k(t) \cdot p_{n-k}(h)$$

But from assumption 3, $p_{n-k}(h) = o(h)$ for $n - k \geq 2$, so for $n \geq 1$

$$p_n(t + h) = o(h) + p_{n-1}(t)[\lambda h + o(h)] + p_n(t)[1 - \lambda h + o(h)]$$
$$= \lambda h p_{n-1}(t) + (1 - \lambda h)p_n(t) + o(h)$$

while

$$p_0(t + h) = (1 - \lambda h)p_0(t) + o(h)^*$$

Subtracting $p_n(t)$, dividing by h, and letting $h \to 0$, we obtain the differential equations

$$p_0'(t) = -\lambda p_0(t)$$
$$p_n'(t) = -\lambda p_n(t) + \lambda p_{n-1}(t) \qquad n \geq 1$$

which must be solved to obtain the desired probability function of X_t.

The first equation may be written

$$(D + \lambda)p_0(t) = 0$$

the solution of which has the form

$$p_0(t) = c_0 e^{-\lambda t}$$

Since $X_0 = 0$, we have $p_0(0) = 1$ so that

$$p_0(t) = e^{-\lambda t}$$

As time passes, the probability of no occurrences decreases, converging to zero as $t \to \infty$.

*Since $o(h)$ denotes a function such that $\lim_{h \to 0} [o(h)/h] = 0$, all such functions can be combined into one.

Each of the remaining equations has the form

$$(D + \lambda)p_n(t) = \lambda p_{n-1}(t)$$

and the equations may be solved successively. Thus

$$(D + \lambda)p_1(t) = \lambda p_0(t) = \lambda e^{-\lambda t}$$

Multiplying by $e^{\lambda t}$ gives

$$D[e^{\lambda t}p_1(t)] = \lambda$$

from which, since $p_1(0) = 0$, we find

$$p_1(t) = \lambda t e^{-\lambda t}$$

Continuing in this manner gives

$$p_n(t) = P(X_t = n) = \frac{(\lambda t)^n e^{-\lambda t}}{n!} \qquad n = 0, 1, 2, \ldots \tag{3}$$

as may be verified by induction.

For each fixed value of t, the random variable X_t of which (3) is the probability function is called a *Poisson* random variable with parameter* λt. The entire process described in assumptions 1–3 is called a *Poisson process* with rate, λ.

Since the series expansion of $e^{\lambda t}$ is

$$e^{\lambda t} = \sum_{n=0}^{\infty} \frac{(\lambda t)^n}{n!}$$

it follows that the probabilities in (3) sum to unity, as they should. The expected value of X_t is

$$E(X_t) = \sum_{n=0}^{\infty} n(\lambda t)^n \frac{e^{-\lambda t}}{n!} = e^{-\lambda t}\lambda t \sum_{n=1}^{\infty} \frac{(\lambda t)^{n-1}}{(n-1)!} = \lambda t$$

In particular, setting $t = 1$, we find that λ represents the expected number of occurrences in a unit time interval, as suggested in the intuitive remarks which followed assumption 3.

Since

$$E(X_t(X_t - 1)) = \sum_{n=0}^{\infty} n(n-1)\frac{e^{-\lambda t}(\lambda t)^n}{n!}$$

$$= (\lambda t)^2 e^{-\lambda t} \sum_{n=2}^{\infty} \frac{(\lambda t)^{n-2}}{(n-2)!}$$

$$= (\lambda t)^2$$

then

$$\text{Var}\,(X_t) = E(X_t(X_t - 1)) + E(X_t) - (E(X_t))^2$$

$$= (\lambda t)^2 + \lambda t - (\lambda t)^2 = \lambda t$$

The variance of a Poisson random variable is equal to its expected value.

*After Siméon D. Poisson, a nineteenth-century French probabilist.

In the above discussion we have interpreted t as a measure of time. However, other interpretations and applications of Poisson processes are possible as the following examples show.

Example 1 (a) Consider a textile loom which is weaving cloth continuously and which from time to time produces a weave with a broken fiber. Suppose that we assume that a broken fiber is as likely to occur in any one small unit of area as another, that numbers of breakages are independent from one area to another, and that the probability of two or more broken fibers per small unit of area is negligible. Then the number of broken fibers per t square units of cloth is a Poisson random variable.

(b) In astronomical studies, it is reasonable to assume that a given region of space is as likely to contain stars as any other region of equal volume, that disjoint regions are independent, and that there is negligible probability of finding two stars in close proximity. It follows that the number of stars per t cubic units of space should be a Poisson random variable.

(c) Other examples of a similar nature concern the distributions of such things as raisins in a cake, misprints in a page of type, and the pattern of hits from mortar rounds fired into a given area. ▶

Example 2 In checking whether the Poisson distribution fits a particular set of data, a common statistical procedure is to compute the average value from the sample data, use this to generate a set of Poisson probabilities, and then compare the theoretical Poisson frequencies with the observed data.

A classic example, due to Bortkiewicz,* concerns the number of deaths in the Prussian army due to kicks by horses. The first two rows of Table 1 show the observed data gathered from the records of 200 corps. The average number of deaths per corps is

$$\lambda = \frac{0(109) + 1(65) + 2(22) + 3(3) + 4(1)}{200} = 0.61$$

TABLE 1

	0	1	2	3	4	Total
Number of deaths k	0	1	2	3	4	
Number of corps with k deaths	109	65	22	3	1	200
Poisson probabilities $p_k = \dfrac{e^{-0.61}(0.61)^k}{k!}$	0.5436	0.3316	0.1011	0.0206	0.0031	
Frequencies $200p_k$	108.7	66.3	20.2	4.1	0.6	199.9

*Bortkiewicz, L. v., "Das Gesetz der kleinen Zahlen," Leipzig, 1898.

The Poisson probability function with expected value 0.61 is shown in row 3 and is multiplied by 200 in row 4 to give Poisson approximations to the actual data. The degree of correspondence is remarkable. ▶

If X is a Poisson random variable with probability function

$$P(X = k) = \frac{e^{-\lambda}\lambda^k}{k!} \qquad k = 0, 1, 2, \ldots$$

the probability generating function of X is

$$g_X(s) = E(s^X) = \sum_{k=0}^{\infty} \frac{s^k e^{-\lambda}\lambda^k}{k!}$$

$$= e^{-\lambda} \sum_{k=0}^{\infty} \frac{(s\lambda)^k}{k!}$$

$$= e^{-\lambda + \lambda s}$$

Let Y be another Poisson random variable, independent of X, having probability function

$$P(Y = k) = \frac{e^{-\mu}\mu^k}{k!} \qquad k = 0, 1, 2, \ldots$$

Then since the generating function of a sum of independent random variables is the product of the individual generating functions, we find

$$g_{X+Y}(s) = g_X(s) \cdot g_Y(s) = e^{-\lambda + \lambda s} e^{-\mu + \mu s}$$

$$= e^{-(\lambda + \mu) + (\lambda + \mu)s}$$

We conclude that the sum of two independent Poisson random variables, one with expectation λ and the other with expectation μ, is another Poisson random variable with parameter $\lambda + \mu$. The next example shows how this same conclusion may be obtained from the basic assumptions for a Poisson process.

Example 3 (a) In a model of individual behavior in a choice situation, Audley* postulates that each of m possible overt choice responses has associated with it an implicit response. Implicit responses of each type are assumed to occur in a Poisson manner, independently of responses of other types.

Let us consider the case $m = 2$, writing X_t and Y_t for the respective numbers of occurrences by time t of the two types r_1 and r_2 of implicit response. Then there are parameters $\alpha = E(X_1)$ and $\beta = E(Y_1)$ such that

$$P(X_t = 0) = 1 - \alpha t + o(t) \qquad P(Y_t = 0) = 1 - \beta t + o(t)$$

*Audley, R. J., "A Stochastic Model for Individual Choice Behavior," *Psychological Review* **67**, 1–15 (1960).

$$P(X_t = 1) = \quad \alpha t + o(t) \qquad P(Y_t = 1) = \quad \beta t + o(t)$$

$$P(X_t > 1) = \quad o(t) \qquad P(Y_t > 1) = \quad o(t)$$

Since X_t and Y_t are independent,

$$P(X_t + Y_t = 0) = P(X_t = 0, Y_t = 0) = [1 - \alpha t + o(t)][1 - \beta t + o(t)]$$

$$= 1 - \alpha t - \beta t + o(t) = 1 - (\alpha + \beta)t + o(t)$$

$$P(X_t + Y_t = 1) = P[X_t = 0, Y_t = 1] + P[X_t = 1, Y_t = 0]$$

$$= [1 - \alpha t + o(t)][\beta t + o(t)] + [\alpha t + o(t)][1 - \beta t + o(t)]$$

$$= \beta t + \alpha t + o(t) = (\alpha + \beta)t + o(t)$$

Similar calculations show that

$$P(X_t + Y_t > 1) = o(t)$$

Hence the random variable $Z_t = X_t + Y_t$ satisfies the Poisson assumptions with intensity $\alpha + \beta$, and it follows that the probability of obtaining exactly k implicit responses in the time interval $(0, t)$ is

$$p_k(t) = P(Z_t = k) = \frac{e^{-(\alpha+\beta)t}(\alpha + \beta)^k t^k}{k!} \qquad k = 0, 1, 2, \ldots \qquad \blacktriangleright$$

PROBLEMS

1. The demand for a certain inventory item is Poisson with an average daily demand of one unit. What is the probability that the number of items requested over a five-day period will be

(a) exactly 2?   (b) less than 2?   (c) at least 4?

2. If voters arrive at the polls in a Poisson fashion at the average rate of 2 per minute, what is the probability that in *each* of two nonoverlapping 5-minute intervals the number of arrivals will be

(a) exactly 2?   (b) 2 or more?   (c) 4 or less?

3. If highway accidents are Poisson distributed at a rate of 0.01 per mile of road per month, find the probability that in one month

(a) at least one accident occurs in each of four nonoverlapping hundred-mile stretches.

(b) at least one accident occurs in at least one of the stretches.

4. Use the procedure of Example 2 to fit a Poisson distribution to the following accident data reported by Mintz and Blum.*

Accidents per man	0	1	2	3	4	5	6	7	8	9	10	≥11	Total
Number of men	201	21	2	1	0	0	0	0	0	0	1	0	226

*Mintz, A., and Blum, M. L., "A Re-examination of the Accident Proneness Concept," *Journal of Applied Psychology* **33**, 195–211 (1949).

5. The Homemade Cookie Company has noted that the number of chips in one of their Chocolate Chip Delights is a Poisson random variable and that the average number of chips per cookie is 2.5.

  (a) What is the probability that a cookie contains at least one chip?

  (b) What is the variance of the number of chips per cookie?

  (c) Determine the probability that a box containing M cookies contains exactly the same number of chips as a separate box of N cookies. (M and N are positive integers.)

  (d) Due to customer complaints, the inspectors have been instructed to dispose of all cookies containing no chips. What is the expected value and variance of the number of chips per cookie for the remaining cookies?

6. (a) In a Poisson process show that for $s < t$,

$$P(X_s = k \mid X_t = n) = \binom{n}{k}\left(\frac{s}{t}\right)^k\left(1 - \frac{s}{t}\right)^{n-k}$$

  Given n occurrences in time t, the conditional distribution of occurrences in time $s < t$ is binomial.

  (b) Argue that the result in (a) is equivalent to the following. Let X and Y be independent Poisson random variables with expected values λ and μ, respectively, and let $Z = X + Y$. Then

$$P(X = k \mid Z = n) = \binom{n}{k}\left(\frac{\lambda}{\lambda + \mu}\right)^k\left(\frac{\mu}{\lambda + \mu}\right)^{n-k}$$

  (c) If X_t and Y_t denote the respective numbers of occurrences in two independent Poisson processes and if $E(X_t) = \lambda t$ and $E(Y_t) = \mu t$, then $Z_t = X_t + Y_t$ is a Poisson random variable with $E(Z_t) = (\lambda + \mu)t$. The conditional probability function of X_t, given $Z_t = n$, is

$$P(X_t = k \mid Z_t = n) = \binom{n}{k}\left(\frac{\lambda}{\lambda + \mu}\right)^k\left(\frac{\mu}{\lambda + \mu}\right)^{n-k}$$

7. (a) Continuing Example 3, prove that the successive occurrences of implicit responses r_1 and r_2 form Bernoulli trials with $p = P(r_1) = \alpha/(\alpha + \beta)$ and $q = 1 - p = P(r_2) = \beta/(\alpha + \beta)$. (*Hint:* Apply Problem 6(c).)

  (b) Audley's model* assumes that an overt response of type R_1 occurs only after two successive occurrences of the implicit response r_1. Examples of such sequences are r_1r_1, $r_2r_1r_1$, $r_1r_2r_1r_1$, ..., with respective probabilities p^2, qp^2, qp^3, ..., where p and q are defined in (a). Show that the probability $P(R_1)$ that an overt response is R_1 is given by

$$P(R_1) = \frac{\alpha^2(\alpha + 2\beta)}{(\alpha + \beta)[(\alpha + \beta)^2 - \alpha\beta]}$$

8. In Problem 7, let V be the number of occurrences of implicit responses of either type prior to the occurrence of an overt response; that is, prior

*Audley, R. J., "A Stochastic Model for Individual Choice Behavior," *Psychological Review* **67**, 1–15 (1960).

to the occurrence of two successive implicit responses of the same kind. For example,

$$V = 0 \quad \text{if } r_1 r_1 \quad \text{or } r_2 r_2$$
$$V = 1 \quad \text{if } r_2 r_1 r_1 \text{ or } r_1 r_2 r_2, \ldots$$

(a) Prove that $E(V) = 3\alpha\beta/[(\alpha + \beta)^2 - \alpha\beta]$ and hence that $E(V)$ depends only on the ratio β/α.

(b) Find the value of $P(R_1)$ for which $E(V)$ is a maximum.

9. Show by induction that the reproductive property of the Poisson distribution extends to any number of random variables. That is, prove that if $X_1, X_2, \ldots, X_n$ are independent Poisson random variables with expected values $\lambda_1, \lambda_2, \ldots, \lambda_n$, then $Z = X_1 + X_2 + \cdots + X_n$ is a Poisson random variable with expectation $\lambda_1 + \lambda_2 + \cdots + \lambda_n$.

10. (a) Suppose that manufacturing errors form a Poisson process in which λ, the average failure rate per hour, is unknown. Moreover, λ varies from time to time, and from past experience it is estimated that $P(\lambda = 1) = \frac{1}{2}$ and $P(\lambda = \frac{1}{2}) = P(\lambda = 2) = \frac{1}{4}$. Given that we observe two errors in a 4-hour period, find the conditional probabilities that $\lambda = \frac{1}{2}, \lambda = 1$, and $\lambda = 2$.

(b) A Bayesian decision rule for estimating λ is one which, for each observed number of errors, chooses that value of λ which has the largest conditional probability. Formulate a Bayesian decision rule for the estimation of λ.

11. Consider a Poisson arrival process, such as voters entering a polling place, which has parameter λ. Let us suppose that each arrival initiates a Bernoulli trial in which success occurs if the arrival is recorded and failure occurs if it is not. Prove that the occurrences of recorded arrivals constitute a Poisson process with parameter $m = p\lambda$. (There are many alternative versions of this problem. The number of eggs which an insect lays may be a Poisson random variable with parameter λ and p is the probability that an individual egg produces a mature insect. The number of mature insects is then Poisson with parameter λp. Or, the number of particles hitting a Geiger counter may be Poisson with constant probability that a particle activates the counter. The number of recorded particles is Poisson, and so forth.)

12. Suppose that a new process is created from the original one by recording only alternate arrivals. Show that this process is *not* Poisson.

13. An inspector monitors the automatic production of plastic castings, looking for tolerance deviations and casting irregularities which occur as independent Poisson processes at the respective rates of two and five per hour. Suppose that tolerance deviations are detected with probability 0.6 and casting irregularities with probability 0.8.

(a) What is the probability that in 2 hours

   (i) no tolerance or casting defects occur?

   (ii) exactly one defect (of either kind) occurs?

   (iii) at most three defects occur?

(b) What is the probability that in 2 hours the inspector detects

   (i) no tolerance or casting defects?

   (ii) exactly one defect?

   (iii) at most, three defects?

(c) Suppose that the inspector wishes to take a 15-minute break. What is the probability that

   (i) no more than two defects will occur?

   (ii) no more than two defects would have been missed had he been present?

(d) Given that six defects have occurred in a half-hour period, what is the probability that

   (i) at least one casting irregularity occurred?

   (ii) exactly one casting irregularity occurred?

   (iii) at most two casting irregularities occurred?

(e) Let Z be the number of tolerance errors in a half-hour period. Find $E(Z)$ and Var (Z).

(f) What is the probability that at least 60% of the casting irregularities are detected, given that six casting irregularities occur in a half-hour period?

(g) Given that six defects occur during an hour interval, what is the probability that exactly three of these defects occur in the first half-hour period and two of the three are detected?

21.2 CONTINUOUS TIME RANDOM PROCESSES

In the Poisson process, the parameter λ represents the expected number of occurrences in a unit interval of time of the phenomenon under study. As such, λ provides a measure of the rate of increase of this number of occurrences. This rate remains constant and is unaffected by the number of occurrences or the times of their occurrence.

In the study of population growth, such assumptions are often not realistic, the rate of increase naturally depending upon population size. Moreover, populations may decrease as well as increase. We are thus led to generalize the approach taken in the Poisson process in the manner indicated in Figure 2. When population size is n, there are opposing tendencies for subsequent increase or decrease of the size of the population. The rates λ_n, of increase, and μ_n, of decrease, may both depend on n. The Poisson process is obtained by choosing $\lambda_n = \lambda$ and $\mu_n = 0$ for all n.

A useful special case (Figure 3) is obtained by choosing $\lambda_n = n\lambda + \alpha$ and $\mu_n = n\mu + \beta$, where α, β, λ, and μ are constants. The rationale behind the choices is as follows: Consider a population subject to rates α, of immigration and β of emigration, which do not depend on population size. In addition, for *each* individual, there is a fixed birth rate λ and death rate μ. Taken together, these assumptions imply that when population size is n, the

FIGURE 2

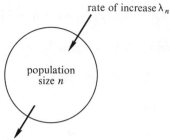

rate of increase λ_n

population size n

rate of decrease μ_n

rate λ_n of increase should have the indicated form $n\lambda + \alpha$ while the cor-responding rate of decrease is $n\mu + \beta$.

Regardless of the particular form of the quantities λ_n and μ_n, a process described in this fashion will be called a *birth and death process*. When population size is n, λ_n and μ_n represent the respective *birth and death rates* of the process.

Proceeding as in the Poisson process, we write X_t for the population size at time t, so that $(X_t = n)$ denotes the event that at time t there are n individuals in the population. We assume that, if $X_t = n$, then

(a) the probability of a "birth" in the time interval $(t, t + h]$ is $\lambda_n h + o(h)$.

(b) the probability of a "death" in $(t, t + h]$ is $\mu_n h + o(h)$.

(c) the probability of more than one change is $o(h)$.

(d) the probability of no change is $1 - \lambda_n h - \mu_n h + o(h)$.

Note that the rates depend only on the present size n of the population and not on the time at which this size is attained or on any other feature of the past history of the process. This is essentially a statement that the process is Markovian in nature, and we use this Markov property to obtain differential equations which describe the process.

Specifically, we have (compare with Equations (1) and (2))

$$X_{t+h} = X_t + (X_{t+h} - X_t)$$

FIGURE 3

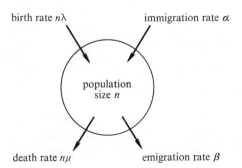

birth rate $n\lambda$

immigration rate α

population size n

death rate $n\mu$

emigration rate β

so that the event $(X_{t+h} = n)$ partitions into

$$(X_{t+h} = n) = \bigcup_{k=0}^{\infty} [X_t = k, X_{t+h} - X_t = n - k]$$

Again writing $p_n(t)$ for the probability $P(X_t = n)$, we have

$$p_n(t + h) = \sum_{k=0}^{\infty} p_k(t) \cdot p_{n-k}(h)$$

Using assumptions (a)–(d) above, this reduces to

$$p_n(t + h) = p_n(t)[1 - \lambda_n h - \mu_n h] + p_{n-1}(t)\lambda_{n-1}h + p_{n+1}(t)\mu_{n+1}h + o(h)$$

Subtracting $p_n(t)$, dividing by h, and letting $h \to 0$ yields the differential equation

$$p_n'(t) = -(\lambda_n + \mu_n)p_n(t) + \lambda_{n-1}p_{n-1}(t) + \mu_{n+1}p_{n+1}(t) \qquad (4)$$

for $n \geq 1$. When $n = 0$, we write $\mu_n = 0$ and $p_{-1}(t) = 0$ to reflect the fact that population size is never negative, obtaining

$$p_0'(t) = -\lambda_0 p_0(t) + \mu_1 p_1(t) \qquad (5)$$

Equations (4) and (5) are the general differential equations for birth and death processes. As an intuitive aid to understanding these equations, we may view λ_n and μ_n as measures of the tendency of the process to *leave* state (population size) n, moving either upward or downward. The rate at which these transitions occur is proportional to the probability $p_n(t)$ of being in state n. Similarly, we *enter* state n, or tend to increase the probability of being in this state, either by being in state $n - 1$ (probability $p_{n-1}(t)$) and moving up (rate λ_{n-1}) or by being in state $n + 1$ (probability $p_{n+1}(t)$) and moving down (rate μ_{n+1}).

Example 1 *A "Pure Death" or "Extinction" Process* Imagine a population whose members can only die (leave the population). Let us assume that members act independently and that during any short interval of time h each member has probability $\mu h + o(h)$ of dying. Then, of n members, the probability that none die in time h is

$$[1 - \mu h + o(h)]^n = 1 - n\mu h + o(h)$$

the probability that exactly one dies is

$$n[1 - \mu h + o(h)]^{n-1}[\mu h + o(h)] = n\mu h + o(h)$$

and the probability that two or more die is

$$1 - [1 - n\mu h + o(h)] - [n\mu h + o(h)] = o(h)$$

To illustrate the calculations, we obtain the second of these equations by noting that the n members act independently and that any one of them may die (this has probability $\mu h + o(h)$), while the others must live (each with probability $1 - \mu h + o(h)$). Expanding the binomial expression and noting that terms involving h^2, h^3, ... are such that we obtain the limit zero when we divide by h and let $h \to 0$ (that is, these terms are all $o(h)$), we find that the only term which is not $o(h)$ is $n\mu h$, as indicated.

Our process is thus described by taking $\lambda_n = 0$ and $\mu_n = n\mu$ in (4) and (5). Thus, we have

$$p_n'(t) = -n\mu p_n(t) + (n + 1)\mu p_{n+1}(t) \qquad n \geq 1$$
$$p_0'(t) = \mu p_1(t)$$

These equations may be solved successively. If when $t = 0$ the size of the population is k (so that $p_n(t) = 0$ for $n > k$), we have

$$p_k'(t) = -k\mu p_k(t)$$

or

$$(D + k\mu)p_k(t) = 0$$

Multiplying both sides by $e^{k\mu t}$ gives (see Theorem 3, Section 15.2)

$$D[e^{k\mu t}p_k(t)] = 0$$

from which

$$p_k(t) = C_k e^{-k\mu t}$$

where the constant $c_k = 1$ since $p_k(0) = 1$.

With $p_k(t)$ known we next solve for $p_{k-1}(t)$ by writing

$$p_{k-1}'(t) = -(k - 1)\mu p_{k-1}(t) + k\mu p_k(t)$$

or

$$[D + (k - 1)\mu]p_{k-1}(t) = k\mu e^{-k\mu t}$$

Multiplying by the integrating factor $e^{(k-1)\mu t}$ yields (see Section 15.2)

$$D[e^{(k-1)\mu t}p_{k-1}(t)] = k\mu e^{-\mu t}$$

which may be solved to give

$$p_{k-1}(t) = ke^{-(k-1)\mu t}(1 - e^{-\mu t})$$

Proceeding by induction, we find for $0 \leq n \leq k$,

$$p_n(t) = \binom{k}{n}e^{-n\mu t}(1 - e^{-\mu t})^{k-n}$$

a binomial probability function with $p = e^{-\mu t}$. Note that as t increases, $p_n(t)$ goes to zero for $n \geq 1$, while $p_0(t) \to 1$. As one would expect, the population eventually dies out.

A typical example of such an extinction process is provided by Ballenger.* Mice were conditioned to exit at the sound of a tone from a shock box with variable exit gates. The parameter μ was estimated from the data and the model found to provide a good approximation to actual behavior. ▶

*Ballenger, W., "A Stochastic Model of Escape Behavior," M. S. thesis, North Carolina State University, 1969 (unpublished).

Example 2 *A "Pure Birth" Process* In contrast to the death process of the preceding example, consider a population whose members create new members but do not die. If we assume that members act independently, each at rate λ, the process is described by (4) and (5) with $\lambda_n = n\lambda$ and $\mu_n = 0$. It is easily verified that if n_0 is the population size at time $t = 0$, then for $n \geq n_0$,

$$p_n(t) = \binom{n-1}{n-n_0} e^{-\lambda t n_0}(1 - e^{-\lambda t})^{n-n_0}$$

which for each t is a negative binomial probability function with

$$p = e^{-\lambda t} \qquad \blacktriangleright$$

Although we shall not attempt a proof here, it can be shown that in any birth and death process, the limits

$$\lim_{t \to \infty} p_n(t) = a_n$$

exist and do not depend on the initial conditions under which the process began. The a_n may be determined from the explicit form of the solution (as in Example 1), but are more often found by solving the equations obtained from (4) and (5) by replacing the derivatives on the left by zero.

Example 3 Consider a system which processes randomly arriving inputs on a first-come, first-served basis. If the processing unit is busy, the new inputs form a waiting line or *queue*. Typical examples of queueing systems are the flow of aircraft in and out of an airport, the flow of cars at a toll booth, arrivals of injured at the emergency room of a hospital, or the arrival of customers at a restaurant or supermarket.

In the simplest case the system contains a single service unit. Inputs are processed individually and remain in the waiting line until processed. Let us assume that inputs arrive according to a Poisson process with rate λ and that individuals depart at rate μ. The system is thus a birth and death process with $\lambda_n = \lambda$ for all n, while $\mu_n = \mu$ for $n \geq 1$ and $\mu_0 = 0$.

Although explicit solutions may be obtained,* these are rather complicated and will not be considered here. On the other hand, limiting values are obtained fairly easily. Replacing the derivatives by zero in (4) and (5), we obtain

$$0 = -(\lambda + \mu)a_n + \lambda a_{n-1} + \mu a_{n+1}$$

for $n \geq 1$, while for $n = 0$,

$$0 = -\lambda a_0 + \mu a_1$$

*See Cox, D. R., and Smith, W. L., *Queues* (John Wiley & Sons, Inc., New York, 1961), pp. 60–64.

Solving recursively we find

$$a_1 = \left(\frac{\lambda}{\mu}\right) a_0$$

$$a_2 = \frac{(\lambda + \mu)a_1 - \lambda a_0}{\mu} = \left(\frac{\lambda}{\mu}\right)^2 a_0$$

and, by induction,

$$a_n = \left(\frac{\lambda}{\mu}\right)^n a_0$$

for all $n \geq 0$.

The series $\sum_{n=0}^{\infty} a_n = a_0 \sum_{n=0}^{\infty} (\lambda/\mu)^n$ converges only when $\lambda < \mu$. In this case, since $\sum_{n=0}^{\infty} a_n = 1$, we have

$$a_0 = 1 - \frac{\lambda}{\mu}$$

and the limiting distribution is geometric with

$$a_n = \left(1 - \frac{\lambda}{\mu}\right)\left(\frac{\lambda}{\mu}\right)^n \qquad n = 0, 1, 2, \ldots$$

If $\lambda \geq \mu$, we must take $a_0 = 0$ since otherwise $\sum_{n=0}^{\infty} a_n$ diverges. In this case, we have

$$a_n = 0 \qquad \text{for all } n$$

and the waiting line must exceed all bounds as $t \to \infty$. ▶

Example 4* Consider the process by which a group of N people formulate their preferences for two opposing political candidates. We assume that if at time t a person favors candidate A, the probability that in the interval $(t, t + h)$ he switches support to candidate B is $\mu h + o(h)$, while if he favors B, the probability of switching is $\lambda h + o(h)$. The group is said to be in state S_n if n members support candidate A.

Assuming group members act independently we have a birth and death process with

$$\lambda_n = (N - n)\lambda \quad \text{and} \quad \mu_n = n\mu \qquad \text{for } 0 \leq n \leq N$$

The basic system of differential equations is

$$p_0'(t) = -N\lambda p_0(t) + \mu p_1(t)$$

$$p_n'(t) = -[(N - n)\lambda + n\mu]p_n(t) + (N - n + 1)\lambda p_{n-1}(t)$$
$$+ (n + 1)\mu p_{n+1}(t) \qquad \text{when } 1 \leq n \leq N - 1$$

and $p_N'(t) = -N\mu p_N(t) + \lambda p_{N-1}(t)$.

*Adapted from Coleman, J. S., *Introduction to Mathematical Sociology* (Free Press, Glencoe, Ill., 1964), pp. 336–343. A wide variety of continuous time models are discussed in this book.

Replacing all derivatives by zero and solving, we obtain

$$a_1 = N\left(\frac{\lambda}{\mu}\right)a_0$$

$$a_2 = \frac{N(N-1)}{2}\left(\frac{\lambda}{\mu}\right)^2 a_0$$

and by induction

$$a_n = \binom{N}{n}\left(\frac{\lambda}{\mu}\right)^n a_0$$

for $0 \le n \le N$. Of course, $a_n = 0$ for $n > N$. Imposing the restriction that

$$1 = \sum_{n=0}^{N} a_n = \sum_{n=0}^{N} \binom{N}{n}\left(\frac{\lambda}{\mu}\right)^n a_0$$

$$= a_0\left(1 + \frac{\lambda}{\mu}\right)^N$$

we find $a_0 = [\mu/(\mu + \lambda)]^N$ and in general

$$a_n = \binom{N}{n}\left(\frac{\lambda}{\mu + \lambda}\right)^n \left(\frac{\mu}{\mu + \lambda}\right)^{N-n} \qquad 0 \le n \le N$$

The limiting probabilities of being in the various states follow a binomial probability function. ▶

PROBLEMS

1. Consider the acquisition and loss of friends as a continuous time random process in which new friends are acquired in accord with a Poisson process with parameter λ. Friendship durations are assumed to be mutually independent, the probability of losing a particular friend during a time interval of length h being $\mu h + o(h)$. The system is said to be in state E_n if a person has n friends. Find the limiting probabilities of being in the various states.

2. In Problem 1, given that at time zero a person has no friends, verify that the probability that he has n friends at time t is

$$p_n(t) = \frac{e^{-\alpha}\alpha^n}{n!} \qquad (n = 0, 1, 2, 3, \ldots)$$

where $\alpha = (\lambda/\mu)(1 - e^{-\mu t})$.

3. Let us suppose that successive automobile purchases by a particular family constitute a continuous time process in which the probability that car A is purchased during a time interval of length h is $\lambda h + o(h)$ and the probability that a different make is purchased is $\mu h + o(h)$. The system is in state $E_n (n \ge 1)$ if the family has purchased car A n times in succession, and in state E_0 if the last purchase was of a different make. Find the limiting probabilities of being in the various states.

4. Consider a "pure birth" process and let X_t denote the number of "births" in time t. Assume that when X_t is odd, the probability of a birth in time interval $(t, t + h)$ is $\lambda_1 h + o(h)$, while if X_t is even, the probability is $\lambda_2 h + o(h)$. Take $X_0 = 0$. Find the probabilities

$$p_1(t) = P(X_t \text{ is odd}) \quad \text{and} \quad p_2(t) = P(X_t \text{ is even})$$

(*Hint:* Derive the differential equations

$$p_1'(t) = -\lambda_1 p_1(t) + \lambda_2 p_2(t) \quad \text{and} \quad p_2'(t) = \lambda_1 p_1(t) - \lambda_2 p_2(t)$$

and solve them.)

5. The following is a "pure birth" process with "contagion." Coleman,* in an application of stochastic models to social data, assumed that record purchases of teen-age girls could be represented as a "contagion" process in that the more records a girl had purchased, the more likely she would be, in a given interval of time, to purchase another. Thus he assumed that if by time t a girl had purchased n records (state E_n), the probability of a new purchase in $(t, t + h)$ was $(\alpha + n\beta)h + o(h)$. The parameter β is called a contagion parameter.

(a) Write the differential equations which govern the process. Use an inductive argument to verify that, assuming the process starts in state E_0, the solution is

$$p_n(t) = \frac{e^{-(\alpha + n\beta)t}(e^{\beta t} - 1)^n \alpha[\alpha + \beta] \cdots [\alpha + (n - 1)\beta]}{n!\beta^n}$$

(b) Prove that $\sum_{n=0}^{\infty} p_n(t) = 1$ for all t.
(*Hint:* Make use of the series expansion for $(1 - x)^{-r}$.)

6. Rework Problem 5 assuming that $\alpha = 0$ and the system starts in state E_1.

7. A social movement gains and loses followers according to a continuous time "contagion" process. Specifically, if n members of an infinite population are followers of the movement at time t, then the probability that in the interval $(t, t + h)$ the movement gains a follower is

$$(\lambda + n\alpha)h + o(h)$$

where α is a "contagion" parameter. Followers of the movement act independently and the probability that any particular one drops out is $\beta h + o(h)$. (That is, $\mu_n = \beta n$.) Find the limiting probability a_n that exactly n members of the population are followers of the movement.

8. Parsons and Shils† postulate that social systems tend to maintain order in their internal structure. Assume a sequence of system states E_0, E_1, E_2, ... arranged according to their degree of disorder. Assume that if the system is in state $E_n (n \geq 1)$ at time t, the probability of moving to E_{n+1} in $(t, t + h)$ is $\lambda h + o(h)$, of moving to E_{n-1} is $\mu h + o(h)$, and of

*Coleman, J. S., *Introduction to Mathematical Sociology* (Free Press, Glencoe, Ill., 1964), pp. 301–304.

†Parsons, T., and Shils, E., Eds., *Toward a General Theory of Action* (Harvard University Press, Cambridge, Mass., 1951).

moving to any other state is $o(h)$. For $n = 0$, these probabilities apply, respectively, to remaining in state E_0 or moving to E_1. Prove that if $\mu > \lambda$, the social system in the long run has greater probability of being in state E_0, the state of least disorder, than in any other state.

9. In Example 4, assume that group members are influenced by the number of persons who support the opposing candidate. Specifically, assume that if at time t a person supports candidate A and there are $N - n$ persons who support B, then the probability that in the interval $(t, t + h)$ he switches his support to B is $[\mu + (N - n)\alpha]h + o(h)$. Under the same circumstances, the probability that a person favoring B switches to A is $[\lambda + n\alpha]h + o(h)$. Group members act independently.

(a) Write the differential equations governing the process.

(b) Show that the limiting probabilities satisfy

$$a_1 = \frac{N\lambda}{\mu + (N - 1)\alpha} a_0$$

$$a_2 = \frac{N(N - 1)\lambda(\lambda + \alpha)}{2 \cdot 1[\mu + (N - 1)\alpha][\mu + (N - 2)\alpha]} a_0$$

and, by induction, that

$$a_n = \frac{\binom{N}{n}\lambda(\lambda + \alpha)(\lambda + 2\alpha)\cdots(\lambda + (n - 1)\alpha)}{[\mu + (N - 1)\alpha][\mu + (N - 2)\alpha]\cdots[\mu + (N - n)\alpha]} a_0$$

(c) Using $\sum_{n=0}^{\infty} a_n = 1$, show that

$$a_0 = \frac{\displaystyle\prod_{j=0}^{n-1}(\mu + j\alpha)}{\displaystyle\prod_{j=0}^{n-1}(\lambda + \mu + j\alpha)}$$

and thus verify that the limiting probability a_n of finding the system in state S_n is the same (if λ, μ, and α are integers) as the probability of drawing n black balls in N draws from a Polya urn which initially contains λ blacks and μ reds and to which α balls are added following each draw.

10. The operator of a car-washing establishment finds that customers arrive at an average rate of one every 5 minutes, and that service takes, on the average, 3 minutes. Assume that arrivals follow a Poisson process and that services are completed at rate $\frac{1}{3}$.

(a) What is the probability that a customer will not have to wait? (*Hint:* Proceed as in Example 3 and compute this as the probability the waiting line is empty.)

(b) Given there is a waiting line, what is the expected waiting time?

11. Assume that the system of Example 3 has M identical processors. If all processors are busy, an incoming element joins a common waiting line and waits until a processor is free.

(a) Argue that the system is a birth and death process with

$$\lambda_n = \lambda \quad \text{and} \quad \mu_n = n\mu \qquad n < M$$

$$= M\mu \qquad n \geq M$$

(b) Find the limiting probability that there is at least one element in the system.

(c) Find the limiting probability that there are n elements in the system.

(d) Under what conditions on λ and μ does a limiting probability function exist?

(e) Find the probability that an input has to wait before being processed.

12. Suppose that inputs in Problem 11 that are not processed immediately leave and do not return. Show that the limiting probability distribution is

$$a_n = \frac{(\lambda/\mu)^n/n!}{\sum_{n=0}^{M} (\lambda/\mu)^n/n!} \qquad (n = 0, 1, 2, \ldots, M)$$

21.3 CONTINUOUS RANDOM VARIABLES

Our discussion thus far has been concerned with discrete random variables. (See Section 19.1.) The graph of the distribution function of such a random variable consists entirely of jumps and level stretches, all probability being concentrated at the jump points. At the other extreme are those random variables whose distribution functions have no jumps at all and thus are continuous. Such random variables are themselves called continuous.

A particularly important continuous random variable is the *exponential* which arises in connection with the Poisson process. Consider the time T that one must wait for the first occurrence. It is obvious that T is greater than the real number t if and only if there are no occurrences by time t. Thus, using the notation of Section 21.1, $(T > t)$ and $(X_t = 0)$ are the same event, and we have

$$P(T > t) = P(X_t = 0) = e^{-\lambda t}$$

For $t \geq 0$, then, the distribution function of T is the continuous function

$$F_T(t) = P(T \leq t) = 1 - e^{-\lambda t} \tag{6}$$

Of course, since T cannot be negative, $F_T(t) = 0$ when $t < 0$. The quantity T is called an exponential random variable because of the form of its distribution function (6).

Example 1 Voters arrive at the polls in a Poisson fashion at the rate of two per minute. If T denotes the waiting time in minutes until the first arrival, then for $t \geq 0$, the distribution function of T is

$$F_T(t) = P(T \leq t) = 1 - e^{-2t}$$

The probability that at least 3 minutes elapse before an arrival is $1 - F_T(3) = e^{-6} \approx 0.0025$, while the probability that no more than 15 seconds are required is $F_T(\frac{1}{4}) = 1 - e^{-1/2} \approx 0.393$. ▶

The derivative of the distribution function of a random variable X is called its *density function*.* For the exponential case (6) the density function is

$$f_T(t) = F'_T(t) = \begin{cases} 0 & \text{when } t < 0 \\ \lambda e^{-\lambda t} & \text{when } t > 0 \end{cases}$$

(See Figure 4.) The distribution function F_T does not have a derivative at $t = 0$, but if we extend the definition of f_T by writing $f_T(0) = 0$, then it follows that for all t

$$F_T(t) = \int_{-\infty}^{t} f_T(x)\, dx$$

FIGURE 4 **(a) Distribution function of T;**
(b) Density function of T.

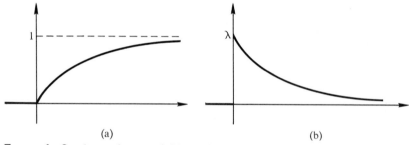

(a)   (b)

Example 2 A random variable U having the density function shown in Figure 5 is called a *uniform* or *rectangular* random variable. Here the density function is

$$f_U(t) = \begin{cases} 0 & \text{if } t < 0 \text{ or } t > 1 \\ 1 & \text{when } 0 \leq t \leq 1 \end{cases}$$

The corresponding distribution function is

$$F_U(t) = \begin{cases} 0 & \text{for } t < 0 \\ t & \text{for } 0 \leq t \leq 1 \\ t & 1 \text{ for } > 1 \end{cases}$$

(See Example 4 of Section 19.1.) ▶

*Although we shall not encounter them in this text, it is possible to find distribution functions which have no derivatives. Thus, in advanced books, it is common to distinguish between a continuous random variable X, one which has a continuous distribution function, and an absolutely continuous random variable Y, one for which there exists a function f_Y such that the distribution function F_Y can be written as
$$F_Y(t) = \int_{-\infty}^{t} f_Y(y)\, dy$$
For our purposes the role of f_Y will be played by the derivative of F_Y.

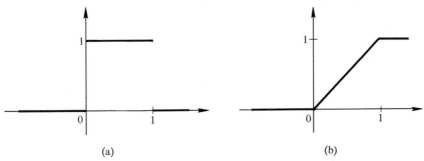

(a)   (b)

Example 3 A random variable X having the density function

$$f_X(x) = \begin{cases} \dfrac{\Gamma(\alpha + \beta)}{\Gamma(\alpha) \cdot \Gamma(\beta)} x^{\alpha-1}(1 - x)^{\beta-1} & 0 < x < 1 \\ 0 & \text{elsewhere} \end{cases}$$

is called a *beta* variable. The Γ-function is defined as

$$\Gamma(\alpha + 1) = \int_0^\infty e^{-t} t^\alpha \, dt$$

For integer values of α, $\Gamma(\alpha + 1) = \alpha!$. The density function has two parameters α and β which must both be greater than zero. When $\alpha = \beta = 1$, the beta variable becomes the uniform variable.

The distribution function, often called the *incomplete beta*, is

$$F_X(x) = \begin{cases} 0 & x \le 0 \\ \displaystyle\int_0^x \frac{\Gamma(\alpha + \beta)}{\Gamma(\alpha)\Gamma(\beta)} t^{\alpha-1}(1 - t)^{\beta-1} \, dt & 0 < x < 1 \\ 1 & x \ge 1 \end{cases}$$

and has been extensively tabulated.*   ▶

For any random variable X, if $b > a$, we must have $P(X \le b) \ge P(X \le a)$. In terms of the distribution function of X then, $F_X(b) \ge F_X(a)$ whenever $b > a$, so that F_X is an increasing function. It follows that the density function $f_X = F_X'$ must be non-negative. In addition, the relation

$$F_X(x) = \int_{-\infty}^x f_X(t) \, dt$$

yields

(1) $\displaystyle\int_{-\infty}^\infty f_X(x) \, dx = 1$

---

*Biometrika Tables of the Incomplete Beta Function, Pearson, K., Ed. (Cambridge University Press, Cambridge, England, 1934).

$$(2) \quad \int_a^b f_X(x)\,dx = \int_{-\infty}^b f_X(x)\,dx - \int_{-\infty}^a f_X(x)\,dx = F(b) - F(a)$$

$$= P(a < X \leq b) \qquad \text{for } a \leq b$$

Thus we see that the total area is unity under the curve representing f_X, and that the area under the curve between the values a and b represents the probability that X takes on a value in the interval $(a, b]$. (See Figure 6.)

FIGURE 6

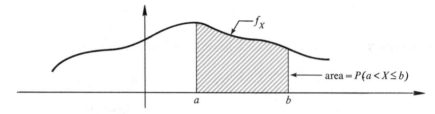

Example 4 Suppose the weekly garbage pickup G (in tons) for a certain city has the density function

$$f(x) = \begin{cases} k[-x^2 + 110x - 1000] & 10 < x < 100 \\ 0 & \text{otherwise} \end{cases}$$

where k is a constant such that

$$\int_{-\infty}^{\infty} f(x)\,dx = 1$$

We compute k by writing

$$\int_{-\infty}^{\infty} f(x)\,dx = \int_{-\infty}^{10} 0\,dx + \int_{10}^{100} k[-x^2 + 110x - 1000]\,dx + \int_{100}^{\infty} 0\,dx$$

or

$$1 = 0 + k \cdot 121{,}500 + 0$$

Hence,

$$k = 1/121{,}500$$

and $f(x) = (1/121{,}500)(-x^2 + 110x - 1000)$.

If the city can handle 40 tons without overtime, then the probability that normal operations will suffice is

$$P(10 < x < 40) = \frac{1}{121{,}500} \int_{10}^{40} (-x^2 + 110x - 1000)\,dx$$

$$= 7/27 \qquad \blacktriangleright$$

The *expected value* of a continuous random variable X with density function f_X is defined by the improper integral

$$E(X) = \int_{-\infty}^{\infty} t f_X(t)\, dt \qquad (7)$$

and the expectation of a function h of X is defined by

$$E(h(X)) = \int_{-\infty}^{\infty} h(t) f_X(t)\, dt \qquad (8)$$

Example 5 The decision as to what stock inventory level to maintain is often determined by the demand anticipated against the inventory. As a simple example, let us assume that demand is a continuous random variable $D \geq 0$ with density function f. Let I denote the inventory level. We assess a unit overage cost c_1 when supply exceeds demand $(I > D)$ and a per unit shortage cost c_2 when demand exceeds supply $(D > I)$. Thus the cost function C given by

$$C = \begin{cases} c_1(I - D) & \text{when } I > D \\ c_2(D - I) & \text{when } D > I \end{cases}$$

is a random variable with expected value

$$E(C) = c_1 \int_0^I (I - t) f(t)\, dt + c_2 \int_I^{\infty} (t - I) f(t)\, dt$$

$$= c_1 I \int_0^I f(t)\, dt - c_1 \int_0^I t f(t)\, dt$$

$$+ c_2 \int_I^{\infty} t f(t)\, dt - c_2 I \int_I^{\infty} f(t)\, dt$$

$E(C)$ is a function of I, and we may minimize expected cost by determining the value I_0 of I for which

$$\frac{d}{dI} E(C) = 0$$

Recalling that

$$\frac{d}{dx} \int_a^x f(t)\, dt = f(x)$$

and writing

$$F(x) = \int_0^x f(t)\, dt$$

for the distribution function of D, we find

$$\frac{d}{dI} E(C) = c_1 I f(I) + c_1 F(I) - c_1 I f(I) - c_2 I f(I)$$

$$- c_2 + c_2 F(I) + c_2 I f(I)$$

Thus the optimum inventory level is the value I_0 for which

$$F(I_0) = \frac{c_2}{c_1 + c_2}$$

or, equivalently, for which

$$\frac{F(I_0)}{1 - F(I_0)} = \frac{c_2}{c_1}$$

In order to minimize expected cost the inventory level should be established at the point where the ratio of the probability of an overage to the probability of a shortage equals the ratio of shortage cost to overage cost. ▶

The *variance* of a continuous random variable X is defined in the same way as that of a discrete variable. That is,

$$\text{Var } (X) = E([X - E(X)]^2)$$

The *factorial moment generating function* is

$$g_X(s) = E(s^X)$$

and the *moment generating function* is

$$m_X(s) = E(e^{sX})$$

As with discrete random variables, means and variances of continuous random variables can be found by differentiating the generating functions. For instance, using the moment generating function

$$m_X(s) = E(e^{sX}) = \int_{-\infty}^{\infty} e^{st} f_X(t)\, dt$$

the derivative is

$$m_X'(s) = \int_{-\infty}^{\infty} t e^{st} f_X(t)\, dt$$

(Here we assume without proof that the order of the operations of differentiation and integration can be interchanged in a manner similar to the differentiation of series expressed in Theorem 4 of Section 16.2.) Setting $s = 0$ gives

$$m_X'(0) = \int_{-\infty}^{\infty} t f_X(t)\, dt = E(X)$$

Similar calculations show that for any k, the kth derivative of m_X, evaluated at zero, is $E(X^k)$.

Example 6 The moment generating function of an exponential random variable T is

$$m_T(s) = E(e^{sT}) = \int_{-\infty}^{\infty} e^{st} f_T(t)\, dt$$

$$= \int_{-\infty}^{\infty} \lambda e^{-\lambda t} e^{st}\, dt = \left. \frac{\lambda}{s-\lambda} e^{-(\lambda-s)t} \right]_0^{\infty}$$

$$= \frac{\lambda}{\lambda - s} \qquad \text{for } s < \lambda$$

Differentiating, we find

$$E(T) = m_T'(0) = \left. \lambda(\lambda - s)^{-2} \right|_{s=0} = \frac{1}{\lambda}$$

$$E(T^2) = m_T''(0) = \left. 2\lambda(\lambda - s)^{-3} \right|_{s=0} = \frac{2}{\lambda^2}$$

Hence

$$\text{Var } (T) = E(T^2) - [E(T)]^2 = \frac{2}{\lambda^2} - \frac{1}{\lambda^2} = \frac{1}{\lambda^2}$$

The value $E(T) = 1/\lambda$ might have been anticipated in an intuitive fashion as follows. The quantity λ represents the average number of occurrences per unit time in the underlying Poisson process. If, on the average, there are λ occurrences in one unit of time, then it is reasonable to find that the average length of time between occurrences is $1/\lambda$. ▶

In working with discrete random variables the relation analogous to (8) was derived from (7) as a theorem. We have used the above approach here to simplify details, essentially accepting without proof the following theorem.

Theorem 1 If X is a continuous random variable with density function f_X and $Y = h(X)$ is also continuous with density function g_Y, then $E(Y)$ is given by either

$$E(Y) = \int_{-\infty}^{\infty} u g_Y(u)\, du$$

or by

$$E(Y) = \int_{-\infty}^{\infty} h(t) f_X(t)\, dt \qquad\qquad ▶$$

While Theorem 1 serves in most cases to eliminate the necessity of finding g_Y if we already have f_X, there are times when g_Y must be found. The procedure is best illustrated through examples.

Example 7 Suppose X has the density

$$f_X(t) = \begin{cases} 2te^{-t^2} & \text{when } t \geq 0 \\ 0 & \text{when } t < 0 \end{cases}$$

and suppose we require the density of $Y = X^2$. Then for $z \geq 0$,

$$P(Y \leq z) = P(X^2 \leq z) = P(X \leq z^{1/2})$$

In terms of distribution functions this says

$$F_Y(z) = F_X(z^{1/2})$$

Using the Chain Rule (Section 12.4) we obtain, for $z \geq 0$,

$$\begin{aligned} f_Y(z) = F'_Y(z) &= \frac{d}{dz} F_X(z^{1/2}) \\ &= f_X(z^{1/2}) \cdot \tfrac{1}{2}z^{-1/2} \\ &= 2z^{1/2}e^{-z}\tfrac{1}{2}z^{-1/2} \\ &= e^{-z} \end{aligned}$$

for the density of Y. Of course, $f_Y(z) = 0$ when $z < 0$. ▶

Example 8 Many behavioral systems which generate responses have an underlying periodic or cyclic component.* Typical examples are heartbeat, periodic variations in population size, and cycles of temperature, capillary level, metabolism, and ovarian activity.

One model proposed by McGill† postulates an underlying process which produces an excitation every τ units of time, where $\tau > 0$ is a constant. From time to time, these excitations elicit responses from the system. It is assumed that at most, one response may occur between two successive excitations and that the time from the first excitation following a response to the next response is an exponential random variable W.

Figure 7 gives a picture of the process. Here R_1 and R_2 denote successive responses, T being the total waiting time between them. No response can

FIGURE 7

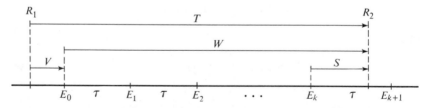

*For an interesting account of many such phenomena, see *Journal of Cycle Research*, published by the Foundation for the Study of Cycles, New York.

†McGill, W. J., "Random Fluctuation of Response Rate," *Psychometrika* **27**, 3–17 (1962).

occur in the time V between R_1 and the next excitation E_0. Following E_0 the waiting time W to the next response is assumed to be an exponential random variable with parameter λ.

Primary interest centers on the random variable S denoting the time between a response and the immediately preceding excitation. Obviously, S takes values only between zero and τ, and the event $(S \leq s)$ occurs if W falls between $k\tau$ and $k\tau + s$ for some value of k. Thus for $0 \leq s \leq \tau$,

$$P(S \leq s) = \sum_{k=0}^{\infty} P(k\tau \leq W \leq k\tau + s)$$

$$= \sum_{k=0}^{\infty} (1 - e^{-\lambda s})e^{-\lambda \tau k}$$

$$= \frac{1 - e^{-\lambda s}}{1 - e^{-\lambda \tau}} \qquad\qquad (9) \quad \blacktriangleright$$

PROBLEMS

1. Subjects arrive at an experimental lab at the average rate of ten per hour in accord with a Poisson process. Suppose that each subject has probability 0.20 of being selected to participate in the experiment. Let T denote the waiting time for the arrival of a suitable experimental subject. Find the mean and variance of T and $P(T \leq 2)$.

2. Let X be a random variable with density f_X given by

$$f_X(t) = \begin{cases} \dfrac{t^3}{4} & \text{when } 0 \leq t \leq 2 \\ 0 & \text{otherwise} \end{cases}$$

   (a) Find the number m such that X is equally likely to be greater than or less than m. (*Note:* such a number is called the *median value* of X.)

   (b) Find a number c such that $P(X > c) = 0.05$.

3. Let the density function of the random variable T be

$$f_T(t) = \begin{cases} 4t^2 e^{-2t} & t > 0 \\ 0 & t \leq 0 \end{cases}$$

   Let the events A_1, A_2, A_3, A_4 be defined by

$$A_1 = (T < 0)$$
$$A_2 = (T \geq 0)$$
$$A_3 = (0 \leq T \leq 2)$$
$$A_4 = (-1 \leq T \leq 0)$$

Find the probability of the following events:

(a) A_1' (b) $A_1' \cup A_2$ (c) $A_3 \cap A_2$

(d) $A_2' \cap A_1$ (e) $(A_1 \cup A_3)'$ (f) $(A_3 \cap A_4) \cap (A_1 \cup A_2)$

(*Hint:* $\lim_{t \to \infty} t^2 e^{-2t} = 0$.)

4. Let X be a random variable with uniform density

$$f_X(x) = \begin{cases} \dfrac{1}{\beta - \alpha} & \text{for } \alpha \le x \le \beta \\ 0 & \text{otherwise} \end{cases}$$

(a) Find the mean and standard deviation of X.

(b) Find the mean and standard deviation of $Y = aX + b$ in terms of those for X (where a and b are constants).

(c) Find the density function of $Z = -\ln X$ when $\alpha = 0$ and $\beta = 1$.

5. Let X be a continuous random variable with distribution function F_X. Define the random variable Y by $Y = F_X(X)$. Show that Y is uniform on the interval $[0, 1]$.

6. Let Z be a continuous random variable with density

$$f_Z(x) = \begin{cases} k_0 x & \text{when } 0 \le x \le 2 \\ 0 & \text{otherwise} \end{cases}$$

(a) Determine k_0.

(b) Find the expected value of $2Z^2 + 1$.

7. Let X be a beta variable (Example 3). Show that $E(X) = \alpha/(\alpha + \beta)$ and Var $(X) = \alpha\beta/(\alpha + \beta)^2(\alpha + \beta + 1)$. (*Hint:* Calculate the moments directly using the relation $\Gamma(\alpha + 1) = \alpha\Gamma(\alpha)$.)

8. Find the optimum inventory stock level in Example 5 when $c_1 = \$2$, $c_2 = \$3$, and demand is an exponential variable with expected value 3.

9. Suppose that on a certain political issue the position P preferred by individual voters is a random variable having the density function

$$f_P(x) = \frac{\lambda}{2} e^{-\lambda|x-m|} \qquad -\infty < x < \infty$$

where $\lambda > 0$ and m are constants. Denote a candidate's position by θ and suppose that if the candidate's position does not coincide with that of an individual voter, he suffers a loss

$$L = (P - \theta)^2$$

If the candidate wishes to minimize his expected loss, what platform position should he adopt? (*Hint:* Write $(P - \theta)^2 = [P - E(P) + E(P) - \theta]^2$.)

10. Let us assume that mortar rounds fall on a plane in accord with a Poisson distribution at an average rate of λ per unit area.* Let D denote the distance from an individual hit H to its nearest neighbor. Prove that $\pi \lambda D^2$ is an exponential random variable with expected value 1. (*Hint:* The distance D exceeds r if and only if there is no other hit in the circle of radius r centered at the hit H.)

11. In Problem 10, find the density function of the distance to the nearest neighbor in a Poisson distribution of points in three-dimensional space.

12. In Example 3 of Section 21.1, suppose implicit responses r_1 and r_2 occur at the respective rates of five and one per minute. What is the probability that the latency period exceeds 3 minutes?

13. The following indicates the *lack of memory* of the exponential distribution. Prove that if T is an exponential random variable then

$$P(T \leq t + s \mid T > s) = P(T \leq t)$$

Intuitively, if the waiting time T is exponential, then having waited s minutes, one has no assurances whatever about the remaining time. In fact, the remaining time has the *same* probability distribution as the original waiting time.†

14. Fill in the details of derivation of Equation (9) by showing that

$$P(k\tau \leq W \leq k\tau + s) = e^{-k\tau\lambda}(1 - e^{-\lambda s})$$

15. (a) Show that the moment generating function of the random variable S in Example 8 is

$$m_S(\theta) = E(e^{\theta S}) = \frac{\lambda}{\lambda - \theta} \frac{1 - e^{-(\lambda-\theta)\tau}}{1 - e^{-\lambda\tau}}$$

   (b) Use (a) to find the mean and variance of S.

16. In Example 8, show that the density function of the random variable V is

$$f_V(t) = \frac{\lambda e^{-\lambda(\tau-t)}}{1 - e^{-\lambda\tau}} \qquad 0 \leq t \leq \tau$$

   (*Hint:* First find the distribution function of V by noting that all S random variables have the same distribution function, as do all V random variables, and that the sum of one S and the *next* V is always τ.)

17. In Example 3 of Section 21.2 show that at equilibrium (that is, in the limit)

   (a) the expected number of customers in the system (including the one being processed) is $\lambda/(\mu - \lambda)$.

*That such an assumption is reasonable whenever the firing distance is large relative to the area considered is confirmed in the analysis of flying-bomb hits on London during World War II. See Clarke, R. D., "An Application of the Poisson Distribution," *Journal of the Institute of Actuaries* **72**, 48 (1946).

†It has been found that lengths of telephone calls within a city closely follow an exponential distribution. (Perhaps you can think of some reasons why.) Exponential laws also describe such random phenomena as the time interval between accidents such as explosions in mines, and the time intervals between successive breakdowns in an electronic system.

(b) the expected number of customers waiting and not being processed is $\lambda^2/[\mu(\mu - \lambda)]$.

(c) the expected service time is $1/\mu$.

(d) the expected waiting time (prior to processing) for a customer is $\lambda/[\mu(\mu - \lambda)]$.

(e) the expected time a customer spends in the system is $1/(\mu - \lambda)$.

18. In Problem 11 in Section 21.2 find, in the limiting case,

(a) the expected number of elements in the system.

(b) the mean length of the common waiting line, *excluding* the elements being processed.

(c) the average waiting time of an input before processing.

(d) the average time that an element spends in the system.

19. Four employment counselors are available to serve the clientele of a private employment agency. Clients arrive in accord with a Poisson process at a mean rate of 24 per 8-hour day. The time each interviewer spends with a client is exponentially distributed with an average interview length of 30 minutes. Clients are seated in a reception room and interviewed on a first-come, first-served basis. The agency manager wishes to know at equilibrium

(a) how long, on the average, a client is kept in the system.

(b) the average length of time a client waits before being interviewed.

(c) the average number of clients in the reception room.

(d) the probability that an interviewer is waiting for a client.

(e) the expected number of idle interviewers.

21.4 THE NORMAL DISTRIBUTION

A continuous random variable X which has a density function of the form

$$f(x) = \frac{1}{\sigma\sqrt{2\pi}} e^{-(x-\mu)^2/2\sigma^2} \qquad -\infty < x < \infty \tag{10}$$

is called a *normal random variable*. Here $\sigma > 0$ and μ are constants. The name "normal" arose in the nineteenth century when it was erroneously thought that most real world situations gave rise to these random variables.

A graph of the normal density function is shown in Figure 8. It is apparent from (10) that f is non-negative and we state without proof the fact that

$$\int_{-\infty}^{\infty} f(x)\, dx = \int_{-\infty}^{\infty} \frac{1}{\sigma\sqrt{2\pi}} e^{-(x-\mu)^2/2\sigma^2}\, dx = 1 \tag{11}$$

regardless of the values chosen for μ and σ.* Hence f is a legitimate density function.

*Except, of course, that σ must be positive. A proof of (11) may be found in Widder, D. V., *Advanced Calculus* (Prentice-Hall, Inc., Englewood Cliffs, N.J., 1961), 2nd ed., p. 371.

FIGURE 8

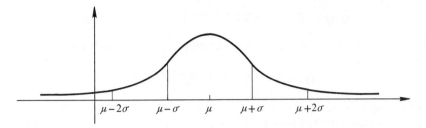

The moment generating function of a normal random variable X is

$$m_X(s) = E(e^{sX}) = \int_{-\infty}^{\infty} e^{sx} \frac{1}{\sigma\sqrt{2\pi}} e^{-(x-\mu)^2/2\sigma^2} \, dx$$

By multiplying out the term $-(x - \mu)^2/2\sigma^2$, combining all terms involving x, and then completing the square in the resulting quadratic, the exponential $e^{sx}e^{-(x-\mu)^2/2\sigma^2}$ may be rewritten as

$$e^{s\mu+\sigma^2 s^2/2} e^{-[x-(\mu+\sigma^2 s)]^2/2\sigma^2}$$

Hence

$$m_X(s) = e^{s\mu+\sigma^2 s^2/2} \int_{-\infty}^{\infty} \frac{1}{\sigma\sqrt{2\pi}} e^{-[x-(\mu+\sigma^2 s)]^2/2\sigma^2}$$

The integral has the *same form* as (11) with constants $(\mu + \sigma^2 s)$ and σ in place of μ and σ. It follows that its value is 1, and the moment generating function of X is

$$m_X(s) = e^{s\mu+\sigma^2 s^2/2} \tag{12}$$

Differentiating m_X gives

$$E(X) = m_X'(0) = (e^{s\mu+\sigma^2 s^2/2})(\mu + \sigma^2 s)\Big]_{s=0} = \mu$$

$$E(X^2) = m_X''(0) = (e^{s\mu+\sigma^2 s^2/2})(\sigma^2 + [\mu + \sigma^2 s]^2)\Big]_{s=0} = \sigma^2 + \mu^2$$

so that

$$\text{Var}\,(X) = E(X^2) - [E(X)]^2 = (\sigma^2 + \mu^2) - \mu^2 = \sigma^2$$

The parameters μ and σ in (10) thus represent the mean and standard deviation of the corresponding random variable.

Henceforth we shall indicate the fact that a random variable X is normal with expected value μ and variance σ^2, by writing X is $N(\mu, \sigma^2)$.

The *standardized random variable* corresponding to a random variable X having mean μ and variance σ^2 is

$$Z = \frac{X - \mu}{\sigma}$$

If X is $N(\mu, \sigma^2)$, the moment generating function of Z is

$$m_Z(s) = E(e^{sZ}) = E(e^{s(X-\mu)/\sigma}) = e^{-\mu s/\sigma} E(e^{sX/\sigma})$$

Since the last term is the generating function (12) of X with s replaced by s/σ, we have

$$m_Z(s) = e^{-\mu s/\sigma} e^{(s/\sigma)\mu + \sigma^2(s/\sigma)^2/2} = e^{s^2/2}$$

Comparing this with (12) we see that Z is a normal random variable with expected value 0 and variance 1. That is, if X is $N(\mu, \sigma^2)$, then $Z = (X - \mu)/\sigma$ is $N(0, 1)$.

Table 2 shows some values of the distribution function of Z.* The fact that the density function of Z is symmetric about zero means that

$$P(Z \leq -z) = P(Z \geq z)$$

for all real z (see Figure 9). For this reason negative values are not tabulated.

TABLE 2. Distribution function of a $N(0, 1)$ random variable

z	$P(Z \leq z)$	z	$P(Z \leq z)$	z	$P(Z \leq z)$
0	0.5000	1.0	0.8413	2.0	0.9772
0.1	0.5398	1.1	0.8643	2.1	0.9821
0.2	0.5793	1.2	0.8849	2.2	0.9861
0.3	0.6179	1.3	0.9032	2.3	0.9893
0.4	0.6554	1.4	0.9192	2.4	0.9918
0.5	0.6915	1.5	0.9332	2.5	0.9938
0.6	0.7257	1.6	0.9452	2.6	0.9953
0.7	0.7580	1.7	0.9554	2.7	0.9965
0.8	0.7881	1.8	0.9641	2.8	0.9974
0.9	0.8159	1.9	0.9713	2.9	0.9981
				3.0	0.9987

Moreover, since

$$P(Z \leq z) = P\left(\frac{X - \mu}{\sigma} \leq z\right) = P(X \leq \mu + \sigma z)$$

*More extensive tables are available. See, for example, *Tables of the Error Function and its Derivative* (National Bureau of Standards Applied Mathematics Series 41, Washington, D.C., 1954), or almost any text on statistics.

FIGURE 9

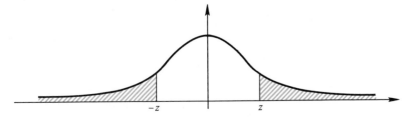

the distribution function of any normal random variable can be determined from that of a standardized variable. For example, if X is $N(2, 9)$, then

$$P(X \leq 5) = P\left(\frac{X - 2}{3} \leq \frac{5 - 2}{3}\right) = P(Z \leq 1) = 0.8413$$

$$P(X \leq 3) = P\left(\frac{X - 2}{3} \leq \frac{3 - 2}{3}\right) = P(Z \leq \tfrac{1}{3})$$

which, interpolating in the table, is approximately 0.6304, and

$$
\begin{aligned}
P(-5.5 \leq X \leq 8) &= P(X \leq 8) - P(X \leq -5.5) \\
&= P(Z \leq 2) - P(Z \leq -2.5) \\
&= P(Z \leq 2) - P(Z \geq 2.5) \\
&= 0.9972 - (1 - 0.9938) \\
&= 0.9910.
\end{aligned}
$$

The probability $P(-k\sigma \leq X - \mu \leq k\sigma)$ that a normal random X takes a value within k standard deviations of its expected value is the same as the probability $P(-k \leq (X - \mu)/\sigma \leq k) = P(-k \leq Z \leq k)$ that Z falls within k units of zero. For $k = 1$, 2, and 3, these probabilities are, respectively, 0.682, 0.954, and 0.998 (see Figure 10).

FIGURE 10

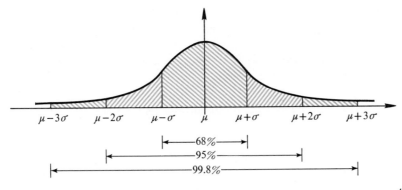

If $X_1, X_2, \ldots, X_n$ are independent normal random variables with respective means $\mu_1, \mu_2, \ldots, \mu_n$ and variances $\sigma_1^2, \sigma_2^2, \ldots, \sigma_n^2$, the moment generating function of the sum

$$Y = a_1 X_1 + a_2 X_2 + \cdots + a_n X_n$$

is

$$m_Y(s) = E(e^{sY}) = E(e^{s(a_1 X_1 + \cdots + a_n X_n)})$$
$$= E(e^{a_1 s X_1} e^{a_2 s X_2} \cdots e^{a_n s X_n})$$

Since the X_n are independent, this may be written as the product of the individual generating functions

$$m_Y(s) = \prod_{i=1}^n E(e^{a_i s X_i}) = \prod_{i=1}^n [e^{a_i \mu_i s + (a_i^2 \sigma_i^2 s^2 / 2)}]$$
$$= \exp\left(s \sum_{i=1}^n a_i \mu_i + \frac{s^2}{2} \sum_{i=1}^n a_i^2 \sigma_i^2\right) \tag{13}$$

(The notation exp (x) means the same thing as e^x.) Thus Y is normal with mean $\sum_{i=1}^n a_i \mu_i$ and variance $\sum_{i=1}^n a_i^2 \sigma_i^2$. In words, any linear combination of independent normal random variables is another normal random variable.

A particularly important case arises if

$$Y = \frac{1}{n}(X_1 + X_2 + \cdots + X_n)$$

is the *average* of n independent normal random variables, each having the same mean μ and variance σ^2. In statistical terms, Y is the mean of a random sample of n observations taken from a normal population $N(\mu, \sigma^2)$. Putting $a_i = 1/n$ in (13), we see that Y itself is normal with mean μ and variance σ^2/n. That is, Y has the same expected value as each of the X_n, but a smaller variance. The fact that Var $(Y) = \sigma^2/n$ goes to zero as $n \to \infty$ provides a basis for the statistical technique of sampling to obtain estimates of population parameters, in this case the mean μ.

Example 1   Let us assume that the amount A of breakfast cereal a filling machine puts into a box is a normal random variable with mean μ and variance σ^2. From time to time the machine may get out of adjustment, in which case the mean amount changes although the variance remains fairly constant. The "net weight" printed on the box constitutes a lower critical level (LCL) of output such that the box is accepted if $A \geq LCL$ and rejected if $A < LCL$.

The process is considered to be "in control" if $P(A < LCL) \leq 0.01$ and out of control otherwise. Since

$$P(A < LCL) = P\left(\frac{A - \mu}{\sigma} < \frac{LCL - \mu}{\sigma}\right) = P\left(Z < \frac{LCL - \mu}{\sigma}\right)$$

we see from Table 2 that the system is in control whenever

$$\frac{LCL - \mu}{\sigma} \leq -2.33$$

or, equivalently, when

$$\mu \geq LCL + 2.33\sigma$$

When the process is found to be out of control, the machine is stopped and adjusted to bring it back within control limits.

Since μ varies and is unknown, the company is forced to draw inferences about the process and to take appropriate action on the basis of a sample of n boxes chosen from the current output. Naturally, because of random fluctuations, the sample will only approximate the actual value of μ and there are two errors the company may make, either of which may be costly:

(I) Decide the process is out of control when it is actually in control.

(II) Decide the process is in control when it is out.

The discussion throughout this example follows the general ideas involved in statistical testing of hypotheses, in this case the hypothesis that the process is in control.

Let $X_1, X_2, \ldots, X_n$ be the weights of the cereal in the n boxes chosen and let

$$\overline{X} = \frac{1}{n}(X_1 + X_2 + \cdots + X_n)$$

be the average of these weights. Then the company reasons as follows:

For a proper decision policy there should be a number c such that the process is assumed to be in control when $\overline{X} \geq c$ and out of control when $\overline{X} < c$. Because of possible sampling errors, there will be a positive probability α of making an error of type I. We shall arbitrarily allow $\alpha = 0.10$ and choose c so that

$$P(\overline{X} < c \mid \mu = LCL + 2.33\sigma) = \alpha$$

To take a specific example of the effect of the company's policy, suppose $LCL = 16$ ounces, $\sigma = 0.1$ ounce, and take $n = 16$ items in a sample. Then $\overline{X}$ is $N(\mu, 0.01/16)$ and

$$P(\overline{X} < c \mid \mu = 16 + 2.33(0.1)) = 0.10$$

means that c must be 16.201.

These values of c, n, σ, and LCL give, corresponding to each value of μ, probabilities of making errors of either type I or type II, as shown in the graphs of Figure 11. (Note carefully the differences in scale on the two graphs.).

Figure 11(a) shows that the probability of a type I error decreases as μ increases and that $\alpha = 0.10$ is, in effect, the maximum probability of such an error. This maximum probability is usually called the *significance level* of

FIGURE 11 **(a) Probability of an error of type I for various values of μ;**
(b) Power curve showing the probability of not making an error of type II for various values of μ.

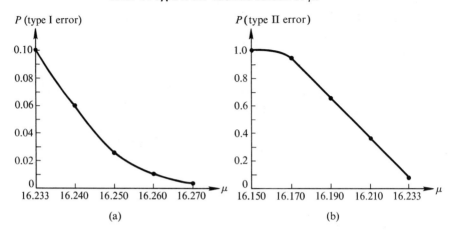

(a)   (b)

the statistical test procedure. Of course, for $\mu < 16.233$, a type I error is impossible.

For a given value of μ, the probability of *not* making a type II error is called the *power* of the test. Figure 11(b) shows the *power curve* of the test procedure. In this case the power decreases as μ increases since as $\mu \to 16.233$, it becomes increasingly more difficult to make the correct decision. For $\mu > 16.233$, no error of type II is possible.

Note that at 16.233, the power is equal to the significance level α. This illustrates a phenomenon typical of statistical tests, that in order to improve (lower) the significance level one must at the same time reduce the power of the test. Usually, the only way to improve both quantities is to increase the size of the sample. ▶

We have seen that a sum of independent normal random variables is itself a normal random variable. This is not a particularly surprising result. However, there is a very surprising result, called the *Central Limit Theorem*, which states that if $X_1, X_2, \ldots, X_n$ are *any* independent random variables having a common distribution function, then for *large* values of n the sums

$$T = X_1 + X_2 + \cdots + X_n$$

$$Y = \frac{1}{n}(X_1 + X_2 + \cdots + X_n)$$

have distribution functions which can be closely approximated by distribution functions of normal random variables. This provides perhaps the most important use of the normal distribution in applications. In practice, n is "large" if it exceeds 30.

A proof of the Central Limit Theorem is rather lengthy so we will content ourselves with a precise statement of the theorem and an example of its use.

Theorem 2   *The Central Limit Theorem*   Let $X_1, X_2, X_3, \ldots$ be a sequence of independent random variables all having the same distribution function and common mean μ and variance σ^2. Define a new sequence of standardized random variables by

$$T_1 = \frac{X_1 - \mu}{\sigma} \qquad T_2 = \frac{X_1 + X_2 - 2\mu}{\sqrt{2}\,\sigma}, \ldots$$

and, in general,

$$T_n = \frac{X_1 + X_2 + \cdots + X_n - n\mu}{\sqrt{n}\,\sigma}$$

Let $F_{T_1}, F_{T_2}, F_{T_3}, \ldots$ be the sequence of distribution functions of the T_n and let Φ denote the distribution function of the standardized normal random variable Z. Then, for each real number x,

$$\lim_{n\to\infty} F_{T_n}(x) = \Phi(x) \qquad\qquad \blacktriangleright$$

In words, the sequence of probabilities $P(T_n \le x)$ that the random variables $T_1, T_2, \ldots$ are less than or equal to x, converges to the number

$$\Phi(x) = P(Z \le x) = \int_{-\infty}^{x} \frac{1}{\sqrt{2\pi}} e^{-y^2/2}\, dy$$

Example 2   A community concert series offers the same program on two successive evenings. Suppose n patrons independently and at random decide which performance they will attend. Then S_n, the number in attendance at the first performance, is the outcome of n Bernoulli trials with probability $\frac{1}{2}$ of "success" (that is, choosing the first evening). We may write $S_n = X_1 + X_2 + \cdots + X_n$ where for $1 \le i \le n$, X_i denotes the number of successes (0 or 1) on the ith trial. The X_i are independent random variables each having the same probability function

$$P(X_i = 0) = P(X_i = 1) = \tfrac{1}{2}$$

It follows that $E(S_n) = n/2$, Var $(S_n) = n/4$ and the Central Limit Theorem states that S_n has a distribution function which is approximately the same as that of a $N(n/2, n/4)$ random variable.

Assuming the auditorium can seat $a < n$ people, the probability $\beta(n)$ that on one of the two nights more patrons attend than can be seated, is given approximately by the integral

$$\beta(n) \approx 1 - \int_{-a}^{a} \frac{1}{\sqrt{n/4}\,\sqrt{2\pi}} e^{-(1/2)(4/n)(x-n/2)^2}\, dx \qquad\qquad \textbf{(14)}$$

The series directors wish to determine the largest number of tickets which can be sold and still keep $\beta(n)$ below some arbitrary risk level α. Making the change of variable $z = (4/n)^{1/2}(x - n/2)$ and writing

$$z_\alpha = (4/n)^{1/2}(a - n/2) \qquad\qquad \textbf{(15)}$$

in (14) gives

$$\beta(n) \approx 1 - \int_{-z_\alpha}^{z_\alpha} \frac{1}{\sqrt{2\pi}} e^{-(1/2)z^2} \, dz$$

If, for example, $a = 6000$ and we want $\beta(n) \leq \alpha = 0.01$, then $z_\alpha = 2.58$. Solving for n in (15), we find $n = 11,718$ as the maximum number of tickets which can be sold to ensure that the seating capacity of the auditorium will be sufficient 99% of the time. ▶

PROBLEMS

1. If X is $N(1, 10)$, use Table 2 to find $P(X > 11)$ and $P(-4 < X < 5)$.

2. Evaluate

$$\int_0^{1.061} e^{-x^2} \, dx$$

3. Suppose X is $N(0, 1)$ and let f_X denote the density function of X. Show that

$$E(X \mid a < X < b) = \frac{f_X(a) - f_X(b)}{P(a < X < b)}$$

4. Determine k so that

$$f(x) = ke^{-x(x-1)} \qquad -\infty < x < \infty$$

is a density function. If X is any random variable having this density, find $E(X)$ and Var (X).

5. (a) In Example 2, suppose 11,900 tickets are sold. What is the probability that attendance exceeds seating capacity on at least one of the two nights?

   (b) What is the probability that capacity is exceeded on the first night?

6. Same as Problem 5 when 12,100 tickets are sold.

7. Show that if X is $N(0, \sigma^2)$, then $Y = X^2$ is a *gamma* random variable with density

$$g(y) = \frac{e^{-y/2\sigma^2} y^{-1/2}}{2^{1/2}\sigma\sqrt{\pi}} \qquad 0 < y < \infty$$

   (*Hint:* First write the distribution function of Y and then differentiate.)

8. Control limits are established on a process with a standard deviation of 100 such that the process is designated "out of control" when the proportion of output greater than 500, the upper critical level, exceeds 0.025. Samples of size 16 are taken and evaluated to determine whether the process is out of control.

   (a) Draw the power curve for this decision scheme.

   (b) Draw the power curve for the case where samples of 36 observations are drawn. Compare with (a).

Section 1. (a) $|2(-1)| = |-2| = 2 = 2 \cdot 1 = |2| \cdot |-1|$. (b) $\left|\dfrac{2}{-1}\right| = |-2| = 2 = \dfrac{2}{1} =$

4.2 $\dfrac{|2|}{|-1|}$. (c) $|2| = 2 = |-2|$. (d) $|2 + (-1)| = |1| = 1$, while $|2| + |-1| =$

$2 + 1 = 3$. (e) $|2 - (-1)| = |3| = 3$, while $|2| - |-1| = 2 - 1 = 1$.
5. In (a), $|a| + |b| > |a + b|$; in (b)$-$(e), $|a| + |b| = |a + b|$. **7.** $|a - b|^2$
$= (a - b)^2 = a^2 - 2ab + b^2$ while $||a| - |b||^2 = a^2 - 2|a||b| + b^2$. Thus,
$|a - b| \geq ||a| - |b||$ is equivalent to $-2ab \geq -2|a||b|$, or to $ab \leq |ab|$. This is
true, by Properties 3 and 10. **10.** $\{x: -3 < x < 9\}$. **12.** $\{x: -8 \leq x \leq 5\}$.
14. Partial answer: In 10, $\{x: |x - 3| < 6\}$ is the set of points x whose distance
from the point 3 is less than 6 units.

Section 1. (a) 3, 3. (b) 1, 1. (c) lub $A = 2$, there is no maximum. **3.** If m and t
4.3 are both maxima of the set A, then the facts that $t \in A$ and $m \geq x$ for all $x \in A$
together mean $m \geq t$. Similarly, $t \geq m$ and thus $t = m$. **6.** b is not a lower
bound for A if there is an element $x \in A$ such that $x < b$. **7.** Follow the
argument in Problem 3. **9.** (a) glb $A = \frac{2}{3}$; lub $A = 1$. (b) $\frac{999}{1000}$.

Chapter 5

Section 1. \$115. **3.** Neither, since $a_2 < a_1$ and $a_3 > a_2$. **5.** Partial answer: Yes.
5.1 **6.** (a) $\frac{1}{3}, \frac{1}{3}, \frac{3}{11}, \frac{2}{9}$. (c) $2, \frac{5}{4}, \frac{10}{9}, \frac{17}{16}$. **7.** (a) Bounded. Lower bound 0. Upper
bound 1. Alternatively, lower bound -10, upper bound $\frac{1}{3}$, and so forth. (c)

Bounded. Lower bound 0. Upper bound 2. **8.** (a) $a_n = \dfrac{n^2 + 9}{(n + 2)^2}$. (d) $a_n =$

$4n - \sqrt{n^2 + 4}$. **10.** (b) \$106,090. (d) $\$100,000 \, (1.03)^{n-1}$. The sequence is
bounded below by $S_1 = \$100,000$, but is not bounded above. **14.** If $|p_n| < M$
for all n, then $-M \leq p_n \leq M$ for all n. Hence M is an upper bound and $-M$ a
lower bound for the sequence.

Section 4. (b) Approximately 70. **13.** In Problem 12, let $a = 1$. Then $2^n \geq 1 + n$ is
5.2 true for all n, from which it follows that $2^n > n$ for all n. **18.** To obtain the
result of Problem 8, let $a = 1$ and $r = 3$; for Problem 10, choose $a = r = 2$; for
Formula (4), let $a = \beta$ and $r = \alpha$.

Section 2. (b) $(1, 2, 3, 4, \ldots)$. (c) $(1, -1, 1, -1, 1, -1, \ldots)$. **4.** The sequence is
5.3 not bounded, hence it does not converge. **8.** The limit is 4. Problem 7 applies.
10. (a) 3. (b) No limit. (e) -3. (g) $\frac{3}{4}$. **15.** Let $A = (1, -1, 1, -1,$
$1, -1, \ldots)$ and $B = (-1, 1, -1, 1, -1, 1, \ldots)$.

Section 2. $\dfrac{1}{1 - a}$. **3.** For $a \leq 1 - r$ and $r < 1$, $\lim\limits_{n \to \infty} (I_n) = 1 - \dfrac{a}{1 - r}$. **6.** $a_n =$
5.4

$a_1 c^{n-1} + k \left(\dfrac{1 - c^{n-1}}{1 - c}\right)$. Use induction. (b) $k/(1 - c)$. (c) Let $c = 1 - \alpha$

and $k = \alpha$. Then $\lim\limits_{t \to \infty} (p_t) = 1$ when $0 < \alpha < 2$. **7.** This follows from the
contrapositive of Property 3. **9.** Parallel the proof of Property 5. **11.** 0.
12. 0. Use Problem 13, Section 5.2. **13.** If $a + b = 0$, $p_n = p_0$ for all n. If

$0 < a + b < 2$, $\lim\limits_{n \to \infty} (p_n) = \dfrac{a}{a + b}$. Otherwise, (p_n) diverges. **14.** Use Example
3 of Section 5.4 with $a = 1 - b$.

Chapter 6

1. (a) 5.　(c) 10.　**3.** (a) $\sum_{k=0}^{5} \frac{t^k}{2^k}$.　(c) $\sum_{k=0}^{4} (a + kd)$.　(e) $\sum_{i=1}^{3} (x_i - \bar{x})^2$.

4. (a) $1 + 2 + 2^2 + 2^3 + 2^4 + 2^5 = 63$.　(c) $k(-2) + k(-1) + k(0) + k(1)$ $= -2k$.　**6.** All sums equal 68.　**7.** $\sum_{k=1}^{12} 2k = 2(78) = 156$.　**9.** \$124,622.73.

11. Apply Problem 11, Section 5.2 with $a = 1, r = n$, and $n = L$.　**13.** $\sum_{i=1}^{n} (x_i - \bar{x})$ $= \sum_{i=1}^{n} x_i - \sum_{i=1}^{n} \bar{x} = n\bar{x} - n\bar{x} = 0$. (Use Rules 1 and 4 for sums.)　**15.** (a) \$5301.95.

(b) \$2246.27.　(c) $\dfrac{400(1.04)^n}{(1.04)^n - 1}$.

1. (a) Approximately 1.05.　**3.** $\frac{1}{2} + \frac{2}{5} + \frac{3}{10} + \frac{4}{17}$.　**5.** $0 + \frac{2}{3} + \frac{16}{9} + \frac{8}{3}$.
9. $0 + \frac{1}{2} - \frac{1}{2} + \frac{3}{8}$.　**11.** $1, \frac{5}{2}, \frac{15}{4}, \frac{37}{8}$.　**13.** $1, \frac{5}{4}, \frac{49}{36}, \frac{205}{144}$.　**15.** (a) $\frac{4}{3}$.
(b) $1/(3 \cdot 4^{a-1})$.　(c) $\frac{3}{5}$.　(d) Series diverges (Why?).　**18.** \$400 per year,

forever; an infinite amount.　**27.** $(1 - c) \sum_{\pi=0}^{\infty} c^\pi y_\pi \leq (1 - c) \sum_{\pi=0}^{\infty} c^\pi Y =$

$(1 - c)Y \sum_{\pi=0}^{\infty} c^\pi = (1 - c)Y \dfrac{1}{1 - c} = Y$.

1. $0.202020 \ldots = \sum_{k=1}^{\infty} \dfrac{2}{3^{2k-1}}$.　**2.** $2202.1111 \ldots, 244.222 \ldots$.　**3.** (b)

$-11010.0100 = -26.25$, the difference of $86\frac{7}{8}$ and $113\frac{1}{8}$.　(c) 11110001101.11 $= 1933.75$, the product of 110.5 and 17.5.　(e) $11.000 \approx 3.0$, the quotient of $\frac{219}{16}$ and $\frac{9}{2}$.　**4.** $0.0100110011001 \ldots + 0.011001100110 \ldots = .10110011001100 \ldots$ $= \frac{7}{10}$.　**5.** (a) 101111100000001 (Binary) and 57401 (Octal).
(d) 1000011100001.000100011 (Binary) and 10341.043 (Octal).　**7.** (a) 421323.
(c) 53333.　(g) 13.770.　**8.** (b) 71 $\delta \alpha \cdot \gamma$.　(d) $.6666 \ldots$.　**10.** 65535.

1. Divergent.　**3.** Convergent.　**5.** Convergent.　**7.** Convergent.　**9.** Convergent.
11. (c) $1/(1 - c + cM)$.　**14.** Converges for $-1 \leq x \leq 1$. (Ratio test fails when $x = \pm 1$.)　**16.** Ratio test fails, but comparison test shows convergence.
18. Alternating series test shows convergence. Second series is a divergent hyperharmonic series.　**19.** Diverges. Sequence of terms does not converge to zero.
21. Converges absolutely. Ratio test applies.　**23.** (b) $d(\lambda - 1)/(1 - \beta\lambda)$.
24. (a) $e_k = (-1)^{k-1}(d/k!)$.　(c) $d/9! = d/362,880$.

1. (a) and (c) $x_{11} + x_{12} + x_{13} + x_{14} + x_{21} + x_{22} + x_{23} + x_{24}$.　**3.** $\frac{4}{7}$.
4. If $|x| < 1$, both sums have the value $1/(1 - x)(1 - x^2)$.

Chapter 7

2. 6; 40,320; 39,916,800.　**4.** 20, 495, 72.　**6.** 480.　**7.** 13.　**9.** 44.
10. (a) 0.　(d) 109.　**11.** (a) 2.　(c) 16.　**13.** $2^{10} = 1024$.　**14.** (b) After four elections, Democrats 25,488,000 and Republicans, 34,512,000.　**15.** $\frac{2}{3}$.

1. (b) 64, 24, 20, 4.　**3.** 1000, 720.　**6.** 120, 10.　**7.** 1, n, 1.　**9.** They are equal.　**14.** 120.　**15.** (b) 60.　**16.** (a) $\binom{20}{5} = 15504$ if sampling is without replacement.　(b) About 20.7%.　**19.** (a) 1260.　(b) 156.　**20.** A $\frac{1}{2}$; B, C, and D $\frac{1}{6}$ each.　**23.** (a) 4, 7, 11.　(b) $\binom{n}{2} + 1$.

Section 7.3

1. $(1 + t)^4 = 1 + 4t + 6t^2 + 4t^3 + t^4$
$(s^2 - 4s)^3 = s^6 - 12s^5 + 48s^4 - 64s^3$.

2. \$2650. 5. (a) $\binom{10}{3} a^7 y^3 = 120a^7 y^3$. (c) $\binom{10}{6} x^6 y^{24} = 210x^6 y^{24}$.

(e) $2x^{12}y^2$. (g) 0.7847. (i) $256y^8 - 1024y^6 + 1792y^4 - 1792y^2 + 1120 - 448y^{-2} + 112y^{-4} - 16y^{-6} + y^{-8}$. 11. Put $a = b = k = n$ in Equation (5).
13. See hint to Problem 12. You should encounter the binomial expansion of $(1 - 1)^{n-1}$. 15. $\binom{10}{3, 4, 3} = 4200$.

Chapter 8

Section 8.1

1. $2 \times 3, 2 \times 1, 1 \times 3, 3 \times 5$. 2. $\binom{n}{2} = \dfrac{n(n - 1)}{2}$. 4. (a) $\begin{pmatrix} 3 & 6 \\ 4 & 7 \end{pmatrix}$.

8. $A + B = \begin{pmatrix} 1 & -1 \\ 5 & 2 \end{pmatrix} = B + A, (A + B) + C = \begin{pmatrix} 2 & 6 \\ 7 & -7 \end{pmatrix} = A + (B + C)$,

$3A - 4B = \begin{pmatrix} 10 & 4 \\ -13 & 13 \end{pmatrix}, (A + B)' = \begin{pmatrix} 1 & 5 \\ -1 & 2 \end{pmatrix} = A' + B', 4A' = \begin{pmatrix} 8 & 4 \\ 0 & 12 \end{pmatrix} = (4A)'$. 11. Already proved for two matrices. Use induction. 15. $A - B = A + (-B)$. Use Problem 11. 17. tr $(A + B) = \sum_{i=1}^{n} (A + B)_{ii} = \sum_{i=1}^{n} (a_{ii} + b_{ii}) = \sum_{i=1}^{n} a_{ii} + \sum_{i=1}^{n} b_{ii} = $ tr $A + $ tr B.

Section 8.2

1. $|V_1| = \sqrt{98}, |V_2| = \sqrt{5}, |V_3| = \sqrt{21}$. 2. (a) 5. (b) 42. (c) 0. 3. \$5870.

11. (a) I. (e) $\begin{pmatrix} 0 & 1 \\ -1 & 0 \end{pmatrix}$. (i) $\begin{pmatrix} 0 & -1 \\ -1 & 0 \end{pmatrix}$. 14. $\begin{pmatrix} 1 & -1 & -1 \\ -b & 1 & 0 \\ -v & 0 & 1 \end{pmatrix}\begin{pmatrix} Y \\ C \\ I \end{pmatrix} = \begin{pmatrix} 0 \\ a \\ u \end{pmatrix}$.

18. Partial answer: $d_{12} = \sqrt{18}$ and $d_{14} = \sqrt{106}$.

Section 8.3

1. $A^2 = \begin{pmatrix} 3 & -2 \\ -4 & 3 \end{pmatrix}$ $A^3 = \begin{pmatrix} 7 & -5 \\ -10 & 7 \end{pmatrix}$ $A^4 = \begin{pmatrix} 17 & -12 \\ -24 & 17 \end{pmatrix}$. 6. (a) If $D = $ diag $(d_1, d_2, \ldots, d_n)$ and $C = $ diag $(c_1, c_2, \ldots, c_n)$, then $CD = DC = $ diag $(c_1d_1, c_2d_2, \ldots, c_nd_n)$. (b) Use induction. 8. Let $I_3 = (1, 1, 1)$. Then $\sum_{i=1}^{3} d_i = I_3 D I_3'$. 9. $BA + 2C'$. 12. (a) $\begin{pmatrix} 1 & 0 \\ \frac{1}{4}(1 + \frac{1}{3}) & (\frac{1}{3})^2 \end{pmatrix}$

(d) $\begin{pmatrix} 1 & 0 \\ \frac{3}{8}[1 - (\frac{1}{3})^n] & (\frac{1}{3})^n \end{pmatrix}$ (e) $\begin{pmatrix} 1 & 0 \\ \frac{3}{8} & 0 \end{pmatrix}$.

Section 8.4

3. Partial answer: The ith diagonal entry is the number of persons who choose person i. 5. col (5.2, 0.4, 0.2).

Chapter 9

Section 9.1

1. (a) $x = \frac{25}{4}, y = \frac{27}{4}, z = \frac{9}{2}$. (b) Inconsistent. (c) If $y = b$, then $x = 13 - b$ and $z = 2b - 9$; b may be any real number. 3. $x = 5, y = \frac{1}{3}$. 5. $x_3 = 0$, $x_1 = -2x_2$, where x_2 can be any real number. 7. $x = \frac{14}{9} - \frac{1}{9}z, y = -\frac{16}{9} + \frac{14}{9}z, z$ is arbitrary. 9. $x = y = z = 0$. 11. 2 type I, 6 type II, 2 type III.
12. The interior equilibrium is not unique. Any output vector which is a positive multiple of the vector (7, 25, 6) is an interior equilibrium.

Section
9.2

2. EB is the matrix obtained from B by multiplying the first row by 3.
4. $\begin{pmatrix} 1 & 0 & 0 \\ 1 & 1 & 0 \\ -2 & 0 & 1 \end{pmatrix}, \begin{pmatrix} 1 & -2 & 0 \\ 0 & -1 & 0 \\ 0 & 6 & 1 \end{pmatrix}, \begin{pmatrix} 1 & 0 & \frac{5}{17} \\ 0 & 1 & \frac{4}{17} \\ 0 & 0 & \frac{1}{17} \end{pmatrix}.$ **7.** No. **9.** Yes. **10.** No.

Section
9.3

3. If $A = 0$, then $AB = 0$ for all B. Hence, there is no matrix B such that $AB = I$.
6. Inverse: $\begin{pmatrix} \frac{1}{3} & \frac{1}{3} \\ \frac{2}{9} & -\frac{1}{9} \end{pmatrix}$ Solution: $x_1 = \frac{10}{3}$, $x_2 = \frac{2}{9}$. **8.** No inverse, no solution.
10. No inverse. Solution: $x = 10z - 6$, $y = -\frac{13}{2}z + \frac{7}{2}$, where z may be any real number. **16.** (a) The corresponding rows are interchanged in the inverse. (c) In the inverse the corresponding row is divided by the same constant.

Section
9.4

7. Solution set empty. **13.** (b) $(0, 0)$ is the only interior equilibrium.
14. Only at $(0, 0)$. Otherwise, no. **15.** $a_{21}a_{12} \leq 1$.

Section
9.5

2. $144, $216, and $0. **3.** (a) In Problem 1, corner points $(0, 0)$, $(0, 3)$, and $(9, 0)$ yield respective functional values 0 (minimum), 18 (maximum), and 18 (maximum). **4.** The maximum value is $17,000 obtained at corner point $(16, 1)$.
6. (a) The profit function $P = 8x + 12y$ attains the maximum value $220 at every point on the line segment between corner points $(\frac{15}{2}, \frac{40}{3})$ and $(\frac{55}{2}, 0)$. (b) The inequality $\frac{5}{2}x + 3y \leq 100$ pertaining to Machine II. **8.** $R = 12,000$ when $P = 2000$, $F = 2000$, $T = 0$. **9.** There are six corner points, in terms of a, c, and h, respectively: $(0, 0, 0)$, $(0, 0, 15)$, $(0, 20, 0)$, $(30, 0, 0)$, $(\frac{45}{4}, 0, 15)$, and $(10, 20, 0)$. The maximum net return is $10,000 from 10 acres of cotton, 20 cows, and no hens.

Chapter 10

Section
10.1

1. 26. **3.** 28. **5.** -27. **7.** $Y = \dfrac{a + I_0 + G_0}{1 - b}$, $C = \dfrac{a + bI_0 + bG_0}{1 - b}$.
9. For instance, col $(-1, 14, 9)$ and col $(\frac{1}{9}, -\frac{14}{9}, -1)$. **12.** If row i is c times row j, subtract c times row j from row i. Apply Theorem 1.

Section
10.2

3. Roots 0, 8; corresponding vectors (c, c) and $(c, -c)$, where c is arbitrary.
5. Roots 15, -3; corresponding vectors $(c, 2c)$ and $(c, -c)$, where c is arbitrary.
7. Roots 1, 2, 3; corresponding vectors $(7c, c, -c)$, $(c, 0, 0)$, and (c, c, c), where c is arbitrary. **9.** Roots 3 and -3 (each a double root); corresponding vectors $(d, 2c + 2d, c, d)$ and $(-2c - d, -2c, c, d)$, where c and d are arbitrary. **10.** (a) 1 and 2. (b) a and b. **13.** Roots $\frac{1}{2}(a + d + \sqrt{(a - d)^2 + 4bc})$ and $\frac{1}{2}(a + d - \sqrt{(a - d)^2 + 4bc})$. (a) $(a - d)^2 + 4bc > 0$. (b) $(a - d)^2 + 4bc = 0$. (c) $(a - d)^2 + 4bc < 0$.
15. $F^* = \begin{pmatrix} .2232 & .4098 & .5964 & .9062 & .9296 \\ .0134 & -.1098 & -.2330 & -.1696 & -.3928 \end{pmatrix}'$. **16.** (a) 2, 2, 3, and 2. (b) 0, 3.

Section
10.3

3. $-2u^2 + 6uv - 10v^2$. **5.** $2y_1y_2 + 4y_1y_3 + 6y_1y_4 + 2y_2y_3 + 4y_2y_4 + 2y_3y_4$. **7.** $(x \ \ y) \begin{pmatrix} 3 & 8 \\ 8 & -2 \end{pmatrix} \begin{pmatrix} x \\ y \end{pmatrix}$. **9.** $(x \ \ y \ \ z) \begin{pmatrix} 1 & 1 & 1 \\ 1 & 2 & 0 \\ 1 & 0 & 1 \end{pmatrix} \begin{pmatrix} x \\ y \\ z \end{pmatrix}$.

11. $D = \text{diag } (0, 2)$, $Q = \dfrac{1}{\sqrt{2}} \begin{pmatrix} 1 & 1 \\ 1 & -1 \end{pmatrix}$. **15.** $D = \text{diag } (0, 0, 9)$, $Q =$
$\begin{pmatrix} 1/\sqrt{2} & 1/\sqrt{18} & \frac{2}{3} \\ -1/\sqrt{2} & 1/\sqrt{18} & \frac{2}{3} \\ 0 & 4/\sqrt{18} & -\frac{1}{3} \end{pmatrix}$.

17. (a) $D = \text{diag } (2, 0.32, 0.68)$, $C = \begin{pmatrix} -5/\sqrt{66} & 1/\sqrt{2} & 2/\sqrt{33} \\ 4/\sqrt{66} & 0 & 5/\sqrt{33} \\ 5/\sqrt{66} & 1/\sqrt{2} & -2/\sqrt{33} \end{pmatrix}$,

$F = \begin{pmatrix} -0.87 & 0.40 & 0.29 \\ 0.70 & 0 & 0.72 \\ 0.87 & 0.40 & 0.29 \end{pmatrix}$.

18. #2 is positive definite, the rest are neither.   **21.** $(P'P)' = P'(P')' = P'P$.
22. Yes, the transpose must have the same dimensions as the original matrix.

Chapter 11

Section 11.1 **4.** If $f(x) = ax + b$, then $8 = f(2) = 2a + b$ and $12 = f(4) = 4a + b$. Solving for a and b gives $a = 2$, $b = 4$, and $f(x) = 2x + 4$.   **6.** (a) $x = -1$ and $x = -2$.   **7.** (a) $f(n^2) = \dfrac{n^2 r}{1 + (n^2 - 1)r}$. Simply replace n by n^2 in $f(n)$.
(b) $nr^2/(1 + nr - 2r + r^2)$. Replace n by $f(n)$ in $f(n)$.   (c) $g(n) = (1 - r)/r$ for all n.

Section 11.2 **1.** Only (c); (a) and (b) contain their limit 2 while in (d) the limit is not 2.   **2.** 5.
3. 2.   **5.** -1.   **7.** -4.   **9.** $-\frac{7}{2}$.   **11.** Only (d) and (e) are correct.
12. Marginal cost at x is $\lim\limits_{y \to x} \dfrac{C(y) - C(x)}{y - x} = \lim\limits_{y \to x} \dfrac{10 + 2y - (10 + 2x)}{y - x} = 2$.
14. (a) $(f + g)(x) = 6x - 4$. Domain and range both the entire set of real numbers.   (c) $(f \cdot g)(x) = (2x - 6)(4x + 2)$. Domain is the set of all real numbers. Range is the set of real numbers greater than or equal to $-\frac{49}{2}$.
(d) $(f/g)(x) = \dfrac{2x - 6}{4x + 2}$. Domain is the set of all numbers except $x = -\frac{1}{2}$. Range is the set of all numbers except $\frac{1}{2}$.

Section 11.3 **1.** ∞.   **3.** (a) ∞.   (c) 0.   (e) 0.   **5.** N.   **7.** Hints: For Example 2, use the result of Example 1 and the fact that if $S > 1$, $S^2 > S$. For Example 3, use Problems 13 and 16 in Section 5.2 with $a = 2$ and $b = e$.   **8.** 0. Hint: Multiply and divide by $(x^2 + 1)^{1/2} + x$.

Section 11.4 **1.** All but (e) are continuous. In (e) the limit is $\frac{8}{7}$, but the function is not defined at $z = 4$.   **3.** (a), (b), (c), and (e) $\lim\limits_{x \to 0} f(x)$ does not exist.   (d) Assign $f(0) = 0$.
5. Use Example 4 in Section 11.2 and the fact that $c = h(p)$.   **10.** Use the Intermediate Value Theorem with $N = 0$.   **11.** Use Problem 20, Section 11.2, to relate (a) and (b). Use Problem 22 in that section to relate (b) and (c). Part (c), of course, is the definition of continuity of f at a.   **14.** (a) If $f(x) = 2x$, then $f(x + y) = 2(x + y) = 2x + 2y = f(x) + f(y)$.   (b) If $f(x) = 2x + 1$, then $f(x + y) = 2(x + y) + 1$ is not equal to $f(x) + f(y) = 2x + 1 + 2y + 1$.
16. (a) Let $F(x) = x$.   (b) Let $F(x) = \log_{10} x$.

Chapter 12

Section 12.1   **3.** $N'(t) = cN(t)$, for some constant c.   **5.** $r'(x) = kx^2$, where k is a constant. **7.** Let $E(t)$ be the amount of energy expended by time t. Then, $E'(t) = k[p(t) - \bar{p}(t)]$, where k is a constant.   **9.** 0.02, 1.   **11.** 0.704008, 35.2004.   **13.** $f'(x) = 2 - 2x$.   **15.** $Du(t) = 3 + 8t$.   **17.** $\dfrac{dv}{dt} = 2t + 1$.   **18.** (a) 4.   (b) 6.

21. (a) The slope of the tangent at $x = 0$ is $f'(0) = \dfrac{aL}{(c + a)^2}$.   (b) 64. (c) $f(25) = 48$.

Section 12.2 $6x^5, 11x^{10}, 3x^2$.   **3.** $f'(x) = 8x$, $g'(t) = -18t^5 + 12t^2 - 2$, $h'(v) = 8v^7 - 1$. **5.** $w'(z) = 4 + z^3$.   **7.** $u'(r) = 3r^2 + 6r + 2$.   **9.** $Dw(z) = 4z^3 + 3z^2 - 4z$. **11.** $z'(x) = 6/(x + 2)^2$.   **13.** $f'(x) = (x^2 + 2x - 10)/(x + 1)^2$.   **15.** $Du(r) = 4 - 2r^{-3} + (2 - 2r)/(r + 2)^4$.   **16.** When $x = 2$, $dy/dx = 36$.   **18.** $u'(a) = \frac{1}{2}a$.   **20.** (a) $\frac{3}{4}$.   (b) 0.

Section 12.3   **1.** At $x = -\frac{2}{3}\sqrt{3} \approx -1.155$ and $x = \frac{2}{3}\sqrt{3} \approx 1.155$.   **3.** $f'(x) = 3x^2$ and $\dfrac{f(1) - f(0)}{1 - 0} = \dfrac{1 - 0}{1} = 1$. If $f'(c) = 1$, then $c = \sqrt{3}/3 \approx 0.577$.   **6.** (a) Increasing for all x.   (c) Increasing if $z \le -4$ or $z > 1$; decreasing if $-4 < z < 1$. **7.** (a) $c = \sqrt{3}/3$.   (b) $c = 1$.   **9.** Apply Rolle's Theorem to the function $g(x) = f(x) - f(a)$.

Section 12.4   **1.** $2(x^2 - 2x + 3)(2x - 2)$.   **3.** $a^2/(a^2 + u^2)^{3/2}$.   **5.** $-3(1 - 2z)^{1/2}$. **7.** $-4t(t^2 - 5)^{-3}$.   **9.** $-\frac{19}{2}(7 - x)^{-1/2}(12 + x)^{-3/2}$.   **11.** $(20t^3 + 43t - 18) \times (6 + 5t^2)^{-3/2}$.   **13.** If $Q(p) = c/p$, then $Q'(p) = -c/p^2$, so $E(p) = -pQ'(p)/Q(p) = 1$.

Section 12.5   **2.** (a) $-2(t - 1)\sin(t - 1)^2$.   (b) $-3x^2\sec^2(-x^3 + 2)$.   (c) $-\frac{3}{2}(x - 1)^{1/2} \times \sin 2x \sin(x - 1)^{3/2} + 2\cos 2x \cos(x - 1)^{3/2}$.   **5.** $Dh(y) = 2y^2\sec^2y^2 + \tan y^2$.   **7.** $g'(x) = 8x\tan 2x^2 \sec^2 2x^2$.   **9.** $f'(x) = -2\sin(2x - 4)$. **11.** $f'(w) = \frac{1}{2}\cos w/(\sin w)^{1/2}$.   **13.** $f'(x) = 2x\tan x\sec^2 x + \tan^2 x$. **15.** $2x\sec^2 x(x\tan x + 1)$.   **17.** $\frac{1}{2}(x + 1)^{-1/2}\sec^2(x + 1)^{1/2}$.

Section 12.6   **1.** For instance, $[0, \pi/2]$ and $[3\pi/2, 5\pi/2]$.   **3.** No.   **6.** Since $f(1) = f(-1) = 1$, f is not $1 - 1$.   **9.** $-1/(y^2 + 1)$.   **11.** Since $\csc^{-1}t^2 = \sin^{-1}(1/t^2)$, $Du(t) = -2/t\sqrt{t^4 - 1}$.   **13.** $2x/(4 + x^2) + \tan^{-1}(x/2)$.   **17.** Let f be any function with domain S and range T. Then $2 \ge 2$, but not $f(2) < f(2)$. That is, $2R2$, but not $f(2)Qf(2)$.

Section 12.7   **1.** $1/(x + 2)$.   **3.** $2^t \ln 2$.   **5.** $2x^x[1 + \ln x]$.   **7.** $2we^{w^2}$.   **9.** $x(2 - x)e^{-x}$. **11.** $-e^{-w}(\sin w + \cos w)$.   **13.** $\log_b \phi(x) = (\log_b e)\ln \phi(x)$. Now use Theorem 11(a).   **16.** About 2109 A.D.

Section 12.8   **1.** $f(1/\sqrt{2}) = -\sqrt{2}$ is a minimum; $f(-1/\sqrt{2}) = \sqrt{2}$ is a maximum.   **3.** $h(-1) = -1$ is a minimum, $h(1) = 1$ is a maximum.   **5.** $f(4) = \frac{1}{4}$ is a maximum. **7.** $h(-\frac{3}{5}) = \frac{16}{7}$ is a maximum; $h(1) = 0$ is a minimum.   **9.** $k(-\sqrt{2}/4) = -\frac{1}{4}$ is a minimum; $k(\sqrt{2}/4) = \frac{1}{4}$ is a maximum.   **11.** $g(t)$ attains its maximum value, zero, at $t = \pm\dfrac{\pi}{2}, \pm\dfrac{3\pi}{2}, \pm\dfrac{5\pi}{2}, \dots$. There is no minimum.   **13.** Maximum and minimum values occur at the points where $\tan b\theta = \dfrac{b}{a}$.   **14.** 1.   **16.** $\frac{1}{2}$.

19. $DAR(x) = \dfrac{xr'(x) - r(x)}{x^2} = 0$ if and only if $r'(x) = \dfrac{r(x)}{x}$. **20.** (a) $x = 2$.
(b) $x = 5$. $MC(5) = AC(5) = \frac{39}{2}$.

Section 12.9 **4.** $f'(x) = 3x^2 + 6x - 2, f''(x) = 6x + 6$. **5.** $Dg(u) = 5u^4 - 6u^2 + \frac{7}{2}$, $D^2g(u) = 20u^3 - 12u$. **7.** $w'(x) = \frac{1}{2}x^{-1/2} + 3x^{-3/2}, w''(x) = -\frac{1}{4}x^{-3/2} - \frac{9}{2}x^{-5/2}$. **9.** $f'(y) = -2/y^{-1/2}(2 + \sqrt{y})^{-2}, f''(y) = (2 + 3\sqrt{y})/y^{3/2}(2 + \sqrt{y})^3$. **11.** $f'(x) = 2e^{-x}(1 - x), f''(x) = -2e^{-x}(2 - x)$. **13.** $g'(x) = -(x^2 + a^2)^{-1/2}$, $g''(x) = x(x^2 + a^2)^{-3/2}$. **14.** (a) $3x^2 - 4x + 1, 6x - 4, 6, 0$. (c) $-e^{-x}$ − $\sin x, e^{-x} - \cos x, -e^{-x} + \sin x, e^{-x} + \cos x$. **17.** At $t = \left[\dfrac{\beta p}{r(\alpha - 1)}\right]^{1/\alpha}$, $D^2RC(t) = p\alpha t^{-3} > 0$. **19.** If $\beta > b$, choose $t = \frac{1}{2}(\beta - b)$. What happens if $\beta < b$?

Chapter 13

Section 13.1 **1.** (a) $0, 0, 0, 0, 0$. (b) $0, 0, 0, 0, 0$. **5.** (a) $0, 2, 4, 6, 2n$. (b) $2, 2, 2, 2, 2$. **7.** (b) no value, $\frac{1}{3}, \frac{1}{12}, \frac{1}{30}, 2/n(n + 1)(n + 2)$. **9.** (a) $-\frac{2}{3}, -\frac{2}{9}, -\frac{2}{27}, -\frac{2}{81}$, $-\dfrac{2}{3^{n+1}}$. (b) $\frac{4}{9}, \frac{4}{27}, \frac{4}{81}, \frac{4}{243}, \dfrac{4}{3^{n+2}}$. **11.** $\Delta^4 f(n) = f(n + 4) - 4f(n + 3) + 6f(n + 2) - 4f(n + 1) + f(n)$. **15.** $20(n)_3, 60(n)_2, 120n, 120, 0$. **16.** (a) $\Delta a^n = a^{n+1} - a^n = a^n(a - 1)$. **18.** (a) $(a - 1)^2 a^n$. (b) 2^n.

Section 13.2 **2.** $(\alpha - \beta)N(t), (\alpha - \beta)^2 N(t), (\alpha - \beta)^r N(t)$. (Use Problem 18, Section 13.1.) **9.** $MC(n) = a - b^{-c(n+1)}(n + 1 - nb^c)$. **11.** $-\dfrac{1}{b}\dfrac{P(t + 1)Q(t) - P(t)Q(t + 1)}{Q(t)Q(t + 1)}$.

Section 13.3 **1.** (a) $k^2 = \Delta[k^3/3 - k^2/2 + k/6] = \Delta[\frac{1}{6}k(k - 1)(2k - 1)]$ so $\displaystyle\sum_{k=1}^{n} k^2 = \frac{1}{6}[(n + 1)(n + 1 - 1)(2(n + 1) - 1)] - \frac{1}{6}(1 - 1)(2 \cdot 1 - 1) = \frac{1}{6}n(n + 1)(2n + 1)$. **2.** 785. **3.** (a) Marginal cost for the tenth unit is $\Delta T(9) = T(10) - T(9) = 1986/5$. (b) $\Delta T(n) = 397 + \frac{1}{5} + \frac{3}{5}n^2 - \frac{27}{5}n$. The minimum is $\Delta T(4) = \Delta T(5) = 385\frac{1}{5}$. (c) Minimum is $T(0) = 10,000$. **5.** (a) $t^3/3 + t^2 + 2t/3$. (b) $t^2 + 2t$. (c) $2 \cdot 2^t + t$.

Section 13.4 **1.** (a) and (b) are first-order linear, (d) is third-order linear, and (c) and (e) are nonlinear. **2.** (a) If $y(n) = 1$ for all n, then $y(n + 1) - y(n) = 1 - 1 = 0$. (d) $\Delta f(n) = f(n + 1) - f(n) = \frac{1}{2}(n + 1)[(n + 1)^2 - 1] + c - [\frac{1}{2}n(n^2 - 1) + c] = 3n(n + 1)/2$. (e) $h(n + 1) - 2h(n) = 2^{n+1} - 2 \cdot 2^n = 0$. **5.** (b) $c_1 = 2$, $c_2 = 4$. **6.** (a) $g(t) = 3t/2 + 1$. (b) $g(t) = 5t^2 + 5t + 10$.

Section 13.5 **2.** $Y(t) = (10 - 4t)2^t$. **3.** $f(t) = (2 - 10t)(-1)^t$. **5.** $h(n) = 6 \cdot 3^n - 15 \cdot 2^n$. **7.** Solution: $Y(t) = (1 + \frac{3}{2}t)2^t$. **9.** $Y(t) = 3 \cdot 2^t$ diverges to infinity as $t \to \infty$. **11.** $g(n) = 4(\frac{1}{3})^n$ converges to zero as $n \to \infty$. **13.** $Y(n) = -(-4)^n$ oscillates and is unbounded as $n \to \infty$.

Section 13.6 **1.** If $y_p(n) = c_0 + c_1 n + c_2 n^2$ is the particular solution, then $y_p(n + 3) - 6y_p(n + 2) + 3y_p(n + 1) + 10y_p(n) = (8c_0 - 6c_1 - 12c_2) + (8c_1 - 12c_2)n + 8c_2 n^2$. Equating this to $8n^2 + 4n$ gives $c_2 = 1, c_1 = 2$, and $c_0 = 3$. **3.** At equilibrium, $D = S$. Thus the equilibrium price P is obtained by solving $a + bP = c + dP$ to get $P = (c - a)/(b - d)$. **5.** $f(n) = a \cdot 3^n + b - 3n$. **7.** $Y(n) = a \cdot 3^n + b(-1)^n - n^2/2 - 3n/2 + \frac{1}{2}$. **9.** $g(n) = a + b \cdot 3^n -$

$2^n + \frac{3}{4}n^2$.   **11.** $Y(n) = a(-\frac{3}{2})^n + b(\frac{1}{2})^n + (\frac{1}{21}n - \frac{19}{441})2^n$.   **13.** $f(t) =$
$(8t - 9)2^t + 10$.   **15.** $Y(t) = 2 \cdot 3^t - 1 - 3t$.   **18.** (a) Setting $S(t) = D(t)$
gives $3P(t - 1) = 10 - 4P(t)$ or $(4\Delta + 1)P(t - 1) = 10$.   (b) $P(t) = c(-\frac{3}{4})^t +$
$\frac{10}{7}$.   (c) $1, \frac{7}{4}, \frac{19}{16}, \frac{103}{64}, \frac{331}{256}, \frac{1567}{1024}$.   **20.** (b) $f(n) = 1/\left[a\left(\dfrac{c_1}{c_2}\right)^n + \dfrac{c_1 c_2}{c_2 - c_1} \right]$,
where a is arbitrary.   **22.** (a) The principal $P(n + 1)$ equals the principal $P(n)$
plus the interest, $0.06\, P(n)$, on $P(n)$, minus the payment R.   (b) $P(n) =$
$[P(0) - 50R/3](1.06)^n + 50R/3$. To make $P(k) = 0$, choose $R = \frac{3}{50}P(0)(1.06)^k/$
$[(1.06)^k - 1]$.   **23.** (a) The basic difference equation is $N(t + 1) = N(t) -$
$\beta N(t) + \alpha[N - N(t)]$.   (b) As $t \to \infty$, $N(t) \to \alpha N/(\alpha + \beta)$.   **25.** (i) $P(t) \to$
$(a + c)/(b + d)$ as $t \to \infty$.   (iii) $P(t)$ oscillates but converges to $(a + c)/(b + d)$
as $t \to \infty$.

Chapter 14

Section 14.1 $\frac{1}{2}$.   **3.** $2(b - a)$.   **5.** $\frac{9}{2}$.   **7.** $\frac{5}{2}$.   **9.** 6.   **10.** The area below $c \cdot f$ is c times the area below f.   **12.** Area above $[a, b]$ is area above $[a, c]$ plus area above $[c, b]$.

Section 14.2 **1.** $\frac{483}{64}, \frac{51}{64}$.   **2.** $(n + 1)/2n$, $(n - 1)/2n$.   **4.** $\frac{1}{2}n$, $-\frac{1}{2}n$.   **6.** $(n + 1) \times (2n + 1)/6n^2$, $(n - 1)(2n - 1)/6n^2$.   **9.** $\frac{1}{2}, 4, 0, \frac{1}{3}, 2, \frac{83}{12}$.   **11.** (a) 12. (c) 1.   (e) $\frac{1}{2}$.

Section 14.3 **1.** $x^7/7 + C$.   **3.** $x^4/4 - 2x^3/3 + 3x + 4x^5 + C$.   **5.** $3z^{2/3}/2 + C$.
7. $(1 + x^2)^{3/2}/3 + C$.   **9.** $(e^x - 2)^3/3 + C$.   **11.** $-\frac{1}{3}\cos^3 x + C$.
13. $x - \ln|x + 2| + C$.   **15.** $\frac{1}{2}\ln(2 + \sin^2 t) + C$.   **17.** $e^{\sin x} + C$.
19. $-e^{1/x} + C$.   **21.** $2\ln(e^x + 1) + C$.

Section 14.4 **1.** $\frac{16}{3}$.   **3.** 456.   **5.** $\sqrt{7} - \sqrt{3}$.   **7.** 0.   **9.** $\frac{3}{2} - 2\ln 2$.   **11.** $2[e^{\sqrt{2}} - e]$.
13. $\frac{5}{2}$.   **15.** $\frac{128}{3} = \displaystyle\int_{-2}^{2}(16 - 4x^2)dx$.

Section 14.5 **1.** $\displaystyle\int_{0}^{1/2}[h(x) - g(x)]dx + \int_{1/2}^{8}[h(x) - |f(x)|]dx =$
$\displaystyle\int_{0}^{1/2}3dx + \int_{1/2}^{8}(4 - \sqrt{2x})dx = \frac{21}{2}$.

Section 14.6 **2.** Since $F'(x) = A\mu^{-x}$, $F(x) = (-A/\ln\mu)\mu^{-x} + C$. Imposing the condition $F(0) = 0$ gives $C = A/\ln\mu$. Hence, $F(x) = (A/\ln\mu)[1 - \mu^{-x}]$.   **4.** Let $s = 100{,}000$ m. The average total repair cost is approximately
$\displaystyle\int_{0}^{400000}A(s/100{,}000)^{1/2}ds = 1{,}600{,}000\,A/3$.
5. (a) $\displaystyle\sum_{t=1}^{50}100tNp(t) = \sum_{t=1}^{50}\frac{100}{3}tN(\frac{2}{3})^t$.
(b) $\dfrac{100N}{3}\displaystyle\int_{0}^{50}t(\frac{2}{3})^t dt \approx \dfrac{100N}{3}(\ln\frac{2}{3})^{-2}$.
7. (a) 7028, 702.8.   (c) 3559, 355.9.
11. $SP/A + \frac{1}{2}(k + 1)S(2S - Ak - A)/(S - A)$.

Section 14.7 **3.** (a) $(x^2/4)(2\ln x - 1) + C$.   (c) $x\sin^{-1}x + (1 - x^2)^{1/2} + C$.
(e) $+(1/x) \times \cos(1/x) - \sin(1/x) + C$.   **4.** (a) $(25e^8 - 1)/2$.   (b) $\pi - 2$.

5. $[a + e^{-aT}(b \sin bT - a \cos bT)]/(a^2 + b^2)$.   **11.** You can replace $\cos^2 x$ in Problem 10 by $1 - \sin^2 x$ to show that the answers in 10 and 11 differ only by a constant.

Section 14.8   **1.** In (19), choose $g(x) = f(x)$ and $f(y) = y^{n+1}/(n + 1)$.   **5.** $2(t + 2)^{3/2}/3 - 4(t + 2)^{1/2} + C$.   **7.** $4(t^2 + t)^{1/2} + C$. Let $u = t^2 + t$.   **9.** $\frac{3}{2} \ln (2 + \sin^2 t) + C$.
11. $\frac{1}{5} \tan^{-1} (x/5) + C$. Substitute $\tan \theta = x/5$.   **13.** $\frac{1}{3} \sin^{-1} (3v/\sqrt{2}) + C$.
15. $\frac{1}{2} \sin^{-1} \sqrt{2} x + C$.   **17.** $-(t^2 + 9)^{1/2}/9t + C$.   **19.** $\sin^{-1} \left(\dfrac{y - 1}{3} \right) -$
$\sqrt{9 - (y - 1)^2} + C$.   **21.** $-(1 + x^2)^{-n+1}/2(n - 1) + C$ if $n \neq 1$.
$\frac{1}{2} \ln |1 + x^2| + C$ if $n = 1$.   **23.** $1 + \sqrt{2}$.   **25.** $(\frac{2}{3}) \tan^{-1} (\frac{1}{3})$.   **27.** $\sqrt{2}/1500$.
29. $\frac{3}{8}x - \frac{1}{4} \sin^3 x \cos x - \frac{3}{8} \sin x \cos x + C$.   **31.** $\frac{1}{128}[\pi + 9\sqrt{3}/16]$.
33. $-\frac{1}{4} \cos 2\theta - \frac{1}{8} \cos 4\theta + C$.   **35.** $\frac{1}{2} \tan^2 u + \ln |\cos u| + C$.

Section 14.9   **1.** $\frac{1}{10} \ln |(x - 5)/(x + 5)| + C$.   **3.** $x + 2 \ln |(x - 4)/(x + 4)| + C$.
5. $\frac{7}{10} \ln 2 + \frac{9}{20} \ln 3 - \frac{13}{40} \ln 11$.   **7.** $\frac{7}{16}$.   **9.** $\ln 5 - \frac{3}{2} \ln 2$.   **11.** $y(t) = (1 - a)y_0 e^{(1-a)kt}/[1 - a - y_0(1 - e^{(1-a)kt})]$.   **13.** Using $x(0) = \alpha + \beta$,
$x(t) = [\beta^2 - \alpha^2 e^{(\beta-\alpha)\gamma t}]/[\beta - \alpha e^{(\beta-\alpha)\gamma t}]$.

Section 14.10   **1.** Divergent.   **3.** 2.   **5.** Divergent.   **7.** Divergent.   **9.** Divergent.
11. $\frac{1}{2}$.   **13.** 1.   **15.** Divergent.   **16.** $\pi/6$.   **18.** $\pi/4$.   **19.** $k < 0$.
21. $b > -1$.   **24.** $3^{1-\alpha} - 4^{1-\alpha}$.   **26.** A/r.

Chapter 15

Section 15.1   In the following answers, C denotes a constant.   **1.** $f(t) = Ce^{3t^2/2}$.   **3.** $y^2 = C(x^2 - 1)^{-1} - 1$.   **5.** $y + \ln |y| = x + \ln |x| + C$.   **7.** $y = C|x|^{-1/2}$.
9. $y = x^3/[Cx^3 + \frac{1}{3}]$.   **11.** $f(x) = 2x^{-1/2}$.   **13.** $S = \ln \tan \left[\dfrac{\pi}{4} - \dfrac{1}{\sqrt{2}} \times \tan^{-1} (\sqrt{2} \cos \theta) \right]$.   **16.** (a) $y = x \tan^{-1} [\ln c|x|]$.   (b) $y = C - x/2$.

Section 15.2   **3.** The operator is $D^2 - b^2 = (D - b)(D + b)$. Use Theorem 4 and Example 3.
4. $y(t) = (6 - 5t)e^{2t}$.   **5.** $x(t) = (1 - k^2/108)e^{-6t/k} + t^2/6 - kt/18 + k^2/108$.
7. $g(x) = (\frac{23}{27} + \frac{47}{9}x)e^{3x} + \frac{4}{27} + \frac{2}{9}x$.   **9.** $(2t - t^2 + t^3/3)e^{2t}$.   **11.** $\frac{1}{1156}(8 \sin 2t + 15 \cos 2t)e^{4t}$.   **14.** (a) $P(t) = [P(0) + (\alpha - a)/(\beta - b)]e^{(\beta-b)t/c} - (\alpha - a)/(\beta - b)$.
(b) If $c > 0$, $P(t) \to \infty$ as $t \to \infty$. If $c < 0$, the equilibrium state is $(a - \alpha)/(\beta - b)$.
17. First, $V_1(t) = V_1(0)e^{\alpha t}$, then
$$V_2(t) = \left[V_2(0) - \frac{R\beta V_1(0)}{\alpha Rc + 1} \right]e^{-t/Rc} + \frac{R\beta V_1(0)}{\alpha Rc + 1} e^{\alpha t}.$$

Chapter 16

Section 16.1   **1.** 0.967; error ≤ 0.0013.   **3.** 0.587792; error ≤ 0.000008.   **5.** $6 + 3x + 4x^2 + 2x^3 + x^4$.   **7.** $1 + x \ln 2 + x^2(\ln 2)^2/2 + x^3(\ln 2)^3/3! + \cdots = \sum_{k=0}^{\infty} x^k(\ln 2)^k/k!$ **9.** $\frac{1}{2} \sum_{k=0}^{\infty} (-\frac{1}{2})^k(x - 2)^k$.   **11.** $\sum_{k=0}^{\infty} (-1)^{k+1} \dfrac{x^{2k}}{(2k + 1)!}$.
13. $x + \dfrac{x^3}{3!} + \dfrac{x^5}{5!} + \dfrac{x^7}{7!} + \cdots$.   **15.** $1 - x + x^2/2 - x^3/6 + x^4/24$; error $-e^{-c}x^5/5!$, where c is some number between 0 and x.   **17.** $x + \dfrac{x^3}{3!}$; error $(1 - c^2)^{-9/2}(9 + 72c^2 + 24c^4)$, where c lies between 0 and x.   **19.** $1 + x/2 -$

$x^2/8 + x^3/16 - 5x^4/128 + 7x^5/256$; error $-21(1 + c)^{-11/2}/1024$, where c lies between 0 and x. **21.** $\sqrt{1.05} \approx 1.0246882577$, correct to 9 decimal places. **23.** $\frac{1}{2} + \frac{1}{48} \approx 0.5208$; $\pi/6 \approx 0.5236$. **25.** 9.025. **27.** 0.669.

1. $a^n \sum\limits_{k=0}^{\infty} \binom{n}{k} (x/a)^k$; radius of convergence is $R = |a|$. **3.** $\sum\limits_{k=0}^{\infty} \binom{\frac{1}{2}}{k} x^k$; $R = 1$.

7. $\sum\limits_{k=0}^{\infty} \binom{3/2}{k} x^k$; $R = 1$. **11.** 1.3176. **13.** 0.4613. **15.** 0.2726. **17.** $(1 - x)^{-2} =$

$D(1 - x)^{-1} = D \sum\limits_{k=0}^{\infty} \binom{-1}{k} (-1)^k x^k = \sum\limits_{k=0}^{\infty} k\binom{-1}{k} (-1)^k x^{k-1}$. (Use Problem 2.)

Chapter 17

1. $\partial S/\partial x = 2xy - y^2$, $\partial S/\partial y = x^2 - 2xy$. **3.** $\partial z/\partial x = x(x^2 + y^2)^{-1/2}$, $\partial z/\partial y = y(x^2 + y^2)^{-1/2}$. **5.** $\partial f/\partial r = s^{-1} + sr^{-2}$, $\partial f/\partial s = -rs^{-2} - r^{-1}$. **7.** $\partial g/\partial x = 3x^2 + 4x - 6y$, $\partial g/\partial y = -6x - 12$. **9.** $\frac{1}{2}, -\sqrt{3}/2$. **11.** $\partial^2 u/\partial x^2 = -y \sin x$, $\partial^2 u/\partial x \partial y = \cos x$, $\partial^2 u/\partial y^2 = 0$. **13.** $\partial^2 g/\partial x^2 = y^2 z^2 e^{xyz}$, $\partial^2 g/\partial y^2 = x^2 z^2 e^{xyz}$, $\partial^2 g/\partial z^2 = x^2 y^2 e^{xyz}$, $\partial^2 g/\partial x \partial y = z(xyz + 1)e^{xyz}$, $\partial^2 g/\partial x \partial z = y(xyz + 1)e^{xyz}$, $\partial^2 g/\partial y \partial z = x(xyz + 1)e^{xyz}$. **18.** (a) $Y = (1 - \beta + \beta\delta)^{-1}(\alpha - \beta\gamma + I + G)$.

1. $-2(t + 1)/t^3 - 2t/(t - 1)^3$. **3.** $\frac{1}{2}t^{3/2}(2 + \ln t) \cos xy + 2t \sin xy$. **5.** $\partial w/\partial r = 2x \sin t - 2y \cos t$, $\partial w/\partial t = 2xr \cos t + 2yr \sin t$. **7.** $\partial w/\partial r = (x - y)/(x^2 + y^2)$, $\partial w/\partial t = a(x \sin t - y \cos t)/(x^2 + y^2)$. **9.** $\partial w/\partial r = (1/w)(xt + y \sin t - zt \sin r)$, $dw/dt = (1/w)(xr + yr \cos t + z \cos r)$. **11.** (a)

$\dfrac{\partial u}{\partial r} = \dfrac{\partial u}{\partial x} \cos \theta + \dfrac{\partial u}{\partial y} \sin \theta, \dfrac{\partial \theta}{\partial u} = -r \dfrac{\partial u}{\partial x} \sin \theta + r \dfrac{\partial u}{\partial y} \cos \theta.$ (b) Solve the equations in (a) for $(\partial u/\partial x)$ and $(\partial u/\partial y)$. **13.** $(2ay + 3x^2)/2(y - ax)$. **15.** $-(e^x \sin y + e^y \cos x)/(e^x \cos y + e^y \sin x)$. **17.** $-y^2 e^x/(1 + ye^x)$. **20.** (a) x/r. (b) $-y$. (c) y/r. (d) x.

1. $\partial f/\partial x = 2x - 2$ and $\partial f/\partial y = 6y$ are zero at $(x, y) = (1, 0)$. The matrix of second partials $\begin{pmatrix} 2 & 0 \\ 0 & 6 \end{pmatrix}$ is positive definite. Thus, $f(1, 0) = -1$ is a relative minimum.

3. $\partial h/\partial x = 2y$ and $\partial h/\partial y = 2x$ are zero at $(x, y) = (0, 0)$. The matrix of second partials $\begin{pmatrix} 0 & 2 \\ 2 & 0 \end{pmatrix}$ is dinefinite. Hence $f(0, 0) = 0$ is neither a maximum nor a minimum. **7.** $\partial h/\partial x_1 = 2x_2$ and $\partial h/\partial x_2 = 2x_1$ are zero at $(x_1, x_2) = (0, 0)$. The matrix of second partials $\begin{pmatrix} 0 & 2 \\ 2 & 0 \end{pmatrix}$ is indefinite. Hence $f(0, 0) = 0$ is neither a maximum nor a minimum. **11.** The function $f(p_1, p_2) = p_1 p_2 (1 - p_1 - p_2)$ has a *relative* maximum at $p_1 = p_2 = \frac{1}{3}$. (In this case $p_3 = 1 - p_1 - p_2 = \frac{1}{3}$, also.) There is no absolute maximum. For instance, take $p_1 = 2k + 1$, $p_2 = -k$, and $p_3 = -k$, where k is any positive integer. **15.** (b) $8k + 16k(c_1 - 1) + 12k(c_2 - 2) + 8k(c_1 - 1)^2 + 24k(c_1 - 1)(c_2 - 2) + 6k(c_2 - 2)^2 + 12k(c_1 - 1)^2(c_2 - 2) + 12k(c_1 - 1)(c_2 - 2)^2 + k(c_2 - 2)^3$. **16.** $y = 195x/58 + 73/58$. **17.** (b) The least squares line is $59w = -44x + 147y + 248z$.

1. $f(\frac{2}{3}, \frac{4}{3}) = \frac{4}{3}$. **3.** $\frac{285}{92}$, when $(x, y, w) = (\frac{52}{46}, \frac{3}{46}, -\frac{9}{46})$. **5.** $\frac{80}{3}\sqrt{3}$, when $u = v = w = 2/\sqrt{3}$. **7.** $\frac{4500}{256}$, when $C = \frac{165}{16}$ and $L = \frac{75}{16}$. **8.** Differentiate the function $F(x, y) - \lambda(px + qy - B)$, where λ is a Lagrange multiplier, to obtain $\lambda = (1/p)(\partial F/\partial x) = (1/q)(\partial F/\partial y)$. **10.** $U(x, y)$ has the maximum value $\frac{59725}{676} \approx 88.35$ when $x = \frac{35}{13}$ and $y = \frac{155}{52}$.

Section 17.5

1. $\frac{3}{2}$. **3.** $121a^7/5 + 20a^5 - 13a^3$. **5.** $4 - \dfrac{1}{a} \ln 3$. **6.** $(3\pi/4) \ln 2$.

9. $\frac{1}{16}\left[\dfrac{\pi}{3} - \dfrac{\sqrt{3}}{2}\right]$. **11.** $\pi t^2/4$, one-fourth the area of a circle with radius t.

13. $\frac{4}{5} = \displaystyle\int_0^2 2x \int_{\sqrt{2}x^{3/2}}^{2x} dy\,dx$. **15.** $\frac{8}{3} + \pi/3 - 3\sqrt{3}/2 = \displaystyle\int_0^1 \int_{x^2}^x \int_0^{\sqrt{4-x^2}} dy\,dz\,dx$.

17. $3\pi/2$.

Chapter 18

Section 18.1

1. Call the men a, b, c and the women x, y, z. (b) $\{xyz\}$ one element. (c) $\{abc, abx, aby, abz, acx, acy, acz, bcx, bcy, bcz\}$ ten elements. **3.** Let N denote on and O denote off. (a) $E \cap F = F = \{NNOO, NONO, NOON, ONON, ONNO, OONN\}$. (c) $E \cup F = E$. (d) $E' = \{OOOO, OOON, OONO, ONOO, NOOO\}$. (e) $E' \cap F = \phi$. **5.** Denote the individuals by A, B, and C and let 1, 2, and 3 denote the respective links AB, AC, and BC. Then each element of Ω is a list of possible links. $\Omega = \{\text{None}, 1, 2, 3, 12, 13, 23, 123\}$. **7.** (b) $\frac{1}{2}\dbinom{2n}{n}$.

Section 18.2

2. (a) $A \cup B \cup C$. (c) $A' \cap B' \cap C'$. (e) $A \cap B \cap C$. (g) $(A \cap B \cap C') \cup (A \cap B' \cap C) \cup (A' \cap B \cap C)$. **4.** (a) 4. (b) 8. (c) 2^n. **7.** The chairman is always pivotal in positions 4 or 5. There are $4! + 4! = 48$ such arrangements. In addition, he is pivotal in position 3 if he is preceded by A. There are 12 such arrangements $(48 + 12 = 60$ total$)$.

Section 18.3

1. 0.464. **3.** $\frac{43877}{47530} \approx 0.923$. **5.** $\dbinom{n}{r}\left(\dfrac{1}{m}\right)^r\left(\dfrac{m-1}{m}\right)^{n-r}$. **7.** (a) $\frac{193}{512}$. (b) 9.

9. (a) $1/(N-1)$. He can choose any one. That person must choose him. (b) $(N-2)/(N-1)^2$. (c) $(N-1)_{k-1}/(N-1)^k = (N-1)(N-2)\cdots(N-k+1)/(N-1)^k$.

Section 18.4

1. $\frac{9}{10}$. **4.** (a) 0.9. (c) 0.9. (e) 0.7. **6.** Let C_r be the event that the rth photograph is correctly matched. Use Theorem 7 to obtain $P\left(\bigcup_{r=1}^k C_r\right)$. For $k = 3, 4, 5$, and 6, the answers are $\frac{2}{3}, \frac{15}{24}, \frac{76}{120}, \frac{455}{720}$. (This problem is equivalent to Example 5.)

Section 18.5

1. $\frac{7}{15}, \frac{8}{15}$. **2.** $P(0–100) = 0.21$, $P(100–300) = 0.41$, and $P(300–500) = 0.38$. Order three lots. **4.** Let M_1 denote the event that the first child is male, etc. Then $P(M_1 \cap M_2 \mid M_1 \cup M_2) = \dfrac{P(M_1 \cap M_2)}{P(M_1 \cup M_2)} = \dfrac{\frac{1}{2}\cdot\frac{1}{2}}{\frac{1}{2}+\frac{1}{2}-\frac{1}{2}\cdot\frac{1}{2}} = \frac{1}{3}$.

6. Partial answer: $P(x_1 \mid y_1) = \frac{48}{63}$; $P(x_1 \mid y_2) = \frac{12}{37}$. **9.** Use Theorems 4 and 6 together with the facts that $F \cap G \subseteq F \subseteq F \cup G$ and $P(\cdot \mid E)$ is a probability function. **11.** Only (b) is true in general.

Section 18.6

1. 0.66. **3.** (a) $\dfrac{1}{r}\left(1 - \dfrac{1}{r}\right)(1 - p)^2$. (c) $(1 - p)^2\left(1 - \dfrac{1}{r}\right)^2$. **4.** $\frac{2}{3}$.

7. (a) $\frac{40}{49}$. (c) $\frac{15}{343}$. **9.** (a) A_2. (b) A_3. (c) A_3. **11.** (a) Plan 1: $\dfrac{(100 - x)(99 - x)}{100 \cdot 99}$ Plan 2: $1 - \dfrac{x(x - 1)}{100 \cdot 99}$ Plan 3: $\dfrac{(98 + 2x)(100 - x)(99 - x)}{100 \cdot 99 \cdot 98}$.

1. As k increases, $P_G \to 1$.   **3.** $1 - [1 - (0.98)(0.97)]^6 \approx 0.999999985$.
5. (a) $P(\text{of } n \text{ contacted, at least one shows}) = 1 - (\frac{2}{3})^n = p_n$. Choose $n = 12$ to make $p_n > 0.99$. (b) $(\frac{1}{3})(\frac{2}{3})^3 = \frac{8}{81}$. **8.** $(m/n)^2$. **9.** 0.766. **15.** $P(E) = (n+1)/2^n$, $P(F) = 1 - 2/2^n$, and $P(E \cap F) = n/2^n$. $P(E \cap F) = P(E)P(F)$ is equivalent to $2^{n-1} = n + 1$, which is true only when $n = 3$.

Chapter 19

1. $\{TTT\}$, $\{HHH, HHT, THH, HTH\}$, $\{TTT\}$, $\{HTT, TTH, THT\}$, Ω. **2.** $P(N = 0) = \frac{1}{10}$, $P(N = 1) = \frac{5}{20}$, $P(N = 2) = \frac{11}{20}$, $P(N = 3) = \frac{1}{10}$, and $P(N = x) = 0$ if x is not 0, 1, 2, or 3. **3.** If $x < 0$, $P(N \le x) = P(\phi) = 0$. If $0 \le x < 1$, $P(N \le x) = P(N = 0) = \frac{1}{10}$. If $1 \le x < 2$, $P(N \le x) = \frac{7}{20}$. If $2 \le x < 3$, $P(N \le x) = \frac{18}{20}$. If $3 \le x$, $P(N \le x) = P(\Omega) = 1$. **5.** $0.4, 0.108, 1$. **7.** (b) $\frac{1}{5}$, $1 - 4e^{-7}/5$, $4(e^{-2} - e^{-5})/5$, $4e^{-2}/5$.

1. $\frac{8}{27}, \frac{4}{27}, \frac{4}{27}, \frac{2}{27}, \frac{4}{27}, \frac{2}{27}, \frac{2}{27}, \frac{1}{27}$. **2.** $P(N = 0) = \frac{1}{27}$, $P(N = 1) = \frac{6}{27}$, $P(N = 2) = \frac{12}{27}$, $P(N = 3) = \frac{8}{27}$. In general, $P(N = k) = \binom{3}{k}(\frac{1}{3})^k(\frac{2}{3})^{3-k}$ for $k = 0, 1, 2, 3$.

9. (a) (i) $\binom{7}{4} (\frac{3}{5})^4(\frac{2}{5})^3$, (ii) $\binom{7}{5} (\frac{3}{5})^5(\frac{2}{5})^2$, (iii) $\sum_{k=4}^{7} \binom{7}{k} (\frac{3}{5})^k(\frac{2}{5})^{7-k}$. **12.** A 4-engine plane is safer if and only if $q < \frac{1}{3}$. **14.** (a) $\frac{32}{243}$. (b) $\frac{80}{243}$. (c) $\frac{80}{243}$. (d) $\frac{51}{243}$. **15.** 8. **18.** $\binom{r-1}{k-1} (p^n)^k(1 - p^n)^{r-k}$. **19.** (a) $\frac{1}{63}$. (b) $\frac{1}{21}$. **20.** Test III gives the strongest evidence in favor. **23.** (a) $\frac{2}{81}$. (c) $\frac{80}{81}$. (e) $\frac{1}{81}$. **28.** For $k = 0, 1, 2, \ldots, t$, $P(X = k) = \binom{t}{k} (p_A p_B)^k(1 - p_A p_B)^{t-k}$.

1. 7. **2.** $P(X = 0) = q$, $P(X = 1) = p$, so $E(X) = p \cdot 1 + q \cdot 0 = p$. **3.** $2p$. **5.** (a) $E(2N^2)$. (c) $E(s^N)$. **6.** (a) 8. (b) 0. (c) 4. **8.** 2000. **9.** $E(C) = \frac{173}{61}$, Var $(C) = \frac{5878}{3721} \approx 1.580$. **12.** (a) \$2.50. (c) \$1.90. **15.** $E(X/n) = p$, Var $(X/n) = pq/n$. **19.** (a) $P(T = k) = \frac{1}{n}\left(\frac{n-1}{n}\right)^{k-1}$, $k = 1, 2, 3, \ldots$; $E(T) = n$, Var $(T) = n(n-1)$. (b) $P(T = k) = 1/n$, $k = 1, 2, \ldots, n$; $E(T) = (n+1)/2$, Var $(T) = \dfrac{(n+1)(n-1)}{12}$. **21.** (a) 3000 (b) 3000. (c) 4000 or less. (d) 2000. (e) 0 or 1000.

1. $P(Y = 1) = 0.425$, $P(Y = 2) = 0.188$, $P(Y = 3) = 0.158$, $P(Y = 4) = 0.109$, $P(Y = 5) = 0.073$, $P(Y = 6) = 0.047$.

4. (a)

k	1	2	3	4	5	6
$P(Y = k \mid X = 2)$	0.627	0	0.158	0.114	0.066	0.035

.

6. (a) $\frac{35}{1536}$. (b)

k	5	6	7
Probability	$\frac{7}{15}$	$\frac{7}{15}$	$\frac{1}{15}$

. **7.** (a) $\binom{8}{5, 2, 1} (\frac{3}{4})^5(\frac{1}{5})^2(\frac{1}{20})$.

(b) $1 - (\frac{3}{4})^8$. (c) $\binom{8}{4} (\frac{3}{4})^4(\frac{1}{4})^4$. **11.** (a) $P(Z = k) = pq^k(2 - q^k - q^{k+1})$, $k = 0, 1, 2, \ldots$. **13.** $P(U_n = r) = \left(\dfrac{N - r + 1}{N}\right)^n - \left(\dfrac{N - r}{N}\right)^n$, $r = 1, 2, \ldots, N$.

Section
19.5
1. 1.4, 2.0, 3.4, 2.7.   **10.** (a) $E(X_n) = P(X_n = 1) = \alpha^n q_0$, from Problem 8, Section 18.6.   (b) $E(T) = \alpha q_0/(1 - \alpha)$, Var $(T) = \alpha q_0(1 + \alpha - \alpha q_0)/(1 - \alpha^2)$.
13. Let $p_i = P(O_i)$, $i = 1, 2, \ldots, m$. Then Cov $(X_i, X_j) = -rp_ip_j$.   **17.** Choose $k = 1$.   **18.** Choose $k = \infty$.   **19.** Choose $k = 3$.

Section
19.6
3. Binomial: rp and $rp(1 - p)$; Poisson: λ and λ; Geometric: q/p and q/p^2; Negative binomial: rq/p and rq/p^2; Uniform: $n/2$ and $n(n + 2)/12$.   **5.** $E(T) = n(k - 1)/2$, Var $(T) = n(k^2 - 1)/12$.   **9.** $g_{X-1}(s) = E(s^{X-1}) = E(s^{-1}s^X) = \frac{1}{s}E(s^X) = \frac{1}{s}g_X(s)$. $g_{3X}(s) = E(s^{3X}) = E[(s^3)^X] = g_X(s^3)$.   **11.** $E(X) = g'(1) = s(1 - s^2)^{-1/2}]_{s=1} = \infty$. The series $\sum_{k=0}^{\infty} kP(X = k)$, representing $E(X)$, diverges.

Chapter 20

Section
20.1
$\pi(2) = \pi(0)P^2 = (\frac{1}{2}, \frac{1}{2})\begin{pmatrix} 0.28 & 0.72 \\ 0.27 & 0.73 \end{pmatrix} = (0.275, 0.725)$. $\pi(3) = (0.2725, 0.7275)$,

$\pi(4) = (0.27275, 0.72725)$.   **7.** (a) $\begin{pmatrix} 1 & 0 & 0 & 0 & 0 \\ p & r & q & 0 & 0 \\ 0 & p & r & q & 0 \\ 0 & 0 & p & r & q \\ 0 & 0 & 0 & 0 & 1 \end{pmatrix}$ **8.** (a) $P = \begin{matrix} U \\ A \end{matrix}\begin{pmatrix} \frac{2}{10} & \frac{8}{10} \\ \frac{1}{10} & \frac{9}{10} \end{pmatrix}$.

(c) $\pi(0) = (1, 0)$, $\pi(1) = (0.2, 0.8)$, $\pi(2) = (0.12, 0.88)$, $\pi(3) = (0.112, 0.888)$, $\pi(4) = (0.1112, 0.8888)$, $\pi(5) = (0.11112, 0.88888)$.   **10.** Partial answer: To go from state ABC to state AB, candidates A and B must hold their delegates, while C loses his. The respective probabilities are $\frac{5}{6}$, $\frac{3}{4}$, and $\frac{2}{3}$, the product of which is $\frac{30}{72}$.
12. (a) $P = \begin{matrix} F \\ U \end{matrix}\begin{pmatrix} 1 & 0 \\ p & 1 - p \end{pmatrix}$.   (b) $P^n = \begin{pmatrix} 1 & 0 \\ 1 - (1 - p)^n & (1 - p)^n \end{pmatrix}$.   **13.** (a) p.   (c) $1 - (1 - p)^n$.   **14.** Let p_i denote the probability of outcome O_i. Then if O_j occurs on one trial the probability of O_i on the next is, since trials are independent, $p_{ji} = p_i$. This gives a transition matrix with identical rows. Now use Problem 3.

Section
20.2
2. $\{S_1, S_4, S_9\}$, $\{S_5\}$, and $\{S_3, S_8\}$ are closed classes; $\{S_2\}$, $\{S_6\}$, and $\{S_7\}$ are open classes.   **9.** (c) The chain remains in the initial class forever.

Section
20.3
1. Only the chain in Problem 4 is regular.   **2.** Each row is $(\frac{2}{24}, \frac{3}{24}, \frac{2}{24}, \frac{3}{24}, \frac{4}{24}, \frac{3}{24}, \frac{2}{24}, \frac{3}{24}, \frac{2}{24})$. The limiting probability of being in a particular compartment is proportional to the number of entries into that compartment.   **3.** $\begin{pmatrix} \frac{1}{9} & \frac{8}{9} \\ \frac{1}{9} & \frac{8}{9} \end{pmatrix}$.

8. Expansion of present facilities in anticipation of eventual average annual sales of $24 million. The eventual total market is $\lim_{t\to\infty} y = \$40$ million, $\frac{3}{5}$ of which will go to Nichols.

Section
20.4
1. $\begin{pmatrix} \frac{21}{8} & 2 \\ \frac{9}{8} & 2 \end{pmatrix}$.   **3.** (a) $\frac{24}{19}$, the expected number of ballots in state ABC. (b) This is the same as the probability that the first transition *out of state ABC* is to state AB, that is, $(\frac{30}{72})/(1 - \frac{15}{72}) = \frac{10}{19}$.   (c) $\frac{1}{19}$.   (d) Either the first transition out of ABC is to A, or the first transition is to BC and then to ϕ. The probability is $\frac{10}{57} + (\frac{1}{19}) \times (\frac{2}{9}) = \frac{32}{171}$.   **5.** $0.0331/(1 - 0.9041) = \frac{331}{959}$.   **8.** $1/(1 - 0.2) = \frac{5}{4}$.   **10.** (a)

$$\tfrac{1}{3}\begin{pmatrix} 4 & 2 & 0 \\ 2 & 4 & 0 \\ 3 & 3 & 3 \end{pmatrix}.$$ (b) 3.   (c) (i) $\tfrac{1}{2}$ (ii) $\tfrac{1}{4}$.   (d) 1.   **11.** (a) 1724, 7.8, $41\tfrac{2}{3}$.

(b) 0.000068.   (c) $\tfrac{8}{29}$.   **14.** Same answer as 12(b), since $p_{ii} = 0$ for all non-absorbing states. Use Problem 13.

Section 20.5 **2.** Solve the equation $V = VP$ as in Section 20.3.   **4.** Two trials, if the chain starts in either D_1 or D_2. Otherwise, zero.   **6.** Proceed as in Problem 2.
7. The probability of transition to S_1 is $\tfrac{1}{2}$, whether the process is in state S_2 or in state S_3. However, the probability of transition to S_2 is not the same from S_1 as from S_3.

Section 20.6 **1.** (a) Proceed as in Problem 25, Section 10.3, to obtain
$$p^n = \begin{pmatrix} \pi & 1-\pi \\ \pi & 1-\pi \end{pmatrix} + (1-\theta)^n \begin{pmatrix} 1-\pi & \pi-1 \\ -\pi & \pi \end{pmatrix}.$$ (b) $\begin{pmatrix} \pi & 1-\pi \\ \pi & 1-\pi \end{pmatrix}$.
2. (b) $U_n = \pi - \pi(1-\theta)^{n-1} + (1-\theta)^{n-1}u_1$.   (c) π.   **6.** vT_1T_2 is closer to Λ_2 while vT_2T_1 is closer to Λ_1.

Chapter 21

Section 21.1 **1.** (a) 0.084.   (c) 0.735.   **2.** (a) 0.000005.   (c) 0.000855.   **3.** (a) 0.160.
(b) 0.982.   **5.** (a) 0.918.   (b) 2.5.   (c) $\displaystyle\sum_{k=0}^{\infty} \frac{e^{-2.5M}(2.5M)^k}{k!} \cdot \frac{e^{-2.5N}(2.5N)^k}{k!}$.
(d) $\lambda/(1-e^{-\lambda})$, $\lambda[1-(\lambda+1)e^{-\lambda}]/(1-e^{-\lambda})$, where $\lambda = 2.5$.   **10.** (a) 0.471, 0.510, 0.019.   **13.** (a) (i) 0.000001 (ii) 0.000012 (iii) 0.000474.   (c) (i) 0.744 (ii) 0.857.   (e) 1, 1.   (f) 0.897.

Section 21.2 **1.** Limiting probability of E_n is $\dfrac{e^{-\lambda/\mu}(\lambda/\mu)^n}{n!}$, $n = 0, 1, 2, \ldots$.   **3.** Limiting probability of E_n is $(\mu/\lambda+\mu)(\lambda/\lambda+\mu)^n$, $n = 0, 1, 2, \ldots$.   **6.** $p_n(t) = e^{-\beta t}(1 - e^{-\beta t})^{n-1}$.   **8.** $a_n = (\lambda/\mu)^n a_0$. Since $\lambda < \mu$, $a_0 > a_n$ for $n \geq 1$.   **10.** (a) $\tfrac{2}{5}$. (b) $\tfrac{15}{4}$.

Section 21.3 **1.** $E(T) = \tfrac{1}{2}$, $\mathrm{Var}(T) = \tfrac{1}{4}$, $P(T \leq 2) = 1 - e^{-4}$.   **2.** (a) 1.682.   (b) 1.975.
4. (a) $E(X) = (\beta+\alpha)/2$, $\mathrm{Var}(X) = (\beta-\alpha)^2/12$.
(c) $f_Z(z) = e^{-z}$, $z \geq 0$.
$\qquad\quad = 0$, $z < 0$.
6. (a) $\tfrac{1}{2}$.   (b) 5.   **8.** $I_0 = 3 \ln\left(\tfrac{5}{2}\right) \approx 2.749$.   **12.** $e^{-18} \approx 0.00000$.   **19.** (a) $\tfrac{466}{905}$.   (b) $\tfrac{27}{1810}$.   (c) $\tfrac{81}{1810}$.   (d) $\tfrac{335}{362}$.   (e) $\tfrac{905}{362}$.

Section 21.4 **1.** 0.0008, 0.8392.   **2.** 0.7679.   **4.** $\sqrt{\pi}e^{1/4}/2$, $\pi e^{1/2}/2$.   **5.** (a) 0.3592.
(b) 0.1796.   **10.** No.   **12.** Snack bar 372, Cafeteria 705.

AUTHOR INDEX

SUBJECT INDEX

laws, for sets, 17
of a set, 10
Complementary:
equation, 382
solution, 382
Composite function. *See*
Function, composite
Conclusion, of an argument, 40
Conditional probability, 531–47
function, 533
Conditional statement, 28
Conjunction, of statements, 28
Consequent. *See* Conditional statement
Contagion model, 436
(*See also* Pure birth process, contagion in)
Contrapositive, 37
Convergence. *See* Sequences, Series
Converse, of an implication, 36
Corner point of a solution set, 232
Correlation, 604
Counting, 152–67
criterion, 153
principles, 152, 154
Covariance, of two random variables, 601
Cramer's Rule, 248

Decimal expansion. *See* Base ten
De Morgan's laws:
for sets, 17
for statements, 35
generalized, 21
Density function. *See* Random variable
Derivative: 297, 503, 504
directional, 471
higher order, 347
of a constant function, 301
of identity function, 301
second, 347
Determinant, 247–52
function, 248, 249
minor, 256
principal minor, 257

Deviations, of numbers from their average, 124
Diagonal matrix. *See* Matrix, diagonal
Difference:
equation, 355, 356, 370–89
backward, 377
forward, 377
initial conditions for, 372
linear, 370
order of, 371
particular solution of, 382
solution of, 370
transient part of solution of, 384
function, 356
of two functions, 280
of two sequences, 95
of two sets, 16
second, 356
table, 357
Differential:
calculus, 294–354
equations, 298, 443–54
linear, 447
Differentiation. *See* Derivative
Diffusion model, 436
Digraph. *See* Directed graph
Directed graph, 205–8
Discrete data, 355
Disjoint sets, 13
Disjunction, of statements, 28
Distance, 76, 82
between vectors, 196
Distribution function. *See* Random variable
Distributive laws:
for sets, 17
for statements, 35
generalized, 20
Divergence. *See* Sequences, Series
Domain of discourse. *See* Set, universal
Dummy variable. *See* Summation notation

e, constant, 117, 333

Economic multiplier, 128
Eigenvalue. *See* Characteristic root
Eigenvector. *See* Characteristic vector
Elasticity of demand, 319, 365
Element:
of a set, 3
of a vector, 187
Elithorn Perceptual Maze Test, 173
Empty set. *See* Set, empty
Equations, linear. *See* Linear equations
Equilibrium:
of a system, 344
state, 384
Equivalence, 34
class, 54
logical, as an equivalence relation, 60
Equivalent statements, 35
Event(s), 514
certain, 514
compound, 514
impossible, 514
independent, 547, 548
occurrence of, 514
pairwise independent, 550
simple, 514
Exclusive or, 28
Expected value. *See* Random variable
Exponential function. *See* Function, exponential
Extinction process, 662
Extreme point, of a set, 232

Factor Analysis, 193
basic equation of, 194
Factorial, 139, 155
moment generating function, 674
Fallacy, 41
Feasible vector, 239
Flow, diagram, 68
Function, 75–83
additive, 293
as a binary relation, 75
as a machine, 80
composite, 315
concave, 348
continuous, 288

Logistic curve, 436
Loop, of a relation, 70

Mapping. *See* Function
Marginal:
    cost, 297, 416
    product, 298
    revenue, 282, 298
    utility, 298
Markov chain, 615–50
    absorbing state of, 624,
        635
    canonical form for, 627
    expansion of a, 634
    flow diagram for, 626
    initial probabilities in a,
        616
    lumpable, 645
    n-step transition proba-
        bilities for, 620
    periodic, 628
    regular, 629
    states of, 616
    transition probabilities
        for, 616
Matching problem, 528
Mathematical induction,
    97–104
Matrices, 180–269
    addition of, 182
    additive inverse of, 183
    blocked, 200
    column, 223
    diagonal, 204
    dimensions of, 180
    elementary column oper-
        ations on, 223
    elementary row, 221
    elementary row opera-
        tions on, 220
    equality of, 181
    inverse, 224–30
        computation of, 228
    main diagonal of, 181
    minor, 256
    multiplication of, 191
    negative of, 183
    nonsingular, 225
    notation for, 180
    orthogonal, 263
    partitioned, 200
    powers of, 198
    preference, 181
    rank of, 259
    singular, 225

skew-symmetric, 268
square, 181
square of, 198
square root of, 265
stochastic, 618
sub-, 200
symmetric, 260
trace of, 187
transpose of, 185
zero, 182, 193
zeroth power of, 198
Maximum:
    of a set, 88
    of an order relation, 73
Maximum and minimum
    value theorem, 289
Maximum likelihood, 571
Mean. *See* Average
Mean value. *See* Random
    variable, expected
    value
Mean Value Theorem, 307,
    309
    for partial derivatives,
        474
Mental tests, 496
Metric, 82
Minimax, of a function,
    487
Minimum:
    of an order relation, 73
    of a set, 87
Modulo arithmetic, 73
Monotone Convergence
    Theorem, 112
Multinomial Expansion
    Theorem, 172
Multinomial trials, 591
Multiplicative inverse, of a
    number. *See* Recipro-
    cal
Multiplier. *See* Economic
    multiplier
Multiplier–acceleration
    principle, 377

Necessary condition, 38
Negation, of a statement,
    27
Normal distribution, 680–
    89
Null set. *See* Set, empty
Number:
    of bridge hands, 164
    of ordered *r*-samples, 160

O.C. curve. *See* Operating
    characteristic curve
Octal numbers. *See* Base
    eight
Operating characteristic
    curve, 546
Operator, 363
Ordered pair, 48
Ordered samples, 159

Partial derivatives, 469–77
Partial fractions, 433–38
Partition, 98
    of a set, 55
    of an integer, 172
Pascal's Triangle, 171
Permanent income, 133
Permutations, 155, 161
PERT technique, 607
Pivotal committee mem-
    ber, 165
Poisson Process, 651–60
    independence assump-
        tion for, 652
    stationarity assumption
        for, 651
Poisson random variable.
    See Random variable
Polya Urn Scheme, 540,
    623
Polynomial, 282
    derivative operator, 447
    difference operator, 362
Population, 519
Power, of a test procedure,
    686
Power curve, 686
Power series, 455–68
Premise(s), 40
    inconsistent, 47
Present value, 418
    of an annuity, 124
Probability, 516
    a posteriori, 543
    a priori, 543
    conditional. *See* Con-
        ditional probability
    multiplication rule for,
        548, 550
    vector, 618
Probability function, 516,
    559
    conditional, 589
    joint, 586, 587

A   0
B   1
C   2
D   3
E   4
F   5
G   6
H   7
I   8
J   9

9. If X and Y are independent normal random variables with respective means μ and ν and variances σ^2 and τ^2, show that $Z = X - Y$ is a normal random variable with mean $\mu - \nu$ and variance $\sigma^2 + \tau^2$.

10. A coin tossed 400 times yields 280 heads. Would you say the coin is fair? (Use the normal approximation.)

11. In testing whether the coin of Problem 10 is fair, you could make one of two errors:

    Type I   Say the coin is not fair when it actually is.

    Type II Say the coin is fair when it is actually biased.

    Draw the power curve for this testing procedure. (See Example 1.)

12. A company employing 1000 workers maintains a cafeteria and a separate snack bar. If each worker chooses independently and has probability $\frac{2}{3}$ of eating in the cafeteria and $\frac{1}{3}$ of eating in the snack bar, how many seats should be provided in each facility if the company wishes seating to be adequate in both areas on 99% of the working days?

SUPPLEMENTARY READING

Feller, W., *An Introduction to Probability Theory and its Applications* (John Wiley & Sons, Inc., New York, 1968), Vol. I, 3rd ed., Chapter XVII.

Parzen, E., *Modern Probability Theory and its Applications* (John Wiley & Sons, Inc., New York, 1960), Chapter 6.

ANSWERS TO SELECTED EXERCISES

Chapter 1

Section 1.1
1. (a) {Russia, Antarctica, U. S., England, France, Spain, Algeria, Mali, Upper Volta, Ghana}. (c) {Alaska, Nevada, Wyoming, Delaware, Vermont}. **2.** (a) $\{x: x$ is a citizen of Canada$\}$. (c) $\{x: x$ is an even integer and $x > 0\}$ or, $\{2, 4, 6, 8, \ldots\}$ or $\{x: x = 2n$ where n is a positive integer$\}$. **3.** (b). **4.** (e) $\{m: m$ is a 26 year old U. S. male who has legally voted in three presidential elections$\} = \phi$. (f) $\{x: x^2 + 3x - 10 = 0\} = \{2, -5\}$. **5.** (a) Set of points on the line with equation $y = 3x - 3$. (c) $\{(2, 3)\}$. The intersection of the lines in (a) and (b). (d) ϕ. (e) Same as (b). **6.** (a) 16. **12.** Take $n^2 + \frac{57}{6}(n - 1)(n - 2)(n - 3)$ for the nth element.

Section 1.2
1. (a) 16. (b) 2^n; in (a) $16 = 2^4$. **3.** The sets are equal in (a) and (c). **4.** Only (b). **5.** $\{1\}, \{1, 2\}, \{1, 3\}$. **6.** (a), (c), (d), and (f) are correct. **7.** 7. **10.** (b) ϕ. **12.** (a) No. (b) No. **14.** (a) Yes. (b) No. **15.** $\{A, B, C, D, AB, AC, AD, BC, BD, CD, ABC, ABD, ACD, BCD, ABCD\}$.

Section 1.3
2. $\{4\}, \{1, 2, 4, 7\}, \phi$. **5.** (a) $R_1 \cup R_2 \cup R_5 \cup R_6 \cup R_7 \cup R_8$. (c) $R_7 \cup R_8$. (d) R_1. (f) $R_1 \cup R_2 \cup R_3 \cup R_5 \cup R_6$. **6.** $\phi, (A \cap B) \cap C, A \cap B, B \cap A$, $B, A \cup B, A \cup (B \cup C), (A \cup B) \cup C, \phi', U$. **10.** (c), (d), (e), (g), (h), and (i) are correct. **13.** (a) 340. (b) 100. (c) 10. (d) 90. **15.** (a) 255. (b) 250. (c) 685.

Section 1.4
1. Partial answer: For the first distributive law
$A \cup (B \cap C) = \{a, f\} \cup \{b, g\} = \{a, b, f, g\}$
$(A \cup B) \cap (A \cup C) = \{a, b, c, f, g\} \cap \{a, b, e, f, g\} = \{a, b, f, g\}$.
3. (a) $M \cap C' \neq \phi$. (c) $F \cap (D \cup C) \subseteq I$. (e) $(M \cap C') \cup (F \cap D') \subseteq I'$. (f) $M \cap D' \cap C \cap I \neq \phi$. **5.** (a) $A \cap B$. (c) A. (d) $A \cap C'$. **7.** (a) A'. (c) U. (d) ϕ. **9.** Using the second distributive law twice gives
$[A \cap (B \cup C)] \cup (B \cap C) = (A \cap B) \cup (A \cap C) \cup (B \cap C)$
$= (A \cap B) \cup [(A \cup B) \cap C]$.
17. $A[M(x)] = A(3x) = 3x + 2$ while $M[A(x)] = M(x + 2) = 3(x + 2) = 3x + 6$.
19. $*$ is both commutative and associative.

Chapter 2

Section 2.1
1. Truth set of $B \subseteq \{a, c\}$ is $\{\phi, \{a\}, \{c\}, \{a, c\}\}$. U is the set of all subsets of A. **2.** $\{1, 4, 6, 10\}$. **5.** (a) P. (c) $P \cap P' = \phi$. **7.** (a) T. (b) T. (c) F. **9.** (a), (b), (c), (e), and (f).

Section **1.** The truth set of $\sim p$ is P'. Hence, by Definition 2 (c), the truth set of $\sim p \Rightarrow q$ is
2.2 $(P')' \cup Q = P \cup Q$.   **2.** No.   **4.** (a), (c), and (f).   **5.** (a), (b), (d), (e), (f),
and (h).   **7.** Try p and q true, and r false. The statements are not equivalent.

Section **5.** Valid.   **7.** Valid.   **9.** Not valid.   **11.** Valid.   **13.** Not valid.   **17.** (a)
2.3 (i) and (iii).

Chapter 3

Section **1.** (a), (c), (d), and (f).   **3.** Let $A = \{a\}$ and $B = \{b\}$.   (a) $\{(\phi, \phi), (\phi, A),$
3.1 $(\phi, B), (\phi, P), (A, A), (A, P), (B, B), (B, P), (P, P)\}$.   **4.** $\{(1, 2), (2, 3)\}$.
8. (a) $\{(1, 2), (1, 4), (2, 4)\}$.   (c) $\{(2, 2)\}$.   **9.** 16.

Section **1.** Let R = reflexive, S = symmetric, and T = transitive.   (a) R, S, and T.
3.2 (c) S only.   (e) S only.   **3.** One possible answer: The sets $\{a, c\}$, $\{b\}$, and
$\{d, e\}$ partition U.   **4.** Partial answer: One equivalence class is $\{$Jefferson,
Monroe, Buchanan, Grant$\}$, the set of presidents born in April.   **6.** (c) and (e)
are equivalence relations; (a) is neither R nor S nor T, (b) is T only, (d) is S only,
(f) is T only, (g) is R and T, (h) is R and T, (i) is R.   **14.** Problem 13 applies
since $U \subseteq U$ and $A \subseteq B$.

Section **3.** Let R, I, S, A, T denote reflexive, irreflexive, etc.   (a) I, S.   (c) I, T.   (d) I,
3.3 S, T.   (e) I, S, A, T (The relation is ϕ, the empty relation.)   (g) I, S, T.   (i) R,
T.   (k) R, S, T.   **5.** (a) "is poorer than".   (c) "is less popular than".
6. (a) $p \Rightarrow p$ for every statement p. Hence I is reflexive. If $p \Rightarrow q$ and $q \Rightarrow r$, then
$p \Rightarrow r$. Hence I is transitive.   (c) $p \Leftrightarrow q$ means that $p \Rightarrow q$ and $q \Rightarrow p$. That is
$pEq \Leftrightarrow pIq \wedge qIp$.   **8.** (a) The relation "has at least as good a racial balance
as".   (c) The relation "is at least as risky as".   **9.** Every equivalence relation is
reflexive and transitive.   **11.** P is asymmetric so we may have either xPy or yPx,
but not both. If neither xPy nor yPx, then xIy holds, by definition of I.

Section **2.** (a) Not consistent. If x lives next to y and y lives in the same precinct as z, it
3.4 does not necessarily follow that x lives next to z.   (c) Consistent.   (d) Not
consistent.   **7.** Not connected. Subject may not prefer either x or y.

Section **3.** (a) $(12, 15) \rightarrow (16, 10) \rightarrow (20, 5) \rightarrow (14, 5) \rightarrow (16, 3) \rightarrow (14, 3) \rightarrow (12, 3)$
3.5 $\rightarrow (14, 1) \rightarrow (12, 1)$.   **4.** Every pair of elements is now a comparable pair.
14. (a) 6.   (b) 0.   (c) 9.

Chapter 4

Section **1.** Domain = $\{$Robert, Sam, James, Teddy$\}$, Range = $\{$James, Teddy, Sam$\}$.
4.1 Domain = Range = set of all real numbers.   **2.** 6, 9.   **3.** $C(A) = D$,
$C(B) = \{a\}$, $C(A \cup B) = C(U) = \phi$, $C(A \cap B) = \{a, b, d\}$. Domain of C is the
set of all subsets of U.   **5.** (a) and (d).   **8.** Assuming the relations consist of
all pairs (x, y) which satisfy the given equations, only (a) is a function.   **10.** (a)
50,000, 25,000, 10,000, 20,000/3.   (b) $pol = \dfrac{g \cdot y}{S_t \cdot L \cdot y_{10}}$.   **11.** $k = 24$,

$\rho = \dfrac{\log_{10} 24 - \log_{10} 5}{\log_{10} 3} = 1.4278$.